AF361027

Piezoelectric Transducers

Piezoelectric Transducers: Materials, Devices and Applications

Editor

Jose Luis Sanchez-Rojas

MDPI • Basel • Beijing • Wuhan • Barcelona • Belgrade • Manchester • Tokyo • Cluj • Tianjin

Editor
Jose Luis Sanchez-Rojas
University of Castilla-La Mancha
Spain

Editorial Office
MDPI
St. Alban-Anlage 66
4052 Basel, Switzerland

This is a reprint of articles from the Special Issue published online in the open access journal *Micromachines* (ISSN 2072-666X) (available at: https://www.mdpi.com/journal/micromachines/special_issues/Piezoelectric_Transducers_Materials_Devices_Applications).

For citation purposes, cite each article independently as indicated on the article page online and as indicated below:

LastName, A.A.; LastName, B.B.; LastName, C.C. Article Title. *Journal Name* **Year**, *Article Number*, Page Range.

ISBN 978-3-03936-856-3 (Hbk)
ISBN 978-3-03936-857-0 (PDF)

Contents

About the Editor

Jose Luis Sanchez-Rojas received the B.E., M.S. and Ph.D. degrees in telecommunication engineering from Universidad Politécnica de Madrid, Spain. In 1996–97, he was an Invited Postdoctoral Scientist with Cornell University and Colorado University, Boulder. He is currently Full Professor of Electronic Technologies with the Universidad Castilla–La Mancha, Spain. Research interests cover optical and electrical characterization and modeling of semiconductor devices, miniaturization of instrumentation, design and applications of MEMS/NEMS, sensors and actuators. He has published more than 200 papers in journals and leaded more than 50 national and international research projects, some of them in cooperation with high-tech companies.

Editorial

Editorial of Special Issue "Piezoelectric Transducers: Materials, Devices and Applications"

Jose Luis Sanchez-Rojas

Microsystems, Actuators and Sensors Lab, Universidad de Castilla-La Mancha, 13071 Ciudad Real, Spain; JoseLuis.SAldavero@uclm.es

Received: 22 June 2020; Accepted: 25 June 2020; Published: 13 July 2020

Advances in miniaturization of sensors, actuators, and smart systems are receiving substantial industrial attention, and a wide variety of transducers are commercially available or possess high potential to impact emerging markets. Substituting existing products based on bulk materials, in fields such as automotive, environment, food, robotics, medicine, biotechnology, communications, Internet of things, and related technologies, with reduced size, lower cost, and higher performance, is now possible with potential for manufacturing using advanced silicon-integrated circuits technology or alternative additive techniques from the milli- to the micro-scale.

This Special Issue has compiled a total of 34 papers focused on piezoelectric transducers, covering a wide range of topics including the design, fabrication, characterization, packaging, and system integration or final applications of milli/micro/nano-electro-mechanical systems based transducers featuring piezoelectric materials and devices.

I would like to take this opportunity to thank all the authors for submitting their papers to this Special Issue. I also want to thank all the reviewers for dedicating their effort and time in assisting to improve the quality of the submitted papers.

In view of the success reached in the number and quality of papers published, we plan to open a second volume where we hope to continue with the latest advances in piezoelectric transducers and their trend to miniaturization, efficiency, and new applications.

Conflicts of Interest: The author declares no conflict of interest.

 *micromachines*

Article

PIN-PMN-PT Single Crystal 1-3 Composite-based 20 MHz Ultrasound Phased Array

Wei Zhou [1,2,3,†], Tao Zhang [1,3,†], Jun Ou-Yang [1], Xiaofei Yang [1], Dawei Wu [4,*] and Benpeng Zhu [1,3,*]

1   School of Optical and Electronic Information, Huazhong University of Science and Technology, Wuhan 430074, China; davidzhou@hubcomm.com (W.Z.); D201880657@hust.edu.cn (T.Z.); oyj@hust.edu.cn (J.O.-Y.); yangxiaofei@hust.edu.cn (X.Y.)
2   Jingzhou Vocational College of Technology, Jingzhou 434000, China
3   State Key Laboratory of Transducer Technology, Chinese Academy of Sciences, Shanghai 200050, China
4   State Key Laboratory of Mechanics and Control of Mechanical Structures, Nanjing University of Aeronautics and Astronautics, Nanjing 210016, China
*   Correspondence: dwu@nuaa.edu.cn (D.W.); benpengzhu@hust.edu.cn (B.Z.)
†   The authors are co-first authors of this paper.

Received: 27 November 2019; Accepted: 20 May 2020; Published: 21 May 2020

Abstract: Based on a modified dice-and-fill technique, a PIN-PMN-PT single crystal 1-3 composite with the kerf of 12 μm and pitch of 50 μm was prepared. The as-made piezoelectric composite material behaved with high piezoelectric constant (d_{33} = 1500 pC/N), high electromechanical coefficient (k_t = 0.81), and low acoustic impedance (16.2 Mrayls). Using lithography and flexible circuit method, a 48-element phased array was successfully fabricated from such a piezoelectric composite. The array element was measured to have a central frequency of 20 MHz and a fractional bandwidth of approximately 77% at −6 dB. Of particular significance was that this PIN-PMN-PT single crystal 1-3 composite-based phased array exhibits a superior insertion loss compared with PMN-PT single crystal and PZT-5H-based 20 MHz phased arrays. The focusing and steering capabilities of the obtained phased array were demonstrated theoretically and experimentally. These promising results indicate that the PIN-PMN-PT single crystal 1-3 composite-based high frequency phased array is a good candidate for ultrasound imaging applications.

Keywords: PIN-PMN-PT; 1-3 composite; high frequency; phased array

---

1. Introduction

Over the past few decades, due to its safety, convenience, and efficiency, ultrasound has attracted significant attention in biomedical research and clinical diagnosis [1–6]. To meet the requirement of high resolution imaging and precise manipulation, the operational frequency of biomedical ultrasound needs to be greater than 20 MHz. For such applications, high frequency ultrasound transducer, which plays an important role in producing and receiving ultrasound signal, is indispensable [7–13]. Currently, to the best of our knowledge, single element transducers and linear arrays are widely used [14–17]. Only a few studies have been carried on the high-frequency ultrasound phased array [18–21], though it is very useful in biomedical imaging by providing electronic-beam-focusing and steering capabilities. Furthermore, there are even fewer researches on composite-material-based high-frequency ultrasound phased arrays, despite the numerous benefits of composite piezoelectric materials (lower acoustic impedance, higher electromechanical coefficient, broader bandwidth, etc.). As is known, the pitch of a phased array is much less than that of a linear array at the same operating frequency, and the kerf of a composite material must be small enough to avoid spurious modes. The above facts give rise to smaller size in both kerf and each element, which poses even greater challenges for piezoelectric material composite preparation and high frequency (≥20 MHz) phased array fabrication.

In recent years, single-crystal piezoelectrics have been developing a revolutionary solution to substitute traditional PZT ceramics for ultrasonic transducer applications [22–24]. Recently, PMN-PT single-crystals have been used to manufacture high frequency (≥20 MHz) phased arrays and some attractive results have been found [19,20]. However, the drawback of PMN-PT single-crystals is their relatively low-temperature usage range [22]. To make the devices maintain a better temperature stability, a high Curie temperature (Tc) ternary piezoelectric single crystal [Pb(In$_{1/2}$Nb$_{1/2}$)O$_3$-Pb(Mg$_{1/3}$Nb$_{2/3}$)O$_3$-PbTiO$_3$, abbreviated as PIN-PMN-PT], which maintains a similar electromechanical and piezoelectric performance (d$_{33}$~1500 pm V^{-1}, k$_{33}$ > 90%) to a PMN-PT single crystal, has been developed [25,26]. In previous studies, the PIN-PMN-PT single crystal and its composite have been proven to be promising candidates for high frequency single element transducer fabrication [7,26,27]. Additionally, compare with piezoelectric single crystal itself and its 2-2 composite, the single crystal 1-3 composite usually exhibits higher coupling coefficient and suitable acoustic impedance. Hence, it is of great interest to demonstrate the feasibility of the development of a PIN-PMN-PT single crystal 1-3 composite-based high frequency (≥20 MHz) ultrasound phased array.

In this work, a modified dice-and-fill technique for PIN-PMN-PT single crystal 1-3 composite preparation is introduced. The design, fabrication, and characterization of a 20 MHz side-looking 48-element high frequency ultrasound phased array are presented. Furthermore, to demonstrate the imaging capability of this obtained device, wire phantom imaging experiments are carried out based on a commercial Verasonics Vantage 128 System.

2. Design and Fabrication

In our fabrication, the [1]-oriented single crystal with a composition of 0.27PIN-0.45PMN-0.28PT was selected. Generally, for the fabrication of phased array with central frequency around 20 MHz, the pitch should be less than λ (wavelength in water, 75 μm). In fact, taking into consideration the technology difficulty, the pitch in our design is 50 μm. Figure 1 describes the 1-3 composite fabrication process, during which a modified dice-and-fill method was adopted to prevent the collapse and cracking of the elements. To obtain a high precision kerf width when dicing the PIN-PMN-PT single crystal, a 10-μm-thick nickel/diamond blade with a DAD323 dicing saw (DISCO, Saitama, Japan) was chosen. First, the PIN-PMN-PT wafer was diced from two perpendicular directions with a 100 μm pitch and 80 μm depth to form periodic rods. The kerf was around 12 μm because of the dicing saw's vibration during dicing process. Then, the kerfs were filled with low viscosity epoxy (Epo-Tek-301, Epoxy Technology, Billerica, MA, USA), and the trapped bubbles were removed in vacuum for 15 min. After epoxy solidification at 40 °C for 10 h, the PIN-PMN-PT rods in the epoxy matrix were further diced in the two previous perpendicular directions with the same dicing pitch and depth to form equal areas. The newly formed kerfs were epoxy filled and cured again. Finally, to grind off the excess single-crystal and epoxy, a PIN-PMN-PT single crystal 1-3 composite with a 50 μm pitch and 80 μm depth was obtained. The main piezoelectric properties of the PIN-PMN-PT single-crystal 1-3 composite are listed in Table 1. With the Archimedes principle, we can know that the density of this kind of composite is 4510 kg/m^3. In addition, the longitude velocity was measured to be 3600 m/s. Consequently, using the equation of Z = ρc, the acoustic impedance can be calculated to be 16.2 MRayl. In addition, the piezoelectric coefficient d$_{33}$ (1500 pC/N) was tested by a d$_{33}$ meter (ZJ-6A, Institute of Acoustics, Chinese Academy of Sciences, Beijing, China), and the loss (0.023) was evaluated by a LCR meter (Agilent, Englewood, CO, USA, 4263B). The electromechanical coupling factor (0.81) was calculated using resonance and anti-resonance frequencies, which were measured by an Agilent 4294A electric impedance analyzer (Agilent Technologies, Englewood, CO, USA). These excellent properties imply that this PIN-PMN-PT single crystal 1-3 composite is competent for ultrasound device application.

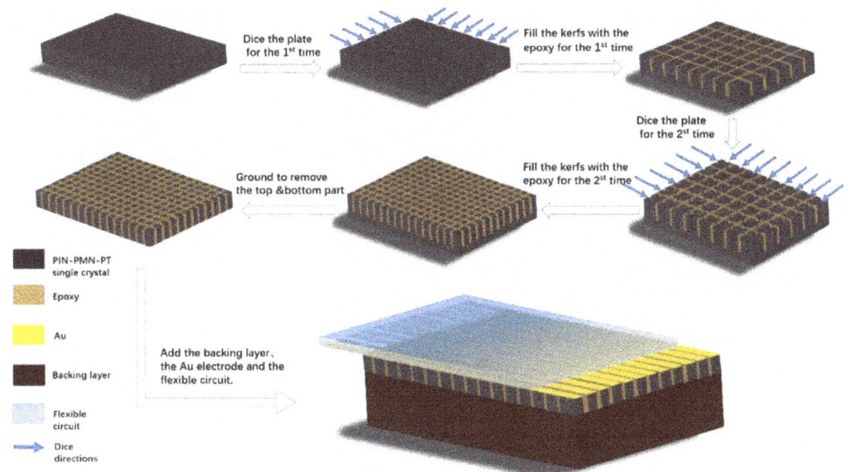

Figure 1. Schematic of the modified dice-and-fill method used for phased array fabrication.

Table 1. Properties of piezoelectric single crystal 1-3 composite.

Piezoelectric material	0.27PIN-0.45PMN-0.28PT
Polymer	Epoxy 301
Longitude velocity	3600 ms^{-1}
Density	4510 kg/m^3
Acoustic impedance	16.2 MRayls
Piezoelectric constant	1500 pC/N
Electromechanical coupling coefficient	0.81
Loss tangent	0.023

After sputtering Au/Cr (150 nm/100 nm) layers on both sides of the 1-3 composite, E-solder 3022 was added as the backing layer, and lithography was used to pattern 48 elements with a 38 μm width, 3 mm length, and 12 μm kerf, as presented in Figure 2A. Then, a piece of 30 μm polyimide-based flexible circuit(Kapton, DuPont, Wilmington, DE, USA) with gold patterns was attached to the front of the array using Epo-Tek 301 epoxy for electrical connections, as shown in Figure 2B,C. Actually, as long as the thickness of Epo-Tek 301 epoxy is in a suitable range, it can have a good adhesion; otherwise, it has no effect on conductivity. Because the flexible circuit (Kapton) has a specific acoustic impedance of 3.4 MRayl, it can also act as a matching layer for the phased array. Lastly, the 48-element side-looking phased array was encapsulated in a 3D-printed housing, as depicted in Figure 2D.

To simulate the acoustic performance of our phased array, the Krimholz, Leedom, and Mattaei (KLM) equivalent circuit-based software package PiezoCAD (Sonic Concepts, Woodinville, WA, USA) was employed. The acoustic design parameters for piezoelectric single crystal 1-3 composite phased array transducer are listed in Table 2. For our design, the thickness of polyimide-based flexible circuit, PIN-PMN-PT single crystal 1-3 composite, and E-solder 3022 are 30 μm, 80 μm, and 2.5 mm, respectively. Figure 3 shows the simulated pulse-echo waveform and frequency spectrum of this phased array. It can be easily seen that the operational frequency is 20.8 MHz and bandwidth at −6 dB is 81.4%.

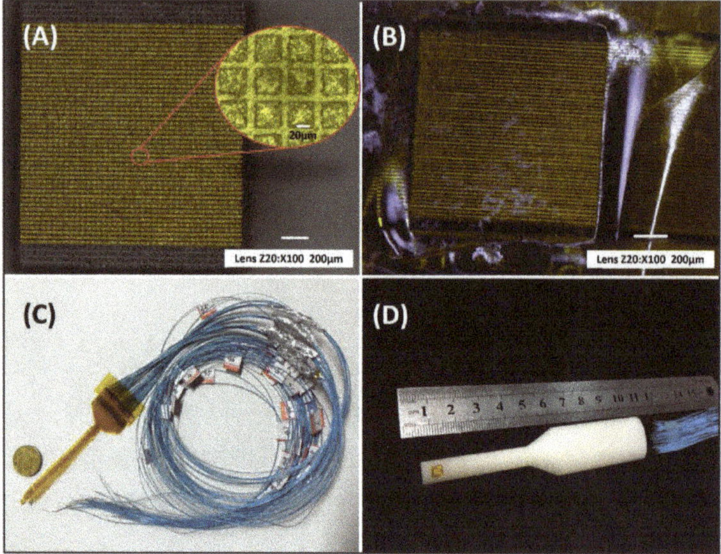

Figure 2. Photographs of (**A**) Au patterned PIN-PMN-PT single crystal 1-3 composite (**B**) with Flexible Circuit and (**C**) electrical connections; (**D**) the side-looking phase array in 3D-printed housing.

Table 2. The acoustic design parameters for piezoelectric single crystal 1-3 composite phased array transducer.

Layer	Material	Acoustic Impendence/MRayl	Thickness/mm
Matching layer	Polyimide-based flexible circuit	3.4	0.03
Piezoelectric layer	Single crystal 1-3 composite	16.2	0.08
Backing layer	E-solder 3.22	5.92	2.5

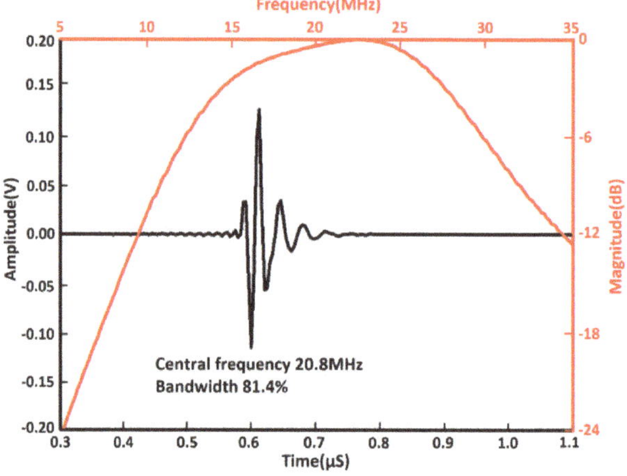

Figure 3. The simulated pulse-echo waveform and frequency spectrum of the transducer using the PiezoCAD software.

3. Characterization and Discussions

Figure 4 shows the measurements of the electrical impedance and phase for the array elements using the impedance analyzer Agilent 4294A. It is clear that no critical differences can be found among the 48 elements, indicating that all the elements are uniform. The peaks in the phase curves are located at 20.4 MHz, suggesting that the operational frequency for the array is approximately 20 MHz. According to the IEEE standard [28], the thickness mode of the electromechanical coupling coefficient (kt) for each array element can be determined from the following equation:

$$k_t^2 = \frac{\pi}{2}\frac{f_r}{f_a}\tan\left(\frac{\pi}{2}\frac{f_a - f_r}{f_a}\right),\tag{1}$$

where f_r and f_a are the resonant and antiresonant frequencies, respectively. Substituting the appropriate values into Equation (1), k_t was calculated to be 0.81. This value is similar to those reported in the literature [26,29].

The pulse-echo response of the representative element of the phased array transducer was acquired in a deionized water tank at room-temperature using a pulser receiver (5900PR, Panametrics, Inc., Waltham, MA, USA). The echo was reflected by an X-cut quartz plate target, the waveform of which was recorded using a 1-GHz oscilloscope (LC534, LeCroy Corp., Chestnut Ridge, NY, USA). As shown in Figure 5A, the random array element exhibits a peak-to-peak echo amplitude of 0.235 V. The measured central frequency and bandwidth at −6 dB are 20 MHz and 77%, respectively, which is closed to the simulated result. It should be noted that, as described in Figure 5B, the array exhibits a good uniformity in acoustic performance (sensitivity: 0.235V ±2.1%, bandwidth: 77% ± 3%).

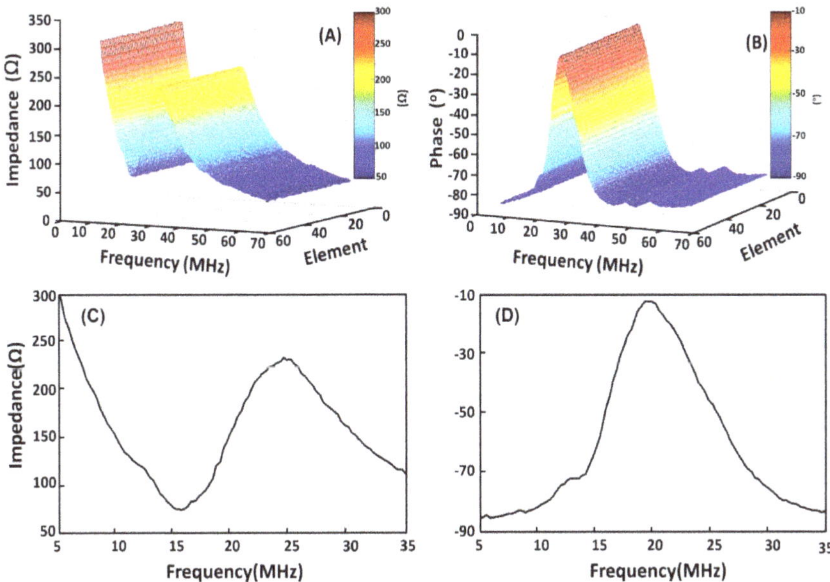

Figure 4. (**A**) Electrical impedance magnitude and (**B**) phase as a function of frequency for 48 elements. (**C**) Electrical impedance magnitude and (**D**) phase as a function of frequency for the 24th element.

The insertion loss (IL) and crosstalk were measured using several cycles of a sinusoid pulse produced by a Sony/Tektronix AFG2020 arbitrary function generator. For the insertion loss (IL) measurement, the amplitude of an echo signal reflected by the quartz target was received by the excited phased-array element. The exciting waveform is a tone burst of 20-cycle sinusoidal wave with the amplitude of the driving signal and the center frequency. An oscilloscope connected with the excited phased-array element was set to 1 MΩ and 50 Ω to test the receiving and transmitting signal, respectively. In consideration of the compensation for the attenuation in water (2.2×10^{-4} dB mm^{-1} × MHz^2) and the loss caused by the imperfect reflection from the quartz target (1.9 dB) [1], the IL was calculated using

$$\text{IL} = 20 \log \frac{V_R}{V_T} + 1.9 + 2.2 \times 10^{-4} \times 2d \times f_c^2, \tag{2}$$

where f_c is the center frequency, V_T and V_R are the transmitting and receiving amplitudes, and d is the distance between the target and transducer. The IL value of the phased array elements was measured to be −19.7 dB, which is superior to those of PMN-PT single crystal- and PZT-5H-based high frequency (≥20 MHz) phased arrays [18–20]. This phenomenon is probably related to the high Curie Temperature of PIN-PMN-PT single crystal. The heat induced by lapping and dicing in fabrication process produces nearly no effluence to its electrical properties. Therefore, PIN-PMN-PT single crystal 1-3 composite array element can sustain a high piezoelectric performance. For the crosstalk measurement, the amplitude of an echo signal reflected by the quartz target was received by the nearest neighbor element of the excited one. According to the method reported in Reference [18], the measured crosstalk between the nearest neighbor array elements at the center frequency was −38 dB.

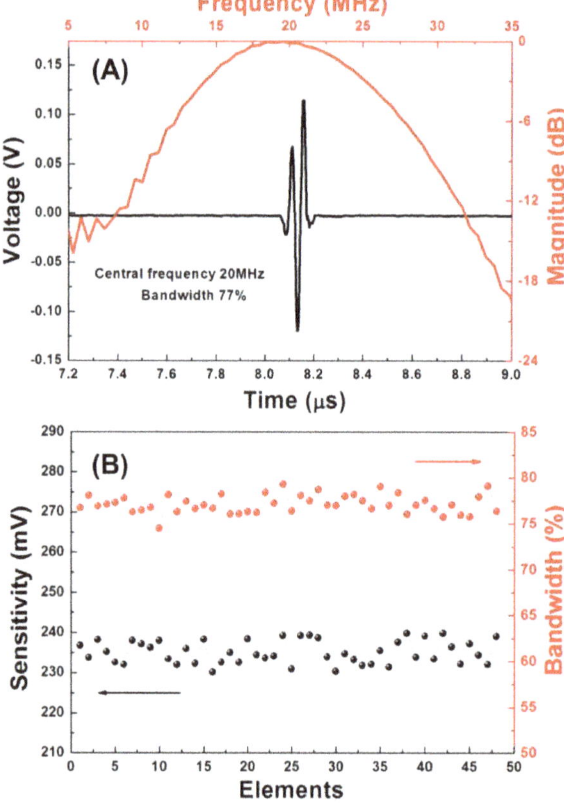

Figure 5. (**A**) Measured pulse-echo response performance of random element; (**B**) sensitivity and bandwidth of the pulse-echo signal for each element.

In order to measure the one-way azimuthal directivity response, a representative array element (24th element) was excited. The amplitude of the time-domain response of the element at discrete angular positions was acquired by a piezoelectric hydrophone with 0.2 mm diameter (Precision Acoustic, Dorset, UK). As can be seen in Figure 6A, for our phased array, the measured −6 dB directivity is approximately ±21°. Meanwhile, the similar result was reported in Reference [20], a −6 dB directivity of approximately ±20° for a 20 MHz phased array. For the evaluation of the acoustic output characteristics of the transducer, both the acoustic pressure of a representative element (24th element) and the entire array in the axis direction were measured by this commercial needle piezoelectric hydrophone. As shown in Figure 6B, the maximum acoustic pressure for an element near the surface is about 323 KPa, and the maximum acoustic pressure of the entire array near the focusing position is 625 KPa. These results indicate that this PIN-PMN-PT single crystal 1-3 composite-based phase array has the potential for ultrasound imaging application.

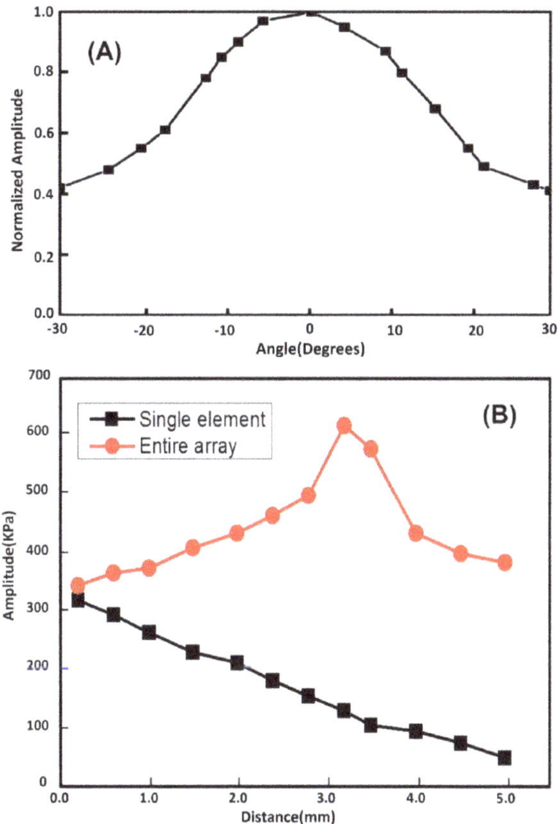

Figure 6. (**A**) Measured one-way azimuthal directivity responses of a representative array element (24th element). (**B**) The acoustic output in the axial direction of a representative array element (24th element) and the entire array.

To predict the imaging performance of the obtained phased array, the Field II Ultrasound Simulation Program was utilized. In the simulation, an aperture consisting of 48 active elements of the array was assumed to test the focusing capability of the phased array. Multiple point targets located evenly between 0–10 mm along the aperture central axis were first created. The simulated result is presented in Figure 7A in a 60 dB dynamic range and the focal distance is about 3.2 mm. The imaging performance can be improved by applying apodization, as shown in Figure 7B. Because the pitch of the fabricated array is 50 μm and the half-wavelength of the 20 MHz sound wave in water is 37.5 μm, the side-lobes thus appear at arcsin(37.5/50), which is 48.6°, and the result is consistent with the simulation result, as shown in Figure 7C. When the beam is steered at the angle of 45°, the grating lobe shares quite a large part of the energy, as Figure 7D. The image results also indicate that the ultrasound energy is concentrated at about 3 mm along the axis. The image resolutions of the points deteriorate when they are off-axis or move away from the 3-mm distance.

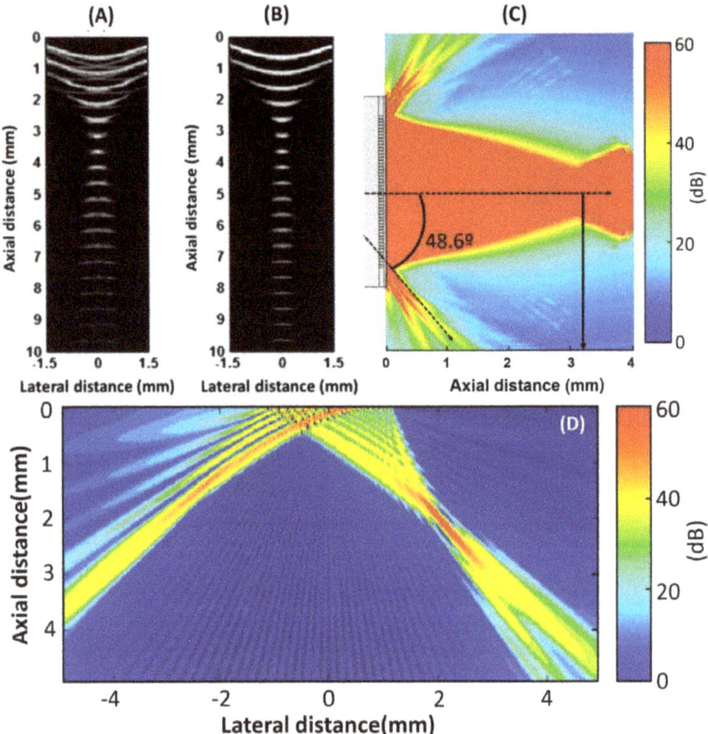

Figure 7. Point spread function phantom-imaged (**A**) without and (**B**) with apodization. (**C**) Simulated phased array acoustic pressure of emission mode when the steering angle is 0°. (**D**) Simulated acoustic beam generated by the 48-element phased array when the steering angle is 45° and the focal distance is 3 mm.

A commercial Verasonics Vantage 128 System (Verasonics, Inc., Kirkland, WA, USA) was utilized to determine the imaging capability of this 20 MHz phased array. In our imaging experiment, two kinds of phantoms were employed. One is 20 μm-diameter tungsten wire in water and the other is 200 μm-diameter copper wire in PDMS (polydimethylsiloxane). In order to determine the axial and lateral spatial resolution of the array transducer, two 20-μm-diameter tungsten wires were positioned at different places, as shown in Figure 8, and immersed in a tank with deionized water. In Figure 9A,B, with gray scale and color scale in 50 dB dynamic range, it can be observed that the two wires are clearly visible. For the second wire, which is located at the focal point, it obviously has a better resolution. We can obtain the theoretical resolutions using the following equations [30]:

$$R_A = PL/2, \tag{3}$$

$$R_L = F\# \times \lambda, \tag{4}$$

where R_A and R_L are the axial and lateral spatial resolutions, respectively. *PL* is the −6 dB spatial pulse length of the received echo(measured PL of 150 μm), *F*# is the F-number of the array transducer (1.25), and λ is the sound wavelength in the transmitting medium (75 μm in the water). Therefore, the theoretical axial resolution is 75 μm and theoretical lateral resolution is 94 μm. Plots of the axial and lateral line spread functions for the second wire are shown in Figure 10A,B. The measured spatial resolutions at −6 dB are approximately 77 μm and 125 μm in the axial and lateral directions, respectively. These values are in accordance with those theoretical values.

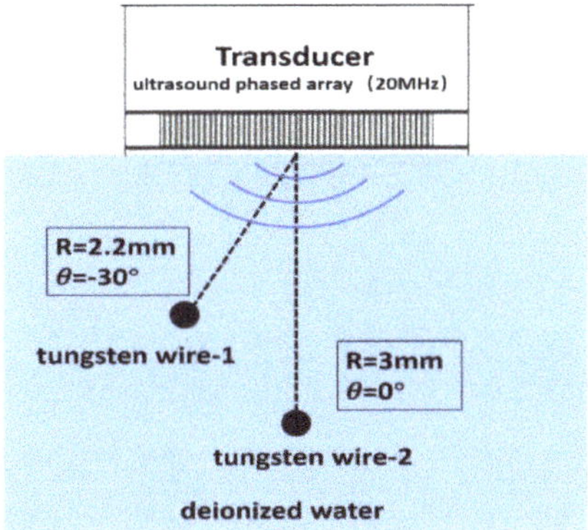

Figure 8. Schematic diagram of tungsten wire phantom in deionized water.

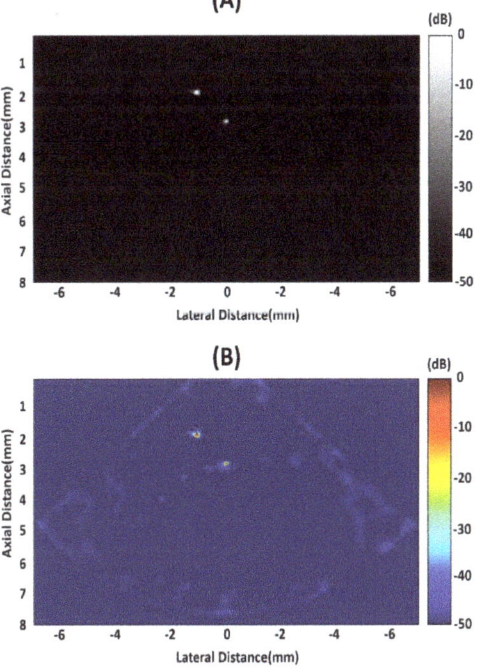

Figure 9. (**A**) Acquired image of custom-made fine-wire phantom. (**B**) Pseudo-color image of custom-made fine-wire phantom.

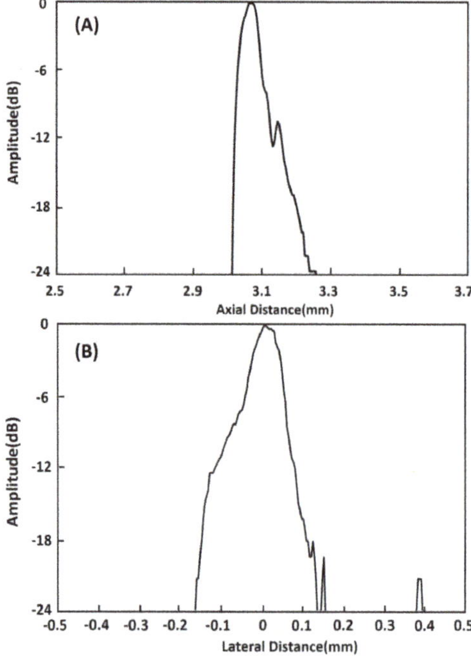

Figure 10. (**A**) Axial and (**B**) lateral line spread functions for the second wire of the wire phantom.

As shown in Figure 11, two 200 µm diameter copper wires are placed in the PDMS at different positions. The purpose is to demonstrate the capability of this PIN-PMN-PT single crystal 1-3 composite-based phase array to detect solid structures embedded in tissue. In Figure 12A,B, with gray scale and color scale in 50 dB dynamic range, it is clear to find that the two wires can be easily observed. Compared with the first wire, the second one's resolution is superior, which demonstrates that this obtained high frequency phased array has an excellent beam steering performance.

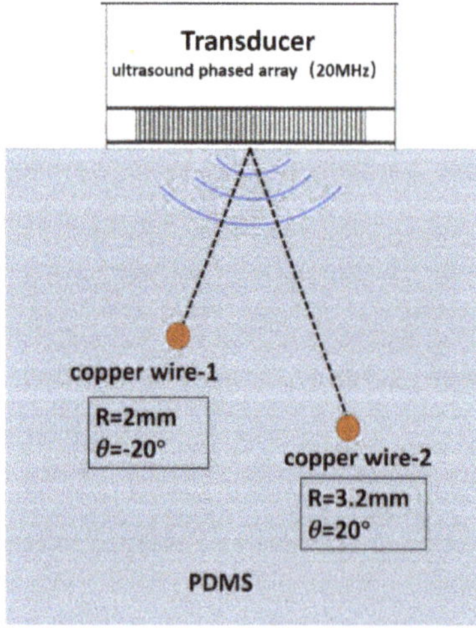

Figure 11. Schematic diagram of copper wire phantom in PDMS.

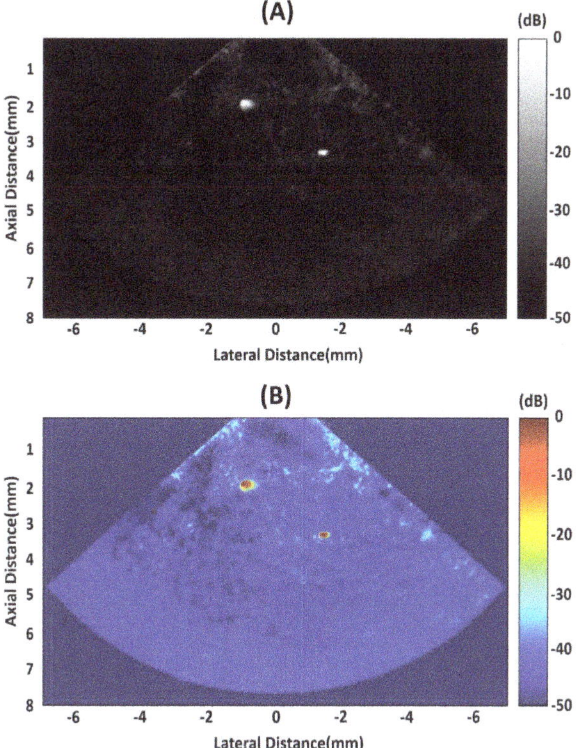

Figure 12. (**A**) Copper wires phantom image. (**B**) Pseudo-color image of copper wires phantom.

4. Conclusions

Based on a modified dice-and-fill technique, a PIN-PMN-PT single crystal 1-3 composite with high piezoelectric constant (d_{33} = 1500 pC/N), high electromechanical coefficient (k_t = 0.81), and low acoustic impedance (16.2 Mrayls) was prepared. Utilizing this kind of composite, a 20 MHz 48-element side-looking high frequency phased array with central frequency of 20 MHz and −6 dB bandwidth of 77% was successfully fabricated, which was confirmed by the electric impedance resonance curve and the pulse-echo response. Of particular significance was that this PIN-PMN-PT single crystal 1-3 composite-based phased array exhibits a superior insertion loss compared with PMN-PT single crystal and PZT-5H-based 20 MHz phased arrays. The focusing and steering capabilities of the obtained phased array were demonstrated theoretically and experimentally. Furthermore, when using such a phased array, wire phantom images in water and PDMS can be achieved. These promising results suggest that the PIN-PMN-PT single crystal 1-3 composite-based high frequency phased array is competent for biomedical ultrasound imaging in the future.

Author Contributions: B.Z. and D.W. conceived and designed the experiments; W.Z. and T.Z. performed the experiments; J.O.-Y. and X.Y. analyzed the data; W.Z. and T.Z. wrote the paper. All authors have read and agreed to the published version of the manuscript.

Funding: This work was supported by the Natural Science Foundation of China (Grant no. 11774117), and the Excellent Youth Foundation of Hubei Province (2018CFA083). We thank the Analytical and Testing Center of Huazhong University of Science & Technology.

Conflicts of Interest: The authors declare no conflict of interest.

References

1. Cannata, J.M.; Ritter, T.A.; Chen, W.H.; Silverman, R.H.; Shung, K.K. Design of efficient, broadband single-element (20–80 MHz) ultrasonic transducers for medical imaging applications. *IEEE Trans. Ultrason. Ferroelectr. Freq. Contr.* **2003**, *50*, 1548–1557. [CrossRef] [PubMed]

2. Zhu, B.P.; Chan, N.Y.; Dai, J.Y.; Shung, K.K.; Takeuchi, S.; Zhou, Q.F. New fabrication of high-frequency (100-MHz) ultrasound PZT film kerfless linear array. *IEEE Trans. Ultrason. Ferro. Freq. Control.* **2013**, *60*, 854–857. [CrossRef] [PubMed]

3. Li, J.P.; Yang, Y.; Chen, Z.; Lei, S.; Shen, M.; Zhang, T.; Lan, X.; Qin, Y.; Ou-Yang, J.; Yang, X.F.; et al. Self-healing: A new skill unlocked for ultrasound transducer. *Nano Energy* **2020**, *68*, 104348. [CrossRef]

4. Fei, C.L.; Liu, X.L.; Zhu, B.P.; Li, D.; Yang, X.F.; Yang, Y.T.; Zhou, Q.F. AlN piezoelectric thin films for energy harvesting and acoustic devices. *Nano Energy* **2018**, *51*, 146–161. [CrossRef]

5. Chen, Z.Y.; Wu, Y.; Yang, Y.; Li, J.P.; Xie, B.S.; Li, X.J.; Lei, S.; Ou-Yang, J.; Yang, X.F.; Zhou, Q.F.; et al. Multilayered carbon nanotube yarn based optoacoustic transducer with high energy conversion efficiency for ultrasound application. *Nano Energy* **2018**, *46*, 314–321. [CrossRef]

6. Li, J.P.; Xu, J.; Liu, X.; Tao, Z.; Lei, S.; Jiang, L.; Ou-Yang, J.; Yang, X.; Zhu, B.P. A novel CNTs array-PDMS composite with anisotropic thermal conductivity for optoacoustic transducer applications. *Compos. Part B* **2020**, *196*, 108073. [CrossRef]

7. Sun, P.; Zhou, Q.F.; Zhu, B.P.; Wu, D.W.; Hu, C.H.; Cannata, J.M.; Tian, J.; Han, P.; Wang, G.; Shung, K.K. Design and fabrication of PIN-PMN-PT single-crystal high-frequency ultrasound transducers. *IEEE Trans. Ultrason. Ferroelectr. Freq. Contr.* **2009**, *56*, 2760–2763.

8. Ou-Yang, J.; Zhu, B.P.; Zhang, Y.; Chen, S.; Yang, X.F.; Wei, W. New KNN-based lead-free piezoelectric ceramic for high-frequency ultrasound transducer applications. *Appl. Phys. A* **2015**, *118*, 1177–1181. [CrossRef]

9. Zhang, L.; Xu, X.; Hu, C.H.; Sun, L.; Yen, J.T.; Cannata, J.M.; Shung, K.K. A high frequency, high frame rate duplex ultrasound linear array imaging system for small animal imaging. *IEEE Trans. Ultrason. Ferroelect. Freq. Contr.* **2010**, *57*, 1548–1557. [CrossRef]

10. Zhu, B.P.; Wu, D.W.; Zhang, Y.; Ou-Yang, J.; Chen, S.; Yang, X.F. Sol-gel derived PMN-PT thick films for high frequency ultrasound linear array applications. *Ceram. Inter.* **2013**, *39*, 8709–8714.

11. Zhu, B.P.; Fei, C.L.; Wang, C.; Zhu, Y.H.; Yang, X.F.; Zheng, H.R.; Zhou, Q.F.; Shung, K.K. Self-focused AlScN film ultrasound transducer for individual cell manipulation. *ACS Sens.* **2017**, *2*, 172–177. [CrossRef] [PubMed]

12. Fei, C.L.; Li, Y.; Zhu, B.P.; Chiu, C.T.; Chen, Z.Y.; Li, D.; Yang, Y.T.; Shung, K.K.; Zhou, Q.F. Contactless microparticle control via ultrahigh frequency needle type single beam acoustic tweezers. *Appl. Phys. Lett.* **2016**, *109*, 173509. [CrossRef] [PubMed]

13. Zhu, B.P.; Xu, J.; Li, Y.; Wang, T.; Xiong, K.; Lee, C.Y.; Yang, X.F.; Shiiba, M.; Takeuchi, S.; Zhou, Q.F.; et al. Micro-particle manipulation by single beam acoustic tweezers based on hydrothermal PZT thick film. *AIP Adv.* **2016**, *6*, 035102. [CrossRef]

14. Zhu, B.P.; Zhang, Z.Q.; Ma, T.; Yang, X.F.; Li, Y.X.; Kirk Shung, K.; Zhou, Q.F. (100)-Textured KNN-based thick film with enhanced piezoelectric property for intravascular ultrasound imaging. *Appl. Phys. Lett.* **2015**, *106*, 173504. [CrossRef] [PubMed]

15. Zhu, B.P.; Zhou, Q.F.; Hu, C.H.; Shung, K.K.; Gorzkowski, E.P.; Pan, M.J. Novel lead zirconate titanate composite via freezing technology for high frequency transducer applications. *J. Adv. Dielect.* **2011**, *1*, 85–89. [CrossRef]

16. Zhu, B.P.; Han, J.X.; Shi, J.; Shung, K.K.; Wei, Q.; Huang, Y.H.; Kosec, M.; Zhou, Q.F. Lift-off PMN-PT thick film for high frequency ultrasonic biomicroscopy. *J. Am. Ceram. Soc.* **2010**, *93*, 2929–2931. [CrossRef]

17. Zhu, B.P.; Zhou, Q.F.; Shi, J.; Shung, K.K.; Irisawa, S.; Takeuchi, S. Self-separated hydrothermal lead zirconate titanate thick for high frequency transducer applications. *Appl. Phys. Lett.* **2009**, *94*, 102901. [CrossRef]

18. Chen, R.; Nestor Cabrera-Munoz, E.; Lam, K.H.; Hsu, H.S.; Zheng, F.; Zhou, Q.F.; Shung, K.K. PMN-PT single crystal high frequency kerfless phased array. *IEEE Trans. Ultrason. Ferroelectr. Freq. Contr.* **2014**, *61*, 1033–1041. [CrossRef]

19. Wong, C.M.; Chen, Y.; Lou, H.S.; Dai, J.Y.; Lam, K.H.; Chan, H.L.W. Development of a 20-MHz wide-bandwidth PMN-PT single crystal phased-array ultrasound transducer. *Ultrasonics* **2017**, *73*, 181–186. [CrossRef]

20. Chiu, C.T.; Kang, B.J.; Eliahoo, P.; Abraham, T.; Shung, K.K. Fabrication and characterization of a 20-MHz microlinear phased-array transducer for intervention guidance. *IEEE Trans. Ultrason. Ferroelectr. Freq. Control* **2017**, *64*, 1261–1268. [CrossRef]

21. Bezanson, A.; Adamson, B.; Brown, J.A. Fabrication and performance of a miniaturized 64-element high-frequency endoscopic phased array. *IEEE Trans. Ultrason. Ferroelect. Freq. Contr.* **2014**, *61*, 33–43. [CrossRef] [PubMed]

22. Zhou, D.; Wang, F.; Luo, L.; Chen, J.; Ge, W.; Zhao, X.; Luo, H. Characterization of complete electromechanical constants of rhombohedral 0.72Pb(Mg$_{1/3}$Nb$_{2/3}$)-0.28PbTiO$_3$ single crystal. *J. Phys. D Appl. Phys.* **2008**, *41*, 185402. [CrossRef]

23. Zhu, B.P.; Zhu, Y.H.; Yang, J.; Ou-Yang, J.; Yang, X.F.; Li, Y.; Wei, W. New potassium sodium niobate single crystal with thickness independent high-performance for photoacoustic angiography of atherosclerotic lesion. *Sci. Rep.* **2016**, *6*, 39679. [CrossRef] [PubMed]

24. Zhang, T.; Ou-Yang, J.; Yang, X.; Wei, W.; Zhu, B.P. High Performance KNN-Based Single Crystal Thick Film for Ultrasound Application. *Electron. Mater. Lett.* **2019**, *15*, 1–6.

25. Zhang, S.; Li, F.; Sherlock, N.P.; Luo, J.; Lee, H.J.; Xia, R.; Meyer, R.J., Jr.; Hackenberger, W.; Shrout, T.R. Recent developments on high Curie temperature PIN–PMN–PT ferroelectric crystals. *J. Cryst. Growth* **2011**, *318*, 846–850. [CrossRef] [PubMed]

26. Li, X.; Ma, T.; Tian, J.; Han, P.; Zhou, Q.; Shung, K.K. Micromachined PIN-PMN-PT crystal composite transducer for high-frequency intravascular ultrasound (IVUS) imaging. *IEEE Trans. Ultrason. Ferroelectr. Freq. Contr.* **2014**, *61*, 1171–1178. [CrossRef]

27. Chen, Y.; Lam, K.H.; Zhou, D.; Cheng, W.F.; Dai, J.Y.; Luo, H.S.; Chan, H.L.W. High-frequency PIN-PMN-PT single crystal ultrasonic transducer for imaging applications. *Appl. Phys. A* **2012**, *108*, 987–991. [CrossRef]

28. *ANSI/IEEE Standard on Piezoelectricity*; ANSI/IEEE Std.: New York, NY, USA, 1987.

29. Li, L.; Xu, Z.; Xia, S.; Li, Z.; Ji, X.; Long, S. PIN-PMN-PT single-crystal-based 1-3 piezoelectric composites for ultrasonic transducer applications. *J. Electron. Mater.* **2013**, *42*, 2564–2569. [CrossRef]

30. Foster, F.S.; Zhang, M.Y.; Zhou, Y.Q.; Liu, G.; Mehi, J.; Cherin, E.; Harasiewicz, K.A.; Starkoski, B.G.; Zan, L.; Knapik, D.A.; et al. A new ultrasound instrument for in vivo microimaging of mice. *Ultrasound Med. Biol.* **2002**, *28*, 1165–1172. [CrossRef]

 *micromachines*

Article

Research on Asymmetric Hysteresis Modeling and Compensation of Piezoelectric Actuators with PMPI Model

Wen Wang [1], Jian Wang [1], Zhanfeng Chen [1,*], Ruijin Wang [1], Keqing Lu [1], Zhiqian Sang [1] and Bingfeng Ju [2]

[1] School of Mechanical Engineering, Hangzhou Dianzi University, Hangzhou 310018, China; wangwn@hdu.edu.cn (W.W.); wangjian_0826@126.com (J.W.); wangrjcn@163.com (R.W.); lkq@hdu.edu.cn (K.L.); sang@hdu.edu.cn (Z.S.)

[2] State Key Laboratory of Fluid Power Transmission and Control, Zhejiang University, Hangzhou 310027, China; mbfju@zju.edu.cn

* Correspondence: czf@hdu.edu.cn

Received: 31 January 2020; Accepted: 27 March 2020; Published: 30 March 2020

Abstract: Because of fast frequency response, high stiffness, and displacement resolution, the piezoelectric actuators (PEAs) are widely used in micro/nano driving field. However, the hysteresis nonlinearity behavior of the PEAs affects seriously the further improvement of manufacturing accuracy. In this paper, we focus on the modeling of asymmetric hysteresis behavior and compensation of PEAs. First, a polynomial-modified Prandtl–Ishlinskii (PMPI) model is proposed for the asymmetric hysteresis behavior. Compared with classical Prandtl–Ishlinskii (PI) model, the PMPI model can be used to describe both symmetric and asymmetric hysteresis. Then, the congruency property of PMPI model is analyzed and verified. Next, based on the PMPI model, the inverse model (I-M) compensator is designed for hysteresis compensation. The stability of the I-M compensator is analyzed. Finally, the simulation and experiment are carried out to verify the accuracy of the PMPI model and the I-M compensator. The results implied that the PMPI model can effectively describe the asymmetric hysteresis, and the I-M compensator can well suppress the hysteresis characteristics of PEAs.

Keywords: piezoelectric actuators (PEAs); asymmetric hysteresis; Prandtl–Ishlinskii (PI) model; polynomial-modified PI (PMPI) model; feedforward hysteresis compensation

---

1. Introduction

With the growth of semiconductor and precision manufacturing industries, the positioning accuracy of micro/nano scale is highly required [1–4]. Because of the advantages of fast frequency response, high stiffness and displacement resolution, piezoelectric actuators (PEAs) based on inverse piezoelectric effect have gradually become one of the most widely used smart material actuators. In addition, PEAs are popularly applied as the actuators in micro/nano scale measurement, micro-electro-mechanical systems (MEMS), flexible electronics manufacturing, and biomedical engineering. However, strong nonlinear hysteresis of piezoelectric materials makes the output of PEAs difficult to predict, and the positioning accuracy and stability of the system are low [5,6]. Specifically, scholars provided plenty of mathematical models to describe the symmetric hysteresis. However, effective mathematical model to describe the asymmetric hysteresis is still lacking, as the asymmetric hysteresis is a more common hysteresis nonlinear phenomenon. Therefore, it is worth exploring the modeling of asymmetric hysteresis behavior and using it to compensate piezoelectric actuators.

To eliminate the influence of hysteresis on accuracy and stability of the system, scholars have proposed numerous hysteresis models and controllers. These hysteresis models can be divided into two categories [7]: mechanistic model and phenomenological model. The former are based on basic physical principles, and derived by energy-displacement or stress–strain methods. Among them, the famous models are Jiles–Atherton model [8,9] and Domain Wall model [10]. The latter uses directly mathematical means to characterize the hysteresis, and ignore the physical meaning behind the hysteresis phenomenon. Because of the high accuracy and flexibility, the phenomenological model is more popular in hysteresis modeling. The phenomenological models include Preisach model [11,12], polynomial model [13,14], Bouc–Wen model [15,16], Duhem model [17,18], neural network model [19,20], Prandtl–Ishlinskii (PI) model [21–23], and etc. Among them, because of simple expression and analytical inverse model, the PI model is the most widely used in hysteresis modeling and compensation. However, the PI model utilizes weighted superposition of the Play operators to describe hysteresis nonlinearity; the Play operator is a basic component of the PI model. Therefore, the symmetric structure of the Play operator determines that PI model can only describe symmetric hysteresis.

Actually, PEAs have slight or severe asymmetric characteristics, as shown in Figure 1. When the application scope of PEAs scales down to micro/nano-meter levels, the gap between symmetric and asymmetric hysteresis modeling approaches results in positioning error. To describe asymmetric hysteresis, scholars [24–26] have tried to modify the PI model, and designed the corresponding controller. Kuhnen [24] presented the dead-zone operator and the modified PI (Ku-PI) model to describe the asymmetric hysteresis. Janaideh et al. [25] provided the generalized PI (GPI) model based on a nonlinear play operator. Different envelope functions make the GPI model have flexible ascending and descending edge. Hence, the GPI model can describe accurately the complex hysteresis phenomenon. By combining memoryless polynomial with PI model, Gu [26] proposed a new modified PI (Gu-PI) model to depict asymmetric hysteresis. These modified PI models and corresponding controller play a good role in the micro/nano positioning compensation. However, these models still have some limitations. The parameter identification of Ku-PI model is complex, which increases the difficulty of compensator design; GPI model is flexible and accurate, but choosing envelope function still depends on the experience and recursive debugging; Gu-PI model is simple in structure and easy to identify, and has good performance with one-side Play operator when the initial loading curve is not considered. But we have tried to extend the application of Gu-PI model with two-side Play operator when the initial loading curve is considered, and its effect is poor. The reason is that the Play operator lacks accuracy in describing displacements near the zero voltage.

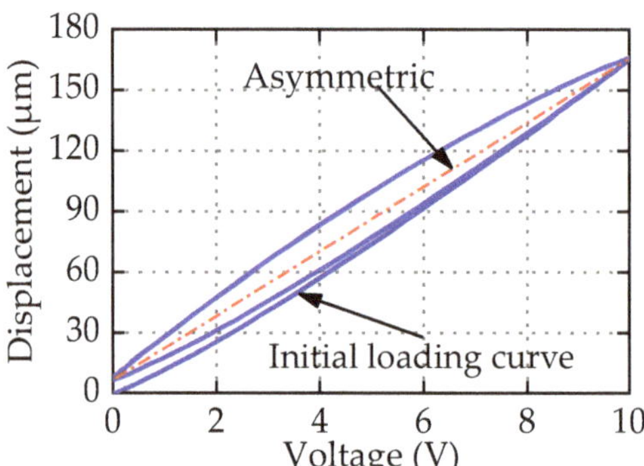

Figure 1. Asymmetric hysteresis loops of piezoelectric actuators (PEAs).

In this paper, a new polynomial-modified PI (PMPI) model is proposed to describe the asymmetric hysteresis of PEAs. Compared with PI model, the most important innovation of PMPI model is the introduction of Modified-Play (M-Play) operator and memoryless polynomial. M-Play operator replaces Play operator as the basic operator in the PMPI model. The memoryless polynomial enables PMPI model to describe the asymmetric hysteresis. The shape of asymmetric hysteresis is determined by both weighted M-Play operators and memoryless polynomial. The main advantages of PMPI model are as following: (1) Whether the initial loading curve is considered or not, in PMPI model, the displacement near zero voltage can be described flexibly by M-Play operator. (2) The inverse of PMPI model can be derived directly from PI model. The feedforward compensation of hysteresis in real-time application can be carried out. To validate the proposed PMPI model and I-M compensator, simulation and experiment are conducted on a piezoelectric micromotion platform.

This paper is organized as follows: Section 2 introduces PMPI model and examines the congruency property. Section 3 develops an I-M compensator and analyzes its stability. Section 4 verifies the PMPI model and I-M compensator on a piezoelectric micromotion platform. Section 5 concludes this paper.

2. Polynomial-Modified Prandtl–Ishlinskii Model

Before introducing the proposed PMPI model, it is necessary to review the PI model in brief.

2.1. Prandtl–Ishlinskii Model

The PI model utilizes weighted superposition of the Play operators to describe hysteresis nonlinearity, so the Play operator is the basic component of PI model. Without special description in the following paper, the Play operator refers to two-side Play operator. For any piecewise monotonic input signal $v(t) \in C_m(0\ t_N)$, the output $w(t) = F_r(v)(t)$ of the Play operator with threshold r is defined as

$$
\begin{aligned}
w(0) &= F_r(v)(0) = f_r(v(0), 0) \\
w(t) &= F_r(v)(t) = f_r(v(t), w(t_{i-1}))
\end{aligned}
\tag{1}
$$

for $t_{i-1} < t < t_i$, $1 < i < N$, with

$$
f_r(v, w) = \max(v - r, \min(v + r, w))
\tag{2}
$$

where $0 < t_1 < t_2 < \cdots < t_N$ is a division of the time domain $(0\ t_N)$ to ensure that the input signal $v(t)$ is monotonic within each subinterval $(t_{i-1}\ t_i)$. The Play operator have partial memory and rate-independent properties. That is, the output of the Play operator depends not only on current input, but also on the input history. However, the input rate does not change the output shape. As an illustration, Figure 2 shows the input–output relationship of play operator. It is easy to notice that there is a symmetry center in the input–output trajectory.

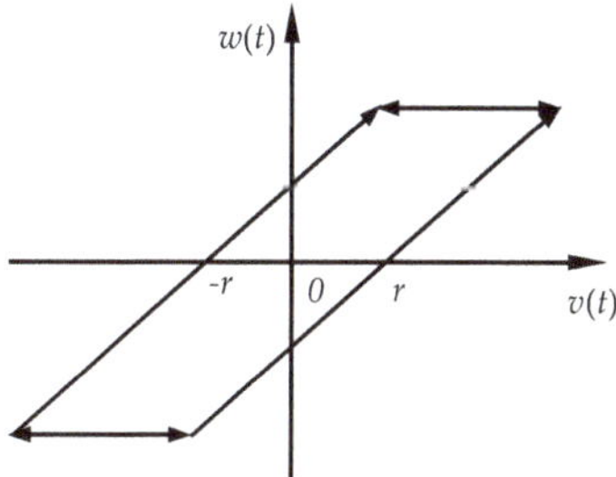

Figure 2. The input–output relationship of the Play operator.

The PI model utilizes the Play operator $F_r(v)(t)$ to describe the hysteresis relationship between input v and output Γ_{PI}:

$$
\begin{cases}
\Gamma_{PI}(v)(k) = p_0 v(k) + \sum_{i=1}^{n} p(r_i) F_{r_i}(v)(k) \\
s.t.\ p(r_i) \geq 0
\end{cases}
\tag{3}
$$

where n is the number of Play operators, p_0 is a positive constant. r_i is the threshold of the Play operator, $p(r_i)$ represents the weighted coefficient for the threshold r_i and approaches 0 as r_i becomes larger. If $p(r_i) < 0$, the PI model will fail to correctly describe hysteresis minor loops.

Observed from Equation (3), the PI model is composed of the weighted superposition of Play operator $F_r(v)(t)$ and linear input signal. The input–output curve is parallelogram, which is in central symmetry, so the PI model can only be used to characterize symmetric hysteresis. As an illustration, Figure 3 shows the hysteresis loop generated by the PI model with $p_0 = 2.1$, $p(r_i) = 2.6e^{-0.25ri}$, $r_i = 0{:}0.5{:}9.5$, and input $v(t) = 5 + 5\sin(2\pi t - \pi/2)$. It is obvious that the hysteresis curve generated by PI model is symmetrical. But PEAs often exhibit more or less asymmetric characteristics as shown in Figure 1. In addition, the max displacement of zero voltage in the PI model is uniquely determined by the initial loading curve, and the relationship can be expressed by Equation (4). In fact, the max displacement of the zero voltage is not necessarily related to the initial loading curve, which reflects that the Play operator lacks accuracy and flexibility in describing displacement near zero voltage. One of the motivations of the paper is to modify the PI model to flexibly and accurately represent the asymmetric hysteresis of PEAs.

$$y(0) = \sum_{i=1}^{n} p(r_i) r_i \tag{4}$$

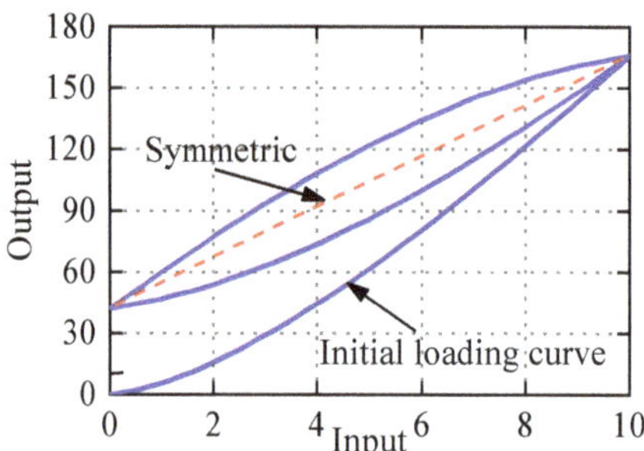

Figure 3. Hysteresis loops generated by the Prandtl–Ishlinskii (PI) model.

2.2. Polynomial-Modified Prandtl–Ishlinskii Model

Utilizing PI model to describe the asymmetric hysteresis of PEAs will produce considerable errors, which cannot meet the needs of precise positioning. To characterize accurately the asymmetric hysteresis, many modified PI models have been proposed, such as Ku-PI model, Gu-PI model, GPI model, and etc. Although these modified PI models can describe the asymmetric hysteresis to some extent, they still have some limitations. To improve the flexibility and accuracy of the model in describing asymmetric hysteresis, we propose Modified-Play (M-Play) operator and polynomial-modified PI model (PMPI).

The M-Play operator is derived from the Play operator, and it is formed by multiplying the threshold value of Play operator on the descending edge by a delay coefficient $\eta > -1$. The M-Play operator can be written as

$$\begin{cases} w(0) = F_{r,\eta_i}(v)(0) = f_{r,\eta_i}(v(0), 0) \\ w(t) = F_{r,\eta_i}(v)(t) = f_{r,\eta_i}(v(0), w(t_i)) \end{cases} \tag{5}$$

where

$$f_{r,\eta_i}(v, w) = \max(v - r, \min(v + \eta_i r, w)) \tag{6}$$

The coefficient η alters the threshold of the descending edge. The larger the η value, the more obvious the delay in the descending state. With proper values of η for every individual M-Play operator, the flexibility and accuracy of the model can be significantly enhanced.

Figure 4 shows the response of M-Play operators with different delay coefficient η. Obviously, one-side Play operator and two-side Play operator are special cases of M-Play operator. When $\eta = 0$, M-Play operator is equivalent to one-side Play operator, and when $\eta = 1$, M-Play operator is equivalent to two-side Play operator.

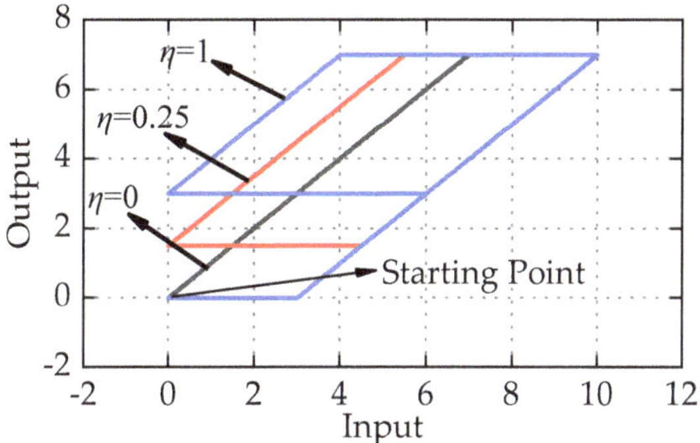

Figure 4. The response of M-Play operators with different delay coefficient η.

Remarks: From Figure 3, although the delay coefficient η is introduced, it can be seen from the input and output trajectories of M-Play operator still exist in the symmetry center. The weighted superposition of M-Play operators alone cannot characterize the asymmetric hysteresis, and the delay coefficient η is only used to improve the description of the displacement near zero voltage.

In this paper, the proposed PMPI model is formulated as:

$$\begin{cases} H(v)(k) = \Gamma_{PI,\eta}(v)(k) + P(v)(k) \\ \Gamma_{PI}(v)(k) = p_0 v(k) + \sum_{i=1}^{n} p_i F_{r_i,\eta_i}(v)(k) \\ P(v)(k) = a_1 v(k)^3 + a_2 v(k)^2 + a_3 \\ s.t.\ p_i \geq 0,\ \eta_i \geq -1 \end{cases} \tag{7}$$

where p_0 and $p(r_i)$ are defined the same as the ones in PI model (3), and $F_{r,\eta i}$ are the M-Play operator. A third-degree memoryless polynomial $P(v)(t)$ is used to characterize asymmetric hysteresis of PEAs.

$$y(0) = \sum_{i=1}^{n} p(r_i)\eta_i r_i + a_3 \tag{8}$$

The proposed PMPI consists of two parts: several weighted M-Play operators (denoted as MPI model) and memoryless polynomial $P(v)(t)$. The introduction of M-Play operator enables the PMPI model to describe accurately the displacement near zero voltage. It can be seen from Equation (8) that, in PMPI model, the max displacement of zero voltage is adjusted flexibly by the delay coefficient η. The combination of M-Play operators and third-order memoryless polynomial can characterize the asymmetric hysteresis of PEAs. To illustrate the advantages of PMPI model, Figure 5 shows the hysteresis loops generated by PMPI model (7) with $P(v)(t) = -0.05v^3 + 1.2v^2 + 0.2$ and $\eta = 0{:}0.025{:}0.475$. It is worth noting that weight function $p(r)$, positive constant p_0, and input $v(t)$ are the same as of the ones used in the hysteresis loops in Figure 3. In contrast, it is the M-Play operator and memoryless polynomial that enabled the PMPI model to characterize flexibly and accurately the serious asymmetric hysteresis.

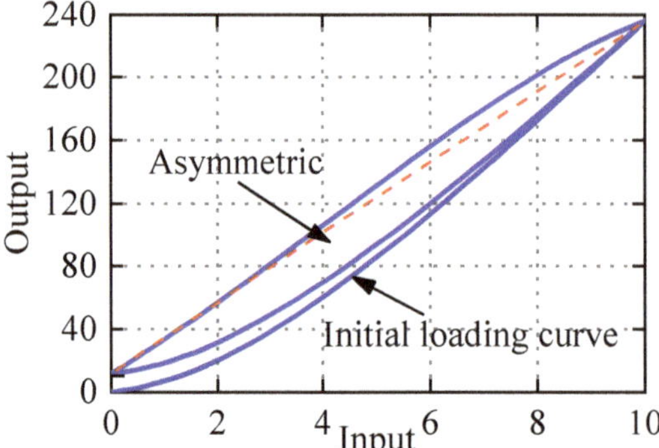

Figure 5. Hysteresis loops generated by polynomial-modified PI (PMPI) model.

2.3. Congruency Property

The congruency property is one of the basic properties of hysteresis model. The congruency property means that minor hysteresis loops with different input history are congruent in the same input range. In this section, we prove the congruency property of PMPI model and establish the minor loop mathematical model.

Because of the introduction of M-Play operator and memoryless polynomial $P(v)(t)$, the congruency property of PMPI model needs to be proven. Memoryless polynomial $P(v)(t)$ is a bijective function and has no partial memory, its output depends only on the current input. Therefore, the congruency property of PMPI model depends on the MPI model.

To illustrate the congruency property of MPI model, Figure 6 shows the hysteresis loops of M-Play operator with different input history in the same input range. Figure 6a,c shows two input signals with different input history in the same range, and Figure 6b,d shows the corresponding hysteresis curves. Combining these four graphs, it can be clearly seen that, although the input history of the two signals is different, the shapes of the corresponding minor loop formed by M-Play operators are same in the same input range ($v_m\ v_M$). This example illustrates the congruency property of the M-Play operator, and further demonstrates that MPI model has congruency property.

Sprekels's monograph [27] has proved that the shape of minor loop is uniquely determined when the vertical height h of minor loop are constant. Therefore, we quantitatively represent the shape of minor loop with its vertical height h. Figure 7 shows the geometric relations of the minor loop of M-Play operator. It can be seen from the figure that the relationship between vertical height h_r and interval length x can be formulated as:

$$h_r(x) = \max(x - (1 + \eta)r, 0) \tag{9}$$

where x is the interval length of input range ($v_m\ v_M$). When $x < (1 + \eta)r$, the M-Play operator cannot form minor loop, that is, $h_r = 0$. Based on Equation (8), the vertical height h_m of the minor loop of MPI model can be expressed as:

$$h_m(x) = p_0x + \sum_{i=1}^{n} p_ih_{r_i}(x) \tag{10}$$

As mentioned above, the output of memoryless polynomial function $P(v)(t)$ only depends on the current input, and there is no partial memory. The PMPI model is based on the MPI model plus $P(v)(t)$, so the vertical height h_{pm} of the minor loop of PMPI model can be expressed as:

$$h_{pm}(x) = p_0x + \sum_{i=1}^{n} p_ih_{r_i}(x) + [P(v_M) - P(v_m)] \tag{11}$$

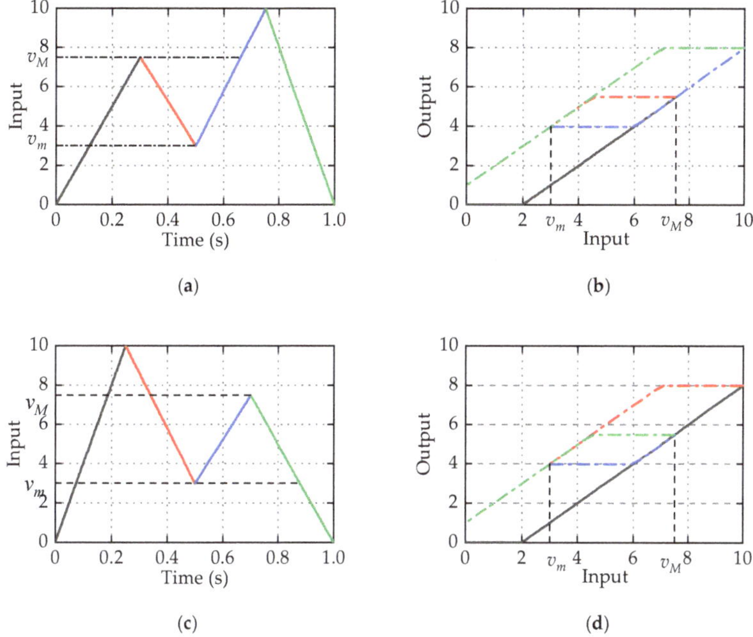

Figure 6. Hysteresis loops of M-Play operator with different input history in the same input range. (**a**) Input history 1; (**b**) hysteresis curve generated by input history 1; (**c**) input history 2; (**d**) hysteresis curve generated by input history 2

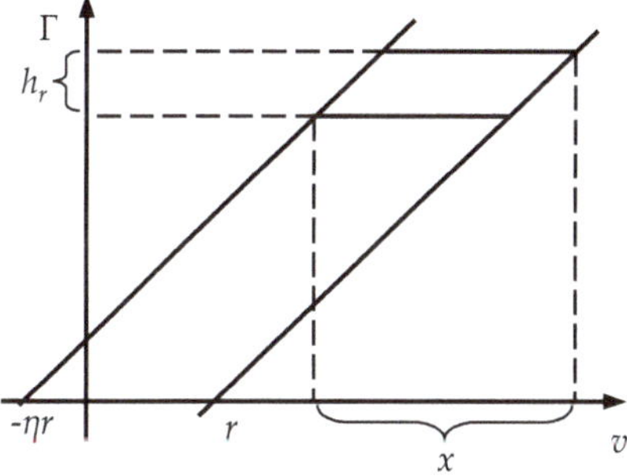

Figure 7. Geometric relations of the minor loop of M-Play operators.

For example, the input $v(t)$ is given to verify the correctness of Equation (11). In this case, the sequence of input maxima and minima is defined as $\{0{\to}7{\to}4{\to}10{\to}4{\to}7{\to}0\}$. Thus, the input voltage has the same input range (4 7). The parameters of the PMPI model are chosen as $r = 0{:}0.5{:}9.5$, $p_0 = 3.93$, $p_i = 2.5e^{-0.44ri}$, $\eta = 0{:}0.05{:}0.95$, $P(v)(t) = -0.02v^3 - 0.48v^2$. For the input voltage shown in Figure 8a, the associated hysteresis loops of PMPI model are given in Figure 8b. It can be seen from the partial enlarged drawing of Figure 8b that the shapes of the minor loops are identical in the same input range and the height of the minor loops are both 19.02 µm. According to the Equation (11) derived above, the vertical height of the minor loops also are 19.02 µm. This example verifies that the PMPI model also has congruency property and the minor loop mathematical model is correct.

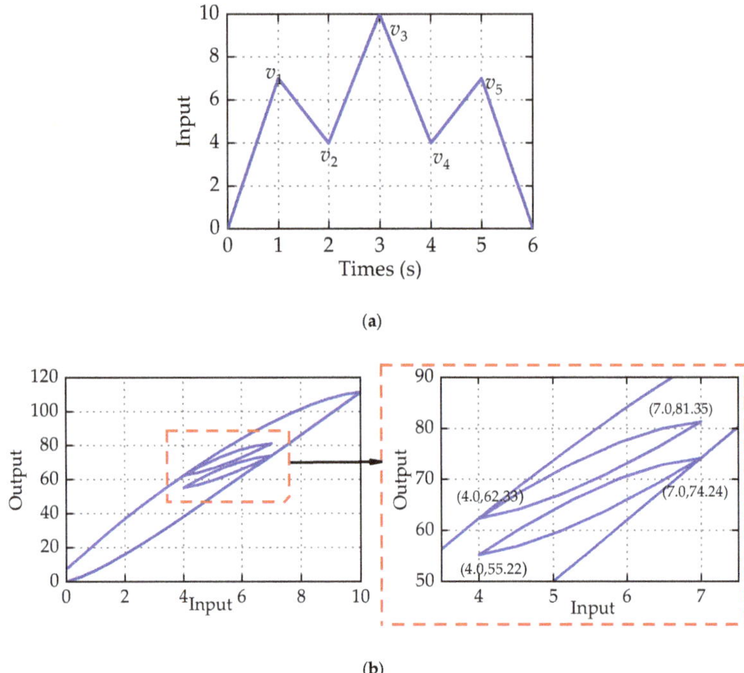

(a)

(b)

Figure 8. Simulation of the congruency property. (**a**) Input voltage. (**b**) Hysteresis loops.

3. The Design and Analysis of the Inverse Model Compensator

Feedforward compensation based on inverse hysteresis model is efficient and practical for reducing hysteresis effect in open-loop systems. The main idea of this method is to construct inverse hysteresis model, which is cascaded to the controlled object for feedforward control. After the cascade, the piezoelectric system can be approximated as a linear system, as shown in Figure 9. In this section, based on PMPI model, we will design the inverse model (I-M) compensator to compensate hysteresis and analyze its stability.

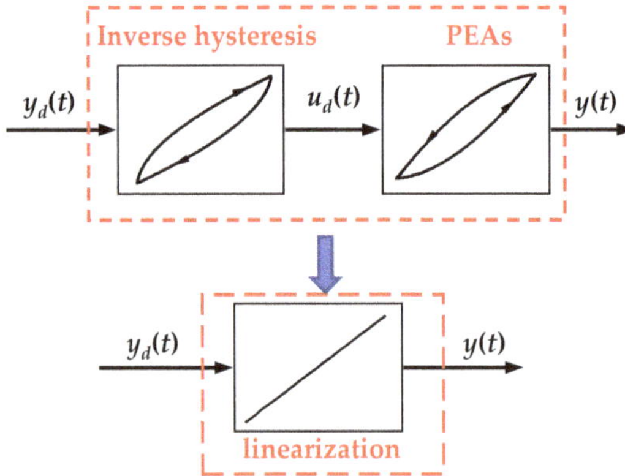

Figure 9. Schematic illustration of feedforward compensation based on inverse hysteresis model.

3.1. Inverse model Compensator Design

The MPI model introduces delay coefficient η on the basis of PI model, but this has no influence on the inverse MPI model, that is, the parameter expression of inverse MPI model is the same as that of inverse PI model. This is because that the parameter expression of inverse PI model is derived by the initial loading curve, the delay coefficient η only has influence on the descending edge but not the initial loading curve. The initial loading curve is still expressed as:

$$\varphi(r) = p_0 r + \int_0^r p(\zeta)(r - \zeta)d\zeta \tag{12}$$

The MPI model has an analytical inverse model, but the PMPI model has one more memoryless polynomial $P[u](k)$. Hence, there is no analytical inverse model. The inverse PMPI model can only be obtained by iterating the memoryless part. In this section we utilize the iterative structure to design a compensator as shown in Figure 10. In practical applications, most compensator systems are discrete, when the sampling frequency is sufficiently high, $u_i \approx u_{i-1}$. So the I-M compensator can be expressed as:

$$u_d(k) = H^{-1}[y_d](k) = \Gamma_{MPI}^{-1}(y_d[k] - P[u_d](k-1)) \tag{13}$$

where $y_d(k)$ is the desired displacement, and $u_d(k)$ is the desired control voltage. $\Gamma_{1\,MPI}^{-}[\cdot](k)$ stands for the inverse MPI model and can be calculated according to the following formula:

$$\Gamma_{MPI}^{-1}[y_d](k) = \hat{p}_0 y_d(k) + \sum_{i=1}^{n} \hat{p}_i F_{\hat{r}_i,\eta_i}[y_d](k) \tag{14}$$

with

$$
\begin{aligned}
\hat{p} &= \tfrac{1}{p_0},\ s.t.\ p_0 > 0 \\
\hat{p}_i &= -\frac{p_i}{(p_0 + \sum_{j=1}^{i} p_j)(p_0 + \sum_{j=1}^{i-1} p_j)} \\
\hat{r}_i &= p_0 r_i + \sum_{j=1}^{i-1} p_j(r_i - r_j)
\end{aligned}
\tag{15}
$$

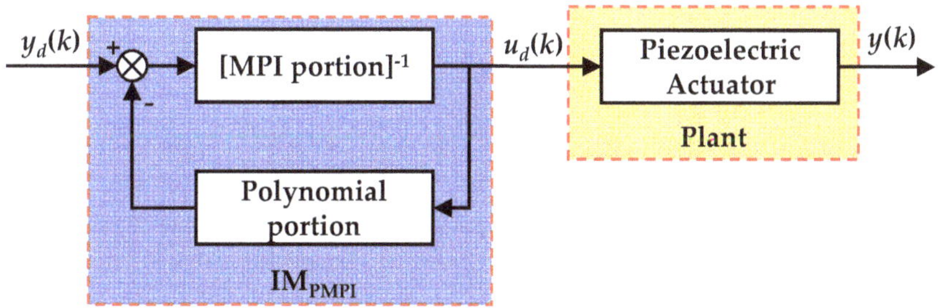

Figure 10. Schematic diagram of inverse model (I-M) compensator.

3.2. Stability Analysis

It can be seen from Figure 10 and Equation (11) that I-M compensator obtains the desired control voltage $u_d(k)$ by iterative solution. The divergence problem may occur during the iterative solution process. Therefore, it is necessary to analyze the stability of the I-M compensator. In this paper, the small gain theorem is used to analyze the stability. The small gain theorem states that when the maximum closed-loop gain of a closed-loop system satisfies |K|<1, then the system is stable in any case. The stable condition can be described as following:

$$\max\left\{K_{inv}(k)|K_{inv}(k) = \left|\frac{d\Gamma_{MPI}^{-1}[y_d](k)}{dy_d} \cdot \frac{dP[u_d](k)}{du_d}\right|\right\} < 1$$
$$\Rightarrow p_0 > \max\left|3a_1u_d(k)^2 + 2a_2u_d(k)\right| \tag{16}$$

where $K_{inv}[u](k)$ denotes the absolute gain of the I-M compensator, and p_0 a_1 and a_2 are the parameters of the PMPI model.

Proof: To satisfy Equation (16), the maximum value of the product of absolute gain $|dP[u_d](k)/du_d(k)|$ and $|d(\Gamma_{1\,MPI}^{-}[y_d](k)/d(y_d(k)))|$ should be less than 1. From Equation (6), we can get

$$0 \le d(F_{\hat{p}_i,\eta_i}[y_d](k)/d(y_d(k))) \le 1 \tag{17}$$

Combining Equations (15), the following relationship can be derived

$$\sum_{i=1}^{n} \hat{p}_i = \sum_{i=1}^{n}\left\{\frac{1}{p_0 + \sum\limits_{j=1}^{i} p_j} - \frac{1}{p_0 + \sum\limits_{j=1}^{i-1} p_j}\right\}$$
$$= \frac{1}{p_0 + \sum\limits_{i=1}^{n} p_i} - \frac{1}{p_0} = \frac{1}{p_0 + \sum\limits_{i=1}^{n} p_i} - \hat{p}_0 \Rightarrow -\hat{p}_0 < \sum_{i=1}^{n} \hat{p}_i < 0 \tag{18}$$

Therefore, the upper limit of the gain $d(\Gamma_{1\,MPI}^{-}[y_d](k)/d(y_d(k)))$ is $1/p_0$. Since the memoryless polynomial $P[u](k)$ is differentiable, its gain can be expressed by the following formula:

$$\left|dP[u_d](k)/du_d\right| = \left|3a_1u_d(k)^2 + 2a_2u_d(k)\right| \tag{19}$$

Combining Equations (18) and (19), it can be concluded that when the identified parameters of PMPI model meet the condition (16), I-M compensator is globally stable.

The I-M compensator using iterative structure has superior performance in accuracy and response speed, but it may appear that the parameter identified by PMPI model does not satisfy Equation (16). Once this happens, the proportional gain $k_e u_d(k)$ addressed in Equation (20) can be introduced to adjust the ratio between the MPI and memoryless polynomial part. In addition, the proportional gain $k_e u_d(k)$ can greatly improve the convergence speed of I-M compensator. Compared with the dichotomy of pure iterative, the I-M compensator utilizes the characteristic of the analytical inverse of the MPI model, and its iterative process is approximately open-loop. Hence, the I-M compensator has fewer iterative steps, faster convergence speed, and higher accuracy.

$$H[u_d](k) = (\Gamma_{MPI}[u_d](k) + k_e u_d(k)) + (P[u_d](k) - k_e u_d(k))$$
$$\Rightarrow p_0 + k_e > \max\left|3a_1u(k)^2 + 2a_2u(k) - k_e\right| \tag{20}$$

4. Experimental Verification and Discussion

In this section, an experimental platform is established. The experiment is conducted to verify the effectiveness of the PMPI model and I-M compensator in hysteresis modeling and compensation.

4.1. Experimental Setup

As shown in Figure 11a, the experimental platform consists of a computer, a USB-6259BNC (from National Instruments, Austin, TX, USA) data acquisition card, a 1-D piezoelectric micro-motion platform, and a piezoelectric servo controller E-625.CR (from Piezomechanik, München, Germany). The P-622.1CD (from Piezomechanik, München, Germany) has a maximum stroke of 200 μm and a built-in capacitive displacement sensor. The E-625.CR has a piezoelectric amplifier and displacement acquisition module. Its voltage amplification factor is 10 and the sensitivity of the displacement acquisition module is 20 μm/V. The USB-6259BNC has multiple 16-bit digital-to-analog converters and 16-bit analog-to-digital converters and cooperates with the host computer to realize the real-time control of the micro-motion platform. Figure 11b shows the process block diagram of the experimental system.

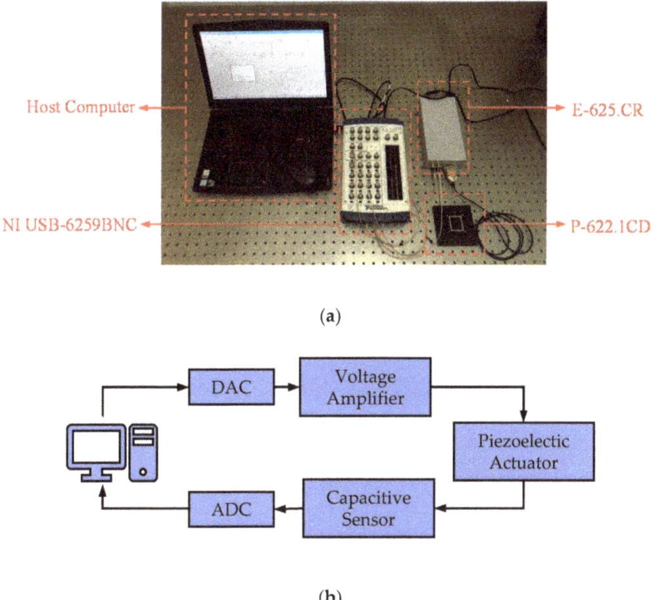

Figure 11. Experimental system. (**a**) Experimental platform; (**b**) process block diagram.

4.2. Asymmetric Hysteresis Description Results and Discussion

To experimentally validate the PMPI model, the first step is to identify the parameters of PMPI model. Model type and parameter identification both affect the accuracy of hysteretic modeling. Many identification algorithms [28–30] have been proposed to obtain model parameters, such as least square method (LSE), particle swarm optimization (PSO), and differential evolution (DE) algorithm. However, ensuring that the identified parameters are the global optimal solutions is a challenging task. In this section, the hybrid algorithm Nelder–Mead differential evolution (NM-DE) [31], based on differential evolution and simplex algorithm, is used to identify the parameters of the PMPI model. The NM-DE algorithm takes into account both global and local search capabilities, and has the advantages of fast convergence and high accuracy.

It should be noted that the larger the number of operators, the more accurately the model can describe the hysteresis in theory. Table 1 shows the relationship between the number of operators n, identification errors, and run time, where the runtime reflects indirectly the computation. From this Table, it can be observed that modest increase in the number of operator can improve the accuracy of the model, but further increase in the number of operator show no significant improvement in the accuracy of model, the identification errors are almost at the same level when n = 10,20,30. In addition, increase in the number of operators will increase the run time (computation) which further affects the real time performance of compensation. We select $n = 10$ for the case studies. As mentioned above, the weighting coefficient $p(r_i)$ approaches 0 as r_i becomes larger. The weighting coefficient $p(r_i)$ can be expressed as $p(r_i) = \alpha_1 e^{-\alpha_2 r_i}$. This form reduces the number of parameters to be identified and greatly reduces the identification burden.

Table 1. The relationship between the number of operators n, identification error, and run time.

Number of Operators n	Identification Error (μm)	Run Time (ms)
5	1.503	48.93
10	0.884	53.30
20	0.831	69.49
30	0.848	87.50

To demonstrate the superiority of PMPI model in characterizing asymmetric hysteresis, comparison of the three models PI, Gu-PI, and PMPI was carried out. The number of operators n is set to be 10, the thresholds are the same, and the parameters of models are the optimal values obtained after repeated identifications. The comparison experiments were carried out respectively in two cases (Case 1 and Case 2) as shown in Figure 12. The input–output curves of the three models appear to coincide because of the small modeling error. To directly reflect the superiority of PMPI model in hysteresis modeling accuracy, Figure 13a,b show respectively the modeling errors of the three models PI, Gu-PI, PMPI in two cases. In order to evaluate the accuracy of hysteresis model and quantify the modelling error, the maximum absolute error (MAE), the mean absolute deviation (MAD), and the root-mean-square error (RMSE) are defined as follows.

$$
\begin{cases}
\text{MAE} = \max_{1 \leq i \leq N} \left| \hat{y}(i) - y(i) \right| \\
\text{MRE} = \frac{MAE}{y_{\max}} \times 100\% \\
\text{MAD} = \frac{1}{N} \sum_{i=1}^{N} \left| \hat{y}(i) - y(i) \right| \\
\text{RMSE} = \sqrt{\frac{1}{N} \sum_{i=1}^{N} \left[\hat{y}(i) - y(i) \right]^2}
\end{cases}
\tag{21}
$$

where N is the number of samples, $y(i)$ is the real measured displacement, $y(i)$ is the model predicted displacement, and $y_{\max}$ is the maximum measured displacement. Among them, the MAE and MRE are used to evaluate local accuracy, and the MAD and RMSE are used to evaluate global accuracy.

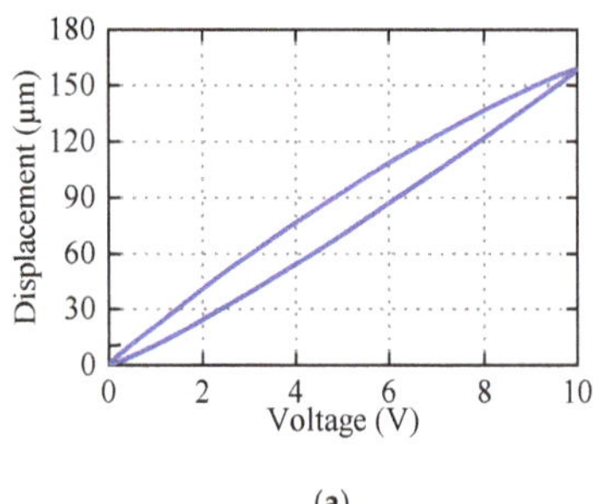

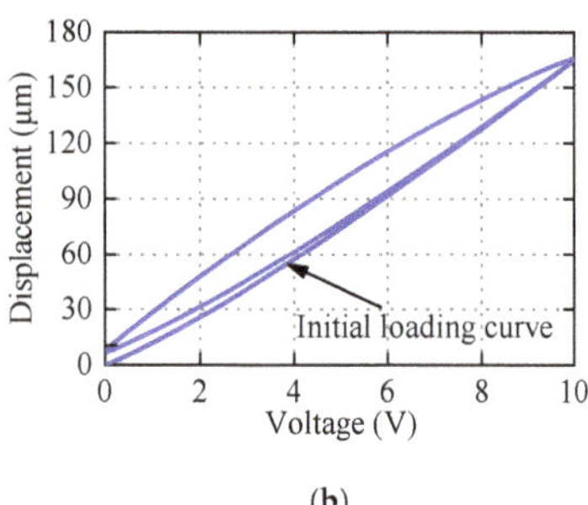

(a)　　　　　　　　　　　　(b)

Figure 12. Two cases of comparative experiment. (**a**) Case 1: the initial loading curve is not considered; (**b**) Case 2: the initial loading curve is considered.

The modeling error evaluation results of the three models in two cases are respectively listed in Tables 2 and 3. It can be seen from Table 2 that the prediction errors of the Gu-PI and PMPI model are significantly lower than that of the PI model in Case 1, and the MRE of prediction are only 0.968% and 0.698%. The result shows that the Gu-PI and PMPI model have obvious advantages in characterizing asymmetric hysteresis in Case 1. However, compared with PI and Gu-PI model, the accuracy of the PMPI model is significantly improved in Case 2, and the MAE of prediction is reduced by 83.3%. This is due to the lack of accuracy of Play operator in describing the displacement near zero voltage on the descending edge. This deficiency of Play operator shows that the local accuracy of Gu-PI model is approximately equal to PI model. The M-Play operator significantly improves the flexibility and accuracy of PMPI model. If the initial loading curve is not considered in hysteresis compensation, the compensator must make PEAs run for a period of time in advance, which will undoubtedly increase the burden of the compensator. In summary, Case 1 has high modeling accuracy, but it will increase the burden of the compensator. Case 2 has slightly low modeling accuracy, but the compensator has no such concern. The proposed PMPI model has superior modeling ability for hysteresis asymmetry in both cases.

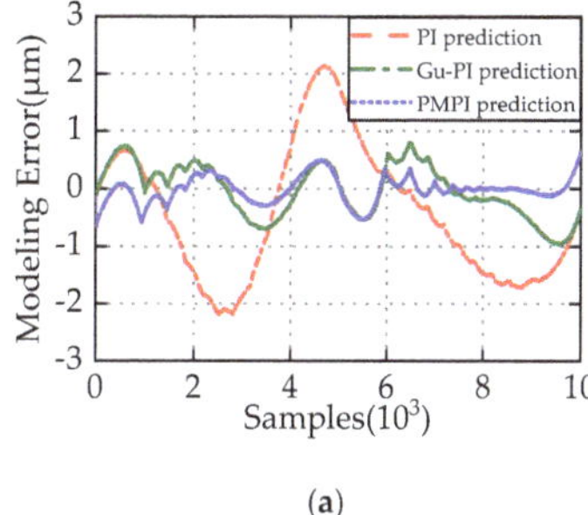

(a)

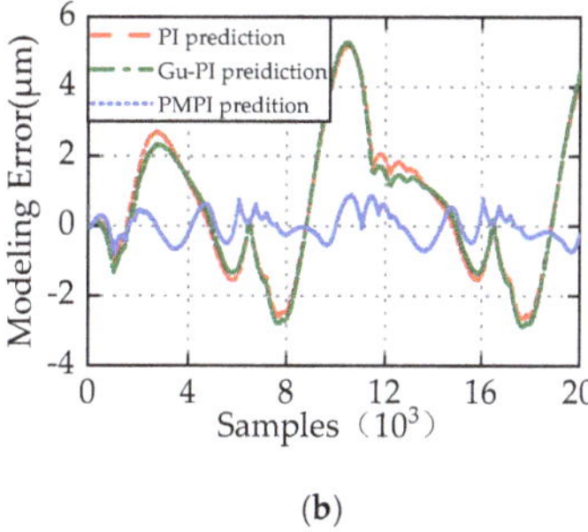

(b)

Figure 13. Performance comparison of three modeling methods in two cases. (**a**) Modeling errors in Case 1; (**b**) modeling errors in Case 2.

Table 2. Comparison of three model errors in Case 1.

Model	MAE (μm)	MRE (%)	MAD (μm)	RMSE (μm)
PI	2.186	1.37	1.058	1.243
Gu-PI	0.968	0.61	0.397	0.463
PMPI	0.698	0.44	0.172	0.232

Table 3. Comparison of three model errors in Case 2.

Model	MAE (μm)	MRE (%)	MAD (μm)	RMSE (μm)
PI	5.193	3.14	1.654	2.049
Gu-PI	5.280	3.19	1.627	2.038
PMPI	0.905	0.55	0.334	0.397

4.3. Hysteresis Compensation Results and Discussion

Table 4 lists the identified parameters of the PMPI model, the parameters of PMPI model satisfy the condition (14) in the range (0 10). Therefore, the I-M compensator is globally stable. To verify the effectiveness of the I-M compensator, the tracking experiment with periodic sinusoidal references with $y_r = 50 + 50\sin(2\pi t - \pi/2)$ is conducted. Figure 14a shows the comparison of the desired and actual trajectory. After compensation, the actual displacement can track the desired trajectory well, and no tracking loss occurs. Figure 14b shows the tracking errors, defined as the difference between the desired and actual trajectory. The MAE is 1.07 μm, the MRE is 1.07%, and the MAD is less than 0.4 μm. It is worth mentioning that, because of the existence of modeling uncertainty, the tracking errors appear periodic in periodic tracking experiments, which can be seen as systematic error, which can be eliminated by closed-loop control. To more intuitively reflect the compensation effect, Figure 14c shows the relationship between the desired and actual displacements. After compensation, the input–output shows an approximate linear relationship. The error is one order of magnitude less than that without any control, which shows that the I-M compensator can well suppress the hysteresis characteristics of PEAs.

Table 4. The identified parameters of PMPI model.

i	r_i	α_i	η_i	a_i
1	0	6.904	0.849	0.037
2	1	0.517	0.043	−0.647
3	2		0.276	0
4	3		0.376	
5	4		0.336	
6	5		0.443	
7	6	-	0.561	-
8	7		0.335	
9	8		0.158	
10	9		0.040	
p_0	4.770		-	

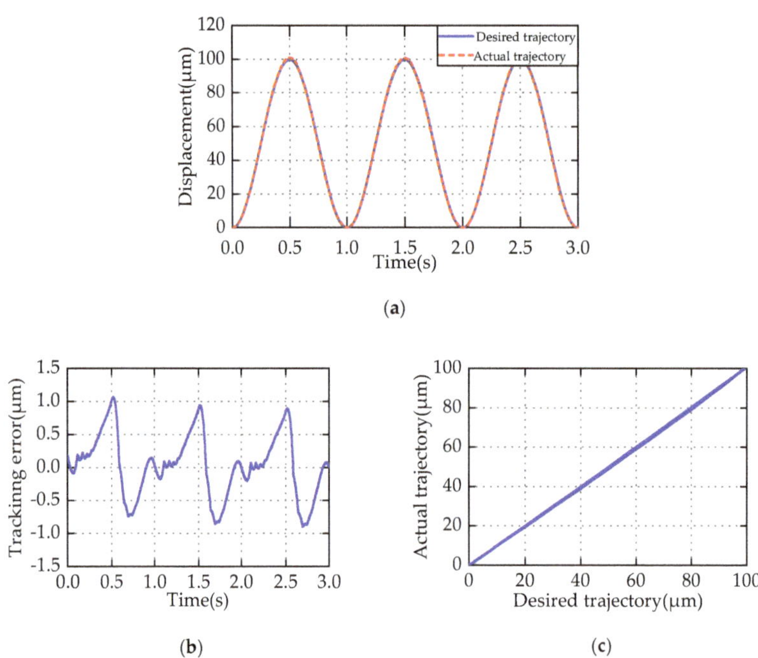

Figure 14. Periodic sinusoidal reference tracking experiment. (**a**) Trajectory tracking; (**b**) tracking error; (**c**) the relationship between desired and actual displacement.

To further verify the effectiveness of I-M compensator, a tracking experiment of frequency conversion attenuated triangular wave is performed. Figure 15 shows the results of this tracking experiment. It can be seen that the I-M compensator still has good tracking performance in tracking complex trajectory. The MRE is 1.18%, which is slightly larger than the ones of periodic sinusoidal. The experimental result further demonstrates the effectiveness of the I-M controller in hysteresis compensation.

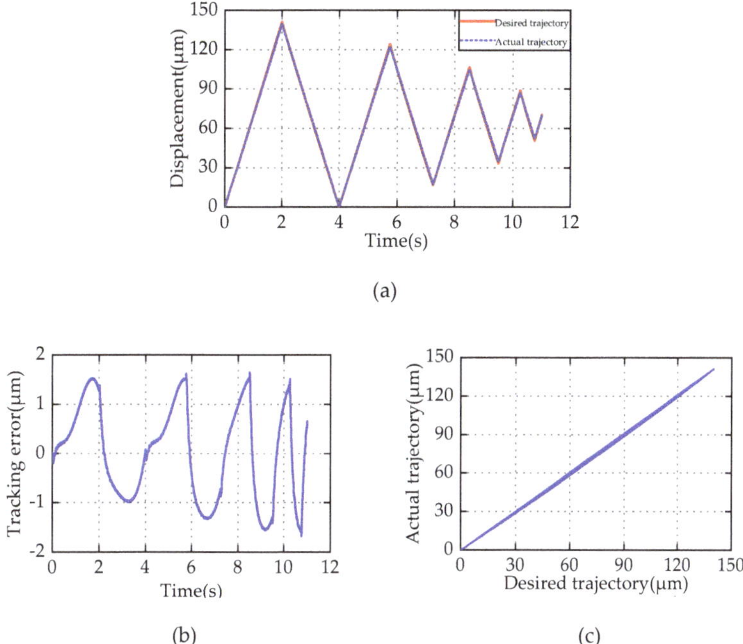

Figure 15. Frequency conversion attenuated triangular wave reference tracking experiment. (a) Trajectory tracking; (b) tracking error; (c) relationship between desired and actual displacement.

5. Conclusions

Because of its simple structure and analytical inverse, the PI model are used widely in hysteretic nonlinear modeling and compensation. However, the PI model can only describe symmetric hysteresis. In this paper, based on PI model, we provide a novel PMPI model to describe and compensate asymmetric hysteresis of PEAs. First, the PMPI model is introduced, which includes the M-Play operator and memoryless polynomial. Then, the congruency property of PMPI model is analyzed and verified. The minor loop mathematical model is also established. Next, the correctness of the PMPI model is proved by simulation. It should be noted that when considering the initial loading curve, the PMPI model can accurately characterize asymmetric hysteresis. Compared with the PI model and the Gu-PI model, the error of PMPI model is reduced by 83.3%. In the end, based on the PMPI model, the I-M compensator is designed for hysteresis compensation. The stability of I-M compensator is analyzed. The experimental results show that the I-M controller has superior tracking performance.

Although the PMPI model has satisfactory results for asymmetric hysteresis, it is not suitable for rate-dependent and load-dependent hysteresis. In the future, further research is to expand the PMPI model to rate-dependent and load-dependent applications.

Author Contributions: Conceptualization, W.W. and J.W.; methodology, J.W.; software, R.W.; validation, K.L. and R.W.; formal analysis, Z.S.; investigation, K.L.; resources, W.W.; data curation, Z.C.; writing—original draft preparation, W.W. and J.W.; writing—review and editing, Z.C. and Z.S.; visualization, J.W.; supervision, W.W.; project administration, W.W.; funding acquisition, W.W. and B.J.; All authors have read and agreed to the published version of the manuscript.

Funding: This research was supported by the Zhejiang Provincial Natural Science Foundation of China under Grant No. LZ16E050001, and the National Natural Science Foundation of China under Grant No. U1709206, No.51275465.

Conflicts of Interest: The authors declare no conflict of interest.

References

1. Devasia, S.; Eleftheriou, E.; Moheimani, S.O.R. A survey of control issues in nanopositioning. *IEEE Trans. Control Syst. Technol.* **2007**, *15*, 802–823. [CrossRef]
2. Lu, H.J.; Shang, W.F.; Xi, H.; Shen, Y.J. Ultrahigh-Precision Rotational Positioning Under a Microscope: Nanorobotic System, Modeling, Control, and Applications. *IEEE Trans. Robot.* **2018**, *34*, 497–507. [CrossRef]
3. Tanaka, Y. A Peristaltic Pump Integrated on a 100% Glass Microchip Using Computer Controlled Piezoelectric Actuators. *Micromachines* **2014**, *5*, 289–299. [CrossRef]
4. Zainal, M.A.; Sahlan, S.; Mohamed Ali, M.S. Micromachined Shape-Memory-Alloy Microactuators and Their Application in Biomedical Devices. *Micromachines* **2015**, *6*, 879–901. [CrossRef]
5. Li, Z.; Zhang, X.Y.; Su, C.Y.; Chai, T.Y. Nonlinear Control of Systems Preceded by Preisach Hysteresis Description: A Prescribed Adaptive Control Approach. *IEEE Trans. Control Syst. Technol.* **2016**, *24*, 451–460. [CrossRef]
6. Xu, Q.S.; Li, Y.M. Micro-/Nanopositioning Using Model Predictive Output Integral Discrete Sliding Mode Control. *IEEE Trans. Ind. Electron.* **2012**, *59*, 1161–1170. [CrossRef]
7. Mehrabi, R.; Kadkhodaei, M. 3D phenomenological constitutive modeling of shape memory alloys based on microplane theory. *Smart Mater. Struct.* **2013**, *22*, 025017. [CrossRef]
8. Hussain, S.; Lowther, D.A. Prediction of Iron Losses Using Jiles–Atherton Model With Interpolated Parameters Under the Conditions of Frequency and Compressive Stress. *IEEE Trans. Magn.* **2016**, *52*, 1–4. [CrossRef]
9. Raghunathan, A.; Melikhov, Y.; Snyder, J.E.; Jiles, D.C. Modeling the Temperature Dependence of Hysteresis Based on Jiles-Atherton Theory. *IEEE Trans. Magn.* **2009**, *45*, 3954–3957. [CrossRef]
10. Antonio, P.P.D.; de Campos, M.F.; Dias, F.M.D.; Campos, M.A.; Capo-Sanchez, J.; Padovese, L.R. Sharp Increase of Hysteresis Area Due to Small Plastic Deformation Studied With Magnetic Barkhausen Noise. *IEEE Trans. Magn.* **2014**, *50*, 1–4. [CrossRef]
11. Tousignant, M.; Sirois, F.; Kedous-Lebouc, A. Identification of the Preisach Model parameters using only the major hysteresis loop and the initial magnetization curve. In Proceedings of the 2016 IEEE Conference on Electromagnetic Field Computation (CEFC), Miami, FL, USA, 13–16 November 2016; p. 1.
12. Bi, S.S.; Wolf, F.; Lerch, R.; Sutor, A. An Inverted Preisach Model With Analytical Weight Function and Its Numerical Discrete Formulation. *IEEE Trans. Magn.* **2014**, *50*, 1–4. [CrossRef]
13. Bashash, S.; Jalili, N. A polynomial-based linear mapping strategy for feedforward compensation of hysteresis in piezoelectric actuators. *J. Dyn. Syst. Meas. Control* **2008**, *130*, 031008. [CrossRef]
14. Ru, C.; Sun, L. Improving positioning accuracy of piezoelectric actuators by feedforward hysteresis compensation based on a new mathematical model. *Rev. Sci. Instrum.* **2005**, *76*, 095111. [CrossRef]
15. Qin, H.; Bu, N.; Chen, W.; Yin, Z. An Asymmetric Hysteresis Model and Parameter Identification Method for Piezoelectric Actuator. *Math. Probl. Eng.* **2014**, *2014*, 1–14. [CrossRef]
16. Ding, B.X.; Li, Y.M. Hysteresis Compensation and Sliding Mode Control with Perturbation Estimation for Piezoelectric Actuators. *Micromachines* **2018**, *9*, 13. [CrossRef]
17. Wang, G.; Chen, G.Q. Identification of piezoelectric hysteresis by a novel Duhem model based neural network. *Sens. Actuator A-Phys.* **2017**, *264*, 282–288. [CrossRef]
18. Lin, C.J.; Lin, P.T. Tracking control of a biaxial piezo-actuated positioning stage using generalized Duhem model. *Comput. Math. Appl.* **2012**, *64*, 766–787. [CrossRef]
19. Tai, N.T.; Ahn, K.K. A hysteresis functional link artificial neural network for identification and model predictive control of SMA actuator. *J. Process Control* **2012**, *22*, 766–777. [CrossRef]
20. Nikdel, N.; Nikdel, P.; Badamchizadeh, M.A.; Hassanzadeh, I. Using Neural Network Model Predictive Control for Controlling Shape Memory Alloy-Based Manipulator. *IEEE Trans. Ind. Electron.* **2014**, *61*, 1394–1401. [CrossRef]
21. Qin, Y.D.; Zhao, X.; Zhou, L. Modeling and Identification of the Rate-Dependent Hysteresis of Piezoelectric Actuator Using a Modified Prandtl-Ishlinskii Model. *Micromachines* **2017**, *8*, 11. [CrossRef]
22. Sun, Z.; Song, B.; Xi, N.; Yang, R.; Hao, L.; Chen, L. Compensating asymmetric hysteresis for nanorobot motion control. In Proceedings of the 2015 IEEE International Conference on Robotics and Automation (ICRA), Seattle, WA, USA, 26–30 May 2015; pp. 3501–3506.
23. Li, Z.; Su, C.Y.; Chen, X. Modeling and inverse adaptive control of asymmetric hysteresis systems with applications to magnetostrictive actuator. *Control Eng. Pract.* **2014**, *33*, 148–160. [CrossRef]

24. Kuhnen, K. Modeling, Identification and Compensation of Complex Hysteretic Nonlinearities: A Modified Prandtl-Ishlinskii Approach. *Eur. J. Control* **2003**, *9*, 407–418. [CrossRef]
25. Janaideh, M.A.; Aljanaideh, O.; Rakheja, S. A Generalized Play Operator for Modeling and Compensation of Hysteresis Nonlinearities. In *Proceedings of Intelligent Robotics and Applications*; Springer: Berlin/Heidelberg, Germany, 2010; pp. 104–113.
26. Gu, G.-Y.; Zhu, L.-M.; Su, C.-Y. Modeling and Compensation of Asymmetric Hysteresis Nonlinearity for Piezoceramic Actuators With a Modified Prandtl–Ishlinskii Model. *IEEE Trans. Ind. Electron.* **2014**, *61*, 1583–1595. [CrossRef]
27. Sprekels, J.; Brokate, M. *Hysteresis and Phase Transitions*; Springer: Berlin/Heidelberg, Germany, 1996; Volume 121.
28. Chan, C.-H.; Liu, G. Hysteresis identification and compensation using a genetic algorithm with adaptive search space. *Mechatronics* **2007**, *17*, 391–402. [CrossRef]
29. Yang, M.; Gu, G.; Zhu, L.-M. Parameter Identification of the Generalized Prandtl-Ishlinskii Model for Piezoelectric Actuators Using Modified Particle Swarm Optimization. *Sens. Actuators A Phys.* **2012**, *189*. [CrossRef]
30. Liu, L.; Tan, K.K.; Teo, C.S.; Chen, S.; Lee, T.H. Development of an Approach Toward Comprehensive Identification of Hysteretic Dynamics in Piezoelectric Actuators. *IEEE Trans. Control Syst. Technol.* **2013**, *21*, 1834–1845. [CrossRef]
31. Lin, H. Hybridizing Differential Evolution and Nelder-Mead Simplex Algorithm for Global Optimization. In Proceedings of the 2016 12th International Conference on Computational Intelligence and Security (CIS), Wuxi, China, 16–19 December 2016; pp. 198–202.

 *micromachines*

Article

Motion of a Legged Bidirectional Miniature Piezoelectric Robot Based on Traveling Wave Generation

Jorge Hernando-García *, Jose Luis García-Caraballo, Víctor Ruiz-Díez and Jose Luis Sánchez-Rojas

Microsystems, Actuators and Sensors Group, Universidad de Castilla-La Mancha, E-13071 Ciudad Real, Spain; JoseLuis.GCaraballo@uclm.es (J.L.G.-C.); Victor.Ruiz@uclm.es (V.R.-D.); JoseLuis.SAldavero@uclm.es (J.L.S.-R.)
* Correspondence: jorge.hernando@uclm.es; Tel.:+34-926-295-300

Received: 21 February 2020; Accepted: 19 March 2020; Published: 20 March 2020

Abstract: This article reports on the locomotion performance of a miniature robot that features 3D-printed rigid legs driven by linear traveling waves (TWs). The robot structure was a millimeter-sized rectangular glass plate with two piezoelectric patches attached, which allowed for traveling wave generation at a frequency between the resonant frequencies of two contiguous flexural modes. As a first goal, the location and size of the piezoelectric patches were calculated to maximize the structural displacement while preserving a standing wave ratio close to 1 (cancellation of wave reflections from the boundaries). The design guidelines were supported by an analytical 1D model of the structure and could be related to the second derivative of the modal shapes without the need to rely on more complex numerical simulations. Additionally, legs were bonded to the glass plate to facilitate the locomotion of the structure; these were fabricated using 3D stereolithography printing, with a range of lengths from 0.5 mm to 1.5 mm. The optimal location of the legs was deduced from the profile of the traveling wave envelope. As a result of integrating both the optimal patch length and the legs, the speed of the robot reached as high as 100 mm/s, equivalent to 5 body lengths per second (BL/s), at a voltage of 65 V_{pp} and a frequency of 168 kHz. The blocking force was also measured and results showed the expected increase with the mass loading. Furthermore, the robot could carry a load that was 40 times its weight, opening the potential for an autonomous version with power and circuits on board for communication, control, sensing, or other applications.

Keywords: robot; piezoelectric; miniature; traveling wave; leg

1. Introduction

Control of locomotion in artificial structures is paramount for development in many disciplines. After the breakthroughs at the macroscale, there is a considerable interest in the scientific community for the development of miniature locomotion systems for multi-functional millimeter-to-centimeter scale robotic platforms capable of performing complex tasks for disaster and emergency relief activities, as well as the inspection of hazardous environments that are inaccessible to larger robotic platforms [1]. Miniaturization in the field of locomotion would result in advantages, such as smaller volume and mass, access to restricted volumes, interaction with same-sized targets, lower cost, profit of scaling laws, and so on [2].

When considering mechanisms of locomotion at the miniature scale, the use of wave-driven structures stands out because of its simplicity, reduced thickness, and low cost. The nature of the waves might be either traveling or standing. Furthermore, the generation of such waves can be easily accomplished with the help of piezoelectric materials, which can be integrated onto the structures during the fabrication process. A clear example of the potentiality of waves for locomotion are ultrasonic motors [3–5]. A history of commercial success was achieved using circular traveling wave-based stators transmitting its energy to a rotor [6,7], with the capability of bidirectional movement. Standing-wave-based motors were also reported, with the requirement of legs to induce the movement of the rotor [8], as well as the proper mixing of two standing waves at the same frequency, with either a bending and a longitudinal mode [9] or two bending modes [10].

In the field of mobile robots based on piezoelectric motors, the literature is more recent. Different reports combined piezoelectric materials with legs to attain locomotion [11–13]. For those robots whose movement relies on the generation of stationary waves, millimeter-sized legged devices have already been reported [14–16]. Furthermore, in the case of traveling waves, the state of the art involves the locomotion of 180-mm long plates actuated by piezoelectric patches, without using legs [17,18]. Taking these as the background, the present work aimed to further develop the traveling-wave-based locomotion of miniature plates with two piezoelectric patches. Miniaturization was accomplished by reducing the size of the structure down to 20 mm long and 3 mm wide. Furthermore, to pursue future untethered applications, the maximum applied voltage was limited to 65 V peak-to-peak (V_{pp}). This voltage limitation resulted in no locomotion of the bare structure, requiring the use of legs to induce movement at a voltage as low as 20 V_{pp}. In addition, the size and location of the patches that maximize the vertical displacement of the plate, while preserving the progressing nature of the generated wave, were calculated with the help of a 1D analytical model of the patches/plate system. This study completes the work begun in a previous study based on numerical finite element analysis [19], where the transversal displacement along the structure was analyzed for five different locations of a fixed-length patch. All these improvements are expected to reduce the miniaturization limits in the field of mobile robots.

2. Device Design

The first part of this section considers the effect of the device design and the excitation signals on the generation of a traveling wave in a plate with two piezoelectric patches. The electromechanical performance of piezoelectric layers on laminates was already well established by Lee and Moon's seminal work [20]. Later reports demonstrated the generation of linear traveling waves on a beam using external actuation forces [21,22], or with two piezoelectric patches on the same beam [23], by combining two vibration modes with appropriate amplitudes and phases. In the latter, the patches were symmetrically situated with respect to the center of the plate, where each of them was actuated with the same sinusoidal signal, but the phase was shifted. Here, we used a similar approach to design the robotic structures to be discussed in the next sections.

Figure 1 shows the schematic of the structure under study. It consisted of a supporting layer of glass with a length of 20 mm, width of 3 mm, and thickness of 1 mm. Piezoelectric patches of electroded lead zirconate titanate (PZT), 0.2 mm thick, covered a length L_{patch}. The thickness of the electrodes was neglected in the model. The patches started at the edges of the glass and both covered a given length to be determined by design, creating a symmetric configuration with respect to the center of the structure.

According to the basic mode superposition [24], the vertical displacement at any time and position along the length of the structure can be expressed as:

$$v(x,t) = \sum_{i=1}^{\infty} \varphi_i(x) \times T_i(t) \tag{1}$$

where φ_i are the shapes of the normalized flexural modes and T_i are the time-dependent modal coefficients. The previous expression can be truncated to the modes nearest to, above, and below the actuation frequency. The time-dependent modal factor can be Fourier-transformed to the following expression for each of the patches with an actuation voltage $Ve^{j(\omega t+\phi)}$ in the complex domain [20,23]:

$$T_i(\omega) = -\frac{Y_p d_{31}\left(\frac{w}{2}\right)t_p\left(t_p + t_s - 2z_n\right)\left(\varphi'_i(l1) - \varphi'_i(l2)\right)}{t_p\left(\omega_i^2 - \omega^2 + 2j\zeta\omega\omega_i\right)}Ve^{j\phi},$$

(2)

which is given by the product of the modal admittance and the complex amplitude of the actuation voltage, where Y_p is the Young modulus of the piezoelectric film; d_{31} is the piezoelectric coefficient; z_n is the neutral axis of the laminate structure; φ' is the first spatial derivative of the modal shape; $l1$ and $l2$ are the initial and final position of the patch, respectively; ω is the frequency of actuation; ω_i is the resonance frequency of mode i; and ζ is the damping ratio. The traveling wave (TW) envelope is taken as the magnitude of $v(x,\omega)$ for each position x at a given frequency. To simplify the calculation, the modal shapes and resonance frequencies were obtained analytically assuming a constant cross-section along the length of the structure, with the PZT covering the entire glass and the electrodes limited to the extension of the patch [25].

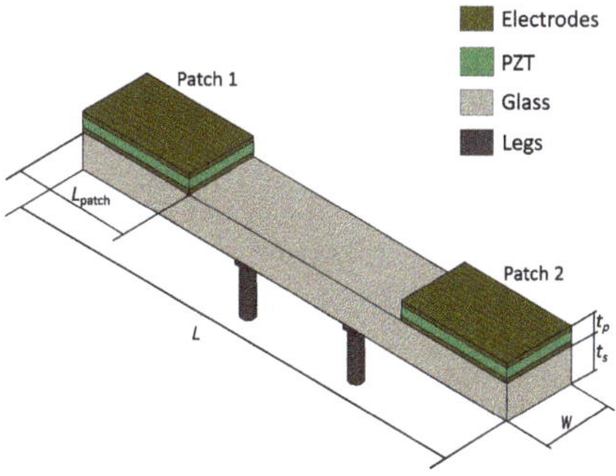

Figure 1. Representation of the fabricated structure. t_p and t_s are the thicknesses of the piezoelectric film and the substrate, respectively, W is the width of the structure, L is the length of the structure, and L_{patch} is the length of the patch.

Regarding the parameters of the model, the amplitude of the voltage applied to the patches was 10 V, the d_{31} piezoelectric coefficient of PZT was 180 pm/V, and the damping factor of the modes was equal to 0.001, corresponding to a quality factor of 500. Table 1 shows the rest of the parameters.

Table 1. Structural properties of the materials.

Material	Thickness (mm)	Young's Modulus (GPa)	Density (kg/m^3)
Glass	1	72.9	3350
PZT	0.2	62	7800

We then considered two figures of merit: (i) standing wave ratio (SWR), defined as the ratio of the maximum to the minimum value of the TW envelope, which is related to the quality of the traveling wave: the closer to 1, the better the traveling wave; and (ii) the average displacement of the TW envelope, named <TW>, which is associated with the speed or energy of the wave. In the rest of the section, SWR and <TW> are taken in a central window of the total length corresponding to 60% to remove the effect of the boundary conditions at the edges.

This approach resulted in three key variables that may be varied to improve the quality and amplitude of the TW: patch length, frequency of actuation, and phase shift between the sinusoidal signals on each of the patches. Malladi et al. [23] reported the effect of the frequency of actuation and the phase shift. Here, we also considered the effect of the patch length on the two figures of merit mentioned before.

First, we studied the dependence of SWR and <TW> on the patch length with a fixed frequency of actuation (average of the two consecutive resonant frequencies) and phase shift (90°). Figure 2 shows the results. Leissa´s nomenclature is used to identify the modes of vibration (the first digit is the number of nodal lines along the length of the plate and the second digit is the number of nodal lines along the width of the plate) [25]. Three different frequencies of actuation were considered depending on the couple of modes to be mixed: half-way between modes (30) and (40), (50) and (60), and (90) and (100); named f_{3-4}, f_{5-6}, and f_{9-10}, respectively. The effect of varying these two variables, namely frequency and phase, is shown below.

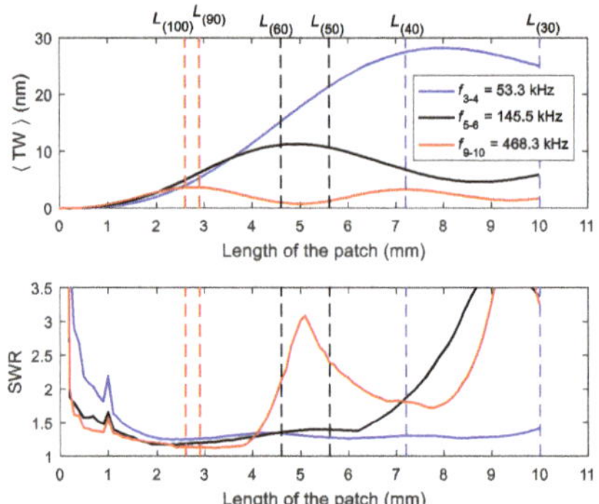

Figure 2. Average displacement of the traveling wave envelope (<TW>) and standing wave ratio (SWR) as a function of the length of the patch for three different frequencies of actuation f_{3-4}, f_{5-6}, and f_{9-10}, and a phase shift of 90° between patches. f_{m-n}: mid-frequency between modes (m0) and (n0). The vertical line $L_{(n0)}$ represents the first zero, excluding the edge of the plate, of the second derivative of the modal shape (n0), where n will vary depending on the mode under consideration.

There was a clear dependence of the figures of merit on the length of the patch, and for each of the frequencies, there was an optimal length of patch L_{m-n}, where m refers to mode (m0) and n to mode (n0), that provided a maximum of <TW>: L_{3-4} = 8.3 mm for f_{3-4}, L_{5-6} = 4.9 mm for f_{5-6}, L_{9-10} = 2.8 mm for f_{9-10}, while maintaining a value for the SWR close to the ideal value of 1.

Furthermore, the information of Figure 2 was correlated with the second derivative of the modal shape along the length of the structure. As already reported, optimal actuation of a given mode can be attained using an electrode distribution that covers only the regions of the surface where the sign of the second derivative of the mode shape (associated with stress on the surface) is either positive or negative [26]. The dashed vertical lines in Figure 2 represent the position of the first zero (excluding the edge of the structure) of the second derivative of the three pairs of modal shapes involved in the calculations. A patch of length $L_{(n0)}$, where n will vary depending on the mode under consideration, would optimally actuate this individual mode (n0). Note that the optimum patch length for the TW generation with modes m–n, L_{m-n}, lays half-way between lengths $L_{(m0)}$ and $L_{(n0)}$ as a balance between the optimal patches for each individual mode contributing the most to the TW: $L_{m-n} \approx (L_{(m0)} + L_{(n0)})/2$. Therefore, as a rule of thumb, the second derivative of the modal shapes can be used to determine the best patch length without requiring any complex simulation approach.

For the optimum patch length deduced before, we then considered whether the figures of merit could be improved by varying the frequency of actuation while maintaining the 90° phase shift. Figure 3 shows the results for the combination of modes (50) and (60) as an example. The best SWR was very close to the mid-frequency f_{5-6}, and the displacement varied only slightly; therefore, there was very little room for improvement by changing the frequency around the mid-frequency under these conditions. This conclusion was also reached for other combinations of modes.

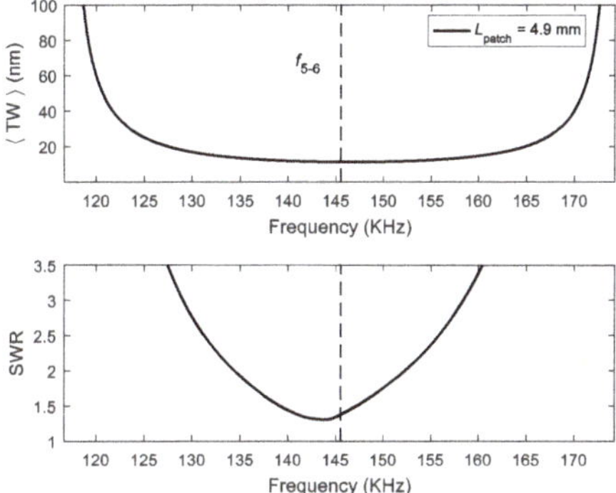

Figure 3. <TW> and SWR as a function of the frequency of actuation for the optimum patch length $L_{5\text{-}6}$ and a phase shift of 90° between the signals applied to the patches.

Now we focus on the effect of both the frequency of actuation and the phase shift for the optimal patch length. Figure 4 shows a surface plot of <TW>, corresponding to the combination of modes (50) and (60), as a function of frequency and phase shift as a color map for different SWR isolines. It can be seen that an increase in <TW> was only possible by deteriorating the SWR and that there were many possible couples of frequency and phase leading to a similar <TW> and SWR, as shown with the isoline corresponding to SWR = 1.4, for example. This is crucial for avoiding undesirable torsional modes that might be in between the two flexural modes under study and hinder the TW generation based on the two-mode-approximation. This was further illustrated by plotting the envelope of the TW for the mixing of modes (50) and (60) under three different combinations of frequency and phase shift (Figure 5).

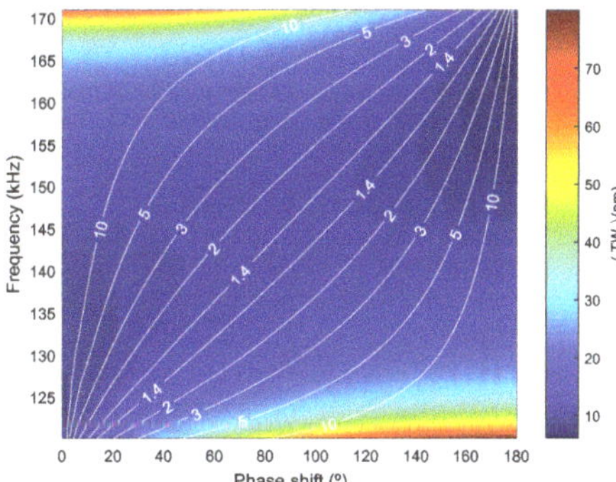

Figure 4. Surface plot representation of <TW> for the combination of modes (50) and (60) as a function of actuation frequency and phase shift, with white isolines corresponding to different values of SWR. The <TW> value is represented by the color scale.

Once the design guidelines for the patch size, the actuation frequency, and the phase shift were decided upon, legs were included in the design. This addition was experimentally required to observe stable locomotion of the structures at voltages below the maximum applied, which was 65 V_{pp}. This differed from previous studies [17,18], where locomotion was attained without the help of legs. We attributed this need for legs to the smaller size and weight in our case, as well as to the limit imposed on the maximum voltage applied. Legs amplify the horizontal displacement [27], and at the same time, confine the point of contact to specific areas of the robot. A pertinent question here is where to locate the legs on the glass plate. A key design goal of propagating-wave-based locomotion is the elliptical trajectory of the surface particles of the elastic body due to the coupling of longitudinal and transverse motions with the appropriate phases to achieve bidirectionality using the 90° shift of the actuation signals. Therefore, the leg positions should coincide with those surface points where elliptical trajectories are realized. To determine those locations on the structure, we note that the vertical displacement $v(x, t)$ along the plate can be expressed as follows:

$$v(x, t) = A(x) \cdot \cos(\omega t + \theta(x)), \tag{3}$$

which resembles a pure TW, but instead of a constant amplitude, we have a position-dependent amplitude $A(x)$ (the envelope of the TW defined above), and instead of a phase term proportional to the position, there is a general function of the phase $\theta(x)$. Furthermore, the horizontal displacement, u, at the bottom face of the plate where the legs are to be placed is [28]:

$$u(x, t) = -h \frac{\partial v(x, t)}{\partial x} = -h \left[\frac{dA(x)}{dx} \cdot \cos(\omega t + \theta(x)) - A(x) \cdot \frac{d\theta(x)}{dx} \cdot \sin(\omega t + \theta(x)) \right], \tag{4}$$

where h is the thickness between the neutral plane of the structure and the bottom face of the plate. For the displacement at the tip of the leg, the same expression holds, but the length of the leg is added to the thickness h.

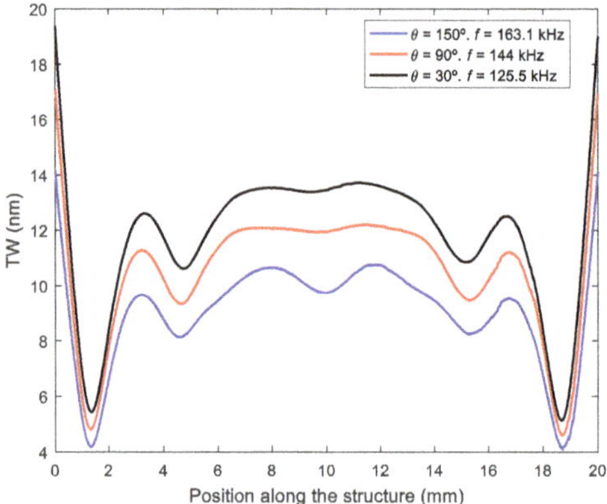

Figure 5. Envelope of the TW along the length of the structure for the combination of modes (50) and (60) under three different conditions of frequency of actuation f and phase shift θ. The optimal patch length $L_{5\text{-}6}$ was used.

According to the two previous expressions, only those positions along the length of the structure where the derivative of $A(x)$ is null (local maximum or minimum) will present an elliptical trajectory. That is to say, at those positions where the TW envelope is almost constant, the surface particles, and hence the tip of the legs, will describe elliptical trajectories. By returning to Figure 5, we notice that the positions to consider were those located at the central plateau of the TW envelope. If the legs are located where the derivative of $A(x)$ is significant, a linear-like displacement is then obtained, as for standing wave linear motors.

3. Materials and Methods

Next, we focus our attention on the fabrication procedure. Figure 6 shows a picture of one of the fabricated structures. A piece of glass with a length of 20 mm, width of 3 mm, and thickness of 1 mm was obtained from a glass slide (VWR International, Radnor, PA, USA) via machine (Buehler, Lake Bluff, IL, USA) drilling. Two PZT patches (PI Ceramic GmbH, Lederhose, Germany) with a thickness of 200 µm and a width of 3.5 mm (slightly larger than the plate to allow for contact to the bottom face) were glued to the glass using a cyanoacrylate adhesive (Loctite, Düsseldorf, Germany).

The actual length of the fabricated patches was 5 mm, close to the optimum value of 4.9 determined in Figure 2, for an efficient actuation of modes (50) and (60). The robots were powered externally using 25-micron wires connected to the piezoelectric patches.

Figure 6. Photograph of a legged structure and U-shaped pairs of legs of 0.5, 1, and 1.5 mm. The numbered ruler marks represent centimeters.

U-shaped pairs of legs were 3D printed with a stereolithography (SLA) B9 Core printer (B9Creations, Rapid City, SD, USA) using a material named Black Resin (B9Creations, Rapid City, SD, USA) (see Figure 6). The pair of legs were glued along the width of the plate. Samples with two and three pairs of legs were fabricated. In the case of the two pairs of legs, these were located at the edges of the TW envelope plateau mentioned earlier. For the three pairs of legs, the third pair was located at the center of the structure. The shape of the legs was cylindrical, 0.6 mm in diameter, with varying lengths from 0.5 mm to 1.5 mm. The mass of the robot was about 240 mg.

4. Results

Figure 7 shows the electrical conductance of a robot with 0.5-mm legs. Two peaks can be clearly identified corresponding to modes (50) and (60). Figure S1, included in the Supplementary Materials, compares the conductance of this device with and without legs. There was almost no difference between the two measurements, which corroborates the negligible impact of the legs on the standing waves corresponding to the modes. Once the resonant frequencies were known, the frequency of actuation was adjusted manually. For this sample, the actuation frequency was set to 161 kHz, close to the mid-frequency between the measured modes (50) and (60). It is important to notice that the frequency of actuation, 161 kHz, differed from the estimated frequency f_{5-6}, by just 10%. This difference might be attributed to the limitations of the 1D model at representing the 3D structure of the robot, as well as to uncertainties in the mechanical parameters of the materials. Table S1 of the Supplementary Materials compares the resonant frequency of different modes, found using both experimentation and calculated using the 1D model and a 3D finite element analysis. It also shows the values for $L_{(nn)}$, which is the first zero of the second derivative of the modal shapes deduced by the 1D and 3D models.

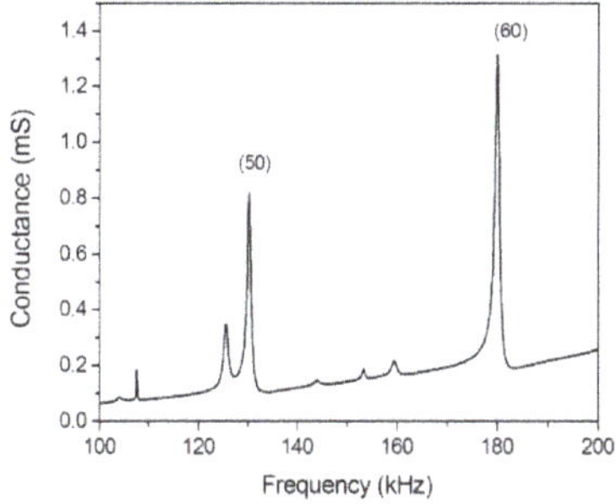

Figure 7. Electrical conductance of a robot with 0.5-mm legs.

Next, we present the characterization of the fabricated robots in terms of speed and force. Figure 8 shows the speed of the robot versus the applied voltage. Ten measurements were taken at each voltage and the standard deviation was about ±3.5 mm/s. The phase shift between patches was fixed to either 90° or −90° to confirm the bidirectional movement. Video S1, included with the Supplementary Materials, shows how the direction was reversed by changing the phase. The set-up for the speed measurement consisted of two infrared LEDs separated by 100 mm, where each was aligned with a photodiode. The set-up allowed for measurement of the time required by the robot to travel 100 mm along a rail on glass by tracking the light interruption events when the robot passed below the infrared LEDs with a frequency counter. Robots with legs of 1.5 mm showed a less uniform speed, with difficulties in maintaining the rectilinear displacement, which might be related to the interference of an intrinsic mode of vibration of the legs. The modes of vibration of 0.5- and 1-mm legs were far away from the frequency of actuation. When comparing robots with 0.5- and 1-mm legs, a better performance was observed for the 1-mm leg, which might be attributed to an enhancement of the horizontal displacement at the tip of the leg, as mentioned previously. For the maximum voltage applied, namely 65 V_{pp}, the velocity for the 1-mm-legged structure reached 60 mm/s, which was equivalent to 3 BL/s (body lengths per second). These results are comparable to the state-of-the-art in miniature soft robotics, with performances similar to arthropods [29]. Furthermore, notice that the minimum voltage required to initiate movement with 1-mm legs was as low as 20 V_{pp}, which might facilitate the implementation of an untethered robot with an integrated driving signal.

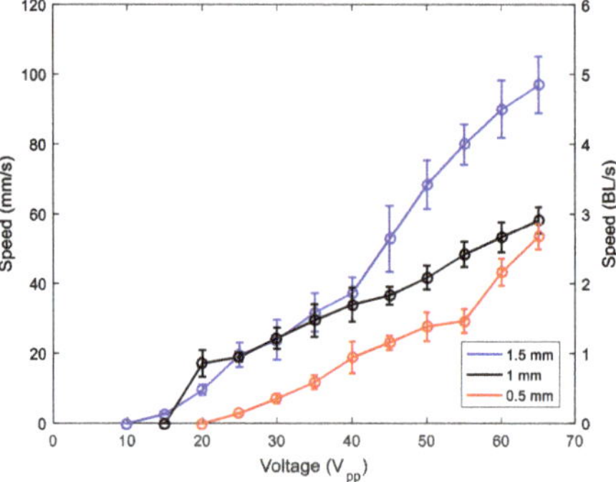

Figure 8. Speed of the robots versus applied voltage for legs with different lengths: 0.5 mm (red), 1 mm (black), and 1.5 mm (blue). Circles represent experimental data and are joined with lines for guidance purposes. BL/s: Body lengths per second.

Furthermore, we investigated the effect of mass loading on the performance of the robot. Figure 9 shows the speed versus applied voltage for different loading masses. The robot carried a mass of 7.5 g, which was 40 times its weight, at a speed of about 40 mm/s at the maximum voltage applied. This result shows the potential to incorporate electronic circuits on board, for communication, control, sensing, or other applications. Video S2 shows the locomotion with a mass of 7.5 g.

To complete the characterization of the robot, Figure 10 displays the blocking force under different mass loadings. The force was measured while the robot contacted a force sensor (Honeywell FSG Series, Morris Plains, NJ, USA) with the actuation voltage applied. As expected, the blocking force increased as the mass loading increased [30].

Finally, Figure 11 shows the comparison between two and three pairs of 1-mm legs. A clear improvement can be seen when using three pairs of legs, with a speed as high as 5 BL/s. Further investigations are in progress to study the effect of increasing the number of legs.

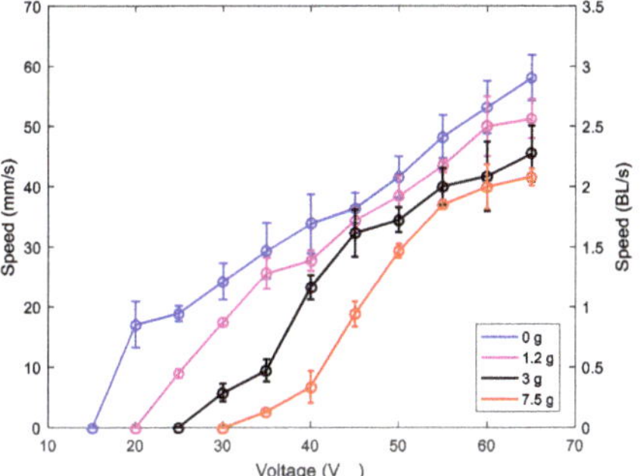

Figure 9. Speed of the 1-mm-legged robot versus applied voltage for different masses: no load (blue), 1.2 g (pink), 3 g (black), and 7.5 g (red).

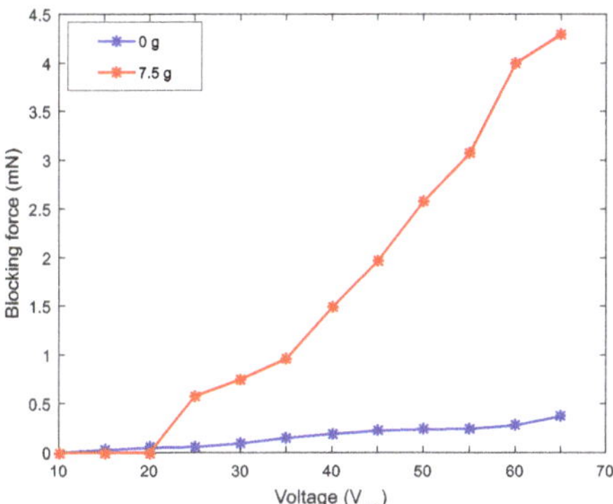

Figure 10. Blocking force of the robot for different masses: no load mass (blue) and 7.5 g (red).

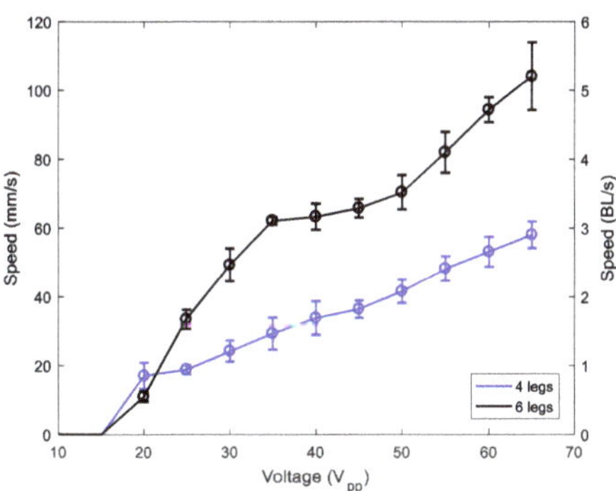

Figure 11. Speed of the 1-mm-legged robot versus applied voltage for two (blue) and three (black) pairs of legs.

5. Conclusions

This paper contributes to the development of miniature mobile robots based on TWs generated by the actuation of symmetrically located piezoelectric patches. Guidelines were proposed for the design of the patches and the driving signals. 3D printed legs were implemented in our devices, which is an approach commonly restricted to standing-wave-based systems. The combination of the optimal patch length and legs resulted in a mobile rigid robot with a speed of 5 BL/s at a voltage of 65 V_{PP}, with the capability of transporting 40 times its weight.

Supplementary Materials: The following are available online at http://www.mdpi.com/2072-666X/11/3/321/s1, Figure S1: Electrical conductance of a 0.5-mm-legged robot compared with the case without legs. Table S1: Resonant frequency and position of the first zero of the second derivative of the shape of modes (40), (50), (60), and (70) for three cases: experiment, 1D analytical model, and 3D finite element analysis. Video S1: Bidirectional locomotion of a 1-mm-legged structure with no load. Video S2: Bidirectional locomotion of a 1-mm-legged structure with a load of 7.5 g.

Author Contributions: Conceptualization, J.L.S.-R.; funding acquisition, J.H.-G. and J.L.S.-R.; investigation, J.L.G.-C.; project administration, J.H.-G. and J.L.S.-R.; software, V.R.-D.; supervision, J.H.-G. and J.L.S.-R.; writing—original draft, J.H.-G.; writing—review and editing, J.H.-G. and J.L.S.-R. All authors have read and agreed to the published version of the manuscript.

Funding: This work was supported by the European Regional Development Fund, the Spanish Ministerio de Ciencia, Innovación y Tecnología project (RTI2018-094960-B-100), and the regional government (JCCLM) project (SBPLY/17/180501/000139).

Conflicts of Interest: The authors declare no conflict of interest.

References

1. Developing Microrobotics for Disaster Recovery and High-Risk Environments. Available online: https://www.darpa.mil/news-events/2018--07--17 (accessed on 17 February 2020).

2. Driesen, W. Concept, Modeling and Experimental Characterization of the Modulated Friction Inertial Drive (MFID) Locomotion Principle: Application to Mobile Microrobots. Ph.D. Thesis, École polytechnique fédérale de Lausanne, Lausanne, Switzerland, June 2008.

3. Uchino, K. Piezoelectric ultrasonic motors: Overview. *Smart Mater. Struct.* **1998**, *7*, 273–285. [CrossRef]

4. Watson, B.; Friend, J.; Yeo, L. Piezoelectric ultrasonic micro/milli-scale actuators. *Sens. Actuators A* **2009**, *152*, 219–233. [CrossRef]

5. Uchino, K. *MicroMechatronics*, 2nd ed.; CRC Press: Boca Raton, FL, USA, 2020.

6. Sashida, T. Motor Device Utilizing Ultrasonic Oscillation. U.S. Patent 4,562,374, 31 December 1985.

7. Hagedorn, P.; Wallaschek, J. Travelling wave ultrasonic motors, Part I: Working principle and mathematical modelling of the stator. *J. Sound Vib.* **1992**, *155*, 31–46. [CrossRef]

8. He, S.; Chen, W.; Tao, X.; Chen, Z. Standing wave bi-directional linearly moving ultrasonic motor. *IEEE Trans. Ultrason. Ferroelectr. Freq. Control* **1998**, *45*, 1133–1139. [CrossRef] [PubMed]

9. Bein, T.; Breitbach, E.J.; Uchino, K. A linear ultrasonic motor using the first longitudinal and the fourth bending mode. *Smart Mater. Struct.* **1997**, *6*, 619–629. [CrossRef]

10. Chen, J.; Chen, Z.; Li, X.; Dong, S. A high-temperature piezoelectric linear actuator operating in two orthogonal first bending modes. *Appl. Phys. Lett.* **2013**, *102*, 052902. [CrossRef]

11. Dharmawan, A.G.; Hariri, H.H.; Soh, G.S.; Foong, S.; Wood, K.L. Design, analysis, and characterization of a two-legged miniature robot With piezoelectric-driven four-bar linkage. *J. Mech. Robot.* **2018**, *10*, 021003. [CrossRef]

12. Shannon, A.; Rios, S.A.; Andrew, J.; Fleming, A.J.; Yong, Y.K. Miniature resonant ambulatory robot. *IEEE Rob. Autom. Lett.* **2017**, *2*, 337–343. [CrossRef]

13. Patel, K.; Qu, J.; Oldham, K.R. Tilted leg design for a rapid-prototyped low-voltage piezoelectric running robot. In Proceedings of the International Conference on Manipulation, Automation and Robotics at Small Scales (MARSS), Nagoya, Japan, 4–8 July 2018. [CrossRef]

14. Ferreira, A.; Fontaine, J.-G. Dynamic modeling and control of a conveyance microrobotic system using active friction drive. *IEEE ASME Trans. Mechatron.* **2003**, *8*, 188–202. [CrossRef]

15. Son, K.J.; Kartik, V.; Wickert, J.A.; Sitti, M. An ultrasonic standing-wave-actuated nanopositioning walking robot: Piezoelectric-metal composite beam modeling. *J. Vib. Control* **2006**, *12*, 1293–1309. [CrossRef]

16. Hariri, H.H.; Prasetya, L.A.; Foong, S.; Soh, G.S.; Otto, K.N.; Wood, K.L. A tether-less legged piezoelectric miniature robot using bounding gait locomotion for bidirectional motion. In Proceedings of the IEEE International Conference on Robotics and Automation (ICRA), Stockholm, Sweden, 16–21 May 2016. [CrossRef]

17. Hariri, H.; Bernard, Y.; Razek, A. A traveling wave piezoelectric beam robot. *Smart Mater. Struct.* **2014**, *23*, 025013. [CrossRef]

18. Hariri, H.; Bernard, Y.; Razek, A. 2-D Traveling wave driven piezoelectric plate robot for planar motion. *IEEE ASME Trans. Mechatron.* **2018**, *23*, 242–251. [CrossRef]

19. Hariri, H.; Bernard, Y.; Razek, A. Dual piezoelectric beam robot: The effect of piezoelectric patches' positions. *J. Intell. Mater. Syst. Struct.* **2015**, *26*, 2577–2590. [CrossRef]

20. Lee, C.-K.; Moon, F.C. Modal sensors/actuators. *J. Appl. Mech.* **1990**, *57*, 434–441. [CrossRef]

21. Loh, B.G.; Ro, P.I. An object transport system using flexural ultrasonic progressive waves generated by two-mode excitation. *IEEE Trans. Ultrason. Ferroelectr. Freq. Control* **2000**, *47*, 994–999. [CrossRef]

22. Gabai, R.; Bucher, I. Excitation and sensing of multiple vibrating traveling waves in one-dimensional structures. *J. Sound Vib.* **2009**, *319*, 406–425. [CrossRef]

23. Malladi, V.V.N.S.; Avirovik, D.; Priya, S.; Tarazaga, P. Characterization and representation of mechanical waves generated in piezoelectric augmented beams. *Smart Mater. Struct.* **2015**, *24*, 105026. [CrossRef]

24. Norton, M.P.; Karczub, D.G. *Fundamentals of Noise and Vibration Analysis for Engineers*, 2nd ed.; Cambridge University Press: The Edinburgh Building, Cambridge, UK, 2003; pp. 85–91.

25. Leissa, A.W. *Vibration of Plates. Scientific and Technical Information Division*; National Aeronautics and Space Administration: Washington, DC, USA, 1969.

26. Ruiz-Díez, V.; Manzaneque, T.; Hernando-García, J.; Ababneh, A.; Kucera, M.; Schmid, U.; Seidel, H.; Sánchez-Rojas, J.L. Design and characterization of AlN-based in-plane microplate resonators. *J. Micromech. Microeng.* **2013**, *23*, 074003. [CrossRef]

27. Roh, Y.; Lee, S.; Han, W. Design and fabrication of a new travelling wave-type ultrasonic linear motor. *Sens. Actuators A* **2001**, *94*, 205–210. [CrossRef]

28. Kuribayashi, M.; Ueha, S.; Mori, E. Excitation conditions of flexural traveling waves for a reversible ultrasonic linear motor. *J. Acoust. Soc. Am.* **1985**, *77*, 1431–1435. [CrossRef]

29. Wu, Y.; Yim, J.K.; Liang, J.; Shao, Z.; Qi, M.; Zhong, J.; Luo, Z.; Yan, X.; Zhang, M.; Wang, X.; et al. Insect-scale fast moving and ultrarobust soft robot. *Sci. Robot.* **2019**, *4*, eaax1594. [CrossRef]

30. Roh, Y.; Kwon, J. Development of a new standing wave type ultrasonic linear motor. *Sens. Actuators A* **2004**, *112*, 196–202. [CrossRef]

Article

System-Level Model and Simulation of a Frequency-Tunable Vibration Energy Harvester

Sofiane Bouhedma [1,*], Yongchen Rao [1,2], Arwed Schütz [2], Chengdong Yuan [1,2], Siyang Hu [1,2], Fred Lange [1], Tamara Bechtold [1,2] and Dennis Hohlfeld [1]

[1] Institute for Electronic Appliances and Circuits, Faculty of Computer Science and Electrical Engineering, University of Rostock, Albert-Einstein-Str. 2, 18059 Rostock, Germany; yongchen.rao@uni-rostock.de (Y.R.); chengdong.yuan@jade-hs.de (C.Y.); siyang.hu@jade-hs.de (S.H.); fred.lange@uni-rostock.de (F.L.); tamara.bechtold@jade-hs.de (T.B.); dennis.hohlfeld@uni-rostock.de (D.H.)

[2] Department of Engineering, Jade University of Applied Sciences, Friedrich-Paffrath-Str. 101, 26389 Wilhelmshaven, Germany; arwed.schuetz@jade-hs.de

* Correspondence: sofiane.bouhedma@uni-rostock.de; Tel.: +49-381-498-7233

Received: 10 December 2019; Accepted: 11 January 2020; Published: 14 January 2020

Abstract: In this paper, we present a macroscale multiresonant vibration-based energy harvester. The device features frequency tunability through magnetostatic actuation on the resonator. The magnetic tuning scheme uses external magnets on linear stages. The system-level model demonstrates autonomous adaptation of resonance frequency to the dominant ambient frequencies. The harvester is designed such that its two fundamental modes appear in the range of (50,100) Hz which is a typical frequency range for vibrations found in industrial applications. The dual-frequency characteristics of the proposed design together with the frequency agility result in an increased operative harvesting frequency range. In order to allow a time-efficient simulation of the model, a reduced order model has been derived from a finite element model. A tuning control algorithm based on maximum-voltage tracking has been implemented in the model. The device was characterized experimentally to deliver a power output of 500 μW at an excitation level of 0.5 g at the respected frequencies of 63.3 and 76.4 Hz. In a design optimization effort, an improved geometry has been derived. It yields more close resonance frequencies and optimized performance.

Keywords: piezoelectricity; vibration-based energy harvesting; multimodal structures; frequency tuning; nonlinear resonator; bistability; magnetostatic force

1. Introduction

The term energy harvesting is the process of capturing or harvesting wasted or unused ambient energy, such as temperature gradients, mechanical vibrations, radio frequencies, etc., and converting it into useable electrical energy. This differs from powering a system using traditional energy sources such as batteries, fuel cells, etc. Energy harvesting has received much attention in the last two decades from various disciplines, mainly due to its potential impact as a key technology enabling autonomous ultra-low-power electronics operating in remote and harsh environments without the need for large energy storage elements. Energy harvesting thus replaces conventional batteries, enhances the environmental friendliness of the system, and lowers the maintenance costs. In other words, it is a promising technology to power various applications ranging from structural health monitoring to medical implants.

Energy harvesting from mechanical sources including industrial machines, human activity, vehicles, building structures, and other environmental sources can be achieved through different conversion methods. Such approaches include piezoelectric, electromagnetic, electrostatic, and magnetostrictive effects. Among the most promising sources for recovering energy are periodic vibrations generated by rotating machinery or environmental sources such as wind or ocean waves.

The present work addresses vibration-based energy harvesting systems based on the piezoelectric effect. Up to now, the piezoelectric effect has been extensively investigated and integrated in different applications, such as sensors and actuators. Vibration energy harvesting converts vibration energy into electrical energy. The large availability of vibrations in industrial environments and the simultaneous need for sensing applications in such locations motivated the research in recent time.

Most often, a mechanical resonator is used to amplify the low vibration levels into usable deflections. Such systems consist usually of a tip-loaded clamped-free cantilever [1–3], which we refer to as 'first generation harvester concepts'. The drawback of such harvesters is that they operate efficiently only if the harvester's resonance frequency coincides with the dominant ambient vibration frequency. Any difference between these frequencies will lead to a decrease in the power output. As realistic ambient vibration spectra exhibit multiple frequencies and vary over time as the vibration source is aging or changing in temperature for instance, 'first generation' energy harvesting schemes fail. In order to overcome this limitation, many research groups are addressing new resonator designs with an optimized active bandwidth, which enables the harvester to collect power on a broader frequency range. This research can be categorized in two groups. On one hand, numerous multimodal resonator designs, which are able to operate resonantly at multiple frequencies, have been introduced in [4–7]. Wu et al. [8] investigated a compact piezoelectric energy harvester, comprising of one main cantilever beam and an inner secondary cantilever beam. The system harvests power at two distinct frequencies. A novel trident (three-pronged spear) shaped piezoelectric energy harvester has been proposed by Upadrashta and Yang in [9] to collect power from wideband, low frequency, and low amplitude ambient vibrations. Lamprecht et al. [10] investigated how multiple vibration harvesters can be combined to a macroscopic array configuration by aiming not on high resonance peak powers, but on close spectral overlaps in lower power regions with a bandwidth 500 Hz. In [11] a multiresonant structure comprising a clamped–clamped piezoelectric fiber composite generator has been proposed by Qi et al., with side mounted cantilevers, which are tuned by added masses to resonate at individual frequencies, resulting in a wider harvesting bandwidth. Other works [12–14] proposed multiple concepts of multiresonant piezoelectric energy-harvesting devices, capable of harvesting power on a wider frequency range using both translational and rotational degrees of freedom.

On the other hand, other groups proved that integrating nonlinearity in the harvesting device broadens the operative bandwidth compared to the standard linear harvesters. The authors of [15–21] demonstrated the usability of bistable resonators for harvesting over a wider operational frequency range, by integrating permanent magnets positioned with respect to another permanent magnet on the resonator. In [22] a compact nonlinear multistable energy harvester array has been presented by Lai et al., for harvesting energy at low frequencies. Hoffmann et al. [23,24] developed a harvesting system capable to autonomously adapt the magnetic field strength, acting on the resonator, by adjusting the orientation of a diametrically polarized permanent magnet. Another bistable rotational energy harvester operating at low frequencies has been proposed by Fu and Yeatman in [25,26]. The harvesting is achieved by magnetic plucking of a piezoelectric cantilever using a driving magnet mounted on a rotating platform. In [27–29] an autonomous tuning mechanism has been developed, allowing a compensation of the hysteresis, as well as maintaining the optimal working point. It allows the use of both coupling modes (attractive and repulsive), which enables the harvester to adapt its operating frequency to the dominant vibration frequency of the environment. Nammari et al. [30] proposed an enhanced novel design of a nonlinear magnetic levitation-based energy harvester, where the tuning effect is achieved by the magnetic and the oblique springs. In [31–34] we studied a dual-frequency piezoelectric energy harvester incorporating permanent magnets for bidirectional frequency tuning and presenting an increased useable frequency range compared to standard harvesters. Other groups used different bandwidth broadening strategies, for example, shunted piezoelectric control systems have been proposed in [35,36]. Zhang and Afzalul [37] analyzed a broadband energy harvester with an array of piezoelectric bimorphs mechanically connected through springs. The operative bandwidth broadening can be achieved by carefully selecting the masses and adjusting the spring stiffness.

In the present paper, we propose a so-called 'folded beam' resonator design for energy harvesting. It consists of an outer tip-loaded beam, mechanically connected to a pair of inner beams, which extend towards the fixed end. In contrast of [8], the resonator is fabricated from a single steel sheet. It uses permanent magnets at the free ends of the cantilevers for higher frequency agility, by making use of the nonlinearity provided for the nonlinear magnetic forces. It resonates at two distinct frequencies, spaced 10 Hz apart, in the range (50,100) Hz. Bidirectional frequency tuning was achieved using a similar approach as presented in [34]. It has been demonstrated that both modes can be tuned independently, together with the dual frequency feature, can provide a superior frequency agility and increase the operational bandwidth of the system compared to existing approaches. Furthermore, an optimized version of the resonator is presented in this paper, where the two power peaks could be nearer to each other and provide the same power levels. This feature further enhances the performances of such a harvester and shows the benefit of such a design compared to an array of simple harvesters. Due to the close resonances enabling the resonator to amplify the deformation of the outer beam and leading to a higher strain distribution on the inner beam compared to the aforementioned simple design, where the second cantilever will be directly connected to the same excitation. A control scheme, which uses an energy efficient maximum- amplitude tracking algorithm, is optimized compared to [34]. A reduced model of the harvester has been developed and enabled us to fully simulate the behavior of the harvester in question. The position-dependent magnetic force obtained from magnetostatic simulations presented in [33] have been implemented together with the reduced model in ANSYS twin builder. The system is able to autonomously choose and tune the closest mode to the dominant vibration frequency to maintain maximum possible oscillation amplitude. Furthermore, we dropped the self-adaptive step size tuning control scheme due to the potential of higher power requirements needed for the fine-tuning operation. Finally, we experimentally characterized the behavior of the first prototype of such a piezoelectric harvester design and the result matched with the simulation results in terms of frequency and expected voltage output.

2. Dual Frequency Piezoelectric Energy Harvester

This section reviews shortly the dual-frequency harvester proposed in [32–34] together with its tuning approach. Continuing our modelling effort from [34] we advance the finite element (FE) model into a system-level model. For this purpose, we derive a reduced order model, integrate nonlinear tuning and damping forces, a control algorithm, and electrical circuitry into a more complex model.

2.1. Design Description

The mechanical resonator design consists of two identical 80 mm long arms (referred to as outer beams), mechanically coupled through a common end to a 60 mm long inner beam, which extends in turn towards the fixed end [34]. The tip masses are of identical weight $m = 7.6$ g. The first two resonance frequencies and mode shapes, as obtained from a harmonic analysis, are shown in Figure 1. The deformation of the piezoelectric layers results in a surface charge distribution and consequently a voltage across the patch electrodes. The maximum power output of the harvester can be derived from this voltage and the patch capacitance as explained in [34]. The phase difference between the inner and outer patch depends on the frequency of operation. We identified the phase difference to be 200 and 30° at the first, respectively the second mode. A connection of both patches shall consider and adapt the polarity of the voltage output. Realistic vibration excitation might excite both modes simultaneously so that individual power processing is mandatory.

In order to increase the frequency agility of the system, we propose magnetic frequency tuning. The approach uses permanent magnets on the resonating structure together with fixed external magnets. The interaction between the fixed and movable magnets creates an attracting or repelling force, which adds to the mechanical restoring force of the structure. The change in stiffness of the structure leads to a frequency up- or down-tuning. In [34] we characterized the resonator and demonstrated bidirectional frequency tuning of 15%. In order to simulate the effect of the magnetic force, the use of the full FE model is not efficient anymore, due to the very long solution time needed for such a simulation. Therefore, in [34] we proposed a compact model of the resonator and we were able to fully simulate the behavior of such a system. However, the integration of the piezoelectric transducer elements is not feasible with a lumped modelling approach. Therefore, in this paper, we derive a reduced order model of the full FE model and implement it in a system-level simulation.

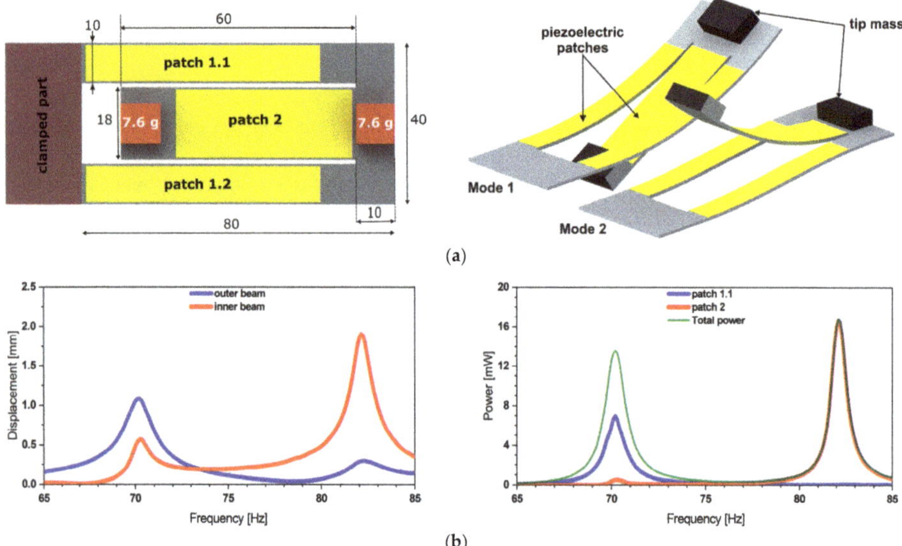

Figure 1. (a) Geometry description of the harvester design together with (b) its simulated displacement, respectively power output. The transfer function illustrates the dual frequency operation of the structure under a base acceleration of 0.5 g. Simultaneously, it demonstrates comparable power output levels.

2.2. Reduced Order Model of the Piezoelectric Energy Harvester

The piezoelectric energy harvester model implemented in ANSYS® Mechanical (V2019, R1) was thoroughly described in [34]. Considering its high computational cost for a transient simulation, Krylov subspace-based model order reduction (MOR) methods, also known as rational interpolation [38–40] were implemented to generate a highly compact but accurate reduced order model (ROM). Furthermore, based on this ROM, a circuit-device co-simulation of the piezoelectric energy harvester became feasible in the system-level simulation.

Model Order Reduction

The finite element method provides a spatially discretized mathematical description of the piezoelectric energy harvester. It is represented by a second order multiple-input multiple-output (MIMO) system of the form:

$$\sum_N \begin{cases} \underbrace{\begin{bmatrix} M_{11} & 0 \\ 0 & 0 \end{bmatrix}}_{M} \begin{bmatrix} \ddot{x}_1 \\ \ddot{x}_2 \end{bmatrix} + \underbrace{\begin{bmatrix} E_{11} & 0 \\ 0 & 0 \end{bmatrix}}_{E} \begin{bmatrix} \dot{x}_1 \\ \dot{x}_2 \end{bmatrix} + \underbrace{\begin{bmatrix} K_{11} & K_{12} \\ K_{21} & K_{22} \end{bmatrix}}_{K} \underbrace{\begin{bmatrix} x_1 \\ x_2 \end{bmatrix}}_{x} = \underbrace{\begin{bmatrix} B_1 \\ B_2 \end{bmatrix}}_{B} u \\ y = \underbrace{\begin{bmatrix} C_1 & C_2 \end{bmatrix}}_{C} \begin{bmatrix} x_1 \\ x_2 \end{bmatrix} \end{cases} \tag{1}$$

where $M, E, K \in \mathbb{R}^{N \times N}$ are the mass, damping, and stiffness matrices, respectively. $B \in \mathbb{R}^{N \times p}$ and $C \in \mathbb{R}^{q \times N}$ are the input and gathering matrices, with $u \in \mathbb{R}^p$ and $y \in \mathbb{C}^q$ being user defined input and output vectors. x_1 and x_2 present the nodal displacement and electrical potentials in the state vector $x \in \mathbb{C}^N$. Considering the large size of system (1) obtained in this work ($N = 166{,}789$ DoF), a highly compact but accurate ROM was generated:

$$\sum_r \begin{cases} \underbrace{V^T M V \cdot \ddot{z}}_{M_r} + \underbrace{V^T E V \cdot \dot{z}}_{E_r} + \underbrace{V^T K V \cdot z}_{K_r} = \underbrace{V^T B}_{B_r} \cdot u \\ y = \underbrace{CV}_{C_r} \cdot z \end{cases} \tag{2}$$

where $V \in \mathbb{R}^{N \times r}$ is the orthonormal basis of the second order Krylov subspace $\mathcal{K}_r\left(-K^{-1}E, -K^{-1}M, -K^{-1}B\right)$ obtained through the Second Order Arnoldi Reduction (SOAR) method [41,42]. The full-scale state vector x is approximated by $x \approx V \cdot z$, where $z \in \mathbb{C}^r$ is the reduced state vector, $r = 35 \ll N$.

In the previous research [43–45], the stability of the reduced system (2) could not be guaranteed by the conventional MOR methods. Several stabilization approaches have been introduced and mathematically proven in [46–48]. In this work, we apply the efficient approach 'Schur after MOR' to generate a stable ROM of the piezoelectric energy harvester model. "Schur after MOR" projects the reduced system (2) onto a sorted orthonormal eigenbasis of matrix M_r:

$$\sum_r \begin{cases} \underbrace{T^T M_r T \cdot \ddot{q}}_{\widetilde{M}_r} + \underbrace{T^T E_r T \cdot \dot{q}}_{\widetilde{E}_r} + \underbrace{T^T K_r T \cdot q}_{\widetilde{K}_r} = \underbrace{T^T B_r}_{\widetilde{B}_r} \cdot u \\ y = \underbrace{C_r T}_{\widetilde{C}_r} \cdot q \end{cases} \tag{3}$$

where $T \in \mathbb{R}^{r \times r}$ and $\widetilde{M}_r$ are the eigenvector matrix and sorted diagonal eigenvalue matrix of M_r, obtained through eigen decomposition of $M_r = T\widetilde{M}_r T^T$. According to the indexes of the relative small eigenvalues in $\widetilde{M}_r$, all the system matrices in (3) can be partitioned:

$$\begin{aligned} \widetilde{M}_r &= \begin{bmatrix} \widetilde{M}_1 & 0 \\ 0 & \widetilde{M}_2 \end{bmatrix} \approx \begin{bmatrix} \widetilde{M}_1 & 0 \\ 0 & 0 \end{bmatrix}; \quad \widetilde{E}_r = \begin{bmatrix} \widetilde{E}_{11} & \widetilde{E}_{12} \\ \widetilde{E}_{21} & \widetilde{E}_{22} \end{bmatrix} \approx \begin{bmatrix} \widetilde{E}_{11} & 0 \\ 0 & 0 \end{bmatrix}; \\ \widetilde{K}_r &= \begin{bmatrix} \widetilde{K}_{11} & \widetilde{K}_{12} \\ \widetilde{K}_{21} & \widetilde{K}_{22} \end{bmatrix}; \quad \widetilde{B}_r = \begin{bmatrix} \widetilde{B}_1 \\ \widetilde{B}_2 \end{bmatrix}; \quad \widetilde{C}_r = \begin{bmatrix} \widetilde{C}_1 & \widetilde{C}_2 \end{bmatrix} \end{aligned} \tag{4}$$

where $\widetilde{M}_1, \widetilde{E}_{11}, \widetilde{K}_{11} \in \mathbb{R}^{I \times I}$, $\widetilde{M}_2, \widetilde{E}_{22}, \widetilde{K}_{22} \in \mathbb{R}^{(r-I) \times (r-I)}$, $\widetilde{E}_{12}, \widetilde{K}_{12} \in \mathbb{R}^{I \times (r-I)}$, $\widetilde{E}_{21}, \widetilde{K}_{21} \in \mathbb{R}^{(r-I) \times I}$, $\widetilde{B}_1 \in \mathbb{R}^{I \times p}$, $\widetilde{B}_2 \in \mathbb{R}^{(r-I) \times p}$, $\widetilde{C}_1 \in \mathbb{R}^{q \times I}$, $\widetilde{C}_2 \in \mathbb{R}^{q \times (r-I)}$, and $I \in [1, r]$. In order to reconstruct the reduced system in an analogous manner as system (1), the relatively small part $\widetilde{M}_2$ and parts $\widetilde{E}_{12}, \widetilde{E}_{21}, \widetilde{E}_{22}$ in (4) are all neglected. In this way, the Schur complement transformation can be performed and the stable piezoelectric energy harvester ROM is obtained:

$$\sum_{r\_Schur} \begin{cases} \widetilde{M}_1 \cdot \ddot{q}_1 + \widetilde{E}_{11} \cdot \dot{q}_1 + \left(\widetilde{K}_{11} - \widetilde{K}_{12}\widetilde{K}_{22}^{-1}\widetilde{K}_{21}\right) \cdot q_1 = \left(\widetilde{B}_1 - \widetilde{K}_{12}\widetilde{K}_{22}^{-1}\widetilde{B}_2\right) \cdot u \\ y = \left(\widetilde{C}_1 - \widetilde{C}_2\widetilde{K}_{22}^{-1}\widetilde{K}_{21}\right) \cdot q_1 + \left(\widetilde{C}_2\widetilde{K}_{22}^{-1}\widetilde{B}_2\right) \cdot u \end{cases} \tag{5}$$

2.3. System-Level Simulation

The system-level model consists of three parts as depicted in Figure 2. The multiphysics part comprises the ROM, which describes the mechanical and piezoelectric behavior of the harvester. The magnetic forces have been derived from magneto-static simulations [33] and implemented in the model as force functions 1 and 2. The magnetic force will alter the effective stiffness of the resonator.

The damping ratio of the mechanical resonator depends on the structure's stiffness. Hence, it also varies while tuning. We address this by adjusting the damping ratio accordingly. The electrical part encompasses the rectification circuitry. A tuning control algorithm, based on maximum amplitude tracking, is included as well.

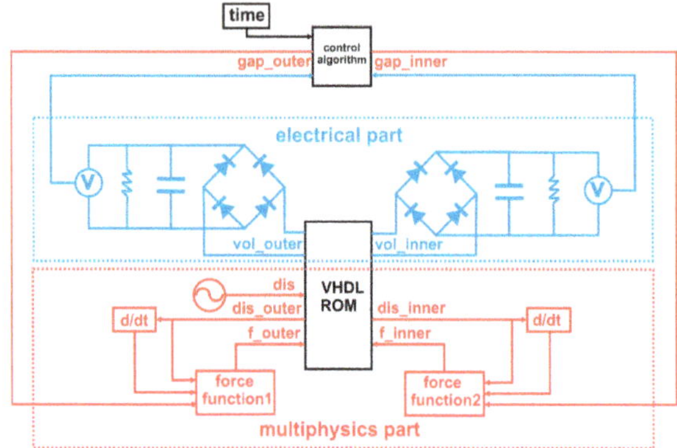

Figure 2. System-level model implemented in ANSYS twin builder, including a reduced order model of the harvester together with tuning actuation, the conditioning circuitry, and the tuning control algorithm.

2.3.1. Mechanical Resonator Reduced Order Model

This subsection presents the validation of the reduced order model by comparison between results of a FE model, the reduced order model and experimental data. The MIMO system of the resonator is illustrated in Figure 3. Both models use three inputs: Base excitation "dis", (displacement amplitude of the ambient vibration), force on outer and inner beam "f_outer" and "f_inner", respectively. The tip displacement of the outer and inner beam "dis_outer" and "dis_inner" are the two outputs. The displacement amplitudes of the outer and inner beam of the FE model, respectively the reduced order model, are shown in Figure 3 for a displacement amplitude of 10 µm at the clamped part.

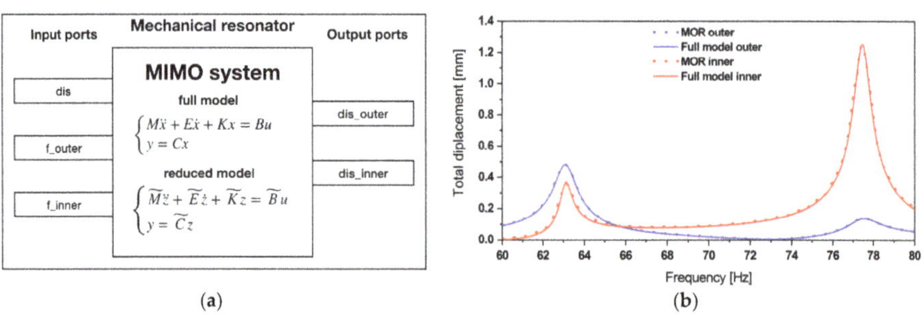

(a)   (b)

Figure 3. (a) The multiple-input multiple-output (MIMO) system of the mechanical resonator together with (b) the resonator reduced model validation through a comparison with the full finite element (FE) model. Both models are subjected to an excitation amplitude of 10 µm.

The obtained results from the reduced order model matched well with the FE model. The two fundamental modes appear at 63.1 and 77.5 Hz. Furthermore, the aforementioned frequencies have been compared to the experimental results obtained in [33], which are 62.6 and 76.1 Hz.

In [32,34] we presented a description of the magneto-static simulations, which enabled us to derive the magnetic forces involved in the frequency tuning. The same forces have been considered in the current system-level model. As shown in Figure 4 the magnetic forces yield a bidirectional frequency shift by up to 18%.

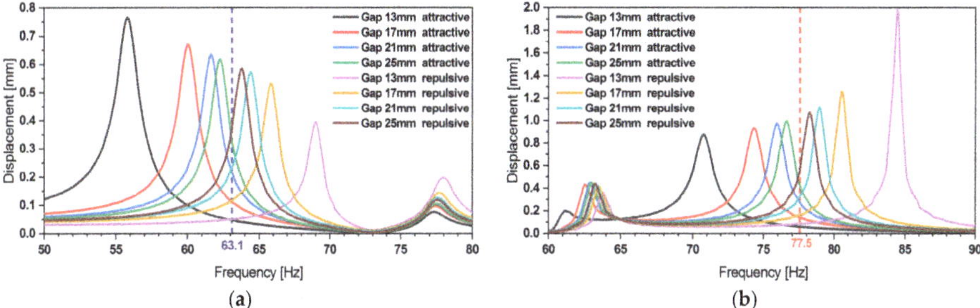

Figure 4. (**a**) Variation of the displacement amplitude of outer and (**b**) inner beam during frequency tuning. An 18% of bidirectional frequency shift can be achieved. The data indicates that frequency tuning does not affect the other resonance frequency.

The frequency tuning system-level simulations showed a relative error of approximately 3.1% and 1.8% for the first, respectively the second mode tuning (see Figure 5) when compared to experimental data. Yet, tuning towards smaller frequencies reveals an increasing discrepancy. We attribute this to some limitations of the underlying magnetostatic model, which neglects the rotation and lateral displacement of the magnet as the resonator undergoes deflection. Furthermore, as displacement amplitude increases while tuning towards smaller frequencies such deviations show a more pronounced effect.

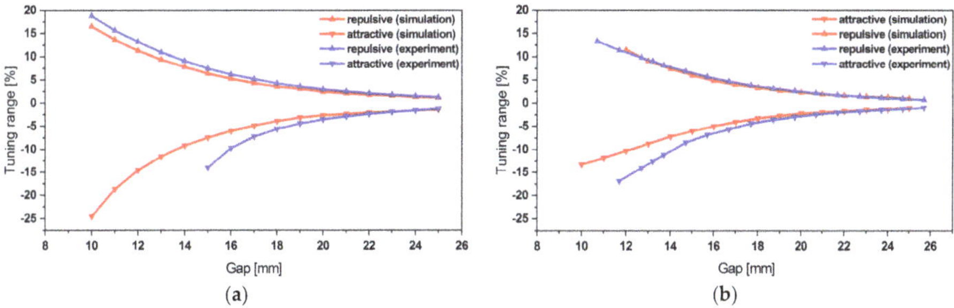

Figure 5. (**a**) Experimental validation of the bidirectional frequency tuning simulation of first and (**b**) second resonance frequency.

2.3.2. Piezoelectric Energy Harvester Reduced Order Model

After the validation frequency-agile resonator model, a reduced order model of the piezoelectric energy harvester has been derived. The corresponding MIMO system is illustrated in Figure 6.

Compared to the mechanical resonator model, the piezoelectric energy harvester model has two additional outputs "vol_outer", "vol_inner", referring to the voltage levels at the piezoelectric patches on the outer and inner beams. The results obtained from the reduced order and the FE model fit well and thus support the applicability of this reduced order modeling approach.

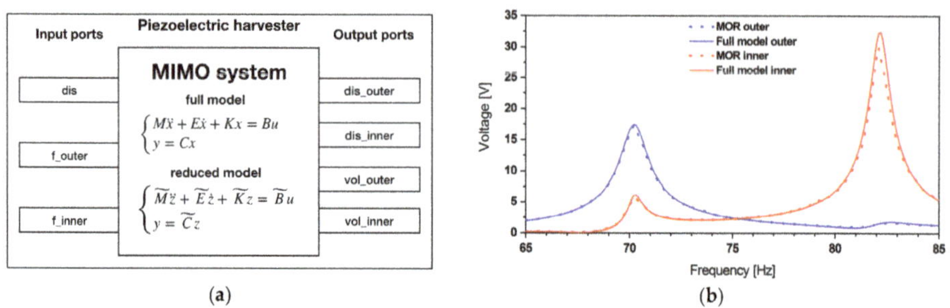

(**a**) (**b**)

Figure 6. (**a**) The MIMO system of the piezoelectric energy harvester and (**b**) the harmonic response of the reduced order and full FE model subjected to a 10 μm excitation amplitude.

2.3.3. Electrical Simulation

The system-level model integrates the reduced order model with electrical circuitry. Here, we considered a diode bridge for full-wave rectification, a capacitor for filtering, a buck converter for voltage regulation, and a resistive load with optimum resistance. The circuitry is connected to the two electrical ports of the reduced order model.

Rectification and Filtering

As presented in Figure 7, after rectification and filtering the AC voltage output, a DC voltage output with small ripple voltage is obtained.

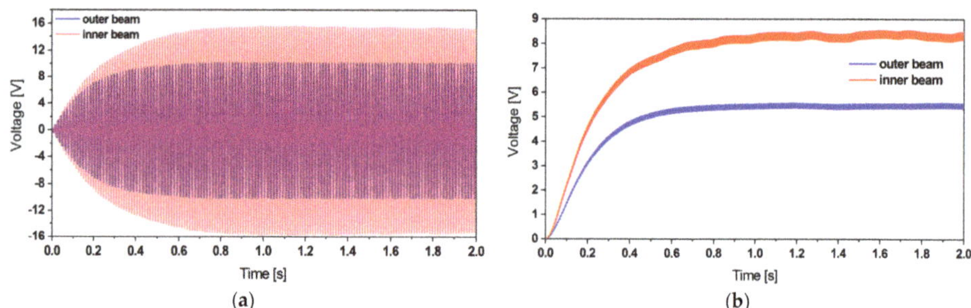

(**a**) (**b**)

Figure 7. (**a**) Simulation results of the AC voltage output before rectification and (**b**) the filtered DC voltage output of the piezoelectric harvester subjected to 0.2 g base excitation.

Optimum Load

Due to the capacitance of the piezoelectric patches, the power output depends on the load resistance. To ensure maximum power delivery, the optimum load has been found to be 100 and 60 kΩ for the outer and inner beam, respectively (see Figure 8). The power in the load is proportional to the square of the excitation amplitude. Therefore, we performed the simulations at several excitation levels.

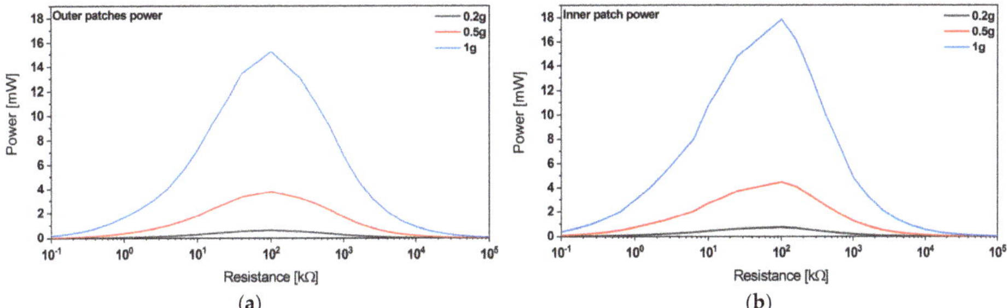

Figure 8. (**a**) Load matching to ensure maximum power output from the outer patches electrically connected in the series, (**b**) respectively the inner patch.

Voltage Regulation

Piezoelectric energy harvesters, which employ PZT as a piezoelectric material, easily generate voltage levels, which exceed the range compatible with electronic circuits and related components, such as microcontrollers or super capacitors. Voltage regulation is required for reliable operation. The voltage regulation is achieved, e.g., by using a buck converter, which includes a MOSFET, a capacitor, an inductor, and a diode as shown in Figure 9. The duty cycle of the MOSFET's state affects the voltage output in such a circuit. A controller alters the duty-cycle to maintain constant output voltage for varying loading situations.

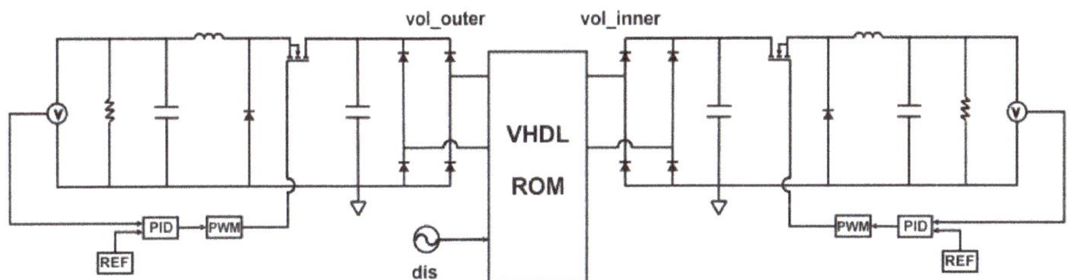

Figure 9. Two independent buck converter circuits for voltage regulation.

Figure 10 shows the simulation results of the voltage regulation. Here, the voltage at the outer and inner beam have been regulated to 5 V within 10 s.

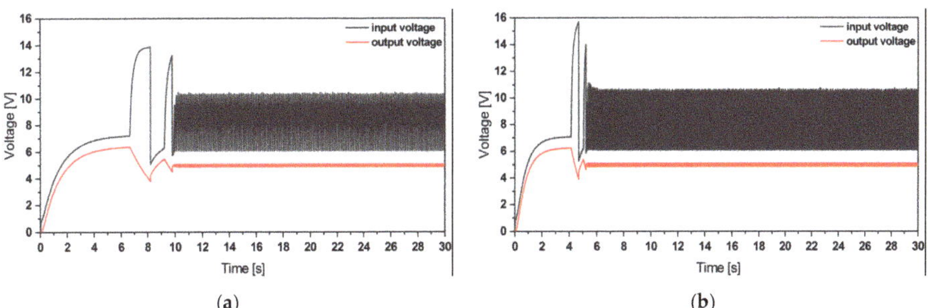

Figure 10. (**a**) Simulation results of the voltage regulation of the outer patches and (**b**) the inner one under 0.2 g base excitation.

2.3.4. Energy Harvester Frequency Tuning

We also studied the effect of our frequency tuning mechanism on the voltage output in analogy to the procedure described in Section 2.3.1. The results of the system-level simulation are shown in Figure 11.

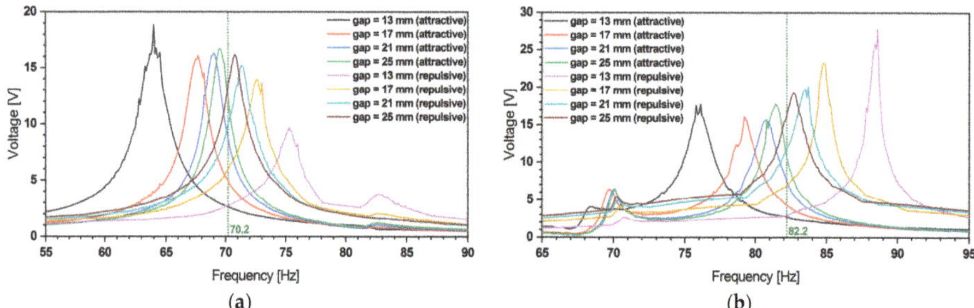

Figure 11. (**a**) The voltage amplitude variation of the piezoelectric energy harvester under bidirectional magnetic frequency tuning of first and (**b**) second resonance frequency at 0.2 g harmonic base excitation.

These results show the voltage amplitude variation as a result of the bidirectional frequency shift of the piezoelectric harvester. These simulations show bidirectional frequency tuning by up to 9% relative to the unaltered resonance frequencies. The reduced tuning range is due to the increased beam stiffness induced by the piezoelectric patches.

2.3.5. Control Algorithm

In this section, we focus on the tuning control as implemented in the system-level model of Figure 2. A control algorithm, as visualized in Figure 12, maintains maximum voltage output even under varying excitation frequency. We consider this as a realistic use case. The time variable and DC voltage output are used as inputs for the control algorithm. The control scheme analyzes the voltage levels at a given frequency and selects one of the two tuning mechanisms.

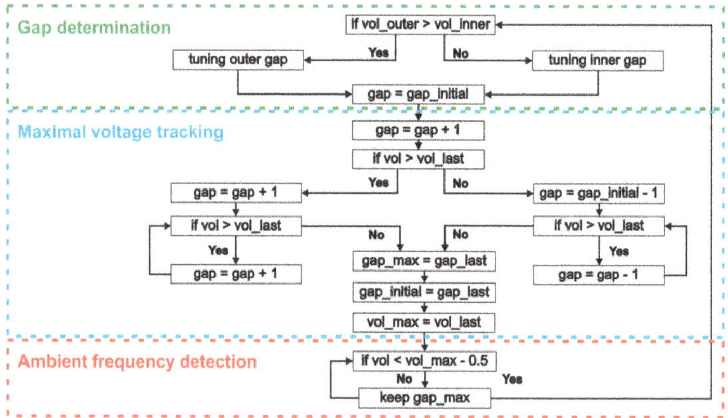

Figure 12. Tuning control algorithm scheme based on maximum-voltage tracking.

In order to demonstrate the self-adaption of our harvester, we excite the structure with a stepwise frequency-varying harmonic excitation, as illustrated in Figure 13. The results demonstrate that the system is able to choose the adequate tuning mechanism. This choice is based on the current voltage level. The maximum voltage is achieved within 15 s. In contrast to our previous work [33], where we implemented an adaptive step size, here the tuning scheme employs a constant step size. This results in less tuning steps, which improve the energy efficiency of the tuning mechanism.

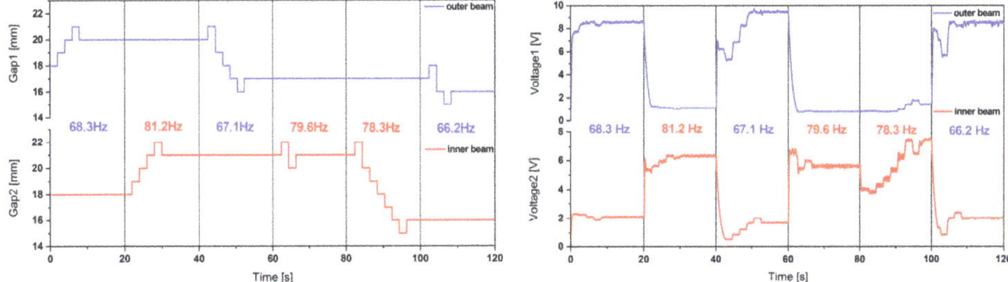

Figure 13. A control algorithm, which is based on maximum-voltage tracking, chooses the most effective tuning actuator.

3. Experimental Investigation

In order to characterize the presented harvester design, we integrate three macro fiber composite (MFC) piezoelectric patches supplied by SMART MATERIAL Corp. (M-8507-P2 on the outer beams and M-8514-P2 on the inner beam) with a resonator fabricated from steel, as depicted in Figure 14. The MFC patches are composed of piezoelectric rods embedded between layers of adhesive, interdigitated electrodes, and encapsulated with polyimide film. The patch dimensions are $60 \times 7 \times 0.18$ mm^3 and $48 \times 14 \times 0.18$ mm^3 for the outer and inner beam, respectively. The material properties of the MFC patches are given in Table 1.

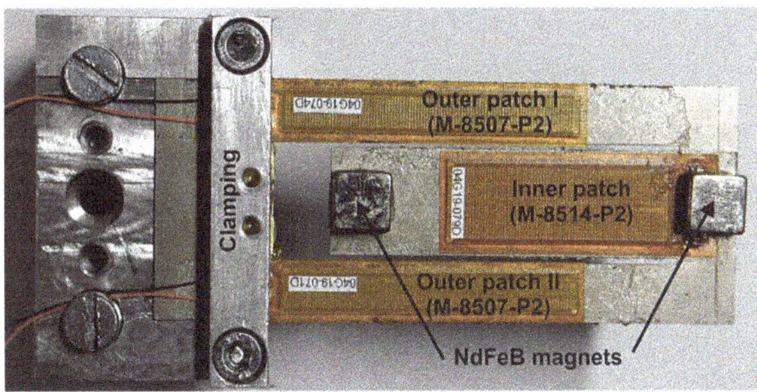

Figure 14. Dual frequency piezoelectric energy harvester with macro fiber composite (MFC) patches on outer and inner beams. Permanent magnets are attached at the beam ends. The base excitation is applied to the clamped part on the left side.

Table 1. Material properties of the active area of the macro fiber composite (MFC) patches.

Material Properties	Value
Mass density (kg/m$^{3)}$)	5440
Tensile modulus, E_1 (rod direction) (GPa)	30.34
Tensile modulus, E_1 (electrode direction) (GPa)	15.86
Poisson's ratio, v_{12}	0.31
Poisson's ratio, v_{21}	0.16
Shear modulus, G_{12} (GPa)	5.515
d_{33} (rod direction) (pC/N)	400
d_{31} (electrode direction) (pC/N)	−170

The harvester has been excited at the acceleration amplitudes of 0.5 and 1.0 g. The results presented on Figure 15 demonstrate the dual frequency feature of the harvester. The experimentally observed resonance frequencies $f_{1\ Exp}$ = 63.27 and $f_{2\ Exp}$ = 76.35 Hz match the simulation results $f_{1\ Sim}$ = 64.30 and $f_{2\ Sim}$ = 77.50 Hz. However, a lower voltage output has been observed. We attribute this to the adhesive tape which attaches the patches to the steel. In our assembly this degrades the strain transfer between the steel resonator and the piezoelectric layers when compared to solid bonding, e.g., using glue. Our simulations considered the adhesive tape as a material of high compliance (E = 450 kPa). The FE model implements a constant damping ratio, which yields correct amplitudes at the first mode and does not describe the damping at the second mode. A mode-specific or even frequency dependent damping ratio shall be applied instead. Furthermore, the slight frequency shift (up to 1.63%) between the model and the experiment results is caused by the additional mass of the solder paste used to electrically connect the inner patch. The patch attachment procedure and the limited reproducibility of the magnets positioning contribute in turn to such a frequency shift.

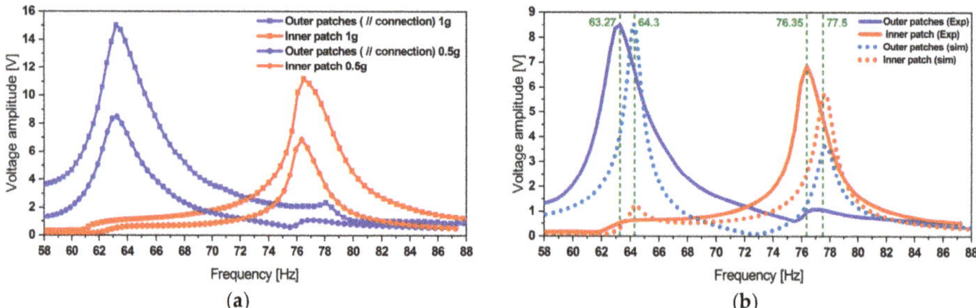

Figure 15. (**a**) Experimentally obtained voltage at the excitation levels of 0.5 and 1.0 g. Comparison of experimental data and (**b**) simulation results for an excitation level of 0.5 g.

We evaluated the harvester's power delivery using the power management board 2151A provided by analog devices (see Figure 16), which also enables battery charging. The board integrates the LTC3331 chip, which provides a regulated voltage from various energy harvesting sources. The circuitry consists of an integrated low-loss full-wave bridge rectifier and a buck converter. The rechargeable coin-cell battery powers a buck-boost converter capable of providing voltages between 1.8 and 5.0 V. Depending on the available power from the harvester the board is either supplied by the harvester or the battery. An internal prioritizer switches between the power sources. If the harvesting source is available, the buck converter is active and the buck-boost is off and vice versa.

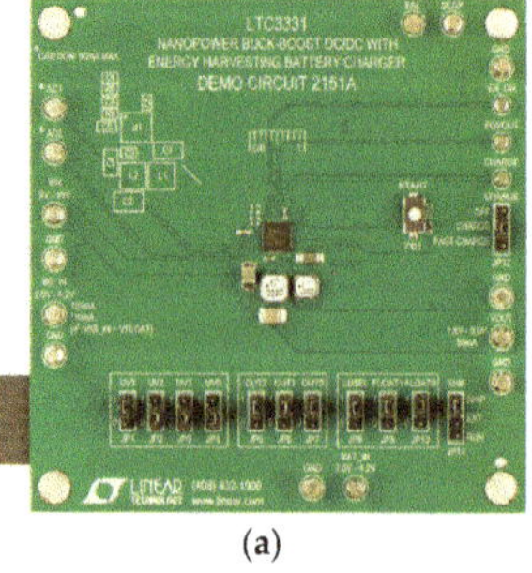

Figure 16. (**a**) Power management boards 2151A from analog devices and (**b**) bq2557OEVM-206 from Texas Instruments used as power management circuits.

The power management board 2151A has been tested under different configurations as presented in Table 2.

Table 2. Harvester characterization at 0.5 g excitation level, using the 2151A power management board.

Patch	f (Hz)	V_{out} (V)	I_{out} (µA)	R (kΩ)	P_{out} (µW)
Outer (// connection)	65.27	2.615	125.0	21.00	653.8
Inner		1.0	0.256	21.00	0.512
Outer (// connection)	78.35	0.381	18.20	22.00	13.87
Inner		2.305	105.1	22.00	484.5

Furthermore, we evaluated the efficiency of different power management boards the 2151A and the bq25570EVM-206, designed for low power applications and providing only 1.8 V output voltage as depicted in Table 3. The efficiency in this case is nothing than the ratio between the output and the input power.

Table 3. Efficiency comparison of the power management boards used as conditioning circuits for the designed harvester.

Board Type	V_{in} (V)	I_{in} (µA)	R (kΩ)	V_{out} (V)	I_{out} (µA)	Efficiency (%)
bq25570EVM-206	3.73	134.5	13.0	1.8	137.5	49.4
2151A	4.42	106.0	14.5	1.8	124.5	47.8

The experiments revealed that a maximum efficiency of approximately 50% can be reached using both boards with our harvester.

4. Parametric Design Optimization

One of the key features of the presented folded beam harvester design is the possibility to enhance the overall performance if the first two resonance frequencies appear closely spaced frequencies (co-resonance) and simultaneously provide the same power levels. This is a unique feature not provided by other multiresonant structures such as an array of two beams. All cantilevers of an array are subjected to the same base excitation, whereas in the case of the coupled resonator, the inner beam is subjected to the maximum tip displacement of the outer one, which is higher than the applied base excitation. This motivated us to investigate the possibility of optimizing the existing design.

The geometry of a vibration energy harvester determines its dynamic properties and thereby its operating frequency and the harvested power. Consequently, optimized dimensions yield higher power and better performance. For this purpose, the reference design was parameterized and optimized for an operating bandwidth centered at 75 Hz. The process of this optimization relied on FE models and is shown in Figure 17.

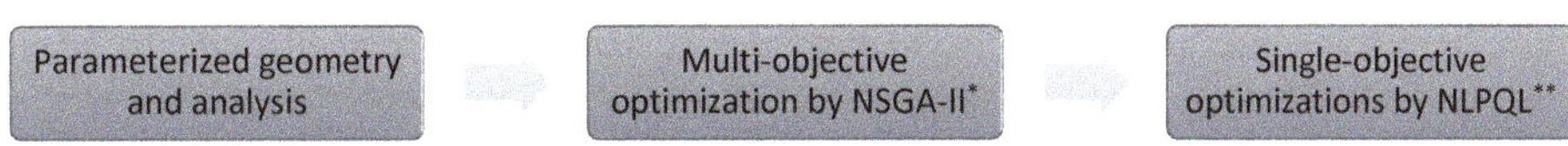

* Non-dominated sorting genetic algorithm II.
** Nonlinear programming by quadratic Lagrangian.

Figure 17. Optimization process.

Firstly, the reference geometry was parameterized and subsequently optimized. Figure 18 presents the parameterized model. Table 4 gives the range of the seven geometry parameters. The size of the magnets and their positions was unchanged during the optimization process. A parameter range of ±50% has been chosen with respect to the reference design. The thickness t is a discrete parameter, because the device is fabricated from a metal sheet, which is available only at certain thickness values.

The parameter L_i has bounds chosen to enable efficient usage of space for all values of L_o. Constraining L_i prevents the inner beam to overlap with the fixed support.

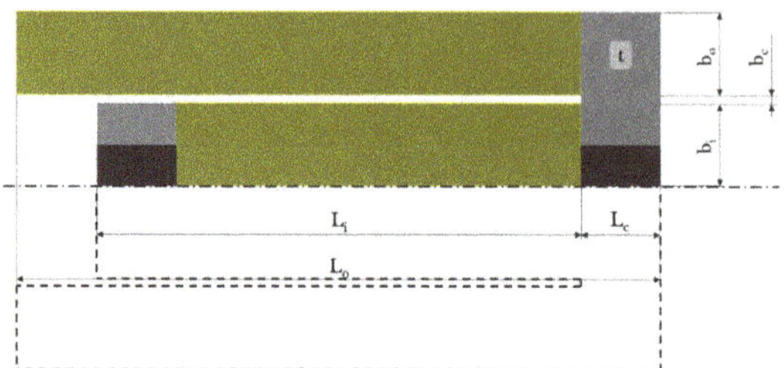

Figure 18. Parameterization of the reference geometry (light grey corresponds to steel, black to NdFeB, and yellow to PIC255).

Table 4. Parameter ranges for the design optimization.

Parameter	Reference Value (mm)	Lower Bound (mm)	Upper Bound (mm)
L_o	80	40	120
b_o	10	5.0	15.0
L_c	10	5.0	15.0
b_c	1.0	0.5	1.50
L_i	60	23	113
b_i	9.0	5.0	15.0
t [1]	1.0	0.5	1.50

[1] Discrete parameter since it is limited to commercial sheet metal; step size 0.5 mm.

A modal analysis and a harmonic analysis have to be performed to compute the objective values of the optimization. A modal analysis determines the eigenfrequencies for the first two modes, while a harmonic analysis computes the electrical power at these modes. The harmonic analysis implements a damping ratio of 0.8% which has been determined experimentally for a similar design. The base excitation acceleration amplitude was 0.01 g. The two outer piezoelectric elements were connected in parallel. The inner element was connected in series in order to obtain maximum power at an optimized load resistance. The results of these analyses yield the parameters and objectives presented in Equations (6)–(13) of which Equations (6)–(9) give the vibrational properties:

$$\overline{f} = \frac{f_1 + f_2}{2} \tag{6}$$

$$obj\, f = \left| 75\,Hz - \overline{f} \right| \tag{7}$$

$$\Delta f_{rel} = \frac{f_1 - f_2}{\overline{f}} \tag{8}$$

$$obj\, \Delta f_{rel} = \left| 0.05 - \Delta f_{rel} \right|. \tag{9}$$

Here, f_1 and f_2 are the first two eigenfrequencies, $\overline{f}$ is their mean value and Δf_{rel} is the relative operating frequency range. The two objectives $obj\, f$ and Δf_{rel} describe the intended operating frequency range. Upper bounds of 0.05 for $obj\, \Delta f_{rel}$ and 5 Hz for $obj\, f$ control the convergence of the optimization algorithm.

Equations (10)–(13) are related to the electrical behavior:

$$V = -g_{31}\, t\, \sigma_1 \tag{10}$$

$$P = 2\pi V^2 \frac{C_o C_i}{2C_o + C_i} f \tag{11}$$

$$obj\ P\ ratio = \frac{\min(PD_1, PD_2)}{\max(PD_1, PD_2)} \tag{12}$$

$$obj\ \overline{PD} = \frac{PD_1 + PD_2}{2}, \tag{13}$$

where V is the approximated voltage of a piezoelectric patch, g_{31} is the piezoelectric voltage coefficient for the 31 mode, t is the thickness of the piezoelectric patch, σ_1 is the normal stress due to bending, P is the electrical power in an attached resistor of at optimum load value, C_o and C_i are the capacitances of the inner and outer piezoelectric patches, f is the frequency and PD_1 and PD_2 are the power densities at the first and second eigenfrequency, respectively. The voltage was obtained analytically from the mechanical model in order to reduce the computational effort. This neglects the electromechanical back coupling. The power objectives $obj\ P\ ratio$ and $obj\ \overline{PD}$ evaluate the frequency spacing of the two maxima and their amplitude ratio.

The optimization follows a two-step procedure: A global multi-objective optimization and subsequent local single-objective optimizations. Methods to decrease the computational effort such as a sensitivity analysis or a metamodel were omitted, as they were suffering from insufficient accuracy. The large design space and the nonlinear objective space require this two-step procedure where the first step searches for promising subspaces. A second, more refined step searches this subspace to find the final candidates. The multi-objective optimization employs the evolutionary algorithm NSGA-II, which iteratively evolves a set of start designs by selection, crossover, and mutation to satisfy the objectives of the optimization. The start population contained 3500 designs; each following generation comprised 100 designs. A crossover probability of 98% and a mutation probability of 1% defined the reproduction. The optimization converged for either 20 generations, a convergence stability of 2% or if 70% of the designs of a generation were Pareto-optimal. The Pareto set provides one start design per thickness for the single-objective optimizations. These start designs were selected to satisfy the two vibrational objectives to guarantee operation at resonance and broaden the harvesting. A subsequent single-objective optimization relied on the algorithm NLPQL for a gradient-based optimization. The local search deployed central differences and a finite difference of 1%. The parameter ranges for each single-objective optimization were ± 10% of the start design. These local optimizations changed the definition of $obj\ \Delta f_{rel}$ to $obj\ \Delta f_{rel} = |0.01 - \Delta f_{rel}|$. The optimization was considered completed if the change for the next iteration fell below 0.1% or if 20 iterations were reached. The local optimization comprised up to three single-objective optimizations.

Figure 19 presents the geometry of the reference design together with the individual optimized designs of all three thicknesses. The optimized design (c) can be compared to the reference design since both have the same thickness. Important differences are the length of the connection and the width of the outer beam, which results in an operating frequency close to 75 Hz.

Figure 19. (a) Reference geometry and (b) the three optimized designs with thicknesses of $t = 0.5$, (c) $t = 1$, and (d) $t = 1.5$ mm.

Figure 20 compares the power densities of the four designs in Figure 19. The power density of the reference design has two dominant peaks. However, the power drops beyond the bandwidth. In contrast, the designs (a) and (b) provide an operational frequency range with a power variation of only 3.5% centered at 75 Hz. Moreover, those designs also have higher peak power densities since their more thin steel structure is more compliant. Up to three local optimizations were performed for each design. Hence, a higher number of local optimizations will further improve the designs.

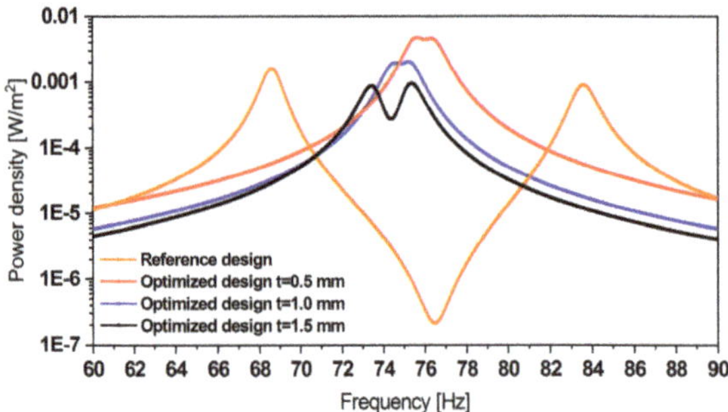

Figure 20. Power density of the reference design and the optimized designs. The co-resonance results in an extended operative bandwidth at comparable power levels.

In summary, an optimized geometry provides equal power at both resonance frequencies at even higher power density, as demonstrated with the 0.5 mm thick design.

5. Conclusions

This work advances the research work presented in [33]. We present a novel self-tunable dual-frequency piezoelectric energy harvester with optimized performances. The dual frequency feature has been thoroughly investigated and we demonstrated that the resonator magnifies the amplitudes at two close fundamental frequencies, enabling simultaneous energy harvesting from both vibration frequencies. The system integrates permanent magnets, whose magnetostatic forces enable the frequency agility of the harvester. In order to simulate the bidirectional frequency tuning effect, we derived a reduced order model of the resonator and the harvester finite element model. The resonator's reduced order model has been experimentally validated and we demonstrated ±18% of bidirectional tuning. Furthermore, the reduced order modelling has been applied to the harvester.

A control algorithm has been developed to drive the tuning mechanism and thereby ensures the self-adaption of the system. The algorithm is based on maximum-voltage tracking and is able to automatically choose the adequate tuning actuator.

Furthermore, we presented experimental results of the piezoelectric harvester. The characterization demonstrated the dual-frequency feature of the harvester and showed that the harvester supplies sufficient voltage and power levels. Additionally, we investigated the efficiency of two commercially available power management systems. Further experiments will be performed to evaluate the harvesting system's performance under realistic applications.

Finally, an optimized version of the harvester design has been proposed. This design presents two modes appearing at two close frequencies and an increased operative bandwidth. Further experiments are planned to verify the characteristics of the optimized version.

Author Contributions: S.B. and D.H. conceived the overall system design and built the finite element models. S.B. was additionally in charge of the experimental characterization, the implementation of the tuning mechanism, and the data processing. D.H. and T.B. were both supervising the project and reviewing the methodology. C.Y. contributed the reduced order model. Y.R. composed the system-level model and implemented the tuning control algorithm. F.L. was responsible for the automation of the tuning mechanism. A.S. and S.H. performed the optimization of the harvester design and derived the power calculations. All authors have read and agreed to the published version of the manuscript.

Funding: This research received no external funding.

Acknowledgments: We gratefully acknowledge the German Academic Exchange Service "DAAD" for a partial financial support through a Ph.D. scholarship.

Conflicts of Interest: The authors declare no conflict of interest.

References

1. Roundy, S.; Leland, E.S.; Baker, J.; Carleton, E.; Reilly, E.; Lai, E.; Otis, B.; Rabaey, J.M.; Sundararajan, V.; Wright, P.K. Improving power output for vibration-based energy scavengers. *IEEE Pervasive Comput.* **2005**, *4*, 28–36. [CrossRef]

2. Erturk, A.; Inman, D.J. A distributed parameter electromechanical model for cantilevered piezoelectric energy harvesters. *J. Vib. Acoust.* **2008**, *130*, 41002. [CrossRef]

3. Liao, Y.; Sodano, H.A. Model of a single mode energy harvester and properties for optimal power generation. *Smart Mater. Struct.* **2008**, *17*, 65026. [CrossRef]

4. Ferrari, M.; Ferrari, V.; Guizzetti, M.; Marioli, D.; Taroni, A. Piezoelectric multifrequency energy converter for power harvesting in autonomous microsystems. *Sensors Actuators A Phys.* **2008**, *142*, 329–335. [CrossRef]

5. Liu, J.-Q.; Fang, H.-B.; Xu, Z.-Y.; Mao, X.-H.; Shen, X.-C.; Chen, D.; Liao, H.; Cai, B.-C. A MEMS-based piezoelectric power generator array for vibration energy harvesting. *Microelectron. J.* **2008**, *39*, 802–806. [CrossRef]

6. Tang, X.; Zuo, L. Enhanced vibration energy harvesting using dual-mass systems. *J. Sound Vib.* **2011**, *330*, 5199–5209. [CrossRef]

7. Tadesse, Y.; Zhang, S.; Priya, S. Multimodal energy harvesting system: Piezoelectric and electromagnetic. *J. Intell. Mater. Syst. Struct.* **2009**, *20*, 625–632. [CrossRef]

8. Wu, H.; Tang, L.; Yang, Y.; Soh, C.K. A novel two-degrees-of-freedom piezoelectric energy harvester. *J. Intell. Mater. Syst. Struct.* **2013**, *24*, 357–368. [CrossRef]

9. Upadrashta, D.; Yang, Y. Trident-shaped multimodal piezoelectric energy harvester. *J. Aerosp. Eng.* **2018**, *31*, 4018070. [CrossRef]

10. Lamprecht, L.; Ehrenpfordt, R.; Lim, C.K.; Zimmermann, A. A 500 Hz-wide kinetic energy harvester: Outperforming macroscopic electrodynamic arrays with piezoelectric arrays. *Mech. Syst. Signal Process.* **2019**, *119*, 222–243. [CrossRef]

11. Qi, S.; Shuttleworth, R.; Olutunde Oyadiji, S.; Wright, J. Design of a multiresonant beam for broadband piezoelectric energy harvesting. *Smart Mater. Struct.* **2010**, *19*, 94009. [CrossRef]

12. Xiao, H.; Wang, X.; John, S. A multi-degree of freedom piezoelectric vibration energy harvester with piezoelectric elements inserted between two nearby oscillators. *Mech. Syst. Signal Process.* **2015**, *6869*, 138–154. [CrossRef]

13. Xiao, Z.; Yang, T.Q.; Dong, Y.; Wang, X.C. Energy harvester array using piezoelectric circular diaphragm for broadband vibration. *Appl. Phys. Lett.* **2014**, *104*, 223904. [CrossRef]

14. Ramírez, J.M.; Gatti, C.D.; Machado, S.P.; Febbo, M. A piezoelectric energy harvester for rotating environment using a linked E-shape multi-beam. *Extreme Mech. Lett.* **2019**, *27*, 8–19. [CrossRef]

15. Challa, V.R.; Prasad, M.G.; Shi, Y.; Fisher, F.T. A vibration energy harvesting device with bidirectional resonance frequency tunability. *Smart Mater. Struct.* **2008**, *17*, 15035. [CrossRef]

16. Spreemann, D.; Manoli, Y. *Electromagnetic Vibration Energy Harvesting Devices*; Springer: Dordrecht, The Netherlands, 2012.

17. Xu, M.; Li, X. Stochastic averaging for bistable vibration energy harvesting system. *Inter. J. Mech. Sci.* **2018**, *141*, 206–212. [CrossRef]

18. Daqaq, M.F. On intentional introduction of stiffness nonlinearities for energy harvesting under white Gaussian excitations. *Nonlinear Dyn.* **2012**, *69*, 1063–1079. [CrossRef]

19. Stanton, S.C.; McGehee, C.C.; Mann, B.P. Nonlinear dynamics for broadband energy harvesting: Investigation of a bistable piezoelectric inertial generator. *Phys. D Nonlinear Phenom.* **2010**, *239*, 640–653. [CrossRef]

20. Zhou, S.; Zuo, L. Nonlinear dynamic analysis of asymmetric tristable energy harvesters for enhanced energy harvesting. *Commun. Nonlinear Sci. Numer. Simul.* **2018**, *61*, 271–284. [CrossRef]

21. Abdelkefi, A.; Barsallo, N. Nonlinear analysis and power improvement of broadband low-frequency piezomagnetoelastic energy harvesters. *Nonlinear Dyn* **2016**, *83*, 41–56. [CrossRef]
22. Lai, S.-K.; Wang, C.; Zhang, L.-H. A nonlinear multi-stable piezomagnetoelastic harvester array for low-intensity, low-frequency, and broadband vibrations. *Mech. Syst. Signal Process.* **2019**, *122*, 87–102. [CrossRef]
23. Hoffmann, D.; Folkmer, B.; Manoli, Y. Experimental Analysis of a Coupled Energy Harvesting System with Monostable and Bistable Configuration. *J. Phys. Conf. Ser.* **2014**, *557*, 12134. [CrossRef]
24. Hoffmann, D.; Willmann, A.; Hehn, T.; Folkmer, B.; Manoli, Y. A self-adaptive energy harvesting system. *Smart Mater. Struct.* **2016**, *25*, 35013. [CrossRef]
25. Fu, H.; Yeatman, E.M. Broadband Rotational Energy Harvesting with Non-linear Oscillator and Piezoelectric Transduction. *J. Phys. Conf. Ser.* **2016**, *773*, 12008. [CrossRef]
26. Fu, H.; Yeatman, E.M. Rotational energy harvesting using bi-stability and frequency up-conversion for low-power sensing applications: Theoretical modelling and experimental validation. *Mech. Syst. Signal Process.* **2018**. [CrossRef]
27. Challa, V.R.; Prasad, M.G.; Fisher, F.T. Towards an autonomous self-tuning vibration energy harvesting device for wireless sensor network applications. *Smart Mater. Struct.* **2011**, *20*, 25004. [CrossRef]
28. Neiss, S.; Goldschmidtboeing, F.; Kroener, M.; Woias, P. Analytical model for nonlinear piezoelectric energy harvesting devices. *Smart Mater. Struct.* **2014**, *23*, 105031. [CrossRef]
29. Neiss, S.; Goldschmidtboeing, F.; Kroener, M.; Woias, P. Tunable nonlinear piezoelectric vibration harvester. *J. Phys. Conf. Ser.* **2014**, *557*, 12113. [CrossRef]
30. Nammari, A.; Caskey, L.; Negrete, J.; Bardaweel, H. Fabrication and characterization of non-resonant magneto-mechanical low-frequency vibration energy harvester. *Mech. Syst. Signal Process.* **2018**, *102*, 298–311. [CrossRef]
31. Bouhedma, S.; Hohlfeld, D. Frequency tunable piezoelectric energy harvester with segmented electrodes for improved power generation. In Proceedings of the MikroSystemTechnik 2017, Munich, Germany, 23–25 October 2017.
32. Bouhedma, S.; Hartwig, H.; Hohlfeld, D. Modeling and Characterization of a Tunable Dual-Frequency Piezoelectric Energy Harvester. In Proceedings of the 2018 IEEE/ASME International Conference on Advanced Intelligent Mechatronics (AIM), Auckland, New Zealand, 9–12 July 2018; pp. 1378–1383.
33. Bouhedma, S.; Zheng, Y.; Hohlfeld, D. Simulation and Characterization of a Nonlinear Dual-Frequency Piezoelectric Energy Harvester. *Proceedings* **2018**, *2*, 908. [CrossRef]
34. Bouhedma, S.; Zheng, Y.; Lange, F.; Hohlfeld, D. Magnetic frequency tuning of a multimodal vibration energy harvester. *Sensors* **2019**, *19*, 1149. [CrossRef] [PubMed]
35. Eichhorn, C.; Tchagsim, R.; Wilhelm, N.; Biancuzzi, G.; Woias, P. An energy-autonomous self-tunable piezoelectric vibration energy harvesting system. In Proceedings of the IEEE 24th International Conference on Micro Electro Mechanical Systems (MEMS), Cancun, Mexico, 23–27 January 2011.
36. Lumentut, M.F.; Howard, I.M. Electromechanical analysis of an adaptive piezoelectric energy harvester controlled by two segmented electrodes with shunt circuit networks. *Acta Mech* **2017**, *228*, 1321–1341. [CrossRef]
37. Zhang, H.; Afzalul, K. Design and analysis of a connected broadband multi-piezoelectric-bimorph- beam energy harvester. *IEEE Trans. Ultrason. Ferroelectr. Freq. Control* **2014**, *61*, 1016–1023. [CrossRef] [PubMed]
38. Freund, R.W. Krylov-subspace methods for reduced-order modeling in circuit simulation. *J. Comput. Appl. Math.* **2000**, *123*, 395–421. [CrossRef]
39. Gugercin, S.; Stykel, T.; Wyatt, S. Model Reduction of Descriptor Systems by Interpolatory Projection Methods. *SIAM J. Sci. Comput.* **2013**, *35*, B1010–B1033. [CrossRef]
40. Beattie, C.A.; Gugercin, S. Model Reduction by Rational Interpolation. In *Model Reduction and Approximation: Theory and Algorithm*; Benner, P., Cohen, A., Ohlberger, M., Willcox, K., Eds.; Society for Industrial and Applied Mathematics: University City, PA, USA, 2014; pp. 297–334.
41. Salimbahrami, B. Structure Preserving Order Reduction of Large Scale Second Order Models. Ph.D. Thesis, Fakultät für Maschinenwesen, Technischen Universität München, Munich, Germany, 2005.
42. Bai, Z.; Su, Y. Dimension Reduction of Large-Scale Second-Order Dynamical Systems via a Second-Order Arnoldi Method. *SIAM J. Sci. Comput.* **2005**, *26*, 1692–1709. [CrossRef]

43. Chen, C.-C.; Kuo, C.-W.; Yang, Y.-J. Generating Passive Compact Models for Piezoelectric Devices. *IEEE Trans. Comput.-Aided Des. Integr. Circuits Syst.* **2011**, *30*, 464–467. [CrossRef]
44. Aftag, T.; Bechtold, T.; Hohlfeld, D.; Rudnyi, E.B.; Korvink, J.G. New Modeling Approach for Micro Energy Harvesting Systems Based on Model Order Reduction Enabling Truly System-level Simulation. In Proceedings of the 23rd Micromechanics and Microsystems Europe Workshop, Ilmenau, Germany, 9–12 September 2012.
45. Kurch, M. Entwicklung einer Simulationsumgebung für die Auslegung piezoelektrischer Energy Harvester. Ph.D. Thesis, Technischen Universität Darmstadt, Darmstadt, Germany, 2014.
46. Kudryavtsev, M.; Rudnyi, E.B.; Korvink, J.G.; Hohlfeld, D.; Bechtold, T. Computationally efficient and stable order reduction methods for a large-scale model of MEMS piezoelectric energy harvester. *Microelectron. Reliab.* **2015**, *55*, 747–757. [CrossRef]
47. Hu, S.; Yuan, C.; Castagnotto, A.; Lohmann, B.; Bouhedma, S.; Hohlfeld, D.; Bechtold, T. Stable reduced order modeling of piezoelectric energy harvesting modules using implicit Schur complement. *Microelectron. Reliab.* **2018**, *85*, 148–155. [CrossRef]
48. Hu, S.; Yuan, C.; Bechtold, T. Quasi-Schur Transformation for the Stable Compact Modeling of Piezoelectric Energy Harvester Devices. In Proceedings of the 12th International Conference on Science Computing in Electrical Engineering (SCEE), Taormina, Sicily, Italy, 23–27 September 2018.

 *micromachines*

Article

Single-Neuron Adaptive Hysteresis Compensation of Piezoelectric Actuator Based on Hebb Learning Rules

Yanding Qin * and Heng Duan

College of Artificial Intelligence; Tianjin Key Laboratory of Intelligent Robotics, Nankai University, Tianjin 300350, China; 2120180411@mail.nankai.edu.cn
* Correspondence: qinyd@nankai.edu.cn

Received: 4 December 2019; Accepted: 11 January 2020; Published: 12 January 2020

Abstract: This paper presents an adaptive hysteresis compensation approach for a piezoelectric actuator (PEA) using single-neuron adaptive control. For a given desired trajectory, the control input to the PEA is dynamically adjusted by the error between the actual and desired trajectories using Hebb learning rules. A single neuron with self-learning and self-adaptive capabilities is a non-linear processing unit, which is ideal for time-variant systems. Based on the single-neuron control, the compensation of the PEA's hysteresis can be regarded as a process of transmitting biological neuron information. Through the error information between the actual and desired trajectories, the control input is adjusted via the weight adjustment method of neuron learning. In addition, this paper also integrates the combination of Hebb learning rules and supervised learning as teacher signals, which can quickly respond to control signals. The weights of the single-neuron controller can be constantly adjusted online to improve the control performance of the system. Experimental results show that the proposed single-neuron adaptive hysteresis compensation method can track continuous and discontinuous trajectories well. The single-neuron adaptive controller has better adaptive and self-learning performance against the rate-dependence of the PEA's hysteresis.

Keywords: piezoelectric actuator; hysteresis compensation; single-neuron adaptive control; Hebb learning rules; supervised learning

---

1. Introduction

As a sub-nanometer-resolution actuation device, piezoelectric actuators (PEAs) have been widely applied in various applications requiring nanometer-accurate motion [1–4]. However, the inherent hysteresis nonlinearity of the PEA greatly degrades its positioning accuracy, thus affecting its applicability and performance in precise operation tasks. The most significant characteristics of the PEA's hysteresis are the rate-dependence and asymmetry [5–7], i.e., the hysteresis loop becomes thicker with the increment in the input rate (or frequency) and the hysteresis loop is not symmetric about the loop center. These characteristics increase the complexity of the system and cause great difficulties in hysteresis modeling and compensation.

To address the above problems, lots of control methods have been proposed to characterize and compensate the hysteresis of the PEA. Physical models can be derived from physical measurement methods, such as magnetization, stress–strain, and energy principles [8,9]. However, the mathematical representations are often complex, making it difficult to obtain the inverse hysteresis model. In the meantime, a phenomenon-based model is also proposed, such as the Preisach model [10], Prandtl–Ishlinskii (PI) model [11,12], and Maxwell model [13]. As the inversion of the classical PI model is analytically available, it has been widely utilized in much research to describe the hysteresis characteristics of the PEA. After the inversion model is obtained, it can be utilized as a feedforward hysteresis compensator. This modeling and inversion approach is widely adopted, and many adaptive methods can be integrated [14–17]. In order to avoid the inversion calculation, the direct inversion method (DIM) is also proposed to identify the inverse hysteresis model directly from the measurements in parameter identification [18–20].

For model-based hysteresis compensation, the performance of the controller is highly dependent on the modeling accuracy of the hysteresis model. However, the PEA's hysteresis is susceptible to many factors, such as the external load and the frequency of the control input. This makes the modeling and compensation of the PEA's hysteresis very case-sensitive. As a result, a high-precision hysteresis model is generally difficult to obtain. Therefore, many intelligent control algorithms have been proposed to achieve higher robustness and adaptability. For instance, sliding mode control has been proposed to improve the accuracy and the robustness against noise and disturbances [14,21]. A linearization control method with feedforward hysteresis compensation and proportional-integral-derivative (PID) feedback has also been proposed [22]. Besides, iterative learning control schemes have been verified to achieve high-performance tracking for PEAs [23].

In the field of intelligent control, the neural network is a highly powerful system identification tool. It has a strong self-learning ability and powerful mapping ability to nonlinear systems, which has been widely used in the control of complex systems [24,25]. In the hysteresis compensation of the PEA, Wang and Chen presented a novel Duhem model based on the neural network to describe the dynamic hysteresis of PEAs [26]. An inversion-free predictive controller was proposed based on a dynamic linearized multilayer feedforward neural network model [27]. A cerebellar model articulation controller neural network PID controller was also proposed [4]. A radial basis function (RBF) network was also used to model and compensate for the PEA's hysteresis [28]. However, the use of the S-type action function increases the calculation difficulty for fast, high-frequency, and fast-response systems such as the PEA.

Among the neural network-based controllers, the single adaptive neuron system retains the advantages of the neural network and can satisfy the requirements of the real-time control of fast processes [29,30]. Therefore, a single-neuron adaptive hysteresis compensation method is proposed in this paper. The controller imitates an adaptive single-neuron system to learn and uses Hebb learning rules and supervised learning to adjust the controller. The controller can respond quickly to time-varying signals, making it suitable for the rate-dependent hysteresis compensation. Positioning and trajectory tracking experiments are carried out to investigate the performance of the proposed method. The performance of the PID control is also investigated for the purpose of comparison. For the positioning control, the proposed method can converge in about 8 ms and the steady-state tracking error can be reduced to the noise level of the system. For trajectory tracking, sinusoidal and triangular trajectories with frequencies up to 50 Hz are utilized. The experimental results show that the proposed method has excellent robustness and adaptability against the rate-dependence of the PEA's hysteresis, and the hysteresis can be successfully compensated.

This paper is organized as follows: Section 2 introduces the properties of the inherent hysteresis of the PEA. Section 3 presents the single-neuron adaptive controller design and analysis. To investigate the efficiency of the proposed method, experimental verifications and performance analyses are provided in Section 4. Section 5 summarizes this paper.

2. The Hysteretic Nonlinearity of the PEA

2.1. Experimental Setup

As shown in Figure 1, in this paper, a standalone PEA (model PZS001 from Thorlabs with integrated strain gauge sensors, Newton, NJ, USA) with a high-voltage amplifier (model ATA-4052 from Aigtek with a bandwidth of DC-500 kHz, Xi'an, China) is selected as the plant. The maximum displacement output of the PEA is measured to be 12.925 μm under the maximum actuation voltage of 10 V, i.e., the actuation gain (displacement/voltage) is 1.2925. According to the datasheet, the resonant frequency of the PEA used in this paper is 69 kHz. A dynamic Wheatstone bridge amplifier (model

SDY2105 from Beidaihe Institute of Practicality Electron Technology with a bandwidth of DC-300 kHz, Qinhuangdao, China) is utilized to measure the strain of the PEA, which is used to calculate the displacement of the PEA. The data acquisition and closed-loop control tasks are implemented on a real-time target (model microlabbox from dSPACE, Paderborn, Germany) with a sampling rate of 10 kHz. The algorithm is programmed in Simulink and implemented in Controldesk. Due to the influence of the strain gauges and the Wheatstone bridge amplifier, the measurement noise of the overall system is found to be ±34 nm.

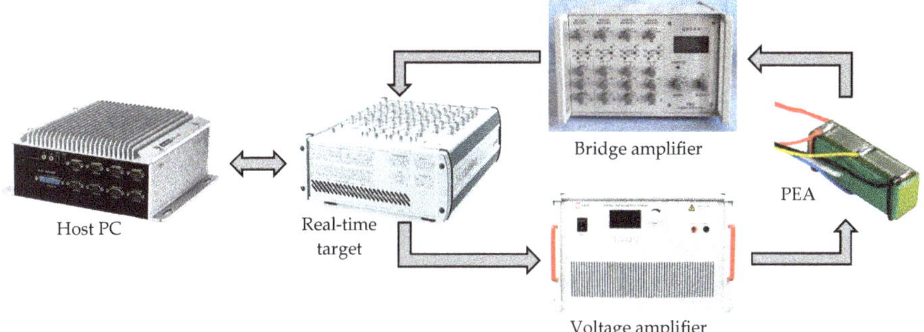

Figure 1. Schematic of the system setup for the standalone piezoelectric actuator (PEA).

2.2. Characteristics of the PEA's Hysteresis

Obvious nonlinearities can be observed in the input–output relationship of the PEA. Generally, the hysteresis is the dominant factor affecting the motion accuracy of the PEA. This paper uses sinusoidal signals of $u(t) = 5\sin(2\pi f t - \pi/2) + 5$ to drive the PEA at different frequencies. By observing the input signal and the measured displacement output of the PEA, hysteresis loops of the PEA can be obtained. Figure 2 depicts the measured input–output loops of the standalone PEA. As the resonant frequency of the PEA is 69 kHz, within the driving frequency of 1–400 Hz, the dynamics of the PEA can be neglected. As a result, the measured input–output loops shown in Figure 2 are totally produced by the hysteretic nonlinearity of the PEA. It can be seen that there is an obvious rate-dependent behavior in the measured hysteresis loops. As the input frequency increases, the hysteresis loop grows bigger and thicker. In addition, the hysteresis loop is not strictly symmetric about the loop center. The above rate-dependence and asymmetry properties increase the model complexity and increase the difficulty in the controller design of the PEA. Therefore, how to compensate the hysteresis and linearize the system have become crucial problems of the PEA.

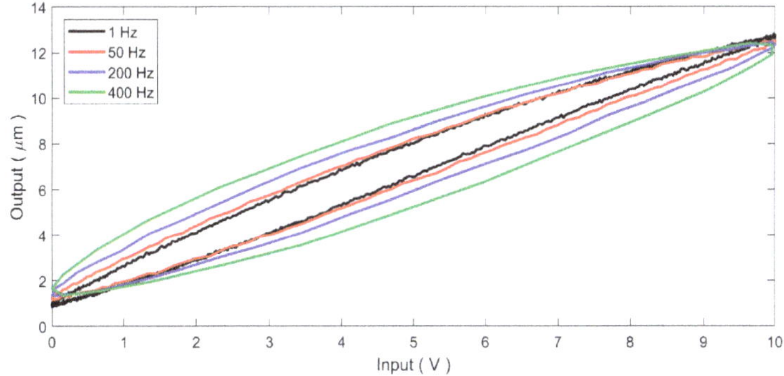

Figure 2. The measured hysteresis loops of the standalone PEA.

3. Single-Neuron Adaptive Controller Design

3.1. Single-Neuron Adaptive Control Algorithm

Aiming to compensate the hysteresis of the PEA, this paper proposes a single-neuron adaptive controller without modelling the hysteresis of the PEA. A single neuron is a non-linear processing unit that has self-learning and self-adaptive capabilities and is applicable for many different control tasks. The input and output of a single-neuron system are expressed as follows:

$$y = K \cdot \sum_{i=1}^{n} w_i x_i + \delta, \tag{1}$$

where K denotes the gain characterizing the response speed of a neuron; x_i, y, and δ are the state variable, output, and threshold, respectively; and w_i represents the weight of x_i that can be adjusted by the learning rules.

Neurons are generally considered to be self-organizing by modifying their synaptic weighting values. Supervised Hebb learning rules are usually used for the adjustment of the weights. Assuming the weight of the neuron $w_i(t)$ during learning is proportional to the signal $p_i(t)$ and decays slowly, the learning rule of the neuron can be expressed as

$$w_i(t+1) = (1-c)w_i(t) + dp_i(t), \tag{2}$$

where c is a positive constant that determines the impact of the last weight value, d is a constant characterizing the learning efficiency, and $p_i(t)$ is the learning rules. To further improve the adaptability of neurons, the following learning rules are employed:

$$p_i(t) = Z(t)S(t)x_i(t), \tag{3}$$

where $S(t)$ indicates that the adaptive neuron adopts the Hebb learning rule, and $Z(t)$ shows supervised learning rules. $Z(t)$ means that external information is self-organized to have a control effect under the guidance of the teacher signal. In this way, the adaptive neuron algorithm combined with Hebb learning rules and supervised learning can perform self-organizing and adaptive control for nonlinear systems.

3.2. Controller Design for the PEA

As shown in Figure 3, the state variables to the controller are calculated by the error between the desired trajectory $r(t)$ and the actual trajectory $y(t)$. The output of the controller is $u(t)$. In order to ensure the convergence and robustness of the learning algorithm, the following modified adaptive learning algorithm is adopted in this paper:

$$\begin{aligned} x_1(t) &= e(t) \\ x_2(t) &= \Delta x_1(t) = e(t) - e(t-1) \\ x_3(t) &= \Delta x_2(t) = e(t) - 2e(t-1) + e(t-2) \end{aligned}, \tag{4}$$

where $e(t) = r(t) - y(t)$ is the error between the desired and actual trajectories, and $x_1(t)$, $x_2(t)$, and $x_3(t)$ are adopted as the state variables to the neuron system.

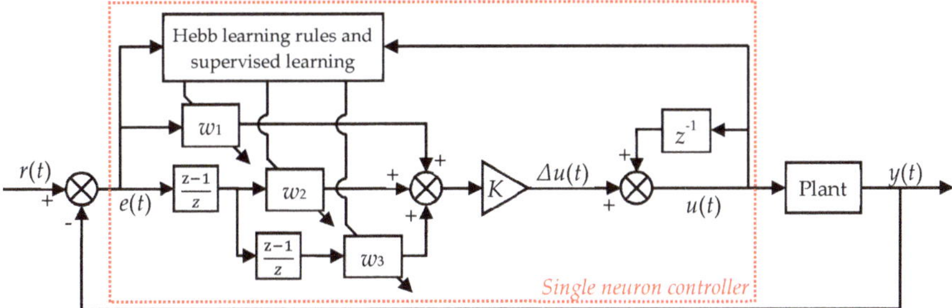

Figure 3. Schematic diagram of the single-neuron adaptive hysteresis compensation method.

The previous controller output $u(t-1)$ can be utilized as the threshold, i.e., $\delta = u(t-1)$. Substituting this into Equation (1), the controller output of the single-neuron adaptive controller can be written as follows:

$$u(t) = K \cdot \sum_{i=1}^{n} w_i x_i + u(t-1), \tag{5}$$

For the adaption of the weights, $Z(t) = e(t)$ is adopted as the supervisory function and $S(t) = u(t)$ is adopted as the Hebb learning rule. Substituting these into Equations (2) and (3), the learning rule of the neuron can be expressed as follows:

$$w_i(t) = w_i(t-1) + d \cdot e(t) \cdot u(t) \cdot x_i(t), \tag{6}$$

where c is set as 0 because $w_i(t)$ will converge to a stable value if c is small enough. According to the common experience of single-neuron adaptive control, d is typically less than 0.5. In this paper, $d = 0.4$ is adopted.

The whole control progress proceeds as follows. After getting the desired trajectory and actual trajectory, the state variable $x_i(t)$ is calculated using Equation (4). Three state variables correspond to three control outputs produced by the neuron, which are the proportional feedback $u_1(t)$, first-order differential feedback $u_2(t)$, and second-order differential feedback $u_3(t)$, respectively. The proportional feedback can quickly reduce the tracking error. The first-order differential feedback can improve the system's transient state performance, i.e., the response speed and overshoot. The second-order differential feedback ensures that the system remains stable during a fast response. The change in the weight reflects the dynamic characteristics of the controlled plant and the response process. The neuron continuously adjusts the weight through its own learning rules, and quickly eliminates the error and enters the steady-state under the correlation of the three feedbacks.

The system response speed is positively proportional to K, but a large overshoot might make the system unstable. On the contrary, if K is too small, the actual trajectory cannot track the desired trajectory. Thus, the tuning of K is very important. In order to determine the proper value for K, we built a mathematical model for the PEA using the Prandtl–Ishlinskii model. Through several simulation tests, the influence of K is computationally investigated. A candidate K is then selected according to the simulation results. Subsequently, this candidate K is adopted as the initial value and it is tuned manually online to achieve improved tracking performance. In this case, only fine tuning within a very small range is necessary.

4. Experimental Verifications

On the basis of the above analyses, the single-neuron adaptive control algorithm is applied to compensate the PEA's hysteresis. Positioning and trajectory tracking experiments are carried out to verify the proposed method's performance in hysteresis compensation.

In order to better compare the performances, this paper also includes the open loop and PID control results for the purpose of comparison. For the open-loop control, the PEA is assumed to be linear and the actuation gain (the ratio between the maximum allowable control input and the maximum displacement output) is utilized to finish the input–output mapping, i.e.,

$$u(t) = r(t) \cdot \frac{U_{\max}}{Y_{\max}}, \tag{7}$$

where $U_{\max}$ and $Y_{\max}$ are the maximum allowable control input and maximum displacement output, respectively. The open-loop control represents the basic characteristics of the system as no controller is utilized.

PID control, a widely utilized controller, has the advantages of simple parameter adjustment and ease of use. However, for nonlinear systems such as the PEA, the tuning of the gains in the PID controller is not an easy task. It might not work properly to systematically adjust the PID gains via strictly following well-developed approaches such as the Ziegler–Nichols method. Further, the behavior of the PEA is susceptible to many factors, making it difficult or impossible for the PID control to maintain the control performance in all scenarios. All these increase the difficulty in PID tuning. In this paper, the critical ratio method is adopted to tune the PID gains. At the beginning, only the proportional gain K_p is tuned with the other gains set to 0. Subsequently, the other gains are adjusted after the K_p is specified. For the PEA, PI control is found to be adequate to achieve satisfactory performance. In fact, a trial and error process is inevitable to finely tune the PID gains to achieve satisfactory results.

4.1. Step Response

The step response is generally utilized to test the system's positioning accuracy. Further, it can also show the system's tracking performance for non-continuous trajectories. For step response, the gains of the proposed single-neuron adaptive controller are tuned to be $K = 0.0002$ and $d = 0.4$ following the method provided in Section 3.2. The gains of the PID controller are manually tuned to be $K_p = 0.8$, $K_i = 1000$, and $K_d = 0$ following the critical ratio method. Step response experiments are carried out and the experimental results are shown in Figure 4. The step response of the open-loop system is also provided for the purpose of comparison.

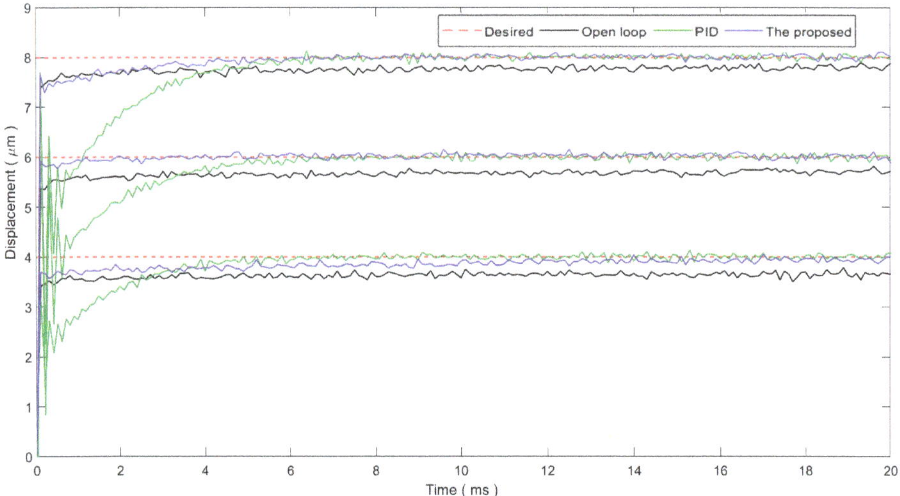

Figure 4. Step response performance.

As the resonant frequency of the PEA is 69 kHz and the sampling rate is set to 10 kHz, the transient state of the open-loop system is not observable. However, the slow creeping of the PEA can be observed and the steady-state positioning error is quite large. As a result, for the PEA, the response is fast, whereas the steady-state positioning accuracy is low. For the PID control, the PEA can converge within 8 ms and the steady-state positioning error is reduced to the noise level, whereas oscillations can be found in the transient state. For the proposed single-neuron adaptive controller, the rise time is on the same level of the open-loop system, indicating a fast response. There are no oscillations in the transient state. The proposed controller can also reduce the steady-state positioning error to the noise level. The convergence time, i.e., the learning time, of the proposed controller is on the level of several milliseconds. The experimental results show that the learning time of the proposed controller has an obvious sensitivity to the magnitude of the step size. For a smaller magnitude step, e.g., 4 μm step, the proposed controller can respond quickly, whereas it will take a relatively longer time (approximately 14 ms) for the controller to converge. For larger magnitudes, e.g., 8 μm step, the proposed controller can respond quickly and converge in about 6 ms.

Based on the above experimental results, both the proposed controller and the PID controller can reduce the steady-state positioning error to the noise level. The proposed controller is superior to the conventional PID controller in that it achieves a fast response and smooth transient state behavior. For the convergence time, the proposed controller converges faster for large step values, whereas the convergence time increases for small step values.

4.2. Tracking of Sinusoidal Trajectories

Sinusoidal trajectories are selected in this paper to verify the tracking performance of the proposed method on continuous trajectories. First, a low-frequency trajectory is adopted to tune the parameters of the controllers. In this case, the rate-dependence of the PEA's hysteresis can be neglected. This helps to ease the tuning of the parameters. The parameters of the controllers are tuned until excellent tracking performance is achieved in this case. These values are then fixed and higher-frequency trajectories are utilized to test the robustness and adaptability of the controllers.

In this paper, a 1 Hz sinusoidal trajectory is utilized to tune the parameters of the proposed method and the PID controller. For the proposed method, the parameters are tuned to $K = 0.002$ and $d = 0.4$. For the PID controller, the gains of K_p, K_i, and K_d are tuned to 1.11, 100, and 0, respectively. Figure 5 shows the tracking performance of the 1 Hz sinusoidal trajectory. Compared to the open-loop system, both the proposed method and the PID controller can successfully compensate the hysteresis of the PEA. The PEA can follow the desired trajectory well. It can be observed that the steady-state tracking error of the proposed method can be reduced to the noise level. The steady-state tracking error of the PID controller is slightly higher than the noise level but is still comparable to the proposed method. In the following experiments, sinusoidal and triangular trajectories with higher frequencies are utilized while the parameters of the two controllers are fixed to the above values. This helps to test the robustness and adaptability of the proposed method against the rate-dependence of the PEA's hysteresis.

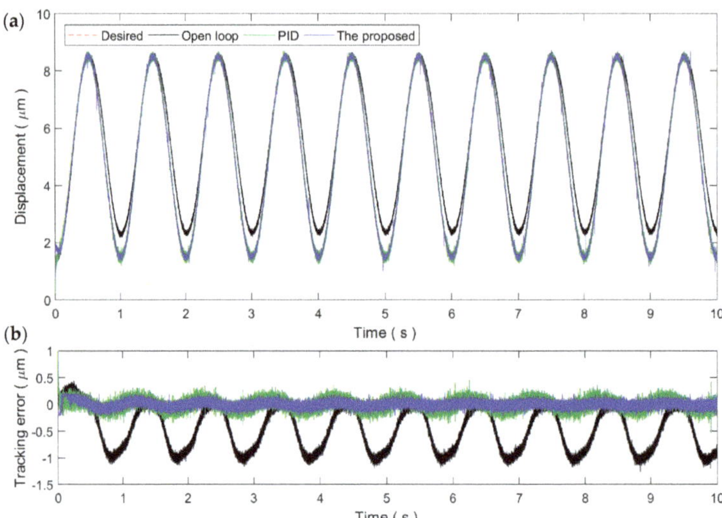

Figure 5. Tracking of the 1 Hz sinusoidal trajectory: (**a**) Time plot and (**b**) tracking error.

Figures 6 and 7 show the sinusoidal trajectory tracking results at 10 and 50 Hz, respectively. As the frequency of the desired trajectory increases, the trajectory tracking error of the proposed method increases slightly but still remains at the same magnitude of the measurement noise, exhibiting high robustness and adaptability against the rate-dependence of the PEA's hysteresis. On the contrary, the tracking performance of the PID controller starts to drop significantly at the frequency of 10 Hz. For the 50 Hz trajectory, the tracking performance of the PID control is even lower than the open-loop control. As a result, the PID gains tuned at the 1 Hz trajectory is only applicable for slow trajectories and might not work properly for fast trajectories.

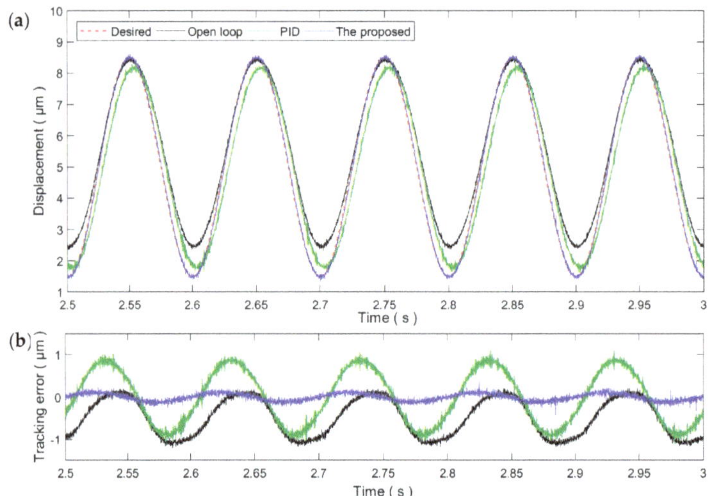

Figure 6. Tracking of the 10 Hz sinusoidal trajectory: (**a**) Time plot and (**b**) tracking error.

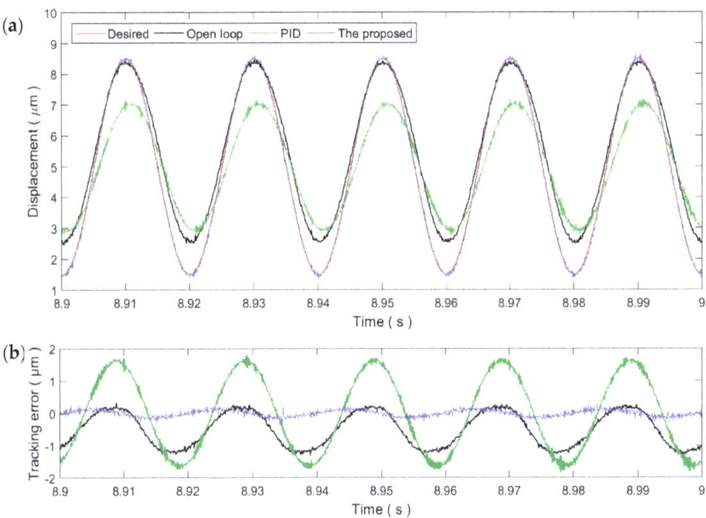

Figure 7. Tracking of the 50 Hz sinusoidal trajectory: (**a**) Time plot and (**b**) tracking error.

More trajectory tracking experiments are performed, whereas not all the experimental results are presented for the conciseness of this paper. In order to quantitatively investigate the tracking performance, the root-mean-square tracking error (RMSE) and relative root-mean-square error (RRMSE) of these sinusoidal trajectories are calculated and presented in Table 1. RMSE and RRMSE are defined in the following equations:

$$RMSE = \sqrt{\sum_{i=1}^{N} (y_i - r_i)^2 / N}, \tag{8}$$

$$RRMSE = \sqrt{\sum_{i=1}^{N} (y_i - r_i)^2 / \sum_{i=1}^{N} r_i} \times 100\%, \tag{9}$$

where y_i and r_i represents the i th values of the actual and desired trajectories, respectively, and N is the length of sampling data.

Table 1. Tracking errors of proportional-integral-derivative (PID) and the proposed approach.

Frequency (Hz)	The Proposed Method		PID	
	RMSE (nm)	RRMSE (%)	RMSE (nm)	RRMSE (%)
1	64.5	0.76	101.9	1.20
5	90.2	1.06	349.3	4.11
10	108.6	1.28	626.1	7.37
20	133.0	1.56	941.5	11.08
50	170.2	2.02	1158.8	13.63

As the parameters of the controllers are tuned at the 1 Hz trajectory, the RMSEs of the proposed method at the PID control are 64.5 and 101.9 nm, corresponding to a 0.76% and 1.2% relative error, respectively. As the frequency increases, the tracking performances of both controllers start to degrade. For the proposed method, the RMSE increases to 170.2 nm, i.e., 2.02% relative error. On the contrary, the RMSE of the PID control increases to 1158.8 nm, an approximately 13.63% relative error. These experimental results demonstrate that the proposed method can successfully compensate the rate-dependent hysteresis of the PEA.

4.3. Tracking of Triangular Trajectories

Sinusoidal trajectories are smooth trajectories that do not contain high-frequency harmonic components. In applications, triangular trajectories are also widely utilized, e.g., the scanning of the sample in an atomic force microscope. Unlike sinusoidal trajectories, triangular trajectories contain high-frequency harmonic components, increasing the difficulty in control. Therefore, the triangular trajectory is also selected in this paper to verify the performance of the proposed method for non-smooth trajectories. The parameters of the PID controller and the proposed method obtained in Section 4.2 are also inherited.

As shown in Figure 8, for the 1 Hz triangular trajectory, the tracking performance is similar to the 1 Hz sinusoidal trajectory. Both the PID control and the proposed method can reduce the tracking error to the noise level. This demonstrates the applicability of the two controllers for slow trajectories. Similarly, the tracking performance of both controllers decreases with the increment in the frequency of the triangular trajectory, which is obvious in Figures 9 and 10. Due to the influence of the high-frequency harmonic components, the tracking performance of the proposed method on the triangular trajectories is slightly lower than that of the sinusoidal trajectories. However, the tracking error still remains at very small ranges, showing strong robustness against the rate-dependence.

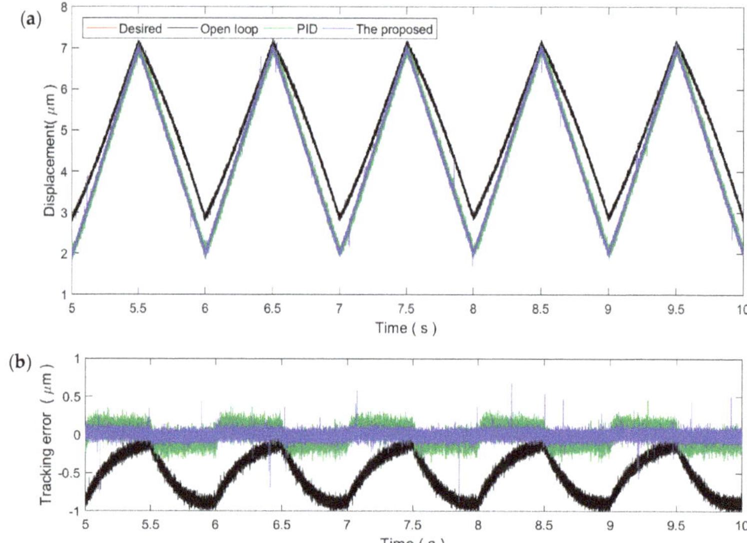

Figure 8. Tracking of the 1 Hz triangular trajectory: (**a**) Time plot and (**b**) tracking error.

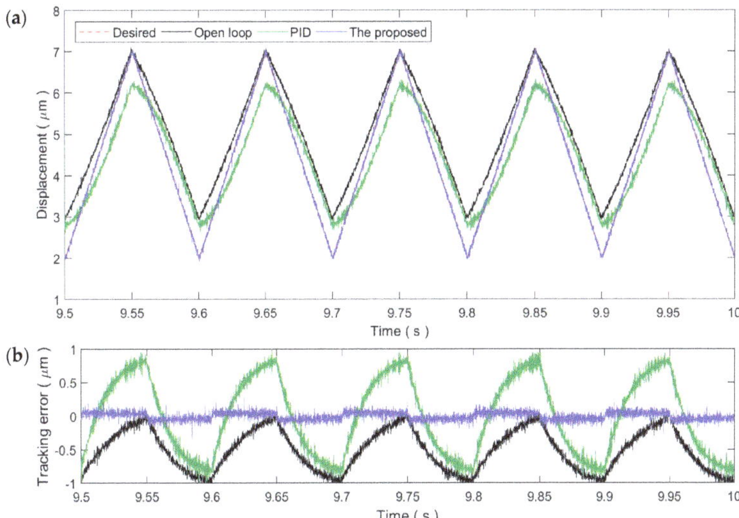

Figure 9. Tracking of the 10 Hz triangular trajectory: (**a**) Time plot and (**b**) tracking error.

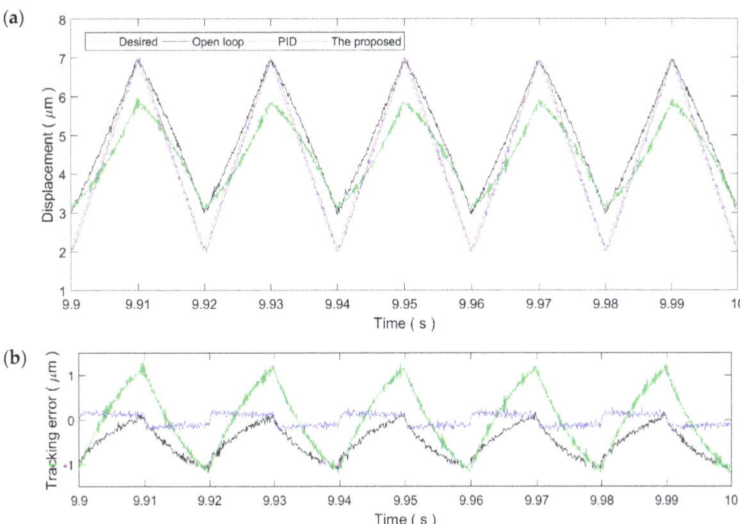

Figure 10. Tracking of the 50 Hz triangular trajectory: (**a**) Time plot and (**b**) tracking error.

4.4. Hysteresis Compensation Efficiency

The trajectory tracking results for sinusoidal trajectories are utilized to analyze the hysteresis compensation efficiency of the proposed method. The hysteresis loops in these experimental results are shown in Figure 11, where a 45° line is included to show the unitary mapping from the desired trajectory to the actual trajectory. For the 1 Hz sinusoidal trajectory, the resultant input–output relationship of both the PID controller and the proposed method coincide with the 45° line. The hysteresis loop is not observable, which means the hysteresis has been efficiently compensated. As the frequency increases, the input–output relationship of the proposed method stays close to the 45° line. On the contrary, for the PID controller, obvious hysteresis loops can still be observed in the input–output relationship for sinusoidal trajectories higher than 5 Hz. From Figure 11, we can conclude that the proposed method can successfully compensate the rate-dependent hysteresis of the PEA when compared to the PID controller.

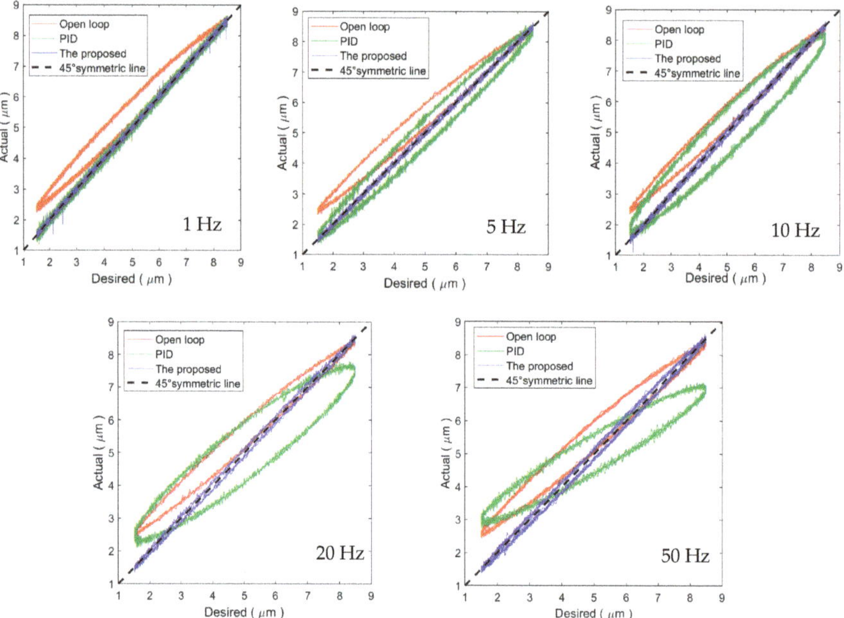

Figure 11. Hysteresis compensation efficiency for the sinusoidal trajectories.

5. Conclusions

The rate-dependence and asymmetry of the PEA's hysteresis increase the difficulty in the hysteresis modeling and compensation. Further, the PEA's hysteresis is susceptible to the system's configurations, making the hysteresis compensation of PEAs very case-sensitive. In this paper, a single-neuron adaptive hysteresis compensation method is proposed. The supervised learning and Hebb learning rules are adopted to dynamically adjust the weights of the neurons according to the error between the actual and desired trajectories and their first-order and second-order differences. As a branch of neural network control, the single-neuron adaptive control simplifies the training process of neural network control while retaining the advantages of neural network control. The learning efficiency and convergence are improved. Positioning control results show that the proposed method can reduce the steady-state tracking error to the noise level, and the transient state performance can be guaranteed. The experimental results of tracking sinusoidal and triangular trajectories with frequencies up to 50 Hz show that the proposed method can successfully compensate the rate-dependent hysteresis of the PEA. The steady-state tracking error can be maintained in a small range, showing great robustness and adaptability against the rate-dependence. Future work will focus on further improving the tracking performance for higher-frequency trajectories.

Author Contributions: Y.Q. conceived and designed the experiments; H.D. performed the experiments and sorted out the experimental results; Y.Q. analyzed the data; Y.Q. and H.D. wrote the paper. All authors have read and agreed to the published version of the manuscript.

Funding: This work is supported in-part by the National Natural Science Foundation of China under Grants 61873133, U1813210 and 61633012, and in part by the Fundamental Research Funds for the Central Universities, Nankai University under Grants 63191717 and 63191739.

Conflicts of Interest: The authors declare no conflict of interest.

References

1. Yang, C.; Li, C.; Xia, F.; Zhu, Y.; Zhao, J.; Youcef-Toumi, K. Charge controller with decoupled and self-compensating configurations for linear operation of piezoelectric actuators in a wide bandwidth. *IEEE Trans. Ind. Electron.* **2019**, *66*, 5392–5402. [CrossRef]
2. Zhang, X.; Xu, Q. Design, fabrication and testing of a novel symmetrical 3-DOF large-stroke parallel micro/nano-positioning stage. *Robot. Comput. Integr. Manuf.* **2018**, *54*, 162–172. [CrossRef]
3. Qin, Y.D.; Zhao, X.; Zhou, L. Modeling and identification of the rate-dependent hysteresis of piezoelectric actuator using a modified Prandtl-Ishlinskii model. *Micromachines* **2017**, *8*, 11. [CrossRef]
4. Lin, C.; Zheng, S.; Li, P.; Shen, Z.; Wang, S. Positioning error analysis and control of a piezo-driven 6-DOF micro-positioner. *Micromachines* **2019**, *10*, 542. [CrossRef] [PubMed]
5. Li, H.; Zhu, B.; Chen, Z.; Zhang, X. Realtime in-plane displacements tracking of the precision positioning stage based on computer micro-vision. *Mech. Syst. Signal Process.* **2019**, *124*, 111–123. [CrossRef]
6. Mishra, J.P.; Xu, Q.; Yu, X.; Jalili, M. Precision Position Tracking for Piezoelectric-Driven Motion System Using Continuous Third-Order Sliding Mode Control. *IEEE/ASME Trans. Mechatron.* **2018**, *23*, 1521–1531. [CrossRef]
7. Li, M.; Wang, Q.; Li, Y.; Jiang, Z. Modeling and discrete-time terminal sliding mode control of a DEAP actuator with rate-dependent hysteresis nonlinearity. *Appl. Sci.* **2019**, *9*, 2625. [CrossRef]
8. Changshi, L. Comprehension of the ferromagnetic hysteresis via an explicit function. *Comput. Mater. Sci.* **2015**, *110*, 295–301. [CrossRef]
9. Wang, W.; Yao, J.E. Modeling and identification of magnetostrictive hysteresis with a modified rate-independent Prandtl-Ishlinskii model. *Chin. Phys. B* **2018**, *27*. [CrossRef]
10. Tao, Y.D.; Li, H.X.; Zhu, L.M. Rate-dependent hysteresis modeling and compensation of piezoelectric actuators using Gaussian process. *Sens. Actuator A Phys.* **2019**, *295*, 357–365. [CrossRef]
11. Oliveri, A.; Maselli, M.; Lodi, M.; Storace, M.; Cianchetti, M. Model-based compensation of rate-dependent hysteresis in a piezoresistive strain sensor. *IEEE Trans. Ind. Electron.* **2019**, *66*, 8205–8213. [CrossRef]
12. Qin, Y.; Shirinzadeh, B.; Tian, Y.; Zhang, D. Design issues in a decoupled XY stage: Static and dynamics modeling, hysteresis compensation, and tracking control. *Sens. Actuators A Phys.* **2013**, *194*, 95–105. [CrossRef]
13. Liu, Y.; Du, D.; Qi, N.; Zhao, J. A Distributed Parameter Maxwell-Slip Model for the Hysteresis in Piezoelectric Actuators. *IEEE Trans. Ind. Electron.* **2019**, *66*, 7150–7158. [CrossRef]
14. Gao, J.; Hao, L.; Cheng, H.; Cao, R.; Sun, Z. Simulation and Experiment Based on FSMLC Method with EUPI Hysteresis Compensation for a Piezo-Driven Micro Position Stage. *J. Syst. Sci. Complex.* **2019**, *32*, 1340–1357. [CrossRef]
15. Qin, Y.; Jia, R. Adaptive hysteresis compensation of piezoelectric actuator using direct inverse modelling approach. *Micro Nano Lett.* **2018**, *13*, 180–183. [CrossRef]
16. Zheng, Y.; Zhang, G. Quantized adaptive tracking control for nonlinear systems with actuator backlash compensation. *J. Frankl. Inst.* **2019**, *356*, 8484–8506. [CrossRef]
17. Liu, X.; Huang, M.; Xiong, R.; Shan, J.; Mao, X. Adaptive inverse control of piezoelectric actuators based on segment similarity. *IEEE Trans. Ind. Electron.* **2019**, *66*, 5403–5411. [CrossRef]
18. Qin, Y.; Tian, Y.; Zhang, D.; Shirinzadeh, B.; Fatikow, S. A novel direct inverse modeling approach for hysteresis compensation of piezoelectric actuator in feedforward applications. *IEEE/ASME Trans. Mechatron.* **2013**, *18*, 981–989. [CrossRef]
19. Gu, G.-Y.; Yang, M.-J.; Zhu, L.-M. Real-time inverse hysteresis compensation of piezoelectric actuators with a modified Prandtl-Ishlinskii model. *Rev. Sci. Instrum.* **2012**, *83*, 065106. [CrossRef]
20. Ryba, Ł.; Dokoupil, J.; Voda, A.; Besançon, G. Adaptive hysteresis compensation on an experimental nanopositioning platform. *Int. J. Control* **2017**, *90*, 765–778. [CrossRef]
21. Ding, B.; Li, Y. Hysteresis compensation and sliding mode control with perturbation estimation for piezoelectric actuators. *Micromachines* **2018**, *9*, 241. [CrossRef] [PubMed]
22. Wang, S.; Chen, Z.; Liu, X.; Jiao, Y. Feedforward Feedback Linearization Linear Quadratic Gaussian with Loop Transfer Recovery Control of Piezoelectric Actuator in Active Vibration Isolation System. *J. Vib. Acoust. Trans. ASME* **2018**, *140*. [CrossRef]

23. Huang, D.; Jian, Y. High-precision tracking of piezoelectric actuator using iterative learning control. *Unmanned Syst.* **2018**, *6*, 175–183. [CrossRef]

24. Liu, D.; Fang, Y.; Wang, H.; Dong, X. Adaptive novel MSGA-RBF neurocontrol for piezo-ceramic actuator suffering rate-dependent hysteresis. *Sens. Actuators A Phys.* **2019**, *297*. [CrossRef]

25. Xu, R.; Zhou, M. A self-adaption compensation control for hysteresis nonlinearity in piezo-actuated stages based on Pi-sigma fuzzy neural network. *Smart Mater. Struct.* **2018**, *27*. [CrossRef]

26. Wang, G.; Chen, G. Identification of piezoelectric hysteresis by a novel Duhem model based neural network. *Sens. Actuators A Phys.* **2017**, *264*, 282–288. [CrossRef]

27. Liu, W.; Cheng, L.; Hou, Z.G.; Yu, J.; Tan, M. An inversion-free predictive controller for piezoelectric actuators based on a dynamic linearized neural network model. *IEEE/ASME Trans. Mechatron.* **2016**, *21*, 214–226. [CrossRef]

28. Wang, S.; Chen, Z.; Jiao, Y.; Mo, W.; Liu, X. Hysteresis modeling and control of piezoelectric actuator. In Proceedings of the ASME 2017 International Mechanical Engineering Congress and Exposition (IMECE), Tampa, FL, USA, 3–9 November 2017.

29. Zhang, H.; Li, R.; Yao, J.; Hu, X.; Gu, C.; Wang, M. Design and implementation of submarine cable way-points tracking control based on single neuron adaptive PID. In Proceedings of the 2019 4th International Conference on Automation, Control and Robotics Engineering (CACRE), Shenzhen, China, 19–22 July 2019; pp. 1–6.

30. Elnady, A.; Alshabi, M. Operation of parallel inverters in microgrid using new adaptive PI controllers based on least mean fourth technique. *Math. Probl. Eng.* **2019**, *2019*. [CrossRef]

 *micromachines*

Article

A Piezo-Electromagnetic Coupling Multi-Directional Vibration Energy Harvester Based on Frequency Up-Conversion Technique

Ge Shi [1,*], Junfu Chen [1], Yansheng Peng [1], Mang Shi [1], Huakang Xia [2], Xiudeng Wang [2], Yidie Ye [2] and Yinshui Xia [2]

[1] College of Mechanical and Electrical Engineering, China Jiliang University, Hangzhou 310018, China; P1801085203@cjlu.edu.cn (J.C.); P1801085240@cjlu.edu.cn (Y.P.); 12b5100080@cjlu.edu.cn (M.S.)

[2] Faculty of Electrical Engineering and Computer Science, Ningbo University, Ningbo 315211, China; xiahuakang@nbu.edu.cn (H.X.); 1611082572@nbu.edu.cn (X.W.); yeyidie@nbu.edu.cn (Y.Y.); xiayinshui@nbu.edu.cn (Y.X.)

* Correspondence: shige@cjlu.edu.cn; Tel.: +86-0571-86914466

Received: 16 December 2019; Accepted: 8 January 2020; Published: 11 January 2020

Abstract: Harvesting vibration energy to power wearable devices has become a hot research topic, while the output power and conversion efficiency of a vibration energy harvester with a single electromechanical conversion mechanism is low and the working frequency band and load range are narrow. In this paper, a new structure of piezoelectric electromagnetic coupling up-conversion multi-directional vibration energy harvester is proposed. Four piezoelectric electromagnetic coupling cantilever beams are installed on the axis of the base along the circumferential direction. Piezoelectric plates are set on the surface of each cantilever beam to harvest energy. The permanent magnet on the beam is placed on the free end of the cantilever beam as a mass block. Four coils for collecting energy are arranged on the base under the permanent magnets on the cantilever beams. A bearing is installed on the central shaft of the base and a rotating mass block is arranged on the outer ring of the bearing. Four permanent magnets are arranged on the rotating mass block and their positions correspond to the permanent magnets on the cantilever beams. The piezoelectric cantilever is induced to vibrate at its natural frequency by the interaction between the magnet on cantilever and the magnets on the rotating mass block. It can collect the nonlinear impact vibration energy of low-frequency motion to meet the energy harvesting of human motion.

Keywords: piezo-electromagnetic coupling; up-conversion; vibration energy harvester; multi-directional vibration; low frequency vibration

---

1. Introduction

Intelligent wearable devices have become a new application hotspot with the rapid development of electronic technology [1]. However, it is difficult for the battery of intelligent wearable devices to meet the requirements of its constantly improving functions and long-time standby [2,3]. With the development of ultra-low power technology and micro-electro-mechanical system (MEMS) technology, it is possible to harvest environmental energy to supply power for intelligent wearable devices [4–6]. At present, the energy that can be used in the environment includes solar energy, heat energy, vibration energy and so forth, and vibration energy widely exists in the human living environment [7–9]. Therefore, researchers have begun to study the technology of harvesting vibration energy in the environment and converting it into electrical energy to power intelligent wearable devices.

At present, the most studied environmental vibration energy harvesting methods are the piezoelectric harvester and the electromagnetic harvester. The output voltage of the piezoelectric vibration energy harvester is higher but the output current is smaller and its output impedance is capacitive [10–12]. The output voltage of electromagnetic vibration energy harvester is lower but the output current is larger and its output impedance is inductive [13–16]. The output power and conversion efficiency of the energy harvester with a single electromechanical conversion mechanism is low and the operating frequency band and load range are narrow [17,18]. Because of the resonance type, a linear vibration energy harvester has higher output power only near its natural resonance frequency point [19]. The smaller the harvester size is, the higher the natural resonant frequency is. At present, the vibration in the range of human application is mostly low-frequency vibration [20,21]. To harvest low-frequency vibration energy, it is necessary to convert the low-frequency vibration to the natural resonant frequency vibration of harvester.

Researchers have also proposed many new structures for vibration energy harvesting. Li et al. [22] designed an enhanced bandwidth nonlinear resonant electromagnetic energy harvester. The inertial mass of the proposed harvester is formed by four stacked ring permanent magnets (PMs), which is suspended axially by two magnetic springs and circumferentially by ferrofluid within a carbon fiber tube. The magnetic springs are made up of two button PMs adhered, respectively, to the end cap at each end of the carbon fiber tube to provide varying repulsive forces to the PM stack, resulting in enhanced resonance frequency band and higher efficiency in energy harvesting. However, this kind of energy harvester can only collect vibration energy in a certain direction [15,22], which has no obvious advantage for human motion or swing. Therefore, many researchers use rotating mass structure to harvest human motion energy [21,23,24]. For example, Smilek et al. [23] present a novel design of a nonlinear kinetic energy harvester for very low excitation frequencies below 10 Hz. The design is based on a proof mass, rolling in a circular cavity in a Tusi couple configuration. This allows for an unconstrained displacement of the proof mass while maintaining the option of keeping the energy transduction element engaged during the whole cycle and thus reducing the required number of transduction elements. Romero et al. [21] present a micro-rotational energy harvester topology for extracting electric energy from human body motion at joint locations. This was accomplished using an inertial-based axial flux machine constructed with multiple permanent magnet poles and stacked micro-fabricated planar coils. An average power of 472 µW was obtained when the 2 cm^3 device was placed on the ankle while walking at 4mph. Some researchers use an up-conversion method to convert external low-frequency excitation to high-frequency vibration of cantilever beam to harvest energy [19,25,26]. For example, Gu et al. [19] present an energy harvesting device in which a low frequency resonator impacts a high frequency energy harvesting resonator, resulting in energy harvesting predominantly at the system's coupled vibration frequency. Pillatsch et al. [25] present an energy harvester specifically targeted at low frequency random excitation as is encountered in human motion. The design uses a frequency up-conversion technique based on magnetic actuation of a piezoelectric beam. The benefits of a rotational design with eccentric proof mass compared to a linear design are discussed. This makes the approach suitable for the random device orientations of the human body. Researchers also try to achieve efficient energy harvesting through the combination of various energy exchange mechanisms [27–30]. For example, Zhu et al. [27] present the design and analysis of a new magnetoelectric energy harvester that uses Terfenol-D/piezoelectric/Terfenol-D laminate to harvest energy from nonlinear vibrations created by magnetic levitation. Due to the high energy density and strong magneto-mechanical coupling effect of magnetostrictive material, the proposed harvester is capable of generating very high voltage and power at low frequency ranges. Ibrahim et al. [29] proposed two hybrid energy harvesters that each employs a combination of piezoelectric (PZT), magnetostrictive (MSM) and electromagnetic technologies. Harvesters employ spiral geometries to allow smaller natural frequencies compared to a straight beam. In particular, the hybrid technology that uses piezoelectric and magnetostrictive has shown significant improvement in the frequency bandwidth of the device.

As mentioned above, most of the methods of human motion energy harvesting adopt a rotating mass structure or up-conversion technology. Up-conversion technology can be divided into collision contact type [19] and non-contact type [25,26]. The non-contact type mainly uses the magnetic force to excite the piezoelectric cantilever beam to generate resonance. However, they also have some defects, for example, they cannot harvest vibration energy in multiple directions; the magnetic method limits the swing amplitude of the cantilever; the driving force of piezoelectric plate vibration only has one direction; the voltage of the electromagnetic coil is too low, which will bring high requirements to the subsequent energy harvesting circuit.

In this paper, a new energy harvester that combines piezoelectric electromagnetic coupling resonance and up-conversion technology is proposed. By composing a multi field nonlinear coupling system of force-electricity-magnetism, a higher working frequency band can be obtained. The up-conversion technology is used to harvest low frequency vibration energy and rotating mass block is used to drive piezoelectric cantilever beam to obtain multi-directional energy harvesting. This paper mainly discusses the piezoelectric electromagnetic coupling model, analyzes the factors that affect the power of the system and tests its energy collection effect on the experimental device.

2. Design of Energy Harvester

It is common to harvest the low-frequency vibration energy of the human body through low-frequency rotating motion. In this paper, a new structure of piezoelectric electromagnetic coupling multi-directional vibration energy harvester based on a frequency up-conversion technique is proposed. The structural design of the energy harvester is shown in Figure 1.

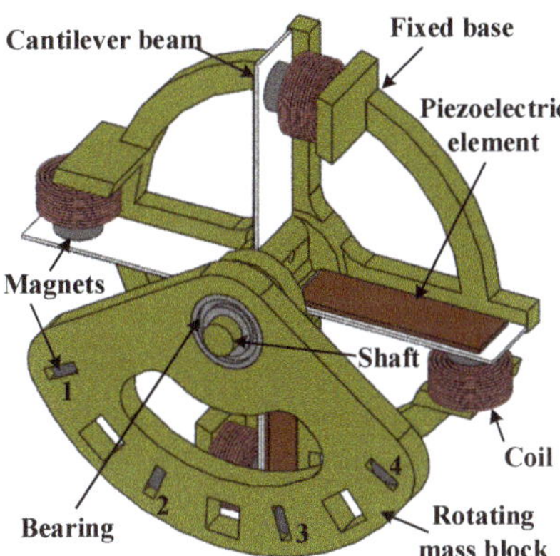

Figure 1. The structural design of the multi-directional vibration energy harvester.

Four piezoelectric electromagnetic coupling cantilever beams were installed on the axis of the base along the circumferential direction. Piezoelectric elements were set on the surface of each cantilever beam to harvest piezoelectric energy. The permanent magnet on the beam was placed on the free end of the cantilever beam as a mass block. Four coils for collecting energy were arranged on the base under the permanent magnets on the cantilever beams to harvest electromagnetic energy. A bearing was installed on the central shaft of the base and a rotating mass block was arranged on the outer ring of the bearing. Four permanent magnets were arranged on the rotating mass block and their positions corresponded to the permanent magnets on the cantilever beams.

The permanent magnets used in the harvester were Neodymium Iron Boron magnets (NdFeB) permanent magnets (Φ6 mm × 3 mm) and were arranged so that they were mutually exclusive as this proved to be advantageous. The bearing model was PNY-686ZZ (SAKAGAMI, Tokyo, Japan), with an inner diameter of 6 mm, an outer diameter of 13 mm and a thickness of 5 mm. It was mounted on the 6 mm central axis of the base. The rotating mass block was made of polylactic acid (PLA) material by way of 3D printing. The diameter of the circular hole matched with the outer ring of the bearing was 6mm, the thickness is 5 mm too and the weight is 30 g. Mass block can be added on it to increase its inertia. The base is also made of PLA material by way of 3D printing. The fixed base was in the shape of a ring and an outer ring was connected by four pillars, with a diameter of 70 mm. It provided four platforms for coils. The macro-fiber composite (MFC) piezoelectric elements (M-2807-P2, Smart Material, Sarasota, FL, USA) were glued to the cantilever beams. The capacitance of the piezoelectric was 14.3 nf and the tested natural frequency of the cantilever beam was about 42 Hz. The small scale device can be realized by reducing the size of the cantilever beam, the area of the piezoelectric, the volume of the coil and so forth.

3. Principle and Model Analysis

In order to further understand the principle and dynamic behaviour of this harvester, the configuration depicted in Figure 2 was studied. A point mass m at a distance r from its axis of rotation was considered under gravity and external excitation. The motion equation of the rotating mass was as shown in Equation (1) [25].

$$\ddot{\gamma}I = F_x r_y + (F_y - mg)r_x, \tag{1}$$

where the distances r_x and r_y describe the rotating mass position and F_x and F_y are the corresponding inertial reaction forces caused by linear external excitation. $I = mr^2/4$ is the mass moment of inertia of the rotor. Gravity g acts in the negative y-direction. The angle γ is the angular deflection of the proof mass in relation to the y-axis.

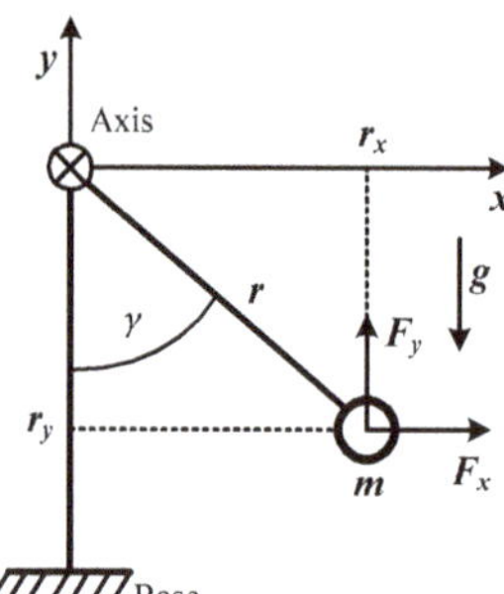

Figure 2. Schematic view of the rotating mass under external excitation.

Through Equation (1), it can be clearly seen that the rotation design can accept the linear and rotation excitation in x and y directions, which makes it more versatile in the case of variable equipment direction and host motion in multiple degrees of freedom at random, just like in the human body. Under the excitation of gravity and rotation, the rotating mass behaves like a pendulum. If the base is rotated, the rotating mass only stays in the negative y-direction, resulting in $\gamma' = -\gamma$. The relative motion between the base and the rotor becomes $\gamma'(t) = -\gamma(t)$. In this scenario the device is behaving purely inertia. In most real applications there will be a mixture of these cases of pendulum and inertial motion. However, the basic analysis method is the same.

The structure diagram of a single piezoelectric electromagnetic coupling cantilever beam is shown in Figure 3. The surface of the cantilever beam was pasted with MFC piezoelectric element and the free end of the cantilever beam is provided with permanent magnet and coil. When the cantilever beam vibrates under external excitation, the MFC piezoelectric element will output alternating voltage because of alternating stress and the closed coil will also produce alternating current because of the change of magnetic flux. The former is based on the positive piezoelectric effect and the latter is based on Faraday's electromagnetic induction law. U is the vibration displacement of cantilever beam. The piezoelectric electromagnetic coupling vibration energy harvester proposed in this paper improves the harvesting efficiency by integrating the two electromechanical conversion mechanisms to a cantilever beam structure.

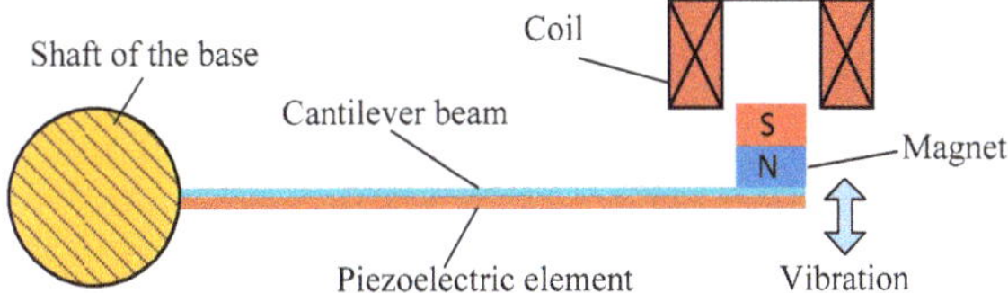

Figure 3. Structural diagram of single piezoelectric electromagnetic coupling cantilever beam.

The equivalent circuit model of a single piezoelectric electromagnetic coupling cantilever beam is shown in Figure 4.

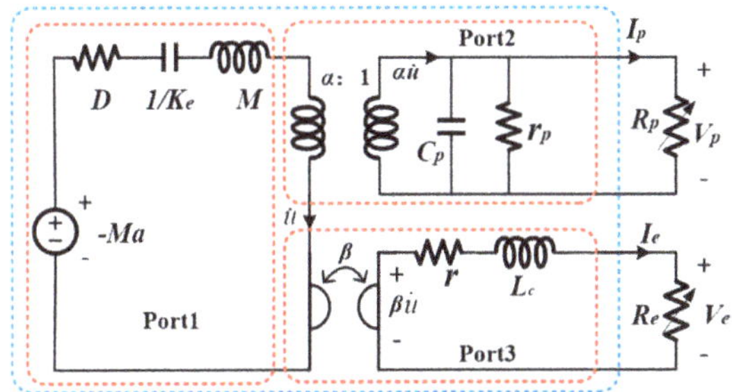

Figure 4. Equivalent circuit model of a single piezoelectric electromagnetic coupling cantilever beam. Port1: mechanical domain, Port2: piezoelectric unit, Port3: electromagnetic unit.

Here, Ma represents the external exciting force in the mechanical domain and the resistance represents the damping D, the capacitance represents the elastic potential energy, which is expressed by $1/K_e$ of the elastic coefficient, the inductance represents the kinetic energy, which is expressed by the magnet mass M. The current of the primary circuit is the vibration velocity u of the mass relative to the base. The equivalent circuit of piezoelectric element is coupled to mechanical domain through transformer, while the equivalent circuit of electromagnetic element is coupled to the mechanical domain through a gyrator [18]. The α represents the piezoelectric force-voltage factor and β the electromagnetic force-current factor. In the piezoelectric equivalent circuit, the C_p and r_p are corresponding clamped capacitor and internal resistance of the piezoelectric element. Because of the large value of r_p, it is often ignored in analysis. In the electromagnetic element circuit, the L_c and r are the corresponding inductance and internal resistance of the electromagnetic coil.

When the output ports of the piezoelectric element and electromagnetic coil are respectively connected to two resistance loads R_p and R_e, V_p and V_e are the corresponding output voltages of the piezoelectric element and electromagnetic coil; I_p and I_e are corresponding to the output currents of the piezoelectric element and electromagnetic coil.

Assuming that the excitation acceleration is single sine wave $a(t) = a_M\sin(\omega t)$ ($0 \le t \le 2\pi/\omega$) for simple analysis, the a_M is the amplitude of acceleration, the ω is angular frequency of rotating mass block. The relative movement displacement between the free end of the cantilever and the base is $u(t)$, the system differential equation can be described as [31]:

$$M\ddot{u}(t) + D\dot{u}(t) + K_e u(t) = -Ma(t) - \alpha V_p(t) - \beta I_e(t), \tag{2}$$

According to Kirchhoff's law, the electrical equations are established as follows:

$$\alpha\dot{u}(t) = C_p\dot{V}_p(t) + \frac{V_p(t)}{R_p}, \tag{3}$$

$$\beta\dot{u}(t) = L_e\dot{I}_e(t) + (R_e + r)I_e(t). \tag{4}$$

The transfer function relationship between the relative movement displacement $U(s)$ of the free end of the cantilever beam and the excitation acceleration $A(s)$ of the foundation can be obtained by the Laplace transformation and simplification under the zero initial condition of Equation (1), Equation (2) and Equation (3):

$$\frac{U(s)}{A(s)} = \frac{-M}{Ms^2 + Ds + K_e + \beta^2 s/(r + R_e + L_c s) + \alpha^2 R_p s/(1 + C_p R_p s)}. \tag{5}$$

By calculating the amplitude of the transfer function, the maximum amplitude of the free end of the cantilever at vibration frequency ω can be obtained:

$$u_M = \frac{Ma_M}{\sqrt{\left[-M\omega^2 + K_e + \frac{\omega^2\beta^2 L_c}{(r+R_e)^2+\omega^2 L_c^2} + \frac{\omega^2 R_p^2\alpha^2 C_p}{1+(\omega C_p R_p)^2}\right]^2 + \left[D\omega + \frac{\beta^2(r+R_e)\omega}{(r+R_e)^2+\omega^2 L_c^2} + \frac{\alpha^2 R_p\omega}{1+(\omega C_p R_p)^2}\right]^2}}. \tag{6}$$

Therefore, each time the magnet of the rotating mass block and the magnet on the cantilever beam pass by, the cantilever will be excited to its maximum amplitude u_M position and then the cantilever will attenuate the vibration with its own natural frequency.

It can be seen from Equation (6) that the electromagnetic element mainly affects the damping of the harvester, while the piezoelectric element also affects the stiffness and damping of the harvester. The output power of the piezoelectric energy harvesting unit in the piezoelectric electromagnetic coupling energy capture device [18,31] is not only affected by the structural parameters and piezoelectric material properties but also by the characteristic parameters of the electromagnetic energy harvesting unit. Similarly, the output power of the electromagnetic energy capture unit is also affected by the characteristic parameters of the piezoelectric unit. Therefore, the appropriate transducer structure size and interface circuits will effectively improve the conversion efficiency of vibration energy harvester.

There are two benefits to using the converters combination. First, since permanent magnets are used in the up-conversion technology of piezoelectric cantilevers, an additional electromagnetic energy harvesting method is added to the existing permanent magnets, which can improve the conversion efficiency of vibration energy. In addition, the optimal load range (set 0.9 peak power band) of a single conversion mechanism is smaller. In the combination of two converters, the corresponding piezoelectric and electromagnetic loads can have more load combinations to obtain the same total power (also 0.9 peak power band) [18]. Therefore, it is easier to achieve maximum power tracking.

4. Experimental Test

In order to test the performance of the vibration energy harvester designed in this paper, two motion platforms were designed to test the experimental prototype.

The first was a single dimension vibration test platform, as shown in Figure 5a, which was composed of a shaker (ZJ-2A, Shanghai Zhurui CO., Shanghai, China), a driver (GF-20W, Shanghai Zhurui CO., Shanghai, China) and a function signal generator (DG3121A, RIGOL CO., Beijing, China). It can simulate a single dimension excitation vibration. Different vibration conditions can be simulated by controlling its vibration amplitude and frequency. By changing different frequency and amplitude, the motion state of the energy harvester under different excitation conditions was obtained and the results under different experimental conditions were obtained.

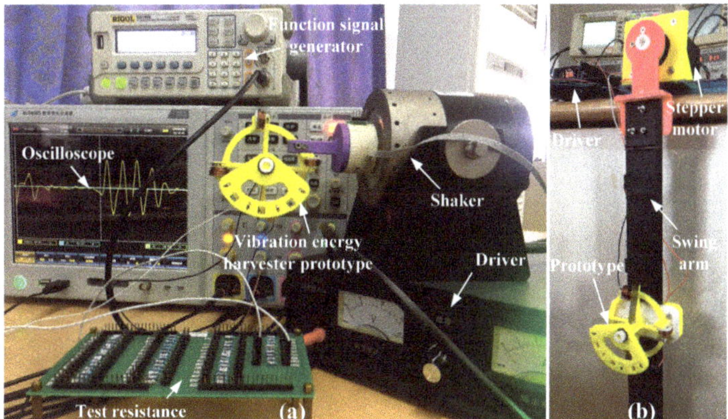

Figure 5. Experimental setup. (**a**) Single dimension vibration test platform. (**b**) Swing device for simulating the swing of human arm.

The second test platform was a swing device for simulating the swing of human arm, as shown in Figure 5b. It is a swing device composed of a stepper motor and a driver, which can simulate the low frequency and large swing characteristics of human arm. It can simulate different motion conditions by controlling its swing amplitude and speed. By changing different acceleration, frequency and swing amplitude, the motion state of the energy harvester under different motion conditions can be obtained and the results under different experimental conditions can be obtained.

The excitation source used in the test was a sinusoidal acceleration signal, the acceleration amplitude was 0.8 g and the excitation frequency was 4.8 Hz. The resistance range used in the test was 1.3 Ω–3.2 MΩ, in which the maximum power output matching resistance of piezoelectric was about 100 KΩ and the maximum power output resistance of coil is 80 Ω.

4.1. Excitation Test of Permanent Magnets

There are four positions (1, 2, 3, 4) on the rotating mass block of the harvester that can be installed with permanent magnets, as shown in Figure 1. The harvester with different number of permanent magnets was tested. The voltage of piezoelectric element just reflects the vibration of cantilever, so the voltage of piezoelectric element was tested.

The test platform used was the swing device for simulating the swing of human arm as shown in Figure 5b. The frequency of swing was 4.8 (Hz). The first case test was to install only one permanent magnet. The permanent magnet was installed at position 1. The maximum output voltage amplitude of a single piezoelectric element can reach 5.4 V. As shown in Figure 6, the cantilever beam moves to the maximum amplitude after being repulsed by the permanent magnet on the rotating mass block. After the permanent magnet on the rotating mass and the permanent magnet on the cantilever beam passing by, the cantilever beam begins to vibrate at the natural frequency of the cantilever beam about 42 (Hz), and the amplitude of the vibration decreases continuously. Every time of the two permanent magnets passing by will cause up-conversion vibration of the cantilever. The low frequency vibration was converted to the inherent high frequency vibration of the cantilever beam.

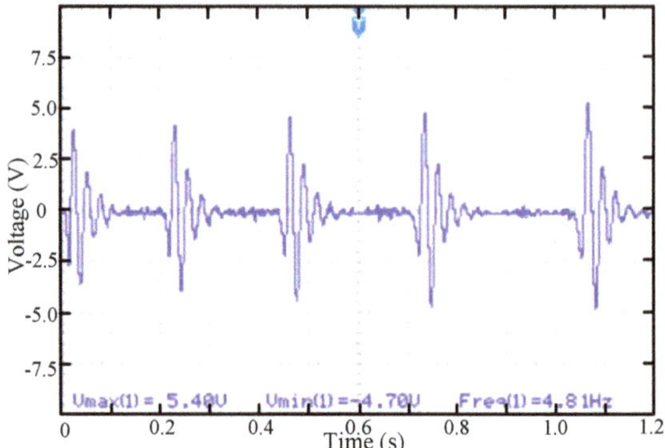

Figure 6. The voltage waveform of the piezoelectric beam while rotating mass block rotates with a permanent magnet (position 1).

In the second case, permanent magnets were installed at positions 1 and 4 of the rotating mass block. Other test conditions remained unchanged from the first test. As shown in Figure 7, there will be two up-conversion processes each time the rotating mass passes passing by the cantilever beam. Because the rotating mass has a fast swing speed, the vibration excited by the first permanent magnet has not been attenuated and the excitation of the second permanent magnet will be added again. The natural vibration frequency of the cantilever is constant and the vibration duration is nearly twice as long.

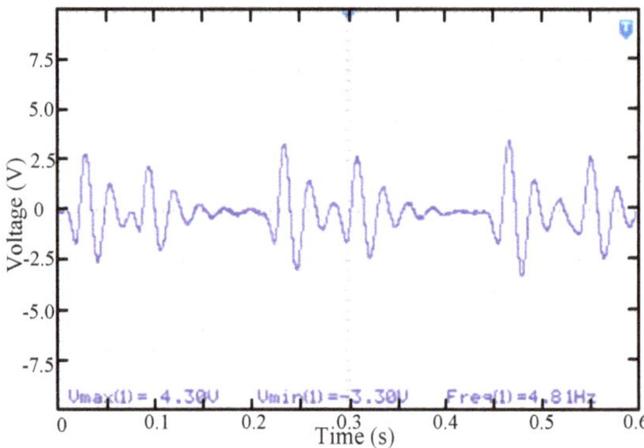

Figure 7. The voltage waveform of the piezoelectric beam while rotating mass block rotates with two permanent magnets (positions 1, 4).

In the third case, the permanent magnet was set at the positions 1, 3 and 4 of the rotating mass block. As shown in Figure 8, since the permanent magnet at position 3 was increased compared with the second case and the distance between 1 and 3 was relatively large, the vibration amplitude was attenuated after the position 1 magnet passed by. However, the distance between the magnets at positions 3 and 4 was relatively close. It can be found that the excitation of the magnet at the position 3 has not yet decayed and the excitation of the position 4 magnet is superimposed to positions 3. So in the wave form, the vibration amplitude of cantilever beam increases again from position 3 and then starts to decay after the position 4 magnet passing by.

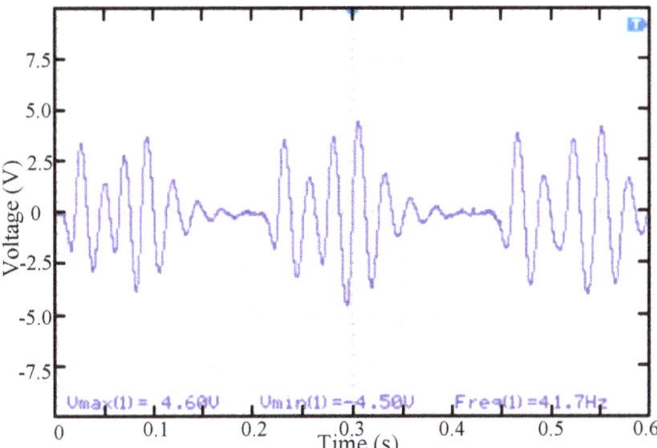

Figure 8. The voltage waveform of the piezoelectric beam while rotating mass block rotates with three permanent magnets (positions 1, 3, 4).

In the fourth case, permanent magnets were installed at all four positions. As shown in Figure 9, the continuous excitation of the four permanent magnets made the cantilever begin to vibrate under the excitation of the first permanent magnet and the vibration amplitude increased continuously. The maximum output amplitude of piezoelectric element can reach 6.7 V and the whole up-conversion vibration process also lasts longer. Therefore, in this case, the harvester can harvest more vibration energy than in the first case.

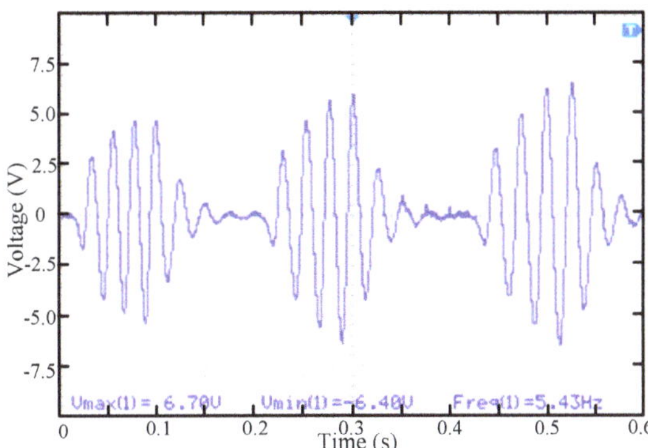

Figure 9. The voltage waveform of the piezoelectric beam while rotating mass block rotates with four permanent magnets (positions 1, 2, 3, 4).

4.2. Single Direction Weak Vibration Test

In many cases, the vibration is weak and the direction of vibration is single. The test device is the single dimension vibration test platform as shown in Figure 5a. Under this excitation condition, only the permanent magnets at positions 2 and 3 can repeatedly excite the piezoelectric cantilever beam. The vibration frequency is 2.5 Hz. The measured voltage waveform is shown in Figure 10. In this case, the cantilever has been in the state of repeated excitation and there is a large amplitude vibration when the permanent magnet passing by and there is a low amplitude vibration when there is no permanent magnet passing by.

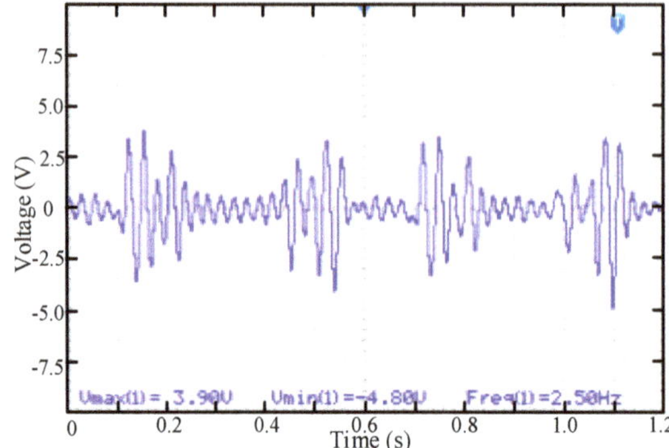

Figure 10. The voltage waveform of the piezoelectric beam while rotating mass block low amplitude reciprocating swing.

In this case, the design of interface circuit is more critical. By harvesting the energy of weak vibration, the conversion efficiency of the whole vibration energy can be improved.

4.3. Piezo-Electromagnetic Coupling Output Test

The harvester adopts the cantilever beam of the piezoelectric and electromagnetic coupling mode. In the process of cantilever vibration, the vibration of the permanent magnet at the end of the cantilever beam will change the magnetic flux of the coil under the permanent magnet, so that the coil can output alternating current. The diameter of the enameled wire used in the coil was 0.1 mm and the number of turns was 200. Due to the limitation of volume, if the thinner enameled wire is used, the internal resistance will be increased and the internal power consumption of the coil will be increased. With a thicker enameled wire, the number of turns of the coil will be fewer and the corresponding output voltage will be lower. According to Faraday electromagnetic induction, the voltage is related to the change rate of magnetic flux. The greater the speed of vibration, the greater the output voltage of the coil. However, the size of cantilever determines the maximum vibration amplitude. As shown in Figure 11, the output voltage of the coil is low and its voltage amplitude is about 100 mV. The frequency of the output voltage of the solenoid is the same as that of the piezoelectric because the coil and piezoelectric cantilever are coupled and resonant. The output frequency of the coil is determined by the vibration frequency of the cantilever. Because of the coupling resonance, the trend of output voltage waveform of the coil is consistent with that of the piezoelectric output. With the up-conversion of the cantilever beam, it first increases and then decreases.

The magnetic field produced by the coil current will react on the permanent magnet of cantilever beam and affect the vibration amplitude of the cantilever beam. It can be seen from Figure 11 that the vibration of the piezoelectric cantilever beam is less affected by the smaller output voltage amplitude of the coil. However, the energy output of the coil is increased. Therefore, the overall energy collection efficiency will be improved by using an efficient low-voltage energy harvesting circuit or using the output energy of the coil as the auxiliary energy for the piezoelectric cantilever energy harvesting.

The maximum power matching test was carried out with different resistors, as shown in Figure 5a. When the piezoelectric device acted alone, the peak power of the energy harvesting was 1.28 mW. When the electromagnetic energy acted alone, the peak power was about 0.03 mW. When piezoelectric and electromagnetic are coupled, the peak power was about 1.31 mW. The power of energy collection was the highest with a better energy conversion effect when there was piezoelectric and electromagnetic coupling resonance.

The permanent magnets and coils used in the prototype were small and the number of turns of as coils is relatively few, while the piezoelectric used in the prototype was relatively large, so the energy extracted by the two transducers was quite different. It can redistribute the harvested energy, such as 50%/50% or other proportions by reducing the size of the piezoelectric elements, thickening the coils to reduce the internal resistance, increasing the number of turns to increase the output voltage and using stronger permanent magnet.

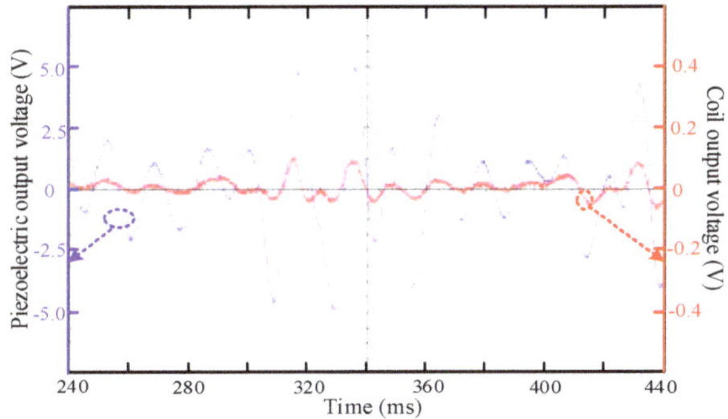

Figure 11. The output voltage wave forms of the piezoelectric beam and coil while rotating mass block random swing.

5. Discussion

At present, there are various kinds of energy harvesters and some novel structures are selected for comparison and discussion. The main characteristics are shown in Table 1. In the structure of Reference [21], printed circuit board (PCB) coil is used to cut the magnetic field line of permanent magnet to generate electricity. It can harvest the energy of the human swing motion process but the energy it harvests is relatively low. In Reference [27], a new kind of vibration energy harvester which combines magnetostrictive material and piezoelectric material is used but it can only harvest single direction vibration and the power density provided in the paper only considers the volume of magnetoelectric laminate composites (MLCs). In Reference [11], the piezoelectric cantilever is excited by permanent magnet but it can only harvest the energy of rotation excitation and the harvesting energy power is low. Compared with the structures proposed in these papers, the harvester in the current paper used four permanent magnets to continuously excite the cantilever beams, which can produce continuous vibration with large amplitude. Four cantilever beams were used to harvest vibration energy in multiple directions. With the coupling of electromagnetic and piezoelectric, more vibration energy can be extracted when the appropriate interface circuit is selected.

Table 1. Comparison of several vibration energy collection structures.

Literature	GTD2019 [11]	TMAGN2012 [27]	MEMS2011 [21]	This Work
Vibration excitation mode	Permanent magnet excitation	Magnetostriction	Permanent magnet excitation	Permanent magnet excitation
Energy harvesting method	Piezoelectricity	Piezoelectricity	Electromagnetism	Piezo-electromagnetic coupling
Maximum Output power	234.47 μW	1.1 mW	472 μW	1.31 mW
Power density [1]	Not available	3.46×10^3 W/m^3 (MLCs)	0.236 W/m^3	11.35 W/m^3
Operating frequency	0.167–1.677 Hz	10 Hz	<10 Hz	1–15 Hz
Directional of vibration	Rotational	Single direction	Swing	Multidirection vibration and swing
Frequency up-conversion	Yes	No	No	Yes

[1] The result is based on the information provided.

6. Conclusions

In this paper, a new structure of piezoelectric electromagnetic coupling up-conversion multi-directional vibration energy harvester is proposed. Four piezoelectric electromagnetic coupling piezoelectric cantilever beams are installed on the axis of base along the circumferential direction. The permanent magnet on the beam is placed on the free end of the cantilever beam as a mass block. Four coils for collecting energy are arranged on the base under the permanent magnets on the cantilever beams. Four permanent magnets arranged on the rotating mass block continuously excite the cantilever beams. The piezoelectric cantilever is induced to vibrate at its natural frequency by the interaction between the magnet on cantilever and the magnets on the rotating mass block. With the coupling of electromagnetic and piezoelectric, more vibration energy can be extracted when the appropriate interface circuit is selected.

Author Contributions: Conceptualization, G.S.; methodology, G.S. and J.C.; software, J.C. and M.S.; validation, H.X., Y.Y. and Y.X.; formal analysis, G.S. and H.X.; investigation, G.S. and J.C.; resources, G.S. and Y.X.; data curation, X.W.; writing—original draft preparation, G.S. and J.C.; writing—review and editing, G.S.; visualization, M.S.; supervision, Y.P.; project administration, G.S.; funding acquisition, G.S., Y.Y., H.X. and Y.X. All authors have read and agreed to the published version of the manuscript.

Funding: This research is supported by the National Natural Science Foundation of China under Grant U1709218, Grant 61971389, Grant 61801253 and Grant 61601429, the Natural Science Foundation of Zhejiang Province Grant LZ20F010006 and Grant LY20F010003 and the Natural Science Foundation of Ningbo Grant 2018A610091,2019A610113.

Conflicts of Interest: The authors declare no conflict of interest.

References

1. Pavelková, R.; Vala, D.; Gecová, K. Energy harvesting systems using human body motion. *IFAC PapersOnLine* **2018**, *51*, 36–41. [CrossRef]

2. Salauddin, M.; Toyabur, R.M.; Maharjan, P.; Rasel, M.S.; Kim, J.W.; Cho, H.; Park, J.Y. Miniaturized springless hybrid nanogenerator for powering portable and wearable electronic devices from human-body-induced vibration. *Nano Energy* **2018**, *51*, 61–72. [CrossRef]

3. Hou, C.; Chen, T.; Li, Y.F.; Huang, M.J.; Shi, Q.F.; Liu, H.C.; Sun, L.N.; Lee, C. A rotational pendulum based electromagnetic/triboelectric hybrid-generator for ultra-low-frequency vibrations aiming at human motion and blue energy applications. *Nano Energy* **2019**, *63*, 103871. [CrossRef]

4. Jackson, N.; Olszewski, O.Z.; O'Murchu, C.; Mathewson, A. Ultralow-frequency PiezoMEMS energy harvester using thin-film silicon and parylene substrates. *J. Micro Nanolithography MEMS MOEMS* **2018**, *17*, 015005. [CrossRef]

5. Xue, T.; Yeo, H.G.; Trolier-McKinstry, S.; Roundy, S. Wearable inertial energy harvester with sputtered bimorph lead zirconate titanate (PZT) thin-film beams. *Smart Mater. Struct.* **2018**, *27*, 085026. [CrossRef]

6. Lockhart, R.; Janphuang, P.; Briand, D.; de Rooij, N.F. A wearable system of micromachined piezoelectric cantilevers coupled to a rotational oscillating mass for on-body energy harvesting. In Proceedings of the 2014 IEEE 27th International Conference on Micro Electro Mechanical Systems (MEMS), San Francisco, CA, USA, 26–30 January 2014; pp. 370–373.

7. Risquez, S.; Woytasik, M.; Coste, P.; Isac, N.; Lefeuvre, E. Additive fabrication of a 3D electrostatic energy harvesting microdevice designed to power a leadless pacemaker. *Microsyst. Technol.Micro Nanosyst. Inf. Storage Process. Syst.* **2018**, *24*, 5017–5026. [CrossRef]

8. Zhang, J.; Fang, Z.; Shu, C.; Zhang, J.; Zhang, Q.; Li, C. A rotational piezoelectric energy harvester for efficient wind energy harvesting. *Sens. Actuators A Phys.* **2017**, *262*, 123–129. [CrossRef]

9. Goudar, V.; Ren, Z.; Brochu, P.; Potkonjak, M.; Pei, Q. Optimizing the output of a human-powered energy harvesting system with miniaturization and integrated control. *IEEE Sens. J.* **2014**, *14*, 2084–2091. [CrossRef]

10. Izadgoshasb, I.; Lim, Y.Y.; Tang, L.; Padilla, R.V.; Tang, Z.S.; Sedighi, M. Improving efficiency of piezoelectric based energy harvesting from human motions using double pendulum system. *Energy Convers. Manag.* **2019**, *184*, 559–570. [CrossRef]

11. Sriyuttakrai, D.; Isarakorn, D.; Boonprasert, S.; Nundrakwang, S. Practical Test of Contactless Rotational Piezoelectric Generator for Low Speed Application. In Proceedings of the 2019 IEEE PES GTD Grand International Conference and Exposition Asia (GTD Asia), Bangkok, Thailand, 19–23 March 2019; pp. 80–85.

12. Song, H.C.; Kumar, P.; Maurya, D.; Kang, M.G.; Reynolds, W.T.; Jeong, D.Y.; Kang, C.Y.; Priya, S. Ultra-Low Resonant Piezoelectric MEMS Energy Harvester with High Power Density. *J. Microelectromechanical Syst.* **2017**, *26*, 1226–1234. [CrossRef]

13. Fan, K.; Cai, M.; Wang, F.; Tang, L.; Liang, J.; Wu, Y.; Qu, H.; Tan, Q. A string-suspended and driven rotor for efficient ultra-low frequency mechanical energy harvesting. *Energy Convers. Manag.* **2019**, *198*, 111820. [CrossRef]

14. Aldawood, G.; Hieu Tri, N.; Bardaweel, H. High power density spring-assisted nonlinear electromagnetic vibration energy harvester for low base-accelerations. *Appl. Energy* **2019**, *253*, 113546. [CrossRef]

15. Fan, K.; Zhang, Y.; Liu, H.; Cai, M.; Tan, Q. A nonlinear two-degree-of-freedom electromagnetic energy harvester for ultra-low frequency vibrations and human body motions. *Renew. Energy* **2019**, *138*, 292–302. [CrossRef]

16. Fan, K.; Cai, M.; Liu, H.; Zhang, Y. Capturing energy from ultra-low frequency vibrations and human motion through a monostable electromagnetic energy harvester. *Energy* **2019**, *169*, 356–368. [CrossRef]

17. Shan, X.; Xu, Z.; Song, R.; Xie, T. A New Mathematical Model for a Piezoelectric-Electromagnetic Hybrid Energy Harvester. *Ferroelectrics* **2013**, *450*, 57–65. [CrossRef]

18. Li, P.; Gao, S.; Niu, S.; Liu, H.; Cai, H. An analysis of the coupling effect for a hybrid piezoelectric and electromagnetic energy harvester. *Smart Mater. Struct.* **2014**, *23*, 065016. [CrossRef]

19. Gu, L.; Livermore, C. Impact-driven frequency up-converting coupled vibration energy harvesting device for low frequency operation. *Smart Mater. Struct.* **2011**, *20*, 045004. [CrossRef]

20. Lin, J.; Liu, H.; Chen, T.; Yang, Z.; Sun, L. A rotational wearable energy harvester for human motion. In Proceedings of the 2017 IEEE 17th International Conference on Nanotechnology (IEEE-NANO), Pittsburgh, PA, USA, 25–27 July 2017; pp. 22–25.

21. Romero, E.; Neuman, M.R.; Warrington, R.O. Rotational energy harvester for body motion. In Proceedings of the 2011 IEEE 24th International Conference on Micro Electro Mechanical Systems, Cancun, Mexico, 23–27 January 2011; pp. 1325–1328.

22. Li, C.; Wu, S.; Luk, P.C.K.; Gu, M.; Jiao, Z. Enhanced Bandwidth Nonlinear Resonance Electromagnetic Human Motion Energy Harvester Using Magnetic Springs and Ferrofluid. *IEEE/ASME Trans. Mechatron.* **2019**, *24*, 710–717. [CrossRef]

23. Smilek, J.; Hadas, Z.; Vetiska, J.; Beeby, S. Rolling mass energy harvester for very low frequency of input vibrations. *Mech. Syst. Signal Process.* **2019**, *125*, 215–228. [CrossRef]

24. Halim, M.A.; Rantz, R.; Zhang, Q.; Gu, L.; Yang, K.; Roundy, S. Modeling and Experimental Analysis of a Wearable Energy Harvester that Exploits Human-Body Motion. In Proceedings of the 2018 IEEE SENSORS, New Delhi, India, 28–31 October 2018; pp. 1–4.

25. Pillatsch, P.; Yeatman, E.M.; Holmes, A.S. A piezoelectric frequency up-converting energy harvester with rotating proof mass for human body applications. *Sens. Actuators A Phys.* **2013**, *206*, 178–185. [CrossRef]

26. Chen, Z.; Zhang, F.; Xu, X.; Chen, Z.; Li, W.; Li, F.; Han, J.; Zhou, H.; Xu, K.; Bu, L. Non-Contact Multiple Plucking Method for Frequency up Conversion Piezoelectric Energy Harvesters in Extremely Low Frequency Applications. In Proceedings of the 2019 20th International Conference on Solid-State Sensors, Actuators and Microsystems & Eurosensors XXXIII (TRANSDUCERS & EUROSENSORS XXXIII), Berlin, Germany, 23–27 June 2019; pp. 1427–1430.

27. Zhu, Y.; Zu, J.W.; Guo, L. A Magnetoelectric Generator for Energy Harvesting from the Vibration of Magnetic Levitation. *IEEE Trans. Magn.* **2012**, *48*, 3344–3347. [CrossRef]

28. Xia, H.; Chen, R.; Ren, L. Analysis of piezoelectric-electromagnetic hybrid vibration energy harvester under different electrical boundary conditions. *Sens. Actuators A Phys.* **2015**, *234*, 87–98. [CrossRef]

29. Ibrahim, M.; Salehian, A. Modeling, fabrication, and experimental validation of hybrid piezo-magnetostrictive and piezomagnetic energy harvesting units. *J. Intell. Mater. Syst. Struct.* **2015**, *26*, 1259–1271. [CrossRef]

30. Xia, H.; Chen, R.; Ren, L. Parameter tuning of piezoelectric-electromagnetic hybrid vibration energy harvester by magnetic force: Modeling and experiment. *Sens. Actuators A Phys.* **2017**, *257*, 73–83. [CrossRef]
31. Yeatman, E.M. Energy harvesting from motion using rotating and gyroscopic proof masses. *J. Mech. Eng. Sci.* **2008**, *222*, 27–36. [CrossRef]

Article

Compensation of Hysteresis in the Piezoelectric Nanopositioning Stage under Reciprocating Linear Voltage Based on a Mark-Segmented PI Model

Dong An [1,2,*], Yixiao Yang [1], Ying Xu [1,3], Meng Shao [1], Jinyang Shi [3] and Guodong Yue [1,*]

[1] School of Mechanical Engineering, Shenyang Jianzhu University, Shenyang 110168, China; peteryang@stu.sjzu.edu.cn (Y.Y.); xuying@sjzu.edu.cn (Y.X.); mshao@sjzu.edu.cn (M.S.)

[2] Research Center for Analysis and Detection Technology, Shenyang Jianzhu University, Shenyang 110168, China

[3] School of Electro-Mechanical Engineering, Guangdong University of Technology, Guangzhou 510006, China; shijy@mail2.gdut.edu.cn

* Correspondence: andong@sjzu.edu.cn (D.A.); ygd@sjzu.edu.cn (G.Y.); Tel.: +86-13504990528 (D.A.); +86-15840363352 (G.Y.)

Received: 4 November 2019; Accepted: 17 December 2019; Published: 19 December 2019

Abstract: The nanopositioning stage with a piezoelectric driver usually compensates for the nonlinear outer-loop hysteresis characteristic of the piezoelectric effect using the Prandtl–Ishlinskii (PI) model under a single-ring linear voltage, but cannot accurately describe the characteristics of the inner-loop hysteresis under the reciprocating linear voltage. In order to improve the accuracy of the nanopositioning, this study designs a nanopositioning stage with a double-parallel guiding mechanism. On the basis of the classical PI model, the study firstly identifies the hysteresis rate tangent slope mark points, then segments and finally proposes a phenomenological model—the mark-segmented Prandtl–Ishlinskii (MSPI) model. The MSPI model, which is fitted together by each segment, can further improve the fitting accuracy of the outer-loop hysteresis nonlinearity, while describing the inner-loop hysteresis nonlinearity perfectly. The experimental results of the inverse model compensation control show that the MSPI model can achieve 99.6% reciprocating linear voltage inner-loop characteristic accuracy. Compared with the classical PI model, the 81.6% accuracy of the hysteresis loop outer loop is improved.

Keywords: nanopositioning stage; piezoelectric hysteresis; mark point recognition; piecewise fitting; compensation control

---

1. Introduction

The nanopositioning stage of the piezoelectric ceramic material driver has the advantages of small volume, high displacement resolution, fast response, large bearing capacity, no noise, and high stability [1]. Hence, it is widely used in modern precision machineries as the core device, such as in atomic force microscopy [2,3] and nanolithography processing [4]. However, the inherent hysteresis nonlinearity of the piezoelectric ceramic materials affects the accuracy of this nanopositioning stage [5]. Thus, it is necessary to model effective compensation for the hysteresis [6,7].

In order to improve the positioning accuracy of the nanopositioning platforms, many scholars at home and abroad have conducted extensive research on piezoelectric ceramic hysteresis. There are two popular approaches. One is to study the hysteresis due to the internal mechanism, where the crystal grains constitute the crystal phase of the piezoelectric ceramic and the electric domains appear in the crystal grains. Polarization treatment enables piezoelectric ceramics to exhibit a piezoelectric effect [8]. However, a small number of grains returns to the original direction after the polarization, which is known as the residual polarization, leading to hysteresis [9]. The second approach is to establish a phenomenological model for the expressed voltage–displacement characteristics, including differential equation models such as the Bouc–Wen model [10], operator models such as the Preisach model [11], the Prandtl–Ishlinskii (PI) model [12–14], and others.

Among them, the PI model is weighted and superimposed by the Play operator [15]. The symmetry and hysteresis characteristics of the Play operator can efficiently describe the appearance characteristics of the hysteresis [16]. Compared with the nonlinear integral of the Preisach model and the nonlinear differential equation of the Bouc–Wen model, the PI model has fewer parameters, a simpler structure, and it is easier to find its inverse model. It has been widely used in modeling and compensation of hysteresis features [17,18]. The classical PI model always has symmetry, which makes it hard to accurately approximate the hysteresis characteristics of the boost phase and buck phase at the same time [19].

Many studies have been conducted on how to improve the accuracy of the nanopositioning stage by improving the traditional PI model. There are ways to change the operator by modifying the classical PI model into a generalized PI model, resetting the initial value that can eliminate the influence of the polarization by changing the hysteresis characteristics of the symmetric bias modeling and piecewise fitting method [20–23]. The above researches greatly improve the hysteresis description accuracy of the nanopositioning stage for single-ring linear voltage.

In a previous study [24], a segmented model was carried out by dividing the boost phase hysteresis characteristics by polarization in single-ring linear voltage. However, it is more likely for an improved method to solve the physical problem than a define-based modeling. For the reciprocating linear voltage hysteresis, the hysteresis characteristics of the inner and outer loops are different and complex. Thus, the segmented modeling method is limited in reciprocating linear voltage due to insufficient theoretical segmentation basis. Therefore, both the classical PI model and the improved model cannot meet the accuracy compensation requirements of the nanopositioning stage under reciprocating linear voltage.

Aiming at the complex hysteresis problem of the piezoelectric stage for the reciprocating linear voltage, this study proposes a mark-segmented PI (MSPI) model that loads the reciprocating linear voltage signal according to the reciprocating linear displacement requirement and compensates using the characteristics of the obtained hysteresis. An approximate hysteresis curve slope can be obtained between the adjacent measurement points of the experimental data, and the segment identification points corresponding to the slope jump segments are found in a threshold manner; thus, the segments are modeled. The results show that the description accuracy of the MSPI model is high, and it has good performance in both the compensation hysteresis inner loop and hysteresis outer loop.

The structure of this paper is as follows. In the second section, a nanopositioning stage with a double-parallel guiding mechanism is designed, and the mechanical characteristics of the platform are analyzed to optimize the design structure. In the third section, the hysteresis characteristics are analyzed and the MSPI model is designed for fitting the reciprocating nonlinear features. After the inversion, the fourth section gives the control comparison between the MSPI model and the classical PI model. Two other verification experiments are also conducted. Finally, the fifth section summarizes this study.

2. Precondition

This section is divided into two parts. The first part designs a nanopositioning stage with a double-parallel guiding mechanism, including its mechanical structure selection and stiffness calculation. The second part is the voltage–displacement experiment process and the conclusion obtained.

2.1. Mechanical Design and Calculation of Nanopositioning Stage

The nanopositioning stage uses a completely flexible mechanism. This flexible mechanism has a microscale range for the elastic deformation motion. It can achieve the transmission of the motion and force without friction [25,26]. Considering the thinness of the flexible hinge, the wire cutting method is adopted. Therefore, the slit that is easier for the machine is selected as a flexible hinge of a rectangular cross-section, as shown in Figure 1a. The flexible hinge, with four rectangular sections as the base members, is the double-parallel guiding mechanism used in the design, as shown in Figure 1b.

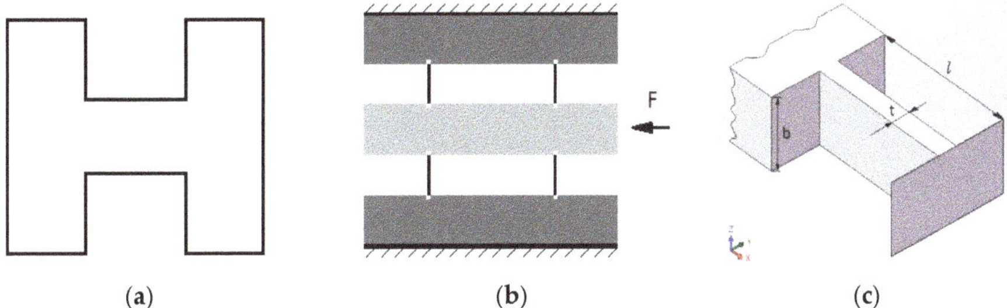

Figure 1. Flexible hinge selection: (**a**) rectangular cut section; (**b**) double-parallel guide mechanism; (**c**) flexible hinge main parameter identification

The mechanical model for the rectangular section of the flexible hinge is shown in Figure 1c. Since the movement causes the flexible hinge to elastically deform, the stiffness in each direction must be calculated. According to the equations of material mechanics [27], the stiffness k_x and torsional stiffness $k_{\theta x}$ motion in the direction x are:

$$k_x = \frac{48EI_{yy}}{l^3} = \frac{4Et^3b}{l^3} \tag{1}$$

$$k_{\theta x} = \frac{Etb^3}{l^3}B^2 \tag{2}$$

where E is the modulus of elasticity, t is the thickness of the flexible hinge, b is the width of the flexible hinge, l is the length of the flexible hinge, I_{yy} is the bending section coefficient of the y-axis, and B is the width of the nanopositioning stage.

The stiffness k_y and torsional stiffness $k_{\theta y}$ in the vertical direction y are:

$$k_y = \frac{12EI_{zz}}{l^3} = \frac{Etb^3}{l^3} \tag{3}$$

$$k_{\theta y} = \frac{Etb}{l}D^2 \tag{4}$$

where I_{zz} is the bending section coefficient of the z-axis and D is the distance between the flexible hinges of the two parallel rectangular sections.

The stiffness k_z and torsional stiffness $k_{\theta z}$ in the vertical direction z are:

$$k_z = \frac{4EA}{l} = \frac{Etb}{l} \tag{5}$$

$$k_{\theta z} = \frac{Etb^3}{3l}\left(\frac{D^2}{2l^2} + \frac{t^2}{5b^2}\right) \tag{6}$$

where A is the flexible hinge cross-sectional area.

The designed flexible hinge has a length of 14.5 mm, a thickness of 0.3 mm, a width of 15 mm, and a material elastic modulus of 72 GPa. The maximum equivalent stress is 7.8 MPa. The stiffness of the double-parallel guiding positioning platform at six degrees of freedom is obtained, as shown in Table 1.

Table 1. Stiffness of double-parallel-oriented positioning platform under six degrees of freedom (unit: k_x, k_y, k_z: N/mm; $k_{\theta x}, k_{\theta y}, k_{\theta z}$: N·mm/rad).

Stiffness	k_x	k_y	k_z	$k_{\theta x}$	$k_{\theta y}$	$k_{\theta z}$
Theoretical value	68.5	99,009.9	22,630.6	21,835,421	100,093,750	7,911,302.6
Calculated value	64.0	106,110.0	29,392.0	23,629,000	107,800,000	7,089,400.0
Error	−6.6%	7.2%	29.9%	8.2%	7.7%	−10.4%

In order to strictly guarantee the accuracy of the nanopositioning stage, a microlevel precision slow wire cutting technology is adopted [28]. At the same time, a material with a small thermal expansion coefficient is selected [29]. In order to prevent the wire mechanism from oxidizing, the surface of the flexible hinge needs to be nickel plated. Figure 2a shows a schematic diagram of the nanopositioning stage. Figure 2b is the actual diagram of the nanopostioning stage with double-parallel guiding mechanism.

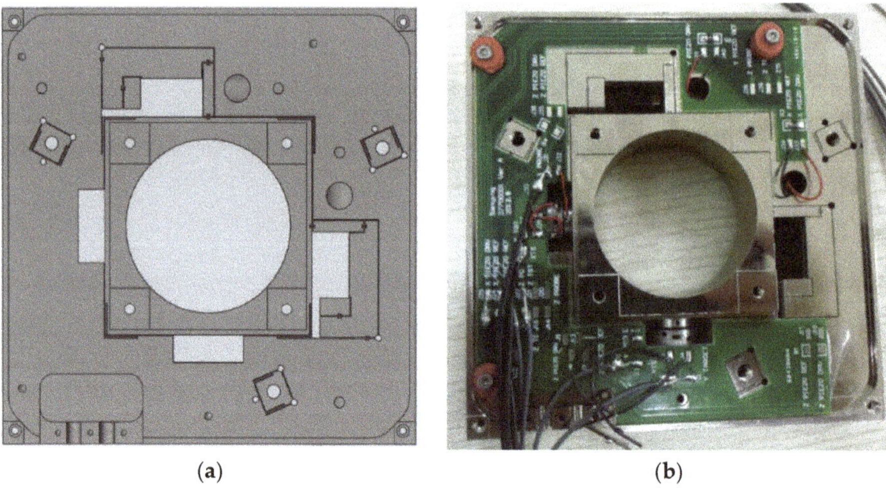

(a)   (b)

Figure 2. Double-parallel guiding mechanism nanopositioning stage: (a) the schematic diagram (b) and the actual diagram.

2.2. Test Results

The experimental system was built using a laser interferometer, nanopositioning platform, reflection mirror, controller, and computer software, as shown in Figure 3a. The laser interferometer used is the Renishaw XL-80 series achieves an accuracy of ± 0.5 ppm. The selected driver was the HVA-150D.A3 instrument from Harbin Xinmingtian Company, whose voltage input variable range is 0 V-150 V. Figure 3b shows the actual experimental system.

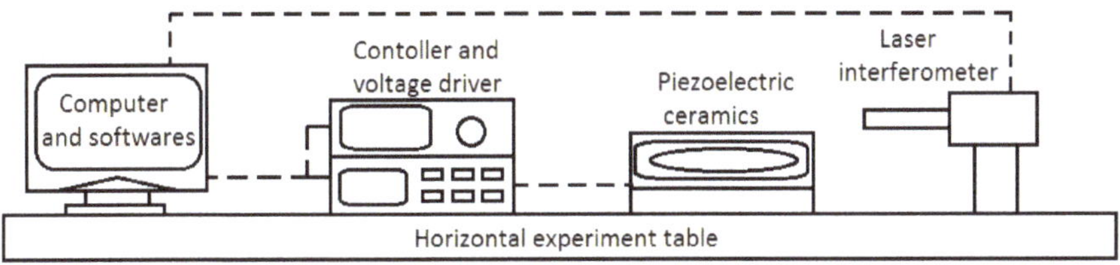

(**a**)

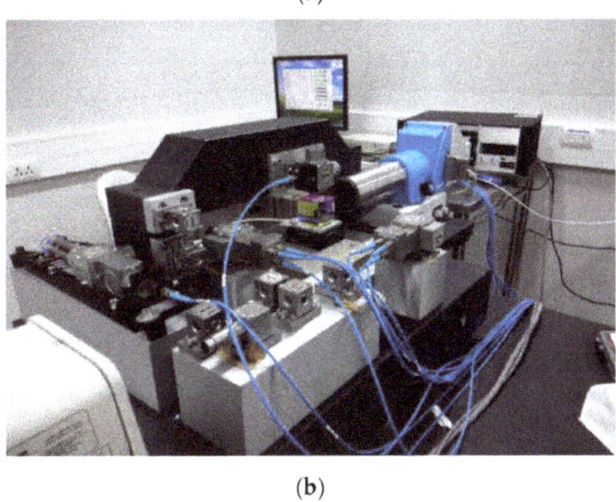

(**b**)

Figure 3. Nanopositioning stage experimental system: (**a**) system schematic (**b**) and actual system diagram.

The user performs the following steps to test the output displacement characteristics of the nanopositioning stage in the x-direction motion, under single-ring linear voltage and under reciprocating linear voltage:

(1) Adjust the laser interferometer so that its light signal intensity is within the confidence range. Connect the computer, controller, and laser interferometer via the USB interface data cable. Data is collected on the computer by the corresponding software of the controller and laser interferometer.

(2) The controller matching software loads the electrical signal to the driving controller. The initial driving voltage is 0 V. Experiment 1 is carried out to give a linear voltage rise signal from 0 V to 150 V. The displacement is measured and recorded by laser interferometer for every 7.5 V. Next, a linear voltage reduction signal is sent from 150 V to 0 V. The displacement is measured and recorded every 7.5 V. A single-ring linear voltage signal is shown in Figure 4a.

(3) Similarly, experiment 2 is carried out. The reciprocating linear voltage signal is recorded, as shown in Figure 4b.

(4) Record several measurements.

(5) Check the instrument and turn it off. Process experimental data.

Through the above experiment, two sets of data can be obtained: a single-ring linear voltage–displacement characteristic curve, as shown in Figure 5a; and a reciprocating linear voltage–displacement characteristic curve, as shown in Figure 5b.

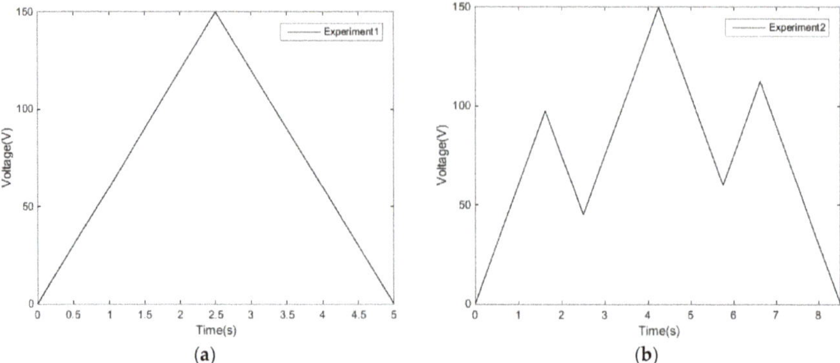

Figure 4. Experimental loading voltage: (**a**) single-ring linear voltage (**b**) and reciprocating linear voltage.

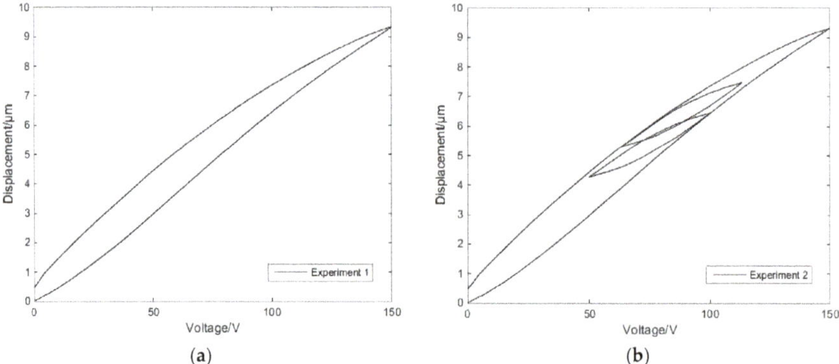

Figure 5. Voltage–displacement experimental curves: (**a**) single-ring linear voltage (**b**) and reciprocating linear voltage.

The expected voltage displacement curves should be linear. Analysis of the experimental data established the nonlinearity of the experimental curve. The voltage displacement curve at the single-ring linear voltage has a hysteresis characteristic. The voltage displacement curve at the reciprocating linear voltage has the same hysteresis characteristics as the single-ring voltage's, while the reciprocating voltage's hysteresis loop conforms to the Madelung principle [30]. Therefore, in order to solve the hysteresis characteristics of the piezoelectric ceramic under single-ring linear voltage and reciprocating linear voltage, an effective compensation method is needed.

3. Modeling

This section describes the process of establishing the MSPI model in three parts. In the first, part the curve of the classical PI model is obtained to describe the hysteresis characteristics. The second part analyzes the specific problem of the classical PI model's inner-loop hysteresis description, and defines the voltage–slope curve corresponding to the hysteresis rate tangent to establish the MSPI model. The third part uses a threshold method to judge whether it is the mark point. The segmentation of the mark points gives the curve described by the MSPI model and its inverse model.

3.1. Play Operator and Classical Prandtl–Ishlinskii Model

The classical PI model is a weighted superposition of a finite number of Play operators. The Play operator is shown in Figure 6a. When the input signal is $x(k)$, the Play operator expression with the threshold r is:

$$p(k) = \max\{x(k) - r, \min[x(k) + r, p(k-1)]\} \tag{7}$$

where $0 = k_0 < k_1 < \ldots < k_s$ is the appropriate division on the input signal interval, $k \in [0, k_s]$. When $k = 0$, $p(-1)$ is the initial value. In the PI model of the piezoelectric effect hysteresis problem, the initial voltage is usually 0 without displacement, so $p(-1) = 0$. Here, $p(k)$ is the output of the input signal.

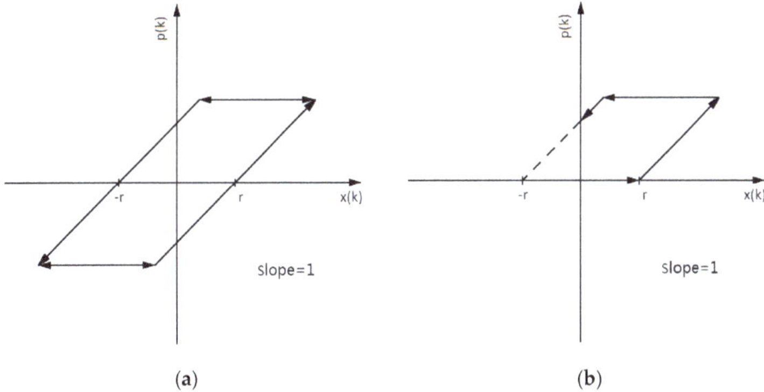

Figure 6. Play operator models: (**a**) full-sided operator model (**b**) and single-sided operator model.

The voltage supplied by the voltage driver is positive, and hence the PI model is usually modeled with a single-sided Play operator. As shown in Figure 6b, when the operator inputs $x(k) \leq r$, the operator outputs $p(k) = 0$; when the operator inputs $r < x(k) \leq x(k_s)$, the unweighted operator has a slope of 1, so the operator outputs $p(k) = x(k) - r$. The operator shown in Figure 6b outputs $p(k) = x(k_s)$ when the input decreases from $x(k)$ to $x(k_s) - 2r$ and outputs $p(k) = x(k) + r$ when the input decreases from $x(k_s) - 2r$ to 0. The operator may have no $p(k) = x(k) + r$ output and a part of $p(k) = x(k_s)$, when the threshold r is increased or the input $x(k_s)$ is decreased. In this case, the specific characteristics of the operator should be considered.

A finite number of Play operators are superimposed according to the weighting of the above output characteristics, and a PI model is obtained to describe the hysteresis of the nanopositioning stage. The equation is:

$$P[x(k)] = \theta_0 \cdot x(k) + \sum_{i=1}^{n} \theta_i \cdot p_i(k) \tag{8}$$

where $P[x(k)]$ is the corresponding PI model output for the operator input $x(k)$. Here, θ_0 is a positive value, $p_i(k)$ is the output of the ith operator that has a threshold r_i and a corresponding weight θ_i.

The more times the PI model is superimposed, the smoother the model contour is and the closer it is to the piezoelectric hysteresis characteristic curve. However, the accuracy of the voltage–displacement characteristics obtained from the experiments is limited, so the number of operators used for the superposition should be realistic.

Figure 7 shows the modelling of the PI model to display the single-ring linear voltage hysteresis characteristic and reciprocating linear voltage hysteresis characteristic of the second section. The modeling results show that the classical PI model describes the hysteresis characteristics well under single-ring linear voltage, but the accuracy under the reciprocating linear voltage is comparatively poor. The main reason is that the hysteresis characteristics of the reciprocating linear voltage are more complicated, and the hysteresis rates between the inner loop and the outer loop are different. Meanwhile, the classical PI model cannot describe the Madelung principle of the hysteresis inner loops.

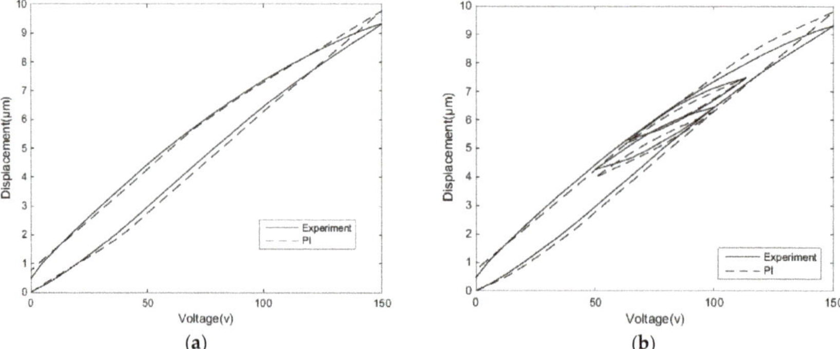

Figure 7. PI modeling: (**a**) modeling of single-ring linear voltage hysteresis characteristics and (**b**) modeling of reciprocating linear voltage hysteresis characteristics.

3.2. Hysteresis Tangent Line and Slope

In order to further improve the description accuracy of hysteresis characteristics, the hysteresis rates must be studied in depth. The hysteresis rates corresponding to the voltage–displacement characteristic curve are the weighted superposition of the Play operators in the PI model at that point. The weight θ_i of the ith operator depends on the angle α between the tangent line of the hysteresis loop at that point and the v-axis. The angle α, as shown in Figure 8a, is not exactly the same in each tangent on the hysteresis loop. The hysteresis rate of the reciprocating linear voltage is even more complicated. As shown in Figure 8b, the tangent at approximately similar positions of the inner loop and outer loop tends to have different hysteresis rates.

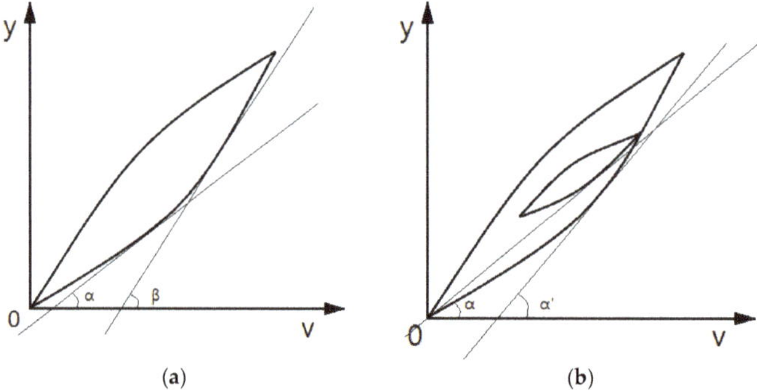

Figure 8. Tangent of the hysteresis loop: (**a**) the hysteresis rate corresponding to each point on the same hysteresis loop is different; (**b**) the hysteresis rates between the inner loop and outer loop of the hysteresis are different. Note: v = voltage; y = displacement.

The voltage–displacement data can approximate the characteristic curve, thereby establishing a v–y coordinate system. If j is the jth data obtained by the experiment, the hysteresis loop passes through the point (v_j, y_j). The equation for the hysteresis rate tangent $l_{\tan}(v)$ at v is defined as:

$$l_{\tan}(v) : y = s(v) \cdot v + t(v) \tag{9}$$

where $s(v)$ is the hysteresis rate tangent slope at v and $t(v)$ is the hysteresis rate's tangent intercept at v.

In a single-ring linear voltage hysteresis characteristic curve, the voltage v corresponds to two hysteresis tangent lines in the linear boost phase and the linear back phase, respectively. Similarly, in the reciprocating linear voltage hysteresis characteristic curve, the v value is likely to correspond to a plurality of hysteresis tangent lines; for example, the hysteresis rate tangent number in Figure 8b corresponding to v is as shown in Figure 9.

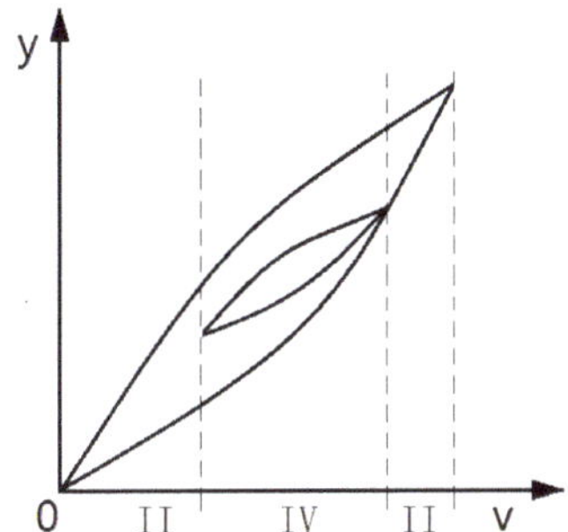

Figure 9. The voltage value corresponding to hysteresis characteristics has more than one tangent. Note: v = voltage; y = displacement.

The slope of the hysteresis tangent can reflect the trend of hysteresis at this point. The hysteresis tangent slope $s(v)$ can be expressed as:

$$s(v) = \frac{\delta y}{\delta v} = \frac{y_{j+1} - y_j}{v_{j+1} - v_j} \tag{10}$$

where $\left(v_j, y_j\right)$ and $\left(v_{j+1}, y_{j+1}\right)$ are adjacent data and satisfy the equation $\min(v_j, v_{j+1}) \leq v < \max(v_j, v_{j+1})$.

The voltage–slope diagram describes the characteristics of the hysteresis rate at any voltage. The different input linear voltage leads to varied hysteresis rate tangent regulation. Figure 10 is the $v - s(v)$ diagram of experimental data for the single-ring linear voltage and the reciprocating linear voltage, respectively. Both groups of data evidently show segmentation in the $v - s(v)$ diagram.

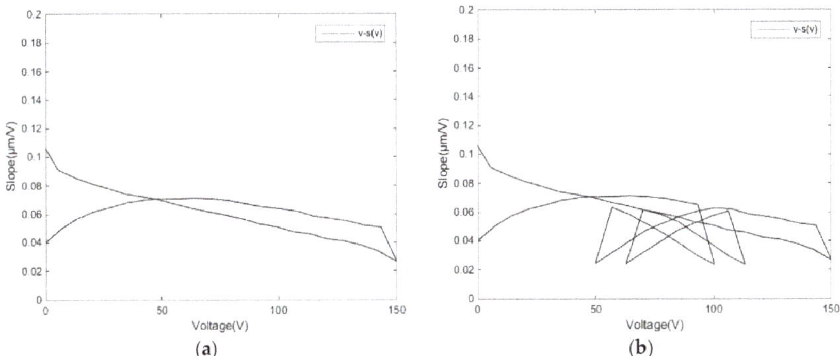

Figure 10. Voltage hysteresis rate tangent slope diagrams: (a) single-ring linear voltage and (b) reciprocating linear voltage.

Due to the fact that the piezoelectric hysteresis characteristic generally has a segmentation variation rule and there are obvious jump points between the segments, a segmented PI model is used to model it.

3.3. Mark-Segmented Prandtl–Ishlinskii Model

The voltage–slope diagram embodies the change of the hysteresis rate. For reciprocating hysteresis, plenty of turning points appear at the critical edge of boost phases and back phases. Compared with the hysteresis under single-ring linear voltage [24], reasonable identification I required for all mark points in order to fulfill the demands of complex hysteresis segmentation. Meanwhile, data with continuous and similar variation laws should be modeled in the same segment. Therefore, a mark-segmented PI (MSPI) model is proposed.

To identify the segmentation mark point, the threshold φ is set in the $v - s(v)$ diagram. The threshold φ is directly proportional to the quantity of experimental data, which of the minimum data amount is always 8–10 times of the average data difference. When the mth hysteresis rate tangent slope value segment satisfies $s(v_{m+1}) - s(v_m)$, then $|s(v_{m+1}) - s(v_m)| \geq \varphi$; $s(v_{m+1}) - s(v_m)$ is defined as the hysteresis rate jump segment. Data (v_{m+1}, y_{m+1}) is defined as the type I mark point.

Therefore, the single-ring linear voltage characteristic experimental curve can be segmented to find one type I mark point, which divides the hysteresis characteristic into 2 segments, as shown in Figure 11a; the reciprocating linear voltage characteristic experimental curve finds five type I mark points, and the data is divided into 6 segments, as shown in Figure 11b.

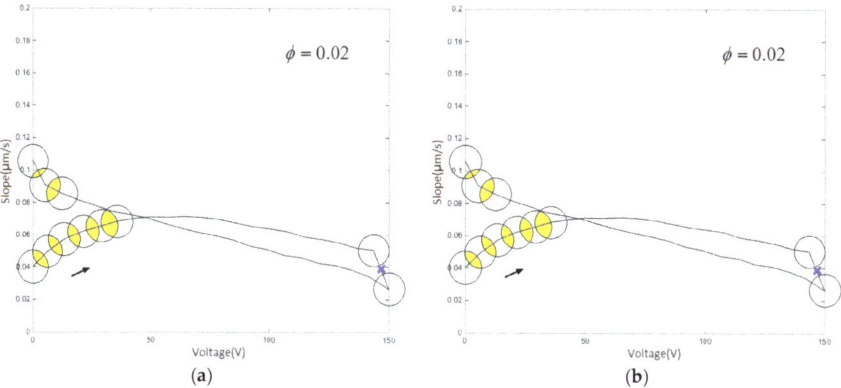

Figure 11. Type I mark points associated with threshold φ: (**a**) single-ring linear voltage $v - s(v)$ characteristic diagram and (**b**) reciprocating linear voltage $v - s(v)$ characteristic diagram.

The segmentation data is required to select an appropriate single-sided Play operator according to its approximate hysteresis characteristics or according to the concavity and convexity. In most cases, the condition for selecting the single-sided Play operator satisfies $s'(v) < 0$ or $s'(v) > 0$, where $s'(v)$ is the differential coefficient of $s(v)$. If there are still some cases where the segmentation data satisfies both the abovementioned conditions at the same time, then it needs to be divided by the segmentation marker point at $s'(v) = 0$, which is defined as a type II mark point. The $v - s(v)$ diagram obtained from the experimental data is not derivable, and the maximum or minimum value can be used as the type II mark point. In the single-ring linear voltage hysteresis $v - s(v)$ diagram shown in Figure 12a, one type II mark point is found, and a total of two segmentation mark points divide the curve into three segments. The reciprocating linear voltage hysteresis characteristic $v - s(v)$ diagram shown in Figure 12b finds one type II mark point, and the total number of segments is 7. Eventually, each segment selects a single-sided Play operator by characteristics.

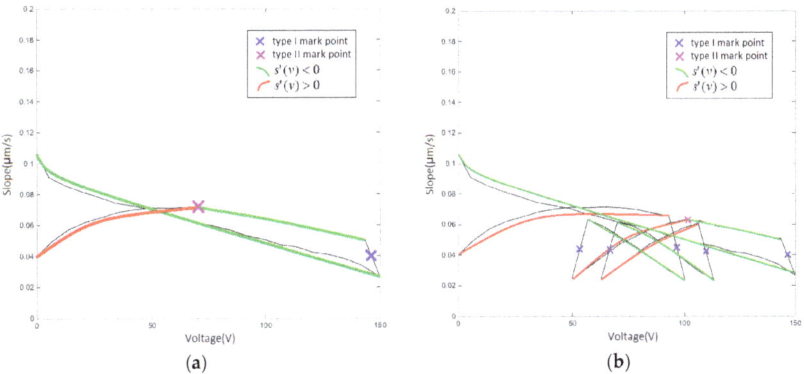

Figure 12. Type II mark point associated with $s'(v)$: (**a**) single-ring linear voltage $v - s(v)$ diagram and (**b**) reciprocating linear voltage $v - s(v)$ diagram.

The placement of the segmentation points is special because they participate in the modeling in both the segments that are divided by themselves. As shown in Figure 13a, the two segmentation points participate in the fitting of the three segments. The MSPI model with single-ring linear voltage hysteresis has good connectivity at the segmentation point. Figure 13b amplifies the MSPI model at one of the mark-segmented points.

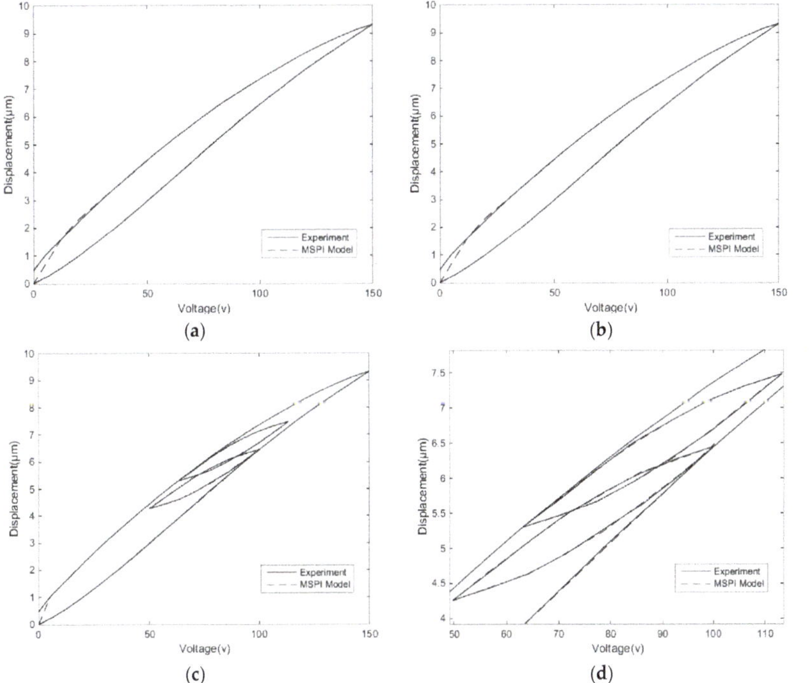

Figure 13. The mark-segemented Prandtl–Ishlinskii (MSPI) model: (**a**) modeling of single-ring linear voltage hysteresis; (**b**) modeling diagram (a) of partial amplification; (**c**) modeling of reciprocating linear voltage hysteresis characteristics; (**d**) modeling diagram (c) of partial amplification.

Similarly, the five segmentation points of the MSPI model of the reciprocating linear voltage shown in Figure 13c participate in the fitting of the six segments. The hysteresis inner-loop MSPI model of the reciprocating linear voltage is enlarged and shown in Figure 13d. The inner-loop hysteresis characteristic can hence be accurately described.

During the modeling process, the slope of the MSPI model at the end is often larger than the tangent slope of the hysteresis rate that is caused by the forced zeroing of the end of the Play operator. This problem can be solved by ignoring the self-property of the superposition end, and by adding end segmentation and modeling according to its specific hysteresis characteristics.

3.4. Inverse Control

The MSPI model obtains an accurate approximation of the voltage–displacement correspondence. In order to achieve accurate compensation of the linear displacement, the displacement–voltage correspondence of the MSPI inverse model is used as a feedforward control. According to the compensation control principle [31], the MSPI inverse model is the inverse function of the hysteresis characteristic curve, as shown in Figure 14.

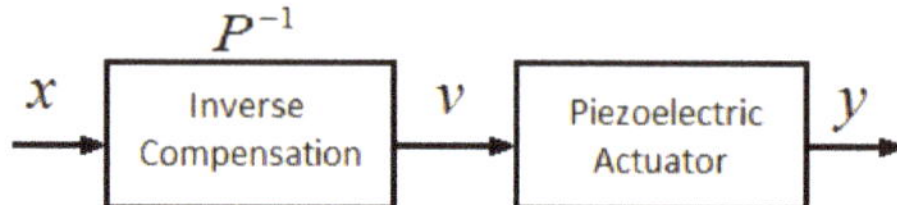

Figure 14. Hysteresis compensation control system diagram.

The classical PI model has an analytical inverse. The MSPI model is composed of separate PI models, so the inverse model equation is consistent with the PI inverse model. The equation is:

$$P^{-1}[p(k)] = \theta'_0 \cdot p(k) + \sum_{i=1}^{n} \theta'_i \cdot x_i(k) = \theta'_0 \cdot p(k) + \sum_{i=1}^{n} \theta'_i \cdot \max\{p(k) - r'_i, \min[p(k) + r'_i, x_i(k-1)]\} \tag{11}$$

where $P^{-1}[p(k)]$ is the output corresponding to the PI inverse model operator input $p(k)$. Here, $x_i(k)$ is the output of the ith operator, $\theta'_0 = \frac{1}{\theta_0}$. The threshold and weight coefficient of the inverse model are:

$$r'_i = \sum_{h=1}^{i} \theta_h \cdot (r_i - r_h) \quad i = 1, 2, \cdots, n \tag{12}$$

$$\theta'_i = -\frac{\theta_i}{\left(\theta_0 + \sum_{h=1}^{i} \theta_h\right)\left(\theta_0 + \sum_{h=1}^{i-1} \theta_h\right)} \quad i = 1, 2, \cdots, n-1 \tag{13}$$

Figure 15 is an MSPI inverse model corresponding to Figure 13a,c. It can be seen that the MSPI inverse model has ideal connectivity between the segments.

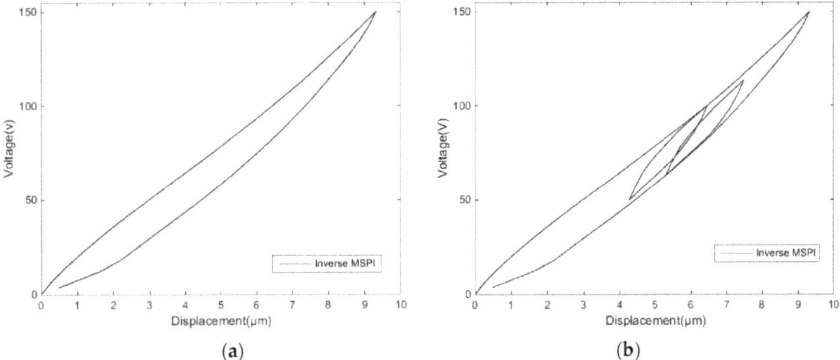

Figure 15. The MSPI inverse models: (**a**) single-ring linear voltage hysteresis inverse model and (**b**) reciprocating linear voltage hysteresis inverse model.

4. Results

This section presents the experimental results of this study. In the first part of this section, the two typical examples of the MSPI model presented in the third section of this study, namely the single-ring linear voltage hysteresis and the reciprocating linear voltage hysteresis, whose compensation control voltages are obtained by the inverse models, are verified and the errors are analyzed. The second part of this section carries out two verification experiments. One of them studies the effects of different frequency voltages on the MSPI model. The other one tests the MSPI model for another type of nanopositioning stage and observes the modeling effect. All of the above experiments demonstrate the contribution of the MSPI model to improving the nanopositioning accuracy of the stage.

4.1. Compensation Results

In order to verify the compensation control effect of the MSPI inverse model, the following experiment was performed on the experimental system from Figure 3, with the inverse model as an input:

(1) Adjust the laser interferometer. Connect computer, controller, and laser interferometer. The related software is turned on and waits for the measurements.

(2) Use the controller-related software to load the control voltage in the inverse model. Experiment 1 is carried out according to the voltage obtained by inverse model of Figure 15a, and the displacement is measured and recorded by the laser interferometer.

(3) Perform the experiment according to the voltage obtained by the inverse model in Figure 15b. Measure and record the displacement data by the laser interferometer. The interval should be the same as (2).

(4) Take several measurements.

(5) Check the equipment and turn it off. Process the experimental data.

Figure 16a is the measured single-ring linear voltage hysteresis feature compensation effect, while Figure 16b is the hysteresis compensation effect of the classical PI inverse model. The mean absolute deviation can be expressed as:

$$e = \frac{1}{\eta} \sum_{\mu=1}^{\eta} (\varepsilon_\mu - \omega_\mu) \tag{14}$$

where η is data quantity, ε_μ are expected results, and ω_μ are experimental results. Hence, the mean absolute deviation of the classical PI inverse model compensation control is 190.2 nm, the mean absolute deviation of the MSPI inverse model compensation control is 35.0 nm, and the nanopositioning accuracy is improved by 81.6%.

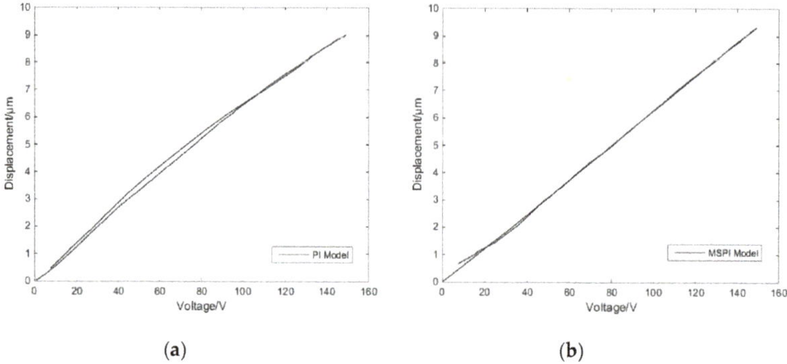

(a)   (b)

Figure 16. Single-ring linear voltage hysteresis compensation control effects: (**a**) classical Prandtl–Ishlinskii (PI) inverse model compensation and (**b**) MSPI inverse model compensation.

The reciprocating linear voltage hysteresis feature compensation effect is shown in Figure 17. The MSPI model in Figure 13c describes the hysteresis characteristics significantly better than the classical PI model description in Figure 7b; hence, the comparison is not made here. The mean absolute deviation of the MSPI inverse model compensation control of reciprocating linear hysteresis is 19.7 nm, and the positioning error is only 0.42%.

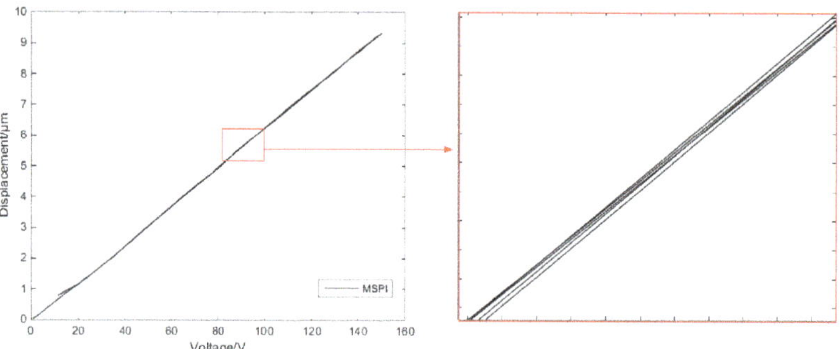

Figure 17. Reciprocating linear voltage hysteresis compensation control effect.

As predicted, the MSPI model is still flawed in its description of the end curve. Therefore, whether it is the reciprocating linear voltage hysteresis MSPI inverse model or the single-ring linear voltage hysteresis MSPI inverse model, the most significant error for both models is at the end of the compensation result. Accuracy can be improved without the end error or with additional segmentation modeling at the end.

In addition, in the two given examples, the MSPI model has more advantages when solving reciprocating linear voltage hysteresis compensation. Compared with the single-ring linear voltage hysteresis MSPI inverse model, the reciprocating linear voltage hysteresis MSPI inverse model has more marks and more segments, and hence the accuracy is improved by 43.7%.

4.2. Verification Tests

In order to study whether the change of the voltage frequency affects the application of the MSPI model, verification experiment 1 is carried out to observe the hysteresis characteristics.

For example, it is observed that for single-ring linear voltage, the elongation speed of the piezoelectric ceramic displacement is a negative value and the shrinkage speed is a positive value. The amplitude of the triangular wave voltage is set to be the same as the amplitude of Figure 4a, which is 150V. Three sets of the speed time diagrams can be obtained by changing the voltage frequency. According to the period–frequency relationship $T = \frac{1}{f}$, the smaller the frequency is, the longer the triangular wave voltage period. The appropriate and easily observed frequency control period time is between 1 and 5 s.

Figure 18 shows the speed time diagrams of 1.0 Hz, 0.4 Hz, and 0.2 Hz voltage frequencies.

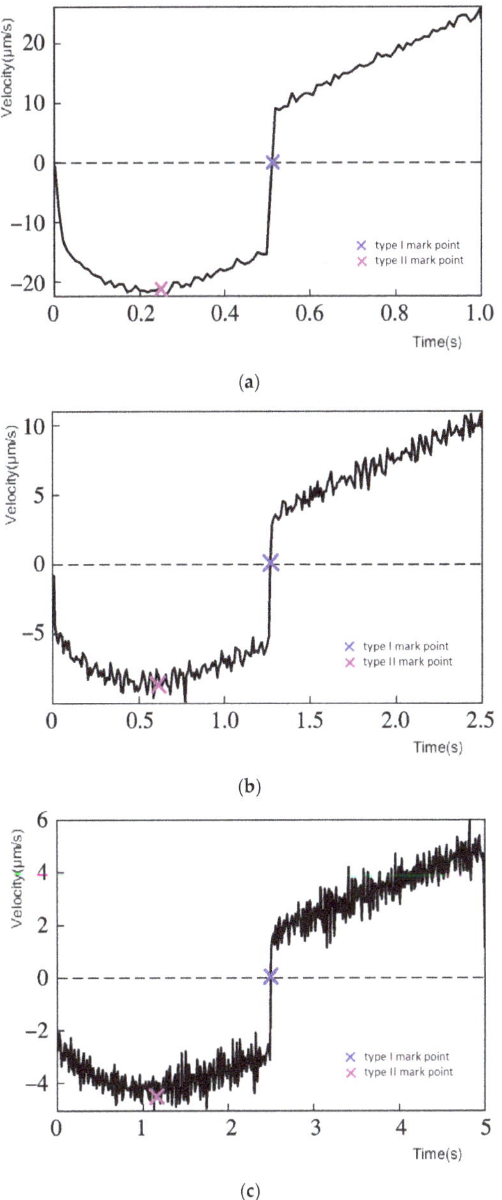

Figure 18. Characteristics of displacement velocity at different frequencies: (**a**) amplitude 150 V, frequency 1.0 Hz; (**b**) amplitude 150 V, frequency 0.4 Hz; (**c**) amplitude 150 V, frequency 0.2 Hz.

Although the output speeds are different at different frequencies and the maximum displacement time is shortened as the frequency increases, the same variation characteristics are maintained. If the speed is taken as an absolute value, the speed–time diagrams can find a similar relationship as with the $v - s(v)$ diagram. Both the type I mark points and the type II mark points have been identified on the map in the same color as the $v - s(v)$ diagram.

The hysteresis characteristic of the reciprocating linear voltage is the same. It can be seen that the different voltage frequencies exhibit the same regularity for hysteresis characteristics, and thus the MSPI model is still effective.

Verification experiment 2 is then carried out. The MSPI model for another type of nanopositioning stage is used to observe the modeling effect.

The equipment is adjusted and connected according to the experimental steps in Section 2.2. The triangular wave voltages of 20 V and 15 V amplitudes are loaded as the reciprocating linear voltage inputs. The experimental data are measured and recorded for every 0.5 V.

Figure 19 shows the hysteresis characteristics of the experimental measurements and the modeling comparison between the classical PI model and MSPI model. The MSPI model has a higher description accuracy, which repeatedly proves that the MSPI model can be applied to different hysteresis characteristics.

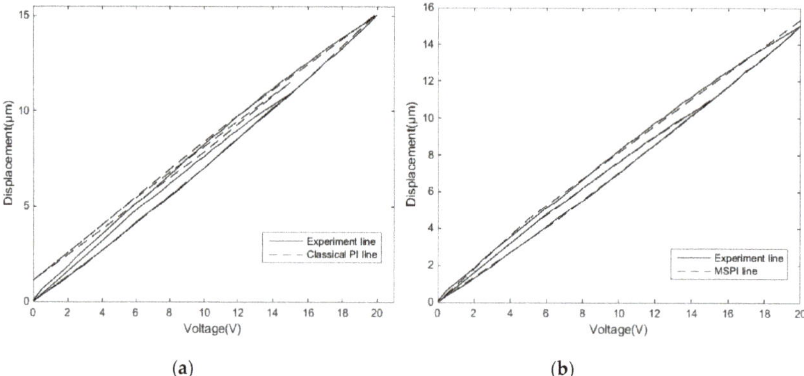

Figure 19. Comparison of the experimental measurements: (**a**) the classical PI model and (**b**) the MSPI model.

5. Conclusions

The MSPI model effectively solves the problem of low description accuracy of reciprocating linear voltage hysteresis of the classical PI model. In this paper, the experimental data can be used to find the hysteresis regulation when the hysteresis rate tangent is defined. With a defined segmented basis, the slope characteristics are analyzed to propose a mark-segmented point, which carries out a theoretical solution to segmented modeling. It has been verified that the further segmented modeling can not only compensate for the nonlinear characteristics of the external hysteresis loop under various linear voltages, but can also effectively identify the segmentation of features generated by the intrinsic microscopic mechanism.

The MSPI model hysteresis compensation method does not introduce a new operator. The hysteresis characteristic parameters are completely based on the experimental data. The two types of mark-segmented point recognition methods are simple and evident. Hence, the segmented basis makes the MSPI model reliable. The model construction is easy to implement as well. This paper also provides sufficient theoretical preparation for further study of the more complex hysteresis characteristics of the nanopositioning stage under the nonlinear voltage.

Author Contributions: Data curation, M.S. and J.S.; methodology, Y.Y. and Y.X.; writing—original draft, D.A.; writing—review and editing, G.Y. All authors have read and agreed to the published version of the manuscript.

Funding: This study was supported by the National Science Foundation of China (51975130, 51705340); Natural Science Foundation of Liaoning Province (20180550002); Basic Research Projects for Colleges and Universities of Liaoning Province (LJZ2017035); and the Key Research and Development Project of Liaoning Province (2017225016).

Conflicts of Interest: The authors declare no conflict of interest.

References

1. Zhang, X.; Xu, Q. Design and development of a new 3-DOF active-type constant-force compliant parallel stage. *Mech. Mach. Theory* **2019**, *140*, 654–665. [CrossRef]
2. Ling, J.; Rakotondrabe, M.; Feng, Z.; Ming, M.; Xiao, X. A Robust Resonant Controller for High-Speed Scanning of Nanopositioners: Design and Implementation. *IEEE Trans. Control Syst. Technol.* **2019**, 1–8. [CrossRef]
3. Raunsmann, C.; Schaffer, T.E. High-speed atomic force microscopy for large scan sizes using small cantilevers. *Nanotechnology* **2010**, *21*, 225705. [CrossRef]
4. Deng, J.; Zhou, H.; Dong, J.; Cohen, P. Three-Dimensional Nanomolds Fabrication for Nanoimprint Lithography. *Procedia Manuf.* **2019**, *34*, 228–232. [CrossRef]
5. Al Janaideh, M.; Rakotondrabe, M.; Al-Darabsah, I.; Aljanaideh, O. Internal model-based feedback control design for inversion-free feedforward rate-dependent hysteresis compensation of piezoelectric cantilever actuator. *Control. Eng. Pract.* **2018**, *72*, 29–41. [CrossRef]
6. Ma, J.; Tian, L.; Li, Y.; Yang, Z.; Cui, Y.; Chu, J. Hysteresis compensation of piezoelectric deformable mirror based on Prandtl–Ishlinskii model. *Opt. Commun.* **2018**, *416*, 94–99. [CrossRef]
7. Chi, Z.; Xu, Q. Precision control of piezoelectric actuator using fuzzy feedback control with inverse hysteresis compensation. In Proceedings of the 10th IEEE International Conference on Nano/Micro Engineered and Molecular Systems, Xi'an, China, 7–11 April 2015; pp. 219–224.
8. Pollack, F.H.; Cardona, M. Piezo-Elector-reflectance in Ge, GA As, and Si. *Phys. Rev.* **1968**, *172*, 816. [CrossRef]
9. Duran, P.; Moure, C. Piezoelectric ceramics. *Mater. Chem. Phys.* **1986**, *15*, 193–211. [CrossRef]
10. Ming, M.; Feng, Z.; Ling, J.; Xiao, X. Hysteresis modelling and feedforward compensation of piezoelectric nanopositioning stage with a modified Bouc-Wen model. *Micro Nano Lett.* **2018**, *13*, 1170–1174. [CrossRef]
11. Chen, Y.; Shen, X.; Li, J.; Chen, J. Nonlinear hysteresis identification and compensation based on the discrete Preisach model of an aircraft morphing wing device manipulated by an SMA actuator. *Chin. J. Aeronaut.* **2019**, *32*, 1040–1050. [CrossRef]
12. Ko, Y.; Hwang, Y.; Chae, M.; Kim, T. Direct identification of generalized Prandtl–Ishlinskii model inversion for asymmetric hysteresis compensation. *ISA Trans.* **2017**, *70*, 209–218. [CrossRef] [PubMed]
13. Stefanski, F.; Minorowicz, B.; Persson, J.; Plummer, A.; Bowen, C. Non-linear control of a hydraulic piezo-valve using a generalised Prandtl–Ishlinskii hysteresis model. *Mech. Syst. Signal Process.* **2017**, *82*, 412–431. [CrossRef]
14. Mokaberi, B.; Requicha, A.A.G. Compensation of Scanner Creep and Hysteresis for AFM Nanomanipulation. *IEEE Trans. Autom. Sci. Eng.* **2008**, *5*, 197–206. [CrossRef]
15. Janaideh, M.A.; Su, C.; Rakheja, S. A generalized asymmetric play hysteresis operator for modeling hysteresis nonlinearities of smart actuators. In Proceedings of the 2008 10th International Conference on Control, Automation, Robotics and Vision, Hanoi, Vietnam, 17–20 December 2008; pp. 240–244.
16. Cao, K.; Li, R.; Du, H.; Ma, J. Modeling and compensation of symmetric hysteresis in piezoceramic actuators. *Results Phys.* **2019**, *13*, 102095. [CrossRef]
17. Sun, Z.; Song, B.; Xi, N.; Yang, R.; Chen, L.; Cheng, Y.; Bi, S.; Li, C.; Hao, L. Systematic Hysteresis Compensator Design based on Extended Unparallel Prandtl–Ishlinskii Model for SPM Imaging Rectification. *IFAC-PapersOnLine* **2017**, *50*, 10901–10906. [CrossRef]
18. Hassani, V.; Tjahjowidodo, T.; Do, T.N. A survey on hysteresis modeling, identification and control. *Mech. Syst. Signal Process.* **2014**, *49*, 209–233. [CrossRef]
19. Gu, G.; Yang, M.; Zhu, L. Real-time inverse hysteresis compensation of piezoelectric actuators with a modified Prandtl–Ishlinskii model. *Rev. Sci. Instrum.* **2012**, *83*, 065106. [CrossRef] [PubMed]

20. Park, J.; Moon, W. Hysteresis compensation of piezoelectric actuators: The modified Rayleigh model. *Ultrasonics* **2010**, *50*, 335–339. [CrossRef] [PubMed]
21. Guo, Z.; Tian, Y.; Liu, X.; Shirinzadeh, B.; Wang, F.; Zhang, D. An inverse Prandtl–Ishlinskii model based decoupling control methodology for a 3-DOF flexure-based mechanism. *Sens. Actuators A Phys.* **2015**, *230*, 52–62. [CrossRef]
22. Yang, M.; Gu, G.; Zhu, L. Parameter identification of the generalized Prandtl–Ishlinskii model for piezoelectric actuators using modified particle swarm optimization. *Sens. Actuators A Phys.* **2013**, *189*, 254–265. [CrossRef]
23. Habineza, D.; Zouari, M.; le Gorrec, Y.; Rakotondrabe, M. Multivariable Compensation of Hysteresis, Creep, Badly Damped Vibration, and Cross Couplings in Multiaxes Piezoelectric Actuators. *IEEE Trans. Autom. Sci. Eng.* **2018**, *15*, 1639–1653. [CrossRef]
24. An, D.; Li, H.; Xu, Y.; Zhang, L. Compensation of hysteresis on piezoelectric actuators based on tripartite PI model. *Micromachines* **2018**, *9*, 44. [CrossRef] [PubMed]
25. Lobontiu, N.; Paine, J.S.N.; Garcia, E.; Goldfarb, M. Design of symmetric conic-section flexure hinges based on closed-form compliance equations. *Mech. Mach. Theory* **2002**, *37*, 477–498. [CrossRef]
26. Yong, Y.K.; Lu, T.F.; Handley, D.C. Review of circular flexure hinge design equations and derivation of empirical formulations. *Precis. Eng.* **2008**, *32*, 63–70. [CrossRef]
27. Gosavi, S.V.; Kelkar, A.G. Passivity-based robust control of Piezo-actuated flexible beam. In Proceedings of the 2001 American Control Conference. (Cat. No.01CH37148), Arlington, VA, USA, 25–27 June 2001; pp. 2492–2497.
28. Petroni, S.; Tegola, C.L.; Caretto, G.; Campa, A.; Passaseo, A.; de Vittorio, M.; Cingolani, R. Aluminum Nitride Piezo-MEMS on polyimide flexible substrates. *Microelectron. Eng.* **2011**, *88*, 2372–2375. [CrossRef]
29. Chu, C.L.; Fan, S.H. A novel long-travel piezoelectric-driven linear nano-positioning stage. *Precis. Eng.* **2006**, *30*, 85–95. [CrossRef]
30. Brokate, M.; Sprekels, J. *Hysteresis and Phase Transaitions*; Springer: Berlin, Germany, 1996.
31. Kuhnen, K. Modeling, identification and compensation of complex hysteretic nonlinearities: A modified Prandtl–Ishlinskii approach. *Eur. J. Control* **2003**, *9*, 407–417. [CrossRef]

Article

Design, Analysis, and Experiment on a Novel Stick-Slip Piezoelectric Actuator with a Lever Mechanism

Weiqing Huang [1,*] **and Mengxin Sun** [2]

[1] School of Mechanical and Electrical Engineering, Guangzhou University, Guangzhou 510006, China

[2] School of Mechanical Engineering, Nanjing Institute of Technology, Nanjing 211167, China; mxsun@njit.edu.cn

* Correspondence: mehwq@nuaa.edu.cn

Received: 30 September 2019; Accepted: 5 December 2019; Published: 8 December 2019

Abstract: A piezoelectric actuator using a lever mechanism is designed, fabricated, and tested with the aim of accomplishing long-travel precision linear driving based on the stick-slip principle. The proposed actuator mainly consists of a stator, an adjustment mechanism, a preload mechanism, a base, and a linear guide. The stator design, comprising a piezoelectric stack and a lever mechanism with a long hinge used to increase the displacement of the driving foot, is described. A simplified model of the stator is created. Its design parameters are determined by an analytical model and confirmed using the finite element method. In a series of experiments, a laser displacement sensor is employed to measure the displacement responses of the actuator under the application of different driving signals. The experiment results demonstrate that the velocity of the actuator rises from 0.05 mm/s to 1.8 mm/s with the frequency increasing from 30 Hz to 150 Hz and the voltage increasing from 30 V to 150 V. It is shown that the minimum step distance of the actuator is 0.875 µm. The proposed actuator features large stroke, a simple structure, fast response, and high resolution.

Keywords: piezoelectric actuator; lever mechanism; analytical model; stick-slip frication

1. Introduction

Though the piezoelectric effect was discovered more than a century ago, research on piezoelectric is still ongoing [1–5] and attracts attention in many areas [6–10]. One of the areas is actuation. Because of their advantages of simple structure, flexible design, high resolution, and low power consumption, many piezoelectric actuators have been developed [11–14]. Actuators can be classified by vibration state into the resonant type and the non-resonant type. The resonant type is the traditional one and is also called the ultrasonic motor. This type is already widely used in precision positioning, nanotechnology, and biomechanics, among other things [15–19]. The main drawback of this type of actuator is that the motion of the motor could be unstable as the system must work under resonance state, which is used to amplify the displacement of the driving element. Non-resonant piezoelectric actuators utilize a piezoelectric stack as the core driving element. Compared to the former type, they would produce enough deformation under the non-resonant state [20,21]. Therefore, the non-resonant actuator can operate more stably and will achieve large stroke and high resolution at the same time.

Non-resonant piezoelectric actuators can be classified into several types, among which the inchworm actuator and the inertial actuator are two important categories. The inchworm motor is a kind of bionic motor with the advantages of a strong loading capacity and high precision accuracy [22–24]. This type of device usually has the limitation of having a complex structure, which is difficult to assemble. In addition, the inchworm type usually requires two or three piezoelectric stacks for operation. The inertial actuator based on the inertial driving principle (stick-slip principle) often has a high pushing force [25–27]. In recent years, many inertia actuators with different mechanisms have been reported. Most inertia actuators are based on the friction drive principle [28–30]. Multilayered piezoelectric stacks (PZTs) are used to increase the amplitude of the mover displacement, and the actuators often have low operation speed because a low input voltage is used [31]. There are also reports presenting piezoelectric actuators that are designed with a flexure mechanism. The flexure mechanisms set in the actuators are not only the connecting joints to implement high precision motion, but also the amplification mechanism of the system. This type of actuator operates with the flexure mechanism to amplify the displacement of the stator per period, which increases the output speed of the mover. Significant efforts have been applied to design and analyze the flexure mechanisms in the actuators. For instance, Meng et al. proposed an approach for analyzing displacement and stiffness characteristics of flexure-based proportion compliant mechanisms based on the principle of virtual work and a rigid body model. The compliant mechanism was designed based on the numerical model and verified via finite element analysis [32]. Tian et al. utilized lever mechanisms to enlarge the working range of a five-bar compliant micro-manipulator. Numerical simulations based on linearization of trigonometric functions and a constant jacobian matrix were carried out to investigate the performance of the system. The experiment results showed that the lever mechanisms can provide the function for displacement amplification [33]. Although the above mentioned methodology can improve both the position accuracy and output displacement of the actuators, it cannot fulfill the requirements of long-stroke applications.

In previous studies, long-travel precision motions have been achieved by setting two or more perpendicular piezoelectric units. Simu et al. introduced dynamic and quasi-static motion mechanisms used in a miniature piezoceramic drive unit. The actuator consisted of six piezoelectric stacks [34]. Chen et al. proposed a piezoelectric actuators based on two groups of orthogonal structures using the friction drive principle. The experimental results indicated that the actuator can stably operate within the scope of 350 to 750 Hz when the step size is about 3.1 μm [35]. Jalili et al. proposed an actuator based on the friction drive principle with two perpendicular vibratory piezoelectric units. Experiment results showed that the maximum mean velocity was about 5.5 mm/sec, and the length of the each step was about 275 nm [36]. In these studies, the vibration of the piezoelectric stacks caused rectangular or elliptical motions at the top of the driving feet. As the driving feet are in contact with the linear guide by appropriate preload force, the micro-vibration at the top of the driving foot is transformed into the macro-linear motion of the guide. Although these actuators have high resolution and fast response, the structure and control mechanisms are complex. Dynamic errors will be caused by non-synchronized driving signals. A novel piezoelectric is proposed in this paper. Based on the mass spring damping system and the Karnopp friction model [37], theoretical analysis is carried out in the frequency domain, and the motion curve of the actuator is obtained. Simulation is carried out in ANSYS (version 18, Canonsburg, PA, USA) to verify the dynamic analysis results. Then the parameters of the flexure hinge are optimized by combining the results of numerical simulation and finite element simulation. Finally, the accuracy of the theoretical model is verified by experiments, and the characteristics of the prototype are tested.

2. Design of the Actuator

2.1. Structure of the Actuator

As shown in Figure 1, a novel stick-slip piezoelectric actuator with a lever mechanism is proposed in this paper. It is comprised of a stator, an adjustment mechanism, a preload mechanism, a base, and a linear guide. As the vibration source of the actuator, a piezoelectric stack is set in the frame structure of the stator preloading by a long flexure hinge. The piezoelectric stack used is a PK4FQP1 from Thorlabs (Newton, NJ, USA). The use of a long flexure hinge has the twin functions of pre-tightening the piezoelectric stack and linking the rigid body. Since the width of the ring in the long flexure hinge is smaller than its outer radius, the hinge flexes more smoothly. Thus, the displacement produced from the piezoelectric stack will deliver to the driving foot with little error. The frame structure of the stator is designed and set on a small guide to adjust the distance between the stator and the linear guide. The preload mechanism consists of a preload spring and a preload screw mechanism to keep the stator and the linear guide in a proper contact state. When the actuator is at work, the application of an appropriate driving signal to the piezoelectric stack causes the guide to perform precision linear motion.

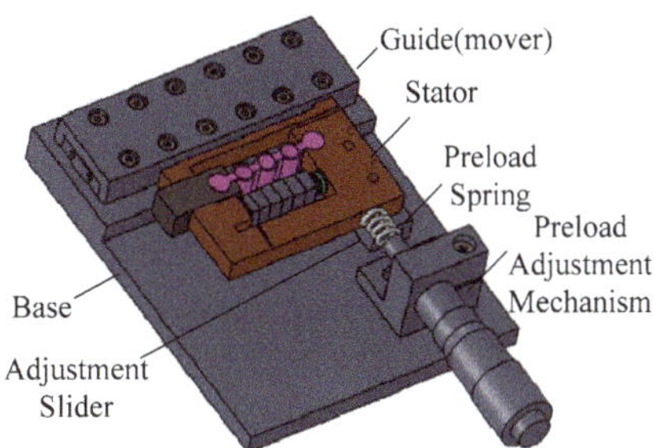

Figure 1. Structure of the stick-slip piezoelectric actuator.

2.2. Working Principle

The proposed stick-slip piezoelectric actuator utilizes the inertia effect to output linear motion. The input signal applied to the piezoelectric stack is shown in Figure 2. Figure 3 illustrates the working state of the stator in a full period.

The working principle of the actuator is performed by the sequence of a stick-phase and a slip-phase. Figures 2 and 3 reveal that when a sawtooth wave is applied to the piezoelectric stack, the inertial skew lines can be produced at the driving foot of the stator. In one period, the proposed actuator is operated as follows: When the sawtooth wave, shown in Figure 2a, is applied to the piezoelectric stack, the piezoelectric stack extends slowly in the voltage up phase (a-b-c in Figure 3, stick-phase). Due to the driving foot being tightly clamped to the mover with the preload force, it pushes the mover to move a distance in the $-x$ direction through static friction force. Then, in the voltage down phase, the piezoelectric stack quickly contracts to its initial length and drives the foot back rapidly to the initial position. In this moment, the mover moves along the $-x$ direction because of its inertia (c–a in Figure 3, slip-phase). By applying the signal in Figure 2b, the movement direction of the mover will change.

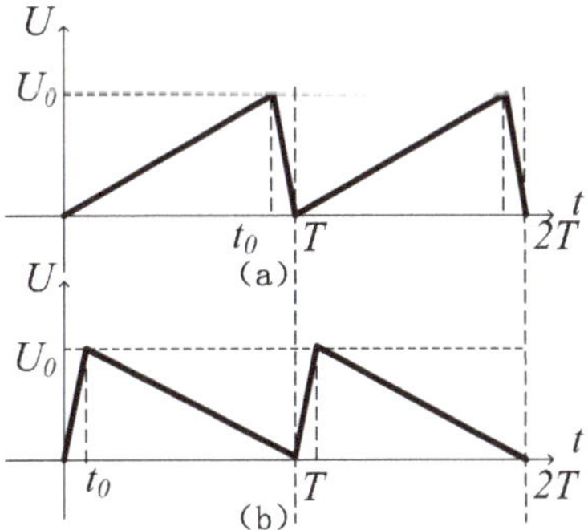

Figure 2. Input signal applied on the piezoelectric stack. (**a**) sawtooth signal with the voltage of slow rise and fast fall; (**b**). sawtooth signal with the voltage of fast rise and slow fall.

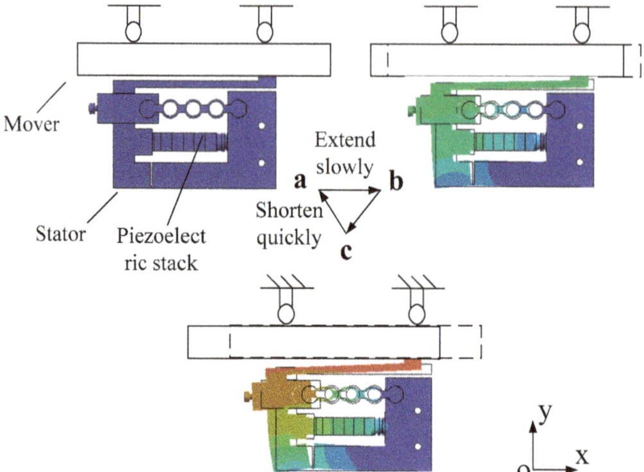

Figure 3. Working principle of the actuator.

2.3. Design and Analysis

The characteristics of the stator determine the motion performance of the actuator. To design and analyze the proposed actuator, modeling analysis is used in this paper. The structure of the stator is shown in Figure 4.

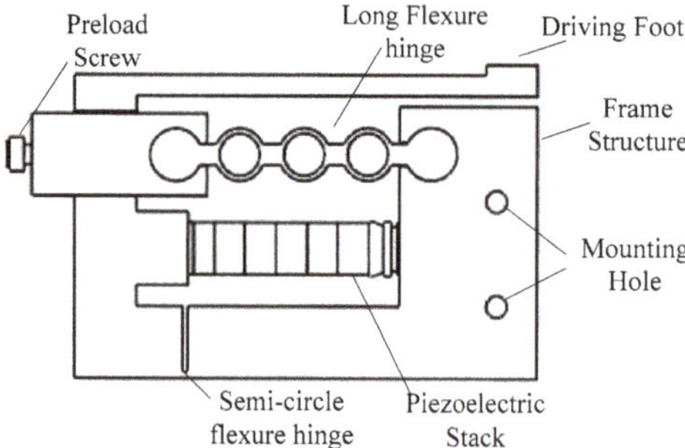

Figure 4. Structure of the stator.

Figure 5 shows the simplified model of the stator. L is the length of the structure, W is the width, h is the distance from the bottom of the structure to the location of the piezoelectric stack, C is the driving foot, and k_1 and k_2 represent the stiffness of the long hinge structure and the flexure hinge in the frame structure, respectively. When the stator works, the input signal applied on the piezoelectric can be written as:

$$U_p(t) = \begin{cases} \frac{U_0}{t_0}(t - kT), & t - kT \leq t_0 \ (k = 0, 1, 2, \ldots) \\ U_0 - \frac{U_0}{T - t_0}(t - kT - t_0), & t - kT > t_0 \ (k = 0, 1, 2, \ldots) \end{cases} \tag{1}$$

where U_0 is the maximum voltage of the input signal, t_0 is the time of the peak point, and T is the cycle time. The output force of the piezoelectric stack is:

$$F_p = nd_{33}U_p(t)k_p. \tag{2}$$

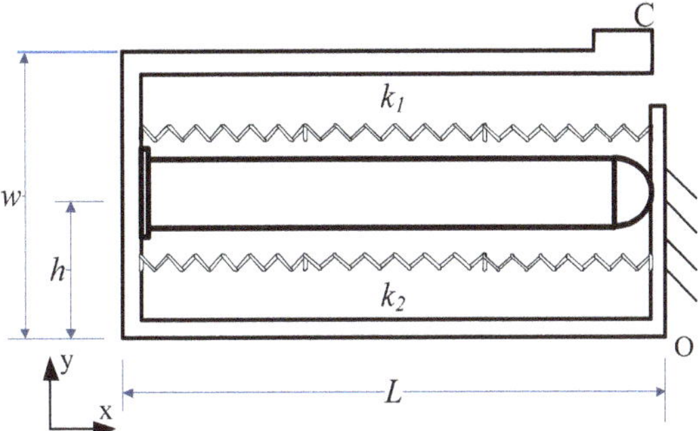

Figure 5. Simplified model of the stator.

The simplified models of the long flexure hinge and the semi-circular flexure hinge are shown in Figures 6 and 7. Dimension parameters of the long flexure hinge are shown in Figure 7. The long flexure hinge consists of three circular structures and four short beams. According to the knowledge of material mechanics, the equivalent tensile rigidity of the circular structure is:

$$k_{11} = \frac{Eb'(R-r)^3}{3\pi R^3}.$$ (3)

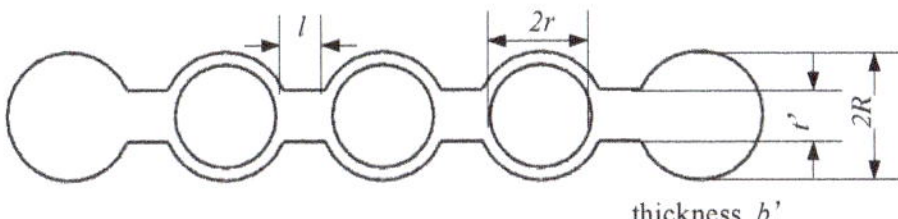

Figure 6. Simplified model of the long flexure hinge.

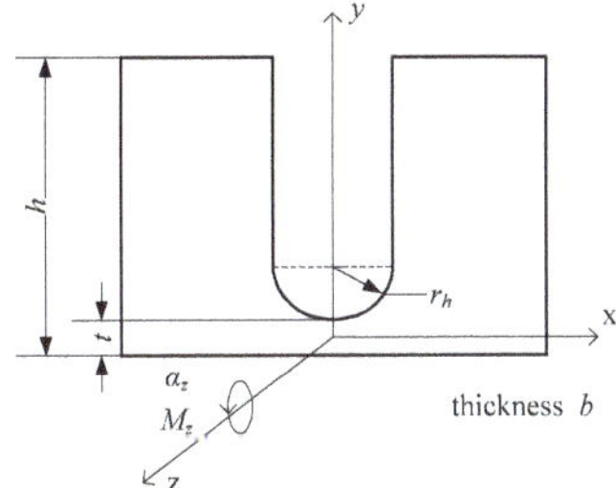

Figure 7. Simplified model of the semi-circular flexure hinge.

The equivalent tensile rigidity of the short beam is:

$$k_{12} = Eb't'/l.$$ (4)

It can be assumed that:

$$k_1 = \frac{k_{11}k_{12}}{3k_{12} + 4k_{11}}.$$ (5)

Another flexure hinge is set in the frame structure, as shown in Figure 7. Take out the small part, whose size is $b \times a \times du$ in the central part. The central angle is β and the equations to describe small part can be written as:

$$a = t + r_h(1 - \cos\beta)$$ (6)

$$du = r_h \cos \beta d\beta. \tag{7}$$

When applying moment, M_z, on the z axis and producing rotate angle, $d\alpha_z$, it can be expressed as:

$$d\alpha_z = \frac{M_z}{EI_z} du \tag{8}$$

where I_z is the inertia moment to the z axis:

$$I_z = \frac{ba^3}{12}. \tag{9}$$

It can be shown that the equivalent stiffness of the semi-circular flexure hinge is [38]:

$$k_2 = \frac{M_z}{\alpha_z} = Eb / \left(12 r_h \int_{-\frac{\pi}{2}}^{\frac{\pi}{2}} \frac{\cos \beta}{\left(\frac{t}{r_h} + 1 - \cos \beta \right)^3} d\beta \right). \tag{10}$$

Based on the stator analysis above, the whole movement between the stator and the mover is under consideration. Figure 8 shows the dynamic model of the actuator.

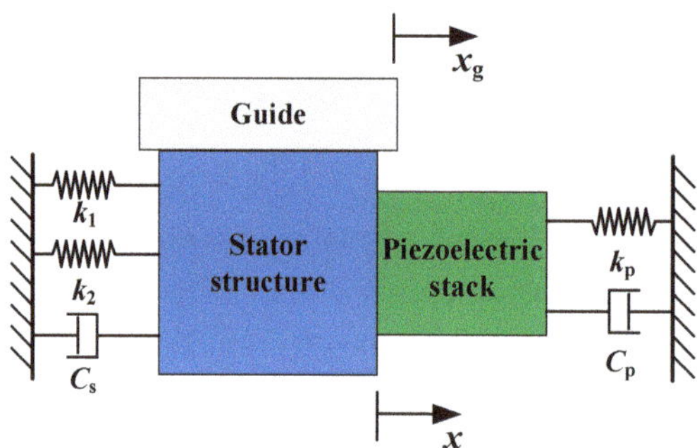

Figure 8. Dynamic model of the actuator.

A dynamic stiffness-damping model of the proposed actuator is established, as shown in Figure 8. The transverse force condition in the dynamic model is shown in Figure 9, where k_p is the stiffness of the piezoelectric stack, C_p is the damping coefficient of the piezoelectric stack, C_s is the damping coefficient of the stator structure, F_a is the internal force of the piezoelectric stack, the stator structure, f, is the friction force, and m_s, m_g, and m_p are the mass of the stator structure, piezoelectric, and guide, respectively.

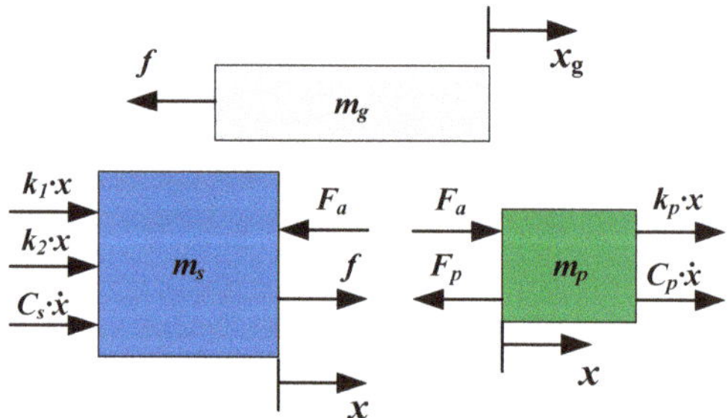

Figure 9. Transverse force conditions in the dynamic model.

The vibration equations can be written as:

$$\begin{bmatrix} m_p \\ m_s \\ m_g \end{bmatrix} \begin{bmatrix} \ddot{x}(t) & \ddot{x}(t) & \ddot{x}_g(t) \end{bmatrix} + \begin{bmatrix} C_p \\ C_s \\ 0 \end{bmatrix} \begin{bmatrix} \dot{x}(t) & \dot{x}(t) & \dot{x}_g(t) \end{bmatrix} + \begin{bmatrix} k_p \\ k_1 + k_2 \\ 0 \end{bmatrix} \begin{bmatrix} x(t) & x(t) & x_g(t) \end{bmatrix} = \begin{bmatrix} F_p - F_a \\ F_a - f \\ f \end{bmatrix} \quad (11)$$

Then the following equation can be derived:

$$(m_p + m_s)\ddot{x}(t) + (C_p + C_s)\dot{x}(t) + (k_p + k_1 + k_2)x(t) = F_p - f. \quad (12)$$

According to the references, the friction force, f, can be ignored when comparing with the output force, F_p [39], and the input voltage of the piezoelectric stack can be obtained as follows:

$$\frac{U_p(s)}{U_0(s)} = k_{amp}\frac{1}{RCs + 1} \quad (13)$$

where R is the resistance of the driving circuit, C is the capacitance of the piezoelectric stack, k_{amp} is the amplification ratio of the input voltage for the piezoelectric stack, and U_0 is the initial input voltage. The following equation for the relationship between $X(s)$ and $U_0(s)$ is derived by Laplace transform:

$$\frac{X(s)}{U_0(s)} = \frac{nd_{33}k_p k_{amp}}{(RCs + 1)[(m_p + m_s)s^2 + (C_p + C_s)s + (k_p + k_1 + k_2)]}. \quad (14)$$

When it comes to the lever mechanism in the stator, the rotate angle is:

$$\theta = \frac{x(t)}{h}. \quad (15)$$

Displacement of the driving foot can be written as:

$$x_C(t) = \frac{x(t)w}{h} \quad (16)$$

$$y_C(t) = \frac{x(t)L}{h}. \quad (17)$$

The longitudinal force condition in the dynamic model is shown in Figure 10, where k_y and C_y are the stiffness and the damping coefficients of the stator structure in the y direction and F_N is the preload force from the preload mechanism. The vertical force acting on the guide can be written as:

$$N = F_N + k_y y(t) + C_y \dot{y}(t) \quad (18)$$

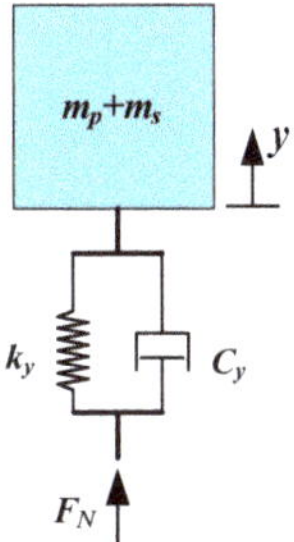

Figure 10. Longitudinal force condition in the dynamic model.

The friction force according to stick-slip can be obtained by applying the Karnopp model [37]. In this model, the friction coefficients are determined from guide velocity, $\dot{x}_g$, and the relative velocity, v_r, which can be derived from:

$$v_r = \dot{x} - \dot{x}_g. \tag{19}$$

The different cases of friction are described as:

$$\begin{cases} f = \frac{m_p + m_s}{m_g + m_p + m_s} \mu_k N \mathrm{sgn}(\dot{x}_g), & |v_r| \le \delta v \,(stick) \\ f = \mu_k N \mathrm{sgn}(\dot{x}_g), & |v_r| > \delta v \,(slip) \end{cases} \tag{20}$$

where, μ_k is the kinetic friction coefficient and δv is the small velocity bound.

The design objective of the stator is to maximize the output displacement of the contact point when it simultaneously meets the requirements of stiffness. The design parameters include the inner diameter, r, and width, t, of the long flexure hinge and the diameter, r_h, of the semi-circular flexure hinge. MATLAB/Simulink (version R2016a) is used to analyze the motion characteristics, and the simulation results for the actuator are shown in Figure 11. It can be illustrated that the output velocity of the actuator is 1.45 mm/s when a sawtooth signal of 100 V and 100 Hz is applied.

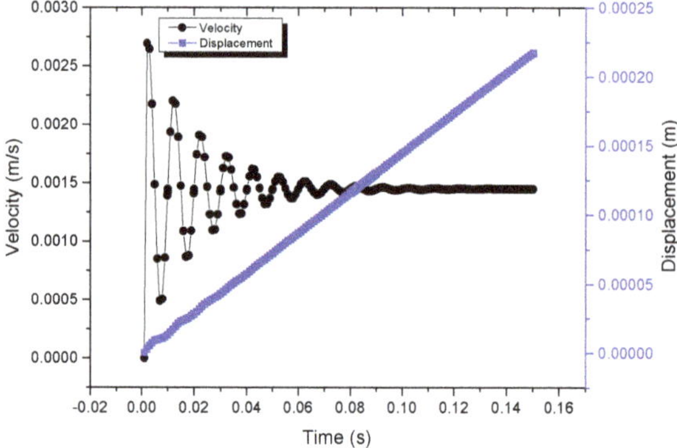

Figure 11. Simulation results of the actuator by MATLAB/Simulink.

According to the analogue simulation, the relationships between the design parameters and the output displacement of the driving foot in the x direction are obtained, as shown in Figure 12.

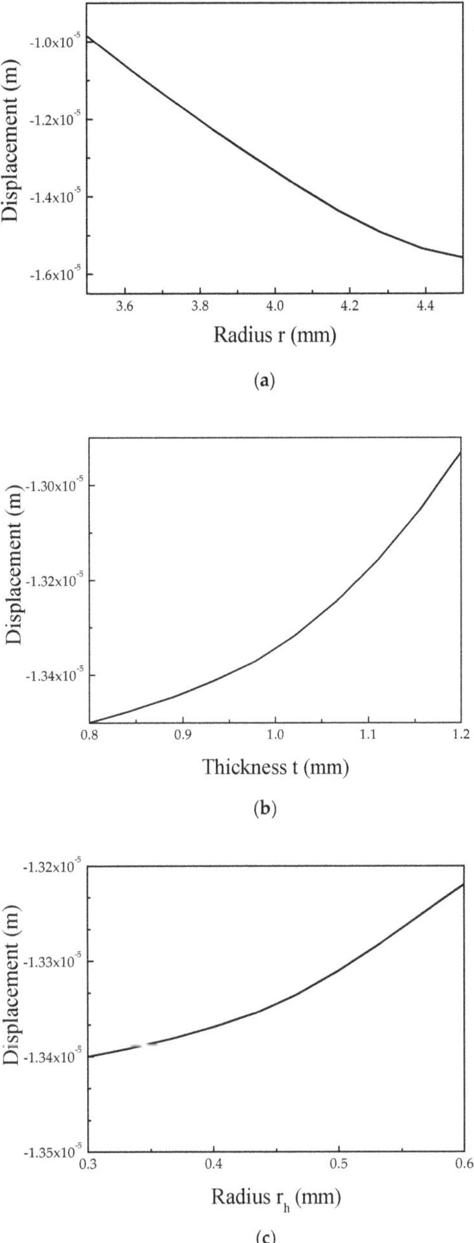

Figure 12. Relationships between the displacement of the driving foot versus the diameter of the long flexure hinge and semi-circular flexure hinge. (**a**) Inner diameter, *r*, and width, *t*, of the long flexure hinge versus the displacement of the driving foot. (**b**)Width, *t*, of the long flexure hinge versus the displacement of the driving foot. (**c**) Diameter, r_h, of the semi-circular flexure hinge versus the displacement of the driving foot.

The finite element method (FEM) of the piezoelectric actuator was performed to calculate the dynamic characteristics of the actuator when the sawtooth signal was applied. The FEM model was made up of a stator structure, a guide, and a piezoelectric stack, as shown in Figure 13. The right part of the stator was rigidly clamped. The mechanical boundary conditions of the model relate to the holding conditions used in the tests. SOLID 95 elements were used to mesh the stator parts while SOLID 98 elements were used to mesh the piezoelectric stack. The model contains 14,598 elements. The stator was made from stainless steel while the piezoelectric stack was made from piezo ceramic. Polarization was aligned in the x direction.

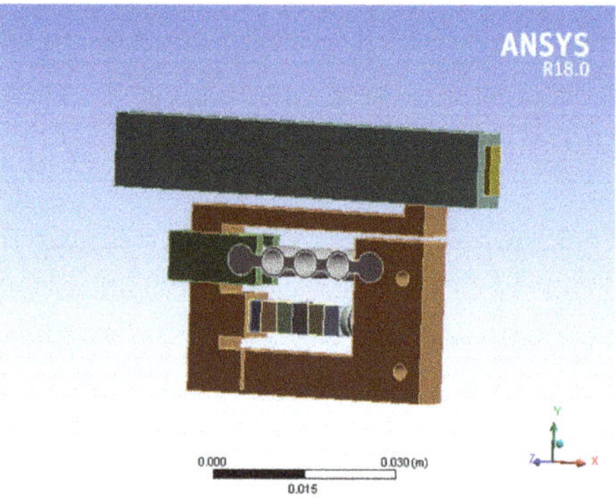

Figure 13. Finite element method (FEM) model.

Modal frequency analysis was first performed to obtain the resonant frequency. Figure 14 shows that the first vibration mode of the stator has a 1223.3 Hz natural frequency. Since the proposed piezoelectric actuator runs on the non-resonant frequency, the working frequency of the system would be less than the resonant frequency.

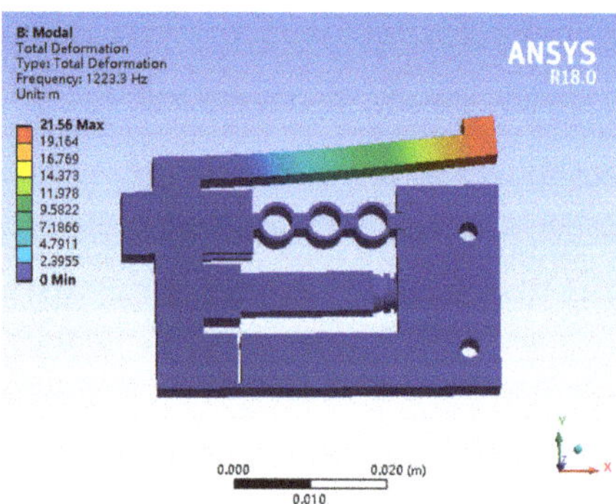

Figure 14. First resonant mode of the stator.

Transient dynamic analysis was carried out next, to calculate the displacement and steady state of the actuator. When a sawtooth voltage signal of 100 Hz and 100 V was applied to the piezoelectric stack, the end of the piezoelectric stack had a displacement of about 10 μm. The output displacement of the driving foot in the x direction was 14.23 μm, as shown in Figure 15. At the same time, the displacement of the guide was obtained, as shown in Figure 16. It was seen that the output velocity of the actuator was 1.4 mm/s.

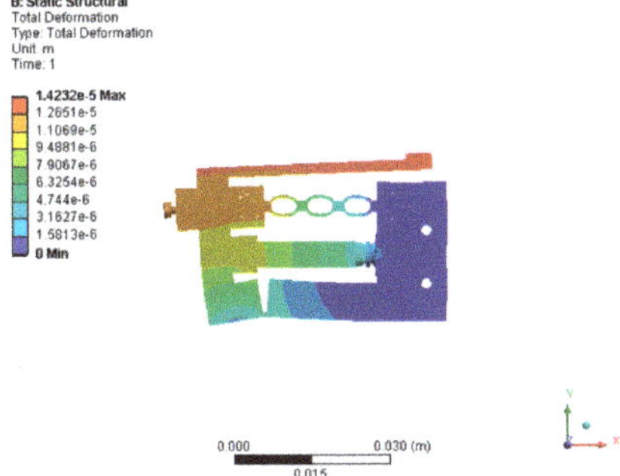

Figure 15. The output displacement of the driving foot.

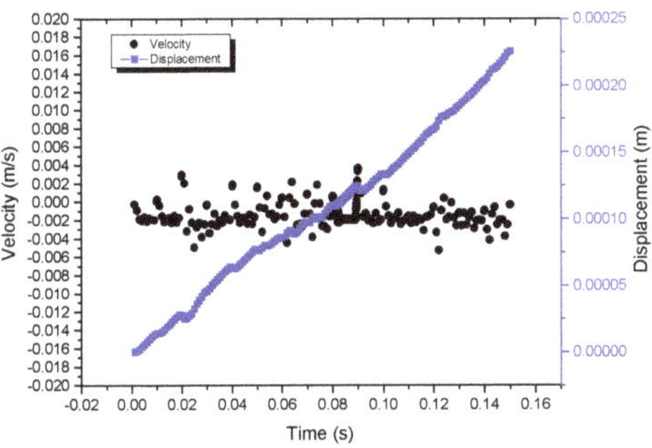

Figure 16. FEM dynamic analysis results.

The FEM analysis results illustrated that the proposed piezoelectric actuator achieved high displacement and response time when a sawtooth voltage signal with low frequency was applied. Considering the simulation results, the stiffness requirement, and the size of the piezoelectric stack, the parameters were determined ($r = 5$, $t = 1$, $r_h = 0.5$).

With the adjustment of the size of the long flexure hinge and the semi-circular flexure hinge, the resolution of the actuator can be changed to adapt to different applications. The lever amplification mechanism can increase the change rate of displacement under different voltage signals.

3. Experiments and Results

Figure 17 shows the established experiment system. Figure 18 shows the prototype of the stick-slip piezoelectric actuator.

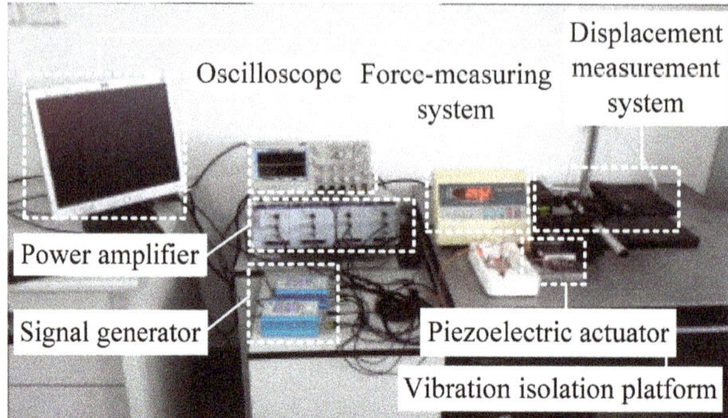

Figure 17. Experiment system.

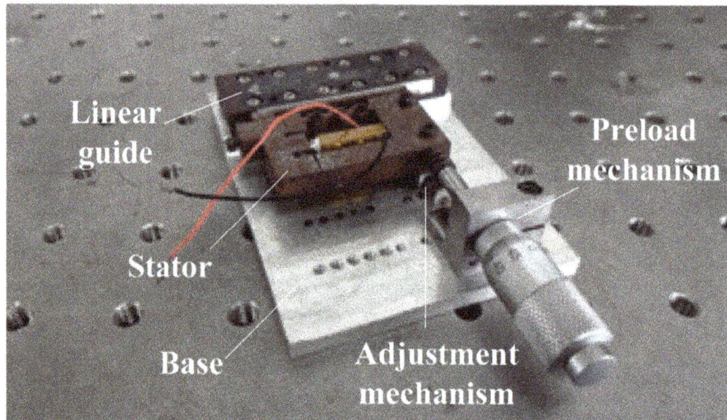

Figure 18. Prototype of the stick-slip piezoelectric actuator.

3.1. Vibration Test of the Stator

To evaluate the effect of amplification, the displacement of the stator driving foot was tested with the help of a laser displacement sensor (KEYENCE LK-HD500). In the test, a sine wave signal with a voltage of 100 V and a frequency of 50 Hz was applied as the driving signal. Figure 19 shows the results of the displacement response measurement of the stator. According to the results, it can be seen that the average amplitude of the driving foot along the x direction and y direction was 16.5 μm and 22.3 μm, respectively, when the elongation of the piezoelectric stack was 11.5 μm.

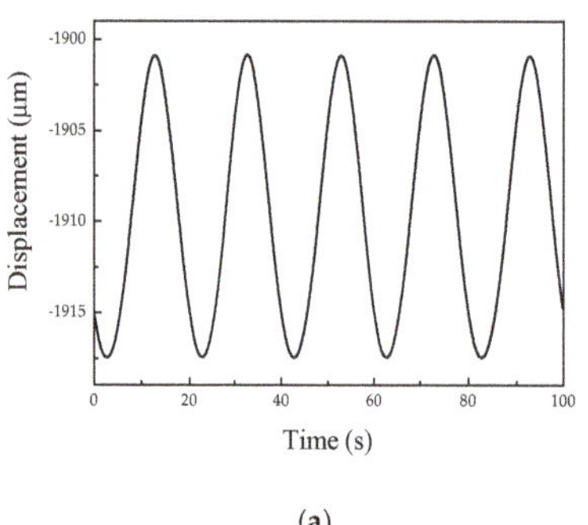

(a)

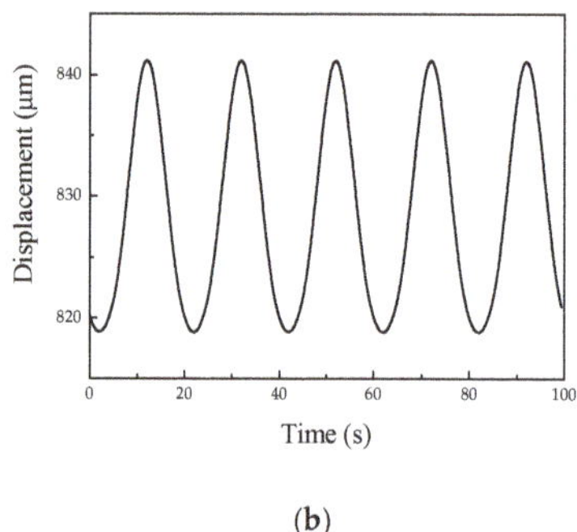

(b)

Figure 19. Vibration test of the driving foot. (**a**) Amplitude of the x direction. (**b**) Amplitude of the y direction.

The theoretical magnification, m_t, and the experimental magnification, m_e, of the x direction can be obtained from:

$$m_t = \frac{14.23}{10} = 1.423 \tag{21}$$

$$m_e = \frac{16.5}{11.5} = 1.434. \tag{22}$$

Obviously, the experimental and theoretical results agree with each other.

3.2. Performance Test

By utilizing a signal generator and power amplification to apply the sawtooth wave signal with different frequency and voltage on the piezoelectric stack, the relationships between the moving velocity and the driving frequency at different driving voltages were obtained and are shown in Figure 20. The results indicate that the velocity of the actuator has a linear relationship with the frequency of the driving signal under different voltages. The errors come from assembly errors and the uneven contact surface between the mover and the stator.

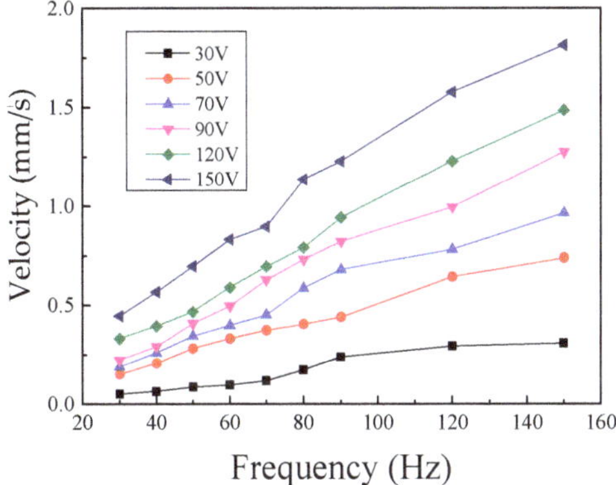

Figure 20. Velocity of the actuator versus the frequency by different voltage.

A driving signal with an input voltage of 100 V and frequency of 100 Hz was sent to the actuator, and the on-off characteristic curve of velocity and displacement was obtained, as shown in Figure 21. It illustrates that the velocity of the actuator is 1.2 mm/s in the stable state, and the response time of startup and shutdown is tens of milliseconds.

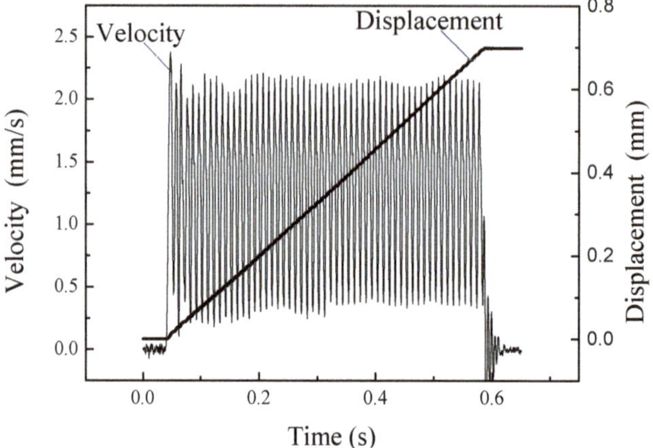

Figure 21. On-off operation characteristics of the actuator.

When the experiment results are compared to the theoretical and FEM results, the motion displacement is slightly smaller than that from the analog results, as shown in Figure 22. Due to manufacture and assembly errors, a difference would exist.

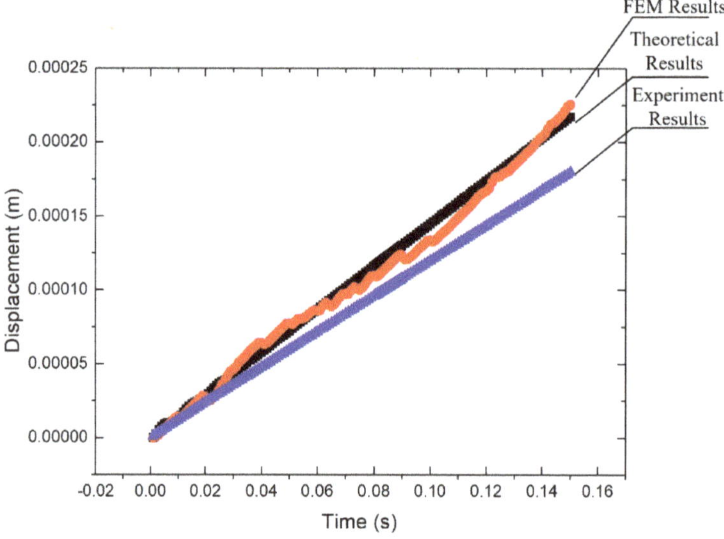

Figure 22. The comparison of theoretical, FEM, and the experiment results.

3.3. Resolution Test

To measure the step distance of an actuator with a low voltage signal, a periodic pulse sawtooth wave signal was used. The experiment results are shown in Figure 23 and indicate that when the voltage of the signal is 30 V, 20 V, and 10 V, the step distance of the actuator is about 3.33 μm, 1.75 μm, and 0.875 μm, respectively. When the voltage of the input signal continues to reduce, there is no clear step distance. This is because the smaller step distance is about the same order of magnitude as the background noise. On the other hand, the movement distance of the mover in the driving phase is not much larger than the possible distance of back off in the return phase.

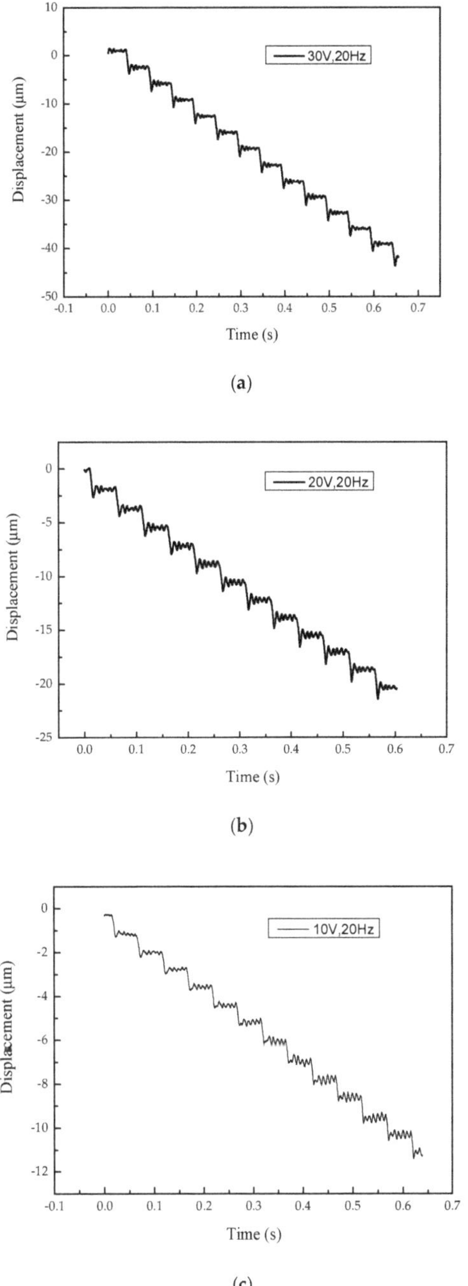

(a)

(b)

(c)

Figure 23. The displacement-time curve with a frequency of 20 Hz and a voltage of (**a**) 30 V, (**b**) 20 V, and (**c**) 10 V.

4. Conclusions

A novel stick-slip piezoelectric actuator with a lever mechanism was designed, fabricated, and tested in this research. A lever mechanism set in the stator was employed to increase the displacement of the actuator by amplifying the displacement of the driving foot. Subsequently, a long flexure hinge was used to pre-tighten the piezoelectric stack and eliminate the lateral offset error of the frame structure in the stator. Based on the modeling analysis and FEM analysis, the working principle was introduced, and the parameters of the stator were designed to meet the requirements. After the fabrication of a prototype, a vibration test of the stator and a performance test were conducted to validate the theoretical results. When a sawtooth wave with a voltage of 150 V and a frequency of 150 Hz was applied, the maximum velocity of the actuator was 1.8 mm/s. The actuator could obtained the minimum step distance of 0.875 μm from the resolution test. The results have confirmed that the design of the frame structure with a lever mechanism ensures that the actuator can undertake long-travel, fast response, and high precision linear motion.

Author Contributions: Conceptualization, Writing—review, Project administration and Funding acquisiton: W.H.; Methodology, Validation, Writing—original draft preparation, editing: M.S.

Funding: This research is funded by the Guangzhou Municipal University research projects (No. 1201610315), The Science and Technology novel Research Team Program in Higher Educational Universities of Guangdong Province (2017KCXTD025), the Research Foundation of the Nanjing Institute of Technology (YKJ201901), and the Produce-Learn-Research projects (No. CXY2013NH09).

Conflicts of Interest: The authors declare no conflict of interest.

Nomenclature

U_0	Initial voltage
T	Cycle time
L	Length of the structure
W	Width of the structure
h	Distance from bottom of the structure to the location of piezoelectric stack
k_1	Stiffness of the long hinge
k_2	Stiffness of the flexure hinge
F_p	Output force of the piezoelectric stack
d_{33}	Piezoelectric coefficient of the piezoelectric stack
U_p	Input voltage
k_p	Stiffness of the piezoelectric stack
k_{11}	Equivalent tensile rigidity of the circle structure in the long hinge
k_{12}	Equivalent tensile rigidity of short beam in the long hinge
E	Young's modulus
x	Displacement of the stator
xg	Displacement of the guide
Cs	Damping coefficients of the stator structure
Cp	Damping coefficients of the piezoelectric stack
Fa	Internal force of the piezoelectric stack and the stator structure
f	Friction force
ms	Mass of the stator structure
mg	Mass of the guide
mp	Mass of the piezoelectric stack
R	Resistance of the driving circuit
C	Capacitance of the piezoelectric stack
k_{amp}	Amplification ratio of input voltage for the piezoelectric stack
k_y	Stiffness of the stator structure in the y direction
C_y	Damping coefficients of the stator structure in the y direction
F_N	Preload force from preload mechanism

N	Vertical force acting on the guide
v_r	Relative velocity
μ_k	Kinetic friction coefficient
δv	Small velocity bound
r	Inner diameter of the long flexure hinge
t	Width of the long flexure hinge
r_h	Diameter of the semi-circular flexure hinge

References

1. Zhang, J.; Huang, Y.; Zheng, Y. A feasibility study on timber damage detection using piezoceramic-transducer-enabled active sensing. *Sensors* **2018**, *18*, 1563. [CrossRef] [PubMed]
2. You, Y.; Liao, W.; Zhao, D.; Ye, H.; Zhang, Y.; Zhou, Q.; Niu, X.; Wang, J.; Li, P.; Fu, D.; et al. An organic-inorganic perovskite ferroelectric with large piezoelectric response. *Science* **2017**, *357*, 306–309. [CrossRef] [PubMed]
3. Gu, G.; Zhu, L.; Su, C.; Ding, H.; Fatikow, S. Modeling and control of piezo-actuated nanopositioning stages: A survey. *IEEE Trans. Autom. Sci. Eng.* **2016**, *13*, 313–332. [CrossRef]
4. Xu, K.; Li, J.; Lv, X.; Wu, J.; Zhang, X.; Xiao, D.; Zhu, J. superior piezoelectric properties in potassium-sodium niobate lead-free ceramics. *Adv. Mater.* **2016**, *28*, 8519–8523. [CrossRef]
5. Du, G.; Li, Z.; Song, G. A PVDF-based sensor for internal stress monitoring of a concrete-filled Steel tubular (CFST) column subject to impact loads. *Sensors* **2018**, *18*, 1682. [CrossRef]
6. Xu, J.; Wang, C.; Li, H.; Zhang, C.; Hao, J.; Fan, S. Health monitoring of bolted spherical joint connection based on active sensing technique using piezoceramic transducers. *Sensors* **2018**, *18*, 1727. [CrossRef]
7. Wang, Y.; Wang, L.; Cheng, T.; Song, Z.; Qin, F. Sealed piezoelectric energy harvester driven by hyperbaric air load. *Appl. Phys. Lett.* **2016**, *108*, 033902. [CrossRef]
8. Ansari, R.; Oskouie, M.F.; Gholami, R.; Sadeghi, F. Thermo-electro-mechanical vibration of postbuckled piezoelectric Timoshenko nanobeams based on the nonlocal elasticity theory. *Compos. Part B Eng.* **2016**, *89*, 316–327. [CrossRef]
9. Wang, Z.; Chen, D.; Zheng, L.; Huo, L.; Song, G. Influence of axial load on electromechanical impedance (EMI) of embedded piezoceramic transducers in steel fiber concrete. *Sensors* **2018**, *18*, 1782. [CrossRef]
10. Xu, K.; Deng, Q.; Cai, L.; Ho, S.; Song, G. Damage detection of a concrete column subject to blast loads using embedded piezoceramic transducers. *Sensors* **2018**, *18*, 1377. [CrossRef]
11. Lu, X.L.; Zhou, S.Q.; Zhao, C.S. Finite element method analyses of ambient temperature effects on characteristics of piezoelectric motors. *J. Intell. Mater. Syst. Struct.* **2014**, *25*, 364–377. [CrossRef]
12. Lu, X.L.; Hu, J.H.; Zhang, Q.; Yang, L.; Zhao, C.S.; Vasiljev, P. An ultrasonic driving principle using friction reduction. *Sens. Actuators A Phys.* **2013**, *199*, 187–193. [CrossRef]
13. Li, X.; Zhou, S. A novel piezoelectric actuator with a screw-coupled stator and rotor for driving an aperture. *Smart Mater. Struct.* **2016**, *25*, 035027. [CrossRef]
14. Liu, Y.; Wang, Y.; Liu, J.; Xu, D.; Li, K.; Shan, X.; Deng, J. A four-feet walking-type rotary piezoelectric actuator with minute step motion. *Sensors* **2018**, *18*, 1471. [CrossRef] [PubMed]
15. Cho, B. Performance evaluation of switching amplifier in micro-positioning systems with piezoelectric Actuator. *Trans. Korean Inst. Power Electron.* **2009**, *14*, 62–71.
16. Huang, H.; Zhao, H.W.; Fan, Z.Q.; Zhang, H.; Ma, Z.C.; Yang, Z.J. Analysis and experiments of a novel and compact 3-DOF precision positioning platform. *J. Mech. Sci. Technol.* **2013**, *27*, 3347–3356. [CrossRef]
17. Yung-Tien, L.; Higuchi, T. Precision positioning device utilizing impact force of combined piezo-pneumatic actuator. *IEEE/ASME Trans. Mechatron.* **2001**, *6*, 467–473. [CrossRef]
18. Presas, A.; Egusquiza, E.; Valero, C.; Valentin, D.; Seidel, U. Feasibility of using PZT actuators to study the dynamic behavior of a rotating disk due to rotor-stator interaction. *Sensors* **2014**, *14*, 11919–11942. [CrossRef]
19. Yu, H.; Quan, Q.; Tian, X.; Li, H. Optimization and analysis of a u-shaped linear piezoelectric ultrasonic motor using longitudinal transducers. *Sensors* **2018**, *18*, 809. [CrossRef]
20. Chen, X.F.; Huang, W.Q.; Luo, J.Q.; Xu, J.J. design of a kind of non-resonant linear piezoelectric motor. *Appl. Mech. Mater.* **2014**, *459*, 418–423. [CrossRef]
21. Li, H.L.; Yin, W.; Qing, H.W.; Liang, Y. A double-foot linear piezoelectric motor. *Chin. Mech. Eng.* **2014**, *25*, 2719–2723.

22. Li, J.; Huang, H.; Zhao, H. A piezoelectric-driven linear actuator by means of coupling motion. *IEEE Trans. Ind. Electron.* **2018**, *65*, 2458–2466. [CrossRef]

23. Wang, S.; Rong, W.; Wang, L.; Pei, Z.; Sun, L. A novel inchworm type piezoelectric rotary actuator with large output torque: Design, analysis and experimental performance. *Precis. Eng.* **2018**, *51*, 545–551. [CrossRef]

24. Tchakui, M.V.; Woafo, P. Dynamics of three unidirectionally coupled autonomous duffing oscillators and application to inchworm piezoelectric motors: effects of the coupling coefficient and delay. *Chaos Interdiscip. J. Nonlinear Sci.* **2016**, *26*, 103118. [CrossRef] [PubMed]

25. Ceponis, A.; Mazeika, D. An inertial piezoelectric plate type rotary motor. *Sensor Actuators A Phys.* **2017**, *263*, 131–139. [CrossRef]

26. Liu, L.; Ge, W.; Meng, W.; Hou, Y.; Zhang, J.; Lu, Q. New design for inertial piezoelectric motors. *Rev. Sci. Instrum.* **2018**, *89*, 033704. [CrossRef]

27. Ma, Y.; Shekhani, H.; Yan, X.; Choi, M.; Uchino, K. Resonant-type inertial impact motor with rectangular pulse drive. *Sensor Actuators A phys.* **2016**, *248*, 29–37. [CrossRef]

28. Chu, C.L.; Fan, S.H. A novel long-travel piezoelectric-driven linear nanopositioning stage. *Precis. Eng.* **2006**, *30*, 85–95. [CrossRef]

29. Nguyen, H.X.; Edeler, C.; Fatikow, S. Modeling of piezo-actuated stick-slip micro-drives: An overview. *Adv. Sci. Technol.* **2012**, *81*, 39–48. [CrossRef]

30. Zhang, Z.M.; An, Q.; Li, J.W.; Zhang, W.J. Piezoelectric friction-inertia actuator-a critical review and future perspective. *Int. J. Adv. Manuf. Technol.* **2012**, *62*, 669–685. [CrossRef]

31. Hunstig, M.; Hemsel, T.; Sextro, W. Stick–slip and slip–slip operation of piezoelectric inertia drives. Part I: Ideal excitation. *Sensors Actuators A Phys.* **2013**, *200*, 90–100. [CrossRef]

32. Meng, Q.; Li, Y.; Xu, J. A novel analytical model for flexure-based proportion compliant mechanisms. *Precis. Eng.* **2014**, *38*, 449–457. [CrossRef]

33. Tian, Y.; Shirinzadeh, B.; Zhang, D.; Liu, X.; Chetwynd, D. Design and forward kinematics of the compliant micro-manipulator with lever mechanisms. *Precis. Eng.* **2009**, *33*, 466–475. [CrossRef]

34. Simu, U.; Johansson, S. Analysis of quasi-static and dynamic motion mechanisms for piezoelectric miniature robots. *Sensor Actuators A Phys.* **2006**, *132*, 632–642. [CrossRef]

35. Chen, X.F.; Wang, Y.; Sun, M.X.; Hunag, W.Q. Linear piezoelectric stepping motor with broad operating frequency. *Trans. Nanjing Univ. Aeronaut. Astronaut.* **2015**, *32*, 137–142.

36. Jalili, H.; Salarieh, H.; Vossoughi, G. Study of a piezo-electric actuated vibratory micro-robot in stick-slip mode and investigating the design parameters. *Nonlinear Dynam.* **2017**, *89*, 1927–1948. [CrossRef]

37. Kim, S.H.; Kim, S. A precision linear actuator using piezoelectrically driven friction force. *Mechatronics* **2001**, *11*, 969–985. [CrossRef]

38. Sun, M.; Huang, W.; Wang, Y.; Lu, Q.; Su, Z. Research on a novel non-resonant piezoelectric linear motor with lever amplification mechanism. *Sensor Actuators A Phys.* **2017**, *261*, 302–310. [CrossRef]

39. Li, J.; Zhou, X.; Zhao, H.; Shao, M.; Li, N.; Zhang, S.; Du, Y. Development of a novel parasitic-type piezoelectric actuator. *IEEE/ASME Trans. Mechatron.* **2017**, *22*, 541–550. [CrossRef]

 *micromachines*

Article

Study on the Critical Wind Speed of a Resonant Cavity Piezoelectric Energy Harvester Driven by Driving Wind Pressure

Xia Li [1,*], Zhiyuan Li [1], Qiang Liu [1] and Xiaobiao Shan [2,*]

[1] School of Mechanical and Power Engineering, Zhengzhou University, Zhengzhou 450001, China; ggyuan11@163.com (Z.L.); 15093367407@163.com (Q.L.)
[2] State Key Laboratory of Robotics and System, Harbin Institute of Technology, Harbin 150001, China
* Correspondences: jennyhit@163.com (X.L.); shanxiaobiao@hit.edu.cn (X.S.)

Received: 30 October 2019; Accepted: 28 November 2019; Published: 1 December 2019

Abstract: In order to solve the problem of continuous and stable power supply for vehicle sensors, a resonant cavity piezoelectric energy harvester driven by driving wind pressure was designed. The harvester has an effective working range of wind speed. According to the energy conservation law, the cut-in (initial) wind speed of the harvester was solved. The pressure distribution law of the elastic beam in the flow field was studied by the Fluent software package, and the results were loaded into a finite element model with a method of partition loading. The relationship between the wind speed and the maximum principal stress of the piezoelectric cantilever beam was analyzed, and the critical stress method was used to study the cut-out wind speed of the energy harvester. The results show that the cut-in wind speed of the piezoelectric energy harvester is 5.29 m/s, and the cut-out wind speed is 24 m/s. Finally, an experiment on the power generation performance of the energy harvester was carried out. The experimental results show that the cut-in and cut-out wind speeds of the piezoelectric energy harvester are 5 m/s and 24 m/s, respectively, and the best matching load is 60 kΩ. The average output power, generated by the harvester when the driving wind speed is 22 m/s, is 0.145 mW, and the corresponding power density is 1.2 mW/cm^3.

Keywords: piezoelectric energy harvester; cut-in wind speed; cut-out wind speed; energy conservation method; critical stress method

1. Introduction

In order to solve the environmental problems caused by automobile exhaust, some automobile companies cooperate with the government to implement the new energy "shared car" services in urban areas, with a maximum speed of no more than 70 km/h. Shared cars have more microsensors than ordinary fuel vehicles. Currently, microsensors are mainly powered by chemical batteries [1,2]. Chemical batteries have the characteristics of a large volume, low energy density, and short service life, and they also need to be replaced regularly [3,4]. These features increase the maintenance cost of shared vehicles. If the renewable energy that exists in the ambition environment can be directly converted into electricity to power microsensors, this problem can be solved very well [5,6]. The process of driving a car produces a lot of energy that can be recycled, such as heat energy [7,8], vibration energy [9–11] and wind energy [12]. Wind-induced vibration energy, which occurs when the car is driving, can be directly converted into electrical energy [13,14]. In recent years, there has been much research on wind-induced vibration energy recovery. There are three main types of electromechanical energy conversions, namely, piezoelectric, electromagnetic, and electrostatic [15,16]. Among them, the piezoelectric energy collecting device has the advantages of having good system compatibility, a simple structure, and a high energy density [17]. This makes piezoelectric energy harvesting a promising approach to energy recovery.

Researchers have developed many effective miniature wind-capacitor energy harvesting devices using electromagnetic or piezoelectric principles. The wind energy harvesters are mainly classified into turbine harvesters [18,19], windmill harvesters [20], reed harvesters [21–27], and other new structures [28–33]. Florian Herrault et al. [18] designed a microturbine electromagnetic generator. The four-pole six-turn/pole NdFeB generator exhibits up to 6.6 mW of AC electrical power across a resistive load at a rotational speed of 392,000 r/min. This milliwatt-scale power generation indicates the feasibility of such ultra-small machines for low-power applications. Sardini [19] et al. designed a turbo-type microgenerator to supply energy to the sensor that measures air temperature and speed. When the wind speed is 9 m/s, the electromagnetic generator matches the resistance of 500 Ω, and electric power with an average power of 45 mW can be obtained. Priya [20] designed a windmill-type piezoelectric energy harvester with a maximum output power of 7.5 mW when the wind speed is 10 mph and the matching resistance is 6.7 kΩ. Tan Y. K. et al. [21] designed a reed-type wind energy recovery device that uses a PZT (lead zirconate titanate) piezoelectric material to power the wind speed sensor. The wind energy harvester can start with a cut-in wind speed of 2 m/s and a cut-off wind speed of 7.5 m/s. When the wind speed is 6.7 m/s and the load is 220 kΩ, the maximum output voltage is 8.8 V, and the maximum output power is 155 μW.

The reed-type wind energy piezoelectric energy harvester can be divided into three types according to the excitation mode, namely, the resonant cavity type [22–24], blunt body spoiler type [25,26], and direct excitation type [27]. D. St. Clair et al. [22] proposed a wind-pressure energy-capturing device based on self-excited airflow oscillation using a single reed and a cavity structure. When the wind speed is 7.5–12.5 m/s, the wind pressure-capture device can obtain an output power of 0.1 mW to 0.8 mW. They found that the resonant cavity-based piezoelectric energy harvester does not require any external vibration sources, thus eliminating the bandwidth problems associated with vibratory energy harvesters. Bibo Amin et al. [23] established an analytical electromechanical model to predict the response characteristics of a wind-pressure energy-capturing device based on self-excited airflow oscillation using a single reed and a cavity structure, and the effectiveness of the model was verified by experiments. In order to improve the vibration response of the wake excitation and increase the output power, L. A. Weinstein et al. [25] added a fin device at the end of the piezoelectric vibrator to make the natural frequency close to the vortex a shedding frequency. Song Rujun et al. [26] proposed a vortex-induced vibration-type piezoelectric energy-trapping device, composed of a piezoelectric cantilever beam and a terminal cylinder. They established its mathematical model and carried out experiments. The results show that this piezoelectric energy harvester can obtain the maximum energy output during vortex-induced resonance.

The wind speed of a piezoelectric energy harvester based on the turbulent vibration of the spoiler cylinder has a narrow adaptive range, and its natural frequency needs to be matched. However, the role of the spoiler cylinder is to increase the degree of chaos in the fluid flow field, and it can increase the pressure of the local fluid. It is not necessary to match the wind speed of the resonant cavity piezoelectric energy harvester. The amplitude response characteristics of the resonant cavity piezoelectric energy harvester are as follows: When the wind speed is low, the fluid kinetic energy cannot overcome the damping of the fluid flowing in the cavity, and the displacement response of the cantilever beam and voltage output is 0. When the wind speed exceeds a certain threshold, the cantilever beam produces self-sustaining and finite period oscillates, and as the wind speed increases, both the amplitude response and the voltage output increase.

In this paper, in order to achieve the purpose of continuous and stable power supply for the vehicle-mounted sensor on the shared car, the traditional resonant cavity piezoelectric energy harvester was improved: A Helmholtz resonant cavity was added to the end of the main cavity, and a spoiler cylinder was added at the entrance of the main cavity. Both the spoiler cylinder and the Helmholtz resonator increase the gas pressure in the main cavity to improve the energy capture efficiency of the harvester. The resonant cavity piezoelectric energy harvester has an effective wind speed range, in which the piezoelectric energy harvester can work effectively, otherwise the energy harvester will not work or even be destroyed. In order to discuss whether the harvester can be installed on a shared car, the effective wind speed range of this harvester has been studied. The energy conservation method was used to solve the cut-in wind speed (threshold wind speed required to start the harvester), and the critical stress method was used to solve the cut-out wind speed (maximum wind speed above which operating could result in harm to the structure) of the energy harvester. Herein, the experimental research that has been carried out is described in detail.

2. Structure and Working Principle

This piezoelectric energy harvester was designed as a composite reed resonator structure, consisting of a structural module and two piezoelectric power generation modules, as shown in Figure 1. The structural module is composed of a main cavity, a spoiler cylinder, a Helmholtz cavity short tube, and a Helmholtz resonator. The two piezoelectric power generation modules are symmetrically distributed on the left and right sides of the main cavity, where one end of which is fixed to the main cavity body, and the other end is free to form an elastic cantilever beam structure. The spoiler cylinder is fixed at the inlet of the main cavity, and the Helmholtz cavity is fixed at the tail of the main cavity. The piezoelectric power generation module is composed of a metal substrate and piezoceramics, and the piezoceramics are attached to the root of the fixed end of the substrate.

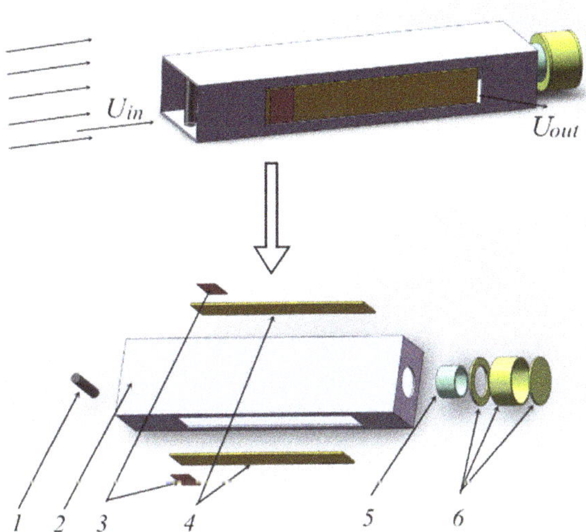

Figure 1. Schematic of the resonant cavity piezoelectric energy harvester: (**1**) Spoiler cylinder; (**2**) main cavity; (**3**) piezoceramics; (**4**) substrate; (**5**) Helmholtz cavity short tube; (**6**) Helmholtz resonator cavity.

When the car is running, the airflow enters the main cavity of the harvester from the air inlet of the main cavity (U_{in}). The air in the main cavity is compressed, then the pressure is increased, the cantilever beam is bent and deformed, and the air outlet of the main cavity (U_{out}) is opened. As a result, the air pressure in the main cavity is reduced, the elastic beam is reset, and the air outlet is closed. Thus, a repeating process is formed. The piezoelectric cantilever beam is periodically and repeatedly deformed to generate charges of opposite polarities in cycles on the upper and lower surfaces of the piezoelectric ceramic sheet. The piezoelectric ceramic sheets and loads form a closed energy output circuit that outputs electrical energy, thereby converting wind pressure energy into electrical energy.

3. Research on Cut-in Wind Speed

The piezoelectric energy harvester has an effective wind speed range, and the minimum wind speed that enables the piezoelectric energy harvester to start capturing energy is called the cut-in wind speed, which is also called the lower critical wind speed. In this paper, the energy conservation method is used to solve the cut-in wind speed.

3.1. Energy Conservation Analysis of the Harvester in one Cycle

The airflow enters the piezoelectric energy harvester from the inlet of the main cavity and flows out from the outlet. In this process, the reduced wind pressure energy is partially consumed by the piezoelectric damper's own damping force, and the other part acts on the piezoelectric harvester converted into electrical energy. Suppose there is a critical wind speed, at which the kinetic energy of the airflow is just consumed by the damping of the piezoelectric trap itself, so that the piezoelectric trap cannot capture energy. This critical velocity of the airflow entering the cavity is the cut-in wind speed.

The damping of the piezoelectric energy harvester consists of three parts, namely, the damping of the material used in the piezoelectric energy harvester (the phenomenon of mechanical energy loss caused by the friction among the internal grains of the material caused by the action of the trap in the wind field), the fluid damping generated by the various structures of the piezoelectric energy harvester after the fluid enters it, and piezoelectric damping produced by piezoelectric ceramic when piezoelectric cantilever beam vibrates. The material of piezoelectric energy harvester produces little damping and can be neglected during the research process. When the wind speed acting on the piezoelectric cantilever beam is equal to the cut-in wind speed, the piezoelectric cantilever beam is not deformed, and the piezoelectric ceramic piece does not generate electric energy. As such, there is no piezoelectric damping.

The cut-in wind speed can be obtained by the energy conservation method. When the energy conservation method is used to calculate the starting wind speed, the work done by the fluid damping force generated by each part of the structure is equal to the kinetic energy of the fluid entering the cavity. The damping of the piezoelectric energy harvester mainly includes the fluid damping force generated by the turbulent cylinder and the fluid damping force generated by the inner surface of the cavity.

The fluid kinetic energy in one cycle is calculated by Equation (1):

$$E_p = \frac{1}{2}mv^2 = \frac{1}{2}\rho(A_0 \cdot vT)v^2 = \frac{1}{2}\rho A_0 v^3 T \tag{1}$$

where E_p is defined as the kinetic energy of the fluid in a cycle, ρ is defined as the density of the air, A_0 is defined as the inlet area of the main cavity, v is defined as the velocity of the fluid at the entrance of the harvester, and T is defined as the vibration period of the piezoelectric cantilever.

The work done by the piezoelectric energy harvester structure damping in one cycle is calculated by Equation (2):

$$W = W_d + W_q \tag{2}$$

where W is defined as the work done by the structural damping force in a cycle, W_d is defined as the work done by the fluid damping force generated by the spoiler cylinder in a cycle, W_q is defined as the work done by the fluid damping force generated on the inner surface of the main cavity and the Helmholtz resonator.

3.2. Work Done by the Spoiler Damping Force

3.2.1. Theoretical Analysis

The average resistance coefficient of the cylinder unit length is [34]:

$$C_d = \frac{F_d}{\frac{1}{2}\rho D v^2} \tag{3}$$

The fluid damping force generated by the spoiler cylinder of length L is:

$$F_d = \frac{1}{2}\rho C_d D L v^2 \tag{4}$$

where C_d is defined as the resistance coefficient of the spoiler cylinder (C_d is 1.11, according to the empirical value [34]), and D and L are respectively defined as the diameter and length of the spoiler cylinder.

The work of the damping force generated by the spoiler cylinder in a cycle is given by Equation (5):

$$W_d = F_d \cdot x = F_d \cdot vT = \frac{1}{2}\rho C_d D L v^3 T \tag{5}$$

where x represents the displacement of the spoiler cylinder relative to the fluid for a cycle.

3.2.2. Simulation Verification of the Value of Resistance Coefficient

The resistance coefficient is a function of the Reynolds number [34]. The Reynolds number is related to the velocity, so the resistance coefficient is related to the speed.

In order to verify the accuracy of the resistance coefficient selected according to the empirical value when solving the starting wind speed, the resistance coefficient of the spoiler cylinder is calculated by the Fluent software package, then compared with the empirical value resistance coefficient selected. If the error is within ±5%, it shows that the resistance coefficient obtained by the empirical value is reasonable when the energy conservation method is used to calculate the wind speed of the resonant cavity piezoelectric harvester.

In order to facilitate the study, this paper neglects the influence of the length of the cylinder. By simplifying the three-dimensional model of the main cavity into two-dimensional model, the influence of the disturbing cylinder on the model is studied. The simplified model is shown in Figure 2.

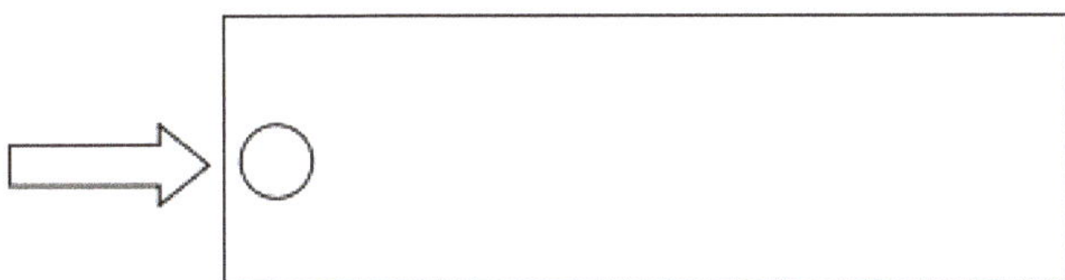

Figure 2. Simplified flow field model of the main cavity.

The Kármán vortex street effect occurs when the fluid passes through a cylinder. At this time, the shedding frequency of the fluid vortex in the flow channel [35] is:

$$f = S_t \frac{v}{\varepsilon D} \tag{6}$$

where

$$\varepsilon = 1 - \frac{2}{\pi}\left[\frac{D}{\Phi}\sqrt{1 - \left(\frac{D}{\Phi}\right)^2} + \arcsin\frac{D}{\Phi}\right] \tag{7}$$

where S_t is the Strouhal number, which indicates the relative relationship between the fluid kinetic energy and the inherent energy of the system. When the fluid flows over the spoiler cylinder, according to the calculation formula of Reynolds number, $Re = \frac{\rho v D}{\mu}$ (v takes the values of 1–30 m/s; ρ is air density; μ is hydrodynamic viscosity; D is the diameter of the spoiler cylinder), where the range of Re is from 336.7 to 10,101.2. It is common practice to assume $S_t = 0.21$ in the sub-critical range, $300 < Re < 1.5 \times 10^5$ [36]; ε represents the ratio of the area of the arcuate faces on both sides of the cylinder to the cross-sectional area of the intake passage; Φ represents the hydraulic diameter of the inlet.

When the flow field dynamic characteristics of the spoiled cylinder are analyzed, the boundary conditions are as follows.

1. The flow field calculation domain is based on Navier–Stokes equations (N-S equation) and is simulated using the realizable k-ε turbulence model. The flowing medium is ideally air at 20 °C, with a density of 1.205 kg/m^3, and a dynamic viscosity coefficient of 1.7894×10^{-5} Pa·s.

2. The inlet and outlet of the model are respectively set as a velocity inlet and a pressure outlet, and the absolute pressure at the outlet is 1.01×10^5 Pa. The wall is set to a fixed wall with no slip and an ambient temperature of 20 °C.

3. In order to accurately obtain the dynamic characteristics of the fluid in the flow field, the basic unit size of the flow field is set to 2 mm, and the meshes of the spoiler cylinder and the two side walls are refined. The number of nodes and grids are 38,383 and 73,990 respectively, and the average mesh quality is 0.97.

4. In order to refine the transient dynamics analysis results of the Kármán vortex generated by the fluid passing through the spoiler cylinder, the time step in the flow field is set to 1/10 of the detuning period of the spoiler cylinder.

The inlet wind speed is set from 2 m/s to 10 m/s, and the step length is 1 m/s. The remaining parameters are not changed. Finally, the resistance coefficient of the spoiler can be calculated. The results from our analysis are shown in Figure 3.

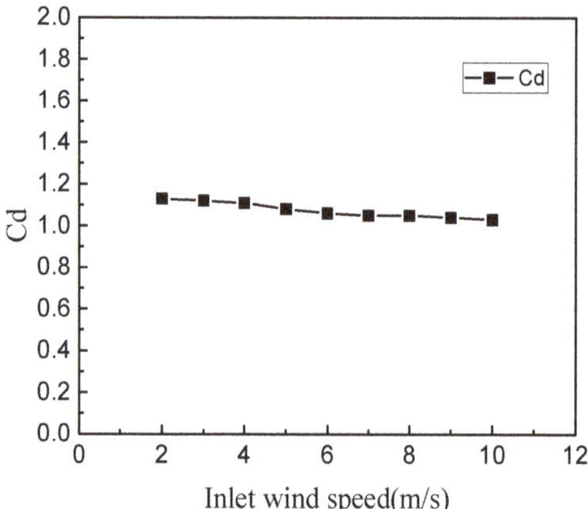

Figure 3. Relationship between inlet wind speed and resistance coefficient.

It can be seen from Figure 3 that when the inlet wind speed is different, the resistance coefficient also changes, but the change is not obvious. Therefore, the influence of wind speed on the resistance coefficient of the turbulent cylinder can be ignored. The average value of the coefficient of resistance of the turbulent cylinder is 1.07. The error is 3.6% when compared with the empirical value of 1.11, and is less than 5%, which is within the allowable range. Therefore, it can be verified that the empirical value of the resistance coefficient selected by the empirical value is reasonable when the energy method is used to solve the cut-in wind speed.

3.3. Work Done by the Damping Force Generated by the Main Cavity and the Inner Surface of the Helmholtz Resonator

The friction of the flat wall surface per unit area [34] is calculated by Equation (8):

$$\tau_f = 0.3232 \sqrt{\frac{\mu \rho v^3}{x}} \tag{8}$$

where μ represents the air viscosity coefficient.

The fluid damping force produced on the upper and lower surfaces of the inner wall of the main cavity is calculated by Equation (9) [34]:

$$F_f = 2 \left(\int_0^{L_B} \tau_f \cdot B \mathrm{d}x - 2 \int_0^{2R} \tau_f \cdot \sqrt{R^2 - (R - x)^2} \mathrm{d}x \right) \tag{9}$$

Equation (8) can be substituted for Equation (9) and then simplified:

$$F_f = 1.2928 \sqrt{\mu \rho v^3} \left(\sqrt{L_B} \cdot B - \frac{2}{3} (2R)^{\frac{3}{2}} \right) \tag{10}$$

where L_B and B denote the length and width of the main cavity, respectively, and R denotes the radius of the spoiler cylinder ($2R = D$).

Similarly, the fluid damping force generated on the left and right surfaces of the inner wall of the main cavity is given by Equation (11) [34]:

$$F_h = 2 \left(\int_0^{L_B} \tau_f \cdot h \mathrm{d}x - 2 \int_{L_2}^{L_3} \tau_f \cdot \Delta h \mathrm{d}x - \int_{L_3}^{L_3 + \Delta b} \tau_f \cdot b \mathrm{d}x \right) \tag{11}$$

where h denotes the height of the main cavity; b denotes the width of the substrate; Δh denotes the gap between the main cavity and the substrate in the width direction; Δb denotes the gap between the main cavity and the substrate in the length direction; L_2 and L_3 denote the distance from the entrance of the main cavity to the front end and the end of the substrate, respectively.

Then, the fluid damping generated on the inner wall surface of the main cavity can be expressed by Equation (12):

$$F_h = 1.2928 \sqrt{\mu \rho v^3} \left(\sqrt{L_B} \cdot h - 2 \left(\sqrt{L_2} - \sqrt{L_3} \right) \cdot \Delta h - \left(\sqrt{L_3 + \Delta b} - \sqrt{L_3} \right) \cdot b \right) \tag{12}$$

Similarly, the fluid damping generated on the inner surface of the Helmholtz resonator at the end of piezoelectric energy harvester can be calculated by Equation (13):

$$F_m = 1.2928\pi \sqrt{\mu \rho v^3} \left(\left(\sqrt{L_4} - \sqrt{L_B} \right) \cdot R_1 + \left(\sqrt{L_5} - \sqrt{L_4} \right) \cdot R_2 \right) \tag{13}$$

where L_4 is defined as the distance from the entrance of the main cavity to the end of the short tube, L_5 is defined as the distance from the entrance of the main cavity to the end of the Helmholtz resonator cavity, R_1 is defined as the radius of the short tube, and R_2 is defined as the radius of the Helmholtz cavity.

3.4. Numerical Calculation of Cut-in Wind Speed

The work done by the fluid damping force of the main cavity in one cycle can be calculated by Equation (14):

$$W_q = (F_f + F_h + F_m) \cdot x = (F_f + F_h + F_m) \cdot vT \tag{14}$$

According to the law of conservation of energy:

$$E_p = W = W_d + W_q \tag{15}$$

Substituting Equations (1), (5), and (14) into Equation (15), Equation (15) can be converted into Equation (16):

$$\frac{1}{2}\rho A_0 v^3 T = F_d \cdot vT + (F_f + F_h + F_m) \cdot vT \tag{16}$$

Substituting Equations (4), (10), (12), and (13) into Equation (16), the cut-in wind speed can be obtained by solving Equation (17).

$$\rho C_d DL + 2.5856 \sqrt{\mu \rho v} \left[\begin{array}{c} (\sqrt{L_B} \cdot B - \frac{2}{3}(2R)^{\frac{3}{2}}) + (\sqrt{L_B} \cdot h - 2(\sqrt{L_3 - L_2}) \cdot \Delta h - (\sqrt{L_3 + \Delta b} \\ - \sqrt{L_3}) \cdot b) + \pi(\sqrt{L_4} - \sqrt{L_B})R_1 + \pi(\sqrt{L_5} - \sqrt{L_4})R_2 \end{array} \right] - \rho A_0 = 0 \tag{17}$$

The values of the parameters in Equation (17) are shown in Table 1.

Table 1. Initial values of each parameter.

Parameters	Values	Parameters	Values
μ/Pa·s	1.78×10^{-5}	Δh/mm	0.5
ρ/kg/m^3	1.225	Δb/mm	6
D/mm	5	L_B/mm	150
B/mm	38	L_2/mm	30
R_1/mm	6	L_3/mm	120
R_2/mm	15	L_4/mm	160
h/mm	23	L_5/mm	175
b/mm	15	-	-

Equation (17) can be solved by mathematics software, and the cut-in wind speed of the piezoelectric energy harvester obtained here was 5.29 m/s.

4. Solution of Wind Cut-off Velocity by Stress Critical Value Method

When the airflow enters the cavity of the piezoelectric harvester, the piezoelectric cantilever beam arranged on both sides of the cavity will bend and deform. When the piezoelectric cantilever beam can work normally and is not damaged, the maximum airflow speed that it can withstand is called the cut-out wind speed, also called the upper critical wind speed. If the inlet wind speed is higher than this critical wind speed, the cantilever beam or the piezoelectric ceramic piece may break and cause a failure. On the contrary, if the airflow velocity is lower than the cut-off wind speed, the piezoelectric cantilever beam will not be destroyed. Therefore, the cut-out wind speed can be obtained by solving the wind speed at the point of critical stress of the piezoelectric beam.

The piezoelectric module is composed of a substrate and a piezoelectric ceramic piece, both of which must be able to work normally within the allowable range of allowable stress. The substrate material used was beryllium bronze, whose strength is 1035 MPa [37]. The piezoelectric ceramic sheet was PZT-5H, where the tensile strength under the d_{31} working mode is 75.845 MPa [38]. Taking a safety factor of 1.3 [39], the allowable stresses for the substrate and the piezoelectric ceramic piece are 796 MPa and 58.345 MPa, respectively. Therefore, it is necessary to ensure that the maximum stress of the substrate and the piezoelectric piece when the piezoelectric cantilever is in operation does not exceed their respective allowable stresses.

The Fluent software package was used to analyze the pressure distribution on the elastic beam of the piezoelectric trap when the airflow acted on the elastic beam in the flow field.

The specific process of fluid simulation was as follows:

1. Model import and establishment of the flow field: The resonant cavity piezoelectric energy harvester structure model was modeled by SolidWorks and saved in the IGS format and then imported into Fluent. The flow field was established according to the size parameters of the energy harvester. Generally, the flow field size was about 5 to 10 times that of the research object.

The length, width, and height of the flow field established in this paper was 1000 mm, 400 mm, and 200 mm, respectively. Finally, the fluid inlet and the fluid outlet were set.

2. Meshing: In this paper, an unstructured tetrahedral mesh was used to divide the flow field, and the piezoelectric trap was partially refined by a local mesh. The number of nodes was 187,126, the number of elements was 1,049,576, and the average mesh quality was 0.904.

3. Physical model selection and boundary condition setting: According to the calculation formula of Re, the Re values at different flow rates were respectively calculated. When Re < 2300, the physical model was set to laminar flow, and when Re > 2300, it was set to turbulent flow. The most commonly used turbulence model is the standard k-ε model. The standard k-ε model introduces the fluid kinetic energy equation (k) and the loss rate equation (ε). Finally, the solutions of k and ε were obtained, and the values of k and ε were used. The fluid viscosity was calculated and the solution of Reynolds stress was obtained by the Boussinesq hypothesis. However, the standard k-ε model has a large error in solving the swirl problem. The RNG k-ε model is an improved model of the Realizable k-ε model. In this paper, the improved RNG k-ε model was selected for numerical simulation.

Boundary condition setting: The inlet wind speed was 15 m/s, outlet pressure was the standard atmospheric pressure, and wall surface was set to no slip. The monitoring was set to monitor the pressure value of the cantilever beam on both sides of the piezoelectric energy harvester.

4. Initialization calculation and postprocessing: Global initialization was required before starting the calculation. The initialization method for this article was standard initialization. The initial pressure was set to the standard atmospheric pressure. Here, the postprocessing mainly extracts and collates the pressure data of the cantilever beams on both sides.

The pressure distribution of each node on the inner surface of the substrate in the wind field was extracted from the analysis results. The pressure distribution law is shown in Figure 4, where the *x*-axis is the length of the substrate, with a range of 30–120 mm, the *y*-axis is the width of the substrate, with a range from −7.5 mm to 7.5 mm, and the *z*-axis is the pressure value, with a range from 0 Pa to 150 Pa.

As can be seen from Figure 4, the pressure in the x-direction was increased first and then decreased, but the change was small. When the y-coordinate range was from −7.5 mm to −2.5 mm, the pressure increased linearly with the increase of the y-coordinate value. When the y-coordinate range was from −2.5 mm to 2.5 mm, the pressure did not obviously change with the increase of the y-coordinate value. When the y-coordinate range was from 2.5 mm to 7.5 mm, the pressure decreased linearly with the increase of the y-coordinate value. According to the distribution of wind pressure on the substrate in the wind field, the substrate can be divided into 12 regions, as shown in Figure 5.

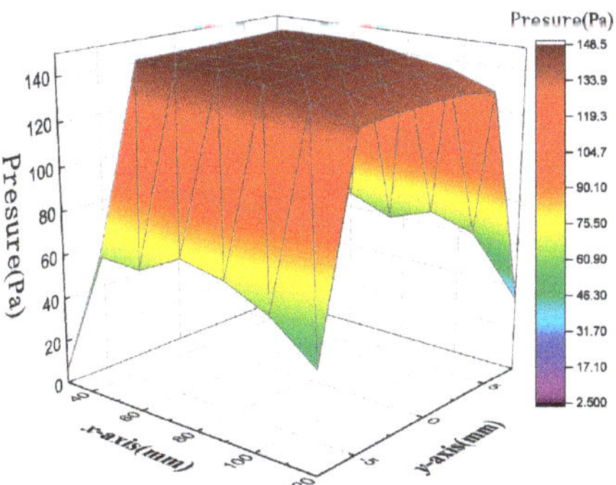

Figure 4. Distribution of the pressure on the substrate in the wind field.

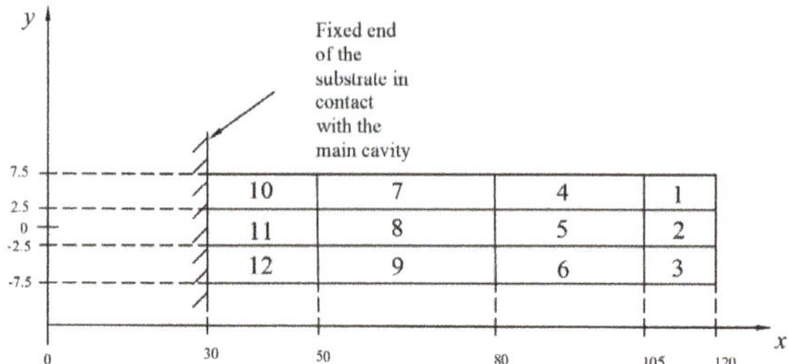

Figure 5. Division of substrates.

As shown in Figure 5, the pressure values of regions 4, 5, 6, 7, 8, and 9 are not much different, and approximate a rectangular area of equal value. The four regions of regions 1, 3, 10, and 12 are approximately triangular. Regions 2 and 11 are approximately rectangular. For the approximate isosurface area, a plurality of values may be averaged, and the triangular region can take the corresponding pressure value at the median line as the average value of the region, and the rectangular region takes the pressure value at the intermediate symmetry axis as the average value of the region.

The pressure of each region at different wind speeds is obtained by changing the inlet wind speed. The static analysis of the piezoelectric cantilever beam was performed using the Ansys software package according to the above loading method. The specific simulation details are described below:

1. Define element type and material parameters: In this paper, the piezoelectric coupling analysis of piezoelectric cantilever beam in piezoelectric energy harvester was carried out by ANSYS 14.5. The piezoelectric ceramic sheet element type was the coupling element Solid226, and the cantilever beam element was the structural element Solid45.

2. Meshing: In this paper, the cantilever beam was meshed by the means of mapping grid method. We used the glue command to combine the nodes on the contact surface of the piezoelectric piece with the substrate. The grid element size of the piezoelectric ceramic sheet, the bonded portion of the substrate, and the piezoelectric ceramic sheet were all set to 0.2 mm. The grid element size of the substrate root was 0.3 mm, and the grid element size of the other portion of the substrate was 0.6 mm.

3. Boundary conditions: A fixed end constraint was added to the left end of the piezoelectric cantilever beam. The upper surface node of the piezoelectric ceramic sheet was coupled to the upper electrode. The lower surface node of the piezoelectric ceramic sheet was coupled to the lower electrode, and the coupling voltage was set to 0 V.

4. Apply load and solve: The pressure was applied to different regions of the piezoelectric sheet by the method of partition loading, described above, and then the solution was calculated. After the calculation was completed, the Mises stress results on the piezoelectric beam were extracted.

After the simulation calculation, since the maximum Mises stress generated on the substrate did not exceed 200 MPa, which is much smaller than its allowable stress, the relationship between the maximum Mises stress on the substrate and the wind speed is not listed here. Figure 6 is a graph showing the relationship between the maximum Mises stress of the piezoelectric piece and the wind speed.

It can be seen from Figure 6 that the maximum Mises stress value of the piezoelectric piece increases with the increase of the wind speed. When the wind speed is lower than 5 m/s, the maximum Mises stress value of the piezoelectric piece approaches 0. When the wind speed is greater than 25 m/s, the maximum Mises stress of the piezoelectric piece exceeds 61.7 MPa, which is greater than the allowable stress of the piezoelectric piece. When the wind speed is between 24 m/s and 25 m/s, there is a wind speed value such that the maximum Mises stress of the piezoelectric piece is exactly equal to the allowable stress of the piezoelectric piece. In order to ensure the safety of the piezoelectric cantilever beam structure, the cut-out wind speed of the piezoelectric energy harvester was determined to be 24 m/s. This speed is greater than the maximum speed that a shared car travels in the city, so the piezoelectric energy harvester installed on the shared car can work properly.

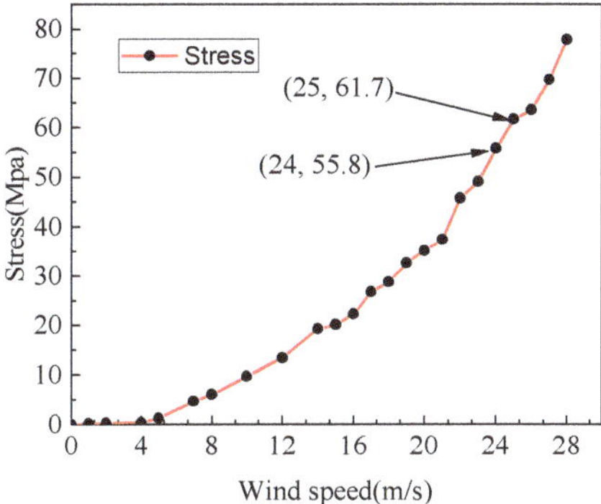

Figure 6. Relationship between wind speed and Mises stress of piezoelectric piece.

5. Experimental Study

In order to determine the output characteristics of the resonant cavity piezoelectric energy harvester and verify the correctness of the critical wind speed calculation results, a prototype of the resonant cavity piezoelectric energy harvester was fabricated, as shown in Figure 7. The thickness of the piezoelectric piece and substrate were both 0.5 mm, and the remaining dimensions were the same as in Table 1.

The resonant cavity piezoelectric energy harvester model is mainly composed of wires, a main cavity, a spoiler cylinder, a bolt, two piezoelectric ceramic pieces, two substrates, and a Helmholtz resonant cavity. The main cavity, the spoiler cylinder, and the Helmholtz resonator were modeled in SolidWorks and then processed separately using 3D printing technology, where the material used was PLA. The left and right sides of the main cavity both have a rectangular groove, with a length of 96 mm and a width of 16 mm, for fixing the piezoelectric cantilever beams. The material of the substrates used was beryllium bronze, which was cut by a cutter. The piezoelectric ceramic pieces used the PZT-5H piezoelectric ceramic pieces produced by the Hebei Hongsheng Acoustic Electronic Equipment Co., Ltd. The size of the piezoelectric piece was 10 mm × 15 mm × 0.4 mm. A through-hole was formed at the entrance of the main cavity and was assembled with the spoiler cylinder. The main cavity has a round hole at the end was assembled with the Helmholtz cavity. The piezoelectric cantilever beam was fixed on both sides of the main cavity by M3 bolts. The piezoelectric cantilever beam was made by bonding the PZT-5H to the substrate by epoxy resin containing copper powder. Two wires were spot-welded on the positive surface of PZT-5H and the substrate.

The experiment platform, as shown in Figure 8, was constructed. The experimental study on the output characteristics of the resonant cavity piezoelectric energy harvester was carried out.

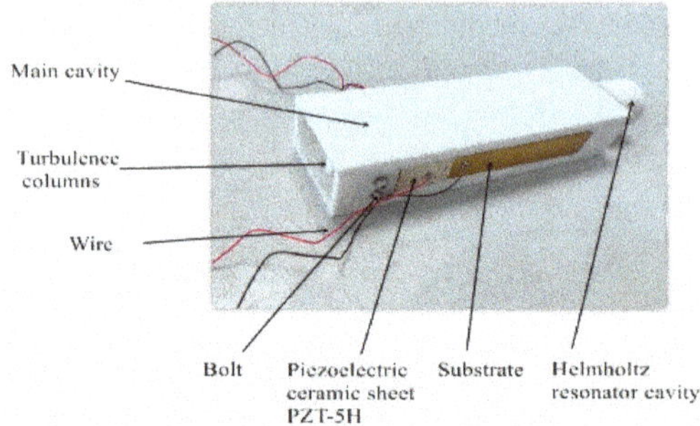

Figure 7. Resonant cavity piezoelectric energy harvester prototype.

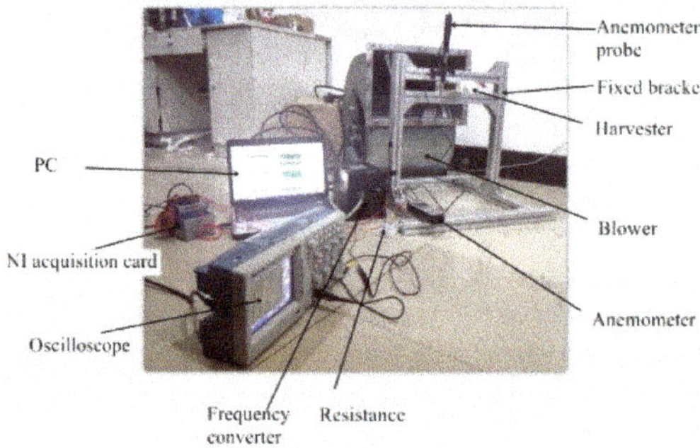

Figure 8. Resonant cavity piezoelectric energy harvester experimental platform.

The equipment used in the experiment were as follows: A blower (LKW315L-4, Shandong Qineng Ventilator Co., Ltd., Zibo, China), with the maximum adjustment frequency of 50 Hz and a maximum speed of 1350 r/min, a NI acquisition card (NI-cDAQ-9174, National Instruments, Austin, TX, USA), an oscilloscope (ADS 1102CAL, Guorui Antai Technology Co., Ltd., Nanjing, China), a frequency converter (VFD 110B 43A Shanghai Xinyu Automation Equipment Co., Ltd., Shanghai, China), an anemometer (CTV100, Beijing Oubai Keyi Instrument Co., Ltd. Beijing, China), with a wind speed measurement range of 0–30 m/s and a sensitivity of 0.1 m/s, and a PC.

Since two piezoelectric cantilever beams were symmetrically mounted on both sides of the piezoelectric energy harvester, only the vibration of one side piezoelectric cantilever beam was studied. Figure 9 shows the average open circuit voltage of the piezoelectric energy harvester at different wind speeds.

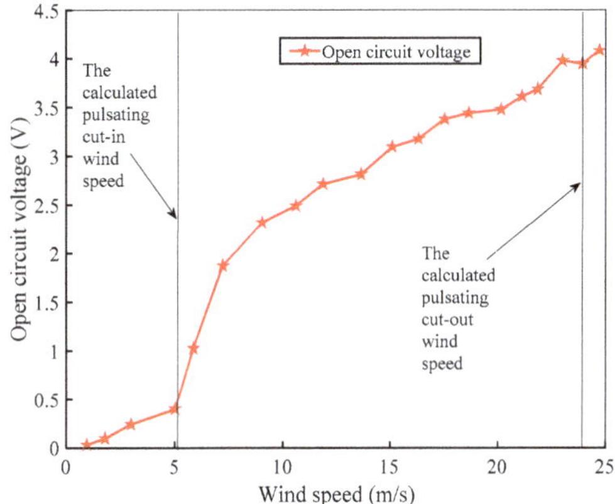

Figure 9. Curve of open circuit voltage with wind speed.

As can be seen from Figure 9, the open circuit voltage of the piezoelectric harvester increases nonlinearly with the increase of the wind speed. When the wind speed increases from 0 m/s to 5 m/s, the open circuit voltage increases slowly. When the wind speed is greater than 5 m/s, the open circuit voltage increases rapidly. In the actual experiment, the external small vibration can also cause the vibration of the energy absorbing device. Considering the influence of the actual factors, the cut-in wind speed of the piezoelectric energy harvester is 5 m/s. The blower can provide a maximum wind speed of 25 m/s, and the trap can still operate at this wind speed. However, in the experiment, it was found that after the harvester worked at a large wind speed for a long time, the piezoelectric cantilever beam would undergo a slight bending deformation. Therefore, the cut-out wind speed of the piezoelectric energy harvester should be 24 m/s. The effective wind speed range of the harvester is 1.58 times that of the micropiezoelectric wind energy harvester with a resonant cavity (with an effective wind speed range of 4–16 m/s), which has a cavity size of 64 mm × 22 mm × 14 mm, with a piezoelectric beam length of 8 mm and a flexible beam length of 30 mm, as described by Du Zhigang [24].

The power generation performance of a piezoelectric energy harvester needs to be measured by a load resistor [40,41]. Figure 10 is a plot of output power as a function of load resistance at the wind speed of 10 m/s.

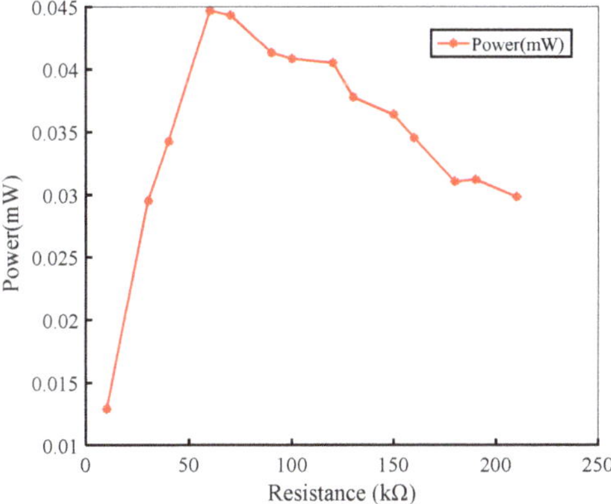

Figure 10. Relationship between output power and load resistance at the wind speed of 10 m/s.

It can be seen from Figure 10 that when the wind speed is 10 m/s, the output power firstly increases and then decreases with the increase of the load resistance. When the load resistance is 60 kΩ, the output power of one side piezoelectric cantilever beam reaches a maximum of 45 μW. In this experiment, the optimal resistance of the piezoelectric energy harvester was 60 kΩ.

Figure 11 shows the relationship of the output voltage with time at a wind speed of 22 m/s when the resistance was 60 kΩ.

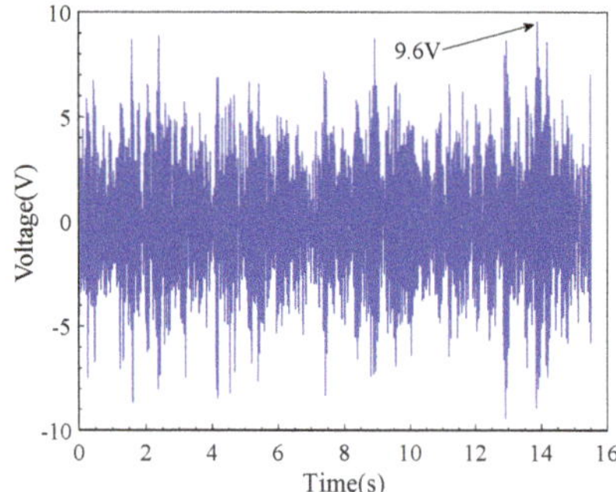

Figure 11. Variation of output voltage at a wind speed of 22 m/s.

It can be seen from Figure 11 that the peak output voltage of the single-sided piezoelectric cantilever beam of the piezoelectric energy harvester can reach 9.6 V when the wind speed is 22 m/s, and the peak output power obtainable at this time is 1.536 mW. The average voltage of one side of the piezoelectric cantilever beam was 2.0877 V, and the corresponding total power and power density here were 0.145 mW and 1.2 mW/cm^3, respectively.

6. Conclusions

In this paper, a wind-pressure resonant cavity piezoelectric energy harvester was designed, and the working wind speed range was studied. Experimental analysis of the harvester was carried out and the following conclusions were obtained:

1. By calculating the work done by the damping generated by the various parts of the harvester in the flow field, which was set to be equal to the initial kinetic energy of the fluid, the pulsating wind speed was obtained. The resistance coefficient of the spoiler cylinder was simulated by the Fluent software package. The results show that the inlet wind speed has little effect on the resistance coefficient. The average simulation result was compared with the empirical value, and the error was within the allowable range, verifying the correctness of the resistance coefficient selected by the empirical value. The resulting cut-in wind speed of the harvester was 5.29 m/s.
2. By solving the wind speed when the piezoelectric modules are not damaged, the cut-out wind speed of the harvester was obtained. The pressure distribution of the piezoelectric cantilever beam in the flow field was obtained by Fluent software simulation. The finite element model of the piezoelectric cantilever beam was loaded by the partition loading pressure method. The relationship between the wind speed and the maximum principal stress of the piezoelectric cantilever beam was studied. The cut-out wind speed of the harvester was 24 m/s.
3. Through experimental research, it was concluded that the cut-in wind speed of the energy harvester was 5 m/s and the cut-out wind speed was 24 m/s. The best external load resistance of the harvester was 60 kΩ. The average voltage of one side piezoelectric cantilever beam was 2.0877 V, and the corresponding total power and power density were 0.145 mW and 1.2m W/cm^3, respectively, at a wind speed of 22 m/s and an external resistance of 60 kΩ.

Author Contributions: Conceptualization, X.L.; Data curation, Z.L. and Q.L.; Formal analysis, Z.L.; Funding acquisition, X.S.; Investigation, Z.L. and Q.L.; Methodology, Z.L.; Project administration, X.L. and Q.L.; Resources, X.L. and X.S.; Software, Q.L.; Supervision, X.L. and X.S.; Writing—original draft, Z.L. and Q.L.; Writing—review and editing, X.L. and X.S.

Funding: This research was funded by National Natural Science Foundation of China, grant number 51107121.

Acknowledgments: The experimental equipment and experimental materials provided by the State Key Laboratory of Robotics and System are gratefully acknowledged.

Conflicts of Interest: The authors declare no conflict of interest.

References

1. Zou, H.X.; Zhao, L.C.; Gao, Q.H.; Lei, Z.; Liu, F.R.; Tan, T.; Wei, K.X.; Zhang, W.M. Mechanical modulations for enhancing energy harvesting: Principles, methods and applications. *Appl. Energy* **2019**, *255*, 1–18. [CrossRef]
2. Radousky, H.B.; Liang, H. Energy harvesting: An integrated view of materials, devices and applications. *Nanotechnology* **2012**, *23*, 502001-1–502001-35. [CrossRef] [PubMed]
3. Vullers, R.J.M.; Schaijk, R.V.; Visser, H.J. Energy harvesting for autonomous wireless sensor networks. *Solid-State Circuits Mag.* **2010**, *2*, 29–38. [CrossRef]
4. Zhou, S.X.; Cao, J.Y.; Inman, D.J.; Lin, J.; Liu, S.S.; Wang, Z.Z. Broadband tristable energy harvester: Modeling and experiment verification. *Appl. Energy* **2014**, *133*, 33–39. [CrossRef]
5. Liu, Y.X.; Yan, J.P.; Xu, D.M.; Chen, W.S.; Yang, X.H.; Tian, X.Q. An I-shape linear piezoelectric actuator using resonant type longitudinal vibration transducers. *Mechatronics* **2016**, *40*, 87–95. [CrossRef]
6. Shan, X.B.; Tian, H.G.; Chen, D.P.; Xie, T. A curved panel energy harvester for aeroelastic vibration. *Appl. Energy* **2019**, *249*, 58–66. [CrossRef]
7. Pancharoen, K.; Zhu, D.; Beeby, S.P. Temperature dependence of a magnetically levitated electromagnetic vibration energy harvester. *Sens. Actuator A-Phys.* **2017**, *256*, 1–11. [CrossRef]
8. Guo, L.; Lu, Q. Potentials of piezoelectric and thermoelectric technologies for harvesting energy from pavements. *Renew. Sustain. Energy Rev.* **2017**, *72*, 761–773. [CrossRef]
9. Abdelkareem, M.A.; Xu, L.; Ali, M.K.A.; Elagouz, A.; Mi, J.; Guo, S. Vibration energy harvesting in automotive suspension system: A detailed review. *Appl. Energy* **2018**, *229*, 672–699. [CrossRef]
10. De Araujo, M.V.V.; Vitoratti, N.R. Electromagnetic harvester for lateral vibration in rotating machines. *Mech. Syst. Sig. Process.* **2015**, *52*, 685–699. [CrossRef]
11. Zhang, Z.; Zhang, X.; Chen, W.; Rasim, Y.; Salman, W.; Pan, H. A high efficiency energy regenerative shock absorber using supercapacitors for renewable energy applications in range extended electric vehicle. *Appl. Energy* **2016**, *178*, 177–188. [CrossRef]
12. Abdelkefi, A. Aeroelastic energy harvesting: A review. *Int. J. Eng. Sci.* **2016**, *100*, 112–135. [CrossRef]
13. Han, N.; Zhao, D.; Schluter, J.U.; Goh, E.S.; Zhao, H.; Jin, X. Performance evaluation of 3D printed miniature electromagnetic energy harvesters driven by air flow. *Appl. Energy* **2016**, *178*, 672–680. [CrossRef]
14. Wang, P.; Pan, L.; Wang, J.; Xu, M.; Dai, G.; Zou, H. An ultra-low-friction triboelectric-electromagnetic hybrid nanogenerator for rotation energy harvesting and self-powered wind speed sensor. *ACS Nano* **2018**, *12*, 9433–9440. [CrossRef]
15. Wei, C.; Jing, X. A comprehensive review on vibration energy harvesting: Modelling and realization. *Renew. Sustain. Energy Rev.* **2017**, *74*, 1–18. [CrossRef]
16. Tran, N.; Ghayesh, M.H.; Arjomandi, M. Ambient vibration energy harvesters: Are view on nonlinear techniques for performance enhancement. *Int. J. Eng. Sci.* **2018**, *127*, 162–185. [CrossRef]
17. Tao, J.X.; Viet, N.V.; Carpinteri, A.; Wang, Q. Energy harvesting from wind by a piezoelectric harvester. *Eng. Struct.* **2017**, *133*, 74–80. [CrossRef]
18. Herrault, F.; Ji, C.H.; Allen, M.G. Ultraminiaturized High-Speed Permanent-Magnet Generators for Milliwatt-Level Power Generation. *J. Microelectromech. Syst.* **2008**, *17*, 1376–1387. [CrossRef]
19. Sardini, E.; Serpelloni, M. Self-powered wireless sensor for air temperature and velocity measurements with energy harvesting capability. *IEEE Trans. Instrum. Meas.* **2011**, *60*, 1838–1844. [CrossRef]
20. Priya, S. Modeling of electric energy harvesting using piezoelectric windmill. *Appl. Phys. Lett.* **2005**, *87*, 184101–184103. [CrossRef]

21. Tan, Y.K.; Panda, S.K. A Novel Piezoelectric Based Wind Energy Harvester for Low-power Autonomous Wind Speed Sensor. In Proceedings of the IECON 2007-33rd Annual Conference of the IEEE Industrial Electronics Society, Taipei, Taiwan, 5–8 November 2007; pp. 2175–2180.

22. Daqaq, M.F.; Bibo, A.; Li, G.; Sennakesavababu, V.R.; Clair, D.S. A scalable concept for micropower generation using flow-induced self-excited oscillations. *Appl. Phys. Lett.* **2010**, *96*, 144103.

23. Bibo, A.; Gang, L.; Daqaq, M.F. Electromechanical Modeling and Normal Form Analysis of an Aeroelastic Micro-Power Generator. *J. Intell. Mater. Syst. Struct.* **2011**, *22*, 577–592. [CrossRef]

24. Du, Z.G.; He, X.F. Micro Piezoelectric Wind Energy Collector with Resonant Cavity. *J. Transduct. Technol.* **2012**, *25*, 748–750.

25. Weinstein, L.A.; Cacan, M.R.; So, P.M.; Wright, P.K. Vortex shedding induced energy harvesting from piezoelectric materials in heating, ventilation and air conditioning flows. *Smart Mater. Struct.* **2012**, *21*, 045003-1–045003-10. [CrossRef]

26. Song, R.J.; Shan, X.B.; Li, J.Z.; Xie, T. Modeling an-d Experimental Study of Vortex-induced Vibration Capacitance of Piezoelectric Energy Harvester. *J. Xi'an Jiaotong Univ.* **2016**, *50*, 55–60.

27. Hobeck, J.D.; Geslain, D.; Inman, D.J. The dual cantilever flutter phenomenon: A novel energy harvesting method. *SPIE Proc.* **2014**, *9061*, 906113.

28. Sirohi, J.; Mahadik, R. Piezoelectric wind energy harvester for low-power sensors. *J. Intell. Mater. Syst. Struct.* **2011**, *22*, 2215–2228. [CrossRef]

29. Shan, X.B.; Song, R.J.; Liu, B.; Xie, T. Novel energy harvesting: A macro fiber composite piezoelectric energy harvester in the water vortex. *Ceram. Int.* **2015**, *41*, S763–S767. [CrossRef]

30. Kwon, S.D. A T-shaped piezoelectric cantilever for fluid energy harvesting. *Appl. Phys. Lett.* **2010**, *97*, 164102. [CrossRef]

31. Wang, J.L.; Zhou, S.X.; Zhang, Z.E.; Yurchenko, D. High-performance piezoelectric wind energy harvester with Y-shaped attachments. *Energy Convers. Manag.* **2019**, *181*, 645–652. [CrossRef]

32. Dias, J.; De Marqui, C.; Erturk, A. Hybrid piezoelectric-inductive flow energy harvesting and dimensionless electroaeroelastic analysis for scaling. *Appl. Phys. Lett.* **2013**, *102*, 044101. [CrossRef]

33. Erturk, A.; Vieira, W.; De Marqui, C.; Inman, D. On the energy harvesting potential of piezoaeroelastic systems. *Appl. Phys. Lett.* **2010**, *96*, 184103. [CrossRef]

34. Jing, S.R.; Zhang, M.Y. *Fluid Mechanics*; Xi'an Jiaotong University Press: Xi'an, China, 2001; pp. 216–231.

35. Xu, F.; Chen, W.L.; Bai, W.F.; Xiao, Y.Q.; Ou, J.P. Flow control of the wake vortex street of a circular cylinder by using a traveling wave wall at low Reynolds number. *Comput. Fluids* **2017**, *145*, 52–67. [CrossRef]

36. Norberg, C. Flow around a Circular Cylinder: Aspects of Fluctuating Lift. *J. Fluids Struct.* **2001**, *15*, 459–469. [CrossRef]

37. Xu, C.P.; Zhao, S.G. *Engineering Materials and Molding Process*; Metallurgical Industry Press: Beijing, China, 2014; pp. 84–89.

38. Kuna, M. Fracture mechanics of piezoelectric materials—Where are we right now. *Eng. Fract. Mech.* **2010**, *77*, 309–326. [CrossRef]

39. Xu, W. *Safety Factor and Allowable Stress*; Mechanical Industry Press: Beijing, China, 1981; pp. 9–14.

40. Guann, M.J.; Liao, W. On the efficiencies of piezoelectric energy harvesting circuits towards storage device voltages. *Smart Mater. Struct.* **2007**, *16*, 498–505. [CrossRef]

41. Chen, Z.S. *Theory and Method of Piezoelectric Vibration Energy Capture*; National Defence Industry Press: Beijing, China, 2017; pp. 25–26.

 *micromachines*

Article

Visual Servo Control System of a Piezoelectric 2-Degree-of-Freedom Nano-Stepping Motor

Cheng-Lung Chen and Shao-Kang Hung *

Department of Mechanical Engineering, National Chiao Tung University, No. 1001, University Road, Hsinchu 30010, Taiwan; bruce.me08g@nctu.edu.tw
* Correspondence: skhung@nctu.edu.tw; Tel.: +886-3571-2121 (ext. 55115)

Received: 9 October 2019; Accepted: 22 November 2019; Published: 25 November 2019

Abstract: A nano-stepping motor can translate or rotate when its piezoelectric element pair is electrically driven in-phase or anti-phase. It offers millimeter-level stroke, sub-micron-level stepping size, and sub-nanometer-level scanning resolution. This article proposes a visual servo system to control the nano-stepping motor, since its stepping size is not consistent due to changing contact friction, using a custom built microscopic instrument and image recognition software. Three kinds of trajectories—straight lines, circles, and pentagrams—are performed successfully. The smallest straightness and roundness ever tested are 0.291 μm and 2.380 μm. Experimental results show that the proposed controller can effectively compensate for the error and precisely navigate the rotor along a desired trajectory.

Keywords: piezoelectricity; visual servo control; stepping motor; nano-positioner; stick-slip

---

1. Introduction

A piezoelectric nano-stepping motor [1] was presented in our previous research work. This article presents its latest enhancements. The device has two degree-of-freedoms (DOF), translation for 6 mm and rotation for endless 360°. Its stepping resolution of translation and rotation are 100 nm and 0.04°, respectively. As it is well-known, piezoelectric elements have inherent nanometer resolution if they work in scanning mode [2–4], i.e., driven by smoothly adjusting its voltage. Adding amplifying mechanisms [5–7] for piezoelectric elements extends the stroke. Switching between scanning and stepping modes [8–10], the proposed device features mm-level working range at nm-level resolution. Referring to Figure 1, the most significant design factor is the parallel arrangement of two piezoelectric elements. Translational or rotational stepping motions are generated when this pair of piezoelectric elements are driven by a sawtooth waveform [11] in-phase or anti-phase, respectively. Our parallel design avoids serial stacking and achieves multiple DOF within a stiffer and more compact structure. This is an important feature for precision nano-positioners, in order to reduce vibration and thermal expansion. In brief, stepping mode offers a theoretically infinite stroke with μm-level resolution; scanning mode offers several μm-level stroke with nm-level resolution. Bridging two modes in 2-DOF is the goal of this study. The proposed system is designed to achieve several mm-level stroke with sub-μm resolution. If a higher accuracy is expected, the device can be switched to scanning mode, which has been well investigated in [2].

The stick-slip principle [12–14] is the foundation of the proposed nano-stepping motor. The periodic high slew-rate sawtooth waveform drives piezoelectric elements to overcome the friction force. Therefore, the inertia motor [15,16], or so-called impact drive [17,18], accumulates a series of steps forward. However, the step size does not remain consistent, because the friction force varies due to the microscopic nature of the contacting surfaces. Since open loop control is impossible, a standard method to control 1-DOF inertia in stepping motors is adding displacement sensors to form a feedback system.

In the case of 2-DOF, crosstalk or coupling may occur between sensors if they are not aligned properly. This article proposes an optically visual servo control system to solve the afore-mentioned problem. Similar piezoelectric motors are used to manipulate a phase plate [19] in a transmission electron microscope (TEM). This requires high resolution, long stroke, and does not allow the electromagnetic field to interfere with electron beams. TEM itself is a perfect 2-DOF motion sensor that can navigate the piezoelectric motor accurately, but only inside TEM itself. This research work extends the nano-stepping motor's applicability beyond the confines of a TEM. The instrumentation, image processing, control method, and validity are discussed in detail in the following sections. The experiments show that the proposed system functions effectively.

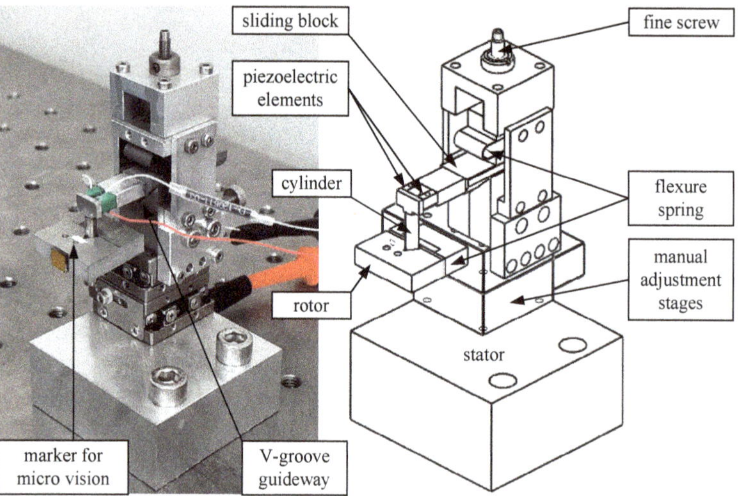

Figure 1. Photograph and schematic of a piezoelectric 2-DOF (degree of freedom) nano-stepping motor, which has a compact size of 5 cm × 5 cm × 3 cm. The fine screw presses the flexure spring and adjusts the clamping force between the V-groove guideway and the sliding block. The rotor can rotate around the cylinder. Translational or rotational stepping motions are generated when the pair of piezoelectric elements are driven in-phase or anti-phase, respectively.

2. Instrumentation and Control

2.1. Microscopic Imaging System

Based on our proposed nano-stepping motor [1], a microscopic imaging system is installed as in Figure 2. The camera (eco655CVGE, SVS-Vistek, Seefeld, Germany) has 2448 × 2050 pixels. Central 1000 × 1000 pixels are cropped for the following study. Two microscopic objectives (Plan 4× and 10×, Olympus) have resolutions of 1 µm/pixel and 0.38 µm/pixel, respectively. The relation between the physical distance and the image pixel count is correlated by an accurate laser displacement sensor (LK-H020, Keyence, Osaka, Japan), which has a 20 nm repeatability. The laser displacement sensor is not used in the following experiments because it only has 1-DOF and cannot measure rotation. The machine vision framework (Precise Eye 1-6044, Navitar, San Ramon, CA, USA) connects the camera and the objective together. The framework's coaxial illuminator provides suitable brightness for acquired images. The video stream is fed into an industrial computer (3.3 GHz, Intel i5 CPU) via Ethernet interface, and is then handled by LabVIEW (National Instrument, Austin, TX, USA). After image processing and decision making, the control signals are generated by a multifunction I/O interface (USB-6341, National Instrument, Austin, TX, USA), which provides two fast analog outputs for generating steep saw-tooth waveform, and at least one digital output for switching the relay. The state of the relay decides whether the sawtooth waveform pair is in-phase for translation, or in anti-phase for rotation.

An atomic force microscope (AFM) [20] probe is small, lightweight, and can serve as a micromachining tool bit [21–23]. Therefore, an AFM probe (Tap300Al-G, BudgetSenors, Sofia, Bulgaria) is glued to the rotor as the marker for micro vision, as labeled in Figure 1. The original image of the AFM probe is acquired as Figure 3a, and then binarized into Figure 3b. Our image processing program detects the edges, which are drawn as green lines in Figure 3c. The intersection of the crossed green lines is defined as the rotor's position, the subject to be controlled in this research work.

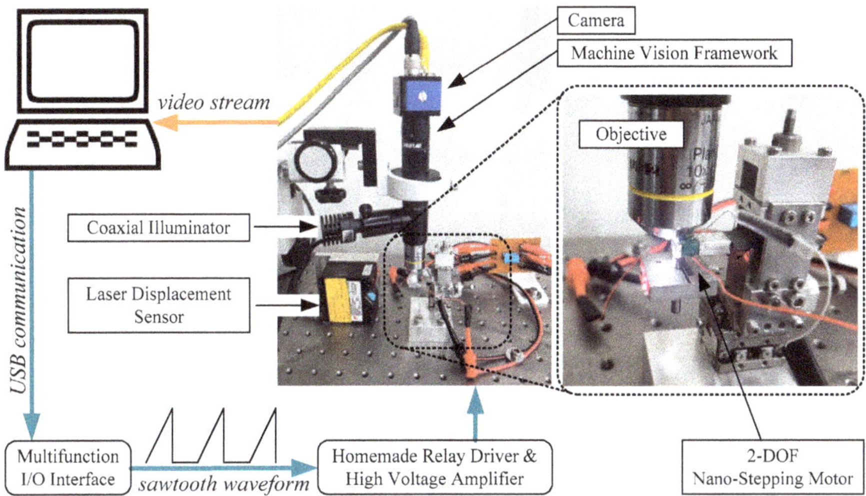

Figure 2. Block diagram and the photograph of the proposed visual servo control system for a piezoelectric 2-DOF nano-stepping motor. The camera acquires microscopic motion video and feeds it into an industrial computer. The image processing and motion control program drive the motor via the multifunction I/O interface and the high voltage amplifier.

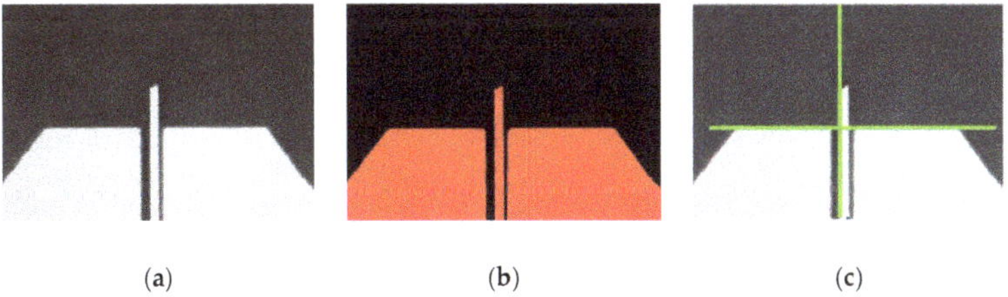

Figure 3. (**a**) The original, (**b**) the binarized, and (**c**) the edge-detected images of an AFM probe, which is treated as a micro marker of the proposed system. The width of above images is 1 mm.

2.2. Control System

As investigated, the proposed nano-stepping motor works up to 300 Hz but its velocity is not linear versus the driving frequency [1]. This phenomenon implies a suitable frequency range. After observing its stepping behavior under a microscope, we find that the device works smoothly between 70–90 Hz. Therefore, the driving frequency 80 Hz (i.e., 12.5 ms per sawtooth waveform) is fixed in the following experiments. For every task, the desired trajectory is separated into a series of checkpoints from start to finish. The amount of checkpoints depends on the intended precision and total length of the desired trajectory.

Referring to Figure 4, we have developed a motion control program to let the rotor's position to sequentially trace the checkpoints. Whenever the rotor's position gets close enough to the current checkpoint, the control program commands the motor to trace the next checkpoint until the whole task is finished. Combining forward/backward with translation/rotation, there are four decisions that can be made. After Cartesian–polar coordinate transformation, the control program calculates its decision related to the current checkpoint, and then moves the motor step by step. If the current position is identical to the previous one, the driving voltage is increased gradually to overcome the local friction force. The driving voltage range is 15 V_{pp} to 30 V_{pp} [1]. The control program dynamically adjusts the "minimum walkable voltage", which achieves precise steps without being stuck. The "closeness" value determines how closely the desired trajectory should be followed. Lower closeness values lead to less errors but take a longer time. The control system has to sacrifice accuracy for speed. The effect will be discussed in the following sections.

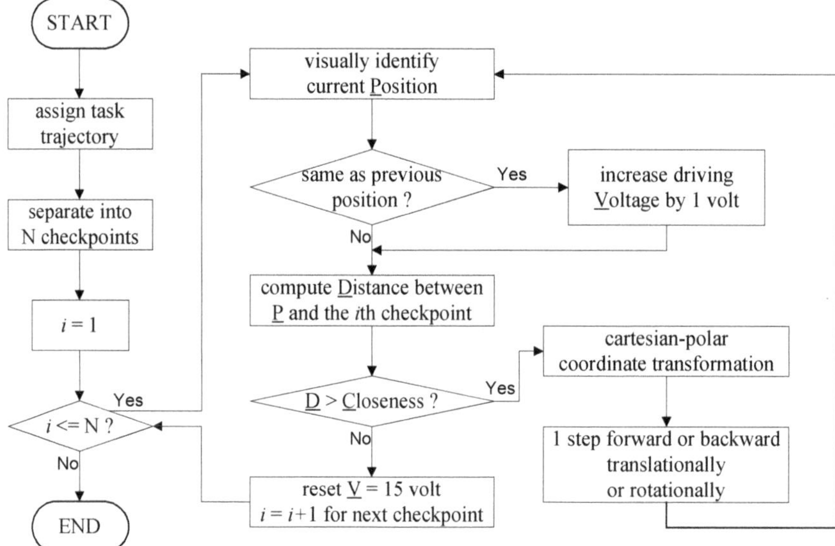

Figure 4. The control flow diagram of the control program. The goal is to trace a series of checkpoints. The driving voltage is increased gradually if the rotor is stuck.

3. Experimental Result

The four-time and 10-time microscope objectives create fields of view (FOV) of 1×1 mm² and 0.38×38 mm², respectively. Three kinds of motion trajectories—straight lines, circles, and pentagrams—are demonstrated in this section.

3.1. Straight Line Trajectory

3.1.1. Objective of Four-Time Magnification

As illustrated in Figure 5, the task trajectory is from the third quadrant to the first quadrant with a distance of 300 μm. Once the imaged position deviates from the desired path, the visual servo system corrects it back. Figure 5a,b shows the closeness values of one pixel (i.e., 1 μm) and five pixels (i.e., 5 μm), respectively. In Figure 5a, the coefficient of determination, the straightness, and the consumed time are 0.9993, 1.184 μm, 165 s, respectively. In Figure 5b, the coefficient of determination (COD), the straightness, and the consumed time are 0.9907, 1.506 μm, 108 s. The results reflect the trade off between accuracy and speed.

Figure 6 shows the experimental results when the walking distance is extended to 500 μm, while other conditions are maintained. In Figure 6a, the coefficient of determination, the straightness, and the consumed time are 0.9995, 1.579 μm, 278 s, respectively. In Figure 6b, COD, the straightness, and the consumed time are 0.9993, 1.812 μm, 117 s. The results present a similar trend to Figure 5.

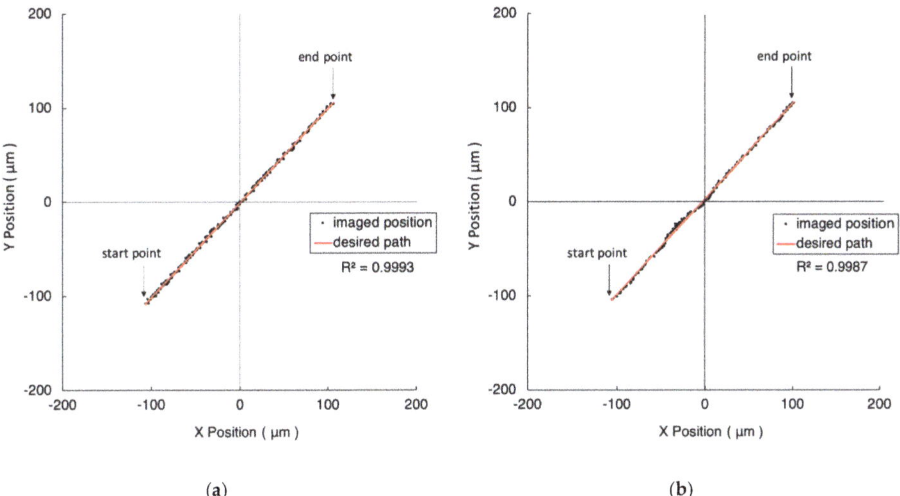

(**a**)

(**b**)

Figure 5. The experiment results of 300 μm straight line trajectories under the FOV of four-time objective. The closeness values are 1 μm and 5 μm in (**a**) and (**b**), respectively. The red line represents the desired path. The black dots are the imaged position of the marker driven by the 2-DOF nano-stepping motor.

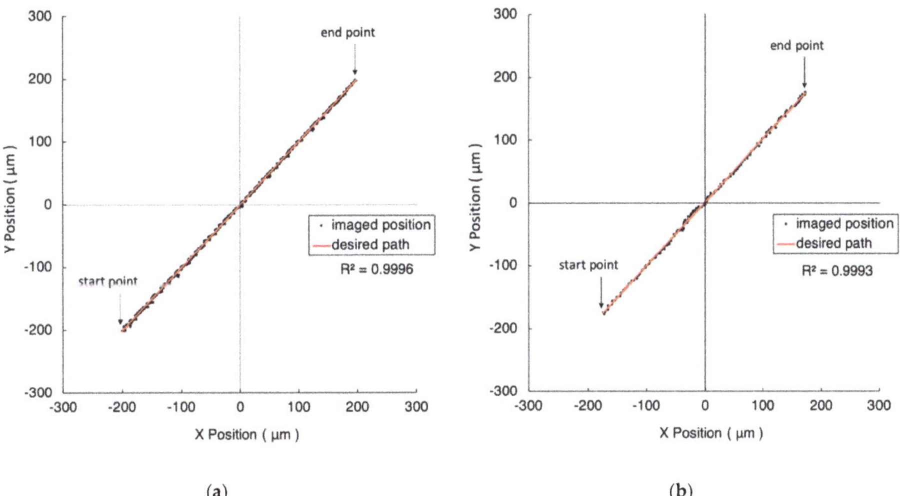

(**a**)

(**b**)

Figure 6. The experiment results of 500 μm straight line trajectories under the FOV of four-time objective. The closeness values are 1 μm and 5 μm in (**a**) and (**b**), respectively. The red line represents the desired path. The black dots are the imaged position of the marker driven by the 2-DOF nano-stepping motor.

3.1.2. Objective of 10-Time Magnification

Replacing the microscope objective with a higher magnification enhances the image resolution and improves the motion accuracy. As illustrated in Figures 7 and 8, the distances from their start points to end points are 100 μm and 200 μm, respectively. Figures 7a and 8a show the results of closeness values of one pixel (i.e., 0.38 μm). The closeness values are five pixels (i.e., 1.9 μm) in Figures 7b and 8b. The detailed experimental results are listed in Table 1.

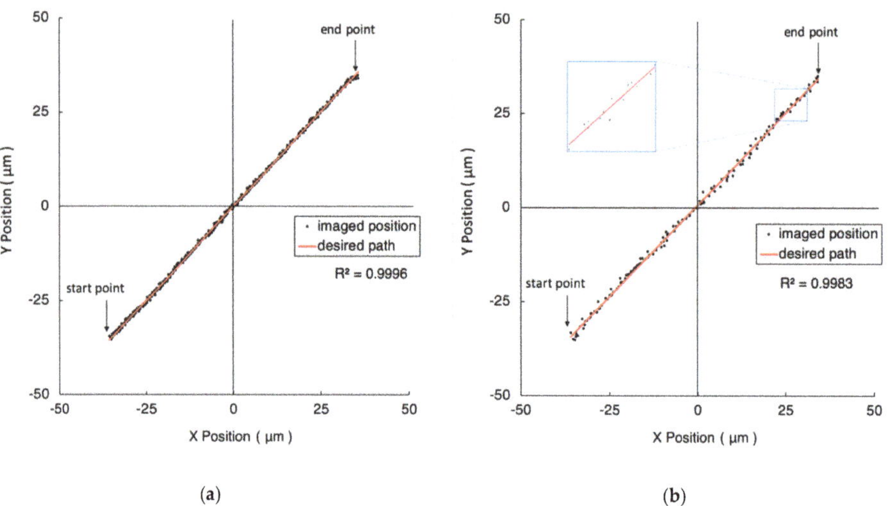

Figure 7. The experiment results of 100 μm straight line trajectories under the FOV of 10-time objective. The closeness values are 0.38 μm and 1.9 μm in (**a**) and (**b**), respectively. The red line represents the desired path. The black dots are the imaged position of the marker.

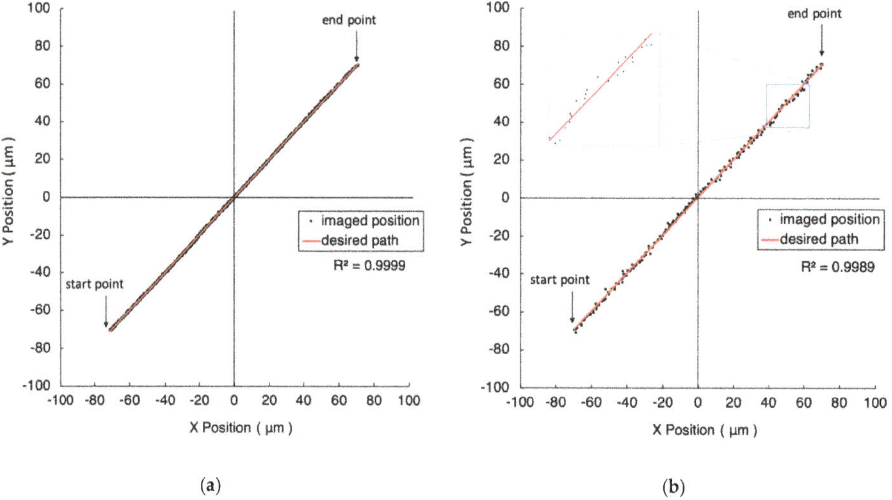

Figure 8. The experiment results of 200 μm straight line trajectories under the FOV of 10-time objective. The closeness values are 0.38 μm and 1.9 μm in (**a**) and (**b**), respectively. The red line represents the desired path. The black dots are the imaged position of the marker.

3.2. Circle Trajectory

3.2.1. Objective of Four-Time Magnification

As illustrated in Figure 9, the desired circular trajectories have diameters of 200 μm, 500 μm, and 800 μm. Figure 9a,b shows the closeness values of one pixel (i.e., 1 μm) and five pixels (i.e., 5 μm), respectively. The roundness varies from 7 μm to 27 μm. The time consumption varies from 174 s to 856 s. The detailed results are listed in Table 2. Obviously, lower closeness values lead to higher accuracy but more time is needed to complete the tasks.

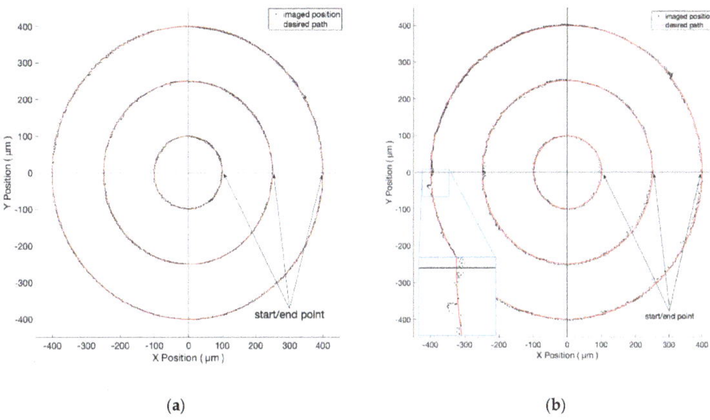

(a)   (b)

Figure 9. The experiment results of circles of different diameters under the FOV of four-time objective. The closeness values are 1 μm and 5 μm in (**a**) and (**b**), respectively.

3.2.2. Objective of 10-Time Magnification

As described in Section 3.1.2, we now change to a higher magnification objective to achieve an improved image resolution. Theoretically, using higher magnification achieves better resolution but smaller FOV, which limits the full working range. Referring to Figure 10, the desired trajectories are circles of diameters at 100 μm, 200 μm, and 240 μm. Figure 10a,b shows the closeness values of one pixel (i.e., 0.38 μm) and five pixels (i.e., 1.9 μm), respectively. Table 2 indicates the detailed results.

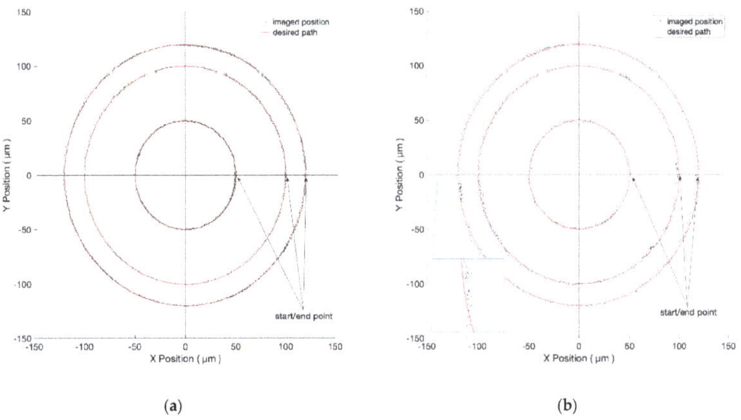

(a)   (b)

Figure 10. The experiment results of circles of different diameters under the FOV of 10-time objective. The closeness values are 0.38 μm and 1.9 μm in (**a**) and (**b**), respectively.

3.3. Pentagram Trajectory

In addition to regular shapes, the proposed visual servo system has the ability to navigate the 2-DOF nano-stepping motor to walk along any other arbitrary trajectory, once it is defined by a series of checkpoints. An experiment of the pentagram trajectory is demonstrated in Figure 11 and recorded in a time-lapse microscopic video (see Supplementary Video) [24]. The diameter of its circumscribed circle is 700 μm. Five-thousand checkpoints are used. The averaged coefficient of determination, straightness, and the consumed times are 0.9994, 1.594 μm, 2310 s.

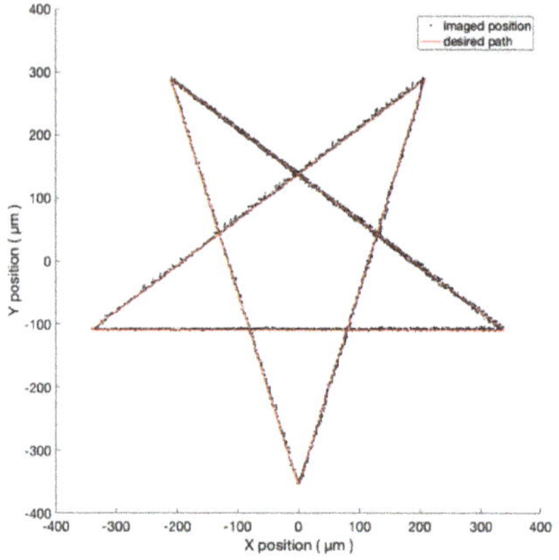

Figure 11. The experiment result of the pentagram trajectory under the FOV of four-time objective.

4. Discussion

The experimental data of straight and circular trajectories are rearranged in Tables 1 and 2, respectively. Higher magnification leads to a better resolution but a smaller working range. The COD, straightness, and roundness all indicate the precision of the controlled motion. All COD values of 45° linear motions are approaching 0.999, which means that the controller can compensate for the crosstalk between two DOFs effectively.

Table 1. Experiment data of straight trajectories.

Objective	Task Distance (μm)	Closeness	COD	Straightness (μm)	Time (s)	AverageSpeed (μm/s)
four-time	300 μm	One pixel	0.9993	1.184	165	1.817
		Five pixels	0.9987	1.506	108	2.780
	500 μm	One pixel	0.9995	1.579	278	1.797
		Five pixels	0.9993	1.812	117	4.274 **
10-time	100 μm	One pixel	0.9996	0.291 *	366	0.273
		Five pixels	0.9983	0.597	99	1.012
	200 μm	One pixel	0.9999	0.303	787	0.254
		Five pixels	0.9989	0.952	131	1.523

* The condition of smallest error; ** the condition of fastest speed.

The smallest straightness 0.291 μm in Table 1 and roundness 2.380 μm in Table 2 both occur at the setting of 10-time objective and one-pixel closeness. The fastest speeds, 4.274 μm/s in Table 1 and 3.607 μm/s in Table 2, both occur at the setting of four-time objective and five-pixel closeness. Referring to Figure 12, plotting straightness/roundness versus the average speed, an obvious correlation can be found between the error and the speed.

Table 2. Experiment data of circular trajectories.

Objective	Task Diameter (μm)	Closeness	Roundness (μm)	Time (s)	Average Speed (μm/s)
four-time	200	One pixel	7.516	475	1.322
		Five pixels	9.178	174	3.607 **
	500	One pixel	8.515	1104	1.423
		Five pixels	18.444	465	3.380
	800	One pixel	11.825	1482	1.696
		Five pixels	26.422	856	2.936
10-time	100	One pixel	2.380 *	659	0.477
		Five pixels	5.558	230	1.369
	200	One pixel	3.000	879	0.714
		Five pixels	6.965	460	1.367
	240	One pixel	3.222	1588	0.475
		Five pixels	7.430	521	1.447

* The condition of smallest error; ** the condition of fastest speed.

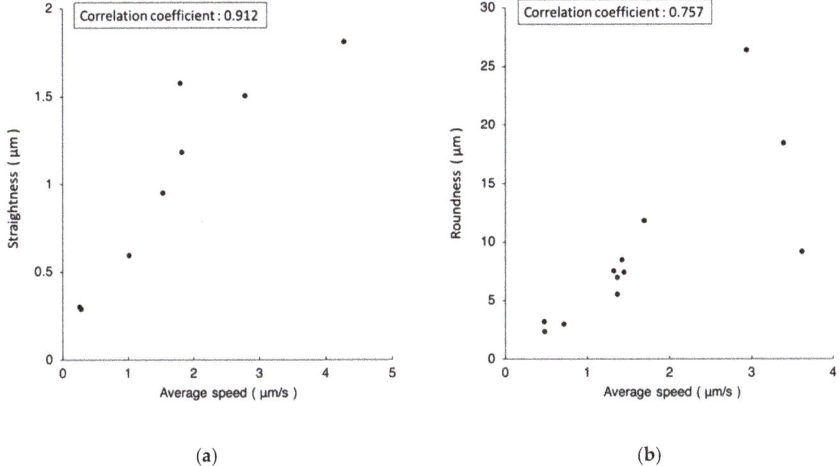

(a)

(b)

Figure 12. (**a**) Relation between the average speed and the straightness of linear motions in Table 1. (**b**) Relation between the average speed and the roundness of circular motions in Table 2. Positive correlation can be found between the error and the speed.

After analyzing the experimental data, we found that "minimum walkable voltage" is position-relative and time-varying. Figure 13 shows the minimum walkable voltage of 700 μm linear translational motion for three repeated tests. Around 420 μm, the local friction force is greater than average; therefore, a higher voltage is needed to keep going forward. The results are similar, however, not identical over three tests because the contact condition had been changed over time. There does not exists a globally consistent minimum walkable voltage value, although the contact surfaces had been carefully polished. In summary, dynamically adjusting the driving voltage is a practical method to deal with the changing friction force.

As the voltage is increased, the rotor suddenly overcomes the maximum static friction and has the chance to step away from the desired trajectory. This phenomenon is illustrated in the zoomed screens of Figures 9b and 10b. At those places with discontinuous friction force, the tracking error becomes greater than average. Eventually, our visual servo system can pull it back and reduces the error effectively.

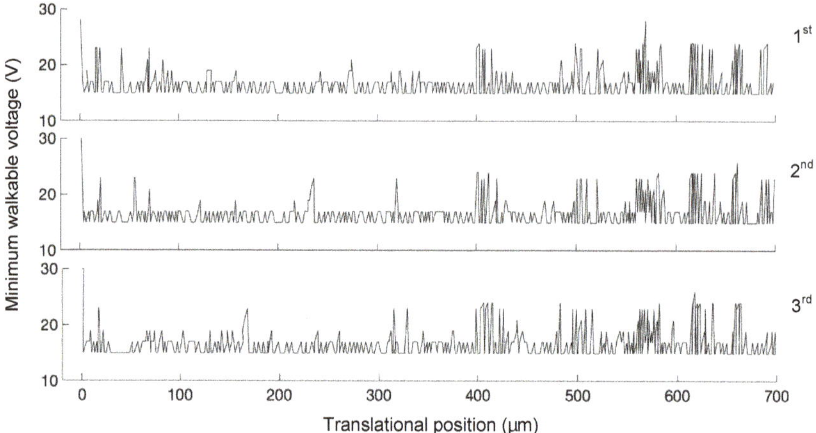

Figure 13. Minimum walkable voltage of 700 μm linear translational motion for three repeated tests.

Depending on the demands of a specific task, a balance between performance and budget is to be expected. Both the camera's pixel density and the objective's magnification affect precision. On the other hand, the camera's frame rate and the computer's image processing ability help to achieve higher speed. The direction of our future research is to build a coaxial multi-camera visual servo system. The large FOV image navigates the coarse motion quickly. The small FOV image controls the fine motion at a relatively low speed. The scheduling method between coarse and fine motions will release the power of the proposed 2-DOF nano-stepping motor. The third DOF, vertical to the image plane, will also be added. This will let the AFM probe fabricate microstructures.

5. Conclusions

The proposed 2-DOF nano-stepping motor offers the advantage of endless rotation and 6-mm translational stroke, at the cost of providing a consistent step size, making it impossible to use an open loop controller. The direct visual measurement is the most practical method to sense 360° rotation precisely and without contact. As a result, this article proposes a visual servo controller for the 2-DOF nano-stepping motor. The experiments show that the proposed controller can precisely navigate the rotor along various trajectories. In the case of following a 45° straight line, the best straightness is 0.291 μm at a speed of 0.273 μm/s. A higher speed could be achieved if a slightly greater tracking error (i.e., closeness) were acceptable. The trade-off between accuracy and speed remains inevitable.

Supplementary Materials: The following are available online at http://www.mdpi.com/2072-666X/10/12/811/s1, **Video:** The time-lapse microscopic video of the pentagram trajectory experiment.

Author Contributions: Conceptualization: S.-K.H.; methodology: S.-K.H. and C.-L.C.; software: C.-L.C.; validation: S.-K.H. and C.-L.C.; formal analysis: C.-L.C.; investigation: S.-K.H. and C.-L.C.; resources: S.-K.H.; data curation: C.-L.C.; writing—original draft preparation: S.-K.H.; writing—review and editing: S.-K.H.; visualization: C.-L.C.; supervision: S.-K.H.; project administration: S.-K.H. funding acquisition: S.-K.H.

Funding: This research was funded by Ministry of Science and Technology, Taiwan, grant number 108-2221-E-009-084-MY2.

Acknowledgments: The authors would like to thank Force Precision Instrument Co. Ltd., Taiwan for technical support.

Conflicts of Interest: The authors declare no conflict of interest.

References

1. Cheng, C.H.; Hung, S.K. A Piezoelectric Two-Degree-of-Freedom Nanostepping Motor With Parallel Design. *IEEE/ASME Trans. Mechatron.* **2016**, *21*, 2197–2199. [CrossRef]

2. Yong, Y.K.; Moheimani, S.O.R.; Kenton, B.J.; Leang, K.K. Invited Review Article: High-speed flexure-guided nanopositioning: Mechanical design and control issues. *Rev. Sci. Instrum.* **2012**, *83*, 121101-1–121101-22. [CrossRef] [PubMed]

3. Kang, D.; Gweon, D. Development of flexure based 6-degrees of freedom parallel nano-positioning system with large displacement. *Rev. Sci. Instrum.* **2012**, *83*, 035003-1–035003-9. [CrossRef] [PubMed]

4. Hung, S.K.; Hwu, E.T.; Hwang, I.S.; Fu, L.C. Postfitting Control Scheme for Periodic Piezoscanner Driving. *Jpn. J. Appl. Phys.* **2006**, *45*, 1917–1921. [CrossRef]

5. Li, Y.; Xu, Q. Development and Assessment of a Novel Decoupled XY Parallel Micropositioning Platform. *IEEE/ASME Trans. Mechatron.* **2010**, *15*, 125–135.

6. Lin, H.R.; Cheng, C.H.; Hung, S.K. Design and quasi-static characteristics study on a planar piezoelectric nanopositioner with ultralow parasitic rotation. *Mechatronics* **2015**, *31*, 180–188. [CrossRef]

7. Li, Y.; Xu, Q. Design and Analysis of a Totally Decoupled Flexure-Based XY Parallel Micromanipulator. *IEEE Trans. Robot.* **2009**, *25*, 645–657.

8. Kleindiek, S.; Kim, H.S.; Kratschmer, E.; Chang, T.H.P. Miniature three-axis micropositioner for scanning proximal probe and other applications. *J. Vacuum Sci. Technol. B* **1995**, *13*, 2653–2656. [CrossRef]

9. Judy, J.W.; Polla, D.L.; ROBBINS, W.P. A Linear Piezoelectric Stepper Motor With Submicrometer Step Size and Centimeter Travel Range. *IEEE Trans. Ultrason.* **1990**, *37*, 428–437. [CrossRef]

10. Yakimov, V.N. Scanning tunneling microscope with a rotary piezoelectric stepping motor. *Rev. Sci. Instrum.* **1996**, *67*, 384–386. [CrossRef]

11. Eng, L.M.; Eng, F.; Seuret, C.; Kündig, A.; Günter, P. Inexpensive, reliable control electronics for stick–slip motion in air and ultrahigh vacuum. *Rev. Sci. Instrum.* **1996**, *67*, 401–405. [CrossRef]

12. Peng, J.Y.; Chen, X.B. Modeling of Piezoelectric-Driven Stick–Slip Actuators. *IEEE/ASME Trans. Mechatron.* **2011**, *16*, 394–399. [CrossRef]

13. Rakotondrabe, M.; Haddab, Y.; Lutz, P. Development, Modeling, and Control of a Micro-/Nanopositioning 2-DOF Stick–Slip Device. *IEEE/ASME Trans. Mechatron.* **2009**, *14*, 733–745. [CrossRef]

14. Van Hoof, J.A.J.M.; Verhaegh, J.L.C.; Fey, R.H.B.; van de Meerakker, P.C.F.; Nijmeijer, H. Experimental Validation of Object Positioning Via Stick–Slip Vibrations. *IEEE/ASME Trans. Mechatron.* **2014**, *19*, 1092–1101. [CrossRef]

15. Xu, D.; Liu, Y.; Shi, S.; Liu, J.; Chen, W.; Wang, L. Development of a Nonresonant Piezoelectric Motor With Nanometer Resolution Driving Ability. *IEEE/ASME Trans. Mechatron.* **2018**, *23*, 444–451. [CrossRef]

16. Howald, L.; Rudin, H.; Gijntherodt, H.J. Piezoelectric inertial stepping motor with spherical rotor. *Rev. Sci. Instrum.* **1992**, *63*, 3909–3912. [CrossRef]

17. Furutani, K.; Higuchi, T.; Yamagata, Y.; Mohri, N. Effect of lubrication on impact drive mechanism. *Precis. Eng.* **1998**, *22*, 78–86. [CrossRef]

18. Morita, T.; Yoshida, R.; Okamoto, Y.; Higuchi, T. Three DOF parallel link mechanism utilizing smooth impact drive mechanism. *Precis. Eng.* **2002**, *26*, 289–295. [CrossRef]

19. Shiue, J.; Hung, S.K. A TEM phase plate loading system with loading monitoring and nano-positioning functions. *Ultramicroscopy* **2010**, *110*, 1238–1242. [CrossRef]

20. Binnig, G.; Quate, C.F. Atomic Force Microscope. *Phys. Rev. Lett.* **1986**, *56*, 930–933. [CrossRef]

21. Matsumoto, K. STM/AFM Nano-Oxidation Process to Room-Temperature-Operated Single-Electron Transistor and Other Devices. *Proc. IEEE* **1997**, *85*, 612–628. [CrossRef]

22. Hu, H.; Kim, H.J.; Somnath, S. Tip-Based Nanofabrication for Scalable Manufacturing. *Micromachines* **2017**, *8*, 90. [CrossRef]

23. Miguel, J.A.; Lechuga, Y.; Martinez, M. AFM-Based Characterization Method of Capacitive MEMS Pressure Sensors for Cardiological Applications. *Micromachines* **2018**, *9*, 342. [CrossRef]

24. PMEL. [PMEL] pentagram trajectory experiment by Piezoelectric 2-Degree-of-Freedom Nano-Stepping Motor. Available online: https://www.youtube.com/watch?v=_MbX6a9ltU8 (accessed on 8 October 2019).

 *micromachines*

Article

Physical Characteristics of and Transient Response from Thin Cylindrical Piezoelectric Transducers Used in a Petroleum Logging Tool

Lin Fa [1,*], Nan Tu [1], Hao Qu [2], Yingrui Wu [1], Ke Sun [1], Yandong Zhang [1], Meng Liang [1], Xiangrong Fang [1] and Meishan Zhao [3]

[1]   School of Electronic Engineering, Xi'an University of Posts and Telecommunications, Xi'an 710121, Shaanxi, China; tn0918@126.com (N.T.); Wu_yingrui@126.com (Y.W.); xayddxsk@163.com (K.S.); zhangyandong@xupt.edu.cn (Y.Z.); focr@xupt.edu.cn (M.L.); fangxr@xiyou.edu.cn (X.F.)

[2]   Baoding Hongsheng Acoustic-electric Equipment Co., Ltd., Baoding 071000, Hebei, China; ydfhcl@163.com

[3]   James Franck Institute and Department of Chemistry, The University of Chicago, Chicago, IL 60637, USA; llcai@uchicago.edu

*   Correspondence: faxiaoxue@126.com

Received: 15 October 2019; Accepted: 19 November 2019; Published: 22 November 2019

Abstract: We report on a transient response model of thin cylindrical piezoelectric transducers used in the petroleum logging tools, parallel to a recently established transient response model of thin spherical-shell transducers. Established on a series of parallel-connected equivalent-circuits, this model provides insightful information on the physical characteristics of the thin cylindrical piezoelectric transducers, i.e., the transient response, center-frequency, and directivity of the transducer. We have developed a measurement system corresponding to the new model to provide a state-of-the-art comparison between theory and experiment. We found that the measured results were in good agreement with those of theoretical calculations.

Keywords: petroleum acoustical-logging; piezoelectric cylindrical-shell transducer; center-frequency; experimental-measurement

---

1. Introduction

Acoustical measurement is ubiquitous in industrial applications, scientific research, and daily life, e.g., mobile and internet communication [1,2], exploration of underground mineral resources (oil, gas, coal, metal ores, etc.) [3], measurement of the in situ stresses of underground rock formation [4], and the inspection of mechanical properties of concrete [5,6], as well as intravascular ultrasound [7], medical imaging [8], biometric recognition [9], implantable microdevices [10], rangefinders [11], nondestructive detection [12–14], experimental verification of acoustic lateral displacement [15], inspection of a specific polarization state of a wave propagating in layered isotropic/anisotropic media [16,17], wave energy devices [18], and more. One of the key factors toward achieving a high-quality acoustic measurement is a good understanding of the properties of the acoustic transducers, e.g., the type of the transducers, the material property, and the geometric structure of the transducers.

A unique characteristic of piezoelectric materials is their electric–mechanical transduction ability, converting mechanical energy to electrical energy and vice versa. This property has been exploited extensively in the construction of acoustic transducers for industrial applications [19], e.g., in electrical engineering, biomedical engineering, and geophysical engineering, among others. Along with technological progress, the quality of piezoelectric transducers has also been improving dramatically, e.g., smaller geometric dimensions, reduced noise level [20], lowered power consumption [21], etc. The typical geometric structures of acoustic transducers in practical applications are cylindrical, schistose, spherical, among others. Thin cylindrical transducers are widely used in industry, e.g., in petroleum logging tools. The physical properties of cylindrical piezoelectric transducers are also widely studied, including radiation, electric–acoustic and acoustic–electric conversions, and more. The radiation of a cylinder transducer with harmonic vibration was reported by Bordoni et al. [22] and Williams et al [23]. Fenlon [24] reported calculations for the acoustic radiation field at the surface of a finite cylinder using the method of weighted residuals. Wu [25] described an application of a variational principle for acoustic radiation from a vibrating finite cylinder. The effect of the length and the radius of a cylinder on its radiation efficiency was discussed and reported by Wang and coworkers [26,27].

We are concerned here with the thin cylindrical piezoelectric transducers used in petroleum logging tools, either as an acoustic source or as a receiver. Conventionally, in studies of acoustic-logging, the simplified analytical-models have been used for measuring acoustic signal waveforms and in processing the measured acoustic signal. Many times, the Tsang wavelet has been used as an acoustic source in acoustic logging [28,29]. Oversimplified acoustic-source functions have been used in forwarding-model research of acoustic logging and/or in inversion analysis and processing of the measured acoustic-logging signal, e.g., Ricker wavelet [30], Gaussian impulse wavelet [31], etc. However, the mathematical expressions of the simplified models have not been able to provide the practical relationship between a driving-voltage signal, geometrical and physical properties of the transducer, and radiated acoustic signal-wavelet.

The transient response of a transducer driven by a sinusoidal electrical signal was reported by Piqtuette [32,33]. However, the reported results were not adequate for practical considerations for either acoustic logging or other acoustic measurements due to the driving-voltage signal of an exciting transducer containing multiple frequency components with different amplitudes and phases. In the process of providing a radiating acoustic signal, the transducer is also counteracted by the acoustic field radiated by itself and creates a radiation resistance and radiation mass, which are functions of the vibration frequency on the transducer's surface.

The complex effect of radiation resistance and radiation mass on the radiated acoustic-signal wavelet was reported by Fa and co-workers, where the deriving-voltage signal contained multi-frequency components. These researchers reported the case of thin spherical-shell transducers polarized in the radial direction, which radiated acoustic waves omnidirectionally [34–37]. Even though these reports are meaningful for modeling, the adopted transducers in the acoustic-logging tools and many other practical applications are mostly cylindrical, with radiation directivities quite different from that of the thin spherical-shell transducers.

It is understood that correct analysis and inversion interpretation of the measured acoustic-logging signal wavelet rely on accurate acoustic measurement instruments, which must be established on a solid physical foundation with a strict engineering mechanism. Based on our current knowledge of the thin cylindrical piezoelectric transducers used in petroleum logging tools, further development is highly desirable to achieve an enhanced understanding to develop applicable measurement instrumentations.

In this paper, we report a study of thin cylindrical transducers widely used in petroleum acoustic-logging tools. By adopting the method describing a thin spherical-shell transducer's transient response [34] for the excited driving-voltage signal wavelet with multi-frequency components, we established the parallel-connected equivalent circuits for the thin cylindrical-shell transducers polarized in the radial direction. By solving the corresponding equations of motion, we analyzed the physical properties related to the transient response. Technically, we employed a measurement system with a high-resolution, high-sampling-rate digitizer to perform an experimental measurement, e.g., with a resolution range from 16-bit to 24-bit and a sampling rate from 500 KS/s to 15,000 KS/s. From this system, we were able to achieve good measurements in a large frequency range, even for weak acoustic signals. Based a proper analysis using the experimentally acquired data, we obtained the relationships between the various physical quantities, e.g., the driving electric-signal wavelet, the electric–acoustic/acoustic–electric conversion factors, the propagation media, and the measured acoustic signal. This, in turn, provided us with the ability to inspect the properties of the transducers, obtain the physical parameters of the measured fluid and solid material, check the quality of the measured objects, and perform a verification of existing and/or on-going scientific research. We report that the calculated results of the physical properties and transient response of the thin cylindrical-shell transducer were in good agreement with those of experimental observations.

2. Theory and Modeling

Let us consider a piezoelectric, thin, cylindrical transducer, with an average radius ρ_0 and a wall thickness l_t, polarized in the radial direction. The electrodes were connected to the inner and outer surfaces, as shown in Figure 1. Because the radius of the thin cylinder is much larger than its thickness ($\rho_0 \gg l_t$), we have the following approximations: $\rho_0 \approx \rho_a \approx \rho_b$ and $\rho_0 = (\rho_a + \rho_b)/2$ From the axis symmetry of the particle displacement, tangential and axial stresses are equal zero, i.e., $T_{\rho z} = T_{\rho \phi} = T_{z\phi} = 0$.

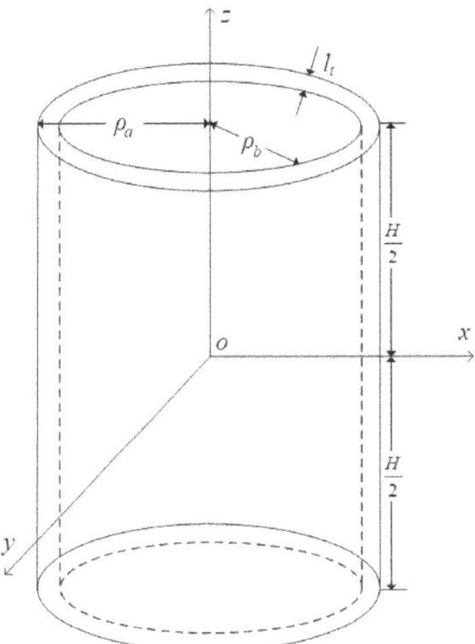

Figure 1. A piezoelectric thin spherical-shell transducer.

If the inner and outer surfaces are free from any other forces, i.e., the normal stress in the radial direction is $T_\rho = 0$, then the equation of motion for the particle vibrations of the transducer can be simplified to:

$$\rho_n \frac{\partial^2 u_\rho}{\partial t^2} = -\frac{T_\varphi}{\rho_0}, \tag{1}$$

$$\rho_n \frac{\partial^2 u_z}{\partial t^2} = -\frac{\partial T_z}{\partial z}, \tag{2}$$

where T_ϕ and T_z are the normal stress in tangential and axial directions, respectively; u_ρ and u_z are the particle displacement in the radial and axial directions, respectively; and ρ_n is the density of the transducer material.

Employing subscripts $\{1, 2, 3\}$ in place of subscripts $\{\varphi, z, \rho\}$, the piezoelectric equations with respect to radial polarization can be expressed as:

$$S_1 = s_{11}^E T_1 + d_{31} E_3, \tag{3}$$

$$D_3 = d_{31} T_1 + \varepsilon_{33}^T E_3, \tag{4}$$

where S_1 and T_1 are the strain and stress in φ-direction, respectively; D_3 and E_3 are the radial components of the electric displacement and electric field vectors, respectively; and s_{11}^E, ε_{33}^T, and d_{31} are the compliance, piezoelectric, and dielectric constants of the piezoelectric material, respectively. Because the height (H) of the transducer is much larger than its thickness, the coupling between the axial and radial vibrations can be neglected and the axial stress T_2 can be ignored. The vibration of the thin cylindrical transducer can be simplified as being one-dimensional in the radial-direction.

Substituting Equation (3) into Equation (1) yields:

$$m\frac{d^2 u_\rho}{dt^2} = -\frac{2\pi H l_t}{s_{11}^E}S_1 + \frac{2\pi H l_t d_{31}}{s_{11}^E}E_3, \tag{5}$$

where $m = 2\pi H l_t \rho_0 \rho_n$ is the mass of the transducer.

We set the transducer to be in coupling fluid, where only its outer surface was in contact with the coupling fluid. The transducer vibrates in the radial direction, where the vibration of the thin cylindrical transducer's outer surface causes the surrounding medium to expand and contract alternately in the radial direction. Consequently, the acoustic waves are radiated outward. Meanwhile, the thin cylindrical transducer is in the acoustic field radiated by itself. Then, it is acted on by a counterforce caused by the acoustic field. On the outer surface, by using a similar method described in References [34,35], we can obtain this counter-force as follows:

$$F_r = -R\left(\frac{k^2\rho_0^2}{1+k^2\rho_0^2} + j\frac{k\rho_0}{1+k^2\rho_0^2}\right)\frac{du_\rho}{dt} = -\left(R_\rho + iX_\rho\right)\frac{du_\rho}{dt} = -Z_\rho\frac{du_\rho}{dt}$$
$$R_\rho = R\frac{k^2\rho_0^2}{1+k^2\rho_0^2}, \; X_\rho = R\frac{k\rho_0}{1+k^2\rho_0^2}, \tag{6}$$

where $R = 2\pi\rho_0 H\rho_m v_m$; ρ_m and v_m are density and acoustic velocity of coupling fluid around the transducer, respectively; i is the unit imaginary number; and the radiation resistance and the radiation reactance are R_ρ.

If the thin cylindrical transducer vibrates harmonically with various frequencies, its radiation resistance and radiation reactance would also be different, where $k = \omega/v_c$.

Due to the viscosity of the coupling liquid, the vibration of the transducer creates a frictional resistance force, which can be expressed as:

$$F_f == -R_m\frac{du_\rho}{dt}, \tag{7}$$

where R_m is the frictional resistance on the surface of the transducer, which is proportional to the viscosity coefficient of the coupling fluid and the outer side wall area of the transducer. The transducer's radiation surface is approximately $A \approx 2\pi\rho_b H$, and the total outer force acted on the transducer is:

$$F = F_r + F_f = -\left(R_\rho + R_m + iX_\rho\right)\frac{du_\rho}{dt}. \tag{8}$$

The axial symmetry of thin-cylindrical transducer leads to:

$$S_1 = \frac{u_\rho}{\rho_0}. \tag{9}$$

For the harmonic vibration, substituting Equations (8) and (9) into Equation (5) yields:

$$u_r = \frac{2\pi H l_t d_{31}/s_{11}^E}{-\omega^2\left(m + m_\rho\right) + j\omega\left(R_\rho + R_m\right) + 1/C_m}E_3, \tag{10}$$

where $C_m = \rho_0 s_{11}^E/(2\pi H l_t)$.

Also, from Equations (3) and (4), we have:

$$D_3 = \frac{d_{31}}{s_{11}^E \rho_0} u_\rho + \varepsilon_{33}^T (1 - K_{31}^2) E_3 \tag{11}$$

where $K_{31} = d_{31} / \sqrt{s_{11}^E \varepsilon_{33}^T}$.

Because the thickness of the cylindrical transducer is small enough, the edge effect of the upper and lower cross-section can be neglected. From the Gauss theorem, the total charge on each electrode (either inner or outer surface) of the thin cylindrical transducer is $Q = 2\pi\rho_0 H D_3$. The instantaneous current into the electrodes is the time derivative of Q. For a harmonic driving electric signal, we have:

$$I = \frac{dQ}{dt} = i\omega C_0 V + N^2 \frac{V}{R_\rho + R_m + i\omega(m + m_\rho) + 1/i\omega C_m}, \tag{12}$$

where $C_0 = 2\pi\rho_0 H \varepsilon_{33}^T (1 - K_p^2)/l_t$; $N = 2\pi H d_{31}/s_{11}^E$, which is the electric–mechanical turn coefficient of the thin cylindrical transducer; and $V = l_t E_r$, which is the voltage across the two electrodes of the thin cylindrical transducer.

Suppose that the driving circuit outputs a sinusoidal signal $U_1(t)$ with an angular frequency ω and output resistance R_o. Based on Equation (12), the electric–acoustic equivalent circuit of the transducer for harmonic vibration is shown in Figure 2a. Its corresponding s-domain network is shown in Figure 2b, where $v_\rho(t)$ and $v_\rho(s)$ are defined as particle vibration speeds at the transducer's surfaces, and m_ρ is defined as the transducer's radiation mass.

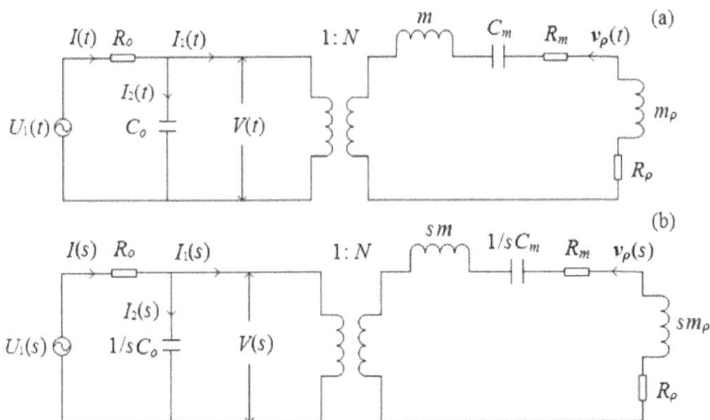

Figure 2. Equivalent circuits of a thin cylindrical transducer for electric–acoustic conversion: (**a**) equivalent circuit in the time domain and (**b**) equivalent circuit in the s-domain, where it is excited by a harmonic sinusoidal electric signal. $U_1(t)$ is the driving voltage source and R_o is its output resistance; $V(t)$ is the voltage signal at the electric terminals of the source; m_ρ, R_ρ, C_m, m, C_o, N, and R_m are the radiation mass, radiation resistance, elastic stiffness, mass, clamped capacitance, mechanical–electric conversion coefficient, and fraction force resistance of transducer, respectively; and $v_\rho(t)$ is the vibration speed at the transducer surface.

At the electric terminals of the s-domain in the network, as shown in Figure 2b, we have:

$$U_1(s) = V(s) + I(s)R_0 \tag{13}$$

and

$$I(s) = sC_0 V(s) + I_1(s) = sC_0 V(s) + Nv_r(s). \tag{14}$$

The transient response process of the thin cylindrical transducer can be held as a zero-state response. In the s-domain, we define the electric–acoustic conversion system function as a ratio of the vibration speed of the transducer's outer surface to the sinusoidal driving voltage signal. In terms of Figure 2b and Equations (13) and (14), this electric–acoustic conversion function can be expressed as:

$$H_1(s) = \frac{v_r(s)}{U(s)} = \frac{ds}{s^3 + as^2 + bs + c},$$
(15)

where:

$$a = \frac{R_m + R_\rho}{m + m_\rho} + \frac{1}{R_0 C_0}, \quad b = \frac{R_m + R_\rho}{(m + m_\rho) R_0 C_0} + \frac{1}{(m + m_\rho) C_m} + \frac{N^2}{(m + m_\rho) C_0},$$
$$c = \frac{1}{(m + m_\rho) C_0 C_m R_0}, \quad d = \frac{N}{(m + m_\rho) C_0 R_0}.$$

By applying the residue theorem to Equation (15), we obtain the electric–acoustic impulse response of the thin cylindrical transducer:

$$h_1(t) = \sum_{i=1}^{L} \mathrm{Res}[H_1(s_i)e^{s_i t}],$$
(16)

where L is the pole number of Equation (15). The denominator of this equation is a cubic polynomial with one unknown variable with the three roots:

$$s_1 = x + y - a/3,$$
(17)

$$s_{2,3} = -(x + y)/2 - a/3 \pm j\sqrt{3}(x - y)/2,$$
(18)

where:

$$x = \sqrt[3]{-q/2 + \sqrt{D}}, \quad y = \sqrt[3]{-q/2 - \sqrt{D}},$$
$$p = b - a^2/3, \quad q = c + 2a^3/27 - ab/3, \quad D = (p/3)^3 + (q/2)^2.$$

In theory, there are three cases with different D parameters—$D > 0$, $D = 0$, and $D < 0$—which correspond to the three motion modes: over-damping, critical-damping, and under-damping (oscillatory), respectively. Practically, the physical properties of a piezoelectric material, i.e. its physical and piezoelectric parameters guarantee that the parameter D is greater than zero, means that the transducer works only in the oscillatory mode and its electric–acoustic impulse response can be written as:

$$h_1(t) = A_3 \exp(-\alpha_1 t) + B_3 \exp(-\beta_1 t) \cos(\omega_1 t + \varphi_1)$$
(19)

where:

$$A = (x + y)/2, \quad B = (x - y)/2, \quad \beta_1 = A + a/3, \quad \alpha_1 = a/3 - 2A, \quad \sigma_1 = \beta_1 - \alpha_1,$$
$$A_3 = \frac{-d\alpha_1}{\sigma_1^2 + 3B^2}, \quad B_3 = -\frac{d(\sigma_1 - \beta_1)}{\sigma_1^2 + 3B^2}, \quad \omega_1 = \sqrt{3}B, \quad \varphi_1 = \arctan\frac{\beta\sigma_1 + 3B^2}{\sqrt{3}B(\sigma_1 - \beta_1)}.$$

The acoustic–electric conversion of the transducer is the inverse process of the electric–acoustic conversion. By repeating the discussed process in reverse order, we obtain the acoustic–electric impulse response of the transducer as follows:

$$h_3(t) = \overline{A}_3 \exp[-\alpha_3 t] + \overline{D}_3 \exp[-\beta_3 t] \cos(\omega_3 t + \varphi_3),$$
(20)

where:

$$\bar{a} = \frac{(m+m_r)+C_oR_o(R_r+R_m)}{C_oR_o(m+m_r)}, \ \bar{b} = \frac{C_oR_o+C_m(R_r+R_m)+C_mR_oN^2}{C_mC_oR_o(m+m_r)}, \ \bar{c} = \frac{1}{C_mC_oR_o(m+m_r)},$$

$$\bar{d} = \frac{\rho_0 v_c N}{C_o(m+m_r)}, \ \bar{x} = \sqrt[3]{-\bar{q}/2+\sqrt{\bar{D}}}, \ \bar{y} = \sqrt[3]{-\bar{q}/2-\sqrt{\bar{D}}}, \ \bar{p} = \bar{b}-\bar{a}^2/3, \ \bar{q} = \bar{c}+2\bar{a}^3/27-\bar{a}\bar{b}/3,$$

$$\bar{D} = (\bar{p}/3)^3 + (\bar{q}/2)^2, \ \bar{A} = (\bar{x}+\bar{y})/2, \ \bar{B} = (\bar{x}-\bar{y})/2, \ \beta_3 = \bar{A}+\bar{a}/3, \ \alpha_3 = \bar{a}/3-2\bar{A},$$

$$\sigma_3 = \beta_3-\alpha_3, \ \bar{A}_3 = \frac{-\bar{d}_3\alpha_3}{\sigma_3^2+3\bar{B}^2}, \ \bar{B}_3 = -\frac{\bar{d}_3(\sigma_3-\beta_3)}{\sigma_3^2+3\bar{B}^2}, \ \bar{D}_3 = \sqrt{\bar{B}_3^2+\bar{C}_3^2}, \ \bar{C}_3 = \frac{-\bar{d}_3(\beta_3\sigma_3+3\bar{B}^2)}{\sqrt{3B}(\sigma_3^2+3\bar{B}^2)},$$

$$\omega_3 = \sqrt{3\bar{B}}, \ \varphi_3 = \arctan\frac{\beta_3\sigma_3+3\bar{B}^2}{\sqrt{3B}(\sigma_3-\beta_3)}.$$

For a harmonic vibration, in the equivalent circuits in Figure 2, the two mechanical components, i.e., radiation resistance and radiation mass, are functions of vibration frequency. Also, for most cases, either the electrical signal of an exciting source transducer or the acoustic signal arriving at the receiver transducer is a signal wavelet with multi-frequency components. Excited by an electric/acoustic signal-wavelet with multi-frequency components, the vibration of the transducer's surface also consists of multiple sinusoidal frequency components.

The Fourier transform of the electric/acoustic signal of an excited transducer can be expressed as a linear superposition of sine-wave components with different frequencies, amplitudes, and phases. The electric–acoustic/acoustic–electric excitation can be processed by the parallel-connected network as shown by parts I and III in Figure 3. Each of these equivalent circuits in the network has its own unique electric–acoustic/acoustic–electric impulse response resulting from its individual radiation resistance and radiation mass.

A continuous driving electric signal $U_1(t)$ with amplitude spectrum $S(\omega)$ and phase spectrum $\phi(\omega)$ can be decomposed into N frequency components using an N-point discrete Fourier transform. Each frequency component can be written as:

$$U_{1j}(t) = |S(\omega_j)|\cos\big[\omega_j t + \phi(\omega_j)\big] \tag{21}$$

where $j = 1, 2, 3, \dots, N$; and $|S(\omega_j)|$ and $\phi(\omega_j)$ are the amplitude and the phase of the jth sinusoidal frequency component. Therefore, a normalized driving electric signal can be expressed as:

$$U_1(t) = \sum_{j=1}^{N} U_{1j}(t)/\max\left[\sum_{j=1}^{N} U_{1j}(t)\right]. \tag{22}$$

The output from the jth circuit in part I of Figure 3 is a convolution of the jth sinusoidal frequency component of the driving electric signal, with the jth electric–acoustic impulse response function being:

$$v_{r1j}(t)\big|_{\omega_j} = [U_{1j}(t) * h_{1j}(t)]\big|_{\omega_j}. \tag{23}$$

Then, the normalized vibration speed of the surface of the thin cylindrical transducer, i.e., the radiated acoustic signal, is defined as:

$$v_{r1}(t) = \sum_{j=1}^{N} v_{r1j}(t)\big|_{\omega_j}/\max[\sum_{j=1}^{N} v_{r1j}(t)\big|_{\omega_j}]. \tag{24}$$

For the jth frequency component of the radiated acoustic signal, if the propagation medium produces an acoustic impulse response $h_{2j}(t)\big|_{\omega_j}$, it would yield the jth frequency component arriving at the receiver transducer as:

$$v_{r3j}(t,\omega_j) = [v_{r1j}(t) * h_{2j}(t)]\big|_{\omega_j, t_{1j}}, \tag{25}$$

where t_{1j} is the propagation time of the *j*th sinusoidal frequency component from the source transducer to the receiver transducer.

The acoustic–electric conversion of a transducer is the inverse of the electric–acoustic conversion. The *j*th frequency component of an acoustic signal arriving at the receiver transducer passing the *j*th circuit (part III of Figure 3) is converted to an electric signal according to:

$$U_{3j}(t) = [v_{r3j}(t) * h_{3j}(t)]\big|_{\omega_j, t_{1j}}. \tag{26}$$

Finally, the measured acoustic signal, i.e., the electric signal at the electric terminals of the receiver transducer, is a collection of the outputs from all circuits in part III of the network (Figure 3), which is normalized as:

$$U_3(t) = \sum_{j=1}^{N} U_{3j}(t)|_{\omega_j} / \max[\sum_{j=1}^{N} U_{3j}(t)|_{\omega_j}]. \tag{27}$$

The above discussion shows that an acoustic-measurement process can be achieved through a parallel-connected transmission network, as shown in Figure 3.

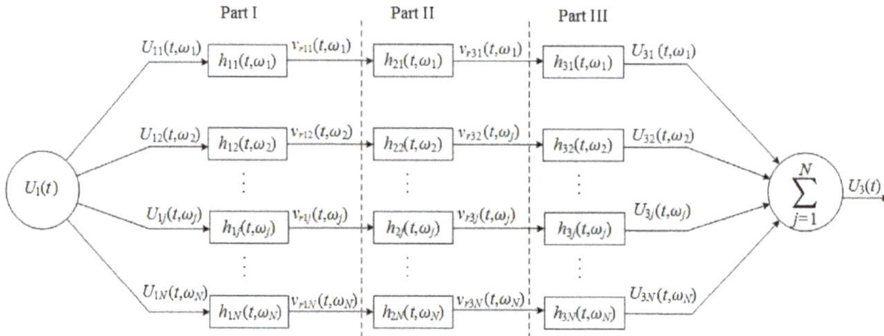

Figure 3. Schematic representation of a transmission network, which shows an acoustic-measurement process. *N* is the total number of components in the frequency spectrum of a driving electric signal. U_{1j} is the *j*th frequency component in the driving electric signal. v_{r1j} is the *j*th sinusoidsal frequency component of the vibration speed on the surface of the transducer. v_{r3j} is the *j*th sinusoidal frequency component in the acoustic signal arriving at the receiver transducer. U_{3j} is the *j*th frequency component of measured acoustic signal (i.e. electric signal at the electric terminals of the receiver transducers) created by the acoustic–electric conversion of the receiver transducer, where $j = 1, 2, \ldots, N$.

3. Calculation

We used a piezoelectric material PZT-5H (Lead Zirconate Titanate-5H, Baoding Hongsheng Acoustic-electric Equipment Co., Ltd., Baoding, Hebei 071000, China) in the construction of the cylindrical transducers. The physical parameters of the piezoelectric material PZT-5H are $\varepsilon_{33}^T = 300.9 \times 10^{-10}$ F/m², $s_{11}^E = 16.5 \times 10^{-12}$ m²/N, $d_{31} = -274 \times 10^{-12}$ m/V, and $\rho_n = 7.5 \times 10^3$ kg/m³. silicone oil was used as the coupling medium around the transducers, where the physical parameters are $v_m = 1424$ m/s and $\rho_m = 856$ kg/m³ (Shandong Longhui Chemical Co., Ltd, Jinan, Shaandong 250131, China).

3.1. Relationship between the Center Frequency versus the Radius for a Thin Cylindrical Transducer

As described above, the transducer always works in an oscillatory mode. As a reference, we defined a free-mechanical-load transducer as a transducer in a vacuum, i.e. $R_\rho = 0$, $m_\rho = 0$, and $R_m = 0$. The center frequency of the piezoelectric PZT-5H thin cylindrical transducer as a function of average radius is presented in Figure 4, along with a case of the mechanical load, i.e., the transducer in the transformer oil, where $R_\rho \neq 0$, $m_\rho \neq 0$, and $R_m \neq 0$.

Figure 4 shows that: (i) the transducer's center frequency decreased with respect to its increased radius, with or without a mechanical load; (ii) the center frequency of the mechanical load was lower than that of the free mechanical load; and (iii) the mechanical load effect on the transducer's center frequency decreased with the increased radius.

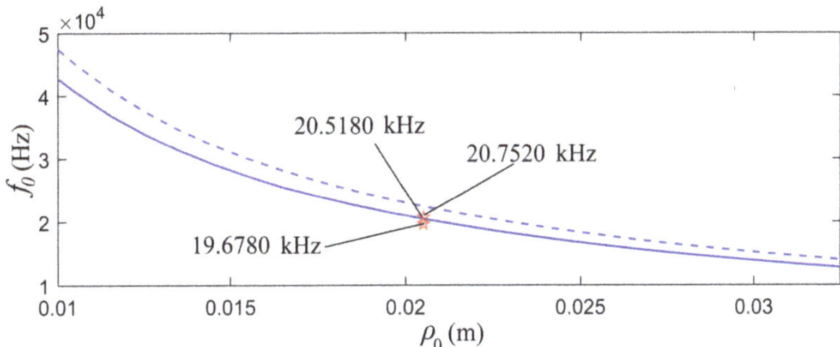

Figure 4. The relationship between the transducer's center frequency versus its radius. The solid line is the center frequency for the case with a mechanical load (i.e. transducer was put in transformer oil) and the dashed line is the center frequency for the case of a free mechanical load (i.e. transducer was put in air or vacuum). The two star signs are the measured values for the case of the transducer with a mechanical load and $l_t = \rho_0/8$.

3.2. Relationship between the Center Frequency versus a Forced Vibrational Frequency

We built two piezoelectric PZT-5H shin-cylindrical transducers polarized in the radial direction, with an average radius $\rho_0 = 20.5$ mm, height $H = 6.0$ mm, and wall thickness $l_t = \rho_0/8$. Figure 4 shows that the center frequency of the transducers with a mechanical load was 20.5180 kHz. We also determined that the center frequency of the free-mechanical-load transducer was at 22.5120 kHz.

For the case of the mechanical load, we selected the parameter $R_m = 0.2R$. When normalized by the transducer's free-load center frequency, for the transducer's forced harmonic vibrational motions at several frequencies, the electric–acoustic/acoustic–electric conversion properties were calculated and are presented in Figure 5.

At the center frequency of 22.5120 kHz for the free mechanical load, both the source and receiver transducers had a maximum transition amplitude, which was the largest for all cases. For the mechanical loaded transducers with harmonic forced vibrations, a lower forced vibration frequency corresponded to a lower center frequency with a larger maximum transition amplitude. With the increased vibration frequency going close to the free-mechanical-load resonance frequency f_0, a mechanical-loaded transducer showed a maximum transition amplitude, but it was much smaller than that of the thin cylindrical transducer free vibration.

The observations above confirmed our understanding that the electric–acoustic/acoustic–electric conversion of the transducer was dependent not only on the physical and geometrical parameters of a transducer and the physical parameters of the medium around the transducer, but also the forced harmonic vibration frequency. The radiation resistance and radiation mass were parameters affecting the forced vibration frequency, which led to the variations of the electric–acoustic/acoustic–electric conversion properties of the transducer.

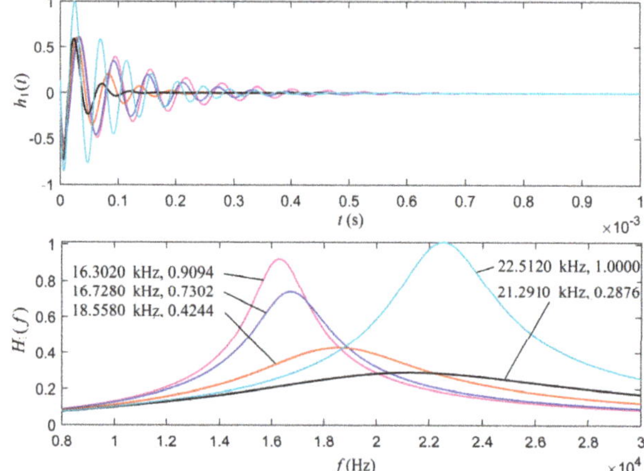

Figure 5. The impulse response and corresponding amplitude spectrum for a thin cylindrical transducer:
(**a**) the impulse response and (**b**) the amplitude spectrum. The cyan line is the case of a free mechanical
load. The other lines are for a mechanical load, where the magenta, blue, red, and black lines stand for
the sinusoidal driving electric signals with frequency $f_s = 0.1f_0$, $0.2f_0$, $0.5f_0$, and $1.5f_0$, respectively,
where $f_0 = 20.5180$ kHz. The first numerical value in Figure 5b is the center frequency and the second
one is the maximal value of the amplitude spectrum.

3.3. Radiation Directivity of a Thin Cylindrical Transducer

Different from the thin-shell spherical transducer, the radiation property of a thin cylindrical
transducer has a determinative directivity, which can be described using [8]:

$$
G(\theta) = \left[\frac{J_0^2(\frac{2\pi\rho_a}{\lambda}\cos\theta) + \cos^2\theta J_1^2(\frac{2\pi\rho_a}{\lambda}\cos\theta)}{J_0^2(\frac{2\pi\rho_a}{\lambda}) + J_1^2(\frac{2\pi\rho_a}{\lambda})} \right]^{1/2} \frac{\sin(\frac{\pi H}{\lambda}\sin\theta)}{\frac{\pi H}{\lambda}\sin\theta},
\tag{28}
$$

where J_0 is the zeroth-order Bessel function; J_1 is the first-order Bessel function; λ is the wavelength of
the acoustic wave in coupling fluid, i.e., transformer oil; and θ is the angle between the propagation
direction of radiated acoustic signal and the normal direction of the thin cylindrical transducer.

From Equation (28), the calculated radiation directivity of a thin cylindrical transducer is shown
in Figure 6 It is shown that the acoustic energy radiated by the transducer was more centralized in the
normal direction with the increased height of the thin cylindrical transducer.

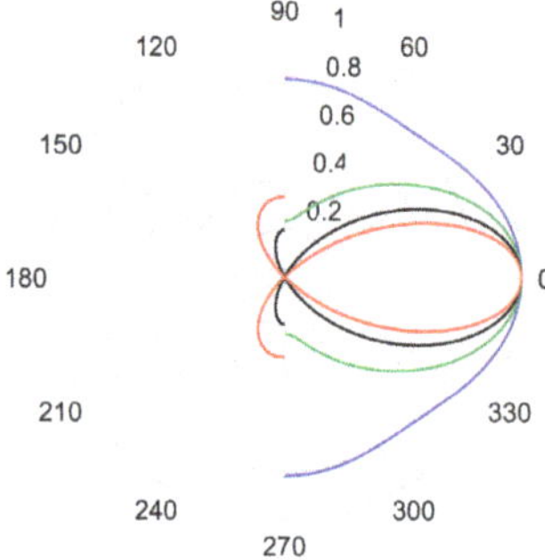

Figure 6. The directivity of thin cylindrical transducer with varied height of the cylindrical transducer,
where $\rho_0 = 20.50$ mm. Red, black, green, and blue colors indicate the directivities of $H = 40$ mm, 60
mm, 80 mm, and 100 mm, respectively.

The effective radiating area of a cylindrical transducer, in the direction of θ, may be defined as the product of the area of the thin cylindrical transducer $S_a (= 2\pi\rho_a H)$ and the directivity $G(\theta)$:

$$ERA = S_a \cdot G(\theta) = 2\pi\rho_a H \cdot G(\theta). \tag{29}$$

For a given cylindrical transducer with a fixed average radius, a normalized effective radiating area is labelled as *NERA*. The relationship of the normalized *NERA* versus H in several different directions are presented in Figure 7. The calculated results indicated that the transducer's effective radiating area did not increase monotonously with respect to its height or its radiation area. This was exactly the effect of the transducer's radiation directivity, which was a function of the radiation direction. With this understanding, by selecting a suitable height of the transducer, we may be able to achieve an increased effective radiation area of the cylindrical transducer for a set special direction.

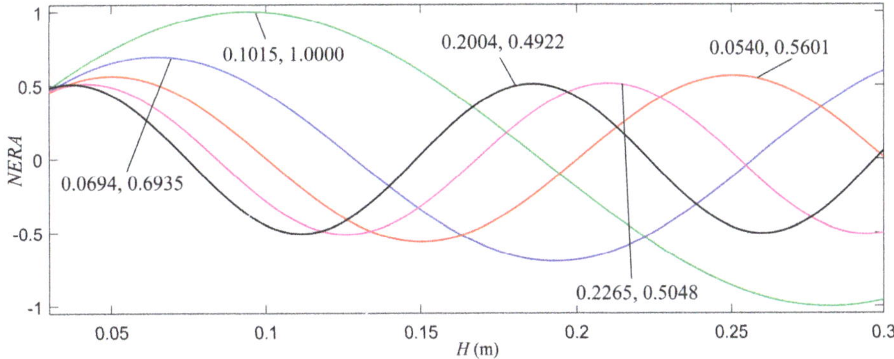

Figure 7. The relationship of the normalized effective radiating area (*NERA*) versus height (*H*) for several different directions, where ρ_0 = 20.5 mm. The green, blue, black, magenta, and red colors designate the cases of radiating direction $\theta = 20°$, $\theta = 30°$, $\theta = 40°$, $\theta = 50°$, and, $\theta = 60°$, respectively. The first numerical value in Figure 8 is the height, *H*, of the transducer and the second one is the maximal value of the normalized effective radiating area. The value of *NERA* is the value of *ERA* normalized using the maximum of *ERA* at $\theta = 20°$.

3.4. Transient Response of a Cylindrical Transducer Excited by a Signal with Multi-Frequency Components

As an example, let us consider a gated sinusoidal electric signal as the source of excitation with multi-frequency components as:

$$U_1(t) = [H(t) - H(t - t_0)] U_0 \sin(\omega_s t), \tag{30}$$

where, U_0, ω_s, and t_0 are the amplitude, angular frequency, and time window of the driving electric voltage signal, respectively; and $H(\cdot)$ is the Heaviside unit step function. A simple Fourier transform yields the source signal in the frequency domain as:

$$S_1(\omega) = U_0\{\omega_s - (\omega_s \cos \omega_s t_0 + j\omega \sin \omega_s t_0) \exp[-j\omega t_0]/(\omega_s^2 - \omega^2) \tag{31}$$

with a corresponding phase spectrum:

$$\phi(\omega) = \text{tg}^{-1}\left[\frac{\text{Im}(S_1)}{\text{Re}(S_1)}\right]. \tag{32}$$

Now, we selected the source signal parameters $U_0 = 1$ V, $t_0 = 4/f_s$, and $f_s = \omega_s/2\pi = 20.5180$ kHz, which was the mechanical-load center frequency of the transducer. The center frequency of this driving electric signal was then 20.1690 kHz, which was slightly smaller than the value of f_s.

Figure 8 shows that the theoretical waveform of the driving electric voltage signal agreed well with the waveform synthesized using a discrete Fourier transform. In turn, it guaranteed the accuracy of our analysis in the following sections.

The gated sinusoidal driving electric signal was expanded on the basis of a series of sine-waves with different frequencies, amplitudes, and phases. Each of these sinusoidal components, as an individual excitation source, was applied to the parallel circuits of Figure 3. The output signal from each parallel circuit in the network (see Figure 3) was calculated and analyzed.

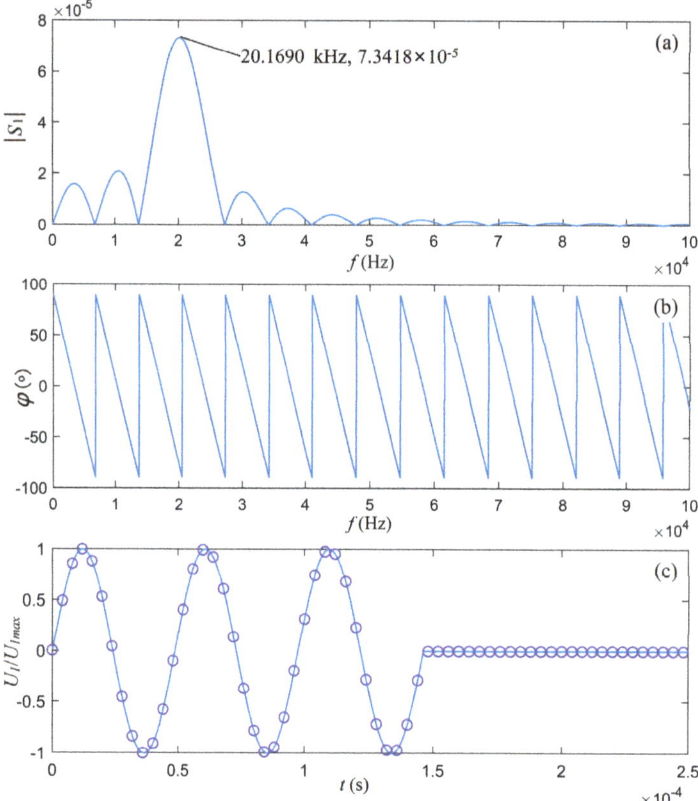

Figure 8. The waveforms of the gated sinusoidal driving electric signal with two cycles and f_s = 20.5180 kHz, where $f_s = \omega_s / 2\pi$. (**a**) The amplitude spectrum, (**b**) the phase spectrum, and (**c**) the waveform. The solid line in (c) was from the theoretical calculation and the cycle line was the synthesized waveform from discretized amplitude and phase spectra.

The cumulative output of the waveform of the parallel circuits in part I of Figure 3 is presented in Figure 9a. The corresponding amplitude spectrum is presented in Figure 9b, which shows that the center frequency of the radiated acoustic signal was 20.2500 kHz. It was smaller than the load center frequency 20.5180 kHz, but larger than that of gated sinusoidal driving electric signal at 20.1690 kHz, which was the result of the joint action of the electric–acoustic conversion through the source transducer and the electric driving signal.

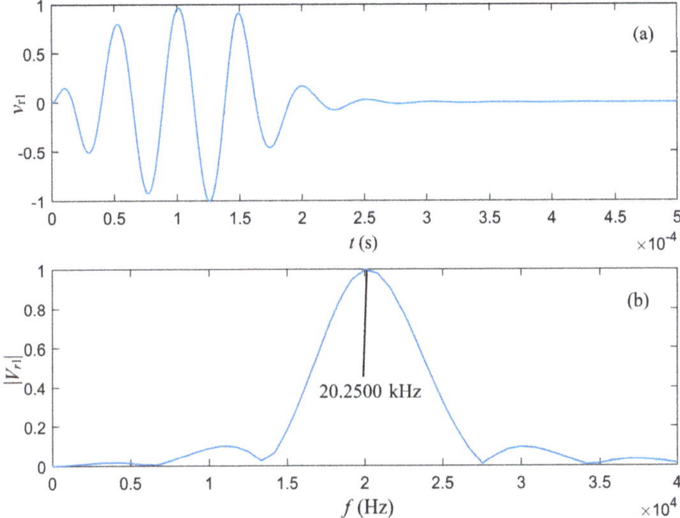

Figure 9. The waveform and amplitude spectrum of the acoustic signal radiated by the source transducer: (**a**) The normalized waveforms, which were the cumulative convolution of all frequency components in the network; and (**b**) the normalized amplitude spectrum of our transducer's transient response model.

The studied source transducer was cylindrical and had a radiation directivity. If we assumed that water is an ideal elastic medium, then the acoustic signal propagating inside the water could have a geometrical attenuation but not a viscous attenuation. Additionally, all frequency components of the acoustic signal would propagate at the same speed. The shapes of the waveform and frequency spectrum would not change. The amplitude would decrease with respect to the increased propagation distance only. Under these conditions, we calculated the signals at the electric terminals of the receiver transducer.

For a thin cylindrical transducer, the shape of its radiation directivity is the same as that of its receiving directivity; therefore, the acoustic impulse response in water can then be written as:

$$h_2(t) = G_1(\theta_1)G_2(\theta_2)\delta(t-t_1)/(1+r), \tag{33}$$

where $h_{21}(t,\omega_1) = \ldots = h_{2j}(t,\omega_j) = \ldots h_{2N}(t,\omega_N) = h_2(t)$, t_1 is the propagation time and r is the distance of the radiated signal in water, θ_1 is the angle between of the propagation direction of the radiated acoustic wave and the normal direction of the side wall of the source-transducer, $G_1(\theta_1)$ is the corresponding radiation directivity at this propagation direction (θ_1), θ_2 is the angle between the direction of the acoustic wave propagating to the receiver transducer and the normal direction of the side-wall of the receiver-transducer, and $G_2(\theta_2)$ is the corresponding receiving directivity.

To determine the quality of a transducer for electric–acoustic/acoustic–electric conversion, the most important piece of information comes from the analysis of the output signal following electric–acoustic–electric transduction. To accomplish this operation, we set $\theta_1 = \theta_2 = 0°$ and the distance from the source transducer to the receiver transducer was 1.0 m. Applying the gated sinusoidal driving electric signal to the source transducer, we calculated the cumulative output signals on the receiver transducer, as shown as solid lines in Figure 10.

Figure 10 shows the cumulative signals of the waveform and amplitude spectrum at the electric terminals of the receiver transducer. From the theoretical calculation, the center frequency of the output signal at the electrical terminals of the receiver transducer was 20.5000 kHz. This center frequency was slightly smaller than the transducer's load center frequency and slightly greater than the center frequency of the acoustic signal radiated by the source transducer. We also believe that this was the result of the joint action of a lower center frequency of the radiated acoustic signal and a higher center frequency of the transducer with a mechanical load.

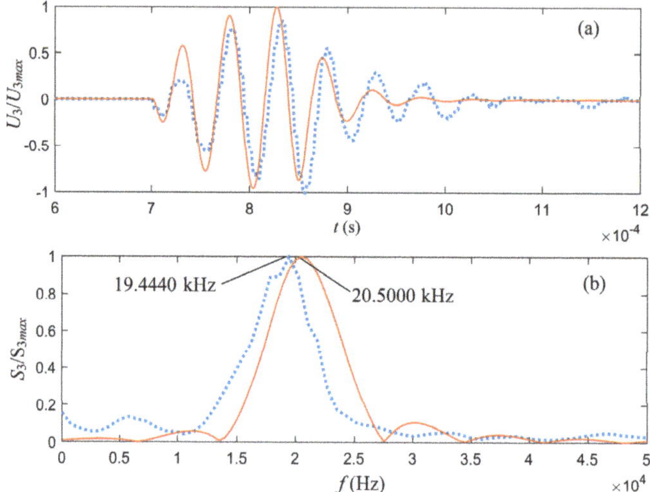

Figure 10. Normalized electric signals at the receiver transducer (part III of Figure 4): (**a**) the waveform and (**b**) the amplitude spectrum. S_3 stands for the amplitude spectrum corresponding to the electric-signals U_3 at the receiver transducer. Solid lines come from the theoretical calculation and dotted lines come from the experimental measurement.

For an average radius of 20.5 mm, Figure 4 shows that the calculated center frequency of the mechanical-loaded transducer was 20.5800 kHz and the measured center frequencies of the two cylindrical transducers with the same size were 19.6780 kHz and 20.7520 kHz. The waveform and amplitude spectrum of the measured acoustic signal, i.e., the measured electric signal at the electric terminals of the receiver transducer, are shown as the dotted lines in Figure 10. Overall, the theoretical calculation had a good agreement with the experimental measurement.

3.5. Influence of the Transducer's Radiating/Receiving Directivity on the Measured Acoustic Signal

Now, let us consider a measurement system, where a cylindrical transducer is used as an acoustic source and four cylindrical transducers as a receiver array, as shown in Figure 11. The transition media of this system is water. The waveforms of the signals at the electric terminals of the four cylindrical transducers in the receiver array were calculated and are presented in Figure 12.

The results of the calculation show that the greater the deflection angle, the greater the influence on the signal at the electric terminals of each cylindrical transducer in the receiver array. This was especially true in the near-field area. The amplitude of the signal at the electric terminals decreased with increased angle θ_j.

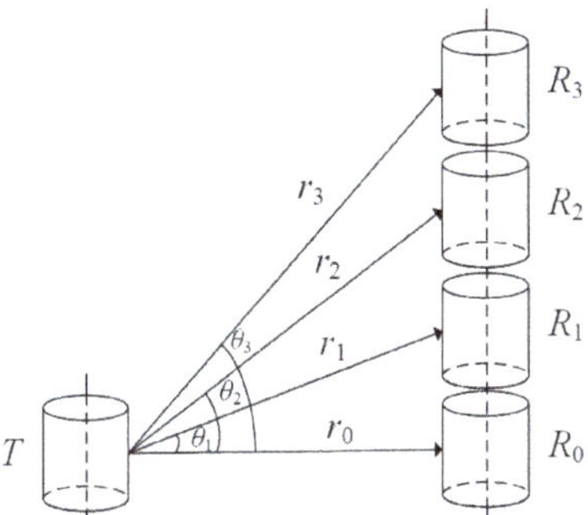

Figure 11. The acoustic-measurement system with a thin cylindrical transducer as a source and an array of four cylindrical transducers as the receivers, where T is the source transducer, R_i is each receiver transducer in the transducers line array, r_i is the distance from the source transducer T to the receiver transducer R_i of the transducer line-array with $\{i\} = \{0, 1, 2, 3\}$, and θ_j is the angle of r_j with respect to r_0 with $\{j\} = \{1, 2, 3\}$.

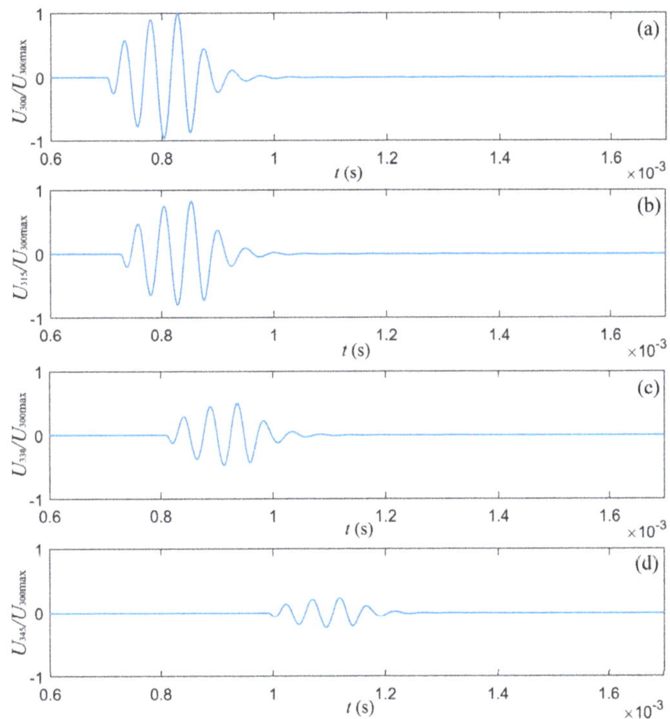

Figure 12. The signal waveforms at the electric-terminals of the receiver-transducer.

4. Devices and Measurements

To rationalize the proposed transient response and physical properties of the cylindrical transducers, we developed an experimental measurement system (see Figure S1 and Figure S2). In particular, this system consisted of a mechanical assembly, an electrical hardware module, and a system software module aimed at system control, measurement, and analysis. A structure flowchart is presented in Figure 13.

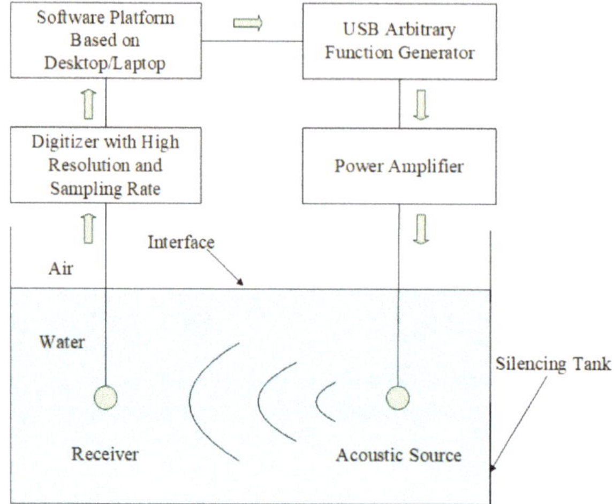

Figure 13. Schematic presentation of the experimental measurement system.

4.1. Mechanical and Electrical Hardware

The mechanical parts include steering engines, stepping motors, and sliding rails. The combined assembly with a microcontroller formed a positioning platform, which was used to slide and/or rotate the source/receiver transducers to the proper positions and directions in a silencing tank filled with water. The silencing tank was used to gauge the physical properties of the transducer, e.g., the electric–acoustic/acoustic–electric transient response, directivity, radiation power, and receiving sensitivity of the transducer.

The electric hardware was composed of a microcontroller, an electric-signal waveform generator, a power amplifier, a high-resolution and high-sampling-rate digitizer (up to 24 bit and 15 MHz), and a desktop/laptop for central system control.

4.2. System Software

The system software was developed on the LabVIEW platform in Graphic Programming Language (G-code). Communications between the computer and the hardware were realized through USB serial ports. Powered by a graphic interface control panel, users have the options of selecting different driving signals, configuring the power amplifier, adjusting the position and rotational angle of the transducers, and attaining data acquisition of the measured acoustic signal.

The developed software consists of four functional modules, as shown in Figure 14. On a graphic computer interface panel, the VISA (Virtual Instrument Software Architecture) library from the LabVIEW platform was employed to achieve the control of the four modules through serial port communications. In the function/selection boxes in the human–computer interface modules, users input the parameters and commands for special operations and acoustic measurements.

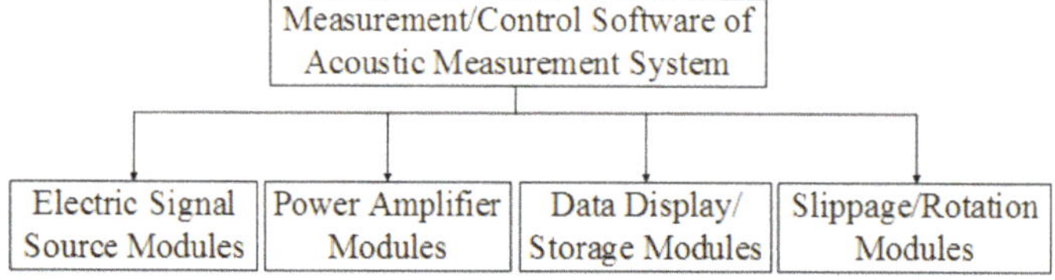

Figure 14. Structure flowchart of the modules for an electric signal source, power amplifier, date display/storage, and slippage/rotation.

4.2.1. Electric Signal Module

In controlling the widgets, knobs, and switches on the software interface panel, this module configures and modulates the types of signal and frequency, amplitude, and the cycle of an electric source signal. The options for the electric source signal are the gated sinusoidal wave, square wave, triangular wave, and sawtooth wave.

4.2.2. Power Amplification Module

This module configures the amplification gain of a power amplifier, which amplifies the electric signal created by the electric signal waveform generator. The amplified electric signal is used to excite the source transducer to radiate the acoustic signal. This is achieved through a rotary button on the interface control panel. For an output voltage signal with a frequency belt from 0.15 MHz to 1.5 MHz, the maximum peak-to-peak value can reach 220 V. Meanwhile, for an output voltage signal with a frequency range from 10 kHz to 150 kHz, the maximum peak-to-peak value can be up to 1600 V. For an electric signal source with general impedance, a suitable input impedance is 50 Ω. For an electric signal source with high impedance, an appropriate input impedance is 50 kΩ.

4.2.3. Display/Storage Module

Using this module, users may configure the acquisition channel, sampling rate, and amplitude range of an acquired signal. Data in the time–frequency domain acquired by the digitizer are sent to the interface panel through a USB port. These data can be displayed on screen in real-time or stored for later analysis. The format of the acquired data can be in either simple text or in a binary form.

4.2.4. Motion/Rotation Module

This module automatically initializes the position and direction of the source/receiver transducers, which means the sliding unite moves to the zero position and the steering engines return to the direction of zero degrees. Users can configure the stepping distance of the sliding unit and the rotation angle of the steering engine by entering specific parameters on the interface control panel. The available stepping distance is 0.0 to 3000.0 mm and the range of the rotating angle of the steering engine is 0 to 360 degrees.

4.3. System Workflow

Through the control panel with a graphic interface, users can manage the operation of the mechanical and electric hardware; complete data acquisition; and perform calculations, analysis, display, and storage of the measured data.

The first step is to perform a system alignment through the control panel operation. Based on the specific need of a given measurement, users may send a specific command to the microcontroller to initialize the system. The stepping motors and steering engines are adjusted on the sliding trails to achieve a specific setting in the vertical and horizontal directions, as well as a set angle relative to the horizontal plane. This operation will make the source/receiver transducers move to a desired position and rotate to a set direction for the acoustic measurement.

Through the control panel, users may send a command to the electric-signal generator to send an electric source signal to the power amplifier, e.g., a gated sinusoidal wave, a square wave, etc. The amplified electric signal excites the source transducer to emit an acoustic signal outward. The acoustic signal then propagates to the receiver through a specific medium or a measured object. Finally, by going through the receiver transducer, the acoustic signal can be converted to an electric signal. The final output is acquired by a high-resolution, high-sampling-rate digitizer and sent back to the computer for display, calculation, analysis, and storage.

The two specially fabricated piezoelectric PZT-5H thin-cylindrical transducers were set to be polarized in the radial direction. They had the same geometrical structure, size, and inner radius. The radius, thickness, and height of the transducers were 20.5 mm, 3 mm, and 50 mm, respectively.

The center frequencies of the thin cylindrical transducers measured using an impedance analyzer PV90A (Bander Electronics Co., Ltd., Beijing, 244021, China) were 20.7520 kHz and 19.6780 kHz with a mechanical load, as shown in Figure 4, as noted by the two five-pointed stars, while the calculated center frequency of the source transducer with a mechanical load was at 20.5180 kHz (see Figure 5). The center frequencies of these two cylindrical transducers were slightly different and close to the center frequency found from the theoretical calculations. We contributed the center frequency difference between the two cylindrical transducers to manufacturing imperfection.

We assembled the cylindrical transducers in a silencing tank filled with water, one as a source transducer and the other as a receiver transducer. The distance between the transducers was set to be 1.0 m. A gated sinusoidal electric signal was generated by a command from the control panel to the waveform generator. Powered by an amplifier, the sinusoidal electric signal provided excitation to the source transducer to emit an acoustic signal. The measured acoustic signal in the time and frequency domains are shown as the dotted lines in Figure 10, i.e., the electric signal at the electric terminals of the receiver transducer. The center frequency of the signal at the electrical terminals of the receiver transducer was analyzed and compared between the theoretical calculation and experimental measurement, which yielded 20.5000 kHz and 19.4440 kHz, respectively. These values are somewhat different from those of the source transducer. This small discrepancy was likely the result of the acoustic–electric filtering effect on acoustic signal arriving at the receiver transducer.

As shown in Figure 10, the shapes of the waveform and the amplitude spectrum of the measured acoustic signal had a noticeable discrepancy from those of the theoretical calculations, even though there was good qualitative agreements in form and shape. Another likely culprit was that only half of the surrounding area of the transducer was receiving the acoustic signal from the source transducer in a nonuniform manner (see Figure 13).

5. Final Remarks

Through modeling, calculation, and experimental measurements, we achieved an enhanced understanding of the transient responses of the electric–acoustic and acoustic–electric conversions based on the piezoelectric cylindrical transducers, which are widely used in petroleum acoustic logging.

For cylindrical transducers, the process of electric–acoustic conversion is the reciprocal process of acoustic–electric conversion. In the processes of these conversions, there are three possible mathematical solutions corresponding to three different physical states, i.e., overdamping, critical-damping, and the oscillatory mode. Practically, the oscillatory mode is the only physically meaningful state for the piezoelectric cylindrical transducers.

For signal conversions, the driving signals applied to stimulate the transducers contained multiple frequency components. This was the reality for both the driving electric signal to the transducer's electric terminals and the acoustic signal to its mechanical terminals. Meanwhile, each frequency component was processed independently and the overall signal transmission was achieved as a superposition of all the individual processes. To achieve this goal, we constructed an equivalent circuits network, where each circuit processed an individual frequency component.

In constructing the circuits for the cylindrical transducers, we noted that radiation resistance and radiation mass were functions of the frequency. One of the simplest processes was to establish a set of equivalent circuits under the harmonic vibrational motions. The acoustic-measurement process could be viewed as a transmission system network with multiple parallel-connected equivalent circuits.

To achieve an enhanced understanding of the electric–acoustic and acoustic–electric conversions, we need to explore many important conversion properties, e.g., the impulse response, amplitude spectrum, center frequency, and more. These properties are influenced not only by the physical and geometrical parameters of the transducer, and the physical parameters of the transition medium, but also by the property and type of the driving electric signal. Obviously, the measured acoustic signal in both the time and frequency domains are dependent on the electric–acoustic and acoustic–electric conversion property and the properties of the driving electric signal.

The directivity of the thin cylindrical transducers can be significant for acoustic measurement. Unlike the spherical shell transducers, which are polarized in the radial direction, the directivity is a unique property of the thin cylindrical transducers, which provides a none-ignorable influence on the measured acoustic signal. It influences the optimal effective radiation area of the thin cylindrical transducers, which varies with the height of the cylindrical transducers.

In summary, for a transient response and relevant physical properties of the piezoelectric, thin cylindrical transducers used in petroleum acoustic logging tools, we established a new transient response model and developed a corresponding measurement system to perform an experimental check on the theoretical calculations. The newly developed experimental system had a high measurement resolution and a high sampling rate, which was able to achieve a considerable measurement precision. It should be able to be applied for both scientific research and for industrial applications.

Supplementary Materials: The following are available online at http://www.mdpi.com/2072-666X/10/12/804/s1, Figure S1: Hardware of acoustic measurement system, Figure S2: Softwave interface of acoustic measurement system.

Author Contributions: L.F. and M.Z. designed the project, performed theoretical derivation, wrote proofread the paper. H.Q. performed the measurement of the center frequency of the cylindrical transducers. N.T., Y.W., and K.S. programmed the software management and calculations. Y.Z., M.L., and X.F. performed experimental measurement and analysis.

Funding: This work was supported in part by the National Natural Science Foundation of China (grant no. 41974130) and by the Physical Sciences Division at The University of Chicago.

Conflicts of Interest: The authors declare no conflict of interest.

References

1. Gubbi, J.; Buyya, R.; Palaniswami, M.S. Internet of Things (IoT): A vision, architectural elements, and future directions. *Future Gener. Comput. Syst.* **2013**, *29*, 1645–1660. [CrossRef]
2. Tadigadapa, S.; Mateti, K. Piezoelectric MEMS sensors: State-of-the-art and perspectives. *Meas. Sci. Technol.* **2009**, *20*, 092001. [CrossRef]
3. Fa, L.; Castagna, J.P.; Suarez-Rivera, R.; Sun, P. An acoustic-logging transmission-network model (continued): Addition and multiplication ALTNs. *J. Acoust. Soc. Am.* **2003**, *113*, 2698–2703. [CrossRef]
4. Fa, L.; Zeng, Z.; Liu, H. A new device for measuring in-situ stresses by using acoustic emissions in rocks. In Proceedings of the 44th US Rock MechanicsSymposium and 5th U.S.-Canada Rock Mechanics Symposium, Salt Lake City, UT, USA, 27–30 June 2010.
5. Ranjith, P.G.; Jasinge, D.; Song, J.Y.; Choi, S.K. A study of the effect of displacement rate and moisture content on the mechanical properties of concrete: Use of acoustic emission. *Mech. Mater.* **2008**, *40*, 453 469. [CrossRef]
6. Scruby, C.B. An introduction to acoustic emission. *J. Phys. E Sci. Instrum.* **1987**, *20*, 945–953. [CrossRef]
7. Dausch, D.E.; Gilchrist, K.H.; Carlson, J.B.; Hall, S.D.; Castellucci, J.B.; von Ramm, O.T. In vivo real-time 3-D intracardiac echo using PMUT arrays. *IEEE Trans. Ultrason. Ferroelectr. Freq. Control* **2014**, *61*, 1754–1764. [CrossRef] [PubMed]
8. Goh, A.S.; Kohn, J.C.; Rootman, D.B.; Lin, J.L.; Goldberg, R.A. Hyaluronic acid gel distribution pattern in periocular area with high-resolution ultrasound imaging. *Aesthet. Surg. J.* **2014**, *34*, 510–515. [CrossRef]
9. Lu, Y.; Tang, H.; Fung, S.; Wang, Q.; Tsai, J.; Daneman, M.; Boser, B.; Horsley, D. Ultrasonic fingerprint sensor using a piezoelectric micromachined ultrasonic transducer array integrated with complementary metal-oxide semiconductor electronics. *Appl. Phys. Lett.* **2015**, *106*, 263503. [CrossRef]
10. He, Q.; Liu, J.; Yang, B.; Wang, X.; Chen, X.; Yang, C. MEMS-based ultrasonic transducer as the receiver for wireless power supply of the implantable microdevices. *Sens. Actuators A Phys.* **2014**, *219*, 65–72. [CrossRef]
11. Przybyla, R.J.; Tang, H.Y.; Guedes, A.; Shelton, S.E.; Horsley, D.A.; Boser, B.E. 3D ultrasonic rangefinder on a chip. *IEEE J. Solid State Circuits* **2015**, *50*, 320–334. [CrossRef]

12. Breazeale, M.A.; Adler, L.; Scott, G.W. Interaction of ultrasonic waves incident at the Rayleigh angle onto a liquid-solid interface. *J. Appl. Phys.* **1977**, *48*, 530–537. [CrossRef]

13. Atlar, A.; Quate, C.F.; Wickramasinghe, H.K. Phase imaging in reflection with the acoustic microscope. *Appl. Phys. Lett.* **1977**, *31*, 791–793. [CrossRef]

14. Briers, R.; Leroy, O.; Shkerdin, G. Bounded beam interaction with thin inclusions. Characterization by phase differences at Rayleigh angle incidence. *J. Acoust. Soc. Am.* **2000**, *108*, 1622–1630. [CrossRef]

15. Fa, L.; Xue, L.; Fa, Y.X.; Han, Y.L.; Zhang, Y.D.; Cheng, H.S.; Ding, P.F.; Li, G.H.; Bai, C.L.; Xi, B.J.; et al. Acoustic Goos-Hänchen effect. *Sci. China Phys. Mech. Astron.* **2017**, *60*, 104311. [CrossRef]

16. Fa, L.; Zhao, J.; Han, Y.L.; Li, G.H.; Ding, P.F.; Zhao, M.S. The influence of rock anisotropy on elliptical-polarization state of inhomogeneously refracted P-wave. *Sci. China Phys. Mech. Astron.* **2016**, *59*, 644301. [CrossRef]

17. Fa, L.; Li, W.Y.; Zhao, J.; Han, Y.L.; Liang, M.; Ding, P.F.; Zhao, M.S. Polarization state of an inhomogeneously refracted compressional-wave induced atinterface between two anisotropic rocks. *J. Acoust. Soc. Am.* **2017**, *141*, 1–6. [CrossRef]

18. Mustapa, M.A.; Yaakob, O.B.; Ahmed, Y.M.; Rheem, C.K.; Koh, K.K.; Adnan, F.A. Wave energy device and breakwater integration: A review. *Renew. Sustain. Energy Rev.* **2017**, *77*, 43. [CrossRef]

19. Arnau, A. *Piezoelectric Transducers and Applications*; Springer: Berlin, Germany, 2008.

20. Wu, J.; Fedder, G.K.; Carley, L.R. A low-noise low-offset capacitive sensing amplifier for a 50-μg/$\sqrt{\text{Hz}}$ monolithic CMOS MEMS accelerometer. *IEEE J. Solid State Circuits* **2004**, *39*, 722–730.

21. Xie, J.; Lee, C.; Feng, H. Design, fabrication, and characterization of CMOS MEMS-based thermoelectric power generators. *J. Micromech. Syst.* **2010**, *19*, 317–324. [CrossRef]

22. Bordoni, P.G.; Gross, W. Sound radiation from a finite cylinder. *J. Math. Phys.* **1949**, *27*, 241–252. [CrossRef]

23. Williams, W.; Parke, N.G.; Moran, D.A.; Sherman, C.H. Acoustic Radiation from a Finite Cylinder. *J. Acoust. Soc. Am.* **1964**, *36*, 2316–2322. [CrossRef]

24. Fenlon, F.H. Calculation of the acoustic radiation field at the surface of a finite cylinder by the method of weighted residuals. *Proc. IEEE* **1969**, *57*, 291–306. [CrossRef]

25. Wu, X.F. Application of a variational principle to acoustic radiation from a vibrating finite cylinder. *J. Acoust. Soc. Am.* **1986**, *79*, 835–836. [CrossRef]

26. Wang, C.; Lai, J.C.S. The sound radiation efficiency of finite length acoustically thick circular cylindrical shells under mechanical excitation I: Theoretical analysis. *J. Sound Vib.* **2000**, *232*, 431–447. [CrossRef]

27. Wang, C.; Lai, J.C.S. The sound radiation efficiency of finite length acoustically thick circular cylindrical shells under mechanical excitation II: Limitations of the infinite length model. *J. Sound Vib.* **2001**, *241*, 825–838. [CrossRef]

28. Tsang, L.; Rader, D. Numerical evaluation of the transient acoustic waveform due to a point source in a fluid-filled borehole. *Geophysics* **1979**, *44*, 1706–1720. [CrossRef]

29. Fa, L.H.; Wang, H.Y.; Ma, H.F. Acoustic array transmitting, receiving sondes and processing method of received full waves. In *Transactions of International Conference on Earth Physics Logging*; Academic Publishing House: Beijing, China, 1990.

30. Ricker, N. Wavelet contraction, wavelet expansion, and the control of seismic resolution. *Geophysics* **1953**, *18*, 769–792. [CrossRef]

31. Gibson, R.L.; Peng, C. Low-and high-frequency radiation from seismic sources in cased boreholes. *Geophysics* **1994**, *59*, 1780–1785. [CrossRef]

32. Piquette, J.C. Method for transducer suppression. I: Theory. *J. Acoust. Soc. Am.* **1992**, *92*, 1203–1213. [CrossRef]

33. Piquette, J.C. Method for transducer suppression. II: Experiment. *J. Acoust. Soc. Am.* **1992**, *92*, 1214–1221. [CrossRef]

34. Fa, L.; Mou, J.P.; Fa, Y.X.; Zhou, X.; Zhang, Y.D.; Liang, M.; Wang, M.M.; Zhang, Q.; Ding, P.F.; Feng, W.T.; et al. On transient response of piezoelectric transducers. *Front. Phys.* **2018**, *6*, 123. [CrossRef]

35. Fa, L.; Zhou, X.; Fa, Y.X.; Zhang, Y.D.; Mou, J.P.; Liang, M.; Wang, M.M.; Zhang, Q.; Ding, P.F.; Feng, W.T.; et al. An innovative model for the transient response of a spherical thin-shell transducer and an experimental confirmation. *Sci. China Phys. Mech. Astron.* **2018**, *61*, 104311. [CrossRef]

36. Fa, L.; Zhao, M. Recent development of an acoustic measurement system. In *Understanding Plane Waves*; Nova Science Publishers: Hauppauge, NY, USA, 2019.

37. Fa, L.; Zhao, M. Recent progress in acoustical theory and applications. In *Understanding Plane Waves*; Nova Science Publishers: Hauppauge, NY, USA, 2019.

 *micromachines*

Article

Simulation of an Adaptive Fluid-Membrane Piezoelectric Lens

Hitesh Gowda Bettaswamy Gowda and Ulrike Wallrabe *

Laboratory for Microactuators, IMTEK - Department of Microsystems Engineering, University of Freiburg, Georges-Koehler-Allee 102, Room 02-081, 79110 Freiburg, Germany; hitesh.gowda@imtek.uni-freiburg.de
* Correspondence: wallrabe@imtek.uni-freiburg.de; Tel.: +49-(0)-761 203-7580

Received: 16 October 2019; Accepted: 16 November 2019; Published: 20 November 2019

Abstract: In this paper, we present a finite-element simulation of an adaptive piezoelectric fluid-membrane lens for which we modelled the fluid-structure interaction and resulting membrane deformation in COMSOL Multiphysics®. Our model shows the explicit coupling of the piezoelectric physics with the fluid dynamics physics to simulate the interaction between the piezoelectric and the fluid forces that contribute to the deformation of a flexible membrane in the adaptive lens. Furthermore, the simulation model is extended to describe the membrane deformation by additional fluid forces from the fluid thermal expansion. Subsequently, the simulation model is used to study the refractive power of the adaptive lens as a function of internal fluid pressure and analyze the effect of the fluid thermal expansion on the refractive power. Finally, the simulation results of the refractive power are compared to the experimental results at different actuation levels and temperatures validating the coupled COMSOL model very well. This is explicitly proven by explaining an observed positive drift of the refractive power at higher temperatures.

Keywords: adaptive lens; piezoelectric devices; fluid-structure interaction; moving mesh; thermal expansion; COMSOL

---

1. Introduction

The expression adaptive optics was initially termed for the technology used in telescopes to deform the mirrors for phase correction of the incoming light [1]. Soon, adaptive optics was implemented in microscopes [2], optical communication systems [3], and optical imaging systems [4]. In conventional imaging systems, the lenses are mechanically moved to focus an image, whereas, with the adaptive optics lens, the surface curvature of the lens is changed to focus an image. The tunable focus lens, also known as the adaptive lens, uses different actuation principles to change the curvature of a deformable surface, thereby changing the focus (refractive power) of the lens. One such adaptive lens using a piezoelectric actuation principle to deform a fluid-membrane interface [5] was developed in the Laboratory for Microactuators, IMTEK - Department of Microsystems Engineering, University of Freiburg, Germany.

The developed adaptive lens consists of a piezoelectric actuator, a fluid chamber, and a transparent flexible membrane, as shown in Figure 1a. The flexible membrane bounds the fluid chamber on one side, and hence any change in the fluid chamber pressure will deform the membrane. An electric field applied on the piezoelectric actuator will deform the fluid chamber and change the fluid chamber pressure. By changing the electric field direction and magnitude, the fluid pressure can be varied to positive or negative pressures resulting in a varied refractive power. The refractive power defined as a function of the applied electric field exhibits piezoelectric hysteresis [6]. At higher temperatures, the fluid expansion will also contribute to the membrane deformation [7]. The piezoelectric hysteresis and thermal expansion contribute to a non-linear response of the adaptive lens. As the membrane deformation is a direct result of the fluid pressure change, the non-linear effects can be addressed by defining the refractive power as a function of the fluid pressure.

To address the non-linear response and hence to compensate for the hysteresis and the temperature effect on the refractive power, it is essential to determine the combined influence of piezoelectric actuation and temperature on the membrane deformation. Hence in this paper, we present a finite-element simulation of the adaptive lens modelled in COMSOL Multiphysics® (5.3a, COMSOL Inc, Burlington, MA, USA) to define the refractive power linearly as a function of both the fluid pressure and temperature.

COMSOL Multiphysics® is based on the finite-element method (FEM), which solves engineering problems such as structural mechanics, fluid dynamics, heat transfer by a numerical approach. In FEM, the complex geometry is divided into simpler domains. These domains are defined with the elementary partial differential equations based on the physics. Then the elementary equations are combined to form a system of global equations, which represent the complex geometry [8]. The system of global equations can be solved using FEM-based simulation software such as ANSYS, ABAQUS, ATILA, and COMSOL [9]. To simulate complex geometry with multiple physics domains, COMSOL Multiphysics® offers a methodological environment to access elementary equations and then couple them with a wide range of available physics modules [10]. Using COMSOL®, authors in [11–13] simulated adaptive lenses using only the piezoelectric physics module, authors in [14,15] simulated micro-pumps using the fluid-structure interaction physics module and authors in [16,17] simulated thermal actuators using the heat transfer physics module. However, the articles [11–17] did not simulate any kind of solid deformation produced by the fluid forces from both the piezoelectric actuation and the thermal expansion. Furthermore, COMSOL® does not provide a direct feature to couple the piezoelectric with the fluid-structure and heat transfer physics modules. Hence, in this paper, we present the explicit coupling of multiple physics modules to simulate the membrane deformation due to the fluid forces from both the piezoelectric actuation and the thermal expansion.

We describe the physical design and working principle of the adaptive lens in Section 2. In Section 3, we describe the simulation model of the adaptive lens and describe the explicit coupling of multiple physics modules using a moving mesh physics module. In Section 4, we present the simulation results of the adaptive lens and compare the simulation results with the experimental results in Section 5. We conclude our paper with results in Section 6.

2. The Fluid-Membrane Piezoelectric Lens

The piezoelectric bi-morph actuator has a circular ring-shaped design with the two piezoelectric ceramic layers glued together in an anti-parallel polarization configuration. The fluid chamber and the flexible membrane are integrated with the actuator using micro-molding techniques to form an active lens chamber [5]. The active lens chamber is glued onto a PCB-based substrate [7] and primed with an optical oil [7] (Figure 1a).

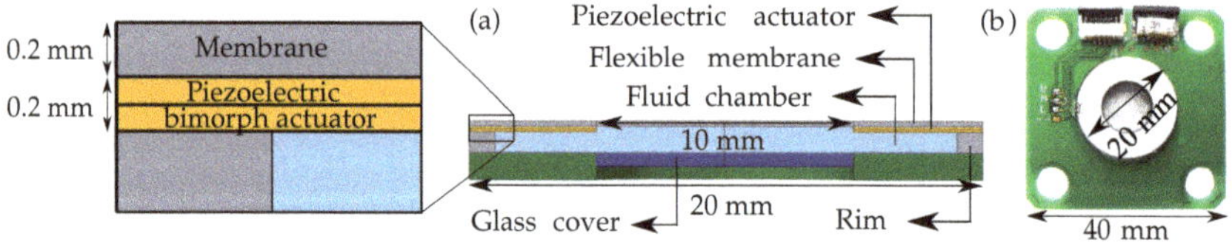

Figure 1. (**a**) 2D cut section of the adaptive lens to show the fluid chamber, flexible membrane, and the integrated actuator. (**b**) The adaptive fluid-membrane piezoelectric lens.

The manufactured adaptive lens with the actuator diameter of 20 mm with an aperture of 10 mm is shown in Figure 1b. The adaptive lens has an overall thickness of around 2.2 mm, with substrate thickness of 1 mm, rim thickness of 0.8 mm, membrane thickness of 0.2 mm and bi-morph actuator thickness of 0.2 mm. Depending on the applied electric field/voltage direction, the actuator deforms the fluid chamber to produce positive or negative fluid chamber pressure. The positive pressure leads to a plano-convex lens (Figure 2a), and the negative pressure leads to a plano-concave lens (Figure 2b).

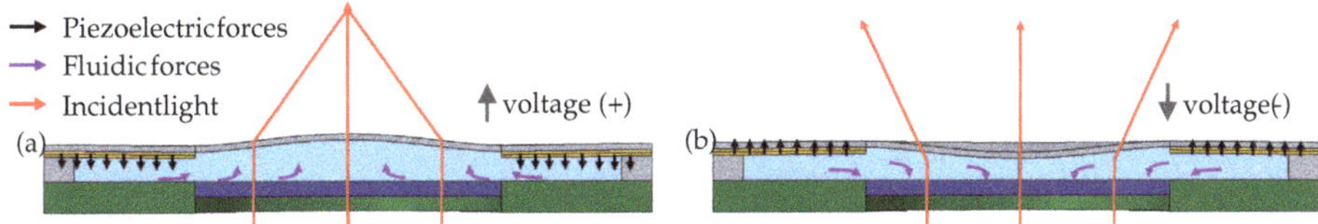

Figure 2. The 2D cross-section of the adaptive lens showing the piezoelectric forces and fluidic forces, which form either (**a**) plano-convex lens or (**b**) plano-concave lens depending on the applied voltage direction.

3. Multiphysics Simulation

To reduce the complexity of a $3D$ geometry and at the same time to decrease the computation time, a $2D$-axisymmetric space dimension is chosen to model the adaptive lens (Figure 3) with the radial axis 'r' and the deformation axis 'z'. The adaptive lens components include a piezoelectric actuator, fluid, membrane, rim, and substrate. The materials for the components are chosen from the COMSOL Multiphysics® inbuilt material library [18]. The material parameters are changed to the equivalent parameters of the materials, which are used in the manufacture of the adaptive lens. The adaptive lens components, along with the modified material parameters used in the simulation, are mentioned in Table 1 and the adaptive lens components thicknesses are mentioned in Table 2. The following section describes the physics modules used in the simulation.

Table 1. The materials of the adaptive lens components used in the simulation model.

Component	Material	Density (kg m^{-3})	Young's Modulus (Pa)	Thermal Expansion Coefficient (K^{-1})
Piezoelectric actuator	Piezo PZT-5H [19]	7500	37×10^9	1×10^{-5}
Flexible membrane, rim	Polydimethylsiloxane [20]	1020	2×10^6	3×10^{-4}
Fluid	Fomblin Y [21,22]	1880	-	2.1×10^{-4}
Substrate	Glass (quartz)	2210	50×10^9	4×10^{-5}

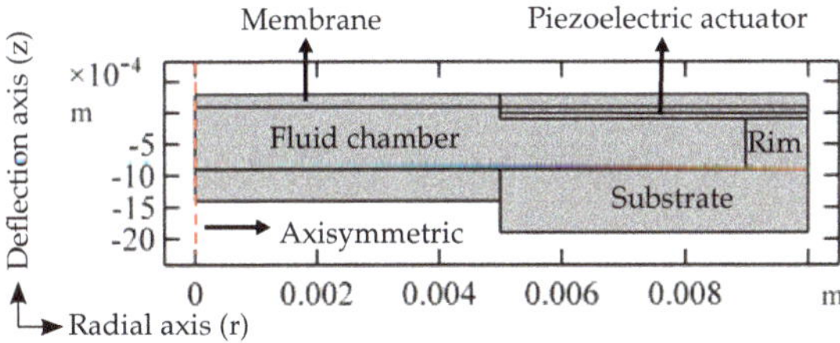

Figure 3. The 2D axisymmetric simulation model of the adaptive lens.

Table 2. The thickness of the adaptive lens components.

Component	Thickness (mm)
Membrane	0.2
Piezoelectric actuator	0.2
Rim	0.8
Substrate	1

3.1. Piezoelectric Devices

The adaptive lens uses the inverse piezoelectric property of the actuator to vary the refractive power. To model the inverse piezoelectric effect, the piezoelectric devices module is used. The module couples the solid mechanics Equation (1) and the electrostatics Equation (2) physics to combine the electrical behavior and the mechanical behavior of the piezoelectric ceramics.

$$\rho \frac{\partial^2 x}{\partial^2 t} = \nabla \cdot s + F_v \tag{1}$$

where ρ is the density, x is the displacement, s is the stress, and F_v is the volume force.

$$E = -\nabla \cdot V \tag{2}$$

where E is the electric field, and V is the electric potential.

The combined behavior is modelled through the coupled Equations (3) and (4) in strain-charge form.

$$S = S_E T + d^T E \tag{3}$$

$$D = dT + \epsilon_0 \epsilon_{rT} E \tag{4}$$

where the solid mechanics parameters are strain S and stress T, the electrostatic parameters are electric field E and electric displacement field D, and the piezoelectric material parameters are compliance coefficient S_E, piezoelectric coefficient d^T, and permittivity ϵ [23]. In the simulation model, the piezoelectric coefficients and compliance coefficients are obtained from the piezo PZT-5H inbuilt material library [18].

3.2. Fluid-Structure Interaction

The piezoelectric actuator in the adaptive lens deforms the fluid chamber and varies the internal fluid pressure. The varied internal fluid pressure results in fluid forces, which act on the flexible membrane. The fluid forces contribute to the deformation of the flexible membrane to an aspherical surface. The fluid-structure interaction (FSI) physics module models the fluid forces acting on the membrane by coupling the solid mechanics Equation (1) and the laminar flow Equation (5) physics modules.

$$\rho \frac{\partial u}{\partial t} + \rho(u \cdot \nabla)u = \mu \nabla^2 u + F + \rho g \tag{5}$$

The Navier–Stokes Equation (5) models the motion of incompressible fluids, where ρ is the fluid density, u is the fluid velocity, F is the external force and g is the gravity. The FSI Multiphysics module couples the fluid inertial forces in Equation (5) with the external forces in Equation (1) [24].

3.3. Heat Transfer in Solids and Fluids

Apart from the piezoelectric actuation that contributes to the deformation of the membrane, the thermal expansion of the fluid at higher temperatures will as well cause the membrane deformation. Hence to model the fluid thermal expansion, heat transfer in solids (Equation (6)) and heat transfer in fluids (Equation (7)) physics modules are used.

$$Q = \alpha T \cdot \frac{dS}{dt} \tag{6}$$

$$Q = \alpha T \left(\frac{\partial \rho}{\partial t} + u \cdot \nabla p \right) \tag{7}$$

where Q is the heat source, T is the temperature, S is the solid stress tensor, α is the coefficient of thermal expansion, p is the fluid pressure, and u is the fluid velocity.

Equations (6) and (7) define the heat source Q that contributes to set the complete adaptive lens domain to the required temperature T. Equation (7) models the thermal expansion that contributes to the fluid pressure p, which acts on the flexible membrane [25].

3.4. Moving Mesh

The adaptive lens working mechanism relies on the transfer of piezoelectric forces to the flexible membrane through the fluid forces. In COMSOL Multiphysics®, the coupling of the piezoelectric effect and the fluid-structure interaction is not possible through a direct Multiphysics feature. Hence, the moving mesh physics module is used to couple the piezoelectric forces with the fluid forces and apply the resultant on the flexible membrane. The explicit coupling of piezoelectric and laminar flow physics is performed in a way such that the solid domain velocities Equations (8) and (9) generated by the deformation of the piezoelectric actuator are applied as the mesh velocities on the walls of the fluid chamber as shown in Figure 4c. The geometric domains with the free deformation mesh and with the fixed mesh are as shown in Figure 4a,b, respectively.

$$V_r = solid \cdot u\_tR \tag{8}$$

$$V_z = solid \cdot u\_tZ \tag{9}$$

Equations (8) and (9) equate the solid velocities u_tR and u_tZ, which are produced by the piezoelectric actuator in radial axis (R) and deformation axis (Z), to the corresponding mesh velocities V_r and V_z. The mesh velocities applied to the fluid chamber wall are implicitly considered to be the external forces in the Navier–Stokes Equation (5) of the fluid-structure interaction physics module Equation (4). In this way, the moving mesh module explicitly couples the piezoelectric actuator force to the fluid force.

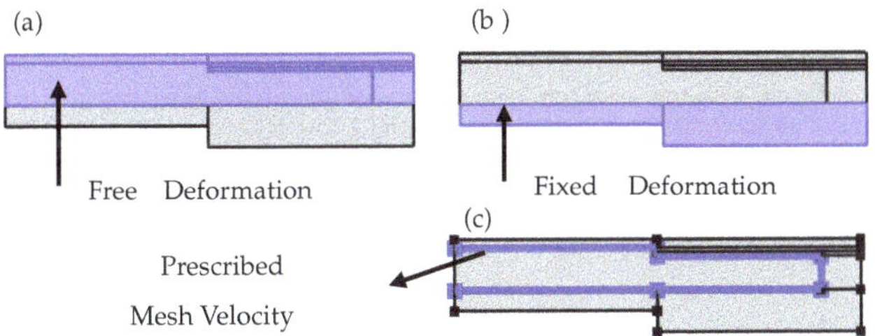

Figure 4. The domains specified in the Moving Mesh module to be (**a**) free mesh and (**b**) fixed mesh. (**c**) The solid domain velocity applied on the walls of the fluid chamber.

3.5. Boundary Condition

In the solid mechanics interface, the glass substrate is selected in the fixed constraint condition, as shown in Figure 5a. The piezo ceramic layers in the bi-morph actuator have an anti-parallel polarization configuration. To adapt for the polarization direction in the simulation, an additional base vector system is created with the base vector 'x3' set to -1 instead of default 1. The default base vector system is selected as the coordinate system for the piezo 1, and the additional base vector system is selected as the coordinate system for the piezo 2 (Figure 5b). In the heat transfer interface, all the boundaries are selected in the temperature constraint, to set the entire domain to the required temperature (Figure 5c). In the laminar flow interface, the walls of the fluid chamber are selected with a no-slip boundary condition (Figure 5d). Since the fluid chamber is a closed domain, there is no inlet, outlet, or open boundary conditions.

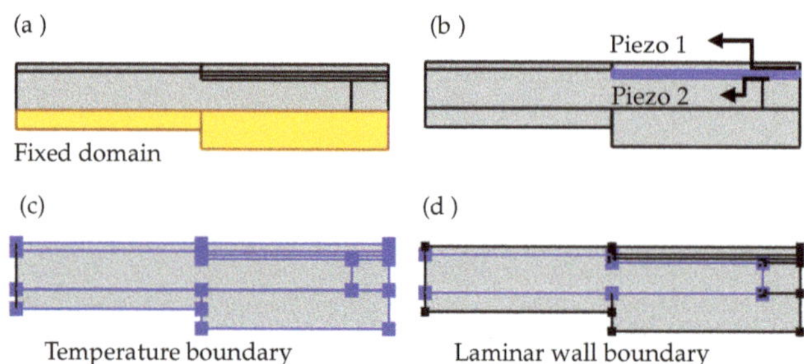

Figure 5. Domains with fixed boundary condition in solid mechanics physics module (**a**), piezo 1 with default base vector system and piezo 2 with modified base vector system (**b**), domains with temperature boundary condition in heat transfer in solids and fluids physics modules (**c**), and domains with wall boundary condition in laminar flow physics module (**d**).

3.6. Mesh

The explicit coupling of the piezoelectric forces with the fluidic forces is achieved through the mesh velocities applied on the walls of the fluid chamber (Figure 4c). Hence the walls of the fluid chamber are finely meshed with the use of corner refinement and boundary layer mesh features (Figure 6). The remaining domains including fluid, piezoelectric actuator, membrane, and the substrate are meshed with the inbuilt physics-controlled mesh.

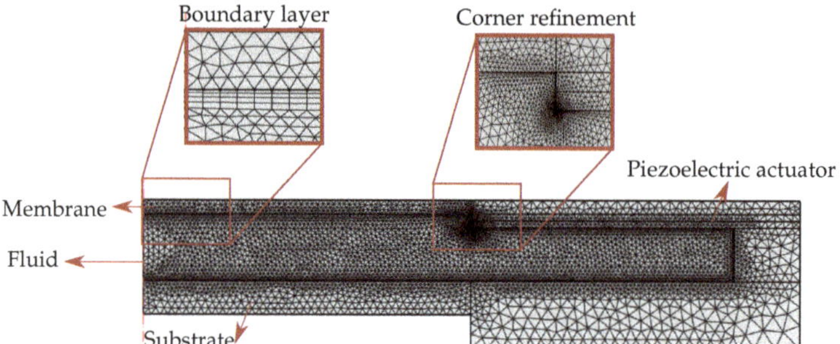

Figure 6. The adaptive lens meshed finely with corner refinement and boundary layer features.

4. Results

The time-dependent study was selected in the simulation as the velocities, which are used in the moving mesh module, are time-dependent values, and as the stationary study does not compute these instantaneous velocities. In the study, the time range was selected from 0 s to 1 s with a step of 0.1 s. The applied voltages were limited to 140 V in positive polarization direction, and −40 V in negative polarization direction to avoid piezo saturation and re-polarization [19]. A sinusoidal voltage within the voltage limits was defined using a piecewise function under global definitions. Figure 7 shows the line plot of voltages applied to the piezoelectric actuator.

Figure 7a shows the top piezo set to positive voltage limit of 140 V, and the bottom piezo set to negative voltage limit of −40 V, the corresponding voltage combination results in a plano-convex lens. In contrast, the voltage combination is reversed in Figure 7b, i.e., the top piezo is set to −40 V and the bottom piezo is set to 140 V, resulting in a plano-concave lens.

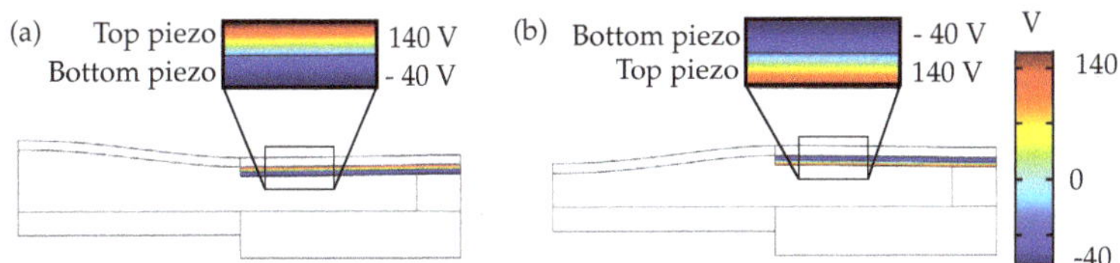

Figure 7. Applied voltages on the piezoelectric actuator with (**a**) convex lens mode and (**b**) concave lens mode.

The volume plots in Figure 8a,b generated by the 2D revolution around the symmetric axis show the adaptive lens in plano-convex and plano-concave lens modes. The revolved plot is used to visualize the aspherical deformation of the membrane.

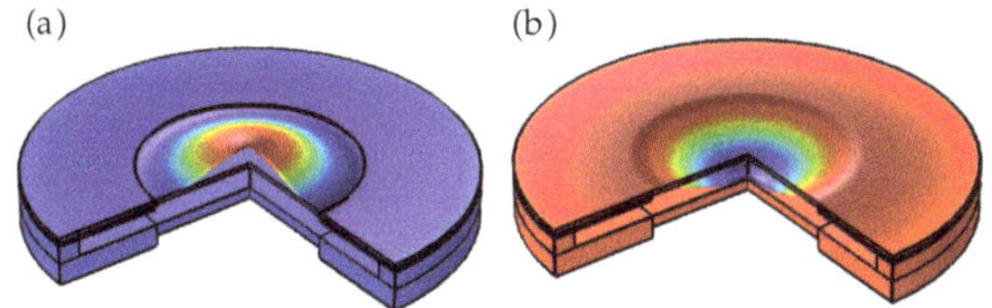

Figure 8. The 2D revolved plots showing (**a**) aspheric convex lens and (**b**) aspheric concave lens mode.

The surface plots in Figure 9a,b show the membrane deformation and the corresponding dynamic internal fluid pressure during actuation. The arrows indicate the fluid velocity direction during actuation. For the plano-convex lens in Figure 9a, in which the piezoelectric actuator is set to a maximum voltage combination results in positive internal pressure of around 270 Pa and a peak deflection of around 300 μm at the center of the membrane. For the plano-concave lens in Figure 9b, in which the voltage combination is reversed compared to the former case, the actuator deforms outwards resulting in negative fluid pressure of around −270 Pa and a peak deflection of around −300 μm at the center of the membrane.

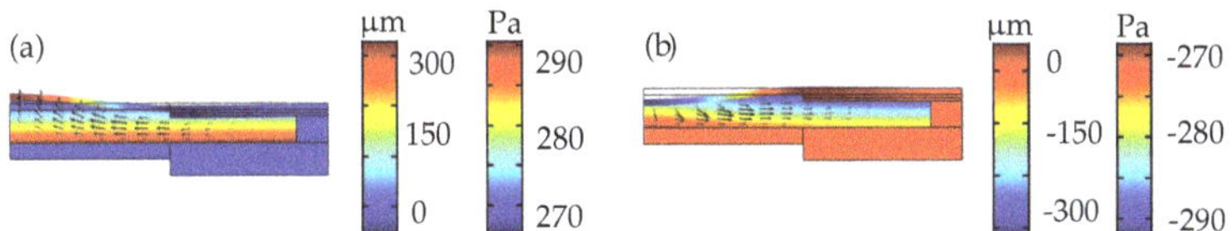

Figure 9. The surface simulation plots of the adaptive lens showing dynamic internal chamber pressure and solid deformation in (**a**) plano-convex and (**b**) plano-concave modes.

The heat transfer interface for solids and fluids was used to set the adaptive lens to a temperature ranging from 20 °C to 80 °C. The surface plot in Figure 10 shows the temperature distribution of the adaptive lens set to 80 °C with the actuator voltage set to 0 V. The expansion of the fluid at 80 °C corresponds to an increase in fluid pressure of around 85 Pa and a peak deflection of around 80 μm at the center of the membrane.

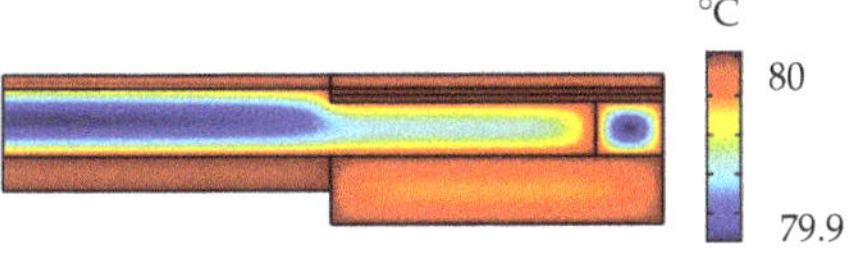

Figure 10. Temperature distribution of the adaptive lens at 80 °C.

The refractive power of the adaptive lens in the simulation (Figure 11) was calculated by double differentiation of the membrane boundary with respect to the deformation component (w) and the radial component (r) Equation (10). Figure 11a shows the refractive power defined as a function of internal fluid pressure. The simulation of the adaptive lens result in a refractive power range of $-16\,\mathrm{m}^{-1}$ to $16\,\mathrm{m}^{-1}$ for an internal fluid pressure range of $-270\,\mathrm{Pa}$ to $270\,\mathrm{Pa}$.

$$Refractive\ Power = \left(\frac{\mathrm{d}^2 w}{\mathrm{d}r^2}\right) \cdot (n-1) \tag{10}$$

where 'w' is the deflection component, 'r' is the radial component and 'n' is the refractive index of the adaptive lens.

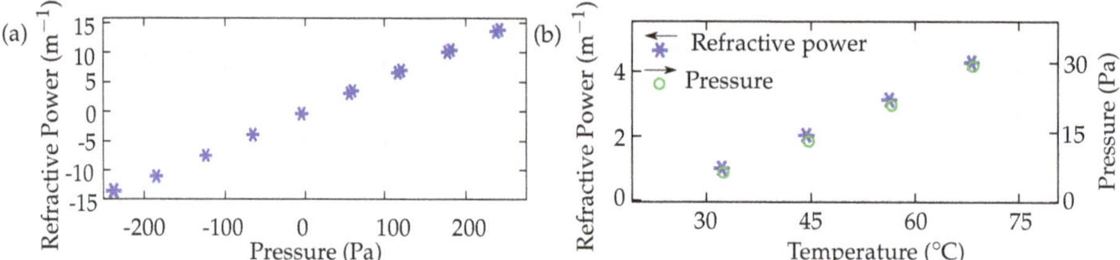

Figure 11. (**a**) The simulated refractive power of the adaptive lens as a function of fluid chamber pressure, and (**b**) the change in refractive power of the adaptive lens due to thermal expansion of the fluid.

5. Experiment

Furthermore, to verify the simulation results, we characterized the adaptive lens at different applied voltages and temperatures [7]. The adaptive lens integrated with a pressure sensor and a temperature sensor, to compensate for the non-linear piezoelectric hysteresis and the fluid thermal expansion effect [7,26], was used in the characterization. Figure 12 shows the block diagram of the experimental setup used to characterize the adaptive lens.

A sinusoidal voltage was applied to the piezoelectric actuator as was assumed in the simulation. A voltage driver was used to limit the negative voltage to $-40\,\mathrm{V}$, and the positive voltage to $140\,\mathrm{V}$. A sensor driver was used to measure the output from the temperature and pressure sensor. The adaptive lens was mounted on a heater, which was used to heat the adaptive lens to the required temperatures. The membrane deformation was measured using a profilometer connected to a confocal displacement sensor providing a resolution of $110\,\mathrm{nm}$. During the characterization, the adaptive lens was actuated and the corresponding applied voltage, membrane deformation, sensor outputs were measured simultaneously. The characterization was repeated with the adaptive lens set to higher temperatures. The refractive power was subsequently calculated from the measured membrane surface and then defined as a function of both the internal fluid pressure and the temperature. The measurements show the refractive power varying from $-16\,\mathrm{m}^{-1}$ to $17\,\mathrm{m}^{-1}$ at $25\,^\circ\mathrm{C}$, and from $-15\,\mathrm{m}^{-1}$ to $28\,\mathrm{m}^{-1}$ at $75\,^\circ\mathrm{C}$.

The simulated and measured refractive power of the adaptive lens at $25\,^\circ\mathrm{C}$, $50\,^\circ\mathrm{C}$ and $75\,^\circ\mathrm{C}$ with different applied voltages are compared in Figure 13. The temperature drift of the refractive power in the positive direction is higher than that in the negative direction because of the superposing effects of the thermal expansion of the fluid, which contribute to positive drift, and the increased actuator deflections at higher temperatures, which contribute to both positive and negative drift [7,19]. Hence, the superposed effect causes a net positive drift.

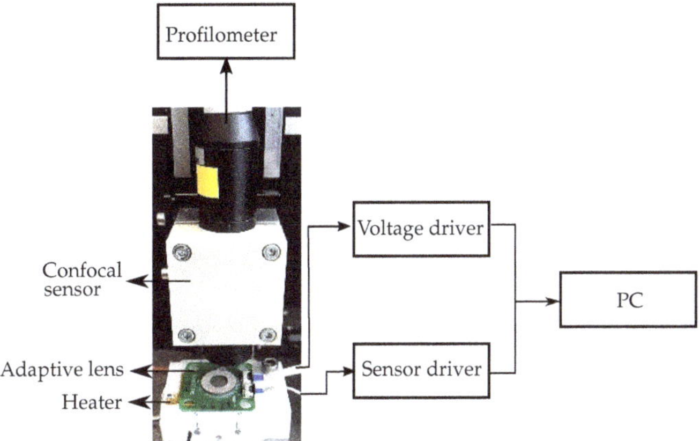

Figure 12. Experimental setup to characterize the adaptive lens at applied voltages and higher temperatures.

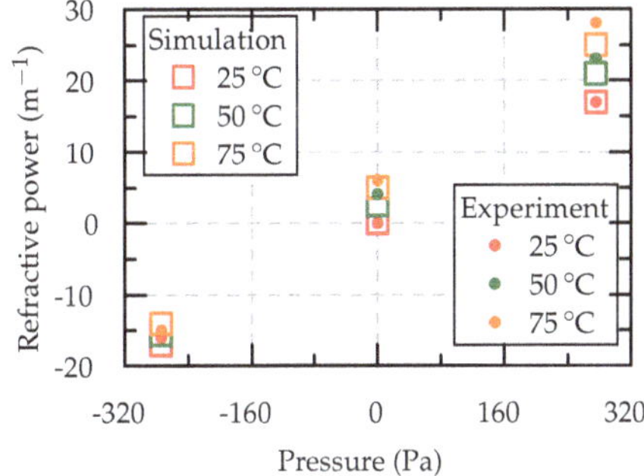

Figure 13. Comparison of the measured and simulated results of refractive power.

6. Conclusions

Our simulation model successfully couples the piezoelectric physics with the laminar flow and heat transfer physics modules. Both prove that the adaptive lens was simulated at different voltages and temperatures to determine the actuator deflection, the fluid pressure, and the refractive power. The simulated results are in close agreement with the experimental results. The adaptive lens can vary the refractive power from $-16\,\mathrm{m}^{-1}$ to $17\,\mathrm{m}^{-1}$ at 25 °C and from $-15\,\mathrm{m}^{-1}$ to $28\,\mathrm{m}^{-1}$ at 75 °C. With this validation, we can now use our model reliably for further geometric optimization of our adaptive lens. Furthermore, the simulation model could be extended to also model the piezoelectric hysteresis and change in piezoelectric coefficients with the temperature.

Author Contributions: Conceptualization, U.W. and H.G.B.G.; methodology, U.W. and H.G.B.G.; software, H.G.B.G.; validation, U.W. and H.G.B.G.; formal analysis, U.W. and H.G.B.G.; investigation, U.W. and H.G.B.G.; resources, U.W. and H.G.B.G.; data curation, H.G.B.G.; writing–original draft preparation, H.G.B.G.; writing–review and editing, U.W.; visualization, H.G.B.G.; supervision, U.W.; project administration, U.W.; funding acquisition, U.W.

Funding: This work was financed by the Baden-Württemberg Stiftung gGmbH under the project VISIR[2] (Variable Intelligent Sensors, Integrated and Robust for Visible and IR light). The article processing charge was funded by the German Research Foundation (DFG) and the University of Freiburg in the funding program Open Access Publishing.

Conflicts of Interest: The authors declare no conflict of interest.

Abbreviations

The following abbreviations are used in this manuscript:

PDMS Polydimethylsiloxane
FSI Fluid-structure interaction
FEM Finite-element method
CNC Computer numerical control
PCB Printed circuit board

References

1. Beckers, J.M. Adaptive Optics for Astronomy: Principles, Performance, and Applications. *Annu. Rev. Astron. Astrophys.* **1993**, *31*, 13–62. [CrossRef]
2. Booth, M.J. Adaptive optics in microscopy. *Philos. Trans. R. Soc. A Math. Eng. Sci.* **2007**, *365*, 2829–2843. [CrossRef] [PubMed]
3. Weyrauch, T.; Vorontsov, M.A.; Gowens, J.; Bifano, T.G. Fiber coupling with adaptive optics for free-space optical communication. In *Free-Space Laser Communication and Laser Imaging*; SPIE: Bellingham, WA, USA, 2002; pp. 177–184. [CrossRef]
4. Schneider, F.; Draheim, J.; Kamberger, R.; Waibel, P.; Wallrabe, U. Optical characterization of adaptive fluidic silicone-membrane lenses. *Opt. Express* **2009**, *17*, 11813. [CrossRef] [PubMed]
5. Draheim, J.; Schneider, F.; Kamberger, R.; Mueller, C.; Wallrabe, U. Fabrication of a fluidic membrane lens system. *J. Micromech. Microeng.* **2009**, *19*, 095013. [CrossRef]
6. Uchino, K. *Advanced Piezoelectric Materials: Science and Technology*; Elsevier: Amsterdam, The Netherlands, 2010; pp. 1–678. [CrossRef]
7. Gowda, H.G.B.; Wallrabe, U. Integrated Sensor System to Control the Temperature Effects and the Hysteresis on Adaptive Fluid-membrane Piezoelectric lenses. In Proceedings of the International Symposium on Optomechatronic Technologies (ISOT 2019), Goa, India, 11–13 November 2019.
8. Norrie, D.H. A first course in the finite element method. *Finite Elem. Anal. Des.* **1987**, *3*, 162–163. [CrossRef]
9. Sahul, R.; Nesvijski, E.; Hackenberger, W. Optimization of piezoelectric transducer design by modeling and simulation. In Proceedings of the 2014 Joint IEEE International Symposium on the Applications of Ferroelectric, International Workshop on Acoustic Transduction Materials and Devices and Workshop on Piezoresponse Force Microscopy, State College, PA, USA, 12–16 May 2014. [CrossRef]
10. The COMSOL® Software Product Suite. Available online: https://www.comsol.com/products (accessed on 30 October 2019).
11. Nguyen, C.H.; Farghaly, M.A.; Akram, M.N.; Hanke, U.; Halvorsen, E. Electrode configurations for layered-plate piezoelectric micro-actuators. *Microelectron. Eng.* **2017**, *174*, 59–63. [CrossRef]
12. Schneider, F.; Müller, C.; Wallrabe, U. A low cost adaptive silicone membrane lens. *J. Opt. Pure Appl. Opt.* **2008**, *10*, 044002. [CrossRef]
13. Lemke, F.; Frey, Y.; Bruno, B.P.; Philipp, K.; Koukourakis, N.; Czarske, J.; Wallrabe, U.; Wapler, M.C. Multiphysics simulation of the aspherical deformation of piezo-glass membrane lenses including hysteresis, fabrication and nonlinear effects. *Smart Mater. Struct.* **2019**, *28*. [CrossRef]
14. Rojas, J.J.; Morales, J.E. Design and Simulation of a Piezoelectric Actuated Valveless Micropump. In Proceedings of the COMSOL Conference 2015, Boston, MA, USA, 7–9 October 2015.
15. Meshkinfam, F.; Rizvi, G. Design and Simulation of MEMS Based Piezoelectric Insulin Micro-Pump. In Proceedings of the COMSOL Conference 2015, Boston, MA, USA, 7–9 October 2015.
16. Crosby, J.V.; Guvench, M.G. Experimentally Matched Finite Element Modeling of Thermally Actuated SOI MEMS Micro-Grippers Using COMSOL Multiphysics. In Proceedings of the COMSOL Conference 2015, Boston, MA, USA, 8–10 October 2015; pp. 1–9.
17. Bansal, V.; Prasad, A.G.B.; Kumar, D. 3-D Design, Electro-Thermal Simulation and Geometrical Optimization of spiral Platinum Micro-heaters for Low Power Gas sensing applications using COMSOL 4.1. In Proceedings of the COMSOL Conference 2011, Boston, MA, USA, 13–15 October 2011.
18. Material Library User's Guide. Available online: www.comsol.com/blogs (accessed on 29 October 2019).

19. Bruno, B.P.; Fahmy, A.R.; Stürmer, M.; Wallrabe, U.; Wapler, M.C. Properties of piezoceramic materials in high electric field actuator applications. *Smart Mater. Struct.* **2019**, *28*, 015029. [CrossRef]

20. Schneider, F.; Fellner, T.; Wilde, J.; Wallrabe, U. Mechanical properties of silicones for MEMS. *J. Micromech. Microeng.* **2008**, *18*. [CrossRef]

21. Fomblin® Y LVAC 06/6, Average mol wt 1800 | Sigma-Aldrich. Available online: https://www.sigmaaldrich.com/catalog/product/aldrich/317926?lang=de{&}region=DE (accessed on 10 October 2019).

22. Lamoreaux, S.K.; Golub, R. Calculation of the ultracold neutron upscattering loss probability in fluid walled storage bottles using experimental measurements of the liquid thermomechanical properties of fomblin. *Phys. Rev. C* **2002**, *66*, 044309. [CrossRef]

23. Ransley, J. Piezoelectric Materials: Understanding the Standards. Available online: https://www.comsol.com/blogs/piezoelectric-materials-understanding-standards/ (accessed on 29 October 2019).

24. What are the Navier-Stokes Equations? Available online: https://www.comsol.de/multiphysics/navier-stokes-equations (accessed on 29 October 2019).

25. Comsol. Heat Transfer Module: User's Guide. Available online: www.comsol.com/blogs (accessed on 29 October 2019).

26. Draheim, J.; Burger, T.; Kamberger, R.; Wallrabe, U. Closed-loop pressure control of an adaptive single chamber membrane lens with integrated actuation. In Proceedings of the 16th International Conference on Optical MEMS and Nanophotonics, Istanbul, Turkey, 8–11 August 2011; pp. 47–48. [CrossRef]

 *micromachines*

 MDPI

Article

The Influence of Piezoelectric Transducer Stimulating Sites on the Performance of Implantable Middle Ear Hearing Devices: A Numerical Analysis

Houguang Liu [1], Yu Zhao [1,*], Jianhua Yang [1] and Zhushi Rao [2]

[1] School of Mechatronic Engineering, China University of Mining and Technology, Xuzhou 221116, China; liuhg@cumt.edu.cn (H.L.); jianhuayang@cumt.edu.cn (J.Y.)

[2] State Key Laboratory of Mechanical System and Vibrations, Shanghai Jiao Tong University, Shanghai 200240, China; zsrao@sjtu.edu.cn

* Correspondence: zhaoy@cumt.edu.cn

Received: 10 October 2019; Accepted: 12 November 2019; Published: 14 November 2019

Abstract: To overcome the inherent deficiencies of hearing aids, implantable middle ear hearing devices (IMEHDs) have emerged as a new treatment for hearing loss. However, clinical results show that the IMEHD performance varies with its transducer's stimulating site. To numerically analyze the influence of the piezoelectric transducer's stimulating sites on its hearing compensation performance, we constructed a human ear finite element model and confirmed its validity. Based on this finite element model, the displacement stimulation, which simulates the piezoelectric transducer's stimulation, was applied to the umbo, the incus long process, the incus body, the stapes, and the round window membrane, respectively. Then, the stimulating site's effect of the piezoelectric transducer was analyzed by comparing the corresponding displacements of the basilar membrane. Besides, the stimulating site's sensitivity to the direction of excitation was also studied. The result of the finite element analysis shows that stimulating the incus body is least efficient for the piezoelectric transducer. Meanwhile, stimulating the round window membrane or the stapes generates a higher basilar membrane displacement than stimulating the eardrum or the incus long process. However, the performance of these two ideal sites' stimulation is sensitive to the changes in the excitation's direction. Thus, the round window membrane and the stapes is the ideal stimulating sites for the piezoelectric transducer regarding the driving efficiency. The direction of the excitation should be guaranteed for these ideal sites.

Keywords: implantable middle ear hearing device; piezoelectric transducer; stimulating site; finite element analysis; hearing compensation

---

1. Introduction

Hearing loss, affecting around 466 million people worldwide, is one of the six leading causes of disease burden in our society [1]. Up to now, there is still no effective medical treatment to sensorineural hearing loss (SNHL), which is the main type of hearing loss taking up approximately 90% of reported hearing loss [2]. The patients with SNHL mainly turn to hearing aids for restoring audibility [3]. Although sophisticated hearing aids have been developed, hearing aids still have a number of inherent shortcomings, such as a limited high-frequency amplification gain, ear canal occlusion, and feedback annoyance [4]. To overcome these problems, many researchers proposed and designed the implantable middle ear hearing devices (IMEHDs), which restores audibility by the mechanical vibration of their implanted transducers [5].

IMEHD primarily comprises four components: the microphone, the sound processor, the transducer, and the battery. A typical schematic illustration of the IMEHD is shown in Figure 1 [6].

Briefly, the microphone, which is implanted closer to the ear canal, receives the outside sound and transmits to the sound processor. Then, the sound processor processes the input signal according to patients' hearing loss and outputs a driving signal to the piezoelectric transducer. The piezoelectric transducer mainly consists of three parts: the piezoelectric stack, the rod, and the support sleeve. One side of the piezoelectric stack is stuck to the rod, which is attached to the incus body. While the other side of the piezoelectric stack is held to the support sleeve, which is fixed to the skull. Under the electrical driving signal's stimulation, the piezoelectric stack, which is a monolithic ceramic construction of many thin piezoelectric ceramic layers, expands and contracts. Finally, the vibration of the piezoelectric stack is transmitted to the incus body by the rod and compensates for hearing loss. All these parts are powered by the battery. Among these IMEHD parts, the transducer is a key component as it is responsible for stimulating the human ear. Based on actuation mechanisms, the IMEHDs' transducers are divided into two types: the electromagnetic transducer and the piezoelectric transducer [5]. Compared with the electromagnetic transducer, the piezoelectric transducer has the advantages of a lower power consumption, compatibility with external magnetic field, and ease of fabrication [4]. Owing to these advantages, piezoelectric transducer have been widely used in IMEHDs, especially the totally implanted type IMEHDs [7]. In terms of the stimulating sites, the transducer can be further classified into five categories: the eardrum driving [8,9], the incus body driving [6], the incus long process driving [10], the stapes driving [11,12], and the round window (RW) membrane driving [13,14].

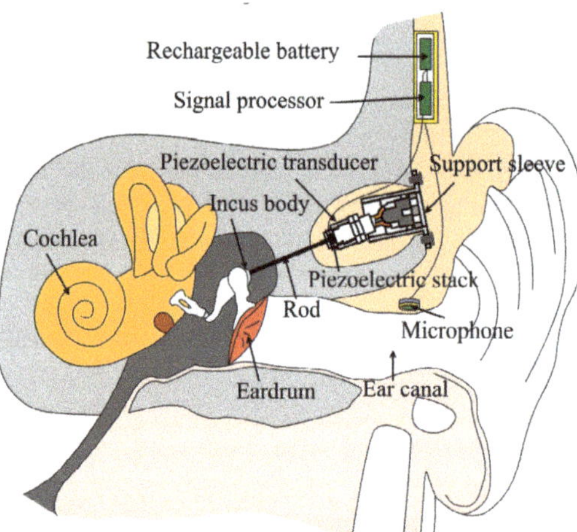

Figure 1. Schematic view of an implantable middle ear hearing device with a piezoelectric transducer attached on the incus body.

Clinical study shows that the stimulating site influence the transducer's hearing compensation performance [15]. To uncover this influence and optimize the transducer's design, some preliminary studies have been carried out. Based on numerical analysis, Zhang et al. found that stimulating the round window membrane is more efficient than stimulating the incus long process [16]. The human temporal bone experiment conducted by Deveze et al. demonstrates that stimulating the stapes is superior to stimulating the incus body and incus long process [17]. However, the above researches only focus on the electromagnetic transducer. A numerical study shows that the stimulating site's influence on the electromagnetic transducer is different from that on the piezoelectric transducer [18]. To investigate the stimulating site's effect on the piezoelectric transducer's performance, Bornitz et al. constructed a human ear finite element (FE) model and compared the stapes displacements under different piezoelectric transducers' stimulation [18]. Their result demonstrates that the incus body is the least effective stimulating site for the piezoelectric transducer. However, auditory response measurements show that the stapes response is unreliable for evaluating round window stimulation [19]. Besides this, stimulating the incus long process, which is widely utilized clinically, was not investigated.

Accordingly, in the present study, we carried out a systematic study on the influence of piezoelectric transducer's stimulating sites. To facilitate this study, we built a human ear FE model and confirmed its validity. Then, the stimulating site's effect was analyzed based on the basilar membrane's displacement, which is reliable for evaluating IMEHD performance. The result could help the surgeon choose a piezoelectric transducer and aid the designer to optimize the piezoelectric transducer.

2. Materials and Methods

2.1. Constructions of the Human Ear FE Model

A 3D FE model of the human ear was built using CT scanning and reverse modelling techniques based on a fresh human temporal bone specimen. Figure 2 shows the constructed model, consisting of the external ear canal, the middle ear (middle ear cavity, ossicular chain, and supporting ligaments and tendons), and the cochlea. The middle ear was separated from the external ear canal by the eardrum. The ossicular chain (malleus, incus, stapes and the joints) was connected to the wall of the middle ear cavity by the ligaments and tendons. The air in the middle ear cavity and the ear canal was meshed by acoustic tetrahedral elements, with a total of 277,863 elements. The eardrum was divided into the eardrum pars tensa and the eardrum pars flaccida. The eardrum pars tensa was established as a three-layer structure [20]. The inner layer and outer layer of the pars tensa was assumed to be isotropic, while the middle layer of the pars tensa was assumed to be orthotropic, with fibers in circumferential and radial directions. The eardrum pars tensa's inner layer, middle layer, and the outer layer had a thickness of 0.017 mm, 0.016 mm, and 0.017 mm, respectively. The thickness of the eardrum annulus ligament and the eardrum pars flaccida were 0.2 mm and 0.1 mm, respectively. A total of 1939 three-noded shell elements were created to mesh the eardrum. The other middle ear structures were meshed by 45,609 four-noded tetrahedral elements.

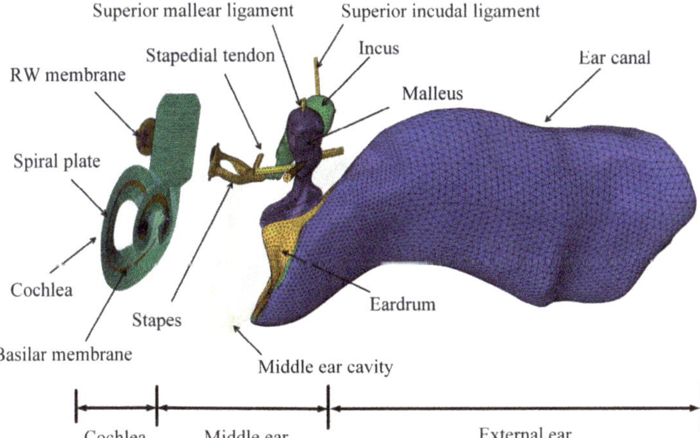

Figure 2. The constructed human ear finite element model.

The middle ear connects to the spiral cochlea with the stapes footplate attached to the oval window. The model's cochlea consists of two fluid-filled chambers: the scala vestibuli (SV) and the scala tympani (ST). These chambers were separated by the basilar membrane (BM) and the bony spiral plate. A total of 361,589 four-noded acoustic tetrahedral elements were created to mesh the fluid in the cochlea. The BM and the bony spiral plate were meshed by 7666 shell elements. The BM thickness and width vary approximately linearly from the base of the cochlea to the apex of the cochlea. The BM length is

34 mm. The thickness of the BM varies from 5.2 μm to 0.6 μm, and the width varies from 0.1 mm to 0.5 mm. The round window membrane was meshed by 851 three-noded tetrahedral elements. The RW membrane has a thickness of 0.1 mm and an area of 2 mm^2, which is close to the size of 2.08 mm^2 reported by Atturo et al. [21].

Considering the ligaments and tendons connect to the bony wall of middle ear cavity, we fixed the end nodes of these components in our FE model. The surfaces of the acoustic elements, which attached to the bony wall in the ear canal, the middle ear cavity, and the cochlea, were defined as rigid walls. The outer edges of the round window membrane and the cochlear spiral plate were set as fixed constraints since they are anchored to the bony wall of the cochlea. Fluid structure interfaces were defined for the surfaces of the acoustic elements attached to the movable structures, i.e., the eardrum, the ossicles, the ligaments, the tendons, the oval window, the BM, and the round window membrane.

2.2. Material Properties

The middle layer of the eardrum pars tensa and the BM were assumed to be orthotropic. Other components of the FE model were assumed to be isotropic. Poisson's ratios were assumed to be 0.3 for all components in the middle ear. The material properties of each component of the middle ear in the FE model were mainly referred to Gentil et al. et al. [20] and Zhang et al. [22], as listed in Table 1.

Table 1. Material properties of the middle ear components.

Components	Layer	Young's Modulus (N/m^2)	Density (kg/m^3)
Eardrum annulus ligament		2.00×10^5	1.20×10^3
Eardrum pars tensa	Outer layer	1.00×10^7	1.20×10^3
	Middle layer	$E_\theta = 2.00 \times 10^7, E_r = 3.20 \times 10^7$	
	Inner layer	1.00×10^7	
Eardrum pars flaccida		1.00×10^7	1.20×10^3
Malleus handle		1.41×10^{10}	3.70×10^3
Malleus neck		1.41×10^{10}	4.53×10^3
Malleus head		1.41×10^{10}	2.55×10^3
Incus body		1.41×10^{10}	2.36×10^3
Incus short process		1.41×10^{10}	2.26×10^3
Incus long process		1.41×10^{10}	5.08×10^3
Stapes		1.41×10^{10}	2.20×10^3
Incudomallear joint		6.00×10^7	3.20×10^3
Incudostapedial joint		2.00×10^6	1.20×10^3
Lateral malleolar ligament		6.70×10^6	2.50×10^3
Superior malleolar ligament		4.90×10^6	2.50×10^3
Anterior malleolar ligament		8.00×10^6	2.50×10^3
Posterior incudal ligament		6.50×10^6	2.50×10^3
Superior incudal ligament		4.90×10^6	1.00×10^3
Tensor tympani tendon		8.00×10^6	2.50×10^3
Stapedial tendon		5.20×10^7	1.00×10^3
Stapedial annulus ligament		1.00×10^4	1.20×10^3

The components of the middle ear and the cochlea were modelled as elastic properties, except for the eardrum, eardrum annulus ligament, incudostapedial joint, incudomallear joint, stapedial annulus ligament, and RW membrane, which were modelled as linear viscoelastic materials. The Rayleigh damping was specified for the elastic components. The Rayleigh damping parameters were taken as $\alpha = 0 \text{ s}^{-1}, \beta = 0.0001 \text{ s}$ [23]. The relaxation modulus of the linear viscoelastic materials was expressed as Equation (1):

$$E(t) = E_0(1 + e_1 \exp(-\frac{t}{\tau_1})) \tag{1}$$

where $E_0, e_1,$ and τ_1 were viscoelastic parameters with constant values for each type of soft tissue, and t is the time. E_0 is the elastic modulus listed in Table 1. e_1, and τ_1 are listed in Table 2. The viscoelastic parameters were referenced to Zhang et al.'s report [24]. These parameters were obtained by dynamic material tests on these components and the cross-calibration method.

The BM is assumed to be the anisotropic membrane with a density of 1200 kg/m^3. The stiffness of BM was decreased from the base to the apex. The BM's longitudinal modulus was assumed as 600 MPa at the base, linearly decreased to 10 MPa at the apex along the BM length. Similarly, the transverse modulus and the vertical modulus of the BM decrease linearly from 6 MPa, and 12 MPa at the base to 0.1 MPa, and 0.2 MPa at the apex, respectively. The density of the RW membrane was set to 1200 kg/m^3 with an elastic modulus of 2.32 MPa [25]. The bulk modulus of the cochlear fluid and the air in the external ear canal and the middle ear cavity were set as 2250 MPa and 0.142 MPa, respectively. The viscosity of the cochlear fluid is 0.001 Ns/m^2 [22].

Table 2. Parameters of linear viscoelastic materials.

Components	e_1	τ_1 (µs)
Eardrum annulus ligament	3.2	28
Eardrum pars flaccida	2.29	25
Eardrum pars tensa	2.8	25
Incudomallear joint	3.0	20
Incudostapedial joint	50	20
Stapedial annulus ligament	2.4	25
RW membrane	3.0	30

2.3. Piezoelectric Transducer Simulation

Since the purpose of this paper is to study the stimulating site's influence rather than the piezoelectric transducer's structural design, we simplified the piezoelectric transducer as an ideal displacement-driven transducer. This idealized representation of the piezoelectric transducer is possible as small displacements and forces are required for hearing compensation in IMEHDs [18]. Based on this simplification, a displacement excitation with the magnitude of 0.1 µm was applied at the commonly used stimulating sites, i.e., the eardrum's umbo, the incus long process, the incus body, the stapes, and the RW membrane, respectively. The magnitude of the applied displacement excitation was ascertained as it can produce a sound pressure level equivalent to 100 dB, which is a design criterion for an IMEHD transducer [4]. The stimulating sites were plotted in Figure 3. When stimulating the eardrum's umbo, the incus long process, the incus body, the stapes, and the direction of the applied displacement excitation was along the longitudinal axis of the stapes, which is efficient for IMEHD stimulation [18]. For stimulating the round window membrane, the excitation's direction was normal to the surface of the round window membrane. Under these forces' stimulation, harmonic analysis was conducted over the frequency range of 0.25–6 kHz using the finite element software package ABAQUS (Dassault Systèmes, Johnston, RI, USA).

The surgical procedure, e.g., the transmastoidal approach for the piezoelectric transducer's implantation, will possibly change the direction of the excitation. To study the stimulating site's sensitivity to the direction changes of their excitations, the excitations were also applied in different directions at each stimulating site with the same magnitude of 0.1 µm. For stimulating the ossicular chain (umbo, incus long process, incus body, stapes), the reference direction was along the stapes' longitudinal axis. The other directions are defined by rotating the direction relative to the reference direction in the plane based on the longitudinal axis and the long axis of the stapes' footplate. The rotation is 20°, 45°, and 60° off the reference direction to crus posterior (20°, 45°, and 60° to CP). For stimulating the RW membrane, the reference direction is the normal direction of the RW membrane. The other directions are rotated 20°, 45°, and 60° off the reference direction.

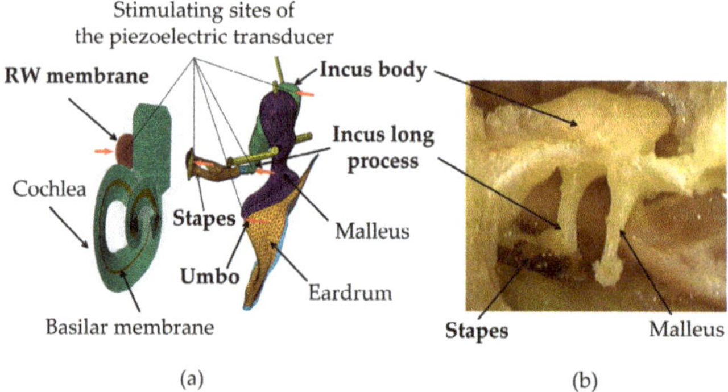

Figure 3. The simulation of the piezoelectric transducer's simulation. (**a**) Stimulating sites on the finite element model; (**b**) the anatomy of the three ossicles (malleus, incus, and stapes).

2.4. Equivalent Sound Pressure Level

The sound transmission property via normal air conduction is different from that by a piezoelectric transducer's stimulation. Considering the basilar membrane inside the cochlea is responsible for sensing the input vibration energy, we used its response to assess the transducer's hearing compensation performance.

The vibration transmitted into the cochlea propagates in the form of a traveling wave from the base to the apex along the basilar membrane. For excitations of different frequencies, the maximum amplitude position of the traveling wave formed on the basilar membrane is different, with high frequencies maximally activating basal regions of the BM and low frequencies maximally activating apical areas of the BM. For a specific frequency excitation, the position of the basilar membrane that is most responsive in the longitudinal direction is referred to as the characteristic place of this frequency. The frequency is called the characteristic frequency of that position on the basilar membrane. The cochlea senses a pure tone sound of a specific frequency through its corresponding characteristic place along the basilar membrane. Therefore, in order to make the sound-perceived effect of the transducer's stimulation of a specific frequency equivalent to that excited by normal sound stimulation (sound pressure applied at the eardrum), the displacements of the BM's characteristic place of the frequency under the two excitations should be equal.

Based on above principle, in the normal sense of sound, when a sound with the frequency of ω and amplitude of P_E is applied at the eardrum, its stimulated BM displacement at the characteristic place x_{CF} is $d_{BM}^{ac}(\omega, x_{CF})$:

$$d_{BM}^{ac}(\omega, x_{CF}) = TF_d^{ac}(\omega) \cdot P_E \tag{2}$$

where $TF_d^{ac}(\omega)$ is the transfer function of the normal human ear sensation from the pressure applied at the eardrum to the displacement of the basilar membrane. The human ear functions as a linear system under the normal acoustic sound pressure excitation [26]. Therefore, based on the model-calculated basilar membrane's displacement under 100 dB SPL sound stimulation applied at the eardrum, we can obtain the transfer function:

$$TF_d^{ac}(\omega) = \frac{d_{BM}^{ac_{100}}(\omega, x_{CF})}{2 \times 10^{-5} \times 10^{\frac{100}{20}}}. \tag{3}$$

Under the excitation of the ideal piezoelectric transducer, its stimulated BM displacement at the characteristic place $d_{BM}^{piezo}(\omega, x_{CF})$ can be calculated by the FE model. Since the basilar membrane vibration is responsible for transmitting the input energy to hair cells, the transducer-stimulated effect is equivalent to that excited by a normal acoustic stimulation $\widehat{P}_E$ applied at the eardrum, which produces the same displacement amplitude $\widehat{d}_{BM}^{ac}(\omega, x_{CF})$ at the characteristic place of the basilar membrane:

$$d_{BM}^{piezo}(\omega, x_{CF}) = \widehat{d}_{BM}^{ac}(\omega, x_{CF}) = TF_d^{ac}(\omega) \cdot \widehat{P}_E = \frac{d_{BM}^{ac_{100}}(\omega, x_{CF})}{2 \times 10^{-5} \times 10^{\frac{100}{20}}} \cdot \widehat{P}_E. \tag{4}$$

Based on Equation (4), the transducer's corresponding equivalent sound pressure $\widehat{P}_E$ applied at the eardrum can be derived as

$$\widehat{P}_E = \frac{d_{BM}^{piezo}(\omega, x_{CF})}{d_{BM}^{ac\_100}(\omega, x_{CF})} \times 2 \times 10^{-5} \times 10^{\frac{100}{20}}. \tag{5}$$

Thus, the performance of the transducer's excitation can be evaluated by L_{EQ}, which is the equivalent sound pressure level (ESPL) of the piezoelectric transducer:

$$L_{EQ} = 20 \log \frac{\widehat{P}_E}{2 \times 10^{-5}} = 100 + 20 \log \left(\frac{d_{BM}^{piezo}(\omega, x_{CF})}{d_{BM}^{ac\_100}(\omega, x_{CF})} \right). \tag{6}$$

3. Results

3.1. Validation of the Human Ear Finite Element Model

To confirm the validity of the established human ear finite element model, three sets of comparisons with the published experimental data were conducted. Since the stapes response is the input of the cochlea, we firstly selected the stapes' footplate displacement to verify our model. Figure 4 shows the mean value of experimental measurements on five temporal bones reported by Gan et al. [27]. In this experiment, a set of pure tone sounds of 90 dB SPL were applied to the eardrum, and the displacement of the stapes footplate was measured using a laser vibrometer. For comparison, we carried out a harmonic analysis across the frequency range of 250–8000 Hz under the same sound pressure applied to the lateral side of the eardrum of our FE model. The model-predicted result was also plotted in Figure 4. It demonstrates that our model-derived displacement of the stapes footplate agrees well with the experimental curve.

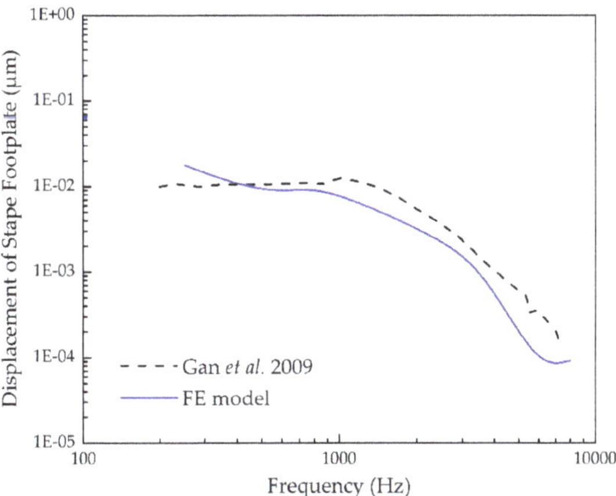

Figure 4. Comparison of the stapes' footplate displacement under 90 dB SPL sound pressure applied at the eardrum.

The BM's response was also selected for our model's validation as it responsible for sensing the cochlear input vibration. Figure 5 displays the experimental curves of the ratio of the BM's velocity at 12 mm from the stapes to the stapes' velocity. The experimental tests were conducted by Gundersen et al. [26] and Stenfelt et al. [28] with a 90 dB SPL input sound pressure applied to the eardrum. Similarly, with a uniform sound pressure applied at the lateral side of the eardrum in our model, a harmonic analysis was conducted across the frequency range of 250–8000 Hz. The model-calculated result was plotted with the experimental curves in Figure 5. It shows that the maximum peak appears at 3500 Hz, which conforms to the experimental data of Gundersen et al. [26]. Besides, our model-predicted result has the same trend as Stenfelt et al.'s [28] data.

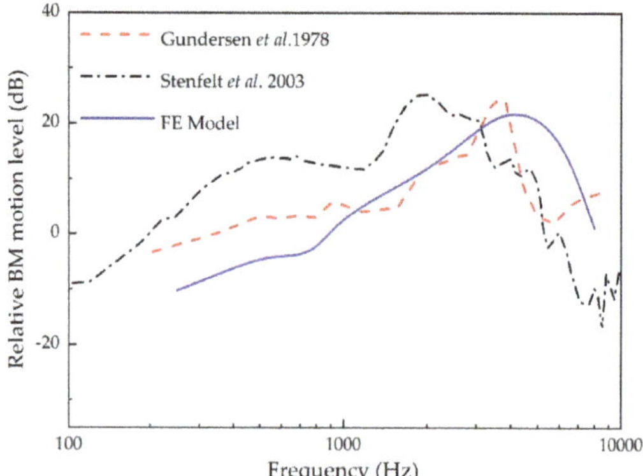

Figure 5. Comparison of the ratio of the basilar membrane's (BM) velocity at 12 mm from the stapes to the stapes' velocity.

Finally, we compared the model-derived cochlear input impedance, which is a measure to represent the cochlear resistance of transmitted vibration from the middle ear, with the experimental data measured by Aibara et al. [29], Puria et al. [30], and Merchant et al. [31], as shown in Figure 6. The cochlear input impedance was calculated from the ratio of the pressure in the SV to the stapes volume velocity (product of the stapes' footplate velocity and the stapes' footplate area). It shows that our predicted result is in the range of these experimental data, and has the same trend with these experimental data, especially the data of Puria et al. [30]. These above comparisons prove that our model can be utilized to predict the sound transmission properties of the human ear.

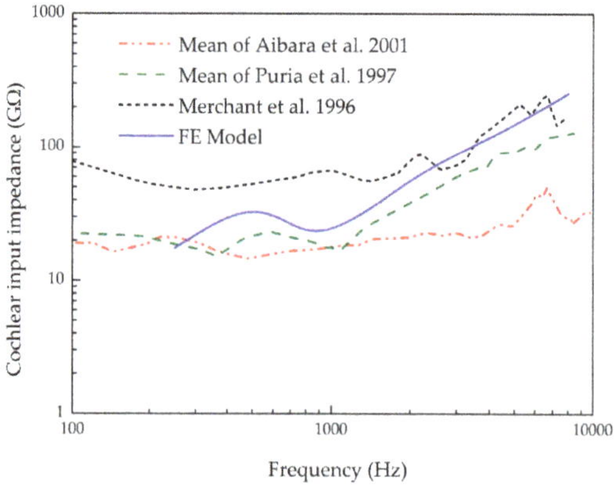

Figure 6. Comparison of the cochlear input impedance.

3.2. The Stimulating Site's Influence on the Piezoelectric Transducer's Performance

Figure 7 shows the influence of the piezoelectric transducer's stimulating sites on its hearing compensation performance. It demonstrates that the piezoelectric transducer can produce high ESPL at high frequency no matter which sites is stimulated. Stimulating the RW membrane as well as stimulating the stapes can produce a more equivalent sound pressure level than stimulating the other sites, especially at a high frequency. Besides, the ESPL under the stimulation applied at the incus-long-process is superior to that generated by the umbo stimulation. The incus body is the worst stimulating site for the piezoelectric transducer.

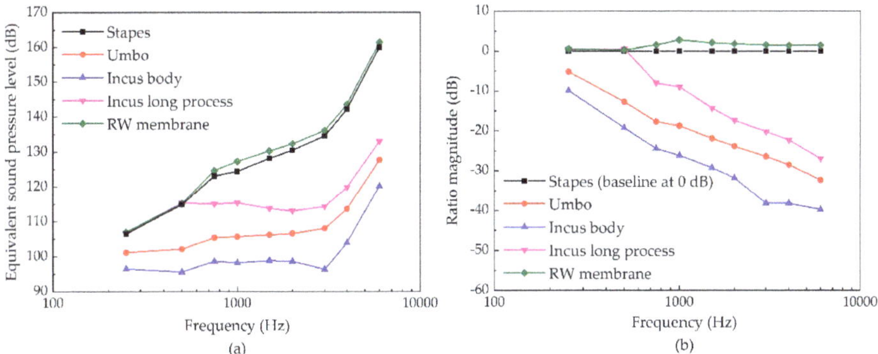

Figure 7. The influence of a piezoelectric transducer's stimulating sites on its hearing compensation performance. (**a**) Equivalent sound pressure level of the piezoelectric transducers stimulating at different sites; (**b**) ratio of equivalent sound pressure of the piezoelectric transducer stimulating at different sites (the reference is the stimulation applied at the stapes along the stapes longitudinal axis).

3.3. The Sensitivities of Each Stimulating Site to the Changes of Excitation's Direction

For stimulating the eardrum's umbo, the influence of a piezoelectric transducer's excitation direction on its hearing compensation performance is shown in Figure 8. It indicates that the transducer's stimulated ESPL decreases with the increase of the angle of the excitation's inclination, especially at the middle frequency. The maximum decrease is found at 1 kHz for 60° to CP, with a reduction of 13 dB.

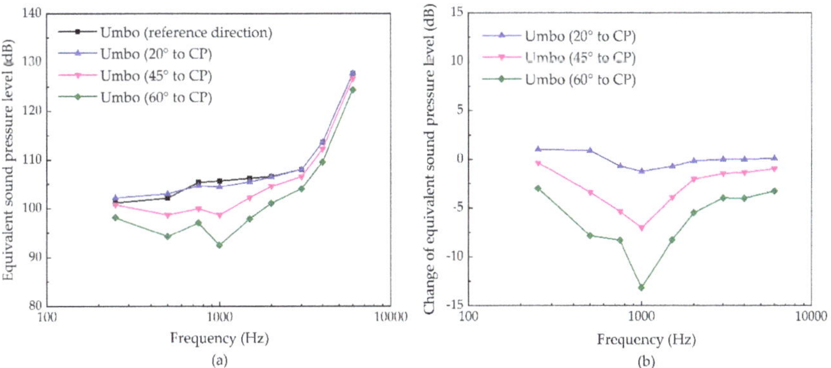

Figure 8. The sensitivity of umbo stimulation due to the change of its excitation's direction. (**a**) Equivalent sound pressure level of the piezoelectric transducer's stimulation; (**b**) change of equivalent sound pressure level.

While the stimulating site is the incus body, the stimulation direction's influence is shown in Figure 9. It demonstrates that the change of the stimulation direction's influence on the transducer's stimulated ESPL is complex in this case. Increasing the angle of the stimulation's inclination decreases the transducer-stimulated ESPL at a lower frequency, but increases the ESPL slightly at a higher frequency. The maximum decrease is at 250 Hz for 45° to CP, with a reduction of 17 dB.

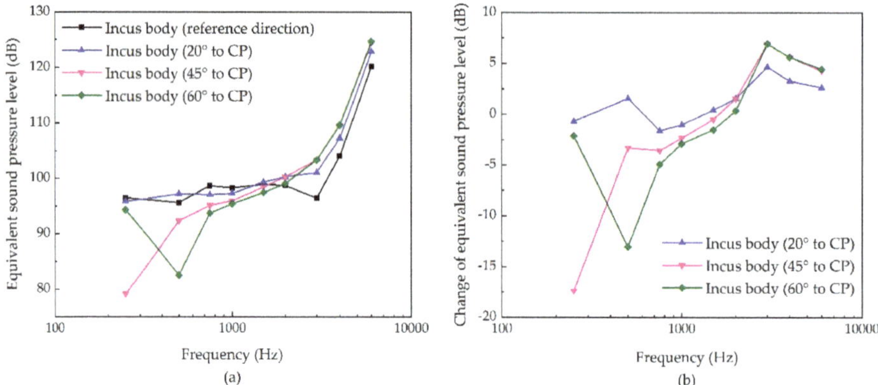

Figure 9. The sensitivity of incus-body stimulation due to the change of its excitation's direction. (**a**) Equivalent sound pressure level of the piezoelectric transducer's stimulation; (**b**) change of equivalent sound pressure level.

As for stimulating the incus long process, the excitation direction's influence is shown in Figure 10. Similar to the influence in stimulating the eardrum's umbo, the boost of the angle of the excitation's inclination reduces the transducer-stimulated ESPL, especially at the middle frequency. The maximum decrease is also at 1 kHz for 60° to CP, with a reduction of 16 dB.

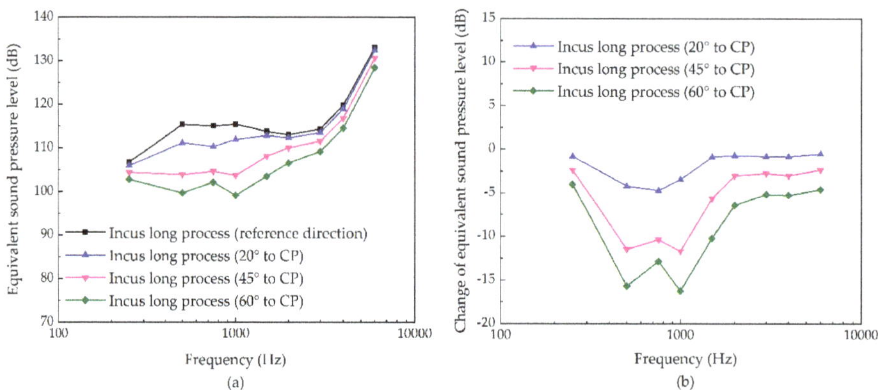

Figure 10. The sensitivity of incus-long-process stimulation due to the change of its excitation's direction. (**a**) Equivalent sound pressure level of the piezoelectric transducer's stimulation; (**b**) change of equivalent sound pressure level.

In terms of stimulating the stapes, the effect of the transducer's stimulation direction is shown in Figure 11. It indicates that the transducer-stimulated ESPL also decreases with the increase of inclination angle. Unlike previous stimulating sites, the ESPL at a high frequency decreases significantly in this case. The maximum decrease is at 400 Hz for 60° to CP, with a reduction of 40 dB. For a high frequency region, the maximum reduction is 36 dB at 4 kHz for 60° off the reference direction.

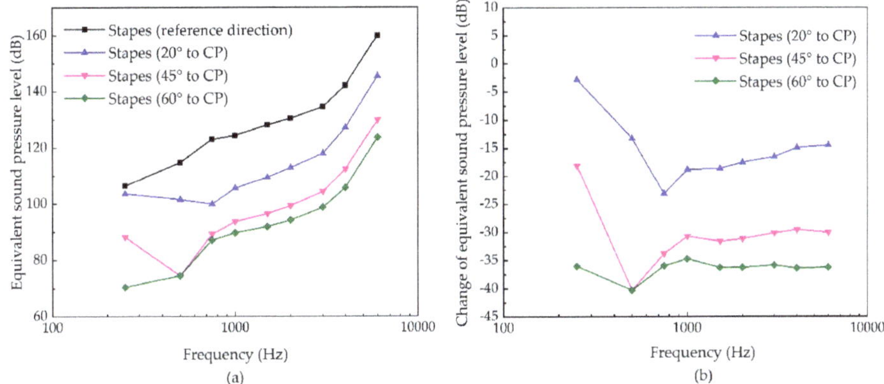

Figure 11. The sensitivity of stapes stimulation due to the change of its excitation's direction. (**a**) Equivalent sound pressure level of the piezoelectric transducer's stimulation; (**b**) change of equivalent sound pressure level.

For stimulating the RW membrane, the influence of the transducer's stimulation direction is shown in Figure 12. It demonstrates that the increase of the inclination angle mainly reduces the transducer's high frequency ESPL. The maximum decrease is at 6 kHz for 60° off the reference direction, with a reduction of 31 dB.

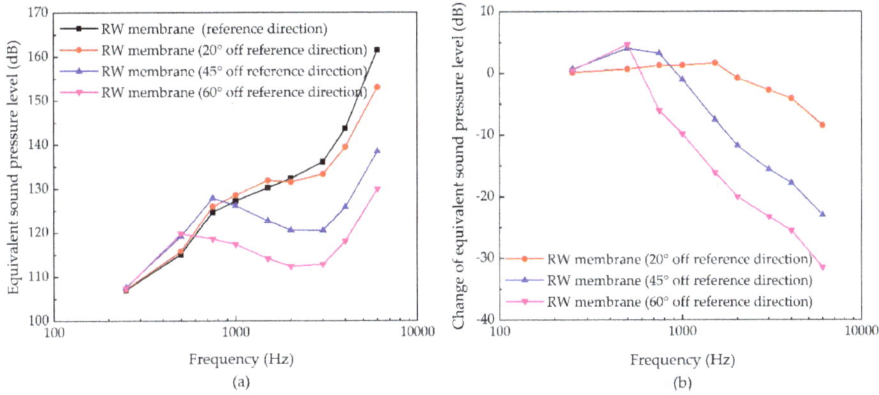

Figure 12. The sensitivity of RW membrane stimulation due to the change of its excitation's direction. (**a**) Equivalent sound pressure level of the piezoelectric transducer's stimulation; (**b**) change of equivalent sound pressure level.

4. Discussion

Since the human ear is a complicated biological system with tiny structures and complex geometry, systematic experimental investigation on it is tough to conduct. Considering the finite element method has the advantage of simulating this complicated biological system, many researchers built a human ear FE model, and used it to study the sound transmission properties of the ear [32–35] and facilitate the design of IMEHDs [14,36,37]. In our human ear FE model, the cochlea was constructed as two fluid-filled channels. This modelling method of the cochlea is widely used in the field of cochlear mechanics [16]. Different individual human ears have similar vibration properties, and to confirm our model's validity, we compared our model-predicted results with the mean value of the experimental results, which were measured on many human ears, to prove that our model can predict the general response of the human ear. This kind of validation has been widely used by scholars in this fields [20,23].

According to the vibrational energy transmission pathway, the implantable middle ear hearing device can be classified as forward stimulation and reverse stimulation [38]. Stimulating the eardrum, the incus body, the incus long process, and the stapes belong to forward stimulation, since their vibration energy are transmitted to the cochlea through the cochlear oval window, which is the same as the normal hearing process. Whereas, stimulating the round window membrane is called reverse stimulation as its vibration energy is transmitted to the cochlea though the cochlear round window, the other opening window of the cochlea. For the forward driving, our results demonstrate that the piezoelectric transducer provides better performance when stimulating the stapes than stimulating the eardrum's umbo or the incus long process. The performance of stimulating the incus long process is superior to that of stimulating the eardrum's umbo. This result can be easily predicted for forward stimulation since the stapes is close to the cochlea and therefore more efficient to transmit vibrational energy into the cochlea. Besides, we found that the superiority of the stapes stimulation is significant at high frequencies. To further analyze this phenomenon, we plot the z direction's (along the longitudinal axis of the stapes) displacement contour plot (Figure 13) of the ossicular chain since the stapes transmits its vibration mainly through its piston motion [39]. Figure 13 shows that the stapes can be efficiently stimulated at a low frequency for all these three stimulating sites, especially for stimulating the stapes and stimulating the incus long process. With an increase in the stimulation's frequency, the vibration cannot be effectively transmitted to the stapes for stimulating the incus long process and the umbo, especially for stimulating the umbo. This may attribute to the incudomallear joint and the incudostapedial joint, whose viscous behavior become significant at a higher frequency and weaken the vibrational energy transmission from the stimulating point to the stapes.

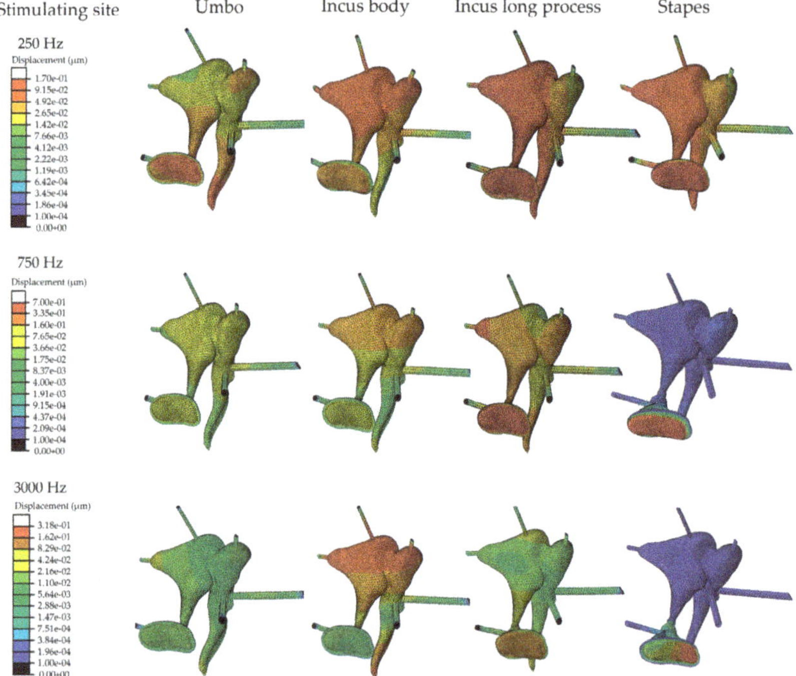

Figure 13. The z-direction (along the longitudinal axis of the stapes) displacement contour plot of the ossicular chain.

For the forward stimulation, the incus body is the worst stimulating site for the transducer. This result is consistent with Bornitz et al.'s report [18] based on stapes displacement. This owing to the fact that the rotation nod of the ossicular chain is close to the incus body [40]; therefore, the incus body is the least efficient point for stimulating the ossicular chain. Figure 13 also shows that most of the stimulated response are restrained around the incus body; the vibration cannot be efficiently transmitted to the stapes footplate under the incus-body stimulation, especially at a high frequency. Thus, compared with other forward stimulation, the high-frequency output should be enhanced for the incus-body simulating-type piezoelectric transducer.

Compared with forward stimulation, it is difficult to estimate the performance of the round window's stimulation, i.e., the reverse stimulation, since its vibration energy transmission pathway is different from that of forward driving. Although comparison of the forward stimulation with the round window's stimulation were reported [16,22], these studies only focus on the electromagnetic transducer, which is a force-driven transducer. Bornitz et al.'s study [18] demonstrates that the stimulating site's influence for the electromagnetic transducer is different from that for the piezoelectric transducer. For the piezoelectric transducer, our study finds that stimulating the round window membrane can produce a similar ESPL as when stimulating the stapes. Regardless of which site is stimulated, the piezoelectric transducer can generate high ESPL at a high frequency. Since most sensorineural hearing loss is severe at a high frequency [41], this characteristic is a crucial advantage for the piezoelectric transducer to compensate for the hearing loss. This better high-frequency performance of the piezoelectric transducer was reported in many experimental studies [6,10,42].

The performance of the RW membrane stimulation, as well as that of the stapes stimulation are susceptible to the change of the excitation's direction. This result for the RW membrane stimulation conforms to Arnold et al.'s temporal bone's study [43], which found that the transducer's direction significantly affects the energy transferred to the cochlea of up to 35 dB. The clinical result also shows that the postoperative performance of the RW stimulation has a high variability [44], which may attribute to the change of the transducer's direction. Meanwhile, this sensitivity for RW membrane stimulation and stapes stimulation to the excitation direction is prominent at a high frequency. Considering the main type of sensorineural hearing loss is the "high-frequency" hearing loss [41], the piezoelectric transducer's orientation for RW stimulation or stapes stimulation should be guaranteed during the surgery. For the design of these two types of transducers, it is recommended to design a fixing part to ensure its orientation after implantation.

The main purpose of this paper is to investigate the stimulating site's influence on the piezoelectric transducer. To facilitate this study, the real structure of the piezoelectric transducer was not considered in this paper; instead, we simplified it as an ideal displacement driven transducer, i.e., a transducer generates a certain displacement without limitations in force. Under this simplification, the retroaction of the human ear system onto the piezoelectric transducer was neglected. This simplification for the IMEHD's piezoelectric transducer is based on the fact that the blocking force of the transducer is much larger than its working force. For instance, the piezoelectric transducer (Model PL-033, Physik Instrumente, Waldbronne, Germany) used in Wang et al.'s study [41] for IMEHD has a blocking force of 300 N, which is much larger than the force (89 μN [45]) required to drive the vibration of ossicles to the equivalent of 100 dB SPL. Thus, the resistant force of the human ear cannot change the piezoelectric displacement output significantly. Based on a coupled FE model of the middle ear and a piezoelectric transducer, which is a 20-layer stack of 3 mm diameter and 4 mm thickness made of PZN-4.5PT, Bornitz et al. [18] also found that there is no retroaction of the human ear onto the piezoelectric transducer. Thus, simplifying the piezoelectric transducer in IMEHD as a displacement-driven transducer is acceptable.

It should be noted that our FE model is constructed only based on one human ear specimen. Based on a numerical study, Daniel et al. [46] found that the human ear's geometrical variation can lead to differences of 4 dB in the lower frequencies and up to 6 dB around 2 kHz, but similar shapes in the calculated response curves. Thus, the patients' individual geometrical differences may alter our results quantitatively at lower frequencies and frequencies around 2 kHz. Nevertheless, the overall trend of our results still holds under different individual human ear geometries.

5. Conclusions

To study the influence of piezoelectric transducer's stimulating sites on its hearing compensation performance, a human ear FE model, including the ear canal, the middle ear, and the cochlea, was constructed. The validity of this model was verified by three sets of comparisons. The results show the piezoelectric transducer provides better performance when simulating the stapes or RW membrane than stimulating other studied sites, especially at a high frequency. The incus body is the worst stimulating site for the piezoelectric transducer. Meanwhile, the performance of the RW membrane stimulation, as well as that of the stapes stimulation, are susceptible to the change of the excitation's direction. Considering most sensorineural hearing loss is severe at high-frequency, the piezoelectric transducer's orientation for RW stimulation or stapes stimulation should be guaranteed during the surgery.

Author Contributions: Conceptualization, H.L. and J.Y.; formal analysis, H.L.; funding acquisition, H.L.; methodology, H.L., Y.Z. and Z.R.; software, Y.Z.; writing—original draft, H.L. and J.Y.; writing—review and editing, Z.R.

Funding: This research was funded by National Natural Science Foundation of China, grant number 51775547, and Top-notch Academic Program Development of Jiangsu Higher Education Institutions, grant number PPZY2015B120.

Acknowledgments: We thank Xinsheng Huang and Lei Zhou for helping us prepare the CT images of the temporal bone.

Conflicts of Interest: The authors declare no conflict of interest.

References

1. World Health Organization. Deafness and Hearing Loss. Key Facts. Available online: https://www.who.int/en/news-room/fact-sheets/detail/deafness-and-hearing-loss (accessed on 30 September 2019).
2. Angeli, S.I.; Yan, D.; Telischi, F.; Balkany, T.J.; Ouyang, X.M.; Du, L.L.; Eshraghi, A.; Goodwin, L.; Liu, X.Z. Etiologic diagnosis of sensorineural hearing loss in adults. *Otolaryngol. Head Neck Surg.* **2005**, *132*, 890–895. [CrossRef] [PubMed]
3. Moore, B.C.J. *Cochlear Hearing Loss: Physiological, Psychological and Technical Issues*, 2nd ed.; Wiley Chichester: John Wiley: Hoboken, NJ, USA, 2007.
4. Hong, E.-P.; Kim, M.-K.; Park, I.-Y.; Lee, S.-H.; Roh, Y.; Cho, J.-H. Vibration Modeling and Design of Piezoelectric Floating Mass Transducer for Implantable Middle Ear Hearing Devices. *IEICE Trans. Fundam.* **2007**, *E90-A*, 1620–1627. [CrossRef]
5. Haynes, D.S.; Young, J.A.; Wanna, G.B.; Glasscock, M.E. Middle ear implantable hearing devices: An overview. *Trends Amplif.* **2009**, *13*, 206–214. [CrossRef] [PubMed]
6. Liu, H.; Cheng, J.; Yang, J.; Rao, Z.; Cheng, G.; Yang, S.; Huang, X.; Wang, M. Concept and Evaluation of a New Piezoelectric Transducer for an Implantable Middle Ear Hearing Device. *Sensors* **2017**, *17*, 2515. [CrossRef]
7. Maurer, J.; Savvas, E. The Esteem System: A totally implantable hearing device. *Adv. Otorhinolaryngol.* **2010**, *69*, 59–71. [CrossRef] [PubMed]
8. Perkins, R.; Fay, J.P.; Rucker, P.; Rosen, M.; Olson, L.; Puria, S. The EarLens system: New sound transduction methods. *Hear. Res.* **2010**, *263*, 104–113. [CrossRef]
9. Song, Y.-L.; Jian, J.-T.; Chen, W.-Z.; Shih, C.-H.; Chou, Y.-F.; Liu, T.-C.; Lee, C.-F. The development of a non-surgical direct drive hearing device with a wireless actuator coupled to the tympanic membrane. *Appl. Acoust.* **2013**, *74*, 1511–1518. [CrossRef]
10. Hong, E.-P.; Park, I.-Y.; Seong, K.-W.; Cho, J.-H. Evaluation of an implantable piezoelectric floating mass transducer for sensorineural hearing loss. *Mechatronics* **2009**, *19*, 965–971. [CrossRef]

11. Wang, Z.; Mills, R.; Luo, H.; Zheng, X.; Hou, W.; Wang, L.; Brown, S.I.; Cuschieri, A. A micropower miniature piezoelectric actuator for implantable middle ear hearing device. *IEEE Trans. Biomed. Eng.* **2011**, *58*, 452–458. [CrossRef]

12. Komori, M.; Yanagihara, N.; Hinohira, Y.; Hato, N.; Gyo, K. Long-term results with the Rion E-type semi-implantable hearing aid. *Otolaryngol. Head Neck Surg.* **2010**, *143*, 422–428. [CrossRef]

13. Shin, D.H.; Cho, J.H. Piezoelectric Actuator with Frequency Characteristics for a Middle-Ear Implant. *Sensors* **2018**, *18*, 1694. [CrossRef] [PubMed]

14. Liu, H.; Wang, H.; Rao, Z.; Yang, J.; Yang, S. Numerical Study and Optimization of a Novel Piezoelectric Transducer for a Round-Window Stimulating Type Middle-Ear Implant. *Micromachines* **2019**, *10*, 40. [CrossRef] [PubMed]

15. Brkic, F.F.; Riss, D.; Auinger, A.; Zoerner, B.; Arnoldner, C.; Baumgartner, W.D.; Gstoettner, W.; Vyskocil, E. Long-Term Outcome of Hearing Rehabilitation With An Active Middle Ear Implant. *Laryngoscope* **2019**, *129*, 477–481. [CrossRef] [PubMed]

16. Zhang, X.; Gan, R.Z. A comprehensive model of human ear for analysis of implantable hearing devices. *IEEE Trans. Biomed. Eng.* **2011**, *58*, 3024–3027. [CrossRef] [PubMed]

17. Deveze, A.; Koka, K.; Tringali, S.; Jenkins, H.A.; Tollin, D.J. Techniques to improve the efficiency of a middle ear implant: Effect of different methods of coupling to the ossicular chain. *Otol. Neurotol.* **2013**, *34*, 158–166. [CrossRef] [PubMed]

18. Bornitz, M.; Hardtke, H.J.; Zahnert, T. Evaluation of implantable actuators by means of a middle ear simulation model. *Hear. Res.* **2010**, *263*, 145–151. [CrossRef]

19. Lee, J.; Seong, K.; Lee, S.-H.; Lee, K.-Y.; Cho, J.-H. Comparison of auditory responses determined by acoustic stimulation and by mechanical round window stimulation at equivalent stapes velocities. *Hear. Res.* **2014**, *314*, 65–71. [CrossRef]

20. Gentil, F.; Parente, M.; Martins, P.; Garbe, C.; Jorge, R.N.; Ferreira, A. The influence of the mechanical behaviour of the middle ear ligaments: A finite element analysis. *Proc. Inst. Mech. Eng. Part H J. Eng. Med.* **2011**, *225*, 68–76. [CrossRef]

21. Francesca, A.; Maurizio, B.; Helge, R.A. Is the human round window really round? An anatomic study with surgical implications. *Otol. Neurotol.* **2014**, *35*, 1354–1360.

22. Zhang, J.; Zou, D.; Tian, J.; Ta, N.; Rao, Z. A comparative finite-element analysis of acoustic transmission in human cochlea during forward and reverse stimulations. *Appl. Acoust.* **2019**, *145*, 278–289. [CrossRef]

23. Sun, Q.; Gan, R.Z.; Chang, K.H.; Dormer, K.J. Computer-integrated finite element modeling of human middle ear. *Biomech. Modeling Mechanobiol.* **2002**, *1*, 109–122. [CrossRef] [PubMed]

24. Zhang, X.; Gan, R.Z. Finite element modeling of energy absorbance in normal and disordered human ears. *Hear. Res.* **2013**, *301*, 146–155. [CrossRef] [PubMed]

25. Zhou, K.; Liu, H.; Yang, J.; Zhao, Y.; Rao, Z.; Yang, S. Influence of middle ear disorder in round-window stimulation using a finite element human ear model. *Acta Bioeng. Biomech.* **2019**, *21*. [CrossRef]

26. Gundersen, T.; Skarstein, O.; Sikkeland, T. A study of the vibration of the basilar membrane in human temporal bone preparations by the use of the Mossbauer effect. *Acta Oto-Laryngologica* **1978**, *86*, 225–232. [CrossRef]

27. Gan, R.Z.; Cheng, T.; Dai, C.; Yang, F.; Wood, M.W. Finite element modeling of sound transmission with perforations of tympanic membrane. *J. Acoust. Soc. Am.* **2009**, *126*, 243–253.

28. Stenfelt, S.; Puria, S.; Hato, N.; Goode, R.L. Basilar membrane and osseous spiral lamina motion in human cadavers with air and bone conduction stimuli. *Hear. Res.* **2003**, *181*, 131–143. [CrossRef]

29. Aibara, R.; Welsh, J.T.; Puria, S.; Goode, R.L. Human middle-ear sound transfer function and cochlear input impedance. *Hear. Res.* **2001**, *152*, 100–109. [CrossRef]

30. Puria, S.; Peake, W.T.; Rosowski, J.J. Sound-pressure measurements in the cochlear vestibule of human-cadaver ears. *J. Acoust. Soc. Am.* **1997**, *101*, 2754–2770. [CrossRef]

31. Merchant, S.N.; Ravicz, M.E.; Rosowski, J.J. Acoustic input impedance of the stapes and cochlea in human temporal bones. *Hear. Res.* **1996**, *97*, 30–45. [CrossRef]

32. Lee, C.F.; Chen, P.R.; Lee, W.J.; Chen, J.H.; Liu, T.C. Three-dimensional reconstruction and modeling of middle ear biomechanics by high-resolution computed tomography and finite element analysis. *Laryngoscope* **2006**, *116*, 711–716. [CrossRef]

33. Liu, H.; Zhang, H.; Yang, J.; Huang, X.; Liu, W.; Xue, L. Influence of ossicular chain malformation on the performance of round-window stimulation: A finite element approach. *Proc. Inst. Mech. Eng. H* **2019**, *233*, 584–594. [CrossRef] [PubMed]

34. Yao, W.; Tang, D.; Chen, Y.; Li, B. Study on vibration characteristics and transmission performance of round window membrane under inverse excitation. *J. Mech. Med. Biol.* **2018**, *18*, 1850033. [CrossRef]

35. Liu, H.; Xu, D.; Yang, J.; Yang, S.; Cheng, G.; Huang, X. Analysis of the influence of the transducer and its coupling layer on round window stimulation. *Acta Bioeng. Biomech.* **2017**, *19*, 103–111. [PubMed]

36. Koch, M.; Essinger, T.M.; Bornitz, M.; Zahnert, T. Examination of a mechanical amplifier in the incudostapedial joint gap: FEM simulation and physical model. *Sensors* **2014**, *14*, 14356–14374. [CrossRef] [PubMed]

37. Wang, X.; Hu, Y.; Wang, Z.; Shi, H. Finite element analysis of the coupling between ossicular chain and mass loading for evaluation of implantable hearing device. *Hear. Res.* **2011**, *280*, 48–57. [CrossRef] [PubMed]

38. Chen, Y.; Guan, X.; Zhang, T.; Gan, R.Z. Measurement of basilar membrane motion during round window stimulation in guinea pigs. *J. Assoc. Res. Otolaryngol.* **2014**, *15*, 933–943. [CrossRef]

39. Edom, E.; Obrist, D.; Henniger, R.; Kleiser, L.; Sim, J.H.; Huber, A.M. The effect of rocking stapes motions on the cochlear fluid flow and on the basilar membrane motion. *J. Acoust. Soc. Am.* **2013**, *134*, 3749–3758. [CrossRef]

40. Koike, T.; Wada, H.; Kobayashi, T. Modeling of the human middle ear using the finite-element method. *J. Acoust. Soc. Am.* **2002**, *111*, 1306–1317. [CrossRef]

41. Wang, Z.G.; Abel, E.W.; Mills, R.P.; Liu, Y. Assessment of multi-layer piezoelectric actuator technology for middle-ear implants. *Mechatronics* **2002**, *12*, 3–17. [CrossRef]

42. Park, I.Y.; Shimizu, Y.; O'Connor, K.N.; Puria, S.; Cho, J.H. Comparisons of electromagnetic and piezoelectric floating-mass transducers in human cadaveric temporal bones. *Hear. Res.* **2011**, *272*, 187–192. [CrossRef]

43. Arnold, A.; Kompis, M.; Candreia, C.; Pfiffner, F.; Häusler, R.; Stieger, C. The floating mass transducer at the round window: Direct transmission or bone conduction? *Hear. Res.* **2010**, *263*, 120–127. [CrossRef] [PubMed]

44. Sprinzl, G.M.; Wolf-Magele, A.; Schnabl, J.; Koci, V. The active middle ear implant for the rehabilitation of sensorineural, mixed and conductive hearing losses. *Laryngorhinootologie* **2011**, *90*, 560–572. [CrossRef] [PubMed]

45. Ko, W.H.; Zhu, W.L.; Kane, M.; Maniglia, A.J. Engineering principles applied to implantable otologic devices. *Otolaryngol. Clin. N. Am.* **2001**, *34*, 299–314. [CrossRef]

46. De Greef, D.; Pires, F.; Dirckx, J.J. Effects of model definitions and parameter values in finite element modeling of human middle ear mechanics. *Hear. Res.* **2017**, *344*, 195–206. [CrossRef] [PubMed]

Article

A Study on the Effects of Bottom Electrode Designs on Aluminum Nitride Contour-Mode Resonators

Soon In Jung [1], Chaehyun Ryu [1], Gianluca Piazza [2] and Hoe Joon Kim [1,*]

[1] Department of Robotics Engineering, Daegu Gyeongbuk Institute of Science & Technology (DGIST), Daegu 42988, Korea; jsi20039@dgist.ac.kr (S.I.J.); chaehyun@dgist.ac.kr (C.R.)

[2] Department of Electrical and Computer Engineering, Carnegie Mellon University, Pittsburgh, PA 15213, USA; piazza@ece.cmu.edu

* Correspondence: joonkim@dgist.ac.kr; Tel.: +82-53-785-6221

Received: 16 September 2019; Accepted: 5 November 2019; Published: 7 November 2019

Abstract: This study presents the effects of bottom electrode designs on the operation of laterally vibrating aluminum nitride (AlN) contour-mode resonators (CMRs). A total of 160 CMRs were analyzed with varying bottom electrode areas at two resonant frequencies (f_0) of about 230 MHz and 1.1 GHz. Specifically, we analyzed the impact of bottom electrode coverage rates on the resonator quality factor (Q) and electromechanical coupling (k^2), which are important parameters for Radio Frequency (RF) and sensing applications. From our experiments, Q exhibited different trends to electrode coverage rates depending on the device resonant frequencies, while k^2 increased with the coverage rate regardless of f_0. Along with experimental measurements, our finite element analysis (FEA) revealed that the bottom electrode coverage rate determines the active (or vibrating) region of the resonator and, thus, directly impacts Q. Additionally, to alleviate thermoelastic damping (TED) and focus on mechanical damping effects, we analyzed the device performance at 10 K. Our findings indicated that a careful design of bottom electrodes could further improve both Q and k^2 of AlN CMRs, which ultimately determines the power budget and noise level of the resonator in integrated oscillators and sensor systems.

Keywords: MEMS; aluminum nitride; resonator; damping; quality factor; electromechanical coupling

---

1. Introduction

Piezoelectric microelectromechanical systems (MEMS) resonators have shown great promise towards fully integrated and high-efficiency RF and wireless communication systems owing to their small footprint and low power budget [1–4]. With the emergence of Internet of Things (IoT), the need for ultra-low-power communication and sensor systems is growing, and recently developed piezoelectric MEMS resonator technologies have been demonstrated to be fundamental building blocks in near-zero power sensors for various application areas [5–7]. Moreover, many piezoelectric resonators are fully compatible with conventional microfabrication processes, enabling monolithic integration for single-chip electronics [8–10]. In contrast to electrostatic resonators [11,12], piezoelectric resonators exhibit better electromechanical coupling and lower motional resistance, which are favorable for 50 Ω RF communication systems.

Among existing piezoelectric MEMS resonators, laterally vibrating aluminum nitride (AlN) contour-mode resonators (CMRs) have drawn much attention as multiple frequency resonators can be fabricated on a single chip [4,13,14]. In more detail, lithographically defined electrodes can realize resonant frequencies (f_0) from MHz to several GHz range, which is a great advantage compared to thin-film bulk acoustic resonators (FBAR) or shear-mode piezoelectric resonators, where resonance is determined by the piezoelectric film thickness [15–17]. In addition, AlN CMRs exhibit low motional resistance of around 50 Ω along with a small footprint, which enable ultra-low-power AlN

CMR-based complementary metal oxide semiconductor (CMOS) oscillators [18,19]. Although AlN CMRs could be building blocks for next generation RF systems, the quality factor (Q) still needs to be improved while maintaining a high electromechanical coupling (k^2) for successful commercialization. The aforementioned properties of the resonator ultimately set the power budget and phase noise performance of the device [20–22], and the figure of merit (FOM) of AlN CMRs is defined as the product of the two.

Previous studies have revealed that damping directly affects both Q and k^2 of piezoelectric MEMs resonators [8,23]. In the case of AlN CMRs, both mechanical and thermal effects induce significant damping on the device and these must be accounted for to ensure stable resonator operation. Mechanical damping (or anchor losses) [24] is more dominant in lower-frequency CMRs, while thermoelastic damping (TED) prevails at higher resonant frequencies around or above 1 GHz [25]. Various designs of electrodes, resonators, and anchors, which connect the resonator to the supporting substrate, have been extensively studied in an effort to mitigate such damping effects and, ultimately, improve device performance. In detail, segmented electrodes [26] and apodization techniques [27] have been applied to improve the FOM and suppress spurious modes. Studies on anchors and the bus region, where the top and bottom electrodes do not overlap near anchors, have also been reported to be important factors on mechanical damping, affecting both Q and k^2 [28]. Since more than 50% of TED arises from the metal electrodes, the type of electrode material and its thickness also affects the electromechanical properties of AlN CMRs [25,29]. A search for the optimal resonator design and metal electrode continues in an effort to build high performance AlN CMRs with outstanding electromechanical properties.

All the aforementioned studies have identified the impact of various design parameters on device performance and improved the current state of AlN CMRs. However, there is a lack of published work on the effects of bottom electrode designs on the electromechanical properties of CMRs. In general, a bottom electrode accounts for more than 50% of the metal in a resonator [25,30,31]. In addition, the size of the bottom electrode ultimately determines the active (or vibrating) region of the resonator. Since the design of the bottom electrode can affect both mechanical damping and TED, it is important to understand its impact on device performance.

To experimentally verify the impact of the bottom electrodes, we fabricated and measured the admittance response of a total of 32 different AlN CMRs with varying anchor designs, bottom electrode coverage rates, and resonant frequencies. In addition, we performed finite element analysis (FEA) to further investigate the impact of electrode designs on mechanical damping in the resonator. To minimize TED, we also measured the resonators at 10K and compared the results to calculations. This work outlines the device design and fabrication, experimental analysis, numerical analysis, and in-depth discussion of the major findings.

2. Materials and Methods

2.1. Resonator Design and Fabrication

The designed AlN CMRs consisted of 100 nm thick Al top electrodes, a 1 μm thick AlN piezoelectric layer, and a 100 nm thick Pt bottom electrode, as shown in Figure 1a. The interdigitated top electrodes work as signal and ground pads while the floating bottom electrode guide the electric field vertically. Such an electric field induces lateral vibrations of the resonator. The resonant frequency of the device can be expressed as

$$f_0 = 1/2w \times \mathrm{sqrt}\,(E_{\mathrm{eq}}/\rho_{\mathrm{eq}}) \tag{1}$$

where w is the pitch of the top electrodes, E_{eq} is the equivalent Young's modulus, and ρ_{eq} is the equivalent mass density of the resonator. Since previous studies have revealed that the type of dominating damping mechanism depends on f_0, we designed a set of CMRs with two different resonant frequencies at 230 MHz and 1 GHz. As anchor losses dominate at lower f_0, we designed the 230 MHz CMRs to have various anchor dimensions. By contrast, the 1.1 GHz CMRs were designed to have full anchors as the impact of anchor losses is rather small compared to TED. In addition, we fabricated CMRs with varying bottom electrode coverage rates ranging from 33% to 120%. Here, 100% was the area where the signal and ground top electrodes overlapped. The bottom electrode coverage rate is defined as the bottom electrode area divided by the area where the top electrodes overlapped, as shown in Figure 1f. For example, a 120% coverage rate means that the bottom electrode covered the entire resonator body including the bus region. For 230 MHz CMRs, the width of the top electrodes were 15 μm with a pitch of 20 μm. For 1.1 GHz CMRs, the top electrodes were 2 μm wide with a pitch of 4 μm. Table 1 summarizes the set of fabricated CMRs with varying design parameters, where L_a, W_a, and L are anchor length, width, and resonator length, respectively. In total, we fabricated CMRs with 32 different designs.

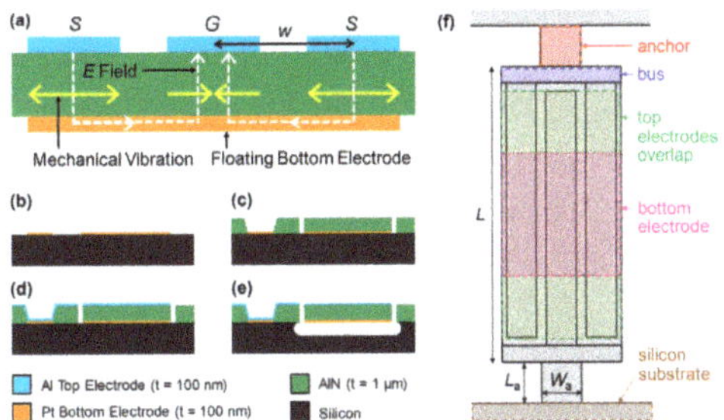

Figure 1. (**a**) Side view of an aluminum nitride (AlN) contour-mode resonator (CMR), which is formed of top electrodes, an AlN layer, and a bottom electrode. (**b**–**e**) The fabrication process consists of a bottom electrode patterning, AlN deposition, top electrode deposition, and the final device release via XeF$_2$ etching of a silicon layer. (**f**) A schematic of 220 MHz CMR consisting of the anchor, bus region, top and bottom electrodes, and silicon substrate. The bottom electrode coverage rate is defined as the bottom electrode area (pink) divided by the area where the top electrodes overlap (green).

Table 1. Parameters of fabricated 230 MHz and 1.1 GHz CMRs with various anchor designs and bottom electrode coverage rates.

| | 230 MHz CMRs | | | 1.1 GHz CMRs | | |
Type	Anchor Length (L_a)	Width (W_a)	Bottom Electrode Coverage	Type	Resonator Length (L)	Bottom Electrode Coverage
A	10 μm	10 μm		G	60 μm	
B	10 μm	20 μm	33%, 50%, 75%,	E	80 μm	50%, 75%,
C	20 μm	10 μm	100%, 120%	F	100 μm	100%, 120%
D	20 μm	20 μm				

Figure 1b shows the overall fabrication process of AlN CMRs. For device fabrication, a 4 inch high resistivity (over 2×10^3 Ω·cm) Si wafer was used to prevent any current leakage from the resonator to the Si substrate. First, a 100 nm thick Pt layer was patterned via liftoff to form the bottom electrodes. Pt was chosen as the bottom electrode material as it allows the growth of c-axis oriented AlN films and can withstand a rather high AlN sputtering temperature of about 400 °C [4]. Then, a 1 μm thick AlN layer was sputter deposited and patterned by photolithography and a reactive ion etching (RIE) process. To form the top electrodes, a 100 nm thick Al layer was patterned and the device was finally released by XeF$_2$ etching of the Si substrate. Throughout the fabrication process, we closely monitored and maintained the intrinsic stress of deposited films to be less than ± 100 MPa to prevent any bending after release. The device yield was over 90% and more than 1000 resonators could be fabricated on a single 4 inch wafer. Here, we defined the yield as the number of devices that gave a measurable Q and k^2 using a vector network analyzer (VNA) out of the total number of released devices. Figure 2 shows the scanning electron microscope (SEM) images of the released devices.

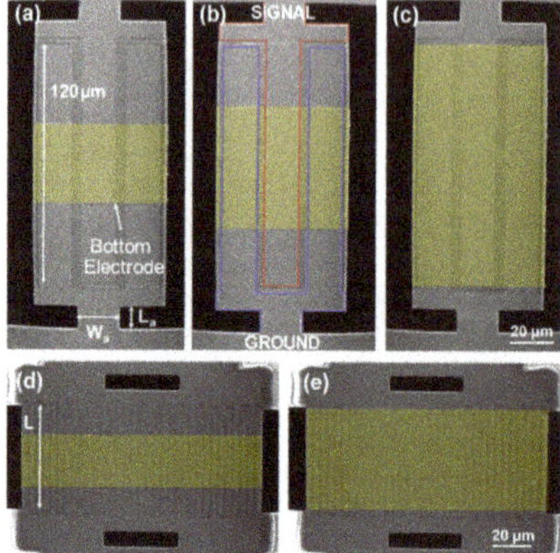

Figure 2. Scanning electron microscope (SEM) micrographs of the released AlN CMRs. (**a–c**) 230 MHz CMRs with bottom electrode coverage rates of 33%, 50%, and 100% (highlighted in yellow). (**d–e**) 1.1 CMRs with 50% and 100% bottom electrode coverage rates.

2.2. Experiment and Finite Element Analysis

The admittance response of AlN CMRs was measured using an RF probing setup and a vector network analyzer. From the measured data, the electromechanical properties, such as k^2 and Q, were extracted using a modified Butterworth–Van Dyke (mBVD) model. In total, we measured 160 devices corresponding to five devices per each resonator design as presented in Table 1. For room temperature analysis, all the devices were measured at about 293 K. We closely monitored and kept a consistent temperature to remove any temperature effects from the experiment. For low-temperature measurements at about 10 K, a cryogenic probe station was used along with the same data extraction method explained above. By comparing the results from the measurements at 293 and 10 K, we can understand how much TED is affecting device performance.

To understand the impact of bottom electrode designs on the mechanical damping of CMRs, we performed an FEA study using COMSOL Multiphysics software. For the purpose of performing quantitative analysis, we implemented a perfectly matched layer (PML) technique that can predict the mechanical damping and, thus, Q of CMRs with reasonable accuracy [23]. However, COMSOL FEA could not be used to predict TED and it was necessary to compare the measurements to predictions when TED is mitigated. Hence, we compared the FEA results to experimental data taken at 10 K, where the effect of any thermoelastic damping were expected to be extremely small in addition to the measurements taken at the ambient temperature of 293 K.

3. Results and Discussions

Figure 3 shows the admittance plots of 230 MHz and 1.1 GHz CMRs with varying bottom electrode coverages. Although the resonator size and top electrode configuration remained the same, the change in bottom electrode dimensions induced a significant shift in the admittance response of the resonators. For both frequency devices, f_0 decreased with increasing coverage rates, while E_{eq} and ρ_{eq} were dependent on the bottom electrode areas. Compared to 230 MHz CMRs, the relative shift in f_0 was smaller for 1.1 GHz devices. The regions with overlapped metal electrodes were substantially larger in terms of acoustic wavelength at 1.1 GHz with respect to the relative coverage at 230 MHz. Thus, changes in the metal overlapping area will have a greater impact on the relative frequency shift of lower-frequency devices than the 1.1 GHz CMRs. For 230 MHz CMRs, the max amplitude of admittance, which directly converts into motional resistance (R_m), also depended on the bottom electrode designs. For 230 MHz CMRs, the shift in R_m increased by as much as 1040%, while the maximum shift in

R_m was much smaller at 180% for the 1.1 GHz resonators. From the admittance plot analysis, it was noticeable that the impact of bottom electrode designs increased for lower-frequency CMRs.

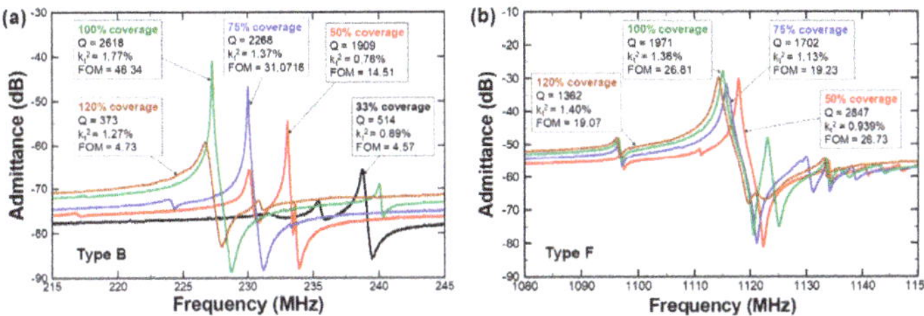

Figure 3. Admittance response of (**a**) 230 MHz and (**b**) 1.1 GHz CMRs. The changes in bottom electrode designs induce significant shift in electromechanical properties of AlN CMRs.

Figure 4 shows the measurement results for Q, k^2, and FOM of 230 MHz and 1.1 GHz CMRs as a function of bottom electrode coverage rate. Each data point represents the measurements from five devices of each design. The standard deviation for each measurement is also indicated on the plot. For 230 MHz CMRs, the measured Q fluctuated over the studied bottom electrode coverage. Such a trend in Q fluctuation was consistent regardless of the differences in anchor designs. By contrast, Q decreased with an increasing bottom electrode coverage rate in the case of 1.1 GHz CMRs, which were fabricated with full anchor designs. k^2 values increased with the bottom electrode coverages for both frequency CMRs. This indicates that the electromechanical transduction was more favorable for the larger active region of CMRs. The resulting FOM of CMRs are shown in Figure 4c,f. The FOM peaks were similar to Q for 230 MHz devices and remained relatively constant for 1.1 GHz CMRs.

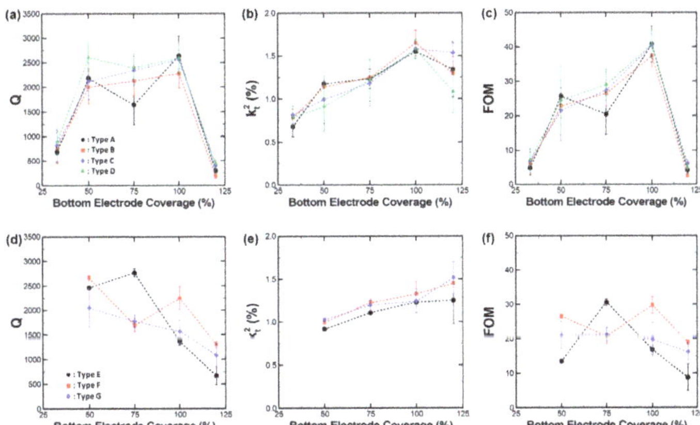

Figure 4. Measured electromechanical properties of (**a–c**) 230 MHz and (**d–f**) 1.1 GHz CMS with varying bottom electrode coverages. It is clear that both the quality factor (Q) and electromechanical coupling (k^2) of CMRs depend on the bottom electrode designs, ultimately affecting the device figure of merit (FOM).

The experiment results suggest that the bottom electrode coverage rates impact the device Q, or the damping inside the resonator. Specifically, the shift in Q was as large as 970% and 392% for 230 MHz and 1.1 GHz CMRs, respectively. To understand the impact of damping on CMRs, we ran FEA using the piezoelectric module and PML methods of the COMSOL Multiphysics software. The PML method accurately predicted the amount of mechanical damping along with the vibration characteristics of the resonators. Figure 5 shows the displacement field along the vibration direction of the resonator at f_0.

The expanded region is in red and the compressed is colored blue, which allows a visualization of CMR's physical motion. The contour images showed that the vibration region of the resonator was strictly limited to the region that was covered with both the top and bottom electrodes. For example, Figure 5a shows that the harmonic vibration of the resonator body was confined in the middle of the bottom electrode covered area. In other words, the significant portion of the region that was covered with the bottom electrode did not vibrate at f_0 for 230 MHz CMRs (Type D) with a 33% bottom coverage rate. However, as the electrode coverage rates increased, the vibrating region fully covered the bottom electrode areas, resulting in improved Q. For the 120% bottom coverage, the bus region also had a floating bottom electrode. Thus, the bus region itself worked as a single-finger CMR with a different resonant frequency compared to the region where the top and bottom electrodes overlapped. Such behavior was likely to hinder harmonic oscillation of the resonator and induce large damping, as shown in Figure 5d.

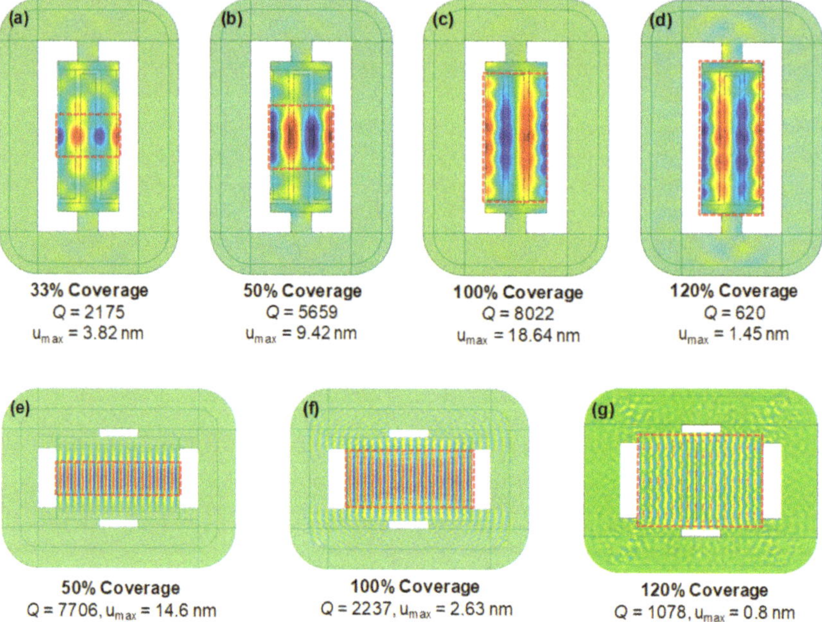

Figure 5. Contour images of vibration displacement fields of (**a**–**d**) Type D 230 MHz CMRs and (**e**–**g**) Type E 1.1 GHz CMRs with varying bottom electrode coverage rates.

Compared to the 230 MHz devices, the 1.1 GHz CMRs exhibited a trend where Q decreased with increasing bottom coverages. Figure 5e shows that the resonator vibration at f_0 was clearly limited to the region covered with a bottom electrode for a 50% coverage. However, an acoustic wave from the resonator body started to propagate towards the supporting Si substrate for higher coverage rates. When the bottom coverage was at maximum (120%), the dissipated wave from the resonator almost reached the PML boundary. Such wave propagations indicate that a significant amount of mechanical energy was being dissipated towards the supporting structures. Although the anchor loss was known to be smaller than other damping mechanisms for high-frequency CMRs [25], our FEA results indicate that the anchor design can still impact damping and such anchor loss should be also be accounted for when considering high-frequency CMRs.

To validate the FEA results, we measured the 230 MHz and 1.1 GHz CMRs at a cryogenic temperature of 10 K, where TED was largely mitigated. Hence, the Q of CMR was mostly limited by anchor losses or other mechanical damping effects. Figure 6 shows the admittance responses of an identical resonator at 10 and 293 K. The device Q improved by about 250% by eliminating TED. A previous study involving TED analysis [25] reported a comparable decrease in unloaded Q at lower operating temperatures. Thus, we can conclude that such large drop in Q is largely due to the mitigated TED at 10 K. In addition, the device f_0 shifted by about 5.7 MHz because the material properties of the resonator were dependent on operating temperatures.

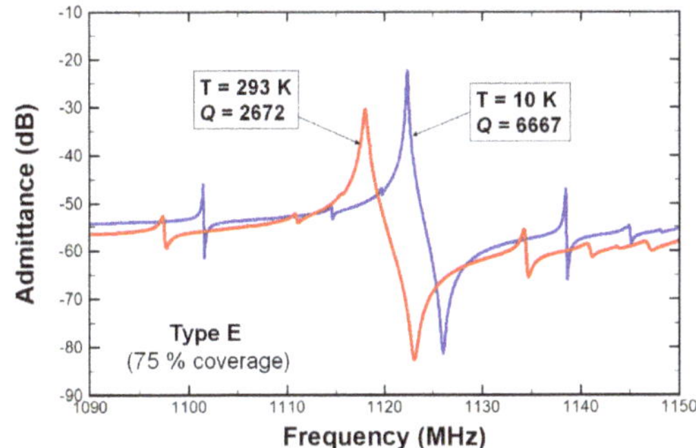

Figure 6. Admittance responses of 1.1 GHz (Type E with 75% bottom electrode coverage) at 293 and 10 K. Both Q and resonant frequencies (f_0) depend on operating temperatures.

In addition to the measurements at 290 K, all 160 CMRs measured at 10 K and Q had significantly improved for both 230 MHz and 1.1 GHz devices, as shown in Figure 7. Quantitatively, Q increased by about 239% and 208% for 230 MHz and 1.1 GHz CMRs, respectively. Such a large enhancement of Q indicates that the impact of TED is significant for CMRs regardless of operating frequencies, as previously reported [25]. Although TED was largely mitigated at 10 K, Q followed the trends observed at 293 K and we can assume that such fluctuations in the electromechanical properties of CMRs were largely affected by mechanical damping effects. The predicted Q of Type D and Type E CMRs were compared to the measured data. Their values followed a similar trend with reasonable accuracy, further validating our FEA approach. A rather large discrepancy between the measurement and the prediction was present for Type D devices with 100% bottom electrode coverage. In such cases, we assumed the actual device experienced additional sources of damping which were neither anchor losses nor TED.

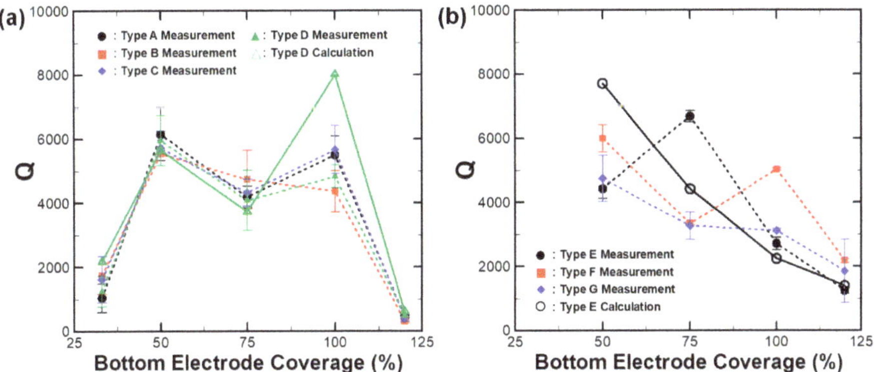

Figure 7. Measured (at 10 K) and predicted Q of (**a**) 230 MHz and (**b**) 1.1 GHz CMRs with varying bottom electrode coverages at 10 K. For all measured resonators, Q drastically improved by eliminating the thermoelastic damping (TED) effect. The prediction follows the general trends of the measurements.

From the measurement and FEA of CMRs with varying bottom electrode coverage rates, it is obvious that the electromechanical properties, such as Q and k^2, were impacted by the bottom electrode configurations. Both parameters were important as they affected both the power budget and noise level of CMR-integrated oscillators. In addition, Q ultimately sets the limit of detection (LOD) of AlN CMR-based sensors. Our results suggest that a shift in the bottom electrode designs induce changes in the amount of mechanical damping or anchor loss. By contrast, the impact of bottom electrode coverage rates on TED was rather small, as the ratio of metals to piezoelectric material in the active regions of the resonator remained constant. Since most CMRs or other types of piezoelectric MEMS resonators operate at temperature ranges where TED is present, it was important to optimize and account for the impact of the bottom electrode designs to alleviate mechanical damping in an effort to enhance the overall device performance.

The findings from this work can contribute to the enhancement of the current state of AlN CMR-based sensors. Such sensors utilize a functional layer (such as a magnetic material or chemically absorbing interface) on top of the resonator to detect a targeted event with outstanding sensitivity and LOD [32–37]. In addition, AlN CMRs with plasmonic nanostructures have been developed for spectrally selective infrared sensing [38]. For the aforementioned sensing applications, a certain portion of the resonator surface is covered or patterned with a functional layer, which cannot effectively work as a metal electrode. Since the sensor in [38] has a floating electrode with partial coverage, similar to the cases covered in this work. As we investigated the impact of varying electrode coverage rates, our results can be extended to determine the optimal ratio between the metal electrodes and a functional layer, or to predict the electromechanical properties of CMR-based sensors. More broadly, the results of this study can be applied to CMRs with interdigitated bottom electrodes and a floating top electrode.

4. Conclusions

We reported the impact of bottom electrode designs on electromechanical properties, such as Q, k^2, and FOM of 230 MHz and 1.1 GHz AlN CMRs. Both measurements and FEA calculation results showed that the change in bottom electrode coverage rates impact CMRs. Such an effect is largely due to the change in the amount of mechanical damping, which was more dominant in lower-frequency CMRs, and led to large fluctuations in Q. For 1.1 GHz CMRs, Q decreased with increasing bottom electrode areas as the damping towards supporting substrate increased. To validate our claims, we measured CMRs at 10 K, confirming that the experimental results match the theoretical predictions. Compared to mechanical damping or anchor loss, the effect of TED remained constant regardless of changes in bottom electrode designs. This is because TED depends on the ratio of metal-to-piezoelectric material coverage only in the active region of the resonator. We believe our findings can contribute to the enhancement of the current state of AlN CMRs and possibly other types of piezoelectric MEMS resonators by improving their electromechanical properties. These finding will directly impact applications in which CMRs are used as the main components, such as in oscillator and sensor systems.

Author Contributions: Conceptualization, G.P. and H.J.K.; Experiment, S.I.J., C.R., and H.J.K.; writing—original draft preparation, S.I.J., C.R. and H.J.K.; writing—review and editing, G.P. and H.J.K.; funding acquisition, G.P. and H.J.K.

Funding: This research was supported by Basic Science Research Program through the National Research Foundation of Korea (NRF) funded by the Ministry of Science, ICT & Future Planning (2018R1C1B6008041), KIMM of the Ministry of Science and ICT of South Korea (2018010152), and DARPA Dynamics Enabled Frequency Sources (DEFYS) program (FA86501217264).

Acknowledgments: Authors would like thank staff members from CMU Nanofab and DGIST CCRF device cleanroom for their assistance in device fabrication and characterization.

Conflicts of Interest: The authors declare no conflict of interest.

References

1. Hung, L.W.; Nguyen, C.T.C. Capacitive-Piezoelectric Transducers for High- Q Micromechanical AlN Resonators. *J. Microelectromech. Syst.* **2014**, *24*, 458–473. [CrossRef]
2. Ruby, R.C.; Bradley, P.; Oshmyansky, Y.; Chien, A.; Larson, J. Thin film bulk wave acoustic resonators (FBAR) for wireless applications. In Proceedings of the 2001 IEEE Ultrasonics Symposium. Proceedings. An International Symposium, Atlanta, GA, USA, 7–10 October 2001; pp. 813–821.
3. Smith, G.L.; Pulskamp, J.S.; Sanchez, L.M.; Potrepka, D.M.; Proie, R.M.; Ivanov, T.G.; Rudy, R.Q.; Nothwang, W.D.; Bedair, S.S.; Meyer, C.D. PZT-based piezoelectric MEMS technology. *J. Am. Ceram. Soc.* **2012**, *95*, 1777–1792.
4. Piazza, G.; Stephanou, P.J.; Pisano, A.P. Piezoelectric aluminum nitride vibrating contour-mode MEMS resonators. *J. Microelectromech. Syst.* **2006**, *15*, 1406–1418. [CrossRef]
5. Qian, Z.; Kang, S.; Rajaram, V.; Cassella, C.; McGruer, N.E.; Rinaldi, M. Zero-power infrared digitizers based on plasmonically enhanced micromechanical photoswitches. *Nat. Nanotechnol.* **2017**, *12*, 969. [CrossRef]
6. Shkel, A.A.; Kim, E.S. Continuous Health Monitoring with Resonant-Microphone-Array-Based Wearable Stethoscope. *IEEE Sens. J.* **2019**, *19*, 4629–4638. [CrossRef]
7. Kochhar, A.; Galanko, M.E.; Soliman, M.; Abdelsalam, H.; Colombo, L.; Lin, Y.-C.; Vidal-Álvarez, G.; Mukherjee, T.; Weldon, J.; Paramesh, J. Resonant Microelectromechanical Receiver. *J. Microelectromech. Syst.* **2019**, *28*, 327–343. [CrossRef]
8. Zhu, H.; Lee, J.E.Y. AlN piezoelectric on silicon MEMS resonator with boosted Q using planar patterned phononic crystals on anchors. In Proceedings of the 2015 28th IEEE International Conference on Micro Electro Mechanical Systems (MEMS), Estoril, Portugal, 19–22 January 2015; pp. 797–800.
9. Lin, C.M.; Chen, Y.Y.; Felmetsger, V.V.; Senesky, D.G.; Pisano, A.P. AlN/3C-SiC composite plate enabling high-frequency and high-Q micromechanical resonators. *Adv. Mater.* **2012**, *24*, 2722–2727. [CrossRef]
10. Karabalin, R.; Matheny, M.; Feng, X.; Defaÿ, E.; Le Rhun, G.; Marcoux, C.; Hentz, S.; Andreucci, P.; Roukes, M. Piezoelectric nanoelectromechanical resonators based on aluminum nitride thin films. *Appl. Phys. Lett.* **2009**, *95*, 103111. [CrossRef]
11. Pourkamali, S.; Hashimura, A.; Abdolvand, R.; Ho, G.K.; Erbil, A.; Ayazi, F. High-Q single crystal silicon HARPSS capacitive beam resonators with self-aligned sub-100-nm transduction gaps. *J. Microelectromech. Syst.* **2003**, *12*, 487–496.
12. Lee, J.-Y.; Seshia, A. 5.4-MHz single-crystal silicon wine glass mode disk resonator with quality factor of 2 million. *Sens. Actuators A-Phys.* **2009**, *156*, 28–35. [CrossRef]
13. Yunhong, H.; Meng, Z.; Guowei, H.; Chaowei, S.; Yongmei, Z.; Jin, N. A review: Aluminum nitride MEMS contour-mode resonator. *J. Semicond.* **2016**, *37*, 101001.
14. Jung, S.I.; Piazza, G.; Kim, H.J. The Impact of Bottom Electrode Coverage Rate on Electromechanical Coupling and Quality Factor of AlN MEMS Contour Mode Resonators. In Proceedings of the 2019 20th International Conference on Solid-State Sensors, Actuators and Microsystems & Eurosensors XXXIII (TRANSDUCERS & EUROSENSORS XXXIII), Berlin, Germany, 23–27 June 2019; pp. 917–920.
15. Qiu, X.; Zhu, J.; Oiler, J.; Yu, C.; Wang, Z.; Yu, H. Film bulk acoustic-wave resonator based ultraviolet sensor. *Appl. Phys. Lett.* **2009**, *94*, 151917. [CrossRef]
16. Lee, J.B.; Kim, H.J.; Kim, S.G.; Hwang, C.S.; Hong, S.H.; Shin, Y.H.; Lee, N.H. Deposition of ZnO thin films by magnetron sputtering for a film bulk acoustic resonator. *Thin Solid Films* **2003**, *435*, 179–185. [CrossRef]
17. Zhang, H.; Marma, M.S.; Kim, E.S.; McKenna, C.E.; Thompson, M.E. A film bulk acoustic resonator in liquid environments. *J. Micromech. Microeng.* **2005**, *15*, 1911. [CrossRef]
18. Wu, X.; Zuo, C.; Zhang, M.; Van der Spiegel, J.; Piazza, G. A 47 µW 204MHz AlN Contour-Mode MEMS based tunable oscillator in 65nm CMOS. In Proceedings of the 2013 IEEE International Symposium on Circuits and Systems (ISCAS2013), Beijing, China, 19–23 May 2013; pp. 1757–1760.
19. Otis, B.P.; Rabaey, J.M. A 300 µmW 1.9 GHz CMOS oscillator utilizing micromachined resonators. In Proceedings of the 28th European Solid-State Circuits Conference, Firenze, Italy, 24–26 September 2002; pp. 151–154.
20. Davis, Z.J.; Svendsen, W.; Boisen, A. Design, fabrication and testing of a novel MEMS resonator for mass sensing applications. *Microelectron. Eng.* **2007**, *84*, 1601–1605. [CrossRef]

21. Vig, J.R.; Kim, Y. Noise in microelectromechanical system resonators. *IEEE Trans. Ultrason. Ferroelectr. Freq. Control* **1999**, *46*, 1558–1565. [CrossRef]

22. Sauvage, G. Phase noise in oscillators: A mathematical analysis of Leeson's model. *IEEE Trans. Instrum. Meas.* **1977**, *26*, 408–410. [CrossRef]

23. Frangi, A.; Cremonesi, M.; Jaakkola, A.; Pensala, T. Analysis of anchor and interface losses in piezoelectric MEMS resonators. *Sens. Actuators A-Phys.* **2013**, *190*, 127–135. [CrossRef]

24. Segovia-Fernandez, J.; Cremonesi, M.; Cassella, C.; Frangi, A.; Piazza, G. Anchor losses in AlN contour mode resonators. *J. Microelectromech. Syst.* **2014**, *24*, 265–275. [CrossRef]

25. Segovia-Fernandez, J.; Piazza, G. Thermoelastic damping in the electrodes determines $ Q $ of AlN contour mode resonators. *J. Microelectromech. Syst.* **2017**, *26*, 550–558. [CrossRef]

26. Cassella, C.; Segovia-Fernandez, J.; Piazza, G. Segmented electrode excitation of aluminum nitride contour mode resonators to optimize the device figure of merit. In Proceedings of the 2013 Transducers & Eurosensors XXVII: The 17th International Conference on Solid-State Sensors, Actuators and Microsystems, Barcelona, Spain, 16–20 June 2013; pp. 506–509.

27. Giovannini, M.; Yazici, S.; Kuo, N.K.; Piazza, G. Apodization technique for spurious mode suppression in AlN contour-mode resonators. *Sens. Actuators A-Phys.* **2014**, *206*, 42–50. [CrossRef]

28. Cassella, C.; Singh, N.; Soon, B.W.; Piazza, G. Quality factor dependence on the inactive regions in AlN contour-mode resonators. *J. Microelectromech. Syst.* **2015**, *24*, 1575–1582. [CrossRef]

29. Kim, H.J.; Jung, S.I.; Segovia-Fernandez, J.; Piazza, G. The impact of electrode materials on 1/f noise in piezoelectric AlN contour mode resonators. *AIP Adv.* **2018**, *8*, 055009. [CrossRef]

30. Ho, G.K.; Abdolvand, R.; Sivapurapu, A.; Humad, S.; Ayazi, F. Piezoelectric-on-silicon lateral bulk acoustic wave micromechanical resonators. *J. Microelectromech. Syst.* **2008**, *17*, 512–520. [CrossRef]

31. Abdolvand, R.; Lavasani, H.M.; Ho, G.K.; Ayazi, F. Thin-film piezoelectric-on-silicon resonators for high-frequency reference oscillator applications. *IEEE Trans. Ultrason. Ferroelectr. Freq. Control* **2008**, *55*, 2596–2606. [CrossRef]

32. Kim, H.J.; Wang, S.; Xu, C.; Laughlin, D.; Zhu, J.; Piazza, G. Piezoelectric/magnetostrictive MEMS resonant sensor array for in-plane multi-axis magnetic field detection. In Proceedings of the 2017 IEEE 30th International Conference on Micro Electro Mechanical Systems (MEMS), Las Vegas, NV, USA, 22–26 January 2017; pp. 109–112.

33. Nan, T.; Hui, Y.; Rinaldi, M.; Sun, N.X. Self-biased 215MHz magnetoelectric NEMS resonator for ultra-sensitive DC magnetic field detection. *Sci. Rep.* **2013**, *3*, 1985. [CrossRef]

34. Rinaldi, M.; Zuniga, C.; Sinha, N.; Taheri, M.; Piazza, G.; Khamis, S.M.; Johnson, A. Gravimetric chemical sensor based on the direct integration of SWNTS on ALN Contour-Mode MEMS resonators. In Proceedings of the 2008 IEEE International Frequency Control Symposium, Honolulu, HI, USA, 18–21 May 2008; pp. 443–448.

35. Qian, Z.; Hui, Y.; Liu, F.; Kar, S.; Rinaldi, M. Chemical sensing based on graphene-aluminum nitride nano plate resonators. In Proceedings of the 2015 IEEE Sensors, Bushan, Korea, 1–4 November 2015; pp. 1–4.

36. Pang, W.; Yan, L.; Zhang, H.; Yu, H.; Kim, E.S.; Tang, W.C. Femtogram mass sensing platform based on lateral extensional mode piezoelectric resonator. *Appl. Phys. Lett.* **2006**, *88*, 243503. [CrossRef]

37. Chen, D.; Wang, J.J.; Xu, Y. Hydrogen sensor based on Pd-functionalized film bulk acoustic resonator. *Sens. Actuators B-Chem.* **2011**, *159*, 234–237. [CrossRef]

38. Hui, Y.; Gomez-Diaz, J.S.; Qian, Z.; Alù, A.; Rinaldi, M. Plasmonic piezoelectric nanomechanical resonator for spectrally selective infrared sensing. *Nat. Commun.* **2016**, *7*, 11249. [CrossRef]

Article

A Novel Capacitance-Based In-Situ Pressure Sensor for Wearable Compression Garments

Steven Lao, Hamza Edher, Utkarsh Saini, Jeffrey Sixt and Armaghan Salehian *

Energy Harvesting and Vibrations Lab, Mechanical and Mechatronics Engineering, University of Waterloo, Waterloo, ON N2L 3G1, Canada; sblao@uwaterloo.ca (S.L.); hedher@uwaterloo.ca (H.E.); utkarsh.saini@edu.uwaterloo.ca (U.S.); jasixt@uwaterloo.ca (J.S.)

* Correspondence: salehian@uwaterloo.ca

Received: 20 September 2019; Accepted: 29 October 2019; Published: 31 October 2019

Abstract: This paper pertains to the development & evaluation of a dielectric electroactive polymer-based tactile pressure sensor and its circuitry. The evaluations conceived target the sensor's use case as an in-situ measurement device assessing load conditions imposed by compression garments in either static form or dynamic pulsations. Several testing protocols are described to evaluate and characterize the sensor's effectiveness for static and dynamic response such as repeatability, linearity, dynamic effectiveness, hysteresis effects of the sensor under static conditions, sensitivity to measurement surface curvature and temperature and humidity effects. Compared to pneumatic sensors in similar physiological applications, this sensor presents several significant advantages including better spatial resolution, compact packaging, manufacturability for smaller footprints and overall simplicity for use in array configurations. The sampling rates and sensitivity are also less prone to variability compared to pneumatic pressure sensors. The presented sensor has a high sampling rate of 285 Hz that can further assist with the physiological applications targeted for improved cardiac performance. An average error of ± 5.0 mmHg with a frequency of 1–2 Hz over a range of 0 to 120 mmHg was achieved when tested cyclically.

Keywords: capacitive pressure sensors; in-situ pressure sensing; sensor characterization; physiological applications; cardiac output

---

1. Introduction

Lower limb compression garments are regularly employed in clinical settings [1] and, more recently, by athletes seeking to improve performance or decrease recovery time [2,3]. In medical settings, compression devices including passive compression socks and active intermittent pneumatic compression devices, are used in the treatment of hypertrophic scarring, venous ulcers, deep vein thrombosis and other diseases [1]. Some examples of physiological testing in [4,5] have shown a significant increase in the mean arterial pressure, cardiac output and cardiac stroke volume, with a significant reduction in the heart rate through applying active compression cyclically at frequencies between 1 to 2 Hz and at pressures of approximately 10 mmHg in the knee region. A major challenge for any compression device, regardless of application, is ensuring the consistency and repeatability of the applied pressure throughout its use [6]. The limited sizing options of most passive compression gear can vary the pressure experienced between subjects due to body shape differences. Further, the consistency of pressure applied by the garment on the same subject can be affected by the donning method [7]. A further challenge is that the shape, size and tissue compliance of the leg can change during compression application due to fluid redistribution, muscle contraction and movement of the garment/device.

Partsch et al. [8] made recommendations on the characteristics that an ideal sensor would need in order to measure the interface pressure between a compression garment and the lower leg skin.

Specifically, they identify the need for sensors that are low cost, low hysteresis, have minimal creep, high sampling rate, high accuracy, flexibility, durability and small sensor surface area. To date, there has yet to be a sensor developed that meets all these requirements.

Burke et al. [9] and Ferguson-Pell et al. [10] examined the use of Tekscan FlexiForce and Interlink Electronics FSR piezoresistive-based sensors for measuring interface pressure in compression garments. Piezoresistive sensors function by measuring the change in resistance under a mechanical load. They are thin and can provide reliable results during dynamic measurements. However, these sensors suffer from significant hysteresis and drift, and thus are not dependable for long time-scale measurements [8]. Additionally, piezoresistive sensors are dramatically affected by base curvatures under 32 mm in radius [10] and require relatively large pressures for reliable measurements [11]. The total error exhibited in the sensors, accounting for errors in repeatability, hysteresis and linearity, is approximately +/− 10 mmHg for a pressure range 0 to 96 mmHg, [9], which limits the utility of these devices for pressure mapping of compression garments. Note, for the remainder of the document, unless otherwise noted, errors are described for the pressure range of interest between 0 and 100 mmHg.

Similar to piezoresistive sensors, strain gauges also respond to a change in resistance with mechanical strain [12,13]. Kraemer et al. [14] and Ghosh et al. [15] utilized strain gauges to measure the compression applied by compression garments/bandages on a subject's thigh and a mannequin leg, respectively. However, since strain gauges exhibit a better sensitivity along their largest dimension, these sensors tend to become thick for measuring interfacial pressures. Additionally, strain gauges tend to exhibit thermal and humidity dependence, and high hysteresis [12].

Pneumatic-based sensors are one of the most prevalent designs for measuring compression garment interfacial pressure [9]. These sensors utilize bladders or compartments that are filled with a small volume of air connected to a pressure transducer through a flexible hose [16]. As pressure is applied, the volume of the bladder decreases, thus increasing the internal pressure. At a constant temperature, it is assumed that the internal pressure measured by the transducer is equivalent to the external applied pressure [16]. Commercially available pneumatic sensors, such as the Salzmann MST MKIV [17], MediGroup Kikuhime [18] and Microlab PicoPress [19] are thin and flexible devices that can be used to measure interfacial pressures between the body and compression device. Furthermore, pneumatic pressure transducers have been found to exhibit repeatable results within +/− 3 mmHg of error [8,19]. A major shortfall of this technology is its temperature dependence [8]. Additionally, their high sensitivity to curvature can result in overestimates of pressure by up to 150% [9]. The sampling rate of the system is affected by the tubing length, which can be important in the evaluation of rapid inflation active compression systems. Additionally, pressure sensing with an array of pneumatic sensors presents multiple inconveniences. The sensors each require a dedicated pressure transducer and have a large surface area (a PicoPress sensor is approximately 50 mm in diameter), resulting in poor spatial resolution. These types of sensors need a great amount of tubing and rely on pressure transducers that are bulky and almost 10 times larger in footprint than the actual sensor, [19]. Sensor multiplexing can reduce the number of transducers in a system, which uses solenoid valves to isolate each bladder for pressure reading (see Figure 1). However, the addition of a manifold increases the system volume, lowering the sampling rate and the transducer's sensitivity to bladder volume change.

Capacitive sensors have also been employed in other textile applications, [20–22], such as muscle activity detection, [20], and pressure mapping for insoles, [21]. These sensors typically consist of a textile used as a dielectric material with conductive yarn woven as electrodes. They have the benefit of being small (good spatial resolution), flexible, and easily used in arrays for pressure distribution measurements [23,24]. Compared to other sensor technologies, they also possess relatively low temperature sensitivities [23,24]. Q. Guo et al. [25] developed a tactile floating electrode capacitive sensor which displayed good linearity up to 350 kPa and improved durability over parallel electrode designs. Capacitive measurements also present their own challenges, often requiring complex circuitry to filter measurement noise [23]. Signal hysteresis can also affect performance, as the dielectric material is often viscoelastic and exhibits nonlinear behaviour when stressed [26]. Additionally, the current-pressure relation is nonlinear [27] which makes it more challenging for proper calibration. Nano-fibers have been employed in the development of pressure sensing technologies for other applications [27,28]. However, due to the high temperature sensitivity of materials used, such as gold, they are not suitable for the application of interest.

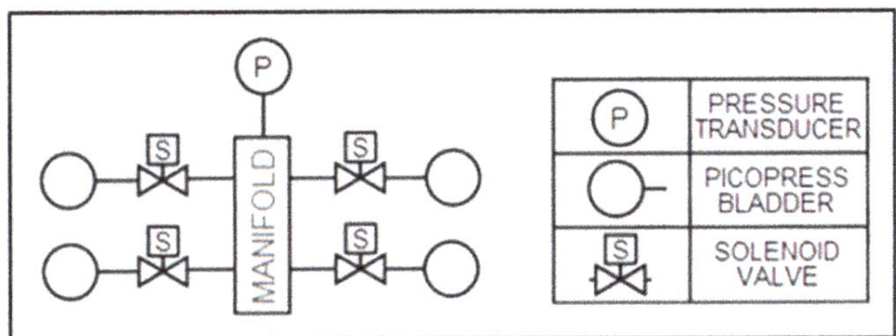

Figure 1. Multiplexing example of four pneumatic sensors.

In this paper, a novel, capacitive-based, dielectric electroactive polymer (DEAP) pressure sensor is developed in collaboration with StretchSense Ltd. for in-situ pressure measurements under textile garments. Specific testing conditions are required when evaluating pressure sensors for compression garments, since the interface is neither flat nor solid. Thus, multiple testing methods are presented in Section 3, with results in Section 4. The sensor's pressure sensitivity, error and durability are characterized. The performance effects due to temperature, humidity, boundary conditions and hysteresis are also examined. This is the first sensor to use DEAP materials measuring low pressure ranges (0 to 100mmHg) for portable and wearable applications, including the fast sampling rate required for many physiological applications. In addition to the sensor fabrication and characterization, the small, portable wireless circuitry for the proposed sensor is one of the novel aspects of this work. The wireless sensing unit communicates data over Bluetooth with the actuation system.

2. Capacitive Sensors

The custom designed sensors developed in collaboration with StretchSense Ltd. for this research are capacitive tactile sensors operating on the principle of a deformable parallel plate capacitor model (shown schematically in Figure 2). The dielectric layer is made of a silicone-based DEAP. The model consists of two compliant electrode layers of length l and width w separated by a dielectric material of thickness d forming a compliant capacitor [29].

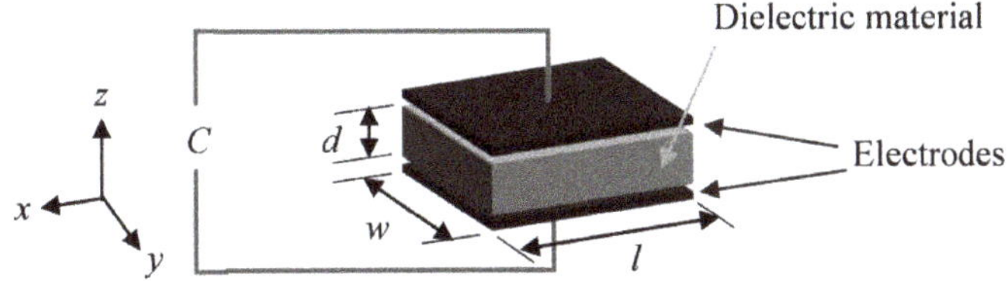

Figure 2. Parallel plate capacitor model of the pressure sensor.

The capacitance C, for a parallel plate capacitor is described by:

$$C = \varepsilon_r \varepsilon_0 \frac{lw}{d} \tag{1}$$

where ε_r and ε_0 are the relative permittivity of the dielectric layer and the vacuum permittivity respectively. Compression of the capacitive sensor along the z-axis, (see Figure 2), causes a decrease in the dielectric layer thickness and the distance between the two electrodes, d. Assuming unconstrained edges and an incompressible material for the dielectric layer, this results in an increase in the area, (lw), [26]. Both changes result in an increase of the capacitance, which can then be correlated to an applied pressure.

A picture of the capacitive sensor used in this study is shown in Figure 3. This is the third generation of the sensor developed by the authors, which has better sensitivity, increased sampling frequency, wider moving average filter, improved manufacturing procedure and improved noise rejection. The system consists of a battery-powered circuit capable of measuring the capacitance of five sensing elements. There is a trade-off with respect to sensor head size, as the change in capacitance with pressure is directly proportional to the sensor area, while the spatial resolution reduces with increasing size. In order to maintain a small footprint, the outer surface of the sensor is electrically insulated and folded onto itself like an accordion to increase the capacitor area. This folding of the sensor effectively creates a multi-layer sensor, where each layer experiences the same pressure, resulting in a higher initial capacitance and improved pressure sensitivity. As evident in Figure 3, the sensor head is perforated with a series of holes. These holes decrease the overall stiffness of the sensor, allowing greater deformation during compression. Figure 4 shows a simplified schematic of the pressure sensor with major dimensions labeled. Table 1 provides the dimensions and elastic modulus of the StretchSense pressure sensor, as well as specifications of the battery used. The elastic modulus was averaged between two sample tensile tests at a strain rate of 12 mm/minute and is applicable for strains below 10%. In comparison to pneumatic sensors, the sensor design used is only ~25% of the PicoPress bladder surface area. This provides a significant advantage in terms of the design's spatial resolution.

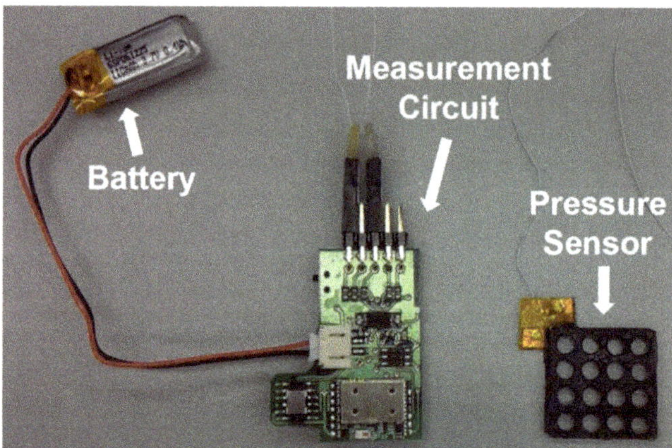

Figure 3. StretchSense pressure sensor, wireless measurement circuit, and battery.

Table 1. StretchSense pressure sensor and measurement equipment details.

Property	Value
Sensor width, a	22 mm
Sensor hole diameter, d	3.75 mm
Sensor thickness, t	2.25 mm
Sensor weight	1.28 g
Sensor elastic modulus	0.78 MPa
Battery weight	3.04 g
Battery capacity	110 mAh
Battery voltage	3.7 V

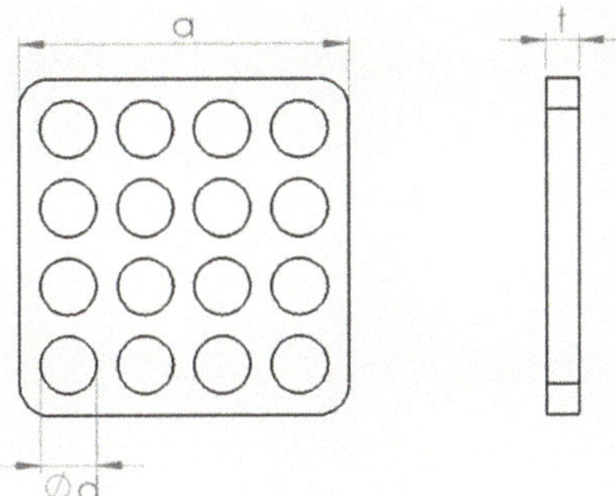

Figure 4. StretchSense pressure sensor schematic.

The capacitance of the sensor is sampled at 660 Hz by a measurement circuit made by StretchSense. The circuit then applies a moving average of 20 data points to generate a final output at 285 Hz. A merit of using capacitive sensing is the stability of sampling rates that are largely independent from experimental setup conditions. Conversely, pneumatic sensor sampling rates are more susceptible to setup conditions as discussed in Section 1. This high sampling rate is particularly useful for the physiological applications where more dynamic pressure applications, such as pulsations tuned at a person's heart rate, are used to increase the cardiac performance [4,5]. The control circuitry used is designed to minimize noise while maximizing the output data rate.

Comparison to Alternative Capacitive Sensors

The unique DEAP construction presents multiple advantages compared to many parallel plate and floating electrode constructions. Since the electrodes are made of a compliant material with similar stiffness to the silicone dielectric layer, interlayer stresses during bending and stretching are minimized. This construction avoids using thin traces that are common for many parallel plate structures, which may break under repeated bending loads, as also noted by M-Y Cheng et al. [30]. Additionally, a major advantage of the DEAP parallel plate construction used is its simplicity and resistance to fail under many cycles of loading. The fabrication of the floating electrode and parallel plate sensors is often more complex, requiring multi-step micromachining processes to create air gaps and metal layers [25,31]. Conversely, the sensor presented avoids such process steps during fabrication. This greatly improves manufacturing scalability and reduces design complexity.

When compared to the DEAP parallel electrode construction used, Q. Guo et al. [25] presented a floating electrode design using a robust construction and good linearity over a wider pressure sensing range (up to 350 kPa, or 2625 mmHg). However, the target application of medical compression garments where pulsations are used requires much lower pressure ranges, (0–100 mmHg). Within the scope of these smaller signals, sensor noise and error reduction becomes increasingly critical for the target design. M-Y Cheng et al. [30] showed that while theoretical sensitivities of parallel plate and floating electrode constructions are the same, parallel plate constructions are predicted to have a better signal-to-noise ratio (SNR). Thus, for the application's low-pressure operating range, the parallel plate construction used is more advantageous to maximize SNR.

3. Experimental Methods

In this section, the test procedures assessing sensor performance in representative configurations are outlined. One of the challenges in assessing in-situ pressure sensor performance is the difficulty in applying a realistic and measurable (known) boundary (loading) condition. Research that includes the measurement of pressures applied to the leg often neglects the measurement errors introduced when moving from laboratory testing facilities to in-situ measurements. For example, sensor calibration is typically performed on a flat, solid surface; the effect of curvature and compliance of the leg are not examined [16]. Therefore, one of the objectives of this section is to study the errors that are introduced when taking in-situ pressure measurements on a leg. For this purpose, a series of representative tests were formulated to examine the behaviour of these capacitance-based sensors in different modes. Figure 5 depicts four test setups used to validate the sensor. The tests in all configurations were repeated several times to gather relevant statistics.

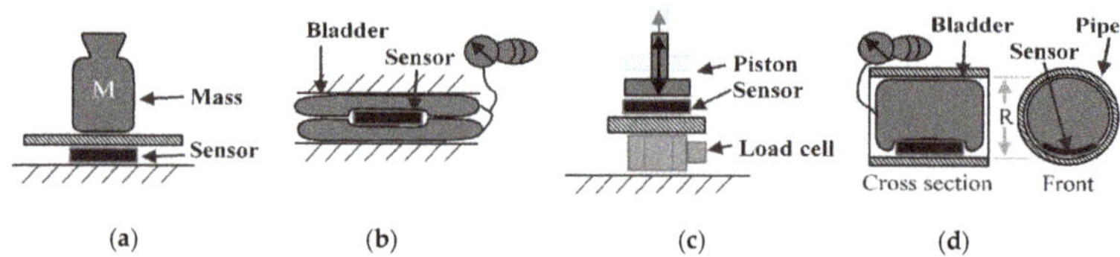

Figure 5. Test configurations for pressure sensor validation. (**a**) Mass Test, (**b**) Bladder Test, (**c**) Piston Test, (**d**) Curvature Test.

The mass test (Figure 5a) is performed on a flat surface where the sensor sits with masses stacked on top of it. The mass applied increases in 50 g increments from 0 g up to 550 g. The applied pressure to the sensor is determined by measuring the total weight placed on the sensor over the area of the sensor. The sensor is subjected to static loads at increments of approximately 7 mmHg that are held for 30 s at a time, up to a maximum pressure of 80 mmHg. The average and standard deviation of the capacitance for a 10 s period is collected for each increment. The 10 s period is taken after the load has been applied and the measurement reading has settled. The loading profile (shown in Figure 6) has four incremental loading and unloading cycles and nine direct loading and unloading cycles. The incremental cycles are used to measure the hysteresis inherent in the sensor. Meanwhile, the direct loading cycles closely match the use-case for these sensors when applied under, for example, intermittent pneumatic compression devices. The results from the first incremental loading and unloading cycle (white section in Figure 5) are discarded as a "bedding-in" cycle, as recommended by StretchSense, to allow the sensor to acclimate to a stable position. The second incremental loading and unloading cycle (light grey section) is then used to calibrate the sensor. A fit is applied to the calibration data and this curve fit is then used to convert the capacitance measurements of the remaining validation data (dark grey section) to pressure estimates. This test is meant to examine repeatability, linearity and hysteresis of the sensor in static conditions.

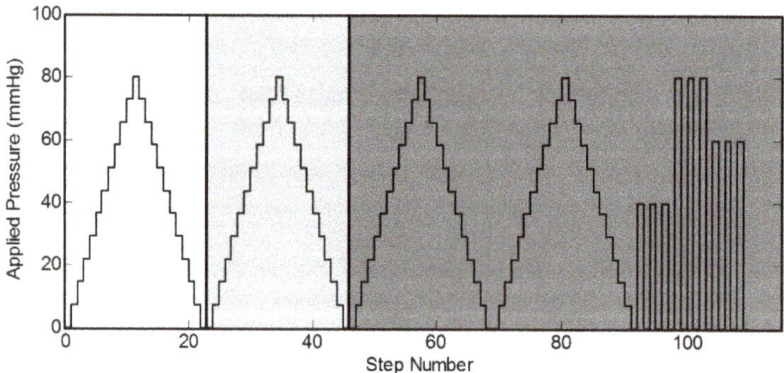

Figure 6. Testing protocol for applied load on sensor in the static tests.

Figure 5b depicts the bladder test, whereby the sensor is placed between two inflatable (18 cm × 38 cm) bladders. The bladders are inflated by a hand pump up to a pressure of 110 mmHg following the same timing procedure as used in the mass test, while the air pressure is measured using a manometer. The assumption is that the internal pressure of the bladder is transmitted to the pressure on the sensor.

The main purpose of this test is to mimic a condition where the sensor rests on a soft surface and is used as additional verification for the previous test. It should be noted that the sensor is extremely lightweight, weighing only 1.28 g, as stated in Section 2. As such, the weight of the sensor per area is negligible compared to the pressures of interest.

In the piston test (Figure 5c), the sensor is placed on top of a dynamic load cell (PCB Piezotronics 208C01, Depew, NY, USA) and an impulse is applied to the unit by a pneumatic cylinder. The pneumatic cylinder is controlled with a pressure regulator and flow restrictor valve to enable dynamic pressure application at 1–2 Hz. In this test, the peak values from the load cell and the sensor measurements in each cycle of the pneumatic cylinder were compared. Similar to the mass test, the pressure being applied to the sensor is simply the force measured by the load cell over the area of the sensor. With this test, the dynamic response and repeatability over 400 loading cycles was tested at peak pressures ranging from 0 to 120 mmHg.

The fourth test (Figure 5d) is used to determine the change in behaviour due to the curvature of the sensor. The curvature is expected to change the response of the sensor since it deforms when it conforms to a curved surface and, therefore, the capacitance changes. In the curvature test, the sensor is placed inside pipes of 19.9 mm and 39.9 mm radii, and a balloon is inserted and inflated to apply pressure in the range of 0 to 110 mmHg. The static pressure of the balloon is compared to the sensor reading, similar to the bladder test. It is noted that the curvature in this test remains constant and one of the surfaces is rigid, which is not the actual use case. However, this configuration was selected to isolate the effect of sensor curvature.

Finally, two additional testing configurations not shown in Figure 5 are employed. First, the effect of temperature and humidity on the no-load capacitance of the sensor was examined. This is done by repeating the mass test at 26 °C and then comparing it to the data from the 29 °C mass test. In addition, the variance in capacitance of the sensor is observed for a range of 17 to 42 °C and 45 to 69 %RH. Second, a curvature test compares the response of the capacitive sensor to that of the PicoPress pneumatic sensor. This test involved a 3D printed cylinder with a radius of 60 mm to serve as a rigid dummy leg (see Figure 7). The two sensors are placed on the cylinder with the StretchSense sensor against the cylinder and the PicoPress directly on top of it. Since the PicoPress sensor is considerably larger in size than the StretchSense sensor, the entirety of the StretchSense sensor contacts the PicoPress bladder. A sphygmomanometer is then wrapped around the cylinder and sensors to apply a known pressure. The pressure in the sphygmomanometer is manually controlled with the hand pump. The test involves a series of 10 inflations and deflations of the sphygmomanometer at three different pressures. The applied pressures are 40 mmHg, 60 mmHg, and 80 mmHg, respectively. During deflation, a minimum pressure of 5 mmHg is kept in the sphygmomanometer to prevent the cuff from moving and/or slipping. In this test, the sensor responses were both compared to the sphygmomanometer pressure and each sensor error were calculated.

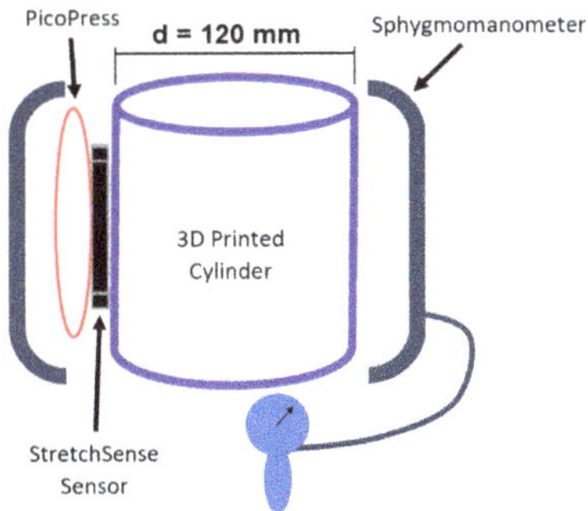

Figure 7. PicoPress and StretchSense pneumatic comparison test.

4. Results and Discussion

The tests described in the previous section were performed on the capacitive pressure sensor presented in Section 2. The results for each test are presented in this section.

4.1. Mass Test

The results of the mass test are shown in Figure 8, where the mean capacitance measurement is plotted along with error bars for each increment. The error bars indicate +/− twice the standard deviation of the measurement at each point, which gives a 95 % confidence that values lie between the bars. The graph is plotted with the capacitance on the x-axis since the interpretation of the sensor would require the reading of the capacitance to determine the pressure applied. The slope of a linear fit through all the data points (loading and unloading) indicates that the sensor sensitivity is 0.0847 pF/mmHg. The average error bar width is +/− 0.12 pF, which corresponds to an average pressure error of +/− 1.4 mmHg when converted using the sensitivity defined above.

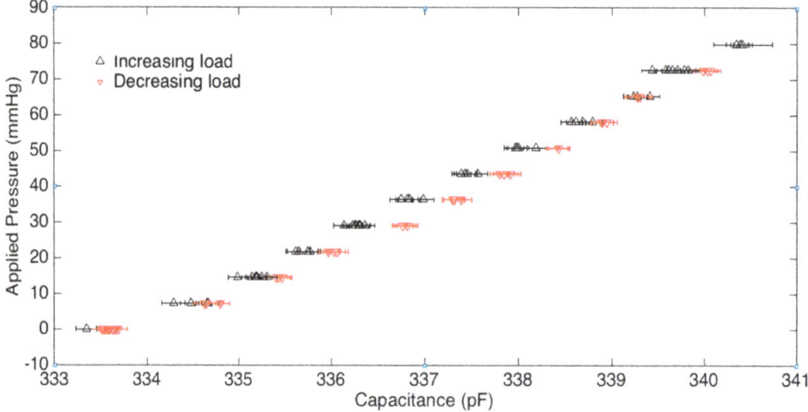

Figure 8. Sensor capacitance at various applied pressures–mass test results.

There is a discrepancy between measurements when the masses are being loaded versus unloaded, indicating that there is hysteresis in the system. The hysteresis error is defined as half of the span of the capacitance at a given pressure from loading versus unloading. The data points with increasing load are shown with triangles pointed upwards, while the decreasing loads are shown with triangles pointed downwards. The average hysteresis error across the applied pressures was computed to be +/− 3.0 mmHg. The overall error of the sensor in this test setup is just outside +/− 5.0 mmHg as also shown in Figure 9 which displays the predicted pressures versus the actual pressures using the aforementioned sensitivity calibration. As shown in this figure, the percentage error is significantly smaller for the upper range of pressure.

4.2. Bladder Test

The bladder test described in Section 3 was repeated three times and the results are shown in Figure 10. To ensure repeatability, the setup was disassembled and reassembled after every test. The pressure was applied using a sphygmomanometer of +/− 3 mmHg accuracy. The test results show an offset of 2 pF in one of the tests for all pressure ranges shown, however, it is noteworthy that the sensor sensitivity for the three tests remains the same. The causes for the offset are explored in the coming section that examines the effect of temperature and humidity on the sensor performance. The sensor sensitivity from the bladder test calibration was found to be 0.100 pF/mmHg, which is approximately 18% higher than the results from the mass test. This is likely due to the softer material in contact with the sensor during these tests that allows for larger expansion along the plane of the sensor, thus increasing its deformation. On a hard, rigid surface, the sensor sticks to the surface and the friction restricts deformation. In the case of the mass test, the rigid surface in contact with the sensor ultimately results in a smaller sensitivity for similar pressure values. The average measurement noise is approximately +/− 0.18 pF (+/− 1.8 mmHg).

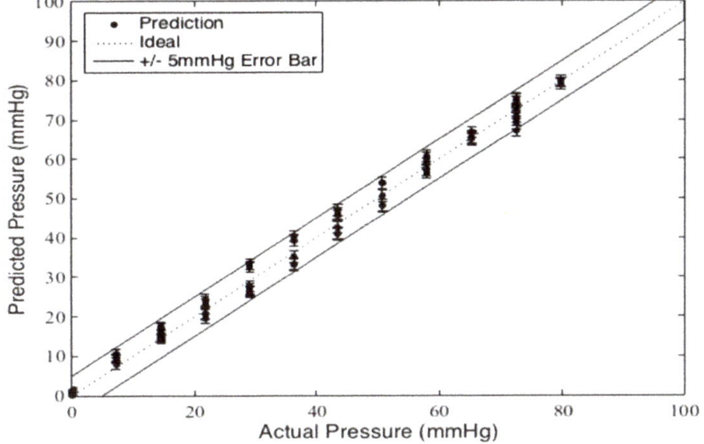

Figure 9. Predicted pressure compared to actual applied pressure in mass test.

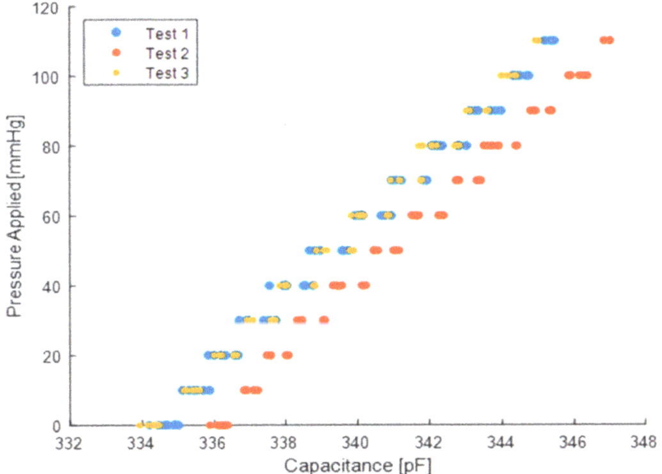

Figure 10. Sensor capacitance at various applied pressures–bladder test results.

4.3. Piston Test

The piston test setup is designed to examine the repeatability of the sensor readings over the course of over 400 loading and unloading cycles. Figure 11 presents the peak values of the load cell pressure readings versus the capacitance of the pressure sensor. The degree of vertical scatter in these data points for a given capacitance represents the repeatability of the sensor measurements, which is considerably smaller at larger pressure values. The hysteresis is not captured during this test since only the maximum load at each cycle is recorded. This results in a smaller overall error of +/− 5.0 mmHg when measuring the vertical scatter in the plot. The best curve fit for Figure 11 corresponds to a sensitivity value of 0.067 pF/mmHg which is on the same order of magnitude as the mass test sensitivity.

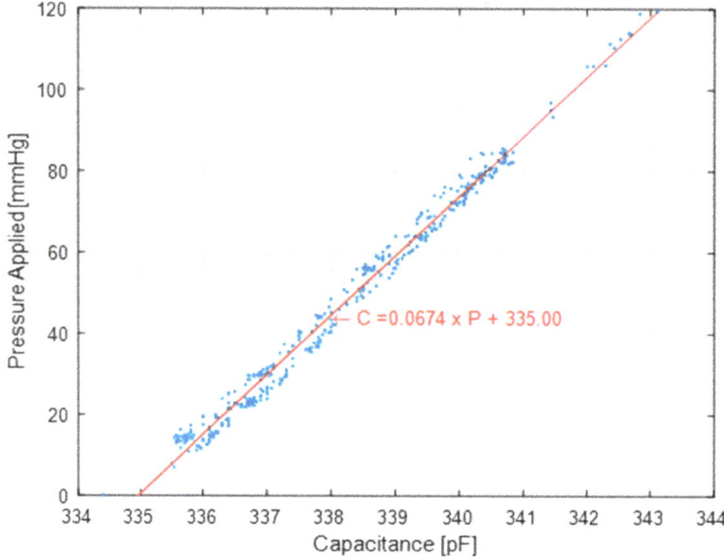

Figure 11. Capacitance change with dynamic pressure application–piston test results.

The temporal signals during one piston stroke are presented for the load cell and capacitive sensor in Figure 12. A sharp peak is noted at the start of the signal, which is attributed to the initial impact of the piston against the sensors. This impact is disregarded in the analysis. In order to allow for a comparison between the load cell and the StretchSense measurements, the capacitance was converted to pressure using a linear curve fit as the calibration. As shown in this figure, the magnitudes of the calibrated StretchSense signal and the load cell are in good agreement. The negligible 10–20 ms latency in the capacitive sensor measurements is expected, given the data smoothing done by the measurement circuit board.

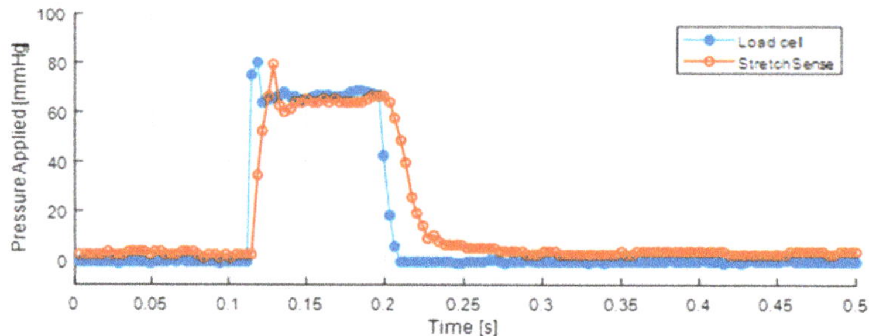

Figure 12. Dynamic response of load cell versus StretchSense sensor during one piston load.

The samples did not show signs of structural damage, including on the electrical connections, after applying more than 400 loading and unloading cycles. This is believed to be due to the perforated geometry and soft sensor material, which allows for compliance in the structure. This allows the sensor to perform better under fatigue loads, in comparison to a solid geometry. Electrical connections are also insulated and reinforced with Kapton tape to improve durability.

4.4. Curvature Test

The sensitivity to curvature of the capacitive sensors is shown in Figure 13 for a variety of pressures. The curvature clearly influences the sensor capacitance when subject to pressure. The sensitivity of the sensor was found to be 0.0812 pF/mmHg and 0.106 pF/mmHg for radii of curvature of 19.9 mm and 39.9 mm, respectively. From the bladder test, the sensitivity on a flat surface (infinite radius) is 0.100 pF/mmHg. As shown in this figure, the sensor sensitivity decreases as the radius of the curvature decreases, i.e., the sensor is more curved. This sensitivity to curvature may potentially impose limitations in the sensor calibration when mounted on a flexible object of variable curvature, such as a calf muscle. This can be mitigated if the sensor is securely placed against a rigid plate of fixed curvature attached to the user. The sensors may then be calibrated where the plate closely fits a person's local body shape during in-situ measurements. Of course, a rigid plate changes the pressure loading experienced by the compression user.

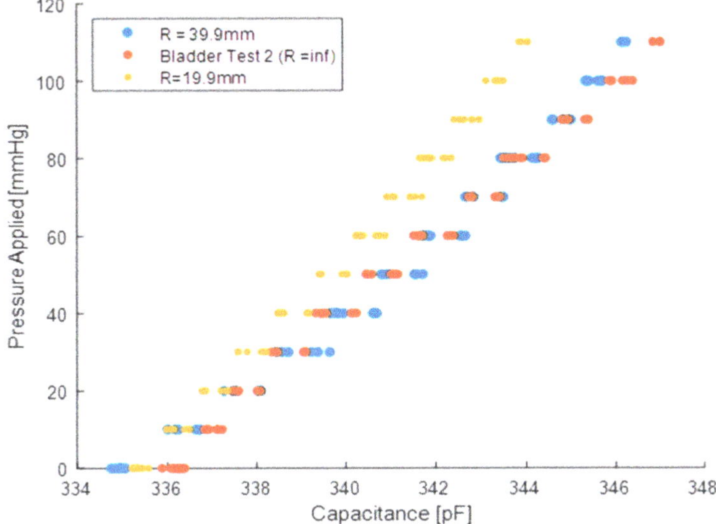

Figure 13. Pressure vs sensor capacitance for various radii of curvature.

4.5. Thermal and Humidity Effects

As shown in Section 4.2, an offset of up to 2 pF was observed between several test runs, which relates to a 20 mmHg variation across the pressure range for these tests. This is a significant source of error, particularly for the smaller pressure values. It is reasoned that the mechanical setup of the fixture should be tolerant to changes in orientation due to the conformity of the air bladders employed in the bladder test. To further examine the source of this offset value shown in Figure 10, the effect of temperature and humidity is studied.

In initial testing in the mass test setup, the temperature of the sensor was allowed to settle at a temperature of 29 °C and the test was performed. The temperature was then dropped to 26 °C and the mass test was repeated twice over an approximately two-hour period. As shown in Figure 14, there is a notable offset with the change in temperature. Like the bladder test results shown in Figure 10, the sensitivity of the sensor is the same despite the offset. This offset is termed the no-load offset.

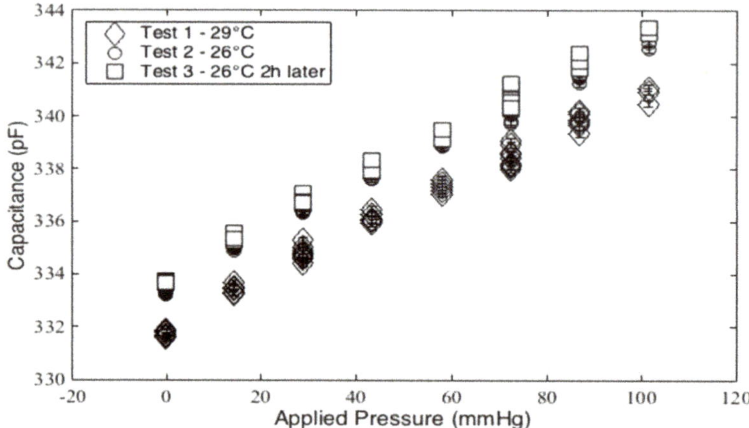

Figure 14. Capacitance vs applied pressure for various temperatures.

Further tests were performed to characterize the offset by measuring the capacitance under the no-load condition while varying temperature and humidity. The sensor was placed on a Peltier module, which acts as a hot/cold plate in order to control the sensor temperature. The capacitance was then measured at several ambient humidity and temperature values, as shown in Figure 15. Using the experimental results shown in this figure, the sensor's temperature sensitivity, or thermal coefficient, was found to be approximately −0.38 pF/°C, and the humidity sensitivity was measured to be −0.345 pF/%RH. This relates to approximate pressure errors of 4 mmHg/°C and 4 mmHg/%RH. These tests indicate that both the temperature and humidity have noticeable effects on the sensor's output. Thus, the variances in the bladder test results are likely attributed to the differences in the temperature and humidity conditions during these tests. It was noted, however, that the ambient conditions were not recorded prior to bladder testing.

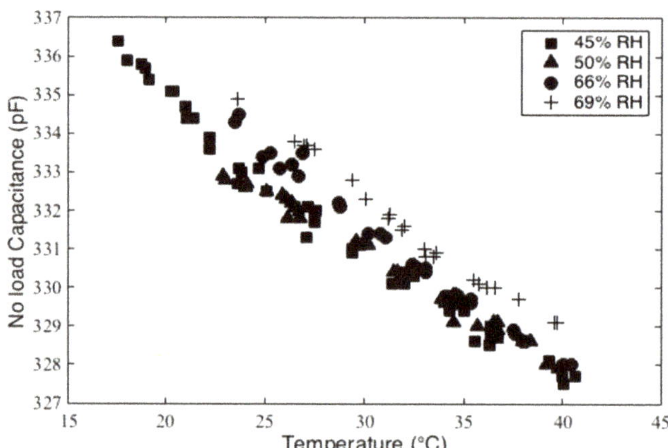

Figure 15. No load capacitance versus temperature and humidity.

The mathematical proof shown below explains the reason behind the offset created by temperature. It should be noted that it may be assumed that the permittivity remains fairly constant for the given temperature range. This is ultimately the reason for the sensitivity (slope of the pressure-capacitance plot) remaining unchanged for various temperatures. Therefore, assuming ε_r remains constant, any change in temperature, T, results in changes of thickness, length and width of the dielectric layer as shown below:

$$C_{T_0} = \varepsilon_r \varepsilon_0 \frac{lw}{d} \rightarrow C_T = \varepsilon_r \varepsilon_0 \frac{l(1 + \alpha\Delta T)w(1 + \alpha\Delta T)}{d(1 + \alpha\Delta T)} = \varepsilon_r \varepsilon_0 \frac{lw(1 + \alpha\Delta T)}{d} = C_{T_0}(1 + \alpha\Delta T) \qquad (2)$$

$$C_T = C_{T_0} + C_{T_0}\alpha\Delta T \qquad (3)$$

Here, C_T is the capacitance for temperature T, and C_{T_0} denotes the capacitance for temperature T_0. The above equation shows that a given temperature change, ΔT, results in a constant offset $C_{T_0}\alpha\Delta T$ compared to its reference value at T_0. Therefore, with permittivity constant, the pressure-capacitance relation has the same slope for all temperature values, and the temperature change only causes a capacitance offset. It should also be noted that the DEAP material has a negative coefficient of thermal expansion which results in a negative offset upon an increase of temperature as shown in Figure 15.

4.6. Pneumatic-Based Sensor Comparison

The results of the pneumatic sensor comparison tests are plotted in Figure 16. The linear sensitivity value obtained during the bladder test was used to calibrate the StretchSense sensor, since those test conditions were most-similar to the cylindrical test bed setup. The results show that the PicoPress exhibits an average error of +/− 6.38 mmHg. Meanwhile, the StretchSense sensor exhibits an average error of +/− 8.03mmHg. The error is calculated by averaging the sum of the sensor calibration error and twice the standard deviation for each pressure value. As described in earlier sections, these errors include the influence of hysteresis, measurement noise, curvature and biases in calibration. Decreasing the capacitive sensor footprint can help to reduce curvature sensitivity.

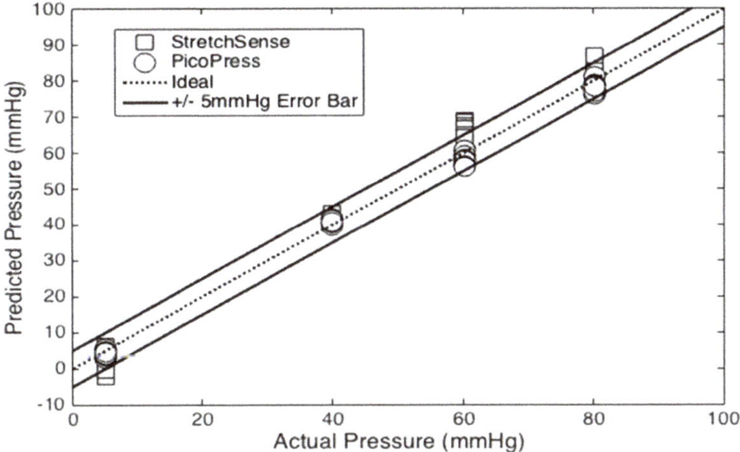

Figure 16. Experimental comparisons of StretchSense and PicoPress sensors.

From the tests performed, recorded sensor sensitivities and errors have been grouped in Table 2 for reference. The sensitivity and error are both seen to vary depending on the testing method implemented. The piston test reported the lowest sensitivity, and the pneumatic sensor comparison test returned the largest error observation. As a result, to perform a conservative characterization of similar sensors in the future, the piston and pneumatic sensor comparison tests should be used to determine sensor sensitivity and error, respectively.

Table 2. Sensor sensitivity and error values for each test.

Test Type	Sensor Sensitivity (pF/mmHg)	Sensor Error (mmHg)
Mass Test	0.085	± 5.0
Bladder Test	0.100	± 1.8
Piston Test	0.067	± 5.0
Curvature Test (19.9 mm radius)	0.081	N/A
Curvature Test (39.9 mm radius)	0.106	N/A
Pneumatic Sensor Comparison–PicoPress Sensor	N/A	± 6.4
Pneumatic Sensor Comparison–StretchSense Sensor	N/A	± 8.0

5. Conclusions

This paper assesses the feasibility of a capacitance-based DEAP sensor technology for tactile pressure measurement in compression garment applications. The sensor is the outcome of a few generations of design iterations between the academic and industry partners. A series of tests were introduced to evaluate the effects of various factors on reliable in-situ pressure sensing. For the wide range of tests presented, the sensor has shown to have errors varying from ±1.8 mmHg for the bladder test to ±8.03 mmHg for the pneumatic sensor comparison test. The variation of the sensor sensitivity in response to mounting curvature was assessed. For sensors of smaller footprints, this becomes less significant. Temperature and humidity were studied and quantified for their impact on a bias error for the no-load capacitance value. The sensor calibration can easily help mitigate this as the sensitivity was shown not to be affected by these variables.

Compared to a pneumatic sensor, some of the advantages of the DEAP capacitive sensor design are its superior spatial resolution, manufacturability for smaller footprints, as well as its overall compact and simple design for use in array configurations measuring pressure distributions. This sensor performs at a high sampling rate of 285 Hz to allow measurements of abrupt pressure changes for physiological applications where a pressure pulsation may need to be tuned to a person's heartrate in order to increase cardiac output. The sensor is thin enough to be worn under a compression garment (thickness of 2.25mm). The durability shown during cyclic testing is attributed to the sensor being made of one solid component and no gaps within the sensor, as well as the compliance of the polymer solution used. Additionally, the perforated design will make the sensor more compliant to help prevent crack initiations during cyclic loading. Future sensor development can potentially improve performance, such as enhanced noise filtering using the electronics. As well, sensor geometry optimization should be further investigated using Finite Element Analysis to optimize the sensor's performance for the desired range of pressures. This includes investigating the trade-off between smaller sensor footprints, which achieves better spatial resolution and smaller curvature errors, but reduces the sensor's overall sensitivity.

Author Contributions: Conceptualization, S.L.; validation, S.L. and H.E.; investigation, S.L., H.E., and J.S.; writing–original draft preparation, S.L.; writing–review and editing, U.S., J.S. and A.S.; visualization, S.L., H.E., U.S., and J.S.; formal analysis, S.L., H.E., and U.S.; supervision, A.S.; funding acquisition, A.S.

Funding: This research was supported by a grant from Lockheed Martin Corporation. The work was also partially supported by NSERC Canada through the discovery grant program [NSERC DG 371472-2009].

Acknowledgments: The authors graciously acknowledge Chekema Prince and Iain Anderson and Sean Peterson for their valuable discussions, feedback, help with testing.

Conflicts of Interest: The authors declare no conflict of interest.

References

1.   Calne, S.; Moffatt, C. *Understanding Compression Therapy*; Position Document of the EWMA; Medical Education Partnership LTD: London, UK, 2003.

2. Troynikov, O.; Ashayeri, E.; Burton, M.; Subic, A.; Alam, F.; Marteau, S. Factors influencing the effectiveness of compression garments used in sports. *Procedia Eng.* **2010**, *2*, 2823–2829.

3. MacRae, B.A.; Cotter, J.D.; Laing, R.M. Compression garments and exercise. *Sports Med.* **2011**, *41*, 815–843. [CrossRef] [PubMed]

4. Zuj, K.A.; Prince, C.N.; Hughson, R.L.; Peterson, S.D. Enhanced muscle blood flow with intermittent pneumatic compression of the lower leg during plantar flexion exercise and recovery. *J. Appl. Physiol.* **2017**, *124*, 302–311. [CrossRef] [PubMed]

5. Zuj, K.A.; Prince, C.N.; Hughson, R.L.; Peterson, S.D. Superficial femoral artery blood flow with intermittent pneumatic compression of the lower leg applied during walking exercise and recovery. *J. Appl. Physiol.* **2019**, *127*, 559–567. [CrossRef] [PubMed]

6. Atiyeh, B.S.; El Khatib, A.M.; Dibo, S.A. Pressure garment therapy (PGT) of burn scars: Evidence-based efficacy. *Ann. Burn. Fire Disasters* **2013**, *26*, 205.

7. Book, J.; Prince, C.N.; Villar, R.; Hughson, R.L.; Peterson, S.D. Investigating the impact of passive external lower limb compression on central and peripheral hemodynamics during exercise. *Eur. J. Appl. Physiol.* **2016**, *116*, 717–727. [CrossRef] [PubMed]

8. Partsch, H.; Clark, M.; Bassez, S.; Benigni, J.-P.; Becker, F.; Blazek, V.; Caprini, J.; Cornu-Thénard, A.; Hafner, J.; Flour, M. Measurement of lower leg compression in vivo: Recommendations for the performance of measurements of interface pressure and stiffness. *Dermatol. Surg.* **2006**, *32*, 224–233. [CrossRef] [PubMed]

9. Burke, M.; Murphy, B.; Geraghty, D. Measurement of sub-bandage pressure during venous compression therapy using flexible force sensors. In Proceedings of the SENSORS, Valencia, Spain, 2–5 November 2014; IEEE: Piscataway, NJ, USA, 2014; pp. 1623–1626.

10. Ferguson-Pell, M.; Hagisawa, S.; Bain, D. Evaluation of a sensor for low interface pressure applications. *Med. Eng. Phys.* **2000**, *22*, 657–663. [CrossRef]

11. Tamez-Duque, J.; Cobian-Ugalde, R.; Kilicarslan, A.; Venkatakrishnan, A.; Soto, R.; Contreras-Vidal, J.L. Real-time strap pressure sensor system for powered exoskeletons. *Sensors* **2015**, *15*, 4550–4563. [CrossRef] [PubMed]

12. Tiwana, M.I.; Redmond, S.J.; Lovell, N.H. A review of tactile sensing technologies with applications in biomedical engineering. *Sens. Actuators A Phys.* **2012**, *179*, 17–31. [CrossRef]

13. Najarian, S.; Dargahi, J.; Mehrizi, A. *Artificial Tactile Sensing in Biomedical Engineering*; McGraw Hill Professional: New York, NY, USA, 2009.

14. Kraemer, W.J.; Bush, J.A.; Bauer, J.A.; Triplett-McBride, N.T.; Paxton, N.J.; Clemson, A.; Koziris, L.P.; Mangino, L.C.; Fry, A.C.; Newton, R.U. Influence of Compression Garments on Vertical Jump Performance in NCAA Division I Volleyball Players. *J. Strength Cond. Res.* **1996**, *10*, 180–183.

15. Ghosh, S.; Mukhopadhyay, A.; Sikka, M.; Nagla, K.S. Pressure mapping and performance of the compression bandage/garment for venous leg ulcer treatment. *J. Tissue Viability* **2008**, *17*, 82–94. [CrossRef] [PubMed]

16. Flaud, P.; Bassez, S.; Counord, J.-L. Comparative in vitro study of three interface pressure sensors used to evaluate medical compression hosiery. *Dermatol. Surg.* **2010**, *36*, 1930–1940. [CrossRef] [PubMed]

17. Salzmann Group. Pressure Measuring Device MST MK IV. Available online: http://www.salzmann-group.ch/4/400/3/130/130/0/index.html (accessed on 2 November 2015).

18. Kikuhime. Available online: http://www.advancis.co.uk/products/other-products/kikuhime (accessed on 2 November 2015).

19. Pico Press. Compression Measurement System. Available online: http://www.microlabitalia.it/case-history.php?azione=show&url=pico-press (accessed on 3 November 2015).

20. Meyer, J.; Lukowicz, P.; Troster, G. Textile pressure sensor for muscle activity and motion detection. In Proceedings of the 10th IEEE International Symposium on Wearable Computers, Montreux, Switzerland, 11–14 October 2006; IEEE: Piscataway, NJ, USA, 2006; pp. 69–72.

21. Holleczek, T.; Rüegg, A.; Harms, H.; Tröster, G. Textile pressure sensors for sports applications. In Proceedings of the SENSORS, Kona, HI, USA, 1–4 November 2010; IEEE: Piscataway, NJ, USA, 2010; pp. 732–737.

22. Sergio, M.; Manaresi, N.; Tartagni, M.; Guerrieri, R.; Canegallo, R. A textile based capacitive pressure sensor. In Proceedings of the SENSORS, Orlando, FL, USA, 12–14 June 2002; IEEE: Piscataway, NJ, USA, 2002; Volume 2, pp. 1625–1630.

23. Yousef, H.; Boukallel, M.; Althoefer, K. Tactile sensing for dexterous in-hand manipulation in robotics—A review. *Sens. Actuators A Phys.* **2011**, *167*, 171–187. [CrossRef]

24. Ko, W.H.; Wang, Q. Touch mode capacitive pressure sensors. *Sens. Actuators A Phys.* **1999**, *75*, 242–251. [CrossRef]

25. Guo, Q.; Mastrangelo, C.; Young, D. High performance MEMS tactile sensor array with robustness and fabrication simplicity. In Proceedings of the IEEE 29th International Conference on Micro Electro Mechanical Systems (MEMS), Shanghai, China, 24–28 January 2016; IEEE: Piscataway, NJ, USA, 2016; pp. 877–880.

26. Siegel, D.; Drucker, S.; Garabieta, I. Performance analysis of a tactile sensor. In Proceedings of the IEEE International Conference on Robotics and Automation, Raleigh, NC, USA, 31 March–3 April 1987; IEEE: Piscataway, NJ, USA, 2003; Volume 4, pp. 1493–1499.

27. Kim, H.; Lee, S.W.; Joh, H.; Seong, M.; Lee, W.S.; Kang, M.S.; Pyo, J.B.; Oh, S.J. Chemically Designed Metallic/Insulating Hybrid Nanostructures with Silver Nanocrystals for Highly Sensitive Wearable Pressure Sensors. *ACS Appl. Mater. Interfaces* **2018**, *10*, 1389–1398. [CrossRef] [PubMed]

28. Gong, S.; Schwalb, W.; Wang, Y.; Chen, Y.; Tang, Y.; Si, J.; Shirinzadeh, B.; Cheng, W. A wearable and highly sensitive pressure sensor with ultrathin gold nanowires. *Nat. Commun.* **2014**, *5*, 3132. [CrossRef] [PubMed]

29. Bar-Cohen, Y. Electroactive polymers as artificial muscles: Capabilities, potentials and challenges. *Robotics* **2000**, 188–196.

30. Cheng, M.-Y.; Huang, X.-H.; Ma, C.-W.; Yang, Y.-J. A flexible capacitive tactile sensing array with floating electrodes. *J. Micromech. Microeng.* **2009**, *19*, 11. [CrossRef]

31. Cheng, M.-Y.; Lin, C.-L.; Yang, Y.-J. Tactile and shear stress sensing array using capacitive mechanisms with floating electrodes. In Proceedings of the 23rd International Conference on Micro Electro Mechanical Systems (MEMS), Hong Kong, China, 24–28 January 2010; IEEE: Piscataway, NJ, USA, 2010; pp. 228–231.

 *micromachines*

Article

Reliability of Protective Coatings for Flexible Piezoelectric Transducers in Aqueous Environments

Massimo Mariello [1,2,*], **Francesco Guido** [1], **Vincenzo Mariano Mastronardi** [1],
Roberto Giannuzzi [1,†], **Luciana Algieri** [3], **Antonio Qualteri** [1], **Alfonso Maffezzoli** [2] and
Massimo De Vittorio [1,2]

1   Istituto Italiano di Tecnologia, Center for Biomolecular Nanotechnologies, 73010 Arnesano (Lecce),
    Italy; francesco.guido@iit.it (F.G.); vincenzo.mastronardi@iit.it (V.M.M.); roberto.giannuzzi@iit.it (R.G.);
    antonio.qualtieri@iit.it (A.Q.); massimo.devittorio@iit.it (M.D.V.)
2   Dipartimento di Ingegneria dell'Innovazione, Università del Salento, 73100 Lecce,
    Italy; alfonso.maffezzoli@unisalento.it
3   Piezoskin s.r.l., 73010 Arnesano (Lecce), Italy; lalgieri@piezoskin.com
*   Correspondence: massimo.mariello@iit.it; Tel.: +39-0832-1816-200
†   Present address: CNR NANOTEC, Institute of Nanotechnology, Via Monteroni, 73100 Lecce, Italy.

Received: 11 October 2019; Accepted: 29 October 2019; Published: 31 October 2019

Abstract: Electronic devices used for marine applications suffer from several issues that can compromise their performance. In particular, water absorption and permeation can lead to the corrosion of metal parts or short-circuits. The added mass due to the absorbed water affects the inertia and durability of the devices, especially for flexible and very thin micro-systems. Furthermore, the employment of such delicate devices underwater is unavoidably subjected to the adhesion of microorganisms and formation of biofilms that limit their reliability. Thus, the demand of waterproofing solutions has increased in recent years, focusing on more conformal, flexible and insulating coatings. This work introduces an evaluation of different polymeric coatings (parylene-C, poly-dimethyl siloxane (PDMS), poly-methyl methacrylate (PMMA), and poly-(vinylidene fluoride) (PVDF)) aimed at increasing the reliability of piezoelectric flexible microdevices used for sensing water motions or for scavenging wave energy. Absorption and corrosion tests showed that Parylene-C, while susceptible to micro-cracking during prolonged oscillating cycles, exhibits the best anti-corrosive behavior. Parylene-C was then treated with oxygen plasma and UV/ozone for modifying the surface morphology in order to evaluate the biofilm formation with different surface conditions. A preliminary characterization through a laser Doppler vibrometer allowed us to detect a reduction in the biofilm mass surface density after 35 days of exposure to seawater.

Keywords: waterproof; coating; reliability; flexible micro-devices; piezoelectric transducers; aqueous environments; seawater

---

1. Introduction

One of the most crucial engineering issues regarding the application of micro and nanoscale electronic devices in liquid environments is the need of being protected from the external harsh surroundings. This is mainly due to the risk of short-circuit caused by the absorption and permeation of water, which is generally also responsible of degradation of devices made of layered functional structures [1–3]. In this respect, delamination is one of the major modes of failure of organic and inorganic layered systems and consists of the weakening or loss of adherence between the different layers, resulting from mechanical strain mismatches or electrochemical reactions at the interfaces [4]. Moreover, the contact between water (especially salty water when dealing with marine applications) and the metal parts inside the micro-systems leads in most cases to spread or localized corrosion of electrodes and wires, degrading the electronic conduction and the device intrinsic performance, as a consequence [5–7]. In addition, an unavoidable phenomenon related to the submersion of devices underwater, is the accumulation of micro-organisms, such as bacteria or unicellular algae, on their surface, namely the biofouling, which is characterized by a gradual and persistent formation of slimy biofilms [8–14].

For applications in aqueous environments, electronic micro-systems must therefore be protected with watertight electrically insulating coatings. Nowadays, the miniaturization of devices, the discovery of new lightweight materials as substrates and the adoption of advanced microfabrication techniques, have led to the design and fabrication of novel flexible devices of cm-size with components of the order of 0.1 mm. They are generally based on stacking sequences of several functional thin-film layers. In this work we focused in particular on the reliability of micro-transducers for harvesting mechanical energy from fluid flows. Two sample categories were selected. The first one embodies transducers made of kapton as substrate, covered by a piezoelectric stack deposited by reactive sputtering and patterned by lithography and etching techniques. The expanded view of the layered structure is reported in Figure 1: here, the active region consists of an aluminum nitride (AlN) piezoelectric layer sandwiched between two thin molybdenum (Mo) electrodes.

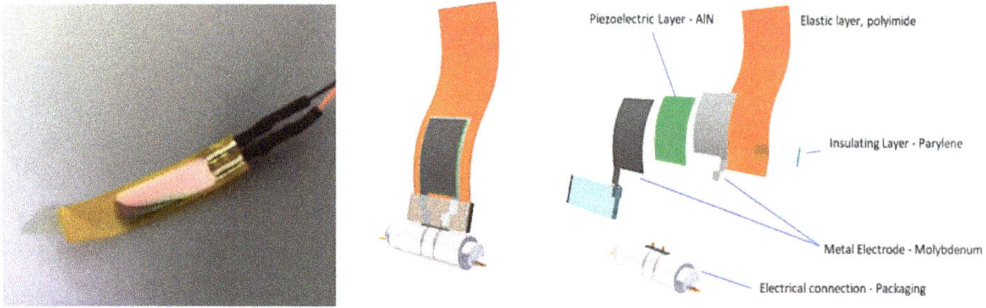

Figure 1. Photo and expanded view of the layered structure of the first group of piezoelectric micro-devices employed for energy harvesting in fluid environments. The thin films were deposited by reactive sputtering, as described in [15,16].

The second sample group includes unimorph devices with a simpler structure made of a poly-(vinylidene fluoride), (PVDF), bi-axially oriented piezoelectric foil sandwiched between two thin-film aluminum (Al) electrodes, as illustrated in Figure 2.

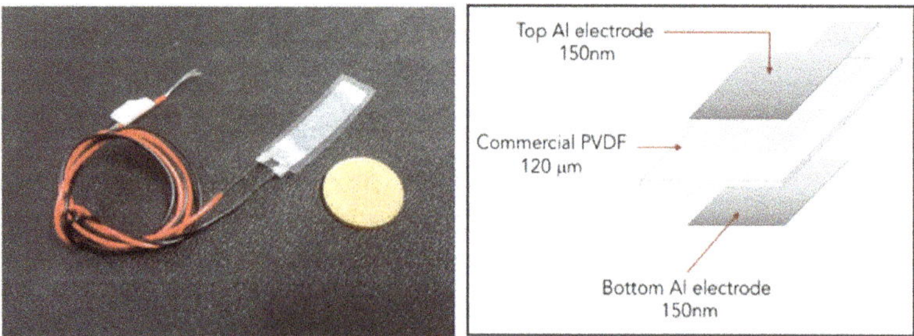

Figure 2. Photo and schematic of the second group of micro-devices. The thin electrodes were deposited by thermal evaporation using shadow masks.

These devices are aimed at undergoing low-frequency fluid-induced oscillations in a single-cantilever-beam configuration. The purpose is to scavenge energy which would be otherwise lost (harvesters) or to sense water motions (sensors), thus they should continuously work for long-lasting periods without decrease in performances [17]. As an example, Figure 3a,b report scanning electron microscope (SEM) images of some defects generated during a long working period of an AlN-based device at the points of crimped regions, whereas Figure 3c shows the general decreasing in output signal after the sudden exposure to seawater: it is worth noticing that after removing and drying the devices, their performances come back unaltered although the presence of defects, such as delaminations grown in short-testing period, was found.

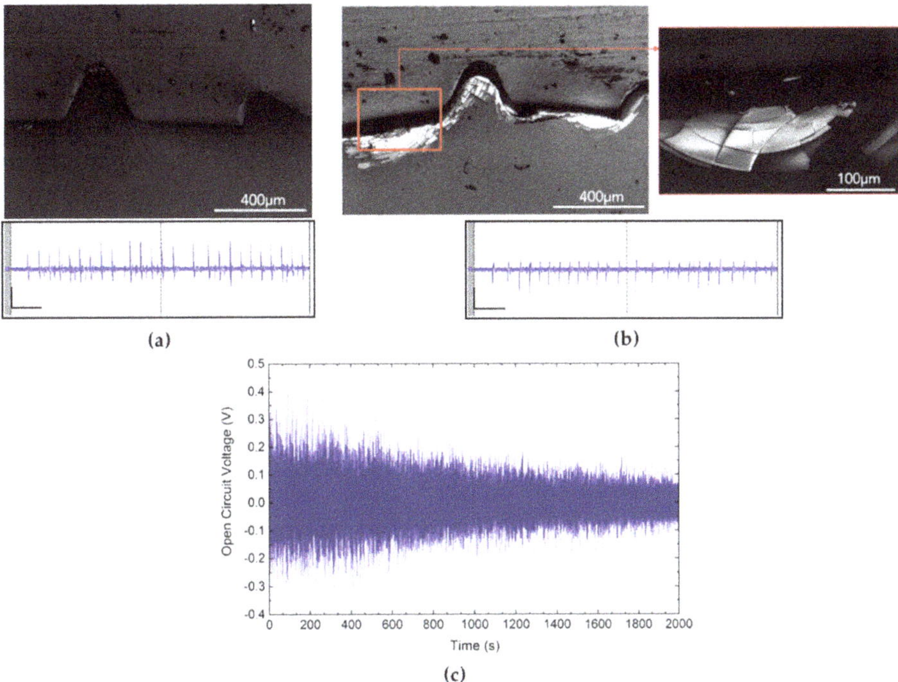

Figure 3. Scanning electron microscope (SEM) images showing crack growth before (**a**) and after (**b**) long-term utilization in correspondence of local defect points, i.e., the areas where the electrical crimp terminals are inserted. The output signals (peak-to-peak voltage) show a decrease in performance due to crack propagation; scale bars: 1.0 s (horizontal), 200 mV (vertical). In short-lasting periods, i.e., after submersion, the decreasing signal amplitude is due to sudden defects, such as delaminations (**c**).

Due to continuous oscillatory movement of the flexible devices, some cracks, growing from surface defects, propagate inside the film and they reach eventually the active region of the device, therefore affecting its performance. Figure 4 shows SEM images of a parylene-coated kapton substrate after a bending deformation. The surface crack may be clearly observed and, although it does not go deeper inside up to the substrate, it could propagate during the submersion period due to fatigue damage, as shown in the cross section inset achieved by focused ion beam (FIB).

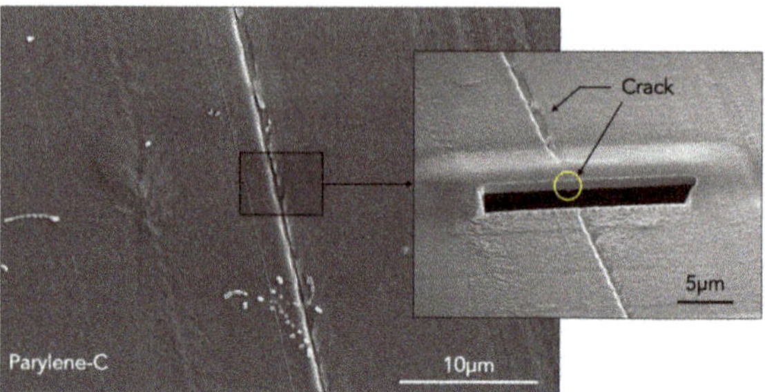

Figure 4. SEM micrograph: flexible kapton substrate coated with thin-film parylene after oscillatory movements in long-term period. The inset shows a cross section made by focused ion beam (FIB) in correspondence of the surface crack. The yellow circle indicates that the crack does not go deep inside up to the substrate.

In this respect, the external coatings must have specific characteristics in order to protect the devices without affecting their flexibility: they should be conformal, lightweight, not fragile, non-porous, insulating and anti-biofouling [18–22].

Materials employed for insulating coatings of microscale electronic devices are mostly polymeric. The most frequently used materials are: (1) elastomers, (2) polyacrylates, (3) fluoropolymers, (4) poly-para-xylylenes (also known as parylenes).

Elastomers are thermosetting polymers with rubber-like properties, such as high stretchability and softness. The most widely used silicone-based elastomer is poly(dimethyl siloxane), PDMS, which has remarkable rheological, optical and non-toxic properties [23]. Generally, it is formed as a viscous bi-component thermosetting mixture between a siloxane-based oligomer and a curing agent, and applied by spin-coating [24], dip-coating [25] or spray-coating [26]. PDMS could also be incorporated, at different contents, into other polymeric coatings, such as polyurethane (PU), in order to enhance their antibiofouling properties, as reported in [27], or to make new composite materials [28].

Poly-(methyl methacrylate), PMMA, is one of the most common polyacrylates: it is a transparent thermoplastic polymer with noticeable mechanical and optical properties. It is generally used as a lightweight alternative to glass, but also for other various applications, such as inks and coatings, or for microfabrication processes as sacrificial layer. Coan et al [29] reported the development of composite coatings of PMMA with hexagonal boron nitride (hBN) as filler, for metal surface protection against corrosion.

Fluoropolymers contain plenty of carbon-fluorine bonds which confer to the backbone a high chemical inertness against many solvents, acids or bases [30]. Some well-known examples are PVDF, or poly(tetrafluoroethylene), (PTFE). In particular, PVDF can be found commercially as extruded foils, pellets and micrometric powder as well, which can be incorporated in other polymeric matrices or dissolved in solvents.

Poly(-para-xylylenes) are thermoplastic semicrystalline polymers discovered by Michael Szwarc in the late 1940s and commercialized in 1965. These polymers are synthesized by chemical vapor deposition (CVD) and have very attractive properties among which low-adhesion coefficient at room temperature, and conformability to different types of substrates. The current method adopted to synthesize parylene is called the Gorham route and is a very efficient polymerization process, in fact it allows complete control of the deposition parameters: the process basically consists of pyrolizing the precursor dimer and polymerizing the resulting monomers during deposition onto the substrate [31–33]. Several different kinds of parylene may be synthesized, depending on the functional groups bonded to the backbone of the precursor (2,2-para-cyclophane): these substituents are not modified during the CVD process, making it possible to tailor chemical, mechanical, electrical and optical properties of parylene thin films and, therefore, to introduce diverse functionalities into the coated surfaces [31].

The range of application fields for parylene is wide: in particular, its insulating and moisture barrier properties make it suitable for protecting implanted biomedical micro-systems or devices in contact with water or wet environments [34–36].

Several works have been published about external polymeric coatings for insulation and barrier purposes [37–39]. Lewis and Weaver provide in [40] a review of thin-film permeation-barrier technologies for flexible organic light-emitting devices. Fredj et al. [41] studied the natural and artificial ageing of marine organic coatings. Deyab et al. [42] prepared a wax coating using waste materials (isolated microcrystalline waxes) to protect petroleum pipelines against corrosion in 0.6 M NaCl solution. The same author analyzed the effect of carbon nanotubes (CNTs) [43], newly synthesized titanium phosphates [44] or M-porphyrins [45] on corrosion protection of carbon steel coated by alkyd resin and tested after immersion in sodium chloride solution. Li et al. [46] performed accelerated soak tests to study the corrosion behavior and failure mechanisms of parylene-metal-parylene thin films. Davies and Evrard [47] studied polyurethanes for marine applications through accelerated tests at high temperatures.

However, the application of protective coatings onto flexible micro-devices continuously moving underwater is a tricky issue, thus finding an optimal solution for guaranteeing at the same time (1) protection from the environment and (2) the best device performances is still an ongoing challenge.

In this work, we have investigated the barrier behavior and surface properties of different polymeric coatings of the flexible piezoelectric transducers described above: parylene-C, poly-methyl methacrylate (PMMA) and poly-dimethyl siloxane (PDMS). These choices were made for the ease of applicability, conformability, low cost and compatibility with the flexible piezoelectric energy harvesters, since the external layers is crucial for the reliability of the device itself. The PDMS coating was used both in the neat form and mixed with a powder of PVDF. The combination of the conformability of the elastomer and the hydrophobicity and chemical inertness of the fluoropolymer was expected to confer higher water repellence to the coating, limiting water permeation in the device [48,49].

Furthermore, parylene was also designated as a suitable surface platform for evaluating the accumulation of microorganisms in a long-term period. Since previous articles in literature reported the surface functionalization of parylene by physico/chemical treatments [19,50,51], besides the pristine parylene-C (pC) coating deposited by CVD, surface-treated pC was taken into account: in particular, two surface treatments were adopted based on oxygen plasma etching and UV/ozonization.

The seawater absorption of coatings was analyzed by impedance spectroscopy (IS) measurements. The anti-corrosion properties of the coatings were tested by dynamic linear scanning voltammetry (LSV) measurements. Additionally, atomic force microscopy (AFM) measurements provided further information in terms of surface morphology. The barrier properties of the coatings and their reliability were therefore correlated to the higher or lower capability of retaining the formed biofilms: this evaluation was possible by estimating the amount of added microbial mass on the exposed surface through a laser Doppler vibrometer (LDV).

2. Materials and Methods

2.1. Materials

Parylene-C was provided by Specialty Coating Systems in form of dimer powders. PDMS (Sylgard 184 Silicone Elastomer) was supplied by Dow Corning Corporation in two compounds: a viscous uncured pre-polymer and a curing agent. PMMA 950 in Anisole e-beam resist (anisole 80–100%, PMMA 1–20%) was purchased from MicroChem Corp. Granular PVDF powder (average Mw ~ 534,000 by GPC) and 2-butanone (MEK) solvent were supplied by Sigma Aldrich.

The piezoelectric devices employed in the present work were grouped into two categories: (i) AlN-based, and (ii) PVDF-based seaweed-shape transducers. The AlN flags were made using kapton HN (by Dupont in the form of 25 µm-thick foils) as substrate for thin film deposition.

The commercial PVDF foils of the second groups of harvesters were provided by TE Connectivity's Measurement Specialties (MEAS).

AISI304 steel samples were selected as substrates for corrosion tests and were provided by RS Components.

2.2. Fabrication of Flexible Transducers

The fabrication of the adopted transducers is reported, as for previous works [15,16].

2.2.1. AlN-Based Transducers

The kapton substrate was attached to a silicon wafer using polydimethyl siloxane (PDMS) spin-coated at 1000 rpm for 30 s and then cured at 90 °C for 15 min. A vacuum step was needed before curing to remove any residual air bubbles. The overlaying layers were deposited by reactive sputtering in a single run in order to minimize contaminations, with the following parameters: an AlN oriented growth-favoring interlayer (120 nm) and the first Mo layer (200 nm) were deposited in a single step and patterned by optical lithography and chemical etching. The AlN interlayer stemmed from a high-purity (99.9995%) Al target in a mixture of Ar (20 sccm) and N_2 (20 sccm) gases with a total pressure of 2.8×10^{-3} mbar and with DC pulsed power supply of 750 W. The Mo electrode was deposited from a pure (99.95%) Mo target at room temperature, with a total pressure of 5×10^{-3} mbar in an Ar atmosphere (66 sccm) and with DC power of 400 W. The pattering of Mo bottom electrode and of AlN interlayer was performed by dry etching with inductively-coupled plasma-reactive ion etching (ICP-RIE) system: the gas mixture was made of BCl_3 (45 sccm) and N_2 (25 sccm) for Mo, and of BCl_3 (100 sccm) and Ar (25 sccm) for the AlN. The applied power was 250 W to the platen and 600 W to the coil. The AlN piezoelectric layer (~1 μm) and the second top Mo electrode (200 nm) were deposited in the same run: for the AlN film a high-purity Al target (99.9995%) was used with a gas mixture of N_2 (20 sccm) and Ar (20 sccm) at a pressure of 2.8×10^{-3} mbar in DC pulsed mode with a frequency of 100 kHz and a power of 1000 W; for the Mo layer the same conditions as for the bottom electrode were applied. The chemical dry etching was performed by the ICP-RIE system under the same conditions as before. The chamber temperature during sputtering increased to ~70 °C and to ~165 °C for the Mo and AlN steps, respectively. Finally, the polymeric substrate was peeled off from the rigid wafer and the wire connections to the electrodes were made with a crimping tool.

2.2.2. PVDF-Based Transducers

The PVDF foil was first cleaned with acetone and isopropanol before depositing the aluminum thin-film electrodes (200 nm) by thermal evaporation. The electrodes were patterned by using shadow masks made of adhesive kapton tape, whose shape was designed properly. Thermal evaporation allows to deposit the electrodes at a temperature of ~60 °C, which is lower than the Curie temperature of PVDF (~100–110 °C) and it doesn't affect the piezoelectric feature of the functional layer. Finally, the wire connections were made with a crimping tool.

2.3. Preparation of Coatings

The methods for preparing the coatings are also reported in [52]. Moreover, in this work, the temperature was kept inside the range allowed by the device materials.

PDMS coating was prepared by mixing the pre-polymer with the cross-linker (10:1 wt), degassed for 30 min and applied on the whole devices by dip-coating. Finally, PDMS was cured in oven at 90 °C for 15 min.

PMMA-based coating was prepared by dip-coating the devices in the PMMA solution, then they were suspended in a heated oven at 90 °C for 1h for solvent evaporation.

Coatings made of blends of PDMS and PVDF were obtained first mixing PDMS uncured pre-polymer with PVDF in a 3:1 (wt) ratio, then the mixture was heated in oven at 200 °C, above the PVDF melting point (177 °C), and stirred every 5 min. After cooling down at room temperature, the blend was mixed adding the curing agent (weight ratio 10:1 with respect to the PDMS). Finally, it was applied onto the devices by dip-coating and cured in oven at 90 °C for 15 min.

Parylene-C deposition process was performed by a RT-CVD equipment (Specialty Coating Systems, PDS 2010 Labcoater system model). The powdered dimer vaporized at a temperature of ~100–150 °C and at a pressure of 1 torr to undergo a pyrolysis and be reduced in monomers; then, the polymerization of the gaseous monomers occurred at ~650–700 °C and 0.5 torr; the gas entered the deposition chamber at 20–25 °C and 0.1 torr and a conformal polymeric coating deposits on the substrate. An amount of 1 g of dimer powder yielded a deposited layer with thickness of 1 μm and the process lasted approximately one hour.

2.4. Characterization and Reliability Tests

With the exception of parylene, the other coatings were applied on the devices by dip-coating so the control of thickness was subjected to uncertainties and non-uniformity. Different portions of the device were selected and the coating thickness was measured by means of a profilometer (Bruker Dektak Xt).

2.4.1. Surface Characterizations

The sample coatings were characterized in terms of wettability, through an OCA 15Pro Contact Angle Tool (DataPhysics), and surface morphology, by means of an atomic force microscope (CSI Nano-Observer AFM) in non-contact (resonant) mode, and a SEM (Helios NanoLab DualBeam, FEI). The reported values for contact angles were averaged over four independent measurements, whereas the roughness values stemmed from the averaging over a scanned area of $5 \times 5 \ \mu m^2$.

2.4.2. Exposure to Seawater

The whole devices and some coated silicon substrates (three for each coating) were submerged in a tank ($440 \times 265 \times 240 \ mm^3$) and filled with seawater which was aimed at reproducing the marine environment (Figure 5a). The natural seawater was collected from the Ionian Sea, with the following physico-chemical parameters: salinity 3.5%, pH 8, metals (Na 10,784 ppm, K 399 ppm, Mg 1294 ppm, Ca 4120 ppm, Si 2.9 ppm), halogenides ($Br^-/Cl^-/F^-$ 19360/67.5/1.3 ppm), carbonates (HCO_3^- 126 ppm), sulphates (SO_4 2712 ppm), and dissolved gases (O_2 7 ppm, N_2 12.5 ppm) [52].

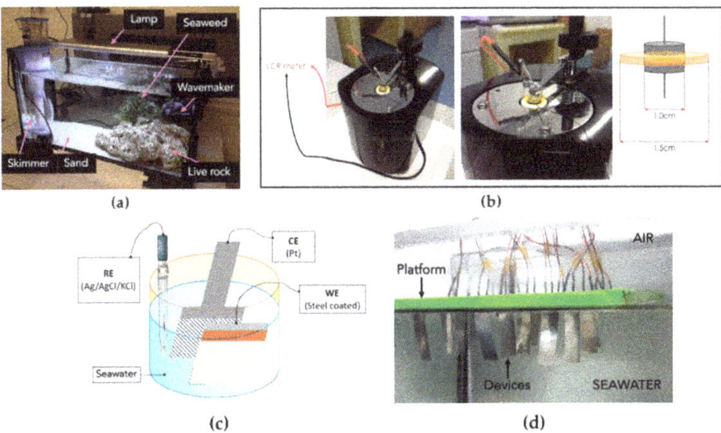

Figure 5. (a) Setup used to mimic the marine environment in the laboratory. (b) Setup used for the impedance spectroscopy (IS) measurements with the schematic of the samples adopted. (c) Schematic of the three-electrode cell used for the corrosion tests. (d) Floating platform for supporting the whole devices while being submerged in seawater for reliability tests.

A wave maker (Jebao RW4 Propeller Pump) was inserted in the tank to create sea waves and currents: in order to avoid the samples tumbling due to the water waves, a metallic grill was used as support; the coated rigid substrates were fixed on Petri dishes and placed in the grill, whereas the whole devices were made floating with a polystyrene supporting slab. A porous live rock was inserted to introduce natural bacteria, algae and invertebrates inside the artificial marine environment. Moreover, it plays the role of biological filter because it harbors aerobic and anaerobic nitrifying bacteria required for the nitrogen cycle. Furthermore, a natural seaweed was added. Its general function is to use waste products as nutrients and perform chlorophyllian photosynthesis, producing oxygen. In this study, it also served as a competing species for the development of bacteria and algae. A LED lighting system (Mini Lumina 30 Marine, Blau-aquaristic; 18 W; 10.000K white and 460 nm blue) was used to provide the solar light radiation, stimulating the growth of microorganisms. Finally, a skimmer (Scuma 0635 Int, Blau-aquaristic; 3.5 W) was installed to purify the environment keeping the water clear and limpid.

2.4.3. Water Absorption Tests

The absorption of water into the devices mainly regards the polymeric layers. The conformity of the coatings is crucial because any possible asperity, uncovered area or inlet could lead to water permeation. Impedance spectroscopy (IS) was carried out to evaluate water absorption: circular samples (1.5 cm radius) of kapton coated with the selected polymers were submerged, then dried with nitrogen flow at regular time intervals. The impedances of the substrate/coating systems were collected by placing the samples between two metallic plates connected to an Agilent E4980A Precision Inductance, Capacitance, Resistance (LCR) meter (applying the same pressure for all samples) and by performing linear frequency sweeps.

2.4.4. Corrosion Tests

Corrosion due to seawater penetration regards all the metallic portions of the devices, thus anti-corrosion tests were performed to evaluate the corrosion resistance of the coatings. The steel panels (4×5 cm^2) were washed with acetone and isopropanol to remove organic contaminants before the application of coatings. The corrosion tests were carried out in a three-electrode cell (Figure 5c) consisting of coated steel samples as working electrode (WE), a 5×5 cm^2 Pt mesh as counter electrode (CE) and an Ag/AgCl/sat-KCl electrode as a reference electrode (RE). Linear scanning voltammetry was performed at 1 mV/s to measure current and voltage using an AUTOLAB PGSTAT302N potentiostat, after 1h of stabilization to reach the equilibrium open circuit potential (OCP).

2.4.5. Piezoelectric Generation in the Long-Term Period

The PVDF-based devices were attached to a polystyrene foam slab and submerged in the tank filled with seawater (Figure 5d): the slab served as a supporting platform to let the devices float firmly on the water. In this way, the supporting platform is always in contact with the water surface although the continuous evaporation of water, keeping the same initial conditions. The measurements were performed by displacing the samples at a proper distance, using a linear micro-actuator (Actuonix Motion Devices Inc., L16-R, 100 mm, 35:1, 6vdc) and the output signal generated by the oscillation of the devices (open circuit voltage) was detected and collected with an oscilloscope (Tektronix MDO 4104-3).

2.4.6. Surface Treatments of Parylene Surface

Wettability and morphology studies were performed on treated parylene C coatings in order to evaluate qualitatively the long-term microbial adhesion. Two different treatments were applied: oxygen plasma and ultraviolet light irradiation with addition of ozone. Oxygen plasma treatment was performed by a low-pressure plasma system NANO (Diener electronic Plasma Surface Technology GmbH & Co. KG), with two different RF powers in order to compare the resulting modified morphologies, i.e., 100 W and 300 W, keeping the process time (20 min) unchanged. UV/ozone treatment was performed by a UV Ozone Cleaner ProCleaner Plus (BioForce Nanosciences) for 20 min. The two treatments were overlapped in order to evaluate the effects of each single treatment. As previously demonstrated [50], the etching effect of oxygen plasma treatment on polymeric substrates induces a structure with nanosized thread-like features with a relatively high roughness. The lower the dimensions of asperities (roughness), compared to the dimensions of air bubbles in water, the less likely these air bubbles remain trapped among the asperities, thus the wider the contact surface/water and the more favored the water absorption. For these reasons, the UV/ozone treatment was superimposed to the oxygen plasma with the intention of making the sample surfaces less rough.

Non-treated samples, samples treated with O_2 plasma, and samples treated with both the O_2 plasma and UV/ozone treatments, were compared. The flexible kapton/pC structures were characterized after submerging them in the tank: the characterization tests were carried out at three different periods of time (15, 25, 35 days), after drying the samples in a desiccator.

The laser Doppler vibrometer (LDV-MSA-500, Polytec), equipped with HeNe laser light source, was employed to investigate the frequency response of the pC-coated kapton samples evaluating the out-of-plane deflections by a non-contact measurement. The experimental setup included also an electromechanical shaker (4819, Bruel & Kjaer) aimed at providing the mechanical vibrations in a range of 0–18kHz. The samples were previously cut and shaped in the form of cantilevers (10×5 mm^2) by an automatic cutting plotter (Graphtec Cutting Plotter CE6000-40).

3. Results

This work was aimed at evaluating the protective and barrier behavior of different selected polymeric coatings for flexible piezoelectric energy harvesting devices in liquid flows, in particular in a seawater environment. The application and the type of micro-devices required specific characteristics for the coatings, i.e., flexibility, conformability, ease of application, besides properties related to the insulating, waterproof and barrier behavior.

As already discussed, parylene-C, PMMA, PDMS and PDMS-PVDF were selected as coating materials to be tested, each characterized by different thicknesses: 2.7 µm for parylene, ~3 µm for PMMA, ~100 µm for PDMS and PDMS-PVDF. Differently from the other coatings applied by dip-coating, parylene C was deposited by CVD, so the thickness of the resulting coating was much more controllable.

3.1. Surface Characterizations

The AFM micrographs and the results of wettability measurements are reported in Figure 6.

Parylene C (pC) exhibited higher roughness (3.71 nm) than PMMA (0.28 nm) and PDMS-based coatings which present comparable values, i.e., 1.19 nm (PDMS), 1.70 nm (PDMS/PVDF). Concerning the wettability characteristics, parylene and PMMA showed contact angles of 89.5° ± 0.6° and 79.3° ± 3.0°, respectively; PDMS-based coatings present c.a. values of 116.6° ± 0.9° (PDMS) and 117.4° ± 0.7° (PDMS/PVDF): these values indicate that PMMA has a moderately hydrophilic character, whereas the other coatings exhibit more hydrophobicity, apart from parylene which may be considered as neutral; the incorporation of PVDF into the PDMS slightly increased the contact angle.

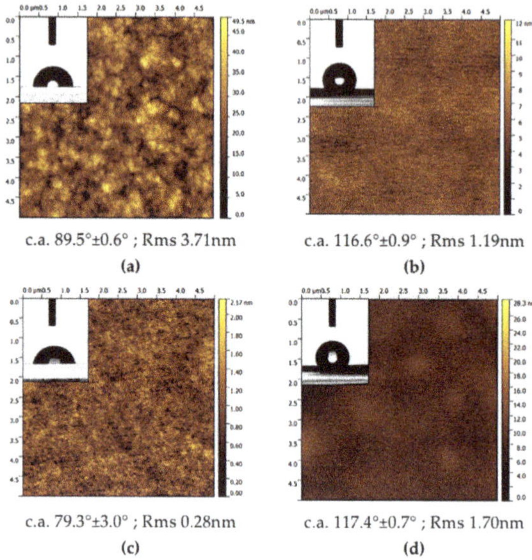

c.a. 89.5°±0.6° ; Rms 3.71nm

(a)

c.a. 116.6°±0.9° ; Rms 1.19nm

(b)

c.a. 79.3°±3.0° ; Rms 0.28nm

(c)

c.a. 117.4°±0.7° ; Rms 1.70nm

(d)

Figure 6. Atomic force microscopy (AFM) topography images (scanned area: 5×5 μm^2) for the examined coatings: (**a**) parylene-C, (**b**) poly-dimethyl siloxane (PDMS), (**c**) poly-methyl methacrylate (PMMA), (**d**) PDMS- and poly-(vinylidene fluoride) (PVDF). The vertical color bar reports the roughness in nm. The insets show the results of contact angle measurements.

3.2. Water Absorption Tests

The IS frequency-sweep analyses performed with the setup shown in Figure 5b on the submerged samples yielded the Nyquist plots reported in Figure 7a–e. The shape of the Nyquist curves is in general semi-circular and the magnitude of the vector connecting the axis origin with each point of the curves gives the values of impedance for different frequencies. Therefore, a curve stretched upward indicates that the imaginary part of impedance (reactance) of the system is higher.

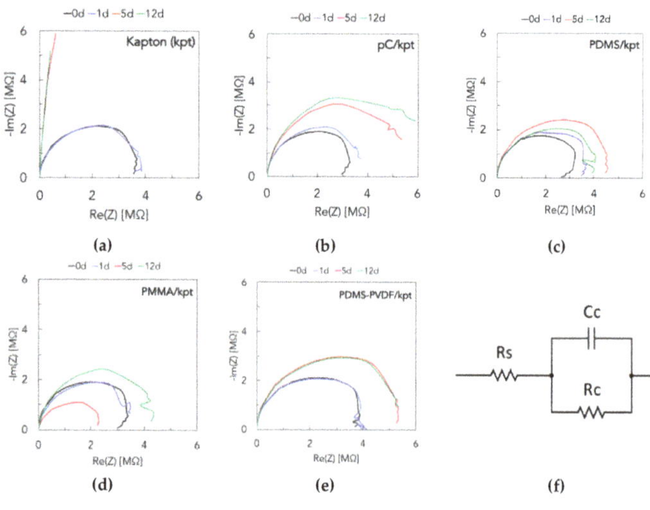

Figure 7. *Cont.*

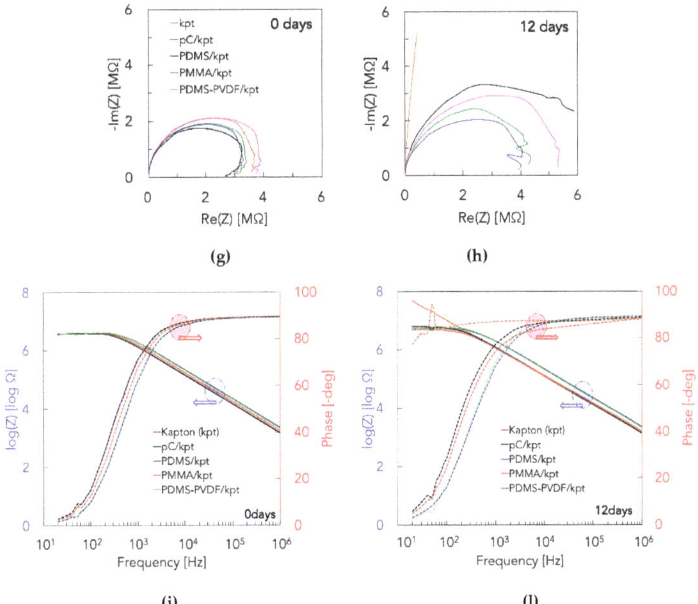

Figure 7. Nyquist plots resulting from IS measurements (frequency sweep in the range 20Hz–1MHz), for the kapton substrate (**a**) and for the coated substrates: (**b**) parylene-C, (**c**) PDMS, (**d**) PMMA, and (**e**) PDMS-PVDF. In each plot, the curves correspond to different periods of exposure to seawater, i.e., 0, 1, 5, 12 days. (**f**) The equivalent circuit model used to fit IS data: R_s is the resistance of wires and electronic parts; R_c and C_c are the resistance and capacitance of the coating. (**g,h**) Nyquist plots for the coated substrates, at 0 (**g**) and 12 (**h**) days of exposure to seawater. (**i,l**) Bode plots for the coated substrates, at 0 (**i**) and 12 (**l**) days of exposure to seawater.

The barrier behavior of the coatings is evident when comparing them with the uncoated substrate. As can be seen from the Nyquist plots, the curves in the latter case become almost linear after 5 and 12 days of exposure to seawater, whereas for all the other samples the curves keep a quasi-circular shape. In other words, with the passing of submersion time, the reactance of the system gets higher because of the absorption of water molecules with high dielectric constant, but to a much lower extent when the substrate is coated. The absorbed water molecules eventually dissociated, and ions are responsible of a change also in the real part with exposure time. The behavior of electrical resistance and reactance of the coating/substrate systems is a result of different phenomena: (1) the water molecules uptake by absorption, (2) the different swelling degree of the coatings, (3) the damage/delamination of the coatings allowing seawater permeation, (4) the dissociation of absorbed water molecules and permeation of solvated ions. The presence of salty water, or dissociated water molecules or solvated ions in the coatings induces a decrease in the electrical resistance, while the absorption of water molecules (with high dielectric constant) and the swelling behavior lead to an increase of the dielectric constant of insulating materials [53–57]. From the analysis of the IS plots it can be deduced that for the selected coatings the latter phenomenon is prevalent. Bode plots reported in Figure 7i,l highlight the frequency behavior of the impedance and phase angle for the tested substrates. The general trend is a decrease in impedance and an increase in phase with increasing frequency due to a more dominant capacitive behavior. In summary, from the Nyquist plots (Figure 7g,h) and the Bode plots (Figure 7i,l), at 0 and 12 days of exposure to seawater, it can be deduced that the protective action of the different coatings is quite comparable, and a further quantitative analysis was necessary: in Table 1, the coating resistance (R_c) and capacitance (C_c) are reported as results of the fitting of IS data in ZView®software, for which the equivalent circuit model in Figure 7f was adopted for the coating/substrate system: generally, the fitted data match the experimental, with an average error of 5.0%, and higher resistances or lower capacitances indicate a more protective ability of the coatings; after 12 days of exposure to seawater, parylene-C shows the best behavior.

Table 1. Impedance parameters resulting from fitting the IS data for the uncoated and coated kapton substrates (1.5 cm diameter).

Submersion Days	Samples	R_c [10^6 Ω]	C_c [10^{-10} F]
	Kapton (kpt)	3.5276	1.0910
	pC/kpt	3.9573	1.0650
0	PDMS/kpt	3.7389	0.7080
	PMMA/kpt	3.6062	0.8885
	PDMS-PVDF/kpt	3.7677	0.7325
	Kapton (kpt)	- (low)	1.3430
	pC/kpt	6.5529	1.2230
12	PDMS/kpt	4.0184	0.7404
	PMMA/kpt	4.3604	1.1510
	PDMS-PVDF/kpt	5.5225	0.7552

3.3. Corrosion Tests

The corrosion tests on the steel samples with the selected coatings yielded the current–voltage curves (Tafel plots) in the Evans diagrams, as reported in Figure 8, which provided more details about the anti-corrosive properties of the coatings: the less porosities and better conformal coverage, the more effective the coating.

Figure 8 indicates that the current passing through the seawater solution, (when the WE is pristine steel), is higher whereas it is much lower when steel is coated. As matter of fact, a coating is more effective if the allowed current level is lower.

As illustrated in Figure 8f, the reason for a non-zero current passing through the coating regards the presence of defects across its thickness: these defects, which include porosities, delamination and cracks, allows the direct contact of seawater with the underlying steel, closing the circuit.

The sparks on the anodic polarization curves are due to noisy current fluctuations which are unavoidable and cannot be eliminated [58]. Since the medium in the electrochemical three-electrode cell consists of seawater (basically water and NaCl) with pH slightly above 8, steels show passive behavior but Cl^- ions continuously disrupt the passivation film which tries to re-form. Another mechanism could involve the formation of OH^* radicals at the metal surface, which attack the polymer leading to deterioration [35]. Therefore, the alternating process of localized formation and breaking of the passive film results in current fluctuations in the Tafel plots: this behavior is more evident for uncoated steel samples whose curves are overall quite noisy. In addition, the sparks at the edge of the curves (i.e., in potentials far from the corrosion potential), even for coated samples, may be caused by bubbling of gases produced by electrolysis of water.

The Tafel plots for the studied coatings were overlapped and compared for the same period of time, i.e., at 0 and 30 days of submersion, as shown in Figure 8g,h. It is worth noticing that:

1. At 0 days the best coatings are parylene-C, PDMS and the combination PDMS-PVDF, whilst the worst is PMMA, in terms of insulation and anti-corrosive behavior;
2. At 30 days the best coatings are parylene-C and PDMS, whereas PDMS-PVDF gets worse perhaps owing to inhomogeneity; PMMA confirms to be the worst coating.

Electrochemical kinetic values (corrosion potential E_{corr} and corrosion current density j_{corr}) were calculated from the intersection of coordinates of Tafel plot [59] and are presented in Table 2.

The protection efficiency ($\eta_j\%$) of the coatings are also reported according to calculations made using the following formula [60,61]:

$$\eta_j\% = [1 - (j_{corr}/j^0_{corr})] \times 100 \tag{1}$$

where j^0_{corr} and j_{corr} are the corrosion current densities in the absence and presence of coatings, respectively.

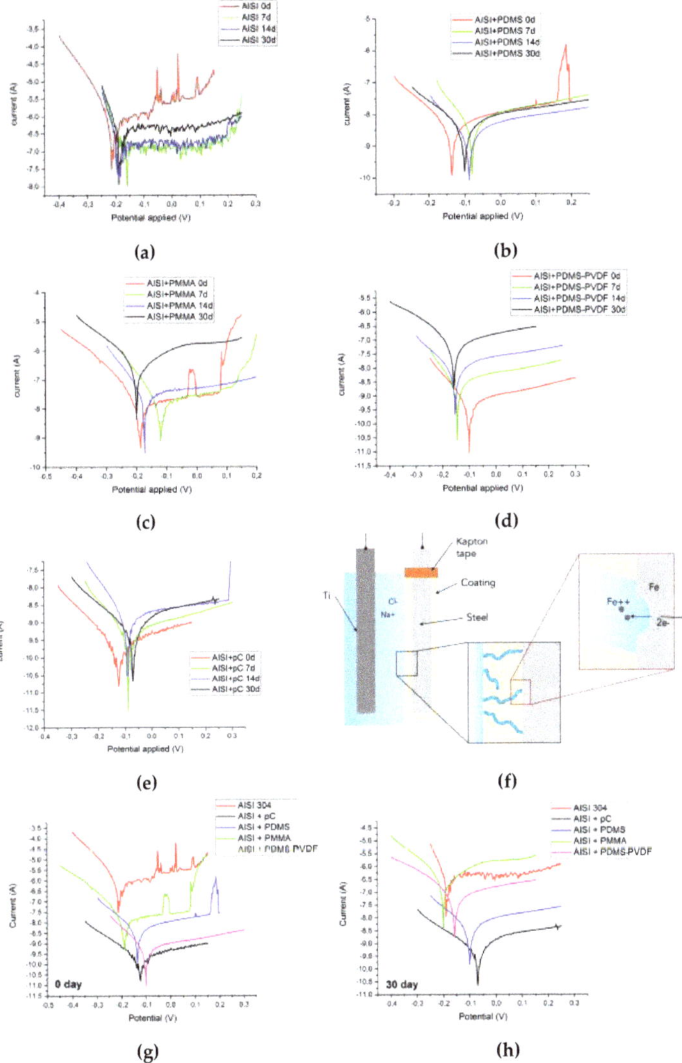

Figure 8. Evan diagrams displaying the Tafel plots resulting from corrosion tests for the pristine steel samples (**a**) and for the coated samples: (**b**) PDMS, (**c**) PMMA, (**d**) PDMS-PVDF, (**e**) parylene-C. The curves in each plot correspond to different periods of submersion, i.e., 0, 7, 14, and 30 days. The *y*-axis in the plots is in logarithmic scale. (**f**) Corrosion mechanism of steel samples due to seawater absorption and permeation. (**g**,**h**) Tafel plots for the differently coated steel samples, at 0 (**g**) and 30 (**h**) days of exposure to seawater.

Table 2. Electrochemical parameters and the corresponding inhibition efficiency for steel coated with the selected coatings.

Submersion Days	Samples	E_{corr} (vs. SCE) [V]	j_{corr} [μAcm^{-2}]	η_j [%]
0	Pristine AISI314 steel	-0.1635 ± 0.0426	$(2.27 \pm 1.14) \times 10^{-2}$	-
	pC	-0.1511 ± 0.0217	$(1.70 \pm 0.65) \times 10^{-5}$	99.9
	PDMS	-0.1677 ± 0.0192	$(3.21 \pm 0.58) \times 10^{-4}$	98.6
	PMMA	-0.1990 ± 0.0191	$(1.49 \pm 0.80) \times 10^{-3}$	93.4
	PDMS-PVDF	-0.1030 ± 0.0202	$(3.67 \pm 0.45) \times 10^{-5}$	99.8
30	Pristine AISI314 steel	-0.2065 ± 0.0084	$(3.10 \pm 0.47) \times 10^{-2}$	-
	pC	-0.2522 ± 0.1398	$(8.84 \pm 4.37) \times 10^{-5}$	99.7
	PDMS	-0.1759 ± 0.0606	$(2.90 \pm 4.43) \times 10^{-2}$	90.7
	PMMA	-0.1542 ± 0.0349	$(2.15 \pm 0.74) \times 10^{-2}$	30.6
	PDMS-PVDF	-0.1635 ± 0.0572	$(6.75 \pm 7.96) \times 10^{-2}$	78.2

The time variation of corrosion potential and current density during submersion is shown in Figure 9 for the pristine steel and the coated samples.

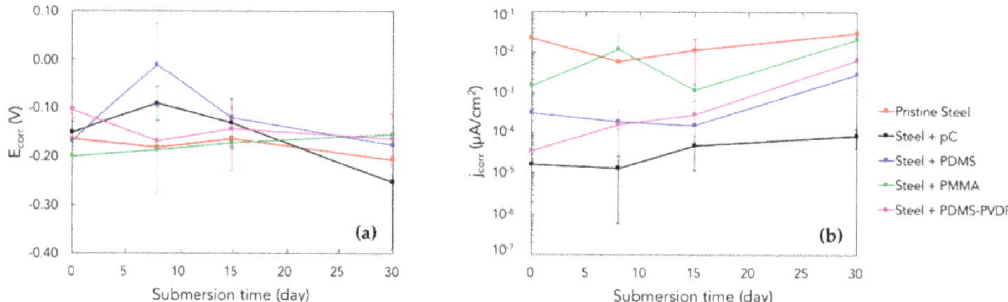

Figure 9. Corrosion potential (**a**) and corrosion current density (**b**) vs. submersion time, for the pristine steel samples and the selected coatings.

From these results it is evident that initially the protection efficiency increases in the following order: PMMA < PDMS < PDMS-PVDF < pC, whereas, for long-term periods: PMMA < PDMS-PVDF < PDMS < pC.

3.4. Piezoelectric Generation in the Long-Term Period

The measurements of the open circuit voltage generated by the submerged devices yielded the plot reported in Figure 10a, as percentage decrease of the output signal. All curves are characterized by a plateau indicating that the piezoelectric performance remains constant in continuous use applications, even if with remarkable differences among the coatings. These results confirm those obtained in corrosion tests, i.e., a complete loss of signal is observed after 7 days with PMMA coating, exhibiting the worst performances (exfoliation can be observed in Figure 10b). PDMS and PDMS/PVDF coatings are substantially equivalent showing a stable voltage loss of 70%, while a voltage loss of less than 20% is observed when pC is used. These results provide a direct correlation among device performance and mass transport properties of the coatings.

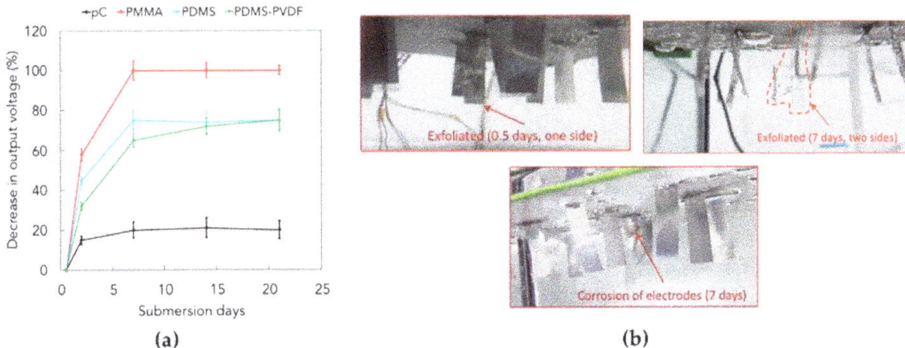

Figure 10. (**a**) Decrease (%) in output voltage of piezoelectric PVDF-based transducers due to prolonged exposure to seawater. (**b**) Exfoliation of PMMA coatings after exposure to seawater.

3.5. Observation of Microbial Adhesion on Pristine and Treated Parylene

This study was also focused on evaluating the possibility of modifying the surface properties of parylene-C thin films in order to use them as protecting materials against marine biofouling in electronic devices. Both the treatments adopted on the polymeric surfaces bring about two kinds of effects due to the simultaneous occurring of different processes [62]: (1) a physical effect, strictly related to the etching and sputtering of the polymer owing to the reactions of oxygen atoms with the surface carbon atoms and to the impingement of plasma ions onto the surface, leading to a change in morphology; (2) a chemical effect, which is provided through the incorporation of hydrophilic and oxygen-rich functional groups on the surface, and through the breakage of organic bonds promoted by the UV radiation emitted by the plasma or by the UV source. Plasma treatments, as well as UV/ozone, are usually used for cleaning, surface activation, deposition and etching, and are typically employed for modifying the chemical and physical surface properties of polymers [63].

Seawater contact angle (WCA) measurements, as shown in Figure 11 (compared to Figure 6), confirmed that the oxygen plasma surface treatment increases the hydrophilic character of the sample surface with respect to non-treated surface ($94.7° \pm 2.8°$), and this also occurs to a larger extent when a higher generator power is set: in fact the recorded WCA is $11.3° \pm 1.4°$ for 100 W, and $3.8° \pm 0.6°$ for 300 W. On the other hand, the UV/ozone treatment, performed subsequently to the oxygen plasma treatment, counter-intuitively results in a slight decrease in hydrophilicity (maintaining however a hydrophilic character, with WCA values of $13.4° \pm 1.2°$ and $19.9° \pm 0.6°$ for 100 W and 300 W powers of oxygen plasma treatment, respectively), revealing a counteraction between the two treatments.

The wettability properties are strictly correlated to the surface roughness (Rms) measured by AFM and reported in Figure 11. The recorded roughness of non-treated samples is 6.17 nm, whereas the oxygen plasma treatment increases the roughness, with the formation of thin threadlike structures, at a higher extent for lower powers (15.82 nm for 100 W, 10.84 nm for 300 W), due to the stronger physical etching provoked by plasma, with a resulting higher etching rate.

The blurry aspect of AFM images of samples treated with oxygen plasma and UV/ozone can be correlated to the chemical modification caused by the UV/ozone treatment. This process increases the amount of oxygen-containing reactive species on the surface, which are very likely to electrostatically interact with the surrounding water vapor molecules. Furthermore, the UV/ozone treatment causes a reduction of roughness (6.00 nm and 5.39 nm for 100 W and 300 W of oxygen plasma treatment, respectively) with less sharp asperities. This result can be explained by a more homogeneous levelling action than that of oxygen plasma. The decrease in roughness is also consistent with the decrease in hydrophilicity, according to Wenzel's wettability model [64].

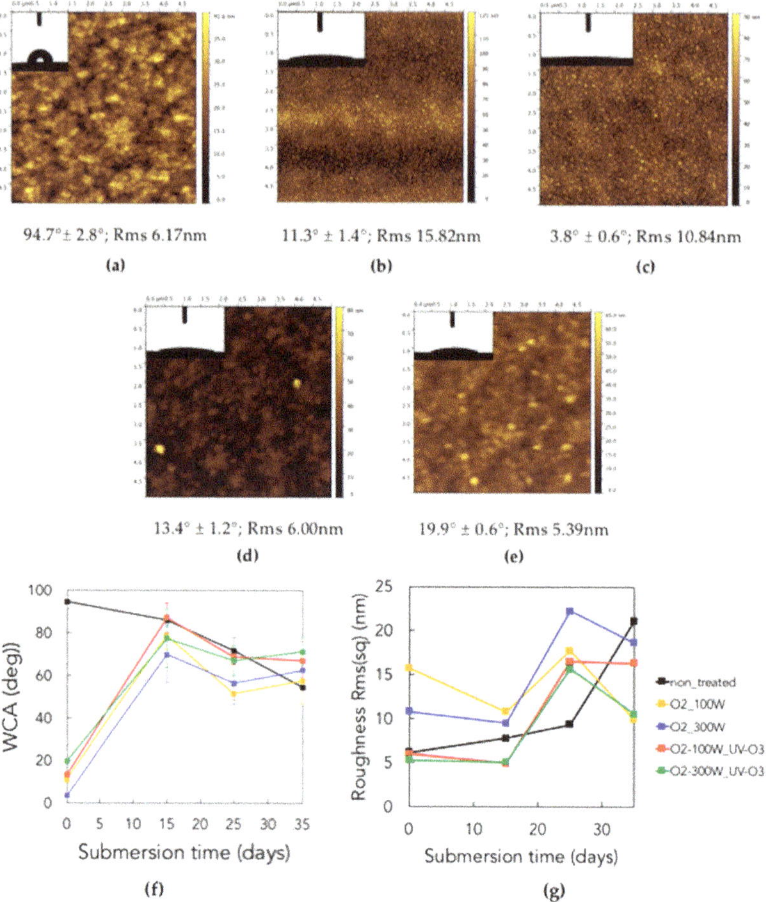

94.7°± 2.8°; Rms 6.17nm — **(a)** · 11.3° ± 1.4°; Rms 15.82nm — **(b)** · 3.8° ± 0.6°; Rms 10.84nm — **(c)**

13.4° ± 1.2°; Rms 6.00nm — **(d)** · 19.9° ± 0.6°; Rms 5.39nm — **(e)**

(f) · **(g)**

Figure 11. **Seawater contact angle** (WCA) measurements and AFM topography micrographs for non-submerged samples: non-treated (**a**), treated with oxygen plasma at 100 W (**b**) and 300 W (**c**), treated with oxygen plasma at 100 W, 300 W and UV/ozone (**d,e**). The scanned area for all AFM images was 5×5 µm². (**f**) Seawater contact angle, WCA, and (**g**) roughness vs. submersion time: each curve corresponds to one specific treatment; each point in the roughness plot is the result of averaging on the scanned area.

During submersion the non-treated sample became more hydrophilic (54.8° ± 7.3° at 35 days) and the others became more hydrophobic (in the range between 57° and 72°). However, the final values of water contact angle were quite similar, as shown in Figure 11f. This is an indirect confirmation that the long-term growth of biofilm induces a common wetting behavior. AFM results for the roughness evolution of the submerged samples are plotted in Figure 11g. As can be seen, microorganisms colonize randomly the devices surface, producing extracellular matrix, thus the roughness values provide information on compactness, homogeneity and regularity of the biofilm. After an initial increase in roughness (up to 25 days), the biofilm becomes more and more populated and composed of microbial aggregates rather than isolated cells, whose effect is a self-levelling action, with a reduction in roughness (after 30 days) [65]. The only exception is given by the non-treated sample, allegedly because the microorganisms attach on the surface but, with the passing of time and the progressive scarcity of nutrients, their bioadhesive strength weakens. Since there are much less asperities than for the other samples, they are washed off more easily, leaving a much more non-homogeneous (thus rougher) surface after 30 days.

AFM and SEM micrographs in Figure 12 highlight the presence of microorganisms on the sample surface, in particular a large number of bacteria (with dimensions in the range of 1 ÷ 2 μm), and diatoms Bacillariophyceae (6 ÷ 10 μm).

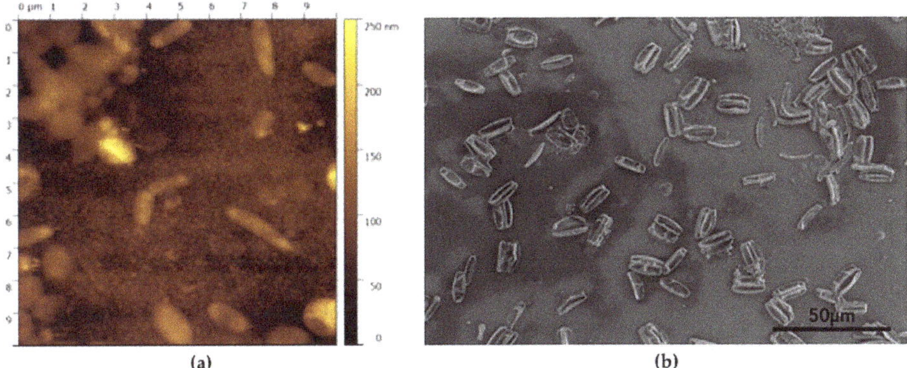

Figure 12. (**a**) AFM image with bacteria (~1 μm). (**b**) SEM micrograph with diatoms (~10 μm).

Finally, as a further proof of the microbial adhesion, LDV analyses highlighted the change of the resonance frequency of the submerged samples with respect to the non-submerged ones (Figure 13a), caused by the growth of biofilm on their surfaces. Finite element method (FEM) simulations (Eigenfrequency analysis, COMSOL multiphysics) were needed to derive the mass surface density of the grown biofilm from the experimental data on resonance frequencies. A simple 3D model was built on (a bi-layer pC/kapton substrate with a variable added biofilm mass on the top of it; see the inset in Figure 13a), showing that the experimental frequencies are in good agreement with the computational values allowing to derive and calculate the biofouling mass added on the sample surfaces: Table 3 summarizes the experimental and computational resonance frequencies, with the corresponding biofilm added mass, for the differently treated samples.

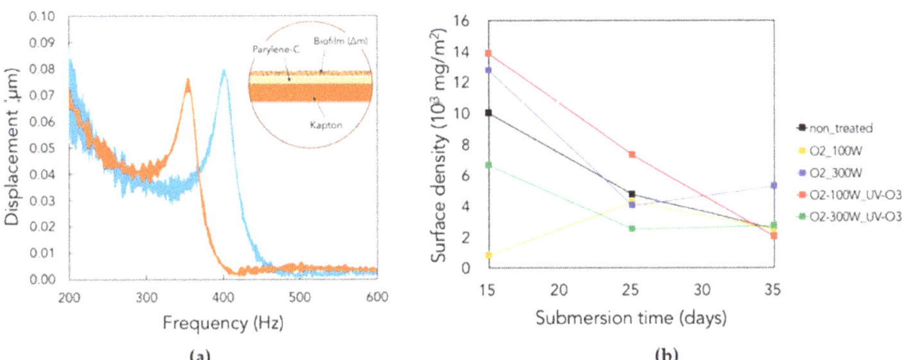

Figure 13. (**a**) Shift in resonance frequency due to the growth of biofilm; the inset shows the model used for the simulations. (**b**) Biofilm surface density vs. submersion days: each curve corresponds to a specific surface treatment.

Table 3. Results of finite element method (FEM) simulations.

Submersion Days	Samples	Length (mm)	Width (mm)	Exp f_{res} (Hz)	FEM f_{res} (Hz)	Biofilm Mass (10^{-5} g)
0	-	5.5	3.0	230.5	232	-
15	NT	6.0	3.0	201.25	201.2	18
	O_2_100W	6.0	3.0	225	225.5	1.5
	O_2_300W	6.0	3.0	208.28	208.4	23
	O_2-100W_UV-O_3	6.0	3.0	196.72	196.8	25
	O_2-300W_UV-O_3	6.0	3.0	226.25	226.1	12
25	NT	4.5	2.2	423.125	423.47	4.75
	O_2_100W	4.5	2.2	400.94	400.17	4.25
	O_2_300W	4.5	2.2	398.75	398.5	4.00
	O_2-100W_UV-O_3	4.5	2.2	354.69	354.9	7.25
	O_2-300W_UV-O_3	4.5	2.2	423.44	424.1	2.50
35	NT	4.5	2.2	408.75	408.9	2.50
	O_2_100W	4.5	2.2	430.94	431.1	2.50
	O_2_300W	4.5	2.2	406.41	406.8	5.25
	O_2-100W_UV-O_3	4.5	2.2	407.81	407.8	2.00
	O_2-300W_UV-O_3	4.5	2.2	427.03	427	2.75

At longer times, a common general decrease of the surface mass density, associated with an increase of resonance frequency, was detected on non-treated and treated substrates due to the degradation of biofilm and the death of microorganisms (Figure 13b).

4. Conclusions

In this study protective, barrier and anti-corrosive properties of different polymeric coatings for applications in marine environments, were compared. The final specific use of these thin films consists of employing them as conformal coatings for insulating flexible micro-devices adopted for harvesting mechanical energy from fluid-induced oscillations.

The desired application poses several challenges and issues to be addressed, i.e., the risk of short-circuit, the corrosion of metal electrodes, the water absorption and permeation into the devices leading to delamination or damages, the adhesion of microorganisms on the device surfaces. Therefore, the selection of the coatings to be tested was made in order to satisfy several requirements, such as flexibility, conformability, ease of application, insulating and barrier properties.

IS analyses, corrosion tests and the measurements of the piezoelectric signal for coated submerged devices revealed that parylene-C provides the best protecting, insulating and adhesion properties among the other coatings, even though it is susceptible to surface crack formation more than elastomers (PDMS-based coatings), because of its thermoplastic semi-crystalline character. Definitely, PMMA is the worst alternative mainly because it easily undergoes exfoliation during exposure to seawater.

Therefore, parylene was also used to observe qualitatively the accumulation of microorganisms and formation of biofilm on the surface of submerged devices: different surface treatments were adopted as well to modify the morphology and wettability of parylene in order to correlate microbial adhesion changes in the long-term period. The surface treatments selected for this purpose comprised an oxygen plasma treatment and a UV/ozone treatment. The reduction in the surface mass density of the grown biofilm, for both treated and non-treated substrates, revealed a loss of adherence for the microorganisms which is ascribed to the local scarcity of nutrients and to the protective action of parylene-C thin film: thus, together with its strong chemical and moisture resistance, this further property of parylene-C makes it more attractive as a conformal passivation coating on polymeric substrate materials and flexible micro-devices.

Author Contributions: M.M. conceived and performed the experiments; F.G., V.M.M. and A.Q. contributed for observation of surface defects; R.G. contributed for corrosion tests; L.A. contributed for AFM imaging of adhered microorganisms; M.M., F.G. and R.G. analyzed the data; F.G., R.G., V.M.M., and A.Q. contributed analysis tools; M.M. wrote the paper; A.M. contributed to mass transport issues, paper writing and supervised the experiments; M.D.V. supervised the experiments.

Funding: This research received no external funding.

Conflicts of Interest: The authors declare no conflict of interest.

References

1. Sacher, E.; Susko, J.R. Water permeation of polymer films. I. Polyimide. *J. Appl. Polym. Sci.* **1979**, *23*, 2355–2364. [CrossRef]
2. Yian, Z.; Keey, S.L.; Boay, C.G. Effects of seawater exposure on mode II fatigue delamination growth of a woven E-glass/bismaleimide composite. *J. Reinf. Plast. Compos.* **2015**, *35*, 138–150. [CrossRef]
3. Chiou, P.; L Bradley, W. Effects of seawater absorption on fatigue crack development in carbon/epoxy EDT specimens. *Composites* **1995**, *26*, 869–876. [CrossRef]
4. Sørensen, P.A.; Dam-Johansen, K.; Weinell, C.E.; Kiil, S. Cathodic delamination of seawater-immersed anticorrosive coatings: Mapping of parameters affecting the rate. *Prog. Org. Coat.* **2010**, *68*, 283–292. [CrossRef]
5. Morales, J.M.H.; Souriau, J.C.; Simon, G. Corrosion Protection of Silicon Micro Systems with Ultra-Thin Barrier Films for Miniaturized Medical Devices. *ECS Trans.* **2015**, *69*, 9–26. [CrossRef]
6. Traverso, P.; Canepa, E. A review of studies on corrosion of metals and alloys in deep-sea environment. *Ocean Eng.* **2014**, *87*, 10–15. [CrossRef]
7. Mohamed, A.M.A.; Abdullah, A.M.; Younan, N.A. Corrosion behavior of superhydrophobic surfaces: A review. *Arab. J. Chem.* **2015**, *8*, 749–765. [CrossRef]
8. Khatoon, Z.; McTiernan, C.D.; Suuronen, E.J.; Mah, T.F.; Alarcon, E.I. Bacterial biofilm formation on implantable devices and approaches to its treatment and prevention. *Heliyon* **2018**, *4*, e01067. [CrossRef]
9. Cao, S.; Wang, J.; Chen, H.; Chen, D. Progress of marine biofouling and antifouling technologies. *Chin. Sci. Bull.* **2011**, *56*, 598–612. [CrossRef]
10. Railkin, A.I. *Marine Biofouling: Colonization Processes and Defenses*, 1st ed.; CRC Press: Boca Raton, FL, USA, 2003.
11. Matsunaga, T.; Nakayama, T.; Wake, H.; Takahashi, M.; Okochi, M.; Nakamura, N. Prevention of marine biofouling using a conductive paint electrode. *Biotechnol. Bioeng.* **1998**, *59*, 374–378. [CrossRef]
12. Bixler, G.D.; Bhushan, B. Biofouling: lessons from nature. *Philos. Transact. A Math. Phys. Eng. Sci.* **2012**, *370*, 2381–2417. [CrossRef] [PubMed]
13. Flemming, H.C. Microbial Biofouling: Unsolved Problems, Insufficient Approaches, and Possible Solutions. In *Biofilm Highlights*, 1st ed.; Flemming, H.C., Wingender, J., Szewzyk, U., Eds.; Springer: Berlin/Heidelberg, Germany, 2011; Volume 5, pp. 81–109.
14. Flemming, H.C. Biofouling in water systems–cases, causes and countermeasures. *Appl. Microbiol. Biotechnol.* **2002**, *59*, 629–640. [CrossRef] [PubMed]
15. Algieri, L.; Todaro, M.T.; Guido, F.; Mastronardi, V.; Desmaële, D.; Qualtieri, A.; Giannini, C.; Sibillano, T.; De Vittorio, M. Flexible Piezoelectric Energy-Harvesting Exploiting Biocompatible AlN Thin Films Grown onto Spin-Coated Polyimide Layers. *ACS Appl. Energy Mater.* **2018**, *1*, 5203–5210. [CrossRef]
16. Petroni, S.; Rizzi, F.; Guido, F.; Cannavale, A.; Donateo, T.; Ingrosso, F.; De Vittorio, M. Flexible AlN flags for efficient wind energy harvesting at ultralow cut-in wind speed. *RSC Adv.* **2015**, *5*, 14047–14052. [CrossRef]
17. Mariello, M.; Guido, F.; Mastronardi, V.M.; Todaro, M.T.; Desmaële, D.; De Vittorio, M. Nanogenerators for harvesting mechanical energy conveyed by liquids. *Nano Energy* **2019**, *57*, 141–156. [CrossRef]
18. Abels, C.; Mastronardi, V.M.; Guido, F.; Dattoma, T.; Qualtieri, A.; Megill, W.M.; De Vittorio, M.; Rizzi, F. Nitride-Based Materials for Flexible MEMS Tactile and Flow Sensors in Robotics. *Sensors* **2017**, *17*, 1080. [CrossRef]
19. Applerot, G.; Abu-Mukh, R.; Irzh, A.; Charmet, J.; Keppner, H.; Laux, E.; Guibert, G.; Gedanken, A. Decorating Parylene-Coated Glass with ZnO Nanoparticles for Antibacterial Applications: A Comparative Study of Sonochemical, Microwave, and Microwave-Plasma Coating Routes. *ACS Appl. Mater. Interfaces* **2010**, *2*, 1052–1059. [CrossRef]

20. Baek, Y.; Kang, J.; Theato, P.; Yoon, J. Measuring hydrophilicity of RO membranes by contact angles via sessile drop and captive bubble method: A comparative study. *Desalination* **2012**, *303*, 23–28. [CrossRef]
21. Jbaily, A.; Yeung, R.W. Piezoelectric devices for ocean energy: a brief survey. *J. Ocean Eng. Mar. Energy* **2014**, *1*, 101–118. [CrossRef]
22. Wang, Z.L.; Jiang, T.; Xu, L. Toward the blue energy dream by triboelectric nanogenerator networks. *Nano Energy* **2017**, *39*, 9–23. [CrossRef]
23. Mukhopadhyay, R. When PDMS isn't the best. *Anal. Chem.* **2007**, *79*, 3248–3253. [CrossRef] [PubMed]
24. Koschwanez, J.H.; Carlson, R.H.; Meldrum, D.R. Thin PDMS films using long spin times or tert-butyl alcohol as a solvent. *PloS One* **2009**, *4*, e4572. [CrossRef] [PubMed]
25. Suleman, M.S.; Lau, K.K.; Yeong, Y.F. Characterization and Performance Evaluation of PDMS/PSF Membrane for CO2/CH4 Separation under the Effect of Swelling. *Procedia Eng.* **2016**, *148*, 176–183. [CrossRef]
26. Choonee, K.; Syms, R.R.A.; Ahmad, M.M.; Zou, H. Post processing of microstructures by PDMS spray deposition. *Sens. Actuators Phys.* **2009**, *155*, 253–262. [CrossRef]
27. Zhang, Z.P.; Song, X.F.; Cui, L.Y.; Qi, Y.H. Synthesis of Polydimethylsiloxane-Modified Polyurethane and the Structure and Properties of Its Antifouling Coatings. *Coatings* **2018**, *8*, 157. [CrossRef]
28. Fischer, S.C.L.; Levy, O.; Kroner, E.; Hensel, R.; Karp, J.M.; Arzt, E. Bioinspired polydimethylsiloxane-based composites with high shear resistance against wet tissue. *J. Mech. Behav. Biomed. Mater.* **2016**, *61*, 87–95. [CrossRef]
29. Coan, T.; Barroso, G.S.; Motz, G.; Bolzán, A.; Machado, R. a. F. Preparation of PMMA/hBN composite coatings for metal surface protection. *Mater. Res.* **2013**, *16*, 1366–1372. [CrossRef]
30. Cui, Z.; Drioli, E.; Lee, Y.M. Recent progress in fluoropolymers for membranes. *Prog. Polym. Sci.* **2014**, *39*, 164–198. [CrossRef]
31. Elkasabi, Y.; Chen, H.-Y.; Lahann, J. Multipotent Polymer Coatings Based on Chemical Vapor Deposition Copolymerization. *Adv. Mater.* **2006**, *18*, 1521–1526. [CrossRef]
32. Fortin, J.B.; Lu, T.M. *Chemical Vapor Deposition Polymerization - The Growth and Properties of Parylene Thin Films*, 1st ed.; Kluwer Academic Publishers: Boston, MA, USA, 2003.
33. Fortin, J.B.; Lu, T.M. A Model for the Chemical Vapor Deposition of Poly(para-xylylene) (Parylene) Thin Films. *Chem. Mater.* **2002**, *14*, 1945–1949. [CrossRef]
34. Chen, T.N.; Wuu, D.S.; Wu, C.C.; Chiang, C.C.; Chen, Y.P.; Horng, R.H. Improvements of Permeation Barrier Coatings Using Encapsulated Parylene Interlayers for Flexible Electronic Applications. *Plasma Process. Polym.* **2007**, *4*, 180–185. [CrossRef]
35. Cieślik, M.; Engvall, K.; Pan, J.; Kotarba, A. Silane–parylene coating for improving corrosion resistance of stainless steel 316L implant material. *Corros. Sci.* **2011**, *53*, 296–301. [CrossRef]
36. Golda-Cepa, M.; Brzychczy-Wloch, M.; Engvall, K.; Aminlashgari, N.; Hakkarainen, M.; Kotarba, A. Microbiological investigations of oxygen plasma treated parylene C surfaces for metal implant coating. *Mater. Sci. Eng. C* **2015**, *52*, 273–281. [CrossRef] [PubMed]
37. Lee, K.N. Current status of environmental barrier coatings for Si-Based ceramics. *Surf. Coat. Technol.* **2000**, *133–134*, 1–7. [CrossRef]
38. Saini, A.K.; Das, D.; Pathak, M.K. Thermal Barrier Coatings -Applications, Stability and Longevity Aspects. *Procedia Eng.* **2012**, *38*, 3173–3179. [CrossRef]
39. Yu, J.; Ruengkajorn, K.; Crivoi, D.G.; Chen, C.; Buffet, J.C.; O'Hare, D. High gas barrier coating using non-toxic nanosheet dispersions for flexible food packaging film. *Nat. Commun.* **2019**, *10*, 1–8. [CrossRef]
40. Lewis, J.S.; Weaver, M.S. Thin-film permeation-barrier technology for flexible organic light-emitting devices. *IEEE J. Sel. Top. Quantum Electron.* **2004**, *10*, 45–57. [CrossRef]
41. Fredj, N.; Cohendoz, S.; Feaugas, X.; Touzain, S. Ageing of marine coating in natural and artificial seawater under mechanical stresses. *Appl. Electrochem. Tech. Org. Coat.* **2012**, *74*, 391–399. [CrossRef]
42. Deyab, M.A.; Mohamed, N.H.; Moustafa, Y.M. Corrosion protection of petroleum pipelines in NaCl solution by microcrystalline waxes from waste materials: Electrochemical studies. *Corros. Sci.* **2017**, *122*, 74–79. [CrossRef]
43. Deyab, M.A. Effect of carbon nano-tubes on the corrosion resistance of alkyd coating immersed in sodium chloride solution. *Prog. Org. Coat.* **2015**, *85*, 146–150. [CrossRef]

44. Deyab, M.A.; Eddahaoui, K.; Essehli, R.; Benmokhtar, S.; Rhadfi, T.; De Riccardis, A.; Mele, G. Influence of newly synthesized titanium phosphates on the corrosion protection properties of alkyd coating. *J. Mol. Liq.* **2016**, *216*, 699–703. [CrossRef]

45. Deyab, M.A.; Mele, G.; Al-Sabagh, A.M.; Bloise, E.; Lomonaco, D.; Mazzetto, S.E.; Clemente, C.D.S. Synthesis and characteristics of alkyd resin/M-Porphyrins nanocomposite for corrosion protection application. *Prog. Org. Coat.* **2017**, *105*, 286–290. [CrossRef]

46. Li, W.; Rodger, D.; Menon, P.; Tai, Y.-C. Corrosion Behavior of Parylene-Metal-Parylene Thin Films in Saline. *ECS Trans.* **2008**, *11*, 1–6.

47. Davies, P.; Evrard, G. Accelerated ageing of polyurethanes for marine applications. *Polym. Degrad. Stab.* **2007**, *92*, 1455–1464. [CrossRef]

48. Pigliacelli, C.; D'Elicio, A.; Milani, R.; Terraneo, G.; Resnati, G.; Baldelli Bombelli, F.; Metrangolo, P. Hydrophobin-stabilized dispersions of PVDF nanoparticles in water. *J. Fluor. Chem.* **2015**, *177*, 62–69. [CrossRef]

49. Kang, G.; Cao, Y. Application and modification of poly(vinylidene fluoride) (PVDF) membranes – A review. *J. Membr. Sci.* **2014**, *463*, 145–165. [CrossRef]

50. Calcagnile, P.; Dattoma, T.; Scarpa, E.; Qualtieri, A.; Blasi, L.; Vittorio, M.D.; Rizzi, F. A 2D approach to surface-tension-confined fluidics on parylene C. *RSC Adv.* **2017**, *7*, 15964–15970. [CrossRef]

51. Calcagnile, P.; Blasi, L.; Rizzi, F.; Qualtieri, A.; Athanassiou, A.; Gogolides, E.; De Vittorio, M. Parylene C Surface Functionalization and Patterning with pH-Responsive Microgels. *ACS Appl. Mater. Interfaces* **2014**, *6*, 15708–15715. [CrossRef]

52. Mariello, M.; Guido, F.; Mastronardi, V.M.; De Donato, F.; Salbini, M.; Brunetti, V.; Qualtieri, A.; Rizzi, F.; De Vittorio, M. Captive-air-bubble aerophobicity measurements of antibiofouling coatings for underwater MEMS devices. *Nanomater. Nanotechnol.* **2019**, *9*, 1847980419862075. [CrossRef]

53. Gonon, P.; Hong, T.; Lesaint, O.; Bourdelais, S.; Debruyne, H. Influence of high levels of water absorption on the resistivity and dielectric permittivity of epoxy composites. *Polym. Test.* **2005**, *24*, 799–804. [CrossRef]

54. Cotinaud, M.; Bonniau, P.; Bunsell, A. The effect of water absorption on the electrical properties of glass-fibre reinforced epoxy composites. *J. Mater. Sci.* **1982**, *17*, 867–877. [CrossRef]

55. Maxwell, I.; Pethrick, R. Dielectric studies of water in epoxy resins. *J. Appl. Polym. Sci.* **1983**, *28*, 2363–2379. [CrossRef]

56. Efros, A.; Shklovskii, B. Critical Behavior of Conductivity and Dielectric-Constant Near Metal-Non-Metal Transition Threshold. *Phys. Status Solidi B* **1976**, *76*, 475–485. [CrossRef]

57. Maffezzoli, A.M.; Peterson, L.; Seferis, J.C.; Kenny, J.; Nicolais, L. Dielectric characterization of water sorption in epoxy resin matrices. *Polym. Eng. Sci.* **1993**, *33*, 75–82. [CrossRef]

58. Videm, K. Phenomena disturbing electrochemical corrosion rate measurements for steel in alkaline environments. *Electrochim. Acta* **2001**, *46*, 3895–3903. [CrossRef]

59. Deyab, M.A.; Keera, S.T.; El Sabagh, S.M. Chlorhexidine digluconate as corrosion inhibitor for carbon steel dissolution in emulsified diesel fuel. *Corros. Sci.* **2011**, *53*, 2592–2597. [CrossRef]

60. Deyab, M.A. Application of nonionic surfactant as a corrosion inhibitor for zinc in alkaline battery solution. *J. Power Sources* **2015**, *292*, 66–71. [CrossRef]

61. Deyab, M.A. Hydroxyethyl cellulose as efficient organic inhibitor of zinc–carbon battery corrosion in ammonium chloride solution: Electrochemical and surface morphology studies. *J. Power Sources* **2015**, *280*, 190–194. [CrossRef]

62. Gaynor, J.F. A heterogeneous model for the chemical vapor polymerization of Poly-P-Xylylenes. *Electrochem. Soc. Proc.* **1997**, *97-98*, 176–185.

63. Rogojevic, S.; Moore, J.A.; Gill, W.N. Modeling vapor deposition of low-K polymers: Parylene and polynaphthalene. *J. Vac. Sci. Technol. Vac. Surf. Films* **1999**, *17*, 266–274. [CrossRef]

64. Wenzel, R.N. Resistance of solid surfaces to wetting by water. *Ind. Eng. Chem.* **1936**, *28*, 988–994. [CrossRef]

65. Chatterjee, S.; Biswas, N.; Datta, A.; Maiti, P.K. Periodicities in the roughness and biofilm growth on glass substrate with etching time: Hydrofluoric acid etchant. *PLoS ONE* **2019**, *14*, e0214192. [CrossRef] [PubMed]

Article

Oscillating U-Shaped Body for Underwater Piezoelectric Energy Harvester Power Optimization

Iñigo Aramendia [1], Aitor Saenz-Aguirre [2], Ana Boyano [3], Unai Fernandez-Gamiz [1],* and Ekaitz Zulueta [2]

[1]   Nuclear Engineering and Fluid Mechanics Department, University of the Basque Country UPV/EHU, 01006 Vitoria-Gasteiz, Spain; inigo.aramendia@ehu.eus
[2]   Automatic Control and System Engineering Department, University of the Basque Country UPV/EHU, 01006 Vitoria-Gasteiz, Spain; asaenz012@ikasle.ehu.eus (A.S.-A.); ekaitz.zulueta@ehu.eus (E.Z.)
[3]   Mechanical Engineering Department, University of the Basque Country UPV/EHU, 01006 Vitoria-Gasteiz, Spain; ana.boyano@ehu.eus
*   Correspondence: unai.fernandez@ehu.eus; Tel.: +34-945-014066

Received: 30 September 2019; Accepted: 29 October 2019; Published: 30 October 2019

Abstract: Vibration energy harvesting (VeH) techniques by means of intentionally designed mechanisms have been used in the last decade for frequency bandwidth improvement under excitation for adequately high-vibration amplitudes. Oil, gas, and water are vital resources that are usually transported by extensive pipe networks. Therefore, wireless self-powered sensors are a sustainable choice to monitor in-pipe system applications. The mechanism, which is intended for water pipes with diameters of 2–5 inches, contains a piezoelectric beam assembled to the oscillating body. A novel U-shaped geometry of an underwater energy harvester has been designed and implemented. Then, the results have been compared with the traditional circular cylinder shape. At first, a numerical study has been carried at Reynolds numbers Re = 3000, 6000, 9000, and 12,000 in order to capture as much as kinetic energy from the water flow. Consequently, unsteady Reynolds Averaged Navier–Stokes (URANS)-based simulations are carried out to investigate the dynamic forces under different conditions. In addition, an Adaptive Differential Evolution (JADE) multivariable optimization algorithm has been implemented for the optimal design of the harvester and the maximization of the power extracted from it. The results show that the U-shaped geometry can extract more power from the kinetic energy of the fluid than the traditional circular cylinder harvester under the same conditions.

Keywords: energy harvesting; piezoelectric; pipelines; underwater networks; wireless sensor networks; control algorithm

---

1. Introduction

Due to industrial development and the improved quality of human life, the economic and social demand for energy is growing. In recent years, the problems arising from the use of coal, petroleum, and other energy resources with high carbon content are becoming almost unbearable. The trend of annual average of atmospheric CO_2 concentration has been increasing non-stop in the last decades. The global average surface temperature was reported to be approximately 1 °C higher than that of the preindustrial period [1]. The transition to an energy system that relies primarily on renewable energy sources has become one of the greatest challenges for alleviating climate change [2,3]. Clean energy sources such as wind, solar, geothermal, biomass, biofuels, waves, tidal, and hydropower can replace fossil fuels. With that in mind, water value is due not only to its agricultural and domestic use, but also to the possibility of being a source of energy, such as hydropower [4,5], ocean energy [6,7], or the energy obtained from its distribution network, for instance [8,9].

In this context, collecting small amounts of energy from the ambient in order to supply power to wireless devices has been investigated for the last decades. Moreover, in some cases where batteries are impractical, such as inaccessible remote systems, health monitoring, or body sensors, energy harvesting technology is very promising [10]. In fact, the power supply is one of the biggest challenges regarding the wireless sensor network applications. Frequently, the lifetime is confined to a battery supply, which is awkward [11]. Several environmental resources that offer enough power can be studied such as vibration, solar irradiation, thermal differences, and hybrid energy sources. Vibration energy harvesting (VeH) transforms mechanical energy obtained from ambient sources to electricity in order to power remote sensors. VeH technologies have been widely studied over the past decade [12–14]. Izadgoshasb et al. [15] proposed a new double pendulum-based piezoelectric system to harvest energy from human movements. They found a high increase in maximum output voltage in comparison to the conventional system and to an analogous system with only one pendulum. Furthermore, they stated that the double pendulum design could be further improved by varying some design parameters regarding dimensions or material. In a more recent work, Izadgoshasb et al. [16] proposed a multi-resonant harvester that consists of a cantilever beam with two triangular branches. They performed a parametric study using the finite element method (FEM) for the design optimization in order to get close resonances at low-frequency spectrum. Experimental results showed that the proposed harvester can harvest broadband energy from ambient vibration sources and is better than analogous piezoelectric energy harvesters with cantilever beams.

Energy harvesting from flow-induced vibrations has also attracted attention in the past years, and energy obtained from fluid–structure interaction (FSI) in different fluids, air, or liquid have been investigated by Elahi et al. [17]. Under flow loads, a structure can suffer different effects, such as limit cycle oscillations, chaotic movements, or internal resonances. From the point of view of aerodynamics, vortex-induced vibrations or vibrations caused by flutter or galloping can occur. Using piezoelectric devices placed in a flow field can convert large oscillations into electrical energy. An airfoil section placed on the end of cantilever piezoelectric beam can be used as a flutter energy harvester. Flutter speed is a critical value, from which the aerodynamic system becomes unstable. The self-excited oscillations that appear on the aeroelastic system once the critical value is overpassed are quite beneficial from the dynamic point of view; see Abdelkefi et al. [18]. Elahi et al. [19] modeled a nonlinear piezoelectric aeroelastic energy harvester that works on a postcritical aerolastic regime. An analytical model was developed taking into account the fluid–structure interaction and electromechanical performance. They concluded that higher electromechanical factor gives better harvesting. Wang et al. [20] conducted a study on a galloping-based piezoelectric energy harvester using isosceles triangle sectioned bluff bodies and applying computational fluid dynamics (CFD) to simulate the aerodynamic forces. They performed a parametric study in order to get the optimum vertex angle of the triangle and provided a guideline for efficient design. They determined that an angle of 130 degrees was the most adequate within the specific electromechanical coupling of their prototype. Dai et al. [21] investigated energy harvesting obtained from wind flow-induced vibrations. They compared experimentally four different test cases of piezoelectric energy harvesters, and based on the results they concluded which was the best orientation of the bluff body to design efficient devices. Jia et al. [22] presented an upright piezoelectric energy harvester (UPEH) that has a cylinder extension along its length. In the case with low speed wind, energy is obtained by vortex-induced vibrations (VIVs) that produce bending deformation. The UPEH can generate energy from low-speed wind by bending deformation produced by vortex-induced vibrations (VIVs). Zulueta et al. [23] developed a new control law for a contactless piezoelectric wind energy harvester, and afterwards, Bouzelata et al. [24] improved this wind energy harvester as a battery charger The simulation results proved that the device could power the battery when the wind speed is v = 2.7 m/s, taking into account that usually the moderate wind speed is considered to be 4 m/s.

Wang and Ko [25] developed a new piezoelectric energy harvester that converted flow energy into electrical energy by means of the oscillation of a piezoelectric film. They concluded that the obtained voltages based on the finite element model they proposed agree adequately with the experiment performed with various pressure differences in the pressure chamber. Therefore, they could use the model to predict the performance of the device, in terms of dimensions, material properties, and pressure loads. Silva-Leon et al. [26] proposed a novel approach to collect wind and solar energy at the same time. After extensive experiments, they founded out that their device was able to generate up to 3–4 mW of total power, which is enough to power remote sensors and small-scale portable electronics. Zhong et al. [27] showed that a type of graphene nanogenerator could generate electricity from the flow of different types of liquid, including water. Qureshi et al. [28] developed an analytical model of a novel and scalable piezoelectric energy harvester, where the kinetic energy from water flow-induced vibration was collected by means of piezoceramic cantilevers. They validated the model by means of a finite element simulation. They concluded that it is possible to install an energy harvester into the Turkey–Cyprus water pipeline with the capability to meet the power requirements of a wireless sensor node to monitor critical parameters.

Another way of collecting energy from water flow is the power generation from fluid flow in pipelines. This energy can be used to power wireless sensors networks; therefore, continuous monitoring of water quality and hydraulic parameters can be performed [29]. In addition, significant leakages can be detected in near real time [30]. Hoffmann et al. [31] presented a radial-flux energy harvester that obtained energy from water flow in water pipelines. Experimental results showed that the obtained energy could be used for powering smart water meter systems when the flow rate was at least 5 L/min. They pointed out that to achieve a power output less dependent on the flow rate, a fluidic bypass could be used. Shan et al. [32] presented a new underwater double piezoelectric energy harvesters system that consists of two harvesters placed in series with same parameters. The results of the experimental work showed that the performance of the double harvester can be improved by equaling the water speed, the specific gravity of the cylinder, and the spacing distance between the two harvesters.

The major novelty of the current research is the development of an innovative U-shaped geometry to be used as the oscillating body in an underwater energy harvester with the aim to increase the power output by the device. Three different geometries for the oscillating body were previously studied by Yao et al. [33]: Circular, triangular, and square. However, they concluded that the circular shape provided the best performance in terms of power output generation. Therefore, we have hypothesized that the proposed U-shaped geometry could improve the power generated by the underwater piezoelectric system. In the current work, we have implemented an Adaptive Differential Evolution (DE)-based (JADE) algorithm for the optimization of the design process of the harvester. JADE is an optimization algorithm based on evolutionary principles that is intended to find the maximum/minimum values of a cost function. In this analysis, a multivariable JADE algorithm intended to maximize the power extracted from the harvester has been designed. The two parameters optimized with the JADE algorithm are the structural spring of the harvester and the constant gain associated to its control algorithm.

2. Differential Evolution-Based Optimization

2.1. Differential Evolution Algorithm

The Differential Evolution (DE) algorithm, first introduced by Storn et al. [34], is an optimization technique based on evolutionary principles. As stated in [34], three main concepts are required for the correct performance of an optimization algorithm such as the DE: The ability to find the global optimum and not get stuck in local optimums, the fast convergence, and simplicity in the configuration parameters.

As it is shown the pipeline in Figure 1, in order to fulfill the previously listed requirements, the DE algorithm proposes the execution of an initial Initialization stage, and the iterative execution of three main steps: Mutation, Crossover, and Selection.

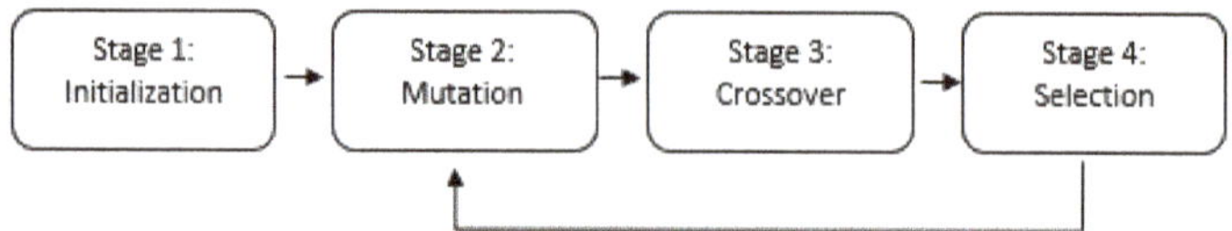

Figure 1. Pipeline of the operating principle of a Differential Evolution (DE) optimization algorithm.

The basic operation of a conventional DE algorithm is explained in detail in the work of Zhang et al. [35]. The first step is the generation (normally at a random process inside certain limit values) of the initial population of the algorithm; see Equation (1). To that purpose, the number of independent variables N and the size of the initial population P must be defined:

$$x_0 = \{x_{i,k,0}\} \tag{1}$$

where $i = 1 : 1 : N$ and $k = 1 : 1 : P$.

After the initialization, the proposed iterative process is an emulation of the natural evolution. First, some individuals suffer from mutations that can improve/worsen their survival capabilities. To that end, different mutation strategies can be implemented in the DE algorithm. One of the most widely used mutation strategies, known as *"DE/rand/1"*, is shown in Equation (2):

$$v_{i,G} = x_{r1,G} + F_i \cdot (x_{r2,G} - x_{r3,G}) \tag{2}$$

Next, in the crossover or recombination step, the characteristics of two different individuals are combined to form a descendant individual, as it is shown in Equation (3):

$$u_{j,i,G} = \left\{ \begin{array}{llll} if\ rand(0,1) \le CR_i\ or\ j = j_{\text{rand}} & then & u_{j,i,G} \\ else & & x_{j,i,G} \end{array} \right\} \tag{3}$$

where $j_{\text{rand}} \in (1, N)$ is defined randomly at each iteration and CR_i is the crossover probability, which is constant in the conventional DE algorithm.

Finally, the fitness of the individuals of the new generation is compared to the value of the fitness of the previous generation, and the best individuals are selected.

$$u_{i,g+1} = \left\{ \begin{array}{llll} if\ f(u_{j,i,G}) < f(x_{j,i,G}) & then & u_{j,i,G} \\ else & x_{j,i,G} \end{array} \right\} \tag{4}$$

This process is repeated during a defined number of generations, and after some iterations the individuals with the best characteristics remain and the evolutionary algorithm converges toward an optimal result.

As stated by Zhang et al. [36], DE algorithms have been widely used due to their simplicity, small number of configuration parameters, and good performance in optimization cases. Nevertheless, one of the problems present in these algorithms is the difficulty associated to the adequate setting of the configuration parameters. As explained in the work of Zhang et al. [35], both theoretical and experimental studies have been proposed for the setting of these parameters (especially the mutational factor F and the crossover probability CR). Nevertheless, the absence of clear guidelines and the necessity for trial and error tuning cause difficulties in achieving a good performance of the algorithm. In order to improve the results of DE algorithms, many variants have been developed and proposed in the literature: Self-adaptive Differential Evolution (SaDE), Self-adaptive Differential Evolution with Neighborhood Search (SaNSDE), Self-adaptive Differential Evolution (jDE), Differential Evolution with Global and Local neighborhoods (DEGL), Adaptive Differential Evolution (JADE), Composite Differential Evolution (CoDE), and Improved Adaptive Differential Evolution (IJADE).

2.2. JADE: Adaptive Differential Evolution

The JADE algorithm is a variation of the DE that belongs to the group of the adaptive parameter control algorithms, as introduced by Zhang et al. [35]. This means that the adaption of the configuration parameters is carried out according to the status of the search process of the algorithm. Some additional examples of adaptive parameter control algorithms are SaDE, jDE, and SaNSDE.

The JADE algorithm was first introduced in the work of Zhang et al. [37]. Later, a second version of the algorithm was presented by Zhang et al. [35]. The objective of the modifications introduced in the JADE strategy with respect to the DE algorithm is the improvement of the convergence of the algorithm and the diversification of the population that is used through the execution of it.

According to the work of Zhang et al. [37], the JADE algorithm is considered to improve the performance of the conventional DE algorithms by implementing a novel mutation strategy referred as *"DE/current-to-p best"* and a new method for the adaption of the mutational factor *F* and the crossover probability *CR*. The recombination and selection steps of the JADE algorithm are the same of the DE algorithm, which were presented in Equations (3) and (4), respectively.

The *"DE/current-to-p best"* mutation method introduced in the JADE algorithm can be expressed as shown in Equation (5):

$$v_{i,G} = x_{i,G} + F_i \cdot \left(x^p_{best,G} - x_{i,G} \right) + F_i \cdot \left(x_{r2,G} - x_{r3,G} \right) \tag{5}$$

where p is the per one value of the number of the best individuals selected for the mutation.

The adaption at each generation of the crossover probability CRi is carried out by random generation according to a normal distribution and a predefined value of a standard deviation, as it is shown in Equation (6):

$$CR_i = randn_i(\mu_{CR}) \tag{6}$$

Similarly, the adaption at each generation of the mutation factor Fi is carried out by random generation according to a Cauchy distribution with predefined parameter values, as shown in Equation (7):

$$F_i = randn_i(\mu_F) \tag{7}$$

According to Islam et al. [38], with the use of the JADE algorithm, a diversity of the population, which avoids a premature convergence of the optimization algorithm, and a more reliable performance of the algorithm are achieved.

3. Harvester Description

The mechanism presented in the current study is connected with the work of Cottone et al. [39] and on the computational fluid dynamics simulations of Aramendia et al. [40]. This device, which is intended to be used inside water pipelines of 2–5 inches of diameter, contains a piezoelectric beam assembled to an oscillating body, as shown in Figure 2.

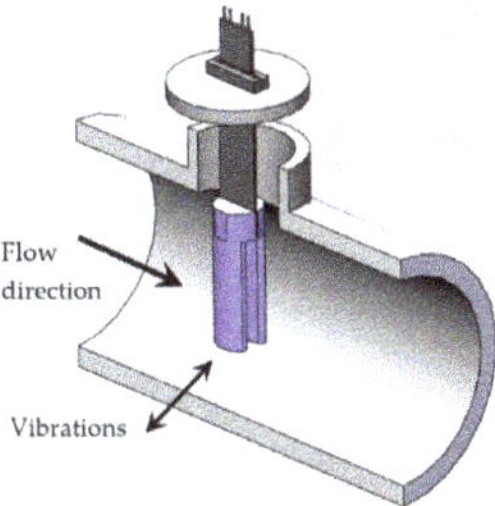

Figure 2. Energy harvester assembled inside a water pipe with the U-shaped geometry as the oscillating body (not to scale).

The impact of the water results in vibrations because of the vortices generated in the region closely behind the oscillating body. Additionally to the cylinder geometry, an innovative U-shaped geometry has been proposed as the oscillating body to optimize the extraction of kinetic energy from the water incoming through the water pipe. Figure 3 shows the dimensions of both geometries used in the present study.

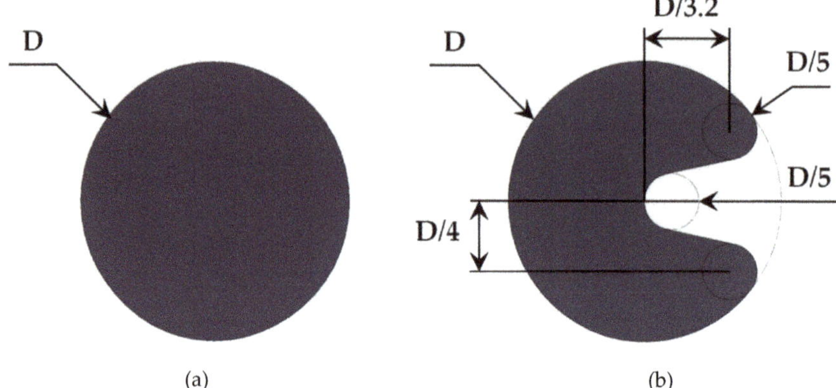

Figure 3. Geometries for the oscillating body. (**a**) Circular cylinder and (**b**) U-shaped geometry.

4. Computational Setup

This section is devoted to provide a detailed description of the numerical model developed to characterize the operation of both oscillating bodies in an underwater harvester. The length of the computational domain consists of 40 times the body diameter (D) behind the oscillating body. This has been considered enough to study accurately the vortices generated by the water passing around the body. Diameters of $D = 10$ mm and $D = 20$ mm have been considered in this study for both geometries, as illustrated in Figure 3.

A velocity inlet and pressure outlet boundary condition has been defined as well as slip condition for the top and bottom boundaries. CFD tools require the subdivision of this computational domain into a number of smaller subdomains to solve the flow physics. Therefore, the mesh generation is a very relevant issue in the pre-process stage. The mesh carried out consists of 2D polyhedral cells; most of them are placed in the area behind the body after the definition of a fully anisotropic wake refinement. Additionally, a volumetric control has been designed to refine the mesh around the body and to keep a y+ value less than 1. The mesh dependency study of the previous work of Aramendia et al. [40] was used to verify the accuracy of the solution.

The Reynolds Averaged Navier–Stokes (RANS) equations have been applied to reach the numerical solution of the unsteady state flow involved. The finite volume method has been used to discretize the integral form of the conservation equations with the CFD commercial software STAR-CCM+ (v. 11.06.011, CD-adapco, Melville, NY, USA) [41]. An upwind scheme [42] was used to discretize the convective terms, ensuring the robustness of the solution. The eddy viscosity models (EVM) with two transport equations have been chosen to define the turbulence modeling by means of the k-ω shear stress transport (SST) turbulence model developed by Menter [43]. A time-step of 0.002 s and 15 inner iterations have been defined as the optimal configuration to capture the vortex shedding. A second-order temporal discretization has been used in all the simulations presented in the current study. The solution obtained with the simulations was considered converged when satisfactory residuals were achieved on pressure, turbulence, and velocity quantities.

5. Computational Results

Four Reynolds numbers have been chosen to study the cross flow (Re = 3000, 6000, 9000, and 12,000). The Re is a dimensionless number based on the oscillating body diameter (D) and was obtained by Equation (8), where ρ_{water} and μ correspond to the density and dynamic viscosity of water at a temperature of 15 °C:

$$Re = \frac{V_{water} \times D \times \rho_{water}}{\mu} \tag{8}$$

The water velocity V_{water} at the inlet has been changed to achieve the different Re numbers. The numerical solution in all cases was simulated for a period of time of 20 seconds. Tables A1 and A2 of the Appendix A show the contour lines of the vorticity at t = 20 s for each Re number studied (Re = 3000, 6000, 9000, and 12,000) and for each geometry proposed as the oscillating body. The oscillating vortex pattern, convection, and diffusion of the vortices are visible. The body lift is calculated by means of CFD tools.

Figures 4 and 5 show the evolution of the lift coefficient C_L at each Re number investigated. This dimensionless coefficient is calculated by Equation (9), where F_L represents the force perpendicular to the flow direction caused by the water in the oscillating body, i.e., the circular cylinder and the U-shaped geometry, respectively.

$$C_L = \frac{F_L}{0.5 \times \rho_{water} \times U_\infty^2 \times D} \tag{9}$$

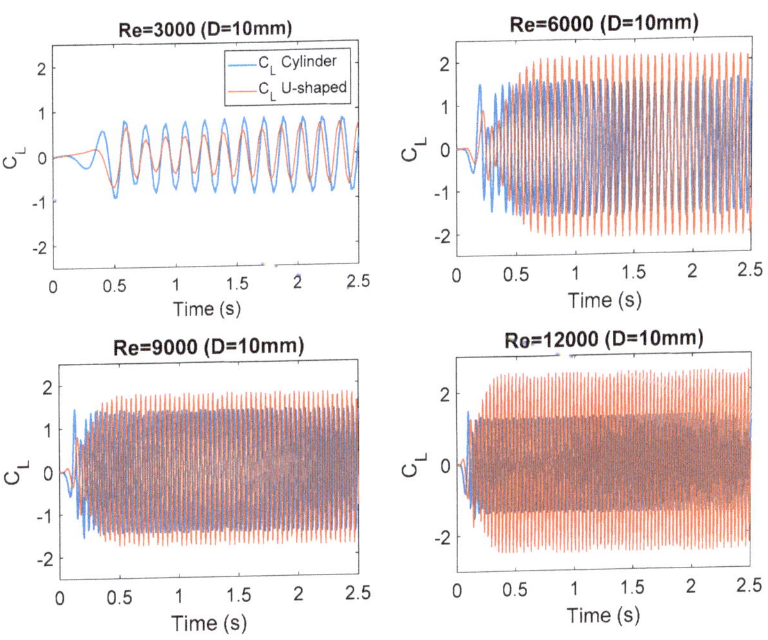

Figure 4. Evolution of the lift coefficient at each Reynolds number (*D* = 10 mm).

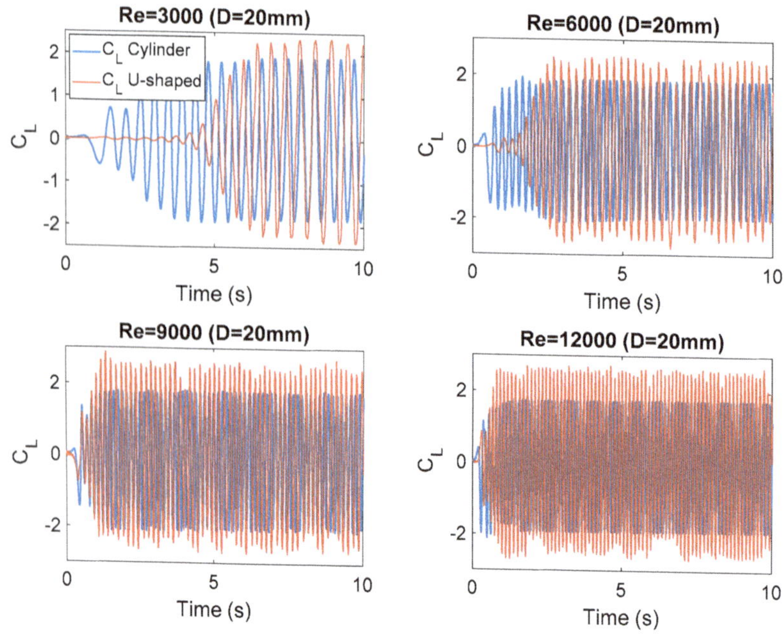

Figure 5. Evolution of the lift coefficient at each Reynolds number ($D = 20$ mm).

6. JADE-Based Underwater Piezoelectric Energy Harvester Optimization

As it was explained in Section 3, a new geometry for an underwater piezoelectric energy harvester is presented in this document. The proposed novel geometry is considered to improve the performance of cylindrical devices used in conventional energy harvesting systems.

A detailed model of the dynamics of an underwater piezoelectric energy harvester system is presented in the work of Aramendia et al. [40]. The selection of the optimal parameters during the design process of the harvester enables maximizing the power generated by the harvesting system. The extracted power has been calculated by means of Equation (10):

$$P = \left(\frac{\alpha \cdot a}{K_{\text{trans}}}\right)^2 \frac{\dot{\theta}^2}{K_p} = \left(\frac{\alpha \cdot a}{K_{\text{trans}}}\right)^2 \frac{1}{K_p} \left(\frac{a_1 w_0 \cdot C_{L \cdot \max}}{\sqrt{(a_4 - a_2 w_0{}^2)^2 + (a_3 w_0)^2}} \sin(w_0 t)\right)^2 \tag{10}$$

Thus, the mean value of the instantaneous power is determined over the period given by an angular pulsation of the lift coefficient w_0; see Equation (11):

$$P_{\text{mean}} = \left(\frac{\alpha \cdot a}{K_{\text{trans}}}\right)^2 \frac{1}{2 \cdot K_p} \left(\frac{a_1 w_0 \cdot C_{L \cdot \max}}{\sqrt{(a_4 - a_2 w_0{}^2)^2 + (a_3 w_0)^2}}\right)^2 \tag{11}$$

Tables 1 and 2 show the system model input parameters and variables used in the control of the energy harvester, respectively.

Table 1. Model parameters.

Name	Definition	Value	Units
ρ_{water}	Fluid density	997.5	kg/m^3
K_{trans}	Transduction gain	2	-
f	Frictional coefficient	0.01	(N·m·s)/rad
a	Force application distance point	0.01	m
α	Voltage induced bending factor	100	A s/m
C	Piezoelectric capacitance	1	nF

Table 2. Model variables.

Name	Definition	Units
$C_{L,\text{max}}$	Maximum lift coefficient	-
V_1	Piezoelectric voltage	V
t	Time	s
θ	Beam angle	rad
$K\text{spring}$	Spring constant	N/m
K_p	Proportional gain	A/V
T_m	Moment generated by the piezoelectric	N·m
T_{Hydro}	Hydro-mechanical torque	N·m
ω_0	Angular pulsation of the lift coefficient	rad/s
J_{wt}	Oscillating body inertia moment	Kg·m^2
u_1	Reference of the piezoelectric deflection	m

In this way, the value of the parameter Kspring and the value of the control parameter Kp stand as two crucial variables for the optimization of the harvester. The Kspring parameter refers to the value of the spring constant of the torsion spring introduced to the hydro-mechanical system of the harvester. The objective of the torsion spring is to maintain the vertical equilibrium of the harvester in the absence of piezoelectric forces. The expression that models the action of the torsion spring in the harvester system is given in Equation (12):

$$J_{wt} \cdot \frac{d^2\theta}{d^2 t} = T_{\text{hydro}} - K_{\text{Spring}} \cdot \theta - f \cdot \frac{d\theta}{dt} - T_m \tag{12}$$

The Kp parameter refers to the proportional gain of the control law proposed for the performance of the underwater piezoelectric energy harvester. The action of the control system of the harvester is given by the following expression in Equation (13):

$$K_P \cdot V_1 = \alpha \cdot \frac{du_1}{dt} - C \cdot \frac{dV_1}{dt} \tag{13}$$

Due to the all the existent possibilities, the optimal setting of both Kspring and Kp parameters could be complicated, which would result in an inefficient operation of the harvester. Consequently, in this paper, a JADE optimization algorithm has been developed and implemented in order to obtain the optimal values of the Kspring and Kp parameters and thus optimize the performance of the system.

The cost function selected for the JADE algorithm is the power generated by the harvester system calculated in Equation (11). The power generated by the harvester is dependent on both Kp and the Kspring parameters and could be expressed as in Equation (14):

$$P = f\left(K_P, K_{\text{Spring}}\right) \tag{14}$$

The configuration parameters defined for execution of the JADE algorithm presented in this paper are listed in Table 3.

Table 3. Configuration parameters of the adaptive differential evolution (JADE) algorithm.

Explanation	Symbol	Value
Number of variables	N	2
Initial population size	P	200
Number of iterations	Niter	2000
Mutation ratio	F	Adaptive
Crossover probability	CR	Adaptive
Mutation ratio adaption parameter	μ_F	0.5
Crossover probability adaption parameter	μ_{CR}	0.5
Kp maximum value	Kpmax	500
Kp minimum value	Kpmin	0
Kspring maximum value	Kspringmax	500
Kspring maximum value	Kspringmin	0
"DE/current-to-p best" mutation parameter	p	0.1

Different scenarios corresponding to various harvester geometries and different Reynolds numbers of the fluid actuating on the harvester, all of them already introduced in Section 3, have been considered and optimized with the application of the JADE algorithm in order to find the best combination of Kp and Kspring parameters that maximizes the energy production of the underwater piezoelectric energy harvester.

An illustration of the progress of the JADE optimization algorithm corresponding to the cylindrical configuration of the harvester with a $D = 10$ mm and a Reynolds number equal to 3000 is presented in Figure 6.

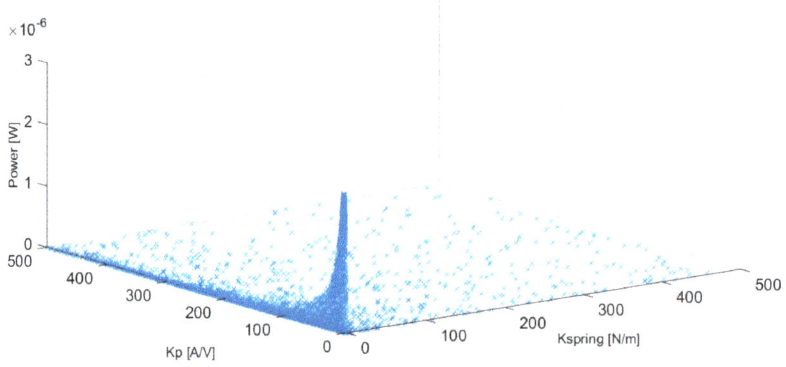

Figure 6. Progress of the JADE optimization algorithm. $D = 10$ mm circular cylinder-based harvester and Re = 3000.

As observed in Figure 6, the search of the optimization algorithm converges toward the optimum combination of the input parameters, Kp and Kspring, until the best solution that maximizes the power generated by the energy harvester is obtained.

7. Results

The search of the optimal combination of the Kp and Kspring parameter values for the maximization of the power generated by the harvester system through the implementation of a JADE algorithm has been proposed in Section 6. The results obtained for each one of the analyzed scenarios are represented in Tables 4 and 5 for $D = 10$ mm and $D = 20$ mm, respectively. The last column of each table represents the increment of the power ΔPower (%) generated by the U-shape underwater energy harvester with respect to the traditional underwater harvester based on a cylindrical oscillating body.

Table 4. Results of the JADE optimization algorithm for the cylinder and U-shaped oscillating bodies with $D = 10$ mm.

| | Cylinder $D = 10$ mm | | | U-shape $D = 10$ mm | | | |
Re	Kspring	Kp	Power [μW]	Kspring	Kp	Power [μW]	ΔPower (%)
3000	9.41×10^{-20}	4.4585	2.25	3.52×10^{-19}	4.4574	1.75	−28.57
6000	2.77×10^{-19}	4.4585	135.76	3.14×10^{-19}	4.4574	237.8	42.91
9000	7.96×10^{-20}	4.4585	560.16	2.58×10^{-20}	4.4574	842.75	33.53
12,000	1.74×10^{-19}	4.4585	1848.3	2.22×10^{-19}	4.4574	5321.7	65.27

Table 5. Results of the JADE optimization algorithm for the cylinder and U-shaped oscillating bodies with $D = 20$ mm.

| | Cylinder $D = 20$mm | | | U-shape $D = 20$mm | | | |
Re	Kspring	Kp	Power [μW]	Kspring	Kp	Power [μW]	ΔPower (%)
3000	2.25×10^{-19}	4.5054	3.03	3.36×10^{-19}	4.4886	4.63	34.56
6000	2.72×10^{-19}	4.5054	49.857	3.76×10^{-19}	4.4886	94.54	47.26
9000	3.03×10^{-19}	4.5054	218.59	3.08×10^{-19}	4.4886	543.02	59.75
12,000	2.38×10^{-19}	4.5054	640.74	3.48×10^{-19}	4.4886	1553	58.74

The power generated by the harvesting system is considerably improved with the application of the proposed U-shaped geometry, especially for cases at higher Reynolds numbers, as observed in Tables 4 and 5. In general, the power achieved by the U-shaped geometry is larger than the cylinder for both diameters considered. However, for the lowest Reynolds number studied, Re = 3000 and $D = 10$ mm, the power achieved by the U-shaped based harvester is lower than the one obtained by the cylinder. The largest power output is achieved at Re = 12,000 and $D = 10$ mm for both geometries. The cylinder oscillating body reaches a power of 1848.3 μW, and the U-shape geometry gets the maximum power with a value of 5321.7 μW, as shown in Table 4. Nevertheless, it must be taken into account that the water velocity associated at this high Reynolds number is difficult to obtain in the water pipes considered in this study from 2 to 5 inches of diameter. A graphical comparison of the optimal power generated by the analyzed four different harvester geometries for different Reynolds number values is presented in Figure 7. The present U-shaped harvester with an oscillating body size of $D = 10$ mm generates up to 5.2 mW at Re = 12,000. This result shows a significant improvement from the literature; please see the model presented on a review on mechanisms for piezoelectric-based energy harvesters for an underwater harvester [17], which is able to produce merely 0.9 mW at the same Re number.

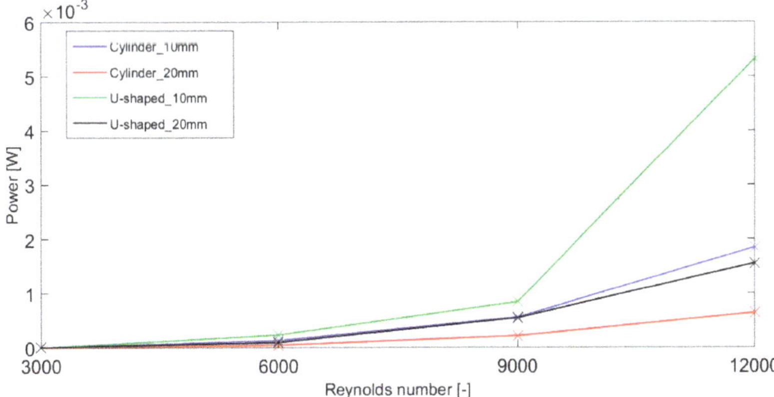

Figure 7. Comparison of the optimal power generated by the proposed four different energy harvesting system geometries.

Similarly, a comparison of the optimal values of the Kp and Kspring parameters for each one of the four analyzed harvester geometries and for different Reynolds number values is presented in Figure 8.

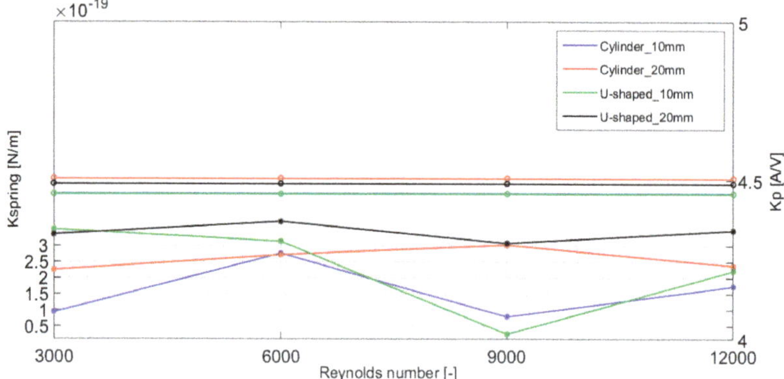

Figure 8. Comparison of the optimal values of the Kp and Kspring parameters for the proposed four different energy harvesting system geometries.

There are slight differences in the optimal value of the Kp and Kspring parameters, especially for the geometries with $D = 10$ mm. This could be translated in a non-optimal performance of the system and in a reduction in the power generation of the harvesting system, with its subsequent decrease of energy yield. These increased power generation of the harvesting system and the differences in the optimal values of the Kp and Kspring parameters prove the correct performance of the proposed harvester geometry and the JADE optimization algorithm presented in this paper.

8. Conclusions

In the current work, a numerical study of an underwater piezoelectric energy harvester has been carried out for the extraction of kinetic energy from the water flow. The mechanism, which is planned to be used inside water pipelines of 2–5 inches of diameter, contains a piezoelectric beam assembled to an oscillating body. Two different geometries for the oscillating body have been considered. The first one is the traditional circular cylinder, and the second one is a novel U-shaped geometry. Both geometries have been studied for two different diameters: $D = 10$ and 20 mm. Thus, 2D numerical simulations have been performed around each proposed geometry at Reynolds numbers Re = 3000, 6000, 9000, and 12,000. Simulations in unsteady-state conditions were made during a period of time of 20 seconds in order to evaluate the vortex shedding generated in the region behind the oscillating bodies. The lift coefficient of the oscillating bodies obtained in the simulations has been used as an input variable in the control system.

Furthermore, a multivariable JADE-based optimization algorithm has been designed to optimize the design process of the harvester and maximize the power extracted from it. The two parameters optimized with the JADE algorithm are the structural spring of the harvester and the constant gain associated to its control algorithm. According to the obtained results, the power generated by the U-shape-based energy harvester is always larger than the one obtained by the circular cylinder for all the Reynolds numbers studied except for Re = 3000 and $D = 10$ mm. The maximum power extracted from the harvester is 5321.7 µW and corresponds to the case with Re = 12,000 and $D = 10$ mm. The results show that thanks to the U-shaped geometry of the oscillating body and to the JADE optimization algorithm, the power output of the harvester has significantly improved.

The power generated by the underwater piezoelectric energy harvester follows an exponential law for all the cases investigated, including the U-shaped geometry. Additionally, the proportional gain of the control law maintains approximately constant at the water speeds studied in the current work.

Author Contributions: I.A. and U.F.-G. conceived and performed the CFD simulations; A.S.-A. and E.Z. developed the new control algorithm, and A.B. analyzed the results and provided constructive instructions in the process of preparing the paper.

Funding: This research was funded by the Government of the Basque Country SAIOTEK (S-PE11UN112) and the University of the Basque Country UPV/EHU EHU12/26 research programs.

Acknowledgments: The authors are grateful to the Government of the Basque Country and the University of the Basque Country UPV/EHU through the SAIOTEK (S-PE11UN112) and EHU12/26 research programs, respectively. The funding of Vital Fundazioa is also acknowledged. Development Agency of the Basque Country (SPRI) is gratefully acknowledged for economic support through the research project "Refrigeración de dispositivos de alto flujo térmico mediante impacto de chorro" (AIRJET), KK-2018/00109, Programa ELKARTEK.

Conflicts of Interest: The authors declare no conflict of interest.

Appendix A

Table A1. Vortex shedding comparison behind the circular cylinder and the U-shaped geometry at different Reynolds numbers where $D = 10$ mm.

Re	Wake Development (Cylinder)	Wake Development (U-Shaped)
3000		
6000		
9000		
12,000		

Table A2. Vortex shedding comparison behind the circular cylinder and the U-shaped geometry at different Reynolds numbers where $D = 20$ mm.

Re	Wake Development (Cylinder)	Wake Development (U-Shaped)
3000		
6000		
9000		
12,000		

References

1. Yildiz, I. Review of climate change issues: A forcing function perspective in agricultural and energy innovation. *Int. J. Energy Res.* **2019**, *43*, 2200–2215. [CrossRef]
2. Zou, C.; Zhao, Q.; Zhang, G.; Xiong, B. Energy revolution: From a fossil energy era to a new energy era. *Nat. Gas Ind. B* **2016**, *3*, 1–11. [CrossRef]
3. Owusu, P.A.; Asumadu-Sarkodie, S. A review of renewable energy sources, sustainability issues and climate change mitigation. *Cogent Eng.* **2016**, *3*. [CrossRef]
4. Bartle, A. Hydropower potential and development activities. *Energy Policy* **2002**, *30*, 1231–1239. [CrossRef]
5. Ranzani, A.; Bonato, M.; Patro, E.R.; Gaudard, L.; De Michele, C. Hydropower Future: Between Climate Change, Renewable Deployment, Carbon and Fuel Prices. *Water* **2018**, *10*, 1197. [CrossRef]
6. Melikoglu, M. Current status and future of ocean energy sources: A global review. *Ocean Eng.* **2018**, *148*, 563–573. [CrossRef]
7. Magagna, D.; Uihlein, A. Ocean energy development in Europe: Current status and future perspectives. *Int. J. Mar. Energy* **2015**, *11*, 84–104. [CrossRef]
8. Telci, I.T.; Aral, M.M. Optimal Energy Recovery from Water Distribution Systems Using Smart Operation Scheduling. *Water* **2018**, *10*, 1464. [CrossRef]
9. Sitzenfrei, R.; Berger, D.; Rauch, W. Design and optimization of small hydropower systems in water distribution networks under consideration of rehabilitation measures. *Urban Water J.* **2018**, *15*, 183–191. [CrossRef]
10. Kiziroglou, M.E.; Yeatman, E.M. Materials and techniques for energy harvesting. In *Functional Materials for Sustainable Energy Applications*; Kilner, J.A., Skinner, S.J., Irvine, S.J.C., Edwards, P.P., Eds.; Woodhead Publishing: Cambridge, UK, 2012; pp. 541–572.
11. Shaikh, F.K.; Zeadally, S. Energy harvesting in wireless sensor networks: A comprehensive review. *Renew. Sustain. Energy Rev.* **2016**, *55*, 1041–1054. [CrossRef]

12. Beeby, S.P.; Tudor, M.J.; White, N.M. Energy harvesting vibration sources for microsystems applications. *Meas. Sci. Technol.* **2006**, *17*, R175–R195. [CrossRef]

13. Wei, C.; Jing, X. A comprehensive review on vibration energy harvesting: Modelling and realization. *Renew. Sustain. Energy Rev.* **2017**, *74*, 1–18. [CrossRef]

14. Ahmed, R.; Mir, F.; Banerjee, S. A review on energy harvesting approaches for renewable energies from ambient vibrations and acoustic waves using piezoelectricity. *Smart Mater. Struct.* **2017**, *26*, 085031. [CrossRef]

15. Izadgoshasb, I.; Lim, Y.Y.; Tang, L.; Padilla, R.V.; Tang, Z.S.; Sedighi, M. Improving efficiency of piezoelectric based energy harvesting from human motions using double pendulum system. *Energy Convers. Manag.* **2019**, *184*, 559–570. [CrossRef]

16. Izadgoshasb, I.; Lim, Y.Y.; Padilla, R.V.; Sedighi, M.; Novak, J.P. Performance Enhancement of a Multiresonant Piezoelectric Energy Harvester for Low Frequency Vibrations. *Energies* **2019**, *12*, 2770. [CrossRef]

17. Elahi, H.; Eugeni, M.; Gaudenzi, P. A Review on Mechanisms for Piezoelectric-Based Energy Harvesters. *Energies* **2018**, *11*, 1850. [CrossRef]

18. Abdelkefi, A. Aeroelastic energy harvesting: A review. *Int. J. Eng. Sci.* **2016**, *100*, 112–135. [CrossRef]

19. Elahi, H.; Eugeni, M.; Gaudenzi, P. Design and performance evaluation of a piezoelectric aeroelastic energy harvester based on the limit cycle oscillation phenomenon. *Acta Astronaut.* **2019**, *157*, 233–240. [CrossRef]

20. Wang, J.; Tang, L.; Zhao, L.; Zhang, Z. Efficiency investigation on energy harvesting from airflows in HVAC system based on galloping of isosceles triangle sectioned bluff bodies. *Energy* **2019**, *172*, 1066–1078. [CrossRef]

21. Dai, H.L.; Abdelkefi, A.; Yang, Y.; Wang, L. Orientation of bluff body for designing efficient energy harvesters from vortex-induced vibrations. *Appl. Phys. Lett.* **2016**, *108*, 053902. [CrossRef]

22. Jia, J.; Shan, X.; Upadrashta, D.; Xie, T.; Yang, Y.; Song, R. Modeling and Analysis of Upright Piezoelectric Energy Harvester under Aerodynamic Vortex-induced Vibration. *Micromachines* **2018**, *9*, 667. [CrossRef] [PubMed]

23. Zulueta, E.; Kurt, E.; Uzun, Y.; Lopez-Guede, J.M. Power control optimization of a new contactless piezoelectric harvester. *Int. J. Hydrogen Energy* **2017**, *42*, 18134–18144. [CrossRef]

24. Bouzelata, Y.; Kurt, E.; Uzun, Y.; Chenni, R. Mitigation of high harmonicity and design of a battery charger for a new piezoelectric wind energy harvester. *Sens. Actuators A Phys.* **2018**, *273*, 72–83. [CrossRef]

25. Wang, D.; Ko, H. Piezoelectric energy harvesting from flow-induced vibration. *J. Micromech. Microeng.* **2010**, *20*, 025019. [CrossRef]

26. Silva-Leon, J.; Cioncolini, A.; Nabawy, M.R.A.; Revell, A.; Kennaugh, A. Simultaneous wind and solar energy harvesting with inverted flags. *Appl. Energy* **2019**, *239*, 846–858. [CrossRef]

27. Zhong, H.; Xia, J.; Wang, F.; Chen, H.; Wu, H.; Lin, S. Graphene-Piezoelectric Material Heterostructure for Harvesting Energy from Water Flow. *Adv. Funct. Mater.* **2017**, *27*, 1604226. [CrossRef]

28. Qureshi, F.U.; Muhtaroglu, A.; Tuncay, K. Near-Optimal Design of Scalable Energy Harvester for Underwater Pipeline Monitoring Applications with Consideration of Impact to Pipeline Performance. *IEEE Sens. J.* **2017**, *17*, 1981–1991. [CrossRef]

29. Mohamed, M.I.; Wu, W.Y.; Moniri, M. Power Harvesting for Smart Sensor Networks in Monitoring Water Distribution System. In Proceedings of the 2011 International Conference on Networking, Sensing and Control, Delft, The Netherlands, 11–13 April 2011; pp. 393–398.

30. Shukla, H.; Desai, H.; Sorber, J.; Piratla, K.R. *Evaluation of Energy Harvesting Potential in Water Pipelines to Power Sustainable Monitoring Systems*; United Engineering Center: New York, NY, USA, 2018; p. 475.

31. Hoffmann, D.; Willmann, A.; Goepfert, R.; Becker, P.; Folkmer, B.; Manoli, Y. Energy Harvesting from Fluid Flow in Water Pipelines for Smart Metering Applications. In Proceedings of the 13th International Conference on Micro and Nanotechnology for Power Generation and Energy Conversion Applications (Powermems 2013), London, UK, 3–6 December 2013; Volume 476, p. 012104. [CrossRef]

32. Shan, X.; Li, H.; Yang, Y.; Feng, J.; Wang, Y.; Xie, T. Enhancing the performance of an underwater piezoelectric energy harvester based on flow-induced vibration. *Energy* **2019**, *172*, 134–140. [CrossRef]

33. Yao, G.; Wang, H.; Yang, C.; Wen, L. Research and design of underwater flow-induced vibration energy harvester based on Karman vortex street. *Mod. Phys. Lett. B* **2017**, *31*, 1750076. [CrossRef]

34. Storn, R.; Price, K. Differential Evolution: A Simple and Efficient Adaptive Scheme for Global Optimization over Continuous Spaces. *J. Glob. Optim.* **1997**, *11*, 341–359. [CrossRef]

35. Zhang, J.; Sanderson, A.C. JADE: Adaptive Differential Evolution with Optional External Archive. *IEEE Trans. Evol. Comput.* **2009**, *13*, 945–958. [CrossRef]

36. Zhang, C.; Gao, L. *An Effective Improvement of JADE for Real-parameter Optimization*; IEEE: New York, NY, USA, 2013; p. 63.
37. Zhang, J.; Sanderson, A.C. *JADE: Self-Adaptive Differential Evolution with Fast and Reliable Convergence Performance*; IEEE: New York, NY, USA, 2007; p. 2258.
38. Islam, S.M.; Das, S.; Ghosh, S.; Roy, S.; Suganthan, P.N. An Adaptive Differential Evolution Algorithm with Novel Mutation and Crossover Strategies for Global Numerical Optimization. *IEEE Trans. Syst. Man. Cybern. Part B Cybern.* **2012**, *42*, 482–500. [CrossRef] [PubMed]
39. Cottone, F.; Goyal, S.; Punch, J. Energy Harvester Apparatus Having Improved Efficiency. U.S. Patent 8,350,394, 8 January 2013.
40. Aramendia, I.; Fernandez-Gamiz, U.; Zulueta Guerrero, E.; Lopez-Guede, J.M.; Sancho, J. Power Control Optimization of an Underwater Piezoelectric Energy Harvester. *Appl. Sci.* **2018**, *8*, 389. [CrossRef]
41. Siemens STAR CCM+ Version 11.06.011. Available online: http://mdx.plm.automation.siemens.com/ (accessed on 1 June 2019).
42. Osher, S.; Chakravarthy, S. Upwind Schemes and Boundary-Conditions with Applications to Euler Equations in General Geometries. *J. Comput. Phys.* **1983**, *50*, 447–481. [CrossRef]
43. Menter, F.R. 2-Equation Eddy-Viscosity Turbulence Models for Engineering Applications. *AIAA J.* **1994**, *32*, 1598–1605. [CrossRef]

Article

Research on the Cascade-Connected Transducer with Multi-Segment Used in the Acoustic Telemetry System while Drilling

Duo Teng [1,2]

[1] School of Marine Science and Technology, Northwestern Polytechnical University, Xi'an 710072, China; tengduo@nwpu.edu.cn; Tel.: +86-29-8849-2599

[2] National Key Laboratory for Underwater Information Processing and Control, Xi'an 710072, China

Received: 22 September 2019; Accepted: 18 October 2019; Published: 21 October 2019

Abstract: The electroacoustic transducer with the performances of low frequency, small size, and high power is desired in the application of the acoustic telemetry system while drilling. In order to fulfill the severe requirements, a novel cascade-connected transducer with multi-segment is developed. The essential framework of such a transducer is to add the cross-beams in the multi-segment cascade-connected arrangement, based on the fundamental configuration of the longitudinal transducer. The flexural vibrations of cross-beams help the transducer to present the appropriate coupling between longitudinal and flexural vibrations, which provide many benefits in keeping the advantages of the longitudinal transducer and lowering the resonance frequency. It is the finite element method to be used for simulating the mode shapes of the cascade-connected transducer, especially the behavior of the cross-beams, and some performances of transducer are also predicted. Several prototypes of cascade-connected transducers with different segments are manufactured. Their related tests show a good agreement with the finite element simulations and analyses. Their characteristics of low frequency, small size, light weight, and high power are attractive for the transmitting or receiving application in the acoustic telemetry system while drilling.

Keywords: cascade-connected transducer; low frequency; small size; finite element; acoustic telemetry; measurement while drilling

---

1. Introduction

During the drilling operations for exploration and development of oil and gas, the measurements play an important role in obtaining the physical parameters of geologic formations and the monitoring of downhole conditions [1]. This valuable information can be derived in different ways. For example, the drill bit can be withdrawn from the borehole and a "wireline logging tool" can be lowered into the borehole to take measurements [2]. Obviously, such a wireline logging after drilling will not be advantageous in obtaining the real-time and reliability data, as well as in the convenient operation of measurements and drilling [3]. Alternatively, the approach called measurement while drilling (MWD) or logging while drilling (LWD) with real-time data transmission increasingly attracts the attention in the oil and gas industry [4].

MWD is a technology of obtaining real-time measurement of various data at the bottom of the well during the drilling process [5]. Compared with conventional wireline logging, MWD is helpful to avoid the shortages in application [6]. This benefits from the measurement without stopping drilling. In the process of MWD, the communication and transmission between a downhole drilling assembly and a surface of a well are the key technology. A wireless system for telemetering data from downhole to surface will be an effective solution. The available approaches include utilizing such as mud pressure pulse, insulated conductor, electromagnetic wave, or acoustic wave [7]. For example, an acoustic telemetry system (ATS) by means of acoustic waves propagating along drill strings was commercialized in 2000 [8,9]. This demonstrates the value of acoustic telemetry. In ATS, the electroacoustic transducer is an indispensable apparatus. Its performances play a determinative role in transmitting or receiving the acoustic waves, and its further impact may be related to the effect of acoustic telemetry [10]. Obviously, the importance of the transducer cannot be ignored.

In the practical applications, considering the special function and severe downhole situation [11,12], the stringent demands for transducers are derived from confined space to assemble in drill collar, adaptability to acoustic channel along drill strings, high downhole pressure and temperature, strong shock and vibration, etc. These limitations facilitate the transducers to implement the performances of small size, light weight, low frequency, high power, pressure-resisting, heat-resisting, reliability, etc.

However, the acoustic telemetry transducers that fulfill all the requirements are difficult to develop. There have been some types of transducers to be developed for ATS while drilling, since the use of telemetry by acoustic waves through the drill strings was suggested in the 1940s due to steel tubular such as drill pipe with good and effective acoustic propagation [13]. In 1961, a magnetostrictive cylinder was invented by Woodworth for generating and/or receiving acoustic waves, and it had a low impedance of about 4000 ohms at the frequencies of 10 to 20 kHz [14]. In 1982, an acoustic well-logging transmitting and receiving transducer was invented by Dennis, which comprised the stacked piezoceramic rings and a resonating metallic plate, and operated at a desirable frequency lower than 15 kHz [15]. In 1989, a rare earth acoustic transducer was utilized by Liu to provide low-frequency acoustic energy in an acoustic well logging apparatus. That was attributed to the low acoustic velocity of the Terfenol-D alloys, so it was possible to produce a lower frequency below 10 kHz without increasing the transducer length [16]. In 1997, one acoustic transducer using PZT-8 type piezoelectric ceramic was provided by Drumheller for use in an acoustic telemetry system. The transducer was capable of delivering 10 W of acoustic energy at the frequency of 1.2 kHz, and its efficiency of energy conversion was approximately 40% [17]. In 2015, a transducer comprised of piezoelectric ceramic rings and thin metallic electrodes was tested in the drill strings for the acoustic telemetry by Zhao. The transducer had the performances of broad bandwidth and low frequency below 3.1 kHz [18].

In addition, there are some other types of transducers to be used in ATS. Their different configurations and characteristics were present in references [6,19–24]. All the studies provide good guidance. In respect of function realization, many of them can be used as a low-frequency acoustic source. While in respect of special requirements of ATS, some types of transducers still have much room for improvement, especially in miniaturizing the transducer to assemble in drill collar. As a practical application, for the most common drill pipe with a diameter of 178 mm, the maximum transverse size will be only about 40 mm to be provided for the electroacoustic transducer. Figure 1 illustrates the limitations subject to a drill collar of 178 mm in diameter. Obviously, it is difficult to develop the low frequency, high power acoustic source in such a restricted size.

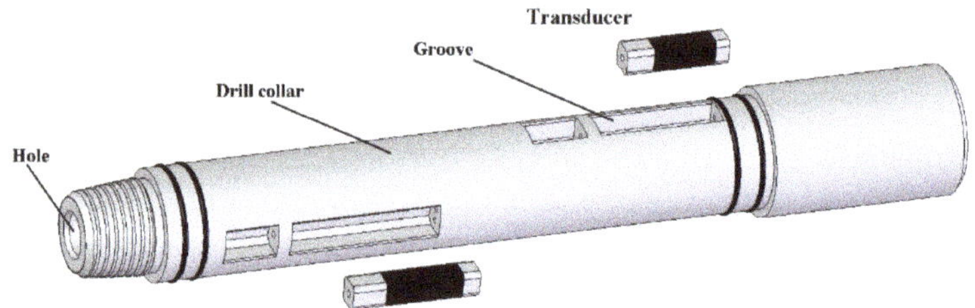

Figure 1. The schematic diagram of a drill collar with a groove to assemble the transducer.

Many types of transducer, such as Tonpilz transducers, flextensional transducers, flexural transducers, Piezoelectric rings, magnetostrictive transducers, piezoelectric ultrasonic transducers [25,26], are not suitable to MWD application because of the requirements of configurations, vibration modes, radiation field, acoustic impedance matching layers, etc. [27,28]. Under these restrictions of MWD application, the available types of transducer used in an acoustic telemetry system while drilling are few. If just considered as a receiver, MEMS may be an available alternative for detecting the acoustic signals propagated in drill strings. While as a transmitter with high power, the Tonpilz transducer is comparatively suitable because of its longitudinal vibration mode [29]. Generally, Tonpilz transducer can work effectively in the frequency range above 10 kHz. According to the advantages of the acoustic channel in drill strings, low frequency of the transducer will be desired in order to achieve a long-distance transmission, especially below 10 kHz and even lower [30]. In order to overcome the limitations in the ATS application, a type of novel longitudinal-flexural complex-mode low-frequency transducers will be presented.

2. Cascade-Connected Transducer with Coupled Longitudinal and Flexural Vibrations

The novel type of low-frequency piezoelectric transducer is called "cascade-connected transducer", which couples the longitudinal and flexural vibrations. Its longitudinal vibration mode is along the direction of drill strings, which is just in accordance with drill collar. Therefore, the acoustic energy can be transmitted forward along drill strings in the close match between the radiating head of the transducer and the drill collar. This characteristic will offer possibilities to take full advantage of the acoustic channel in drill strings. Its flexural vibration mode plays an important role in lowering the resonance frequency of the transducer. Hence, the above coupling makes it possible to implement the characteristics of low frequency, miniaturization, and high power.

The essential framework of the cascade-connected transducer is the multi-segment cascade-connected arrangement connected by the cross-beams, based on the conventional longitudinal transducer. The roles of the beam lie in not only its connection purpose as a structural part but also its flexural vibration as a functional component. Therefore, the Tonpilz configuration is fundamental, the multi-segment cascade-connected arrangement is essential, while the bending beam is the key. Figure 2 illustrates the configuration of the cascade-connected transducer. Its main components include piezoelectric stacks, bending beams, radiating head, tail mass, prestressed bolts, and other appurtenances. The radiating head is the region of transmitting the acoustic energy. Generally, its material should be similar to the drill collar, because the more approximate the characteristic impedances of materials are, the more smoothly the acoustic energy will propagate [31]. Typically, the tail mass should be comparatively heavy metal. A large tail-to-head mass ratio is desirable because it will yield a large head velocity. As the derivation, the more acoustic energy will be transmitted from the head. Every piezoelectric stack is glued closely together in series and wired in parallel. The configuration of multi-segment in a cascade-connected arrangement will be designed to match the length of the groove in the drill collar. There are eight segments in Figure 2. Every segment includes two columns of piezoelectric stacks, fastened to the bending beams alongside each other by a prestressed bolt. A type of cross-beam will be accepted to ensure the cross-connection between the adjacent segments. Compared with the other types of bending beams, this type of cross-beam will provide sufficient benefits to optimize the performances of the cascade-connected transducer, especially to lower the resonance frequency.

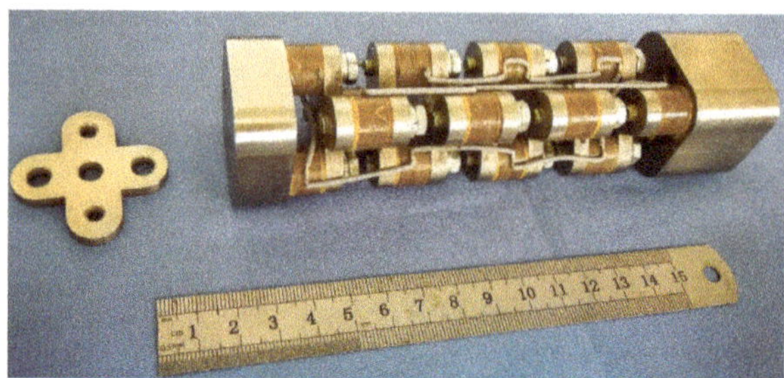

Figure 2. The cascade-connected transducer with eight segments.

3. Finite Element Analysis of the Cascade-Connected Transducer

The models and methods used in transducer analysis and design are always desired. However, it is difficult to develop an ideal solution because the piezoelectric transducer is an integrated system, which needs to be described in three different domains. A piezoelectric transducer is part acoustical at its moving surface in contact with the acoustic medium, part mechanical as a moving body controlled by forces, and part electrical as a current controlled by voltage [32]. So far, the equivalent circuit method (ECM) and finite element method (FEM) are comparatively comprehensive and effective, especially FEM is the prevailing method in the engineering development of transducer.

As a numerical method, the mathematical fundamentals of FEM are variational principle, subdivision, and interpolation [33]. After the whole piezoelectric transducer system is divided into finite elements connected at nodes, the matrix equations of the whole system will be formed. The key to these matrix equations are the governing equations, which can describe the behavior of piezoelectric coupled field. The computer solution of the whole system will be obtained, and the response of any position in the system will be calculated by interpolating. Then, a comprehensive explanation of how the system acts as a whole will be provided. Generally, FEM can model a complicated transducer without large-scale assumptions [34]. The other advantages of convenient modeling, rapid solution, accuracy result, and intuitive illustration are also attractive. Anyway, FEM has become one of the most effective methods to design or simulate the piezoelecteic transducers, especially for the sophisticated configurations, or the complex boundary conditions.

For the piezoelectric finite elements of the transducer model, their inclusion is the governing equation, which can describe the electric structure coupled field problem, written as [35]

$$\begin{bmatrix} \mathbf{M} & 0 \\ 0 & 0 \end{bmatrix} \begin{bmatrix} \ddot{\xi} \\ \ddot{\mathbf{U}} \end{bmatrix} + \begin{bmatrix} \mathbf{C} & 0 \\ 0 & 0 \end{bmatrix} \begin{bmatrix} \dot{\xi} \\ \dot{\mathbf{U}} \end{bmatrix} + \begin{bmatrix} \mathbf{K} & -\mathbf{K}^Z \\ \mathbf{K}^Z & \mathbf{K}^d \end{bmatrix} \begin{bmatrix} \xi \\ \mathbf{U} \end{bmatrix} = \begin{bmatrix} \mathbf{F} \\ \mathbf{q} \end{bmatrix} \tag{1}$$

where $[\xi]$ is vector of nodal displacements, $[\mathbf{U}]$ is the vector of nodal electrical potential, $[\mathbf{M}]$ is the mass matrix, $[\mathbf{C}]$ is the damping matrix, $[\mathbf{K}]$ is the stiffness matrix, $[\mathbf{K}^Z]$ is the piezoelectric coupling matrix, $[\mathbf{K}^d]$ is the dielectric conductivity matrix, $[\mathbf{F}]$ is the nodal force vector, $[\mathbf{q}]$ is the electrical load vector.

For the finite element program, the complete material properties should be specified. Table 1 lists the details as follows.

During the typical process of FEM solution, it is the most important step to build the finite element model. In some ways, the closer to the transducer prototype the model is, the more accurate the solution will be. However, in practice, in order to reduce the modeling difficulties or save the calculation time, simplifying the model without influencing the accuracy of the solution will be helpful. For our cascade-connected transducer, a 1/4 symmetrical finite element model will be built according to some assumptions. The following will present two results of finite element analysis, based on two different configurations of cascade-connected transducers including twelve segments and ten segments respectively.

Table 1. Material properties of cascade-connected transducer.

Components	Material	Quantity	Material Property [32]
Piezoelectric stack	PZT-4 piezoelectric ceramic	Density (kg/m^3)	7600
		Stiffness coefficients matrix ($\times 10^{10}$ N/m^2)	$c^E = \begin{bmatrix} 13.9 & 7.78 & 7.43 & 0 & 0 & 0 \\ 7.78 & 13.9 & 7.43 & 0 & 0 & 0 \\ 7.43 & 7.43 & 11.5 & 0 & 0 & 0 \\ 0 & 0 & 0 & 3.06 & 0 & 0 \\ 0 & 0 & 0 & 0 & 2.56 & 0 \\ 0 & 0 & 0 & 0 & 0 & 2.56 \end{bmatrix}$
		Piezoelectric stress matrix (C/m^2)	$e = \begin{bmatrix} 0 & 0 & -5.2 \\ 0 & 0 & -5.2 \\ 0 & 0 & 15.1 \\ 0 & 0 & 0 \\ 0 & 12.7 & 0 \\ 12.7 & 0 & 0 \end{bmatrix}$
		Relative permittivity matrix	$\dfrac{\varepsilon^S}{\varepsilon_0} = \begin{bmatrix} 730 & 0 & 0 \\ 0 & 730 & 0 \\ 0 & 0 & 635 \end{bmatrix}$
Radiating head Tail mass Cross-beam	Steel	Density (kg/m^3) Young's modulus (N/m^2) Poisson's ratio	7840 2.16×10^{11} 0.28

3.1. Cascade-Connected Transducer with Twelve Segments

Figure 3 illustrates the finite element model of the cascade-connected transducer, which is symmetrical to the XOZ and YOZ plane. The model includes 45,499 nodes and 22,519 elements. The different colors show the different components of the transducer. Its detailed sizes are as follows. The radiating head is 38 × 38 × 15 mm, the tail mass is 38 × 38 × 30 mm, the thickness of cross-beam is 5 mm and the piezoelectric ceramic ring is Φ 14 × 4 mm with a hole of Φ 6 mm. The cascade-connected transducer includes twelve segments, every segment includes two columns of piezoelectric stacks, and every piezoelectric stack includes four piezoelectric ceramic rings.

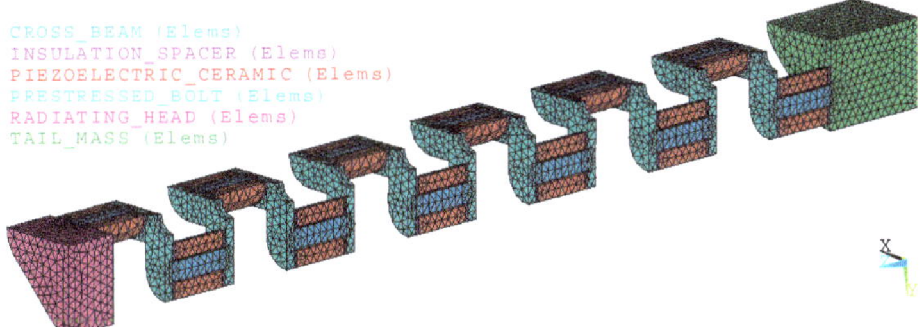

Figure 3. Finite element model of cascade-connected transducers with twelve segments.

When the finite element model is assumed to be free at both the head and tail end, Figure 4 illustrates the mode of vibration. The modal frequency is near 957 Hz, which is the resonance frequency of the cascade-connected transducer with twelve segments. On the whole viewpoint, the mode shape is longitudinal. Essentially, the piezoelectric stacks, the head, the tail, and the bolts are all longitudinal, while the cross-beams are flexural (their mode shape will be shown in Section 4). The obvious vibrations occur at the ends of the transducer, and comparatively the radiating head section is stronger than the tail mass section. This indicates that more energy is transmitted from the head.

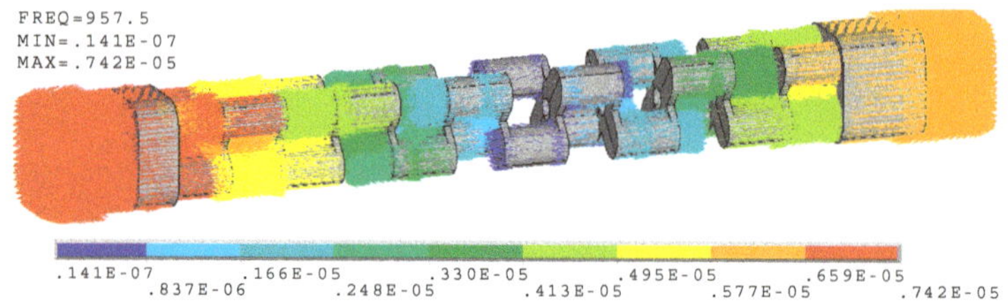

Figure 4. Vector illustration of vibration at 957.5 Hz (cascade-connected transducer with twelve segments).

Figure 5 presents the admittance performances of the cascade-connected transducer. The curves show that the peak of the conductance curve appears at 957.5 Hz, which is near the resonance frequency derived from the coupled longitudinal and flexural vibrations. The sharp peak of the conductance curve also predicts that the bandwidth of the cascade-connected transducer is narrow. Therefore, this type of transducer cannot be used as a broadband transducer individually.

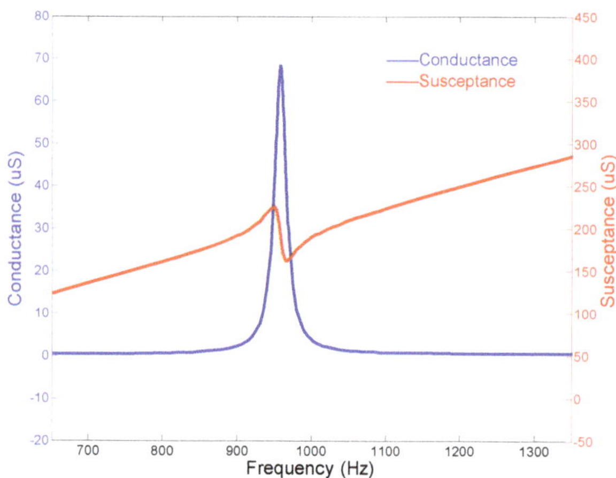

Figure 5. Admittance curves of the cascade-connected transducer with twelve segments.

3.2. Cascade-Connected Transducers with Ten Segments

A similar cascade-connected transducer, which includes ten segments, is also presented. All the configurations are the same except the number of segments. The finite element model of the cascade-connected transducer with ten segments is the same as the above. The results are also similar. Figure 6 shows the mode of vibration and Figure 7 shows the admittance curves. They predict that the resonance frequency is near 1092 Hz.

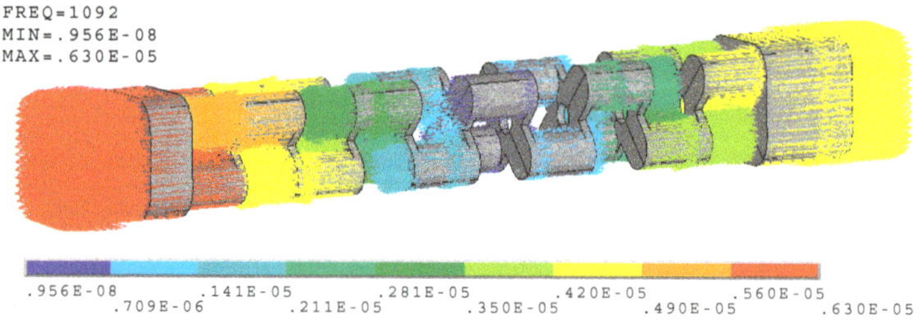

Figure 6. Vector illustration of vibration at 1092 Hz (cascade-connected transducer with ten segments).

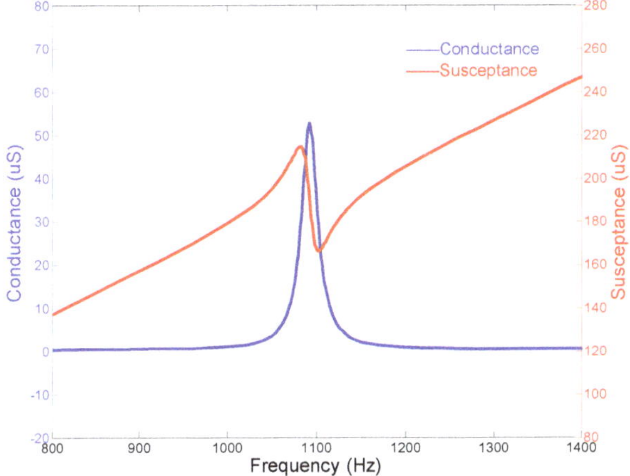

Figure 7. Admittance curves of the cascade-connected transducer with ten segments.

4. Cross-Beams

The bending beams are one of the main differences between the cascade-connected transducer and the conventional Tonpilz transducer. The beams should be helpful for enhancing the longitudinal vibration, and also for lowering the resonance frequency of the transducer. Therefore, the bending beams need to be designed and optimized. The cross-beam is a good alternative. The material and the thickness of the beam are the major factors to control the flexural mode of cross-beam. The flexural mode, as an indispensable vibration of the whole, plays an important role in the cascade-connected transducer. Hence, its performances need to be comprehended.

For the cascade-connected transducer with twelve segments, there need to be eleven cross-beams. The serial number of each beam is shown in Figure 8, where the beam near the tail is assigned as No. 1 and the beam near the head is assigned as No. 11. According to the above analysis, a vector illustration of vibration of cross-beams is shown in Figure 8, which is separated from the whole system in Figure 4. The result shows that the move of the cross-beams near the ends of the transducer is significant, while the move of the middle cross-beam is low. The vibrations of both ends are opposite. All of the performances agree with the longitudinal vibration of the cascade-connected transducer. Although the longitudinal move of the cross-beam is real, it disguises the fact of flexural vibration. The reason is that each cross-beam is attached to a constant move of longitudinal vibration, which is provided by the whole system. This constant longitudinal move of each beam is different, but every one is larger than the scope of its other vibrations. If the constant longitudinal move of each beam is separated, only the flexural vibration will be left.

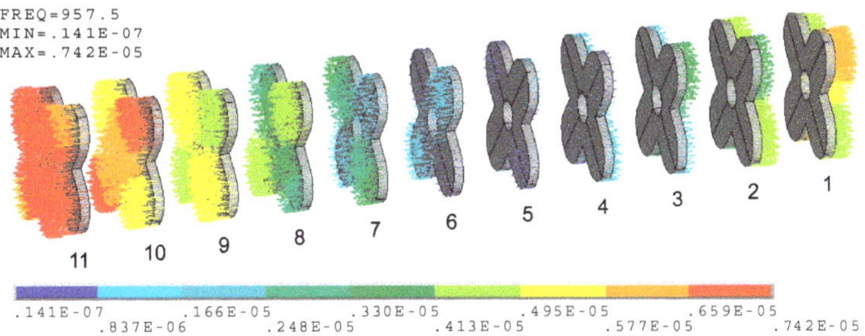

Figure 8. Vector illustration of vibration at 957.5 Hz (cross-beams).

Figure 9 illustrates that the clear flexural vibration happens to each cross-beam. Their distinctions are just the vibration intensity. Unlike the before-mentioned longitudinal move of the cross-beams, the flexural vibrations of the cross-beams near the ends of the transducer are weak, while those of the middle cross-beam are strong. The strongest flexural vibration occurs at No. 5 beam. Figure 9 illustrates its mode shape. If the undeformed edge is assigned as the reference, it is clear that the two orthogonal branches of the cross-beam vibrate respectively along opposite directions. This mode shape is propitious to enhance the longitudinal vibration of the cascade-connected transducer, and also to achieve the low resonance frequency. This is the advantage of cross-beam.

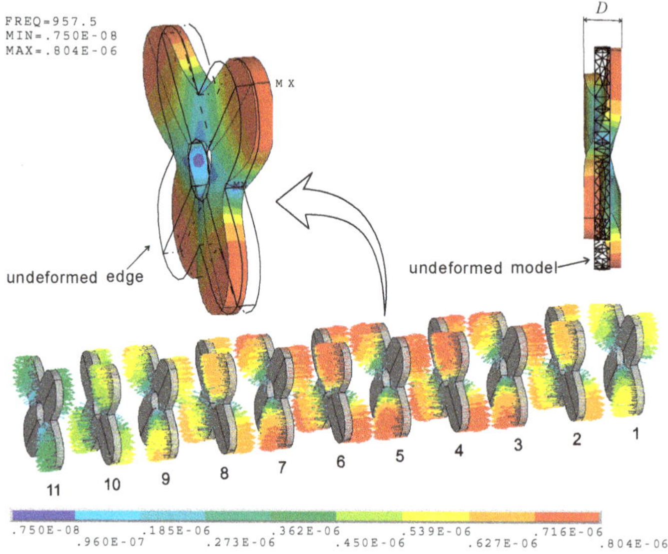

Figure 9. Vector illustration of flexural vibration at 957.5 Hz (cross-beams).

If the maximum relative displacement between the two orthogonal branches of each cross-beam is defined as D_i, shown in Figure 9, and the subscript "i" stands for the serial number of each cross-beam, the value of D_5 is maximum. Figure 10 illustrates the relative displacements of all the cross-beams. Its Y-axis label is the normalized relative displacement $\frac{D_i}{D_5}$, and its X-axis is the serial number of beams. D_i of the beams at the two ends are dissymmetric because the masses of the head and tail are different.

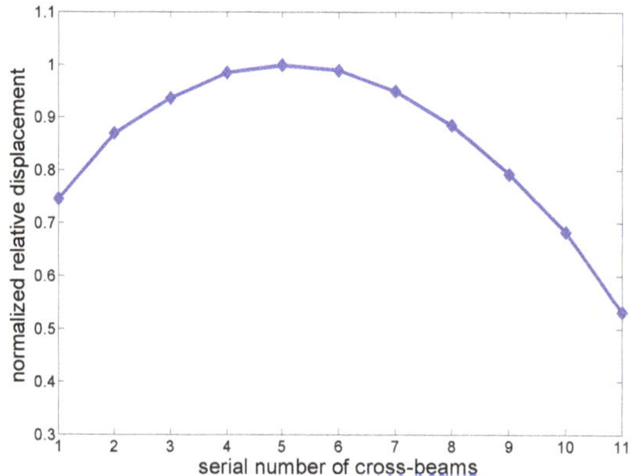

Figure 10. Normalized relative displacements of all the cross-beams.

5. Test and Discussion

Some cascade-connected transducer prototypes have been manufactured according to the above designs. Two kinds of cascade-connected transducer have the external size of 38 × 38 mm. The prototype with twelve segments is 310 mm long and weighs 1.23 kg, while the one with ten segments is 266 mm long and weighs 0.88 kg. Figure 11 shows the prototypes of cascade-connected transducers, which are encapsulated in polyurethane rubber.

Figure 11. Prototypes of cascade-connected transducers.

The admittance curves are obtained from the precision impedance analyzer Agilent 4294A (Agilent, Santa Clara, CA, USA). Figure 12 shows the curves of the cascade-connected transducer with twelve segments. The curves show that the resonance frequency is near 985 Hz, which is slightly higher than the result of FEM. The admittance curves of the cascade-connected transducer with ten segments are shown in Figure 13. Its resonance frequency is near 1112 Hz. The comparison between the measurements and the simulations illustrated respectively in Figure 5 or Figure 7 shows consistency. In practice, the two kinds of cascade-connected transducers can be wired together in parallel to broaden the bandwidth. Figure 14 illustrates the admittance curves of four wired cascade-connected transducers (two for twelve segments and two for ten segments, shown in Figure 11). There are two adjacent peaks. Their coupling can optimize the performances of transducers, which will provide higher power and wider bandwidth to achieve a better effect in the acoustic telemetry system while drilling.

Based on the measurements of resonance frequency, f_r, and anti-resonance frequency, f_a, the transducer's effective electromechanical coupling coefficient, k_{eff}, can be calculated. The two kinds of cascade-connected transducer have k_{eff} of 10.11% and 11.93%, respectively.

For a transmitting transducer, the performance of high power is always desired. The power consumption of two kinds of cascade-connected transducers is monitored from a small excitation signal to high driving voltage. The increasing response is basically linear. A safe operating state is as follows. The power consumption of single cascade-connected transducer with twelve segments is 30.9 watts at the frequency of 970 Hz, where the driving voltage is 1188 V p–p (peak to peak), the current is 243 mA p–p, and the current leads the voltage by 64.6 degrees. The power consumption of a single cascade-connected transducer with ten segments is 31.3 Watts at the frequency of 1110 Hz, where the driving voltage is 1188 V p–p, the current is 234 mA p–p, and the current leads the voltage by 63.2 degrees. These is not the maximum power consumption for the transducers because of the limitation of the power amplifier.

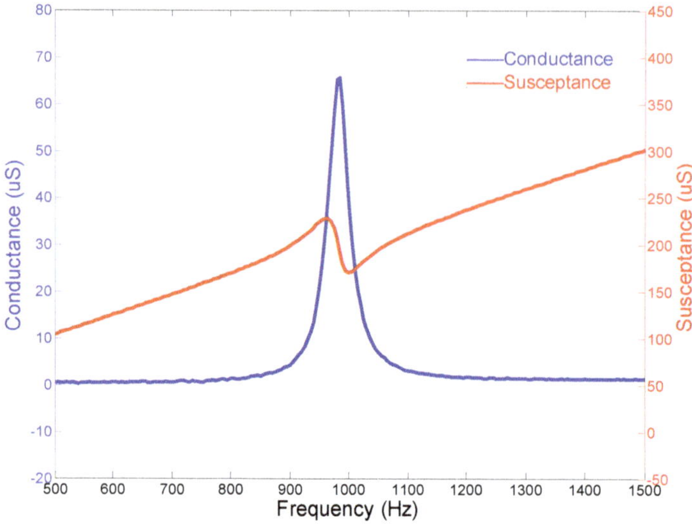

Figure 12. Admittance curves of cascade-connected transducers with twelve segments (obtained from Agilent 4294A).

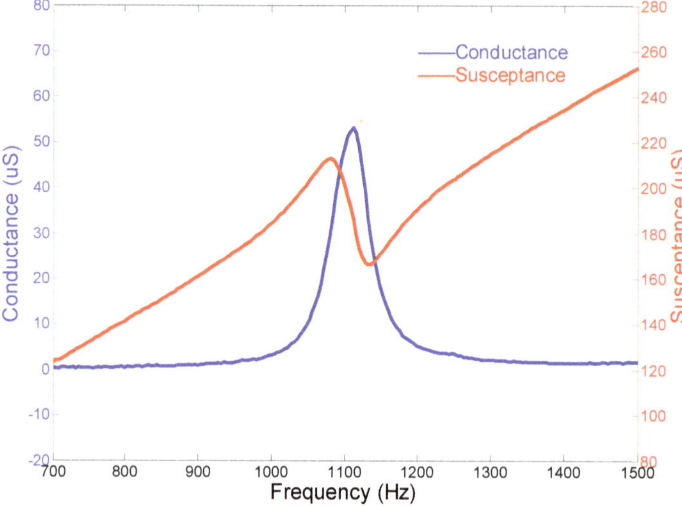

Figure 13. Admittance curves of cascade-connected transducers with ten segments (obtained from Agilent 4294A).

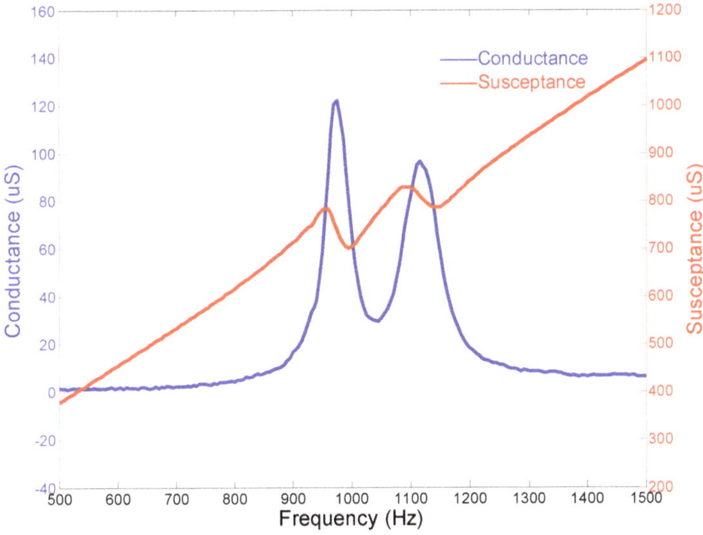

Figure 14. Admittance curves of four wired cascade-connected transducers (two for twelve segments and two for ten segments).

6. Conclusions

A novel type of cascade-connected transducer is present. Its performances are predicted through finite element method, especially the behavior of cross-beams. Several prototypes with different segments are manufactured. The relevant tests show that the cascade-connected transducer can satisfy the requirements for the transmitting or receiving application in the acoustic telemetry system while drilling. In summary, the following conclusions can be obtained.

1. The major configuration of the proposed transducer is the multi-segment cascade-connected arrangement connected by the cross-beams, based on the conventional longitudinal transducer. Such design not only keeps the advantages of longitudinal radiation and high power, but also provides the characteristics of low frequency and small size.
2. The cascade-connected transducer utilizes longitudinal and flexural complex mode. The coupling of two vibration modes is the key to achieve the desired electroacoustic performances. During this process, the cross-beam plays a pivotal role based on its flexural vibration. Therefore, the design and optimization of the bending beams is an important research point.
3. The simulation and test show that the cascade-connected transducer has a resonance frequency of approximately 1 kHz. If the number of segments in the cascade-connected transducer increases, the operating frequency will be lowered. Hence, besides the bending beam, the number of segments is also the deciding factor to lower the resonance frequency.
4. The simulation based on the finite element method shows that cascade-connected transducer can lower the resonance frequency by approximately 80% more than the conventional Tonpilz type transducer when the two transducers have the same length.
5. The relevant tests show that the effective electromechanical coupling coefficient, k_{eff}, of cascade-connected transducer is approximately 11%. It is less than the value of conventional Tonpilz type transducer. There are many factors which can affect k_{eff} in practice. One possible reason is that the small radiation surface leads to the small radiation resistance at low frequency.

Funding: This research was funded by the National Key Laboratory, grant number JCKY2018207CD04.

Acknowledgments: The author acknowledges the National Key Research and Development Program of China, grant number 2016YFF0200902, for the experiment support.

Conflicts of Interest: The author declares no conflict of interest.

References

1. Gravley, W. Review of downhole measurement-while-drilling systems. *J. Pet. Technol.* **1983**, *35*, 1439–1445. [CrossRef]
2. Snyder, D.D.; Fleming, D.B. Well logging—A 25-year perspective. *Geophysics* **1985**, *50*, 2504–2529. [CrossRef]
3. Johnson, H.M. A history of well logging. *Geophysics* **1962**, *27*, 507–527. [CrossRef]
4. Su, Y.N.; Dou, X.R. Measurement while drilling, logging while drilling and logging instrument. *Oil Drill. Prod. Technol.* **2005**, *27*, 74–78.
5. Inglis, T.A. Measurement while drilling (MWD). In *Directional Drilling*, 1st ed.; Inglis, T.A., Ed.; Springer: Dordrecht, The Netherlands, 1987; Volume 2, pp. 130–154.
6. Shah, V.V.; Linyaev, E.J.; Kyle, D.G.; Gardner, W.R.; Moore, J.L. Acoustic telemetry transceiver. U.S. Patent 7,339,494B2, 4 March 2008.
7. Spies, B.R. Electrical and electromagnetic borehole measurements: A review. *Surv. Geophys.* **1996**, *17*, 517–556. [CrossRef]
8. Shah, V.V.; Gardner, W.R.; Johnson, D.H.; Sinanovic, S. Design considerations for a new high data rate LWD acoustic telemetry system. In Proceedings of the SPE Asia Pacific Oil and Gas Conference and Exhibition, Perth, Australia, 1 January 2004.
9. Gardner, W.R.; Shah, V.V. High Data Rate Acoustic Telemetry System. U.S. Patent 6,320,820B1, 20 November 2001.
10. Squire, W.D.; Whitehouse, H.J. A new approach to drill-string acoustic telemetry. In Proceedings of the SPE Annual Technical Conference and Exhibition, Las Vegas, NV, USA, 23–26 September 1979.
11. Drumheller, D.S. An overview of acoustic telemetry. In Proceedings of the Office of Scientific & Technical Information Technical Reports, Albuquerque, NM, USA, 1 January 1992.
12. Baggeroer, A. An overview of acoustic telemetry: 2001–2011. *J. Acoust. Soc. Am.* **2012**, *131*, 3237. [CrossRef]
13. Drumheller, D.S. Acoustical properties of drill strings. *J. Acoust. Soc. Am.* **1989**, *85*, 1048–1064. [CrossRef]
14. Woodworth, J.H. Acoustic Logging Transducer. U.S Patent 3,009,131A, 14 November 1961.
15. Dennis, J.R. Acoustic Well-Logging Transmitting and Receiving Transducers. U.S. Patent 4,319,345A, 9 March 1982.
16. Liu, O.Y. Acoustic Well Logging Method and Apparatus. U.S. Patent 4,813,028A, 14 March 1989.
17. Drumheller, D.S. Acoustic Transducer. U.S. Patent 5,703,836A, 30 December 1997.
18. Zhao, G.S.; Wang, B.B.; Du, Z.C.; Huang, M.Q.; Zhai, W.T. Experimental study on the transmission characteristics of transducer acoustic signal within drill string. *Sci. Technol. Eng.* **2015**, *15*, 14–17.
19. Kent, W.H. Resonant Acoustic Transducer System for a Well Drilling String. U.S. Patent 4,302,826A, 24 November 1981.
20. Drumheller, D.S. Electromechanical Transducer for Acoustic Telemetry System. U.S. Patent 5,222,049A, 22 June 1993.
21. Tochikawa, T.; Sakai, T.; Taniguchi, R.; Shimada, T. Acoustic telemetry: The new MWD system. In Proceedings of the SPE Annual Technical Conference and Exhibition, Denver, CO, USA, 6–9 October 1996.
22. Birchak, J.R. Acoustic Transducer for LWD tool. U.S. Patent 5,644,186A, 1 July 1997.
23. Drumheller, D.S. Acoustic Transducer. U.S. Patent 6,147,932A, 14 November 2000.
24. Wisniewski, L.T.; Arian, A.; Varsamis, G.L. Transducer for Acoustic Logging. U.S. Patent 6,213,250B1, 10 April 2001.
25. Mo, X.P.; Zhu, H.Q. Thirty years' progress of underwater sound projectors in China. *AIP Conf. Proc.* **2012**, *1495*, 94–104.
26. Meng, X.D.; Lin, S.Y. Analysis of a cascaded piezoelectric ultrasonic transducer with three sets of piezoelectric ceramic stacks. *Sensors* **2019**, *19*, 580. [CrossRef] [PubMed]
27. Gao, L.; Gardner, W.; Robbins, C.; Memarzadeh, M.; Johnson, D. Limits on data communication along the drillstring using acoustic waves. *SPE Reserv. Eval. Eng.* **2008**, *11*, 141–146. [CrossRef]

28. Qiao, W.X.; Ju, X.D.; Che, X.H.; Lu, J.Q. Progress in acoustic well logging technology. *Well Logging Technol.* **2011**, *35*, 14–19.

29. Massa, D.P. An Overview of Electroacoustic Transducers. Available online: https://www.massa.com/wp-content/uploads/DPM_Overview_of_Electroacoustic_Transducers.pdf (accessed on 3 September 2019).

30. Drumheller, D.S. Attenuation of sound waves in drill strings. *J. Acoust. Soc. Am.* **1993**, *94*, 2387–2396. [CrossRef]

31. Buckingham, M.J. Sound Propagation. In *Applied Underwater Acoustics*, 1st ed.; Neighbors, T.H., Bradley, D., Eds.; Elsevier: Amsterdam, The Netherlands, 2017; pp. 85–95.

32. Butler, J.L.; Sherman, C.H. *Transducers and Arrays for Underwater Sound*, 2nd ed.; Springer: Cham, Switzerland, 2016; pp. 207–219.

33. Hladky-Hennion, A.C.; Dubus, B. Finite Element Analysis of Piezoelectric Transducers. In *Piezoelectric and Acoustic Materials for Transducer Applications*; Springer: New York, NY, USA, 2008; pp. 241–258.

34. Wang, J.J.; Qin, L.; Li, W.J.; Song, W.B. Parametric Analysis and Optimization of Radially Layered Cylindrical Piezoceramic/Epoxy Composite Transducers. *Micromachines* **2019**, *9*, 585. [CrossRef] [PubMed]

35. Allik, H.; Hughes, T.J.R. Finite element method for piezoelectric vibration. *Int. J. Number Meth. Eng.* **1970**, *2*, 151–157. [CrossRef]

 *micromachines*

Article

A Frequency Up-Converted Hybrid Energy Harvester Using Transverse Impact-Driven Piezoelectric Bimorph for Human-Limb Motion

Miah Abdul Halim [1,*], **M. Humayun Kabir** [2], **Hyunok Cho** [3] **and Jae Yeong Park** [3]

[1] Department of Electrical and Computer Engineering, University of Florida, Gainesville, FL 32601, USA
[2] Department of Electrical and Electronic Engineering, Islamic University, Kushtia 7003, Bangladesh; h.2011.kabir@gmail.com
[3] Department of Electronic Engineering, Kwangwoon University, Seoul 01897, Korea; whgusdhr@gmail.com (H.C.); jaepark@kw.ac.kr (J.Y.P.)
[*] Correspondence: md.miah@ufl.edu

Received: 29 September 2019; Accepted: 14 October 2019; Published: 15 October 2019

Abstract: Energy harvesting from human-body-induced motion is mostly challenging due to the low-frequency, high-amplitude nature of the motion, which makes the use of conventional cantilevered spring-mass oscillators unrealizable. Frequency up-conversion by mechanical impact is an effective way to overcome the challenge. However, direct impact on the transducer element (especially, piezoelectric) increases the risk of damaging it and raises questions on the reliability of the energy harvester. In order to overcome this shortcoming, we proposed a transverse mechanical impact driven frequency up-converted hybrid energy harvester for human-limb motion. It utilizes the integration of both piezoelectric and electromagnetic transducers in a given size that allows more energy to be harvested from a single mechanical motion, which, in turn, further improves the power density. While excited by human-limb motion, a freely-movable non-magnetic sphere exerts transverse impact by periodically sliding over a seismic mass attached to a double-clamped piezoelectric bimorph beam. This allows the beam to vibrate at its resonant frequency and generates power by means of the piezoelectric effect. A magnet attached to the beam also takes part in generating power by inducing voltage in a coil adjacent to it. A mathematical model has been developed and experimentally corroborated. At a periodic limb-motion of 5.2 Hz, maximum 93 μW and 61 μW average powers (overall 8 μW·cm^{-3} average power density) were generated by the piezoelectric and the electromagnetic transducers, respectively. Moreover, the prototype successfully demonstrated the application of low-power electronics via suitable AC-DC converters.

Keywords: transverse impact; frequency up-conversion; piezoelectric bimorph; human-limb motion; hybrid energy harvester

---

1. Introduction

With the ongoing development of microelectronic technologies, multiple low-power consuming wireless sensor devices are being embedded within hand-held and wearable consumer electronics. These devices are mainly powered by an external power source (e.g., electrochemical batteries) of the consumer electronics and their continuous use allows it to run out of power quickly. Compared to device technologies, the development of the power sources (i.e., batteries) are still slower, even though the devices require less power to operate. Electrochemical batteries have a limited lifespan, and require periodic charging that is inconvenient or sometimes impossible. Moreover, since most batteries contain toxic chemicals, disposal of the expired batteries produces hazardous waste that enhances environmental pollution and poses threats to human and animal health. Therefore, there is great interest in developing self-powered electronics for uninterruptible and long-lasting operation by eliminating the need for recharging or replacing the power source. In recent years, energy harvesting from surrounding energy sources (e.g., light, heat, sound, vibration, etc.) has drawn much attraction to address these circumstances [1–3]. Among these sources, vibration is the most attractive physical energy source due to its versatility, incorruptibility, and abundance in nature [4]. However, different vibration sources (e.g., human and machine motion, water and wind flow, rotary motion etc.) generate vibrations of different frequencies and amplitudes, and mostly exhibit low-frequency, large-amplitude characteristics with various cyclic movements in different directions [5–7]. These vibrations, in the form of kinetic energy, can effectively be converted into electrical energy by employing compatible electromechanical transduction mechanisms that include piezoelectric [8], electromagnetic [9], electrostatic [10], magnetostrictive/magnetoelectric [11], and triboelectric [12] mechanisms.

The performance of a vibration energy harvester greatly depends on the characteristics of vibration, the type of transducer, and how the transducer is coupled to the mechanical system. Generally, vibration energy harvesters utilize an inertial mechanism employed by a cantilevered spring-mass system, having a specific resonant frequency. Harvested energy (power) is at its maximum when the harvester's resonant frequency matches the applied vibration frequency. Unfortunately, the power output decreases dramatically as the frequency of excitation (i.e., the resonant frequency of the harvester) decreases [13]. Moreover, employing a cantilevered spring-mass system for low-frequency (<10 Hz) energy harvesting is quite challenging due to the size constraints for specific application. Human-body-induced motion (e.g., walking, running, shaking limbs, etc.) also generates low-frequency (<6 Hz) vibrations, which do not allow the cantilever structure to be employed conveniently [14]. Hence, efficient energy harvesting from human-body-induced motion for hand-held and wearable smart devices requires clever design choices. Micro/nano-structured triboelectric nanogenerators [15,16], flexible piezo-composite based piezoelectric nanogenerators [17,18] etc. have shown great application potential in wearable biomechanical energy harvesting and motion sensing. However, they require huge efforts in material development, which was not of our interest. Our primary interest was to design and develop inertial based, low-frequency (e.g., human-body-induced motion) energy harvesters.

The mechanical frequency up-conversion mechanism [19], among numerous design approaches over the past few years, has become the mainstream approach for human-motion based energy harvesting. It allows the transducer element (in the form of a spring-mass system) to actuate at its own resonant frequency (considerably high) by a low-frequency oscillatory or rotary system that responds to the external low-frequency vibration generated by human-motion. Commonly used methods of mechanical frequency up-conversion include mechanical impact and plucking [20–23]. Impact excitation transfers an instantaneous momentum into the transducer element whereas plucking excitation implies a slow deflection of the transducer element followed by its sudden release. In general, these methods exert direct force straight to the transducer element that could potentially lead to damage, especially in the case of piezoelectric devices. In order to overcome these issues with piezoelectric energy harvesters, we introduced the transverse impact-based frequency up-conversion mechanism in a human handy-motion driven electromagnetic energy harvester by employing a double-clamped FR4 cantilever beam as a high-frequency oscillator and a freely movable sphere as a low-frequency oscillator [24]. The transverse impact mechanism meets the reliability challenge and the freely-movable sphere allows the device to operate efficiently at extremely low frequencies (with sufficiently large amplitudes) of handy-motion vibration, meaning its non-resonant behavior [25]. However, the device generates low power and its average power density is poor.

In order to improve its performance, we attempted to hybridize our previous work by incorporating a piezoelectric transducer without cost to the harvester volume. A hybrid energy harvesting technology combines two or more types of transducers that simultaneously capture energy from the same excitation [26–28]. In this paper, we present the theoretical modeling and experimental characterization of a piezoelectric (PE) and electromagnetic (EM) hybrid energy harvester for human-limb motion by utilizing the transverse mechanical impact-based frequency up-conversion strategy. Transverse impact, created by a sliding sphere over the parabolic tip of a mass attached to a clamped–clamped piezoelectric beam, eliminates the reliability issue from rapid damage of the piezoelectric cantilever due to direct impact. Moreover, simultaneous power generation from both PE and EM transducers offers a higher power density. A theoretical model for the hybrid generator under transverse impact was developed and experimentally validated with a prototype device. The proposed approach has the potential of reliable operation under low-frequency and high-amplitude excitation of human-body-induced motion toward the development of self-powered portable and wearable smart devices.

2. Design and Modeling

2.1. Harvester Structure and Its Operation

Figure 1a shows the schematic structure of the proposed transverse-impact driven hybrid energy harvester for human-limb motion. A PE transducer in the form of a clamped–clamped lead zirconate titanate (PZT) bimorph beam and an EM transducer consisting of one cylindrical magnet attached to the center of the piezo-beam and a hollow cylindrical multi-turn copper coil fixed to the housing constituted the hybrid generator structure. An additional mass with a parabolic-top was attached to the piezo-beam, opposite to the magnet. A hollow rectangular channel that contained a freely-movable spherical ball was placed on top of the piezo-beam in parallel. The channel had an opening at the center of its bottom wall that allowed the parabolic-top of the mass to be positioned through it, so that the ball was able to slide over the parabolic-top while the device was operated.

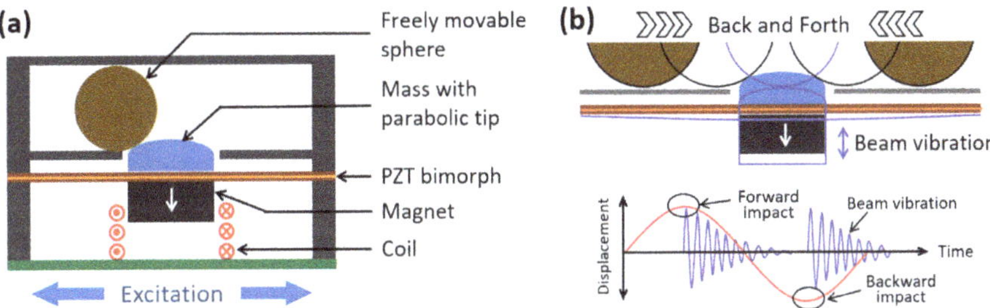

Figure 1. Schematics of the proposed transverse-impact driven frequency up-converted hybrid energy harvester (**a**) and its operation principle (**b**).

The principle of frequency up-conversion by transverse impact in the proposed hybrid energy harvester is illustrated in Figure 1b. When the device is excited by a low-frequency vibration with a sufficiently large amplitude (i.e., human-limb motion), the ball moves back and forth along the length of the channel. In its back and forth motion, the ball slides over the parabolic-top of the mass and produces a transverse impact on it that pushes mass as well as the piezo-beam downward, allowing it to vibrate freely at its own resonant frequency in a direction perpendicular to the direction of the ball movement. As a result, stresses are generated on the surfaces of the piezo-beam that generate voltage by virtue of the piezoelectric effect. Simultaneously, the magnet attached to the piezo-beam vibrates with respect to the adjacent coil and an electromotive force (e.m.f) voltage is generated by electromagnetic induction between them. The frequency (resonant) at which the beam vibrates is much higher than that of the excitation applied by human-limb motion and can be determined by the material and structural parameters of the beam. As shown in Figure 1b, the ball exerts transverse impact twice in one cycle of its back and forth motion; each time the system undergoes an impulse excitation, resulting in an exponentially decayed oscillatory motion between two consecutive impacts, so the output responses from both piezoelectric and electromagnetic transducers will be.

2.2. Electromechanical Modeling

The system is considered as a single-degree-of-freedom (SDOF) forced spring-mass-damper system excited by a periodic force $F(t)$. When the ball slides over the parabolic-top of the cantilevered proof-mass, the collision between them is stated as the low-velocity transverse impact of a rigid body on a flexible element [20]. Accordingly, when the bodies (ball and proof-mass) come into contact, they tend to interpenetrate each other, and a local compression force develops in their interface, which increases as the ball slides over the parabolic-top of the proof-mass, resulting in bending of the beam. When the compression force is large enough, the ball slides over the parabolic-top before the force becomes large enough. However, the curvatures of both the ball and parabolic-top as well as the overlap between them plays crucial roles on the transverse impact mechanism. Finally, the bodies are separated and each vibrates independently until the next collision occurs. According to the force diagram of the transverse impact mechanism [20], the force experienced on the proof-mass in the transverse direction is $F(L, t) = \mu_k F(t) \sin\theta$ and the governing equation of motion of the proposed system can be expressed as

$$m\ddot{y}(t) + c\dot{y}(t) + ky(t) = \mu_k F(t) \sin\theta \int_0^{L/2} \varphi(x)dx \tag{1}$$

where m is the mass (including the masses of the attached proof-mass and magnet); $y(t)$ is the mass displacement; c is the equivalent damping coefficient; k is the stiffness of the beam; L is the length of the beam; μ_k is the coefficient of kinetic friction while the ball slides over the parabolic-top; and $\varphi(x)$ is the mass normalized eigenfunction of the first vibration mode for the boundary condition $x = L/2$, which is [29]

$$\varphi(x) = \sqrt{\frac{2}{m_l L}}\left[\cos h\frac{2\lambda}{L}x - \cos\frac{2\lambda}{L}x - \varsigma\left(\sin h\frac{2\lambda}{L}x - \sin\frac{2\lambda}{L}x\right)\right] \tag{2}$$

where m_l is the mass per unit length; λ is the dimensionless frequency parameter for the first mode; and $\varsigma = (\sin h\lambda - \sin\lambda)/(\cos h\lambda + \cos\lambda)$.

The beam is a piezoelectric bimorph and generates voltage when lateral stress is generated on the surfaces of the piezoelectric material due to the transverse impact. The cross-section of the beam with a metallic shim sandwiched between two piezoelectric layers is shown in Figure 2. Each piezo-material is poled along its thickness direction and are connected in parallel. During bending of the beam, the stresses in the top and bottom piezoelectric layers will be in opposite directions: one is in tension and the other is in compression. An equivalent moment of inertia of the bimorph beam is defined as [30]

$$I_{eq} = 2\left(\frac{wh_p^3}{12} + wh_p h_{eq}^2\right) + \frac{E_s wh_s^3}{12E_p} \tag{3}$$

where E_s and E_p are the Young's modulus of the shim and piezoelectric materials; h_s and h_p are the thickness of the shim and piezoelectric layers; and w and $h_{eq}\left(=\left(h_s + h_p\right)/2\right)$ are the width and the equivalent thickness of the beam, respectively. The maximum stress on the piezoelectric surface due to the transverse impact at the center of the beam $(x = L/2)$ can be calculated as [31]

$$\sigma_{max} = \frac{M(x)\left(h_p + \frac{h_s}{2}\right)}{I_{eq}} = \frac{\mu_k F_{max} L\left(h_p + \frac{h_s}{2}\right)}{8I_{eq}} \tag{4}$$

where $M(x)$ is the bending moment and $F_{max}(= ky_{max})$ is the magnitude of the transverse force determined by Hook's law where k is the stiffness and y_{max} is the maximum displacement of the bimorph beam. Now, the generated peak open circuit voltage can be determined as

$$V_{oc} = \frac{-d_{31} h_p \sigma_{max}}{\varepsilon_0 \varepsilon_r} \tag{5}$$

where $-d_{31}$ and ε_0 are the piezoelectric charge constant and dielectric constant of the piezoelectric material, respectively and ε_r is the permittivity of free space. According to the dynamics of the transverse mechanical impact by the freely movable spherical ball described earlier, the output voltage from the piezoelectric transducer can be written as a function of time t as

$$V_{PE}(t) = V_{oc}e^{-\zeta_m \omega_r t}\sin(\omega_d t); \quad n\frac{2\pi}{\omega_d} < t < (n+1)\frac{2\pi}{\omega_d}, \quad (n = 0,1,2,3,\ldots) \tag{6}$$

where n is the number of impacts; ω_r, ω_d, and ζ_m are the resonant frequency, damped resonant frequency, and mechanical damping ratio, respectively, which are defined as [31,32]

$$\omega_r = \sqrt{\frac{k}{m}} = \frac{\lambda}{L^2}\sqrt{\frac{E_p I_{eq}}{m}}; \quad \omega_d = \omega_r\sqrt{\left(1 - \zeta_m^2\right)}; \quad \zeta_m = \frac{c_m}{2m\omega_r} \tag{7}$$

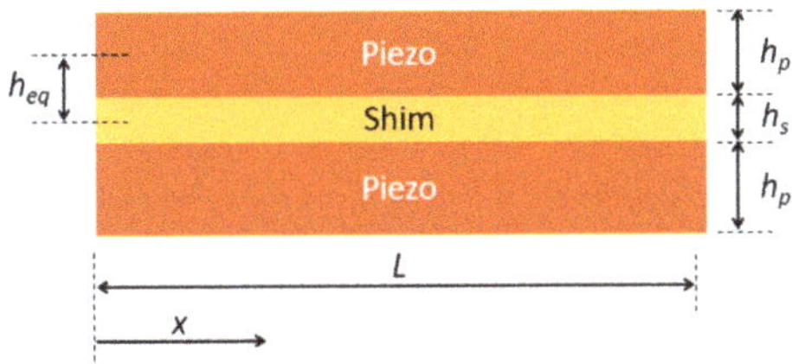

Figure 2. Cross-section of the piezoelectric bimorph beam.

As the magnet attached to the piezoelectric beam also vibrates simultaneously, it induces voltage in the coil due to relative motion between them. According to Faraday's law of electromagnetic induction, the induced open circuit e.m.f voltage generated by the electromagnetic transducer is [33]

$$V_{EM}(t) = -N\frac{d}{dt}\left[\int \vec{B}.d\vec{A}\right] = -NBl\dot{y}(t) \tag{8}$$

where N is the number of coil turns and $\int \vec{B}.d\vec{A}$ indicates the net magnetic flux through the differential element area dA of the magnet-coil assembly. B is the magnetic flux density; l is the coil length across the magnetic flux lines; and $\dot{y}(t)$ is the relative velocity between the magnet and coil, which is determined by solving Equation (1) as

$$\dot{y}(t) = -\frac{\mu_k F_{max}\omega_r\left[\int_0^{L/2}\varphi(x)dx\right]}{k\sqrt{1-\zeta_m^2}}e^{-\zeta_m\omega_r t}\sin(\omega_d t) \tag{9}$$

In the case of both transducers, the instantaneous power delivered to corresponding load resistance R_l can be expressed as

$$P(t) = \frac{1}{T}\int_0^T \frac{V(t)^2}{R_l}dt \tag{10}$$

It is to be noted that the damping (ζ_T) for each standalone (either piezoelectric or electromagnetic) transducer includes the mechanical damping ζ_m and the electrical damping ζ_e of the corresponding transducer: $\zeta_T = \zeta_m + \zeta_{e\,(PE)}$ for the piezoelectric transducer and $\zeta_T = \zeta_m + \zeta_{e\,(EM)}$ for the electro-magnetic transducer. However, for coupled transducers (when both transducers are terminated to corresponding loads simultaneously), the damping values are the same: $\zeta_T = \zeta_m + \zeta_{e\,(PE)} + \zeta_{e\,(EM)}$ for both transducers.

2.3. Simulation

Based on the above discussion on the electromechanical modeling of the proposed energy harvester, we performed time domain simulations using an appropriate simulation tool (MATLAB) to predict the output voltage generated by both PE and EM transducers simultaneously while operated by low-frequency excitation (i.e., human-limb motion). The parameters used in the simulation were calculated from the geometry and material parameters of the device components and will be discussed in the following sections. In the simulation, it was considered that the system was excited periodically in the horizontal direction at 5 Hz frequency and 2 g (g = 9.8 ms^{-2}) peak acceleration. It was also assumed that the sphere started moving from the left end of the rectangular channel and moved back and forth periodically in response to the applied excitation.

Figure 3 shows the simulated open circuit voltage waveforms generated by the PE and EM transducers. In each voltage waveform, two consecutive maximum peaks occurred due to the transverse impact when the sphere slid over the proof-mass during its forward and backward motion in one cycle and the process continued as long as the excitation existed. The positive half cycle of the acceleration waveform indicates the forward motion and the negative half cycle indicates the backward motion of the sphere. Since both transducers generated voltage simultaneously, the frequency of both open circuit voltage waveforms was the same, which was the resonant frequency of the vibrating piezoelectric beam. As seen in the figure, the amplitudes of the instantaneous voltage waveforms (in both cases) decayed exponentially with time due to mechanical damping and became almost zero until the next impact occurred.

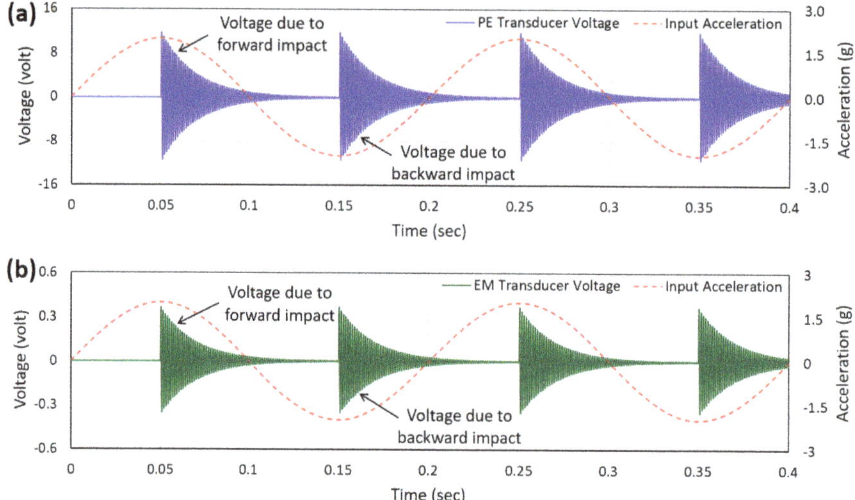

Figure 3. Simulated open circuit voltage waveforms of the piezoelectric (**a**) and electromagnetic (**b**) transducers at 5 Hz excitation frequency and 2 g peak acceleration.

3. Prototype and Test Setup

3.1. Prototype Fabrication

A macro-scale prototype of the proposed hybrid energy harvester was fabricated and tested. The PE transducer of the prototype comprised a piezoelectric (PZT) parallel bimorph (SMBA4510T05M, STEMiNC, Davenport, FL, USA), a neodymium (N52) cylinder magnet, and a cubic iron mass of 6 mm length with a 1 mm high parabolic top, both glued to the middle of either sides of the bimorph beam. A suitable assembly of a cylinder magnet and a 1000-turn coil (0.1 mm diameter laminated copper wire) attached to a printed circuit board (PCB) constituted the EM transducer. A 316 stainless steel ball was enclosed in a rectangular shaped aluminum channel (inner area 10.5×10.5 mm^2) with a square (7×7 mm^2) opening at the middle of its bottom wall, which was assembled on top of the PE transducer. The channel opening was occupied by the parabolic-top of the cubic mass with a 0.4 mm overlap with the ball. Figure 4 shows a photograph of the fabricated prototype device along with the schematics of the electrical connection of the piezoelectric bimorph and the magnet-coil assembly. The geometric parameters and material properties of the components are tabulated in Table 1.

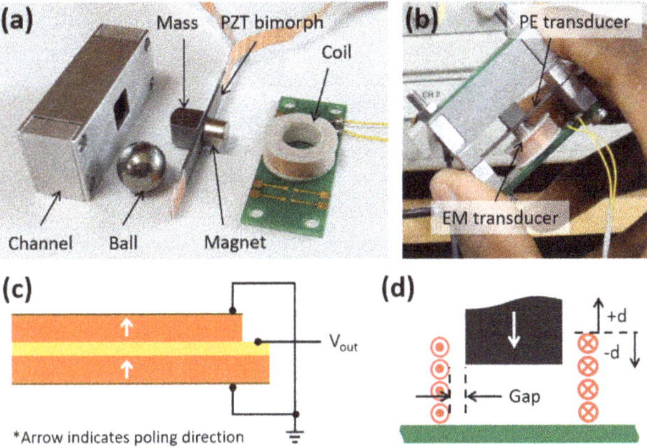

Figure 4. Photographs of the prototype components (**a**), fabricated prototype (**b**), schematics of the parallel bimorph connection (**c**), and the magnet-coil assembly (**d**).

Table 1. Geometric parameters and materials properties of the harvester components.

Parameter	Value
Dimension of the Piezoelectric bimorph	$40 \times 6 \times 0.5$ mm^3
Thickness of each piezoelectric (PZT) layer	0.2 mm
Thickness of middle shim (copper) layer	0.1 mm
Young's modulus of PZT	72 GPa
Dimension of the cylinder magnet	$\varnothing 6 \times 5$ mm^2
Remnant flux density of the magnet	1.18 T
Mass of the magnet	1 g
Mass of the attached proof-mass	1.73 g
Diameter of the sphere	10.3 mm
Mass of the sphere	4.36 gm
Length of the channel	30 mm
Inner diameter of the coil	8 mm
Outer diameter of the coil	10 mm
Number of coil turns	1000
Height of the coil	5 mm
Resistance of the coil	84 Ω
Dimension of the fabricated prototype	$40 \times 30 \times 16$ mm^3

3.2. Human-Limb Motion Test Setup

Our fabricated energy harvester was tested by human-limb motion to observe its power generation capability under a real-world situation. In order to achieve a robust test setup, it required convenient (small and portable) measuring equipment to record the characteristics (frequency and amplitude) of the excitation generated by human-limb motion. An EVAL-ADXL326Z (Analog Devices Inc., Norwood, MA, USA) tri-axial MEMS accelerometer kit (mounted on the harvester prototype) in conjunction with a XR5-SE (Pace Scientific Inc., Mooresville, NC, USA) data logger was used to record the excitation profile of human-limb motion for further analysis. The outputs of both PE and EM transducers were connected to a digital storage oscilloscope (TDS 5052B, Tektronix Inc., Beaverton, OR, USA) to observe and record the output responses. Furthermore, benchtop tests using an electrodynamic shaker were conducted to observe the damping behavior of both transducers and to determine the optimal overlap (±d) between the magnet and coil, as described in our previous work [24].

4. Experimental Results and Discussion

4.1. Optimal Overlap and Damping Measurements

In order to generate maximum possible voltage and power from the prototype, it was important to determine the optimum overlaps between the magnet and coil as well as between the freely-movable sphere and parabolic-top of the proof-mass. The optimum magnet-coil overlap was determined by a benchtop test setup [24] using an electrodynamic shaker whereas the overlap between the sphere and the mass-top was determined by the human-limb vibration test setup (due to the limitation of the shaker to generate low-frequency, large-amplitude excitation). As seen from Figure 5a, the optimum magnet-coil overlap was −1 mm. The lateral gap between the magnet and coil was also 1 mm. Since the absolute values were not primarily of interest in determining the optimum magnet-coil overlap, normalized values were used. Figure 5b shows the change in the open circuit voltages generated by both the PE and EM transducers with the change in the overlap between the sphere and the mass-top. As seen from the figure, the sphere could not make significant contact with the parabolic top when the overlap was 0.2 mm as the clearance between the ball and inner surface of the channel was 0.2 mm. On the other hand, the sphere could not slide over the mass-top and was captured in the middle when the overlap was 0.5 mm because the speed/force of the sphere was not sufficient to pass through. The 0.4 mm overlap between the sphere and mass-top was considered as the optimum value since the open circuit voltages were the maximum for both the PE and EM transducers. The error bars in Figure 5b indicate the range of voltages generated for multiple attempts as the characteristics of the excitation (frequency and amplitude) applied by human-limb were not always the same.

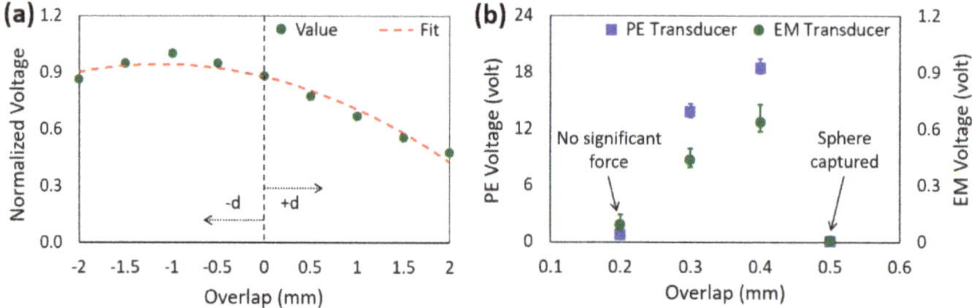

Figure 5. Normalized open circuit voltages for different magnet-coil overlaps (**a**) and open circuit voltages of the transducers for various overlaps between the freely movable sphere and the parabolic top of the proof-mass (**b**).

The damping behavior of both PE and EM transducers were determined by the impulse response test using an electrodynamic shaker [24]. A high amplitude impulse (30.3 ms^{-2} with 50 ms pulse period and 500 µs pulse width) was applied to the harvester. Then, the mechanical damping ratio (ζ_m) and total damping ratio (ζ_T) of both transducers were estimated from the open circuit and loaded impulse response signals, respectively. The logarithmic decrement method was used to calculate the damping ratio as

$$\zeta = \frac{1}{2\pi} \ln\left(\frac{a_1}{a_2}\right) \tag{11}$$

where a_1 and a_2 are the amplitudes of two consecutive peaks in the impulse response plot of the transducer. Subtraction of the mechanical damping ratio (ζ_m) from the total damping ratio (ζ_T) gives the electrical damping ratio (ζ_e). By conducting this experiment, the mechanical damping ratio was found to be 0.011. On the other hand, the electrical damping ratio for the piezoelectric transducer and electromagnetic transducer were 0.017 and 0.016, respectively. It should be noted that the electrical damping values were determined by the impulse response across the corresponding optimum load resistances of the transducers, which were determined by measuring the voltage across various load resistors and calculating the power delivered to them. The power is experimentally equal to $V_{p-p}^2/4R_l$, where V_{p-p} is the peak–peak value of the measured voltage across each load resistance R_l.

4.2. Transducer Outputs

Figure 6 illustrates the measured peak–peak voltages and peak powers delivered to the load resistances connected to the PE and EM transducers, while the prototype was excited by human-limb motion. Error bars indicate the range of voltage and power values measured for multiple attempts. On average, the maximum of 0.98 mW and 0.64 mW peak power were delivered to 40 kΩ and 85 Ω load resistances connected to the PE and EM transducers, respectively. Note that the optimum load resistances for the PE and EM transducers were in the range of kΩ and Ω, respectively. The resistance of the coil was measured as 84 Ω, which closely matches that of the measured optimum load of 85 Ω. On the other hand, the source resistance (R_{source}) of a piezoelectric material depends on its vibration frequency (f) and capacitance (C), according to $R_{source} = 1/(2\pi f C)$. The capacitance of the piezoelectric beam (doubly clamped parallel bimorph) was measured as 16 nF and the frequency of its vibration was calculated as 815 Hz (measured as 818 Hz). This gives the calculated R_{source} as 38.3 kΩ, which closely matches the measured optimal load resistance of 40 kΩ. Figure 7 shows the instantaneous voltage and power waveforms across 40 kΩ optimum load resistances of the PE transducer. The voltage and power waveforms generated by the EM transducer also followed the same trend. The maximum peak–peak voltages across the corresponding optimum load resistances generated by the PE and EM transducers were 12.53 V and 0.47 V, respectively. However, the peaks of both voltage and power waveforms decayed exponentially with time due to the damping, which, in turn, reduced their rms (1.92 V for PE and 72 mV for the EM transducers) and average values (93 µW for the PE and 61 µW EM transducers), respectively. Peak amplitudes of the instantaneous power were reduced to almost zero as the time passed, and before the next impact occurred. As a result, the values of average power reduced dramatically. It is to be noted that the waveforms were collected simultaneously, therefore, the overall damping was composed of mechanical damping and the electrical damping of both transducers, as discussed earlier. As seen from the figure, the amplitude decays were not perfectly exponential due to process variation in assembling the harvester components. Two consecutive maximum peaks were generated in one cycle of the applied excitation since the sphere exerted transverse impact on the mass-top twice during its back and forth movement in one cycle. It should be noted that there was no significant change in the peak values of the voltage and power with the change in the frequency of human-limb motion as the variation in the acceleration amplitude was small, however, the values of the rms voltage and average power output changed with the change in the frequency of excitation [34]. This occurred because of the change in the time interval between two consecutive impacts with the change in the frequency and was also due to the exponentially decaying behavior of the voltage waveform generated by the transducer.

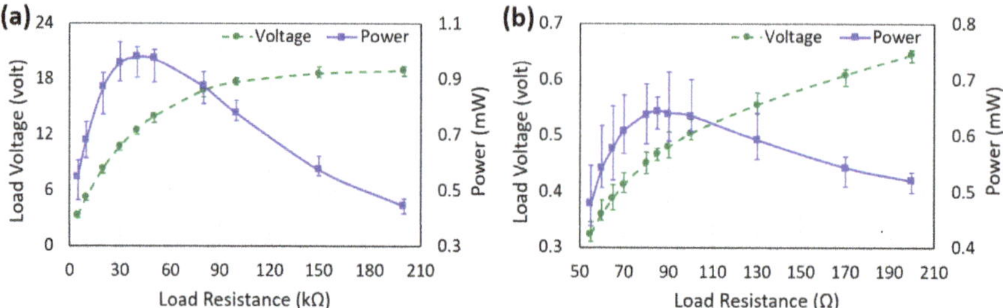

Figure 6. Measured voltage and power vs. load resistances connected to the piezoelectric (**a**) and electromagnetic (**b**) transducers while excited by human-limb motion.

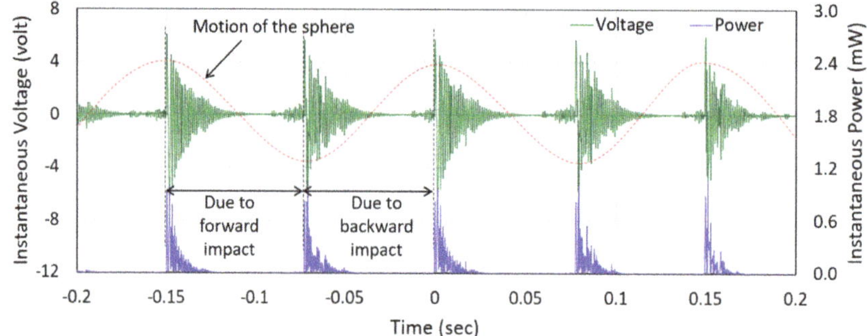

Figure 7. Instantaneous voltage and power waveforms measured across the 40 kΩ optimum load resistance of the piezoelectric transducer during the human-limb motion test.

The input excitation characteristics (frequency and amplitude of human-limb motion) were measured along each axis of the accelerometer mounted on the prototype during the test. As the harvester prototype was driven along the accelerometer's Y-axis, the peak acceleration amplitude was maximum in this direction (~2 g), whilst those in other directions were relatively low (~0.95 g along X-axis and ~0.75 g along Z-axis). Data were collected at the 50 Hz sampling rate. The frequency components of both applied acceleration and the generated voltage waveforms were determined by Fast Fourier Transform (FFT) analysis. Figure 8 shows that the frequency of the applied excitation was 5.2 Hz whereas the frequency of the voltage waveform generated by the PE transducer (same for the EM transducer) was 818 Hz, indicating the frequency up-conversion behavior of the harvester.

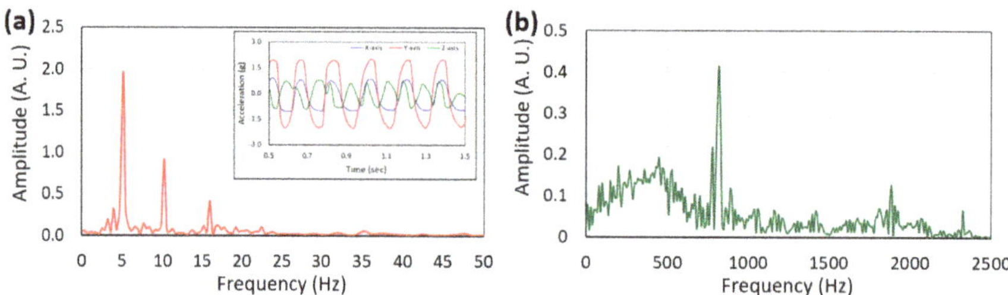

Figure 8. Frequency components (FFT) of the applied excitation (along Y-axis) obtained from the accelerometer data (inset) during the test (**a**) and the voltage waveform generated by the piezoelectric transducer (**b**).

4.3. AC–DC Conversion

The voltage generated by the proposed harvester has alternating (AC) characteristics due to the time-varying characteristic of the input excitation. Most electronic devices are driven by DC voltage source. Therefore, AC–DC conversion is necessary before using the harvested energy. Generally, a full bridge rectifier using four diodes is used to rectify the ac voltage generated by the harvester unit. In our prototype harvester, the voltage generated by the EM transducer was very low when compared to that of the PE transducer. Therefore, a conventional bridge rectifier cannot satisfy the need for rectification and significant voltage generation to drive an electronic load. This is why, a 4-stage Villard's voltage multiplier circuit was used with the EM transducer whereas a bridge rectifier, on the other hand, was used with the PE transducer for AC–DC conversion, as shown in Figure 9a. The voltage multiplier rectifies the voltage output with voltage multiplication based on the number of stages used [35]. The bridge rectifier used four Schottky barrier diodes whereas the voltage multiplier circuit used four pairs of Schottky barrier diodes (HSMS-2852-BLKG, Broadcom Inc., San Jose, CA, USA) and 10 µF, 50 V capacitors, soldered on a printed circuit board (PCB) designed by a professional PCB design tool (Proteus 8.0). The outputs of both bridge rectifier and multiplier circuit were connected to a 33 µF, 50 V storage capacitor (C_s) to accumulate the rectified and multiplied DC electrical energy that was used to power a number of parallelly connected LEDs that demonstrated its application potential, as shown in Figure 9b.

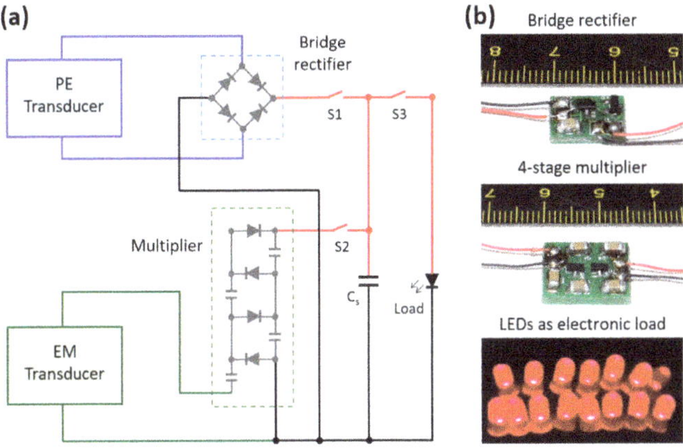

Figure 9. Schematic of the hybrid energy harvester circuit diagram (**a**), photographs of a bridge rectifier, 4-stage voltage multiplier as the AC–DC converters and LEDs powered by the harvester as electronic load (**b**).

Figure 10a shows the output AC voltage waveforms of the PE and EM transducers of the prototype harvester (with the rectifier and multiplier connected) while excited by human-limb motion, to be converted to DC and stored in the storage capacitor (C_s). The charging characteristics of the storage capacitor (C_s) was also observed at the same time, as presented in Figure 10b. The charging behavior is influenced by the inherent output characteristics (voltage and current) of the piezoelectric and electromagnetic transducers where the voltage determines the maximum limit of charging and the current determines the charging speed. As a result, the high output current and low output voltage of the electromagnetic transducer charges the capacitor relatively faster than the low output current and high output voltage of the piezoelectric transducer. When the DC outputs from both transducers were coupled together, the storage capacitor was charged even faster and reached over 2 V DC voltage and was able to turn on the LEDs used as the electronic load.

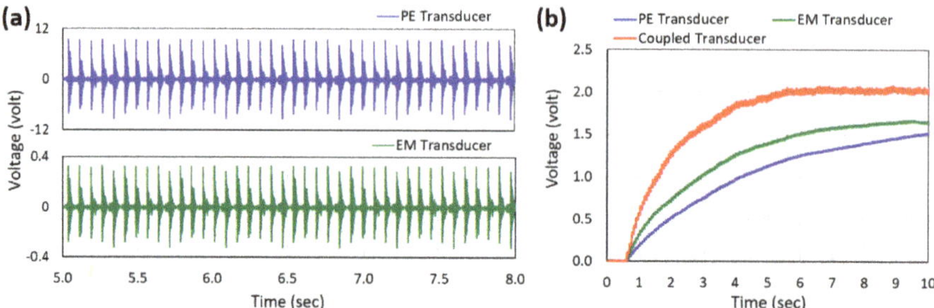

Figure 10. Voltage waveforms generated by each transducer (**a**) and accumulated rectified and multiplied voltages across the storage capacitor (**b**) as a function of time while excited by human-limb motion.

5. Conclusions and Future Works

This paper presents a human-limb motion driven, piezoelectric and electromagnetic hybrid energy harvester that utilized the frequency up-conversion technique by the transverse impact mechanism. Instead of using any resonant structure (e.g., compliant cantilever beam), a freely movable non-magnetic metallic sphere was used as the low-frequency oscillator, which overcomes the limitations of designing energy harvesters for human-body-induced motion. Use of two transducers allows simultaneous power generation from a single mechanical excitation, which increases the power density of the harvester. The theoretical model was derived based on its working principle, and then a macroscale prototype was fabricated and tested. A series of tests were carried out to partially optimize its parameters and to observe its output performances. The piezoelectric and electromagnetic transducers of the prototype energy harvester simultaneously generated maximum 93 μW and 61 μW average powers, respectively, while excited by human-limb motion at ~2 g peak acceleration. Analysis of the measured voltage and acceleration data shows that the frequency was up-converted to 818 Hz from 5.2 Hz human-limb motion. In order to utilize the harvested energy for practical low-power electronics applications, suitable AC–DC converters (rectifier for PE and voltage multiplier for EM transducers) were constructed and demonstrated. For a functional volume of 19.2 cm^3, the average power density of the hybrid energy harvester prototype was 8 μW cm^{-3}, which is ~1.5× higher than its electromagnetic only counterpart (5.4 μW cm^{-3}). However, the generated power and the power density of the harvester was still low as the size, mass, and diameter of the ball, height and curvature of the attached mass-top, stiffness of the piezoelectric beam, etc. were chosen arbitrarily, which are all significantly related to power generation. Further optimization of these parameters would be able to deliver higher power within a reduced volume. A more portable design and lighter packaging material should be adopted for its intended use. Our future work will include further optimization of the design parameters (e.g., spring stiffness, mechanical and electrical damping, transverse impact, magnet-coil assembly, etc.) through finite element analysis (FEA) tools, and to fabricate a compact and smaller device with improved output performances to be efficiently used in powering portable and wearable smart devices from human-body-induced motion.

Author Contributions: Conceptualization, M.A.H.; Formal analysis, M.A.H. and M.H.K.; Funding acquisition, M.A.H. and J.Y.P.; Methodology, M.A.H.; Project administration, J.Y.P.; Resources, H.C.; Software, M.H.K.; Supervision, J.Y.P.; Validation, M.A.H. and H.C.; Writing—original draft, M.A.H.; Writing—review & editing, M.A.H., M.H.K., H.C., and J.Y.P.

Funding: The authors are grateful to acknowledge the Basic Science Research Program (2013R1A1A2A 10064810) and the Pioneer Research Centre Program (2010-0019313) through the National Research Foundation of Korea (NRF) funded by the Ministry of Science, ICT and Future Planning, Korea.

Conflicts of Interest: The authors declare no conflicts of interest.

References

1. Prauzek, M.; Konecny, J.; Borova, M.; Janosova, K.; Hlavica, J.; Musilek, P. Energy harvesting sources, storage devices and system topologies for environmental wireless sensor networks: A review. *Sensors* **2018**, *18*, 2446. [CrossRef] [PubMed]

2. Cook-Chennault, K.A.; Thambi, N.; Sastry, A. Powering MEMS portable devices—A review of non-regenerative and regenerative power supply system with special emphasis on piezoelectric energy harvesting systems. *Smart Mater. Struct.* **2008**, *17*, 043001. [CrossRef]

3. Ahmed, R.; Mir, F.; Banerjee, S. A review on energy harvesting approaches for renewable energies from ambient vibrations and acoustic waves using piezoelectricity. *Smart Mater. Struct.* **2017**, *26*, 085031. [CrossRef]

4. Beeby, S.P.; Tudor, M.J.; White, N.M. Energy harvesting vibration sources for microsystems applications. *Meas. Sci. Technol.* **2006**, *17*, R175–R195. [CrossRef]

5. Mitcheson, P.D.; Yeatman, E.M.; Rao, G.K.; Holmes, A.S.; Green, T.C. Energy harvesting from human and machine motion for wireless electronic devices. *Proc. IEEE* **2008**, *96*, 1457–1486. [CrossRef]

6. Wang, D.-A.; Ko, H.H. Piezoelectric energy harvesting from flow-induced vibration. *J. Micromech. Microeng.* **2010**, *20*, 025019. [CrossRef]

7. Khameneifar, F.; Moallem, M.; Arzanpour, S. Modeling and analysis of a piezoelectric energy scavenger for rotary motion applications. *J. Vib. Acoust.* **2011**, *133*, 011005. [CrossRef]

8. Priya, S.; Song, H.-C.; Zhou, Y.; Varghese, R.; Chopra, A.; Kim, S.-G.; Kanno, I.; Wu, L.; Ha, D.S.; Ryu, J.; et al. A review on piezoelectric energy harvesting: materials, methods, and circuits. *Energy Harvest. Syst.* **2017**, *4*, 1–37. [CrossRef]

9. Tan, Y.; Dong, Y.; Wang, X. Review of MEMS electromagnetic vibration energy harvester. *J. Microelectromech. Syst.* **2017**, *26*, 1–16. [CrossRef]

10. Khan, F.U.; Qadir, M.U. State-of-the-art in vibration-based electrostatic energy harvesting. *J. Micromech. Microeng.* **2016**, *26*, 103001. [CrossRef]

11. Deng, Z.; Dapino, M.J. Review of magnetostrictive vibration energy harvesters. *Smart Mater. Struct.* **2017**, *26*, 103001. [CrossRef]

12. Wang, Z.L.; Chen, J.; Lin, L. Progress in triboelectric nanogenerators as a new energy technology and self-powered sensors. *Energy Environ. Sci.* **2015**, *8*, 2250–2282. [CrossRef]

13. Williams, C.B.; Yates, R.B. Analysis of a micro-electric generator for microsystems. *Sens. Actuators A Phys.* **1996**, *52*, 8–11. [CrossRef]

14. Ju, S.; Chae, S.H.; Choi, Y.; Lee, S.; Lee, H.W.; Ji, C.-H. A low frequency vibration energy harvester using magnetoelectric laminate composite. *Smart Mater. Struct.* **2013**, *22*, 115037. [CrossRef]

15. Gogurla, N.; Roy, B.; Park, J.-Y.; Kim, S. Skin-contact actuated single-electrode protein triboelectric nanogenerator and strain sensor for biomechanical energy harvesting and motion sensing. *Nano Energy* **2019**, *62*, 674–681. [CrossRef]

16. Lee, S.; Ko, W.; Oh, Y.; Lee, J.; Baek, G.; Lee, Y.; Sohn, J.; Cha, S.; Kim, J.; Park, J.; et al. Triboelectric energy harvester based on wearable textile platforms employing various surface morphologies. *Nano Energy* **2015**, *12*, 410–418. [CrossRef]

17. Alluri, N.R.; Vivekananthan, V.; Chandrasekhar, A.; Kim, S.-J. Adaptable piezoelectric hemispherical composite strips using a scalable groove technique for a self-powered muscle monitoring system. *Nanoscale* **2018**, *10*, 907–913. [CrossRef]

18. Wu, Y.; Qu, J.; Daoud, W.A.; Wang, L.; Qi, T. Flexible composite-nanofiber based piezo-triboelectric nanogenerators for wearable electronics. *J. Mater. Chem. A* **2019**, *7*, 13347–13355. [CrossRef]

19. Halim, M.A.; Park, J.Y. Theoretical modeling and analysis of mechanical impact driven and frequency up-converted piezoelectric energy harvester for low-frequency and wide-bandwidth operation. *Sens. Actuators A Phys.* **2014**, *208*, 56–65. [CrossRef]

20. Renaud, M.; Fiorini, P.; van Schaijk, R.; van Hoof, C. Harvesting energy from the motion of human limbs: the design and analysis of an impact-based piezoelectric generator. *Smart Mater. Struct.* **2009**, *18*, 035001. [CrossRef]

21. Halim, M.A.; Cho, H.; Salauddin, M.; Park, J.Y. A miniaturized electromagnetic vibration energy harvester using flux-guided magnet stacks for human-body-induced motion. *Sens. Actuators A Phys.* **2016**, *249*, 23–31. [CrossRef]

22. Pozzi, M.; Zhu, M. Characterization of a rotary piezoelectric energy harvester based on plucking excitation for knee-joint wearable applications. *Smart Mater. Struct.* **2012**, *21*, 055004. [CrossRef]

23. Xue, T.; Yeo, H.G.; Trolier-McKinstry, S.; Roundy, S. Wearable inertial energy harvester with sputtered bimorph lead zirconate titanate (PZT) thin-film beams. *Smart Mater. Struct.* **2018**, *27*, 085026. [CrossRef]

24. Halim, M.A.; Park, J.Y. Modeling and experiment of a handy motion driven, frequency up-converting electromagnetic energy harvester using transverse impact by spherical ball. *Sens. Actuators A Phys.* **2015**, *229*, 50–58. [CrossRef]

25. Halim, M.A.; Park, J.Y. A non-resonant, frequency up-converted electromagnetic energy harvester from human-body-induced vibration for hand-held smart system applications. *J. Appl. Phys.* **2014**, *115*, 094901. [CrossRef]

26. Fan, K.; Liu, S.; Liu, Y.; Zhu, Y.; Wang, W.; Zhang, D. Scavenging energy from ultra-low frequency mechanical excitations through a bi-directional hybrid energy harvester. *Appl. Energy* **2018**, *216*, 8–20. [CrossRef]

27. Salauddin, M.; Toyabur, R.M.; Maharjan, P.; Rasel, M.S.; Kim, J.W.; Cho, H.; Park, J.Y. Miniaturized springless hybrid nanogenerator for powering portable and wearable electronic devices from human-body-induced vibration. *Nano Energy* **2018**, *51*, 61–72. [CrossRef]

28. Wang, X.; Yang, B.; Liu, J.; Zhu, Y.; Yang, C.; He, Q. A flexible triboelectric-piezoelectric hybrid nanogenerator based on P(VDF-TrFE) nanofibers and PDMS/MWCNT for wearable devices. *Sci. Rep.* **2016**, *6*, 36409. [CrossRef]

29. Erturk, A. Electromechanical Modeling of Piezoelectric Energy Harvesters. Ph.D Thesis, Virginia Polytechnic Institute and State University, Blacksburg, VA, USA, 2009.

30. Gu, L. Low-frequency piezoelectric energy harvesting prototype suitable for the MEMS implementation. *Microelectron. J.* **2011**, *42*, 277–282. [CrossRef]

31. Budynas, R.G.; Nisbett, J.K. *Shigley's Mechanical Engineering Design*, 8th ed.; McGraw-Hill: New York, NY, USA, 2006; ISBN 0-390-76487-6.

32. Wang, Q.-M.; Cross, L.E. Performance analysis of piezoelectric cantilever bending actuators. *Ferroelectrics* **1998**, *215*, 187–213. [CrossRef]

33. Mizuno, M.; Chetwynd, D.G. Investigation of a resonance micro generator. *J. Micromech. Microeng.* **2003**, *13*, 209–216. [CrossRef]

34. Halim, M.A.; Park, J.Y. Piezoelectric energy harvester using impact-driven flexible side-walls for human-limb motion. *Microsyst. Technol.* **2018**, *24*, 2099–2107. [CrossRef]

35. Lui, H.; Ji, Z.; Chen, T.; Sun, L.; Menon, S.C.; Lee, C. An intermittent self-powered energy harvesting system from low-frequency hand shaking. *IEEE Sens. J.* **2015**, *15*, 4782–4790.

Article

Development of a Class-C Power Amplifier with Diode Expander Architecture for Point-of-Care Ultrasound Systems

Hojong Choi

Department of Medical IT Convergence Engineering, Kumoh National Institute of Technology, Gumi 39253, Korea; hojongch@kumoh.ac.kr; Tel.: +82-54-478-7782

Received: 8 September 2019; Accepted: 11 October 2019; Published: 14 October 2019

Abstract: Point-of-care ultrasound systems are widely used in ambulances and emergency rooms. However, the excessive heat generated from ultrasound transmitters has an impact on the implementation of piezoelectric transducer elements and on battery consumption, thereby affecting the system's sensitivity and resolution. Non-linear power amplifiers, such as class-C amplifiers, could substitute linear power amplifiers, such as class-A amplifiers, which are currently used in point-of-care ultrasound systems. However, class-C power amplifiers generate less output power, resulting in a reduction of system sensitivity. To overcome this issue, we propose a new diode expander architecture dedicated to power amplifiers to reduce the effects of sinusoidal pulses toward the power supply. Thus, the proposed architecture could increase the input pulse amplitudes applied to the main transistors in the power amplifiers, hence increasing the output voltage of such amplifiers. To verify the proposed concept, pulse-echo responses from an ultrasonic transducer were tested with the developed class-C power amplifier using a resistor divider and the designed diode expander architecture. The peak-to-peak amplitude of the echo signals of the ultrasonic transducers when using a class-C power amplifier with a diode expander architecture (2.98 V_{p-p}) was higher than that for the class-C power amplifier with a resistor divider architecture (2.51 V_{p-p}). Therefore, the proposed class-C power amplifier with diode expander architecture is a potential candidate for improving the sensitivity performance of piezoelectric transducers for point-of-care ultrasound systems.

Keywords: class-C power amplifier; diode expander; piezoelectric transducers; point-of-care ultrasound systems

1. Introduction

Ultrasound has been widely used for patient diagnosis because it is a non-ionizing, real-time, non-invasive, and inexpensive imaging modality [1]. Especially, point-of-care ultrasound systems have been used to diagnose acute diseases for immediate treatment in ambulance vehicles and emergency rooms with remote help from physicians [2,3]. However, the performance of point-of-care ultrasound systems typically suffers from a limited number of ultrasonic array transducer elements, overheating, and battery issues, thereby affecting the sensitivity and resolution of such ultrasound systems [3,4]. To overcome overheating and battery issues, manufacturers need to use cooling systems such as cooling fans and aluminum heat pipe structures even though the internal structures of point-of-care ultrasound systems are much smaller than conventional bench-top ultrasound machines [3].

Conventional ultrasound systems are composed of ultrasound transmitters, ultrasonic (piezoelectric) transducers, and ultrasound receivers [5–7]. Most of the overheating issues come from the power amplifiers in the ultrasound transmitters and by the analog-to-digital converter or digital-to-analog converter in the ultrasound receivers [8]. However, the excessive heat could be resolved by employing non-linear power amplifiers in the ultrasound transmitters owing to their different transition period between voltage and current [9]. Among non-linear power amplifiers, class-C power amplifiers could achieve relatively low DC current consumption while lowering the gain [10]. Therefore, the scheme to increase the gain would be beneficial for class-C power amplifiers in low-sensitivity ultrasonic transducers.

Several non-linear power amplifiers have been developed using different types of ultrasonic transducers to improve the performance of ultrasound instruments. A class-E inverter was proposed for inductive piezoelectric transducers to determine the optimal resonance frequency based on the equivalent circuit model [11]. A class-D half-bridge driving power amplifier was proposed for piezoelectric transducers to find the optimal conversion frequency operation [12]. A class-E power amplifier was proposed to reduce power dissipation in Langevin transducers [13].

Figure 1 describes how input signals in power amplifiers affect the bias circuit and how the power supply controls the power amplifiers through the bias circuit. Positive and negative pulse signals can pass through the bias circuit and get into the power supply while the power supply is applied to DC voltages through the bias circuit to the power amplifier [10]. In the output, the amplified positive and negative pulse signals trigger the ultrasonic transducer. The bias circuit needs to be stabilized during signal amplification through control of the power supply. Therefore, the pulse signals should not affect the bias circuit operation. Otherwise, it could generate output signals with high noise and harmonic distortion or reduce the output signal amplitudes. These scenarios directly affect the resolution and sensitivity performance of ultrasound systems [2,14].

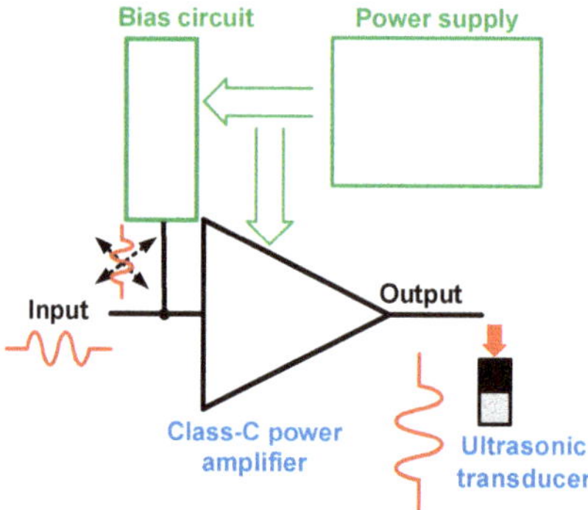

Figure 1. Concept for class-C power amplifiers with input pulse signals and bias circuit.

To reduce the effects of pulse signals on bias circuits, different architecture using an analog-to-digital converter, a digital-to-analog converter, and a look-up table have been used for wireless applications [15]. This method needs to regulate several DC bias voltages to properly operate the transistors. This might not be suitable for point-of-care ultrasound systems because of their higher DC current consumption. Additionally, a number of ultrasonic transducers are required for combination with each transmitter channel to achieve a limited size of point-of-care ultrasound systems. However, the proposed method is a much simpler approach for easy regulation of a single power amplifier, thereby increasing the output voltages of the power amplifiers for point-of-care ultrasound systems.7

A power metal-oxide semiconductor field-effect transistor (MOSFET) simulation model has been developed to estimate the expected performance in a high-voltage environment [9]. Given that a power MOSFET simulation library for a high-voltage environment provides inaccurate performance for sub-decibel level operation, the performance of power amplifiers needs to be verified through real measurements [15,16]. Therefore, in Section 2, we report the detailed experimental circuit design analysis in DC and AC levels rather than calculated and simulated results. The equivalent circuit model with measured parameters for power amplifiers is provided. Section 3 shows the measured performance results for the power amplifier including pulse-echo response with ultrasonic transducers. Section 4 concludes the paper.

2. Materials and Methods

Diode expanders placed after power amplifiers have originally been used to reduce unwanted ring down signals generated from class-A power amplifiers in ultrasound instruments. Class-A amplifiers produce a higher gain compared to class-C power amplifiers while consuming a higher DC current. Therefore, class-A power amplifiers can use relatively lower input voltage signals compared to class-C power amplifiers. In ultrasound instruments, the maximum allowable input voltages are generally limited to 1 V_{p-p} [17,18]. Therefore, the bias circuits in class-A power amplifiers would be less affected by input pulse signals. However, class-C power amplifiers produce a relatively lower voltage gain compared to class-A power amplifiers and also consume a lower DC current. Consequently, higher input voltage signals could affect the bias circuit operation for class-C power amplifiers. Overall, a diode expander architecture is proposed for bias circuits of class-C power amplifiers because higher input pulse signals can pass through the bias circuits while the power supply controls the bias circuits and the amplified pulse signal reaches the ultrasonic transducers. The proposed diode expander architecture is intended to suppress unwanted input pulse signals into the bias circuits, resulting in stabilization of the DC current and improvement of the output voltages of class-C power amplifiers.

Figure 2 shows a schematic diagram and implemented printed circuit board of the developed class-C power amplifier with a resistor divider and with a diode expander architecture. The printed circuit board was fabricated from a manufacturing service (ExpressPCB LLC, Mulino, OR, USA). The mounting holes were tied to reduce parasitic effects. The signal traces were not perpendicularly cross the ground plane. Given that high-voltage pulse signals are needed to trigger ultrasonic transducers, power film resistors, capacitors, inductors, choke inductors, and power transistors are usually selected because they tolerate high voltages. The selected power transistors (T_1, PD57018-E, STMicroelectronics Inc., Geneva, Switzerland) were lateral-diffusion metal-oxide semiconductor field-effect transistors (LDMOSFETs) that can be used for high voltages up to 65 V and high currents up to 2.5 A operations. A small-size (1 cm length and width) heat-sink was attached on the top of the LDMOSFET devices. Electrolytic capacitors (C_{E1} = 10 µF and C_{E2} = 220 µF) including additional ceramic capacitors (C_{L1} = 0.1 µF, C_{L2} = 1 nF, and C_{L3} = 47 pF) admitting up to 200 V were used to stabilize the DC voltages from the power supplies. A radio frequency (RF) choke inductor with a self-resonant frequency at 190 MHz (L_c = 1 µH) admitting high currents up to 2 A was selected. The power film resistors (R_1 to R_5) operated up to 250 V. The input and output matching circuits were used in the input and output sides, respectively, to match the operating frequency of the power amplifier. In the input matching circuit, one resistor, two capacitors, and one inductor were used. In the output matching circuit, three capacitors and two inductors were employed. In both matching circuits, the resistors, capacitors, and inductors also admitted high voltages, up to 100 V. Cooling fans were used for accurate measurement, though class-C power amplifiers generate less heat than class-A power amplifiers. Table 1 provides component values for the power amplifiers.

Table 1. Component values for class-C power amplifiers.

Components	Values	Components	Values	Components	Values	Components	Values
R_{r1}	5 kΩ	C_{E2}	220 µF	C_5	150 pF	C_6	100 pF
R_{r2}	0.5 kΩ	C_{L1}	0.1 µF	C_{L4}	0.1 µF	L_c	1 µH
R_1	56 Ω	C_{L2}	1 nF	C_{L5}	1 nF	L_2	560 nH
C_{E1}	10 µF	C_{L3}	47 pF	C_{L6}	47 pF	L_3	220 nH

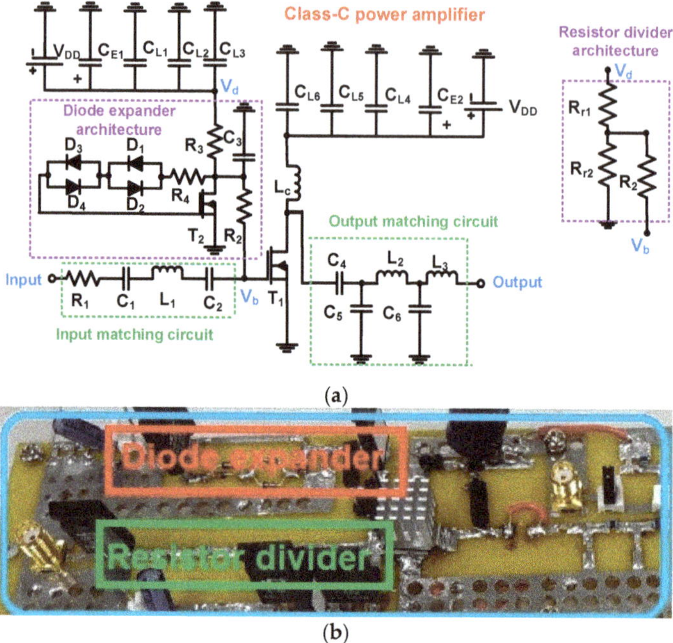

(**a**)

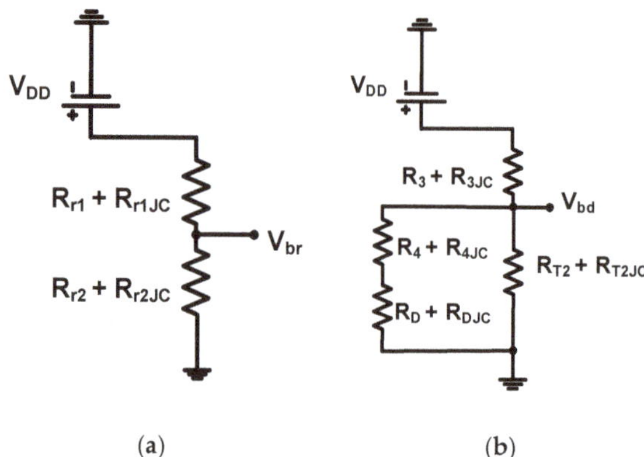

(**b**)

Figure 2. (**a**) Schematic diagram and (**b**) implemented printed circuit board of the developed class-C power amplifier with a resistor divider and diode expander architecture.

Figure 3 shows the operating mechanisms of the resistor divider and diode expander architecture including thermal resistances. Although the manufacturers did not provide the thermal model parameters of the power film resistors, diodes, and LDMOSFETs, their thermal resistances were included to estimate thermal effects in the resistor divider and diode expander architecture. In the resistor divide architecture, the DC bias voltages were obtained by the ratio of two power film resistors (R_{r1} = 5 kΩ and R_{r2} = 500 Ω) with their corresponding thermal resistances (R_{r1JC} and R_{r2JC}). In the diode expander architecture, the DC bias voltages were obtained by two power film resistors, diode equivalent resistance (R_D), and drain-source resistance of one of the LDMOSFET (T_2) including their thermal resistances (R_{3JC}, R_{4JC}, R_{DJC}, and R_{T2JC}).

(**a**)   (**b**)

Figure 3. DC Operating mechanisms of the (**a**) resistor divider and (**b**) diode expander architecture.

In Figure 3a, the resistor divider architecture used two power film resistors (R_{r1} and R_{r2}). The bias voltage of the resistor divider (V_{br}) is expressed as:

$$V_{br} = \frac{R_{r2} + R_{r2JC}}{R_{r1} + R_{r1JC} + R_{r2} + R_{r2JC}} V_{dd} = \frac{1}{1 + \frac{R_{r1} + R_{r1JC}}{R_{r2} + R_{r2JC}}} V_{dd} \qquad (1)$$

Here, V_{dd} is the supply voltage from the DC power supply and R_{r1JC}, R_{r2JC}, and R_{r3JC} are the thermal resistances of the power film resistors (R_{r1}, R_{r2}, and R_{r3}, respectively).

The power MOSFET equivalent circuit model has been used to estimate the performance of diode expander architecture [19,20]. The diode expander architecture in Figure 3b used double cross-coupled diodes. The bias voltage of this architecture (V_{bias2}) is expressed as:

$$V_{bd} = \frac{1}{1 + \frac{R_3 + R_{3JC}}{(R_{T2} + R_{T2JC})//(R_4 + R_{4JC} + R_D + R_{DJC})}} V_{dd} \approx \frac{1}{1 + \frac{R_3 + R_{3JC}}{R_{T2} + R_{T2JC}}} V_{dd} \qquad (2)$$

where R_{3JC}, R_{T2JC}, R_{4JC}, and R_{DJC} are the thermal resistances of the power film resistors (R_3 and R_4), the diode equivalent resistance (R_D), and the drain-source resistance of LDMOSFET T_2.

The thermal effect on the bias voltages for the resistor divider architecture is dependent on the two power film resistors (R_{r1} and R_{r2}) with their corresponding thermal resistances (R_{r1JC} and R_{r2JC}). Conversely, the thermal effect on the bias voltages for the diode expander architecture is dependent on a single power film resistor (R_3) and the drain-source resistance of LDMOSFET R_{T2}, with their corresponding thermal resistances (R_{3JC} and R_{T2JC}). Therefore, the drain-source resistance of LDMOSFET R_{T2} is less insensitive to sudden temperature changes compared to power film resistors because temperature changes are dependent on resistance values [21]. To confirm these temperature dependences, the bias voltages for each architecture were measured after 1 min and 1 hr. At 1 min, the measured bias voltage of the class-C power amplifier when using the resistor divider and diode expander architecture was the same, i.e., 2.3 V. After 1 h, the measured bias voltage of the class-C power amplifier with the resistor divider and diode expander architecture did not coincide anymore; they were 2.25 V and 2.29 V, respectively. Therefore, we confirm that class-C power amplifiers with diode expander architecture are less dependent on temperature variations.

Figure 4 shows the AC operating mechanisms to predict the expected input signal paths of class-C power amplifiers with the resistor divider and diode expander architecture. As shown in Figure 4a, the pulse signals from the input port (path 1) could be transmitted to two power film resistors (R_{r1} and R_{r2}, paths 3 and 4) through one power film resistor (R_2). Likewise, Figure 4b shows that the pulse signals from the same input port (path 1) could be split into one capacitor (C_3, path 3), one power film resistor with cross-coupled diode pairs (R_4, D_1 to D_4, path 4), and one LDMOSFET (T_2, path 5). We estimated that there are more paths to alleviate the burden into the DC power caused by unwanted AC signals. Therefore, we expect that more AC input signals could get to LDMOSFET T_1 for a class-C power amplifier with a diode expander architecture. To confirm this idea, the input voltages at the gate of the LDMOSFET (V_b) were measured as depicted in Figure 4c. The arbitrary function generator produced 5 V_{p-p} pulsed signals at 25 MHz for class-C power amplifiers with the resistor divider and diode expander architecture using a 10:1 voltage probe (TPP0200, Tecktronics Inc., Beaverton, OR, USA) from a digital oscilloscope (MDO4104C, Tecktronics Inc., Santa Clara, CA, USA). The measured bias voltages of the class-C power amplifier with resistor divider and diode expander architecture were 0.21 V and 0.24 V_{p-p}, respectively. Therefore, we can confirm that the class-C power amplifier with a diode expander architecture would be more suitable to reduce the voltage attenuation, resulting in higher output voltage amplification.

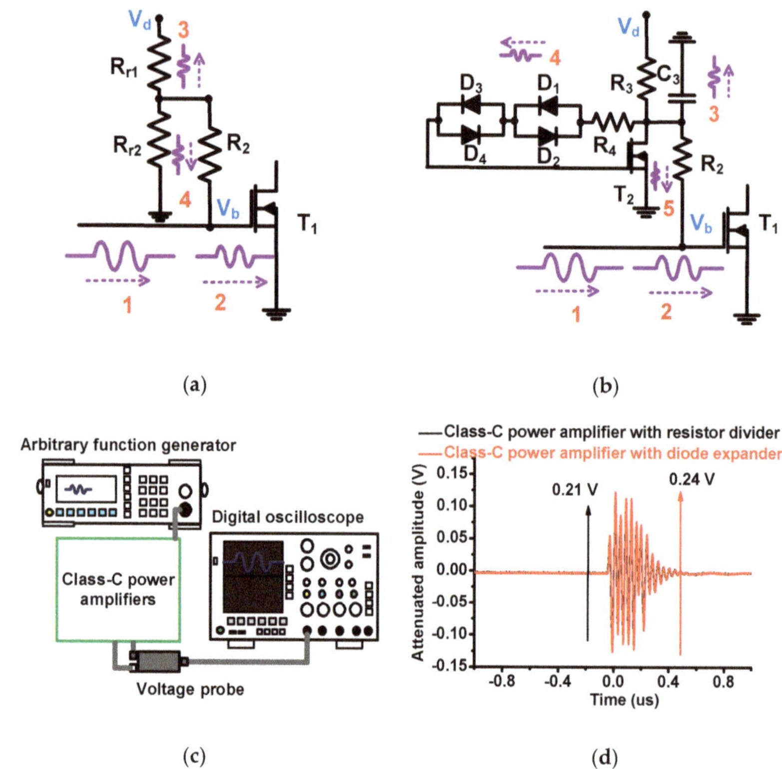

Figure 4. The AC operating mechanisms of the (**a**) resistor divider and (**b**) diode expander architecture; (**c**) measurement setup for input voltage attenuation in the class-C power amplifier; and the (**d**) measured AC input voltages at the gate of the LDMOSFET T_1.

The gain (V_G) and DC current consumption (P_D) of the class-C power amplifiers can be expressed [9,22] as:

$$V_G = 20 \cdot \mathrm{Log}_{10}\left(i_o \cdot \frac{R_L}{2\pi \cdot V_{in}}(2\theta_c - \sin 2\theta_c)\right) \tag{3}$$

$$I_D = \frac{i_o}{\pi}(\sin\theta_c - \theta_c\cos\theta_c) \tag{4}$$

Here, i_o, R_L, V_{in}, and θ_c are the output current, load resistance, input voltage, and conduction angle of the class-C power amplifiers.

The gain and DC current consumption equations for class-C power amplifiers with resistor divider and diode expander architecture coincide [22]. However, the output current (i_o) and conduction angles (θ_c) are different given that those values are dependent on the DC bias voltages [22]. The LDMOSFET T_1 cannot provide the expected parameter for low DC bias voltage in class-C-type power amplifiers. There are no simulation model parameters either for power film resistors (CADDOCK Electronics Inc., Roseburg, OR, USA) and LDMOSFET T_2 in the diode expander. Therefore, modeled equations of the gain and DC current consumption are provided without simulation data. The measured performance of the class-C power amplifier with resistor divider and diode expander architecture is characterized next.

3. Results and Discussion

To estimate the performance of the class-C power amplifier with a resistor divider and diode expander architecture, the measurement setup shown in Figure 5a for voltage gain versus input voltage of power amplifiers was employed. A 25 MHz five-cycle sinusoidal pulse signal with up to 5 V_{p-p} was generated from an arbitrary function generator (DG5701, Rigol Inc., Beijing, China) and applied to the designed class-C power amplifiers to measure the output peak-to-peak voltage amplitude in a digital oscilloscope (MDO4104C). One DC power supply (2231A-3-30, Keithley Instruments, Cleveland, OH, USA) provided gate-source DC voltage and another power supply (E3647A, Agilent Technologies, Santa Clara, CA, USA) provided drain-source DC voltage for the class-C power amplifiers. The voltage gain in each case was obtained by dividing the measured output peak-to-peak voltage by the corresponding measured input peak-to-peak voltage. The pulse-echo mode is a standard method to evaluate the performance of the developed electronic components or ultrasonic transducers [23]. In the pulse-echo mode, the voltage amplitude of the echo signal generated by ultrasonic transducers is typically measured in terms of peak-to-peak voltage [24]. Therefore, voltage gain was measured through peak-to-peak voltages for ultrasound systems.

Figure 5b,c show the measured voltage gain versus input voltages of a class-C power amplifier with resistor divider and diode expander architecture. The voltage gain of the amplifier with diode expander architecture (14.96 dB) was higher than that of the amplifier with resistor architecture (12.04 dB) at 5 V_{p-p} input. Additionally, the gain deviation of the amplifier with diode expander architecture (16.17 dB) was lower than that of the amplifier with resistor architecture (24.43 dB) at 5 V_{p-p} input. Therefore, the class-C power amplifier with a diode expander architecture exhibited higher voltage gain compared to that with a resistor divider architecture. Figure 5e,f show the measured voltage gain versus frequencies of the class-C power amplifier with the resistor divider and diode expander architecture. The voltage gain of the class-C power amplifier with the diode expander architecture (1.81 dB) was higher than that of the amplifier with the resistor architecture (−52.39 dB) at 50 MHz. Additionally, gain deviation of the amplifier with diode expander architecture (−0.38 dB) is lower than that of the amplifier with resistor architecture (−20.69 dB) at 50 MHz. Therefore, the class-C power amplifier with the diode expander architecture exhibited a relatively wider bandwidth compared to that with the resistor divider architecture.

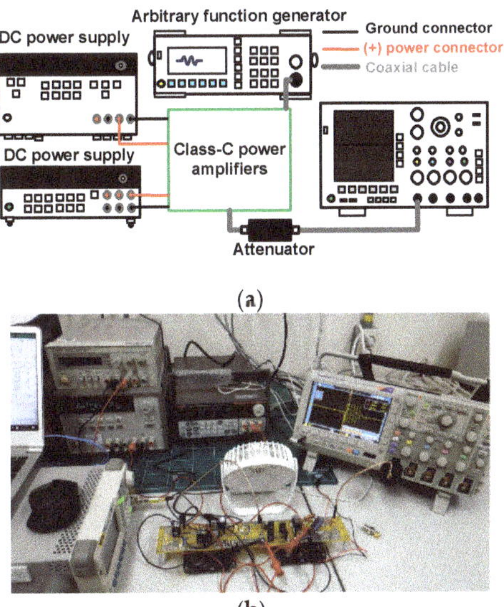

(a)

(b)

Figure 5. *Cont.*

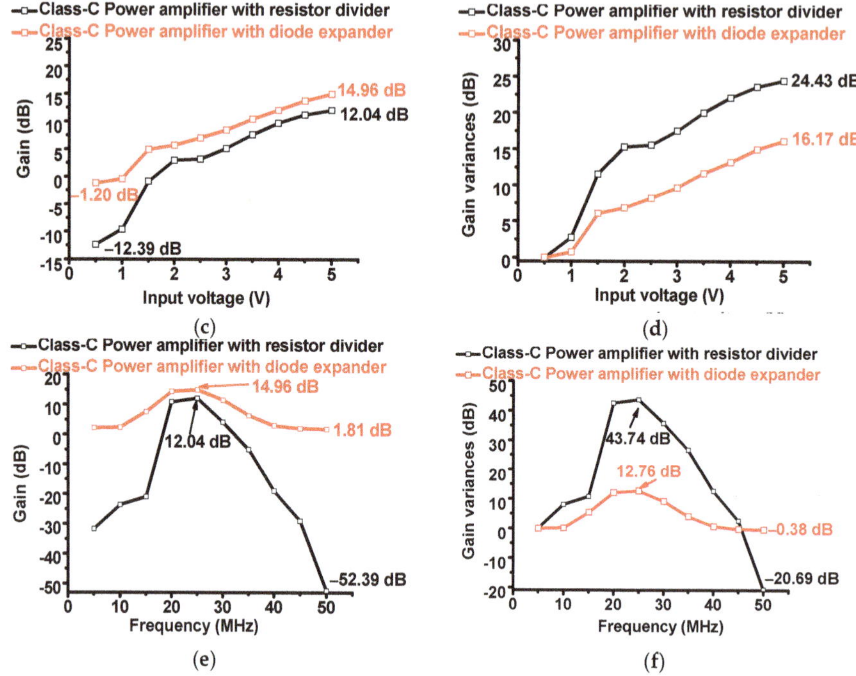

Figure 5. (**a**) Measurement setup sketch and (**b**) an image of its actual realization for measurement of voltage gain and DC current consumption of the power amplifiers; measured voltage gain of the class-C power amplifier versus input voltage with the (**c**) resistor divider and diode expander architecture; (**d**) measured voltage gain deviation of the class-C power amplifier versus input voltage with resistor divider and diode expander architecture; measured voltage gain of the class-C power amplifier versus frequency with the (**e**) resistor divider and (**f**) diode expander architecture.

Measured data for DC current consumption versus input voltages and frequencies of the class-C power amplifiers with the resistor divider and diode expander architecture are depicted in Figure 6a,b. In Figure 6a, a 25 MHz sinusoidal pulse signal with up to 5 V_{p-p} was applied to the designed class-C power amplifier to measure the current consumption. In Figure 6b, a 5 V_{p-p} sinusoidal pulse signal up to 50 MHz was applied to the designed class-C power amplifiers to measure the current consumption. The measured current consumption of the class-C power amplifier with resistor diode and diode expander architecture was less than 1 A. The measured current consumption of the amplifier with the diode expander architecture (0.52 A) was a bit higher than that of the amplifier with the resistor divider architecture (0.36 A). Therefore, both class-C power amplifiers still generated low current consumption such that it can produce relatively low heat, resulting in lower battery consumption for point-of-care ultrasound systems.

The currently designed class-C power amplifier with the diode expander architecture had a reasonable gain between 15 MHz and 35 MHz. To re-design the amplifier to cover the bandwidth between 1 MHz and 25 MHz, the input and output matching circuits need to be removed and series capacitors may be used. For such a case, the bandwidth between 1 MHz and 25 MHz could be covered in the class-C power amplifiers. However, the voltage gain could be decreased or increased as the frequency increased. To flat the bandwidth, the linearizer circuit would be helpful [25]. Therefore, the diode expander architecture needs to be replaced with another linearizer circuit. Therefore, each architecture has some trade-offs in terms of performance, such as voltage gain and bandwidth.

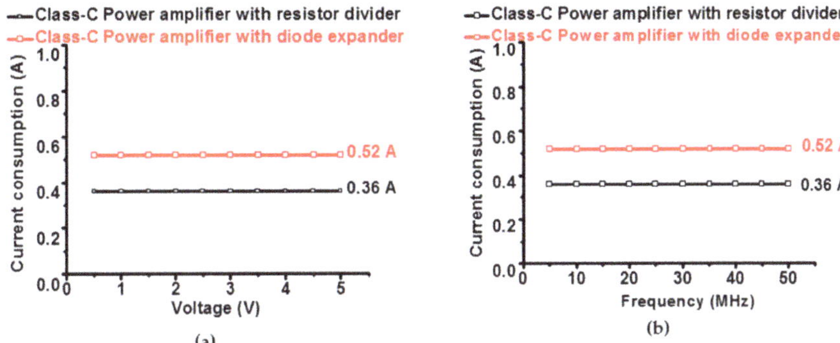

Figure 6. (a) Current consumptions of the class-C power amplifier versus input voltage and (b) frequency with the resistor divider and diode expander architecture.

There is more research that needs to be done for the use of higher frequencies for a variety of applications, such as skin, eye, and intravascular imaging [23,26]. The future direction of point-of-care ultrasound systems could utilize a frequency range higher than 25 MHz, since current commercial bench-top ultrasound machines are already doing so [2]. Higher frequency operation in ultrasound machines will be more demanding due to the higher spatial resolution while scarifying the penetration depth [2].

Figure 7a shows a typical setup for pulse-echo response measurement using class-C power amplifiers with a resistor divider and diode expander architecture [18]. A 25 MHz five-cycle 5 V_{p-p} sinusoidal waveform from a generator (DG5701) was the input of the designed class-C power amplifiers. This input triggered the ultrasonic transducer (V324, Olympus Scientific Solutions Americas Inc., Waltham, MA, USA). Ultrasonic transducers transmitted acoustic waveforms into a circular quartz target. The reflected weak acoustic waveforms were converted into electrical waveforms by ultrasonic transducers. The resulting waveforms were then amplified by a 36 dB preamplifier (AU-1526, L3 Narda-MITEQ Inc., Hauppauge, NY, USA) to be displayed on a 1 GHz digital oscilloscope with a 5 GS/s sampling rate (MDO3102). Figure 7b,c show echo signal amplitudes and their normalized spectrum when using the class-C power amplifier with a resistor divider and diode expander architecture. The echo signal amplitude is related with the sensitivity of ultrasonic transducers [27–29]. In Figure 7b, echo signal amplitudes when using the class-C power amplifiers with resistor divider (2.51 V_{p-p}) is a little bit smaller than that when using the amplifiers with diode expander architecture (2.98 V_{p-p}). The −6 dB bandwidth is related with the image resolution of ultrasonic transducers in point-of-care ultrasound systems [30–32]. In Figure 7c, the echo signal spectrum when using class-C power amplifiers with a resistor divider (17.51%) was smaller than that when using the amplifiers with a diode expander architecture (18.52%). Therefore, the class-C power amplifier with a diode expander architecture outperformed the class-C power amplifier with a resistor divider architecture when it came to improving the sensitivity of point-of-care ultrasound systems.

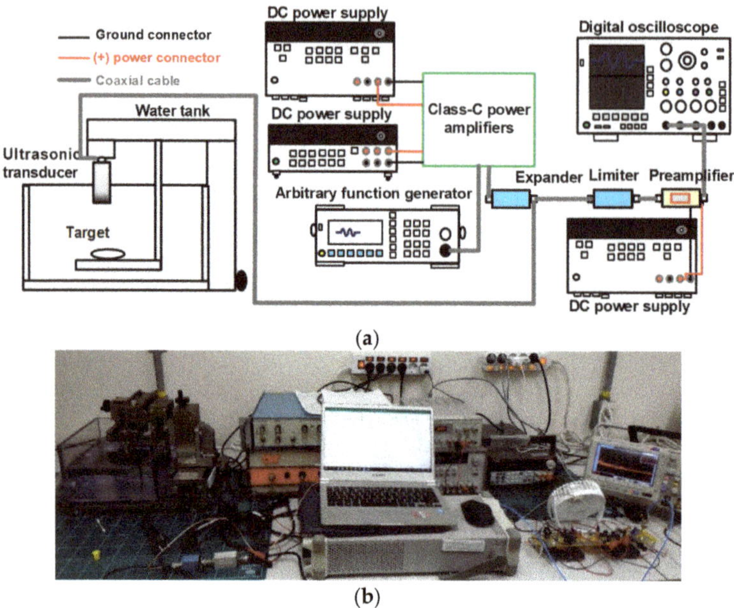

(a)

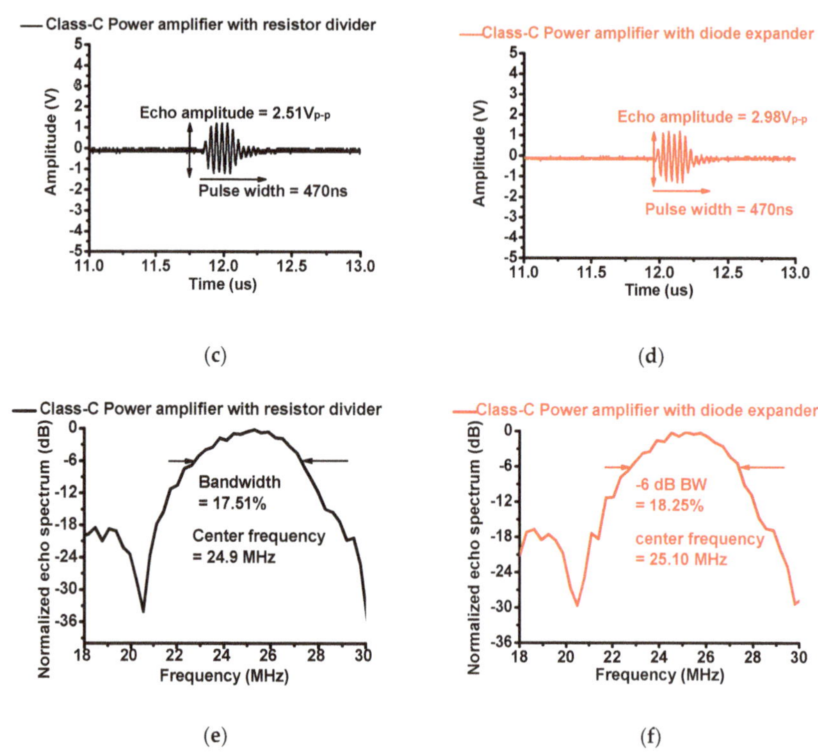

(b)

(c)
(d)

(e)
(f)

Figure 7. (**a**) Schematic diagram and (**b**) pictures during pulse-echo response measurement; echo signal amplitudes when using the class-C power amplifier with a (**c**) resistor divider and (**d**) diode expander architecture; echo signal spectrums when using the class-C power amplifier with a (**e**) resistor divider and (**f**) diode expander architecture.

4. Conclusions

In point-of-care ultrasound systems, excessive overheating critically reduces the performance of piezoelectric transducers because class-A power amplifiers in ultrasound transmitters generate unwanted heat during the entire operation. To reduce excessive overheating, non-linear power amplifiers such as class-C power amplifiers can be used given that these amplifiers have on and off transition periods during its entire operation. Because of pulse signals affecting bias circuits, class-C power amplifiers could be affected by bias circuits, resulting in the generation of sensitive outputs for ultrasonic transducers. Therefore, a new diode expander architecture dedicated to improving input signal conditions for class-C power amplifiers was proposed. The diode expander architecture could reduce the effects of unwanted input pulse signals toward bias circuits, thus reducing the attenuation of the input pulse signals for class-C power amplifiers. As a result, higher input signals could be transferred to the class-C power amplifiers. We have shown that the gain of a class-C power amplifier with a diode expander (14.96 dB) was higher than that with a resistor diode expander (12.04 dB) for a 5 V_{p-p} input. However, the current consumption of a class-C power amplifier with a diode expander architecture (1.02 W) was a little bit higher than that with a resistor divider architecture (0.75 W).

To confirm the proposed idea, typical one-way pulse-echo response measurements were taken. The echo signal amplitude and its −6 dB bandwidth when using a class-C power amplifier with diode expander architecture (2.98 V_{p-p} and 18.25%) was higher and wider than those of a class-C power amplifier with resistor divider architecture (2.51 V_{p-p} and 17.51%). The limitation of the developed class-C power amplifier with a resistor divider architecture made it difficult to block the unwanted pulse input signals. The limitation of the developed class-C power amplifier with a diode expander architecture is that the maximum voltage of the LDMOSFET in the diode expander needs to be much higher than the power supply with unwanted pulse signals. However, a class-C power amplifier with diode expander architecture can be a useful way to improve the output voltage amplitude. In the future, the developed architecture combined with a multiplexer/de-multiplexer will be applied to the array transducers because the sensitivity in point-of-care ultrasound systems is one of the critical performance issues.

Funding: This research was supported by the Basic Science Research Program through the National Research Foundation of Korea (NRF) funded by the Ministry of Science, ICT and Future Planning (NRF-2017R1C1B1003606).

Conflicts of Interest: The author declares no conflict of interest.

References

1. Shung, K.K. *Diagnostic Ultrasound: Imaging and Blood Flow Measurements*; Taylor Francis: Boca Raton, FL, USA, 2015.
2. Szabo, T.L. *Diagnostic Ultrasound Imaging: Inside Out*; Elsevier Academic Press: London, UK, 2013.
3. Daniels, J.M.; Hoppmann, R.A. *Practical Point-of-Care Medical Ultrasound*; Springer: New York, NY, USA, 2016.
4. Choi, H.; Choe, S.W. Therapeutic Effect Enhancement by Dual-bias High-voltage Circuit of Transmit Amplifier for Immersion Ultrasound Transducer Applications. *Sensors* **2018**, *18*, 4210. [CrossRef] [PubMed]
5. Choi, H. Prelinearized Class-B Power Amplifier for Piezoelectric Transducers and Portable Ultrasound Systems. *Sensors* **2019**, *19*, 287. [CrossRef] [PubMed]
6. Shin, S.-H.; Yoo, W.-S.; Choi, H. Development of Public Key Cryptographic Algorithm Using Matrix Pattern for Tele-Ultrasound Applications. *Mathematics* **2019**, *7*, 752. [CrossRef]
7. Zhu, B.; Fei, C.; Wang, C.; Zhu, Y.; Yang, X.; Zheng, H.; Zhou, Q.; Shung, K.K. Self-focused AlScN film ultrasound transducer for individual cell manipulation. *ACS Sens.* **2017**, *2*, 172–177. [CrossRef] [PubMed]
8. Choi, H.; Yoon, C.; Yeom, J.-Y. A Wideband High-Voltage Power Amplifier Post-Linearizer for Medical Ultrasound Transducers. *Appl. Sci.* **2017**, *7*, 354. [CrossRef]
9. Kazimierczuk, M.K. *RF Power Amplifier*; John Wiley Sons: Hoboken, NJ, USA, 2014.
10. Albulet, M. *RF Power Amplifiers*; SciTech Publishing: London, UK, 2001.

11. Yuan, T.; Dong, X.; Shekhani, H.; Li, C.; Maida, Y.; Tou, T.; Uchino, K. Driving an inductive piezoelectric transducer with class E inverter. *Sens. Actuators A* **2017**, *261*, 219–227. [CrossRef]
12. Mortimer, B.; Du Bruyn, T.; Davies, J.; Tapson, J. High power resonant tracking amplifier using admittance locking. *Ultrasonics* **2001**, *39*, 257–261. [CrossRef]
13. Dong, X.; Yuan, T.; Hu, M.; Shekhani, H.; Maida, Y.; Tou, T.; Uchino, K. Driving frequency optimization of a piezoelectric transducer and the power supply development. *Rev. Sci. Instrum.* **2016**, *87*, 105003. [CrossRef]
14. Choi, H.; Park, C.; Kim, J.; Jung, H. Bias-Voltage Stabilizer for HVHF Amplifiers in VHF Pulse-Echo Measurement Systems. *Sensors* **2017**, *17*, 2425. [CrossRef]
15. Kenington, P.B. *High Linearity RF Amplifier Design*; Artech House, Inc.: Norwood, MA, USA, 2000.
16. Vuolevi, J.; Rahkonen, T. *Distortion in RF Power Amplifiers*; Artech House: London, UK, 2003.
17. Choi, H.; Woo, P.C.; Yeom, J.-Y.; Yoon, C. Power MOSFET Linearizer of a High-Voltage Power Amplifier for High-Frequency Pulse-Echo Instrumentation. *Sensors* **2017**, *17*, 764. [CrossRef]
18. Jeong, J.J.; Choi, H. An impedance measurement system for piezoelectric array element transducers. *Measurement* **2017**, *97*, 138–144. [CrossRef]
19. Minasian, R.A. Power MOSFET dynamic large-signal model. *IEE Proc. I Solid-State Electron Device* **1983**, *130*, 73–79. [CrossRef]
20. Rohde, U.L.; Rudolph, M. *RF/Microwave Circuit Design for Wireless Applications*; John Wiley Sons: Hoboken, NJ, USA, 2013.
21. Razavi, B. *RF Microelectronics*; Prentice Hall: Upper Saddel River, NJ, USA, 2011.
22. Lee, T.H. *The Design of CMOS Radio-Frequency Integrated Circuits*; Cambridge University Press: Cambridge, UK, 2006.
23. Zhou, Q.; Lam, K.H.; Zheng, H.; Qiu, W.; Shung, K.K. Piezoelectric single crystal ultrasonic transducers for biomedical applications. *Prog. Mater. Sci.* **2014**, *66*, 87–111. [CrossRef] [PubMed]
24. Choi, H.; Yeom, J.Y.; Ryu, J.M. Development of a Multiwavelength Visible-Range-Supported Opto–Ultrasound Instrument Using a Light-Emitting Diode and Ultrasound Transducer. *Sensors* **2018**, *18*, 3324. [CrossRef] [PubMed]
25. Choi, H.; Jung, H.; Shung, K.K. Power Amplifier Linearizer for High Frequency Medical Ultrasound Applications. *J. Med. Biol. Eng.* **2015**, *35*, 226–235. [CrossRef] [PubMed]
26. Wang, X.; Seetohul, V.; Chen, R.; Zhang, Z.; Qian, M.; Shi, Z.; Yang, G.; Mu, P.; Wang, C.; Huang, Z. Development of a mechanical scanning device with high-frequency ultrasound transducer for ultrasonic capsule endoscopy. *IEEE Trans. Med. Imaging* **2017**, *36*, 1922–1929. [CrossRef] [PubMed]
27. Choi, H.; Choe, S.W. Acoustic Stimulation by Shunt-Diode Pre-Linearizer Using Very High Frequency Piezoelectric Transducer for Cancer Therapeutics. *Sensors* **2019**, *19*, 357. [CrossRef] [PubMed]
28. Qiao, Y.; Zhang, X.; Zhang, G. Acoustic radiation force on a fluid cylindrical particle immersed in water near an impedance boundary. *J. Acoust. Soc. Am.* **2017**, *141*, 4633–4641. [CrossRef]
29. Choi, H.; Ryu, J.; Kim, J. A Novel Fisheye-Lens-Based Photoacoustic System. *Sensors* **2016**, *16*, 2185. [CrossRef]
30. Choi, H. Class-C Linearized Amplifier for Portable Ultrasound Instruments. *Sensors* **2019**, *19*, 898. [CrossRef]
31. Choe, S.W.; Choi, H. Suppression Technique of HeLa Cell Proliferation Using Ultrasonic Power Amplifiers Integrated with a Series-Diode Linearizer. *Sensors* **2018**, *18*, 4248. [CrossRef] [PubMed]
32. Choi, H.; Ryu, J.M.; Yeom, J.Y. Development of a Double-Gauss Lens Based Setup for Optoacoustic Applications. *Sensors* **2017**, *17*, 496. [CrossRef] [PubMed]

Article

A Modified Duhem Model for Rate-Dependent Hysteresis Behaviors

Jinqiang Gan, Zhen Mei, Xiaoli Chen, Ye Zhou and Ming-Feng Ge *

School of Mechanical Engineering and Electronic Information, China University of Geosciences, Wuhan 430074, China; ganjq@cug.edu.cn (J.G.); mz@cug.edu.cn (Z.M.); gysjdd@163.com (X.C.); yezhou72@163.com (Y.Z.)
* Correspondence: fmgabc@163.com; Tel.: +86-027-6788-3273

Received: 18 September 2019; Accepted: 7 October 2019; Published: 9 October 2019

Abstract: Hysteresis behaviors are inherent characteristics of piezoelectric ceramic actuators. The classical Duhem model (CDM) as a popular hysteresis model has been widely used, but cannot precisely describe rate-dependent hysteresis behaviors at high-frequency and high-amplitude excitations. To describe such behaviors more precisely, this paper presents a modified Duhem model (MDM) by introducing trigonometric functions based on the analysis of the existing experimental data. The MDM parameters are also identified by using the nonlinear least squares method. Six groups of experiments with different frequencies or amplitudes are conducted to evaluate the MDM performance. The research results demonstrate that the MDM can more precisely characterize the rate-dependent hysteresis behaviors comparing with the CDM at high-frequency and high-amplitude excitations.

Keywords: piezoelectric ceramic materials; Duhem model; hysteresis model

---

1. Introduction

Piezoelectric ceramic, a new type of functional material, plays an important role in many real-world applications due to its superior performance in converting electrical energy into mechanical energy [1–4]. Piezoelectric ceramic actuators (PCAs) are ones of the most important applications of the piezoelectric ceramic material owing to their small size, high accuracy, and fast response. Hysteresis behavior is an inherent characteristic of PCAs and has already become a bottleneck in developing the applications of the PCAs. Therefore, it is of great significance to develop more precise hysteresis models for characterizing hysteresis behaviors.

The existing models of PCAs can be generally divided into the rate-independent and rate-dependent hysteresis models [5]. The rate-independent hysteresis models include the Preisach model [6,7], Prandtle–Ishlinskii model [8,9], Maxwell-slip model [10,11] and polynomial-based hysteresis model [12,13], and can be used to describe the nonlinear relationship between the input voltage and the output displacement of PCAs. However, the input rates of these models are generally lower than that of the rate-dependent hysteresis models, such as the Bouc–Wen model [14,15] and Dahl model [16–18]. That is mainly because, unlike the rate-independent hysteresis model, the rate-dependent one can describe the dynamic relationship between the input rate and the output. However, the existing rate-dependent hysteresis models generally have low prediction precision and high complexity of model equations. Therefore, how to construct a new, simpler rate-dependent hysteresis model is an urgent and challenging issue of great significance.

Due to its differential equations, the Duhem model has been used to describe and compensate piezoelectric hysteresis behaviors [19–22]. For example, C.-J. Lin and P.-T. Lin [23] combined the Bouc–Wen model, Dahl model and Duhem model as a modified Duhem model and presented a feedforward controller. Wang et al. [24] identified the Duhem model by neural network methods and designed a robust adaptive controller to compensate hysteresis behaviors. Xie et al. [25] presented an observer-based adaptive controller based on the Duhem model for piezoelectric actuators.

The classical Duhem model only characterizes symmetrical hysteresis loops while the actual hysteresis loops of piezoelectric actuators are non-symmetrical. It is worth mentioning the fact that the higher the frequency or the amplitude of the input excitation is, the more serious the hysteresis behaviors are [26,27]. When the frequency or amplitude of input excitation signal is increasing, the non-symmetrical of hysteresis loops is more serious. Therefore, the classical Duhem model already cannot precisely describe rate-dependent hysteresis behaviors at high-frequency and high-amplitude excitations. Thus, Oh and Bernstein [28] proposed the rate-independent and rate-dependent semilinear Duhem models without analyzing the modeling errors in detail at high-frequency and high-amplitude excitations by using a complex model. So far, few efforts have been devoted to developing new hysteresis models based on Duhem model.

Motivated by the aforementioned discussions, this paper proposes a modified Duhem model to describe rate-dependent hysteresis behaviors by introducing trigonometric functions. The proposed model has a simple expression and can detailly characterize rate-dependent hysteresis behaviors precisely at high-frequencies and high-amplitude input excitations. The parameters of models can be easily identified by the nonlinear least squares method. The validity of the proposed model is demonstrated via simulation experiments. The rest of this article is organized as follows: In Section 2, the hysteresis system is constructed to introduce the expression of the classical Duhem model. Section 3 introduces the proposed model and the identification of corresponding parameters. Section 4 aims to verify the validity of the established model, and compares it with the classical Duhem model. The results of the analysis are obtained immediately. The conclusion of this paper is placed in Section 5.

2. Classical Duhem Model (CDM)

In 1986, Coleman and Hodgdon [29] proposed a hysteresis model for ferromagnetic materials, which describes the relationship between the magnetic field $H(t)$ and magnetic flux $B(t)$ as follows:

$$\dot{B}(t) = \alpha \left| \dot{H}(t) \right| \cdot [\beta(H(t)) - B(t)] + \gamma \dot{H}(t) \tag{1}$$

where α, β and γ are the parameters controlling the shape and size of the hysteresis loop. According to the relationship between single-input and single-output [28], the hysteresis system is given by

$$\dot{x}(t) = f(x(t), u(t), \dot{u}(t)), x(0) = x_0, t \geq 0, \tag{2}$$

$$y(t) = h(x(t), u(t)) \tag{3}$$

where $u(t)$ is the input, $y(t)$ is the output and $x(t)$ is a part of it. When this hysteresis model is used to describe hysteresis system of PCAs, the classical Duhem model (CDM) is proposed and expressed as follows:

$$\begin{cases} Y(t) = X(t) - h(t) \\ X(t) = k \cdot u(t) \\ \dot{h}(t) = \alpha \dot{u}(t) - \beta \left| \dot{u}(t) \right| h(t) + \gamma \left| \dot{u}(t) \right| u(t) \end{cases} \tag{4}$$

where $X(t)$ is the linear component and $h(t)$ is the hysteretic component, $u(t)$ is the input voltage and $\dot{u}(t)$ is the derivative of voltage, and k, α, β and γ are the model parameters.

To evaluate the performance of the CDM, two groups of experiments were conducted. First of all, a sinusoidal input signal $u_a(t) = \sin(2\pi \cdot t) + 1$ with the frequency of 1 Hz was taken as a reference signal to identify the CDM parameters. The corresponding CDM parameters were identified by utilizing the nonlinear least squares method as $k = 0.261$, $\alpha = 0.062$, $\beta = 0.131$ and $\gamma = 0.001$. The first group of experiments, called Exp-a, adopted an input excitation signal $u_a(t) = 8\sin(2\pi \cdot f_1 t) + 6\sin(2\pi \cdot f_2 t) + 14$ with $f_1 = 15$ Hz and $f_2 = 40$ Hz. Figure 1 shows the corresponding comparison between the experimental and simulation results. The results reveal that the error of one point of the CDM (the dotted line) is nearly 2 μm (20% of the displacement range). The maximum modeling error is about 2 μm, which is undoubtedly big. In the second group of experiments, called Exp-b, an input excitation signal

$u_a(t) = 20\sin(2\pi \cdot t) + 20$ was used. Figure 2 shows the corresponding comparison between the experimental and simulation results of the CDM. The maximum error is about 0.5 µm. It should also be noticed that the point with the maximum modeling error is the point with $\dot{u}(t) \to 0$. These actual experimental results demonstrate that the CDM cannot precisely describe the rate-dependent hysteresis behaviors at high-frequency and high-amplitude excitation signals.

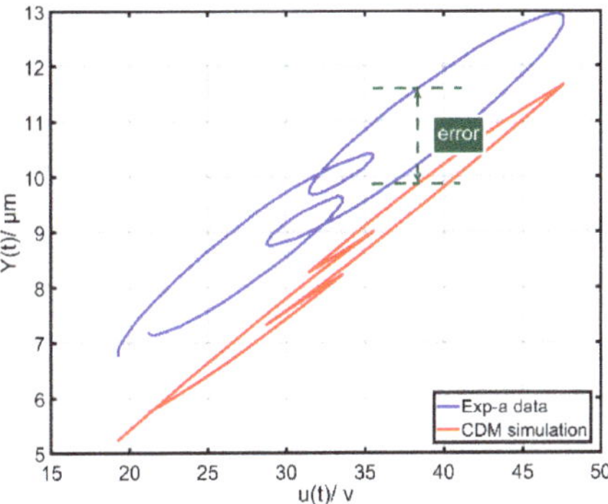

Figure 1. Comparison of the output displacements between experimental data (Exp-a) and simulation of the classical Duhem model (CDM).

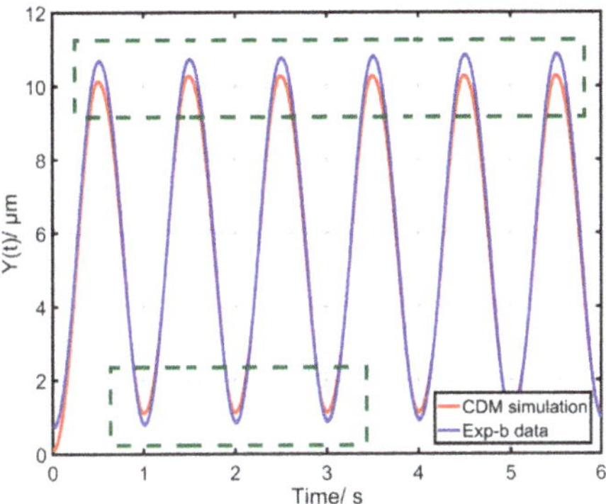

Figure 2. Comparison of the output displacements between experimental data (Exp-b) and simulation of CDM.

3. Modified Duhem Model (MDM)

The main components of CDM are linear component $X(t)$ and hysteretic component $h(t)$, the former having large influence on the output. Thus, the optimization based on the linear component $X(t)$ is an important research hotspot [30,31]. It should be noted that the main structures of CDM are kept, which can still describe the fundamental characteristics of hysteresis behaviors. With respect to the direct relationship between the input and output, there is an important variation $\dot{u}(t)$, which has large influences on the whole model when the input frequency is high. The special points $\dot{u}(t) = 0$ are the demarcation points where the input voltage curves go up and down. These important points decide the final shape of hysteresis loops, which has been demonstrated in the previous literature [12,13]. The CDM only characterizes symmetrical hysteresis loops while the actual hysteresis loops of piezoelectric actuators are non-symmetrical. When the frequency or amplitude of input excitation signal is increasing, the non-symmetrical of hysteresis loops is more serious, Therefore, the corresponding errors of the CDM are higher at high-frequency excitations, especially the special points

$\dot{u}(t) = 0$. Figure 2 also demonstrates that the points with $\dot{u}(t) \to 0$ have bigger errors. The actual output displacement of PCA varies little when the value of $\dot{u}$ varies greatly at high-frequency excitations. The trigonometric function also has the similar characteristics. Its output varies little when the input varies greatly. Therefore, the trigonometric function as a periodic function has the special points where their derivatives are zero, which can be easily used to compensate the bigger errors of the special points $\dot{u}(t) = 0$. Furthermore, it has a simple expression. Thus, it is a good try to introduce the trigonometric function based on the CDM. Lastly, a modified Duhem model (MDM) based on CDM is proposed and expressed as follows:

$$\begin{cases} Y(t) = X(t) - h(t) \\ X(t) = k \cdot u(t) + p \cdot u(t) \cdot \cos\big[|\dot{u}(t)|\big] + q \cdot \dot{u}(t) \\ \dot{h}(t) = \alpha \cdot \dot{u}(t) - \beta \cdot |\dot{u}(t)| \cdot h(t) + \gamma \cdot u(t) \cdot |\dot{u}(t)| + \varepsilon \cdot \dot{u}(t) \cdot \sin\big[|\dot{u}(t)|\big] \end{cases} \tag{5}$$

where p, q, ε, k, α, β and γ are constants. It must be noticed that when $\dot{u}(t) \to 0$, there is $\cos\big[|\dot{u}(t)|\big] \to 1$ and $p \cdot u(t) \cdot \cos\big[|\dot{u}(t)|\big] \to p \cdot u(t)$, which can be perfectly used to compensate for the bigger errors of the special points $\dot{u}(t) \to 0$.

Over the past decade, several methods for parameters identification of models [19,24,32] have been developed, but their identification processes are generally complex. In our previous work [15,33], the nonlinear least squares method is proposed to identify Bouc–Wen model. The nonlinear least squares method adopts the trust-region-reflective algorithm and take the nonlinear least squares function for optimization through the MATLAB/Simulink Optimization Toolbox. Compared with the previous methods, the method is much simpler and can be more easily applied to identify other models. In this paper, the nonlinear least squares method is adopted to identify the MDM and CDM. The objective function F is defined as follows:

$$F = Min \sum_{i=1}^{n} f^2(u) \tag{6}$$

$$f(u) = Y_i - Y_i^{HM} \tag{7}$$

In CDM, there is

$$\begin{cases} Y_i^{HM}(iT) = X(iT) - h(iT) \\ X(iT) = k \cdot u(iT) \\ \dot{h}(iT) = \alpha \cdot \dot{u}(iT) - \beta \cdot |\dot{u}(iT)| \cdot h(iT) + \gamma \cdot u(iT) \cdot |\dot{u}(iT)| \end{cases} \tag{8}$$

In MDM, there is

$$\begin{cases} Y_i^{HM}(iT) = X(iT) - h(iT) \\ X(iT) = k \cdot u(iT) + p \cdot u(iT) \cdot \cos\big[|\dot{u}(iT)|\big] + q \cdot \dot{u}(iT) \\ \dot{h}(iT) = \alpha \cdot \dot{u}(iT) - \beta \cdot |\dot{u}(iT)| \cdot h(iT) + \gamma \cdot u(iT) \cdot |\dot{u}(iT)| + \varepsilon \cdot \dot{u}(iT) \cdot \sin\big[|\dot{u}(iT)|\big] \end{cases} \tag{9}$$

where $i = 1, 2, 3, \cdots, n$ is the number of sample experiments, T is the period of a sample, Y_i is the i-th output displacement of the PCAs obtained from experiments, and Y_i^{HM} is the i-th output simulated by the hysteresis model. The corresponding identification steps of the nonlinear least squares method were carried out offline as follows:

(1) Data collection: Experimental data including output displacements and input voltages for piezoelectric actuators were obtained and recorded.

(2) Model implementation: Classical and modified Duhem models were implemented using the MATLAB/Simulink blocks as shown in Figures 3 and 4, respectively. In these figures, block In1 represents the input voltage $u(iT)$ and block Out1 represents the output displacement

$Y_i^{HM}(iT)$ predicted by the CDM or MDM. Equations (8) and (9) are expressed using MATLAB/Simulink blocks.

(3) Parameter estimation: The trust-region-reflective algorithm was used to identify the parameters of hysteresis models based on experimental data.

(4) Validation: Comparison of the measured and simulation results predicted by hysteresis models were shown, and the corresponding modeling errors were obtained.

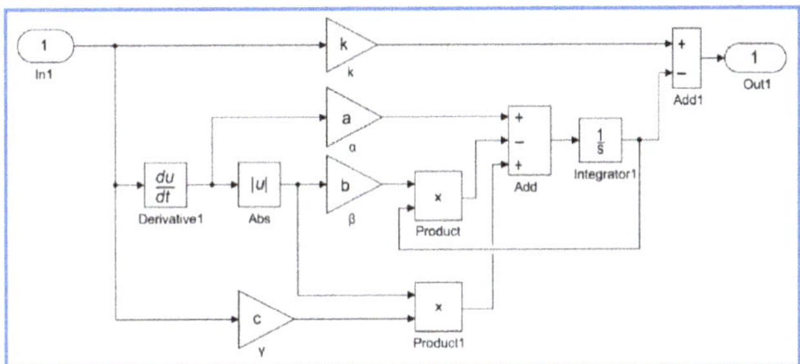

Figure 3. Classical Duhem model implemented with Matlab/Simulink.

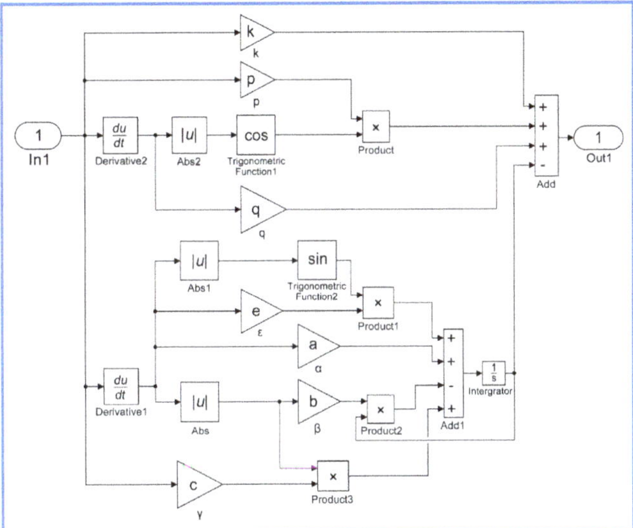

Figure 4. Modified Duhem model implemented with Matlab/Simulink.

4. Experimental Results

4.1. Experiment Setup

To demonstrate the validity of the proposed model, there are some groups of experiments conducted. Figure 5 shows the experimental setup, where a 1-DOF compliant mechanism stage was actuated by a stack piezoelectric ceramic actuator (PST 150/7/60VS12, Coremorrow, Harbin, China). Its nominal displacement was 60 μm for the maximum input voltage of 150 V. This piezoelectric ceramic actuator was made of PZT (Pb-based Lanthanum-doped Zirconate Titanates), whose detailed information is shown in Table 1. The strain gauge position sensor (SGS) included in the piezoelectric ceramic actuator measured the output displacement. dSPACE-DS1104 rapid prototyping controller board equipped with a 16-bit analogue-to-digital converter (ADC) and 16-bit digital-to-analogue converter (DAC) was used to control this 1-DOF compliant mechanism. In addition, an XE-500 controller (Coremorrow, Harbin, China) equipped with an amplifier with 15 times and a signal conditioner was also adopted. A computer with Control Desk 5.0-dSPACE and MATLAB/Simulink was used to conduct all experiments and obtain all experimental data. The detailed experimental steps were carried out as follow:

(1) Building the experimental platform: Connecting all the devices as shown in Figure 5;
(2) Model implementation: Constructing the input excitation signals and designing the control system for the piezoelectric ceramic actuators by using the MATLAB/Simulink blocks;
(3) Experiment start: Actuating piezoelectric ceramic actuators by using control desk 5.0-dSPACE;
(4) Data collection: Obtaining and recording experimental data including output displacements and input voltages for piezoelectric actuators.

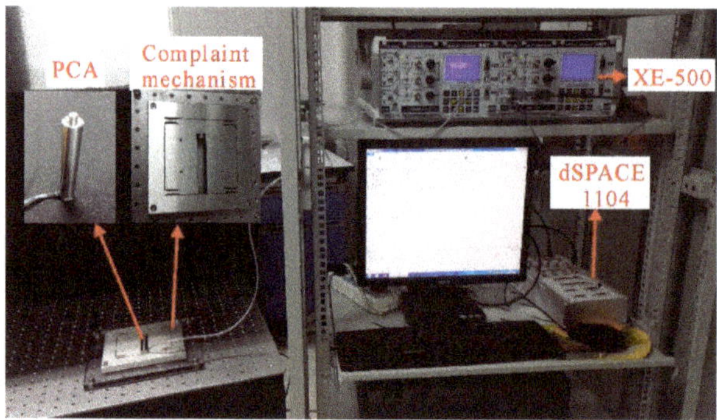

Figure 5. Experimental setup.

Table 1. Information about the piezoelectric ceramics actuator (PCA).

Material	PZT
Nominal stroke (μm) ±15%	60
Stiffness (N/μm) ±20%	15
Length (mm) ±0.3	64
Nominal thrust/tension (N)	1800/300
Electrical capacitance (μF) ±20%	5.4
Resonant frequency (kHz)	15
Stiffness (N/μm) ±20%	15

4.2. Experiment Results and Discussion

In order to evaluate the MDM performance comprehensively, six groups of experiments with different frequencies or amplitudes were conducted. The first and second groups of experiments adopted high-frequency and high-amplitude excitation signals. These two groups had the same amplitudes, but different frequencies. The third and fourth groups adopted low-frequency and low-amplitude excitation signals. The above four groups used multi-frequency signals. The fifth and sixth groups adopted high-frequency and low-amplitude excitation signals. The last two groups used single-frequency signals. The piezoelectric actuator used in our experiments was made of encapsulated stacked piezoelectric ceramics. It must be noted that its input frequency was controlled under 150 Hz to avoid a high dynamic force for security protection. The maximum input voltage of the actuator was 150 V. In the product manuals, it is recommended to control the input voltage within 120 V to guarantee the service life. Therefore, in the experiments, the six groups needed to follow this rule and its input frequencies and amplitudes were generally not high. Finally, the input frequencies including 1, 5, 10, 15, 20, 30, 40 and 50 Hz, and the amplitudes including 3, 5, 6, 8 and 10, were selected randomly following the rule above. In each group of experiments, the chosen excitation signal was used to actuate the PCA in the experimental setup and the experimental output displacements would be recorded and obtained. Subsequently, the displacements predicted by the CDM and MDM were obtained by using the MATLAB/Simulink. Lastly, the comparison results were obtained and drawn.

In the first group of experiments, the excitation signal was $u_1(t) = 10\sin(2\pi \cdot t) + 8\sin(2\pi \cdot 10t) + 6\sin(2\pi \cdot 20t) + 24$ with multi-frequency 1, 10 and 20 Hz. The excitation signal in the second group of experiments was $u_2(t) = 10\sin(2\pi \cdot t) + 8\sin(2\pi \cdot 15t) + 6\sin(2\pi \cdot 40t) + 24$ with multi-frequency 1, 15 and 40 Hz. The excitation signal in the third group of experiments was $u_3(t) = 10\sin(2\pi \cdot 5t) + 6\sin(2\pi \cdot 10t) + 16$ with multi-frequency 5 and 10 Hz. In the fourth group of experiments, the excitation signal was $u_4(t) = 5\sin(2\pi \cdot t) + 3\sin(2\pi \cdot 5t) + 8$ with multi-frequency 1 and 5 Hz. The fifth group of experiments took the excitation signal $u_5(t) = 5\sin(2\pi \cdot 30t) + 5$ at a frequency of 30 Hz to actuate the piezoelectric actuator. The last group of experiments took another excitation signal $u_6(t) = 5\sin(2\pi \cdot 50t) + 5$ at a frequency of 50 Hz.

To further demonstrate the effectiveness of the MDM, the CDM is set as a comparison. In addition, the modeling errors between the predicted output displacements of the two models and experimental displacements were drawn. In theory, any experimental signals can be used to identify the parameters of the MDM using the nonlinear least squares method. To get better prediction performances of hysteresis models including both the CDM and MDM, the more complex signals were generally adopted for identification, which is acceptable. In this work, the signals of both the first and second group of experiments, which were more complex than the others, were adopted to identify the parameters of the MDM and CDM. The detailed identified parameters of MDM and CDM are shown in Table 2.

Table 2. Identified parameters of CDM and modified Duhem model (MDM).

Parameters	CDM	MDM
k	0.39854	0.46992
α	0.18695	0.24599
β	0.049939	0.016074
γ	0.0056835	0.0027751
ε	\	−0.030389
p	\	−0.00072364
q	\	−0.00035258

Figure 6 shows the comparison of the experimental and simulation results of the CDM and MDM. Figure 6a gives the input voltage at each moment, Figure 6b shows the simulation and experimental results and Figure 6c presents the final modeling errors of the CDM and MDM. The blue dotted line represents the predicted results of the CDM. Meanwhile, the red solid line represents the predicted results of the MDM. It can clearly be seen that the simulation results of the MDM are closer to the experimental data. The modeling errors of the MDM are obviously smaller than that of the CDM. Figures 7–11 show the experimental results of the last five groups of experiments. These results further reveal that the MDM simulation results are much closer to the experimental output displacement than the CDM simulation results. The corresponding modeling errors of MDM are much smaller than that of the CDM. In addition, it should be also noted that though the second group of experiments has higher frequency and higher amplitude compared with the other three groups, the MDM still maintains better stability and accuracy compared with the CDM.

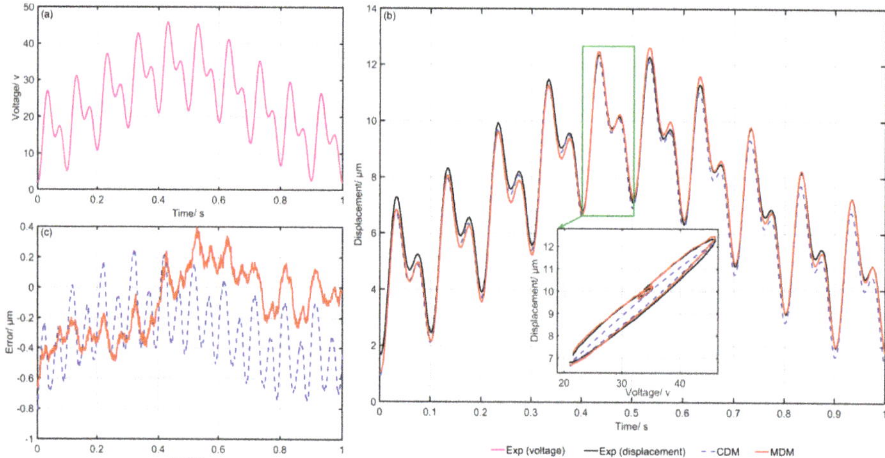

Figure 6. Exp1: Comparison of the experimental and simulation results of the CDM and MDM under $u_1(t) = 10\sin(2\pi \cdot t) + 8\sin(2\pi \cdot 10t) + 6\sin(2\pi \cdot 20t) + 24$: (**a**) Time histories of input voltage, (**b**) time histories of output displacements and (**c**) time histories of errors of the CDM and MDM.

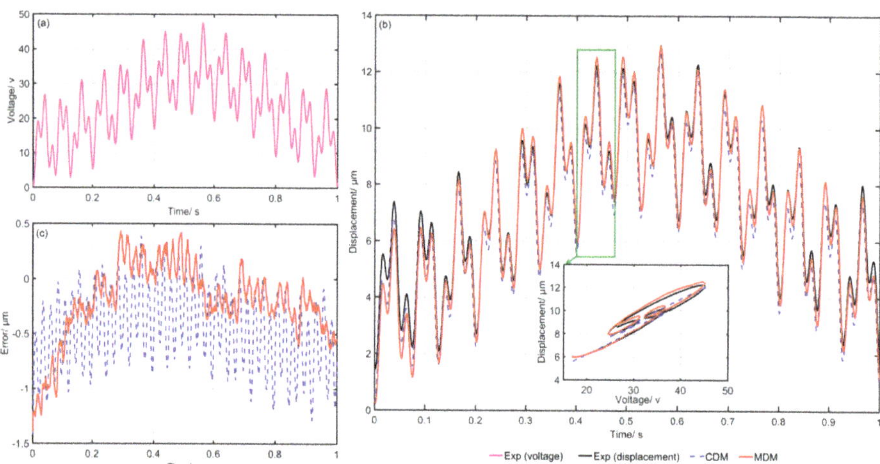

Figure 7. Exp2: Comparison of the experimental and simulation results of the CDM and MDM under $u_2(t) = 10\sin(2\pi \cdot t) + 8\sin(2\pi \cdot 15t) + 6\sin(2\pi \cdot 40t) + 24$: (**a**) Time histories of input voltage, (**b**) time histories of output displacements and (**c**) time histories of errors of the CDM and MDM.

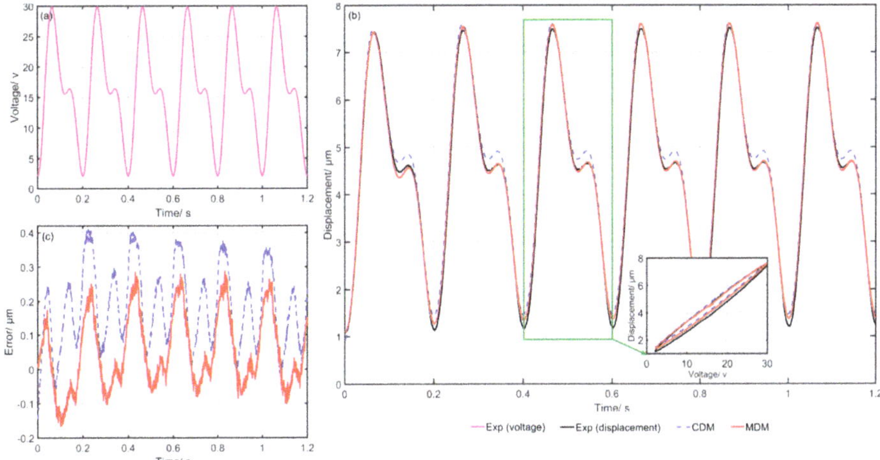

Figure 8. Exp3: Comparison of the experimental and simulation results of the CDM and MDM under $u_3(t) = 10\sin(2\pi \cdot 5t) + 6\sin(2\pi \cdot 10t) + 16$: (**a**) Time histories of input voltage, (**b**) time histories of output displacements and (**c**) time histories of errors of the CDM and MDM.

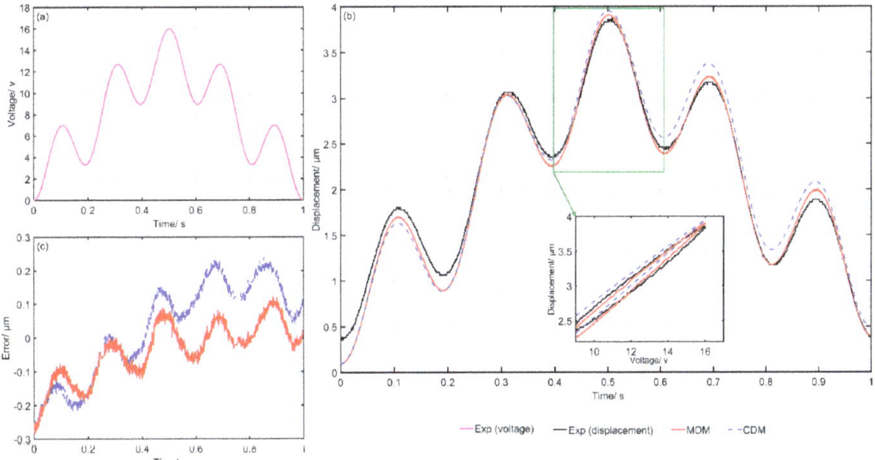

Figure 9. Exp4: Comparison of the experimental and simulation results of the CDM and MDM under $u_4(t) = 5\sin(2\pi \cdot t) + 3\sin(2\pi \cdot 5t) + 8$: (**a**) Time histories of input voltage, (**b**) time histories of output displacements and (**c**) time histories of errors of the CDM and MDM.

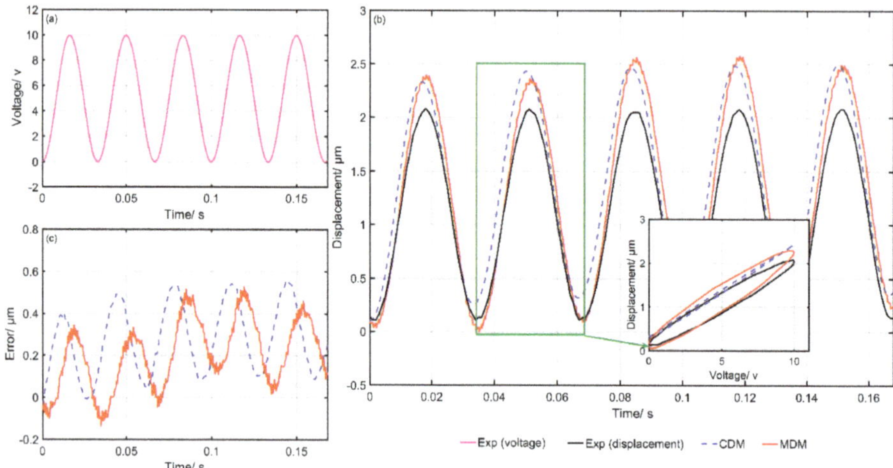

Figure 10. Exp5: Comparison of the experimental and simulation results of the CDM and MDM under $u_5(t) = 5\sin(2\pi \cdot 30t) + 5$: (**a**) Time histories of input voltage, (**b**) time histories of output displacements and (**c**) time histories of errors of the CDM and MDM.

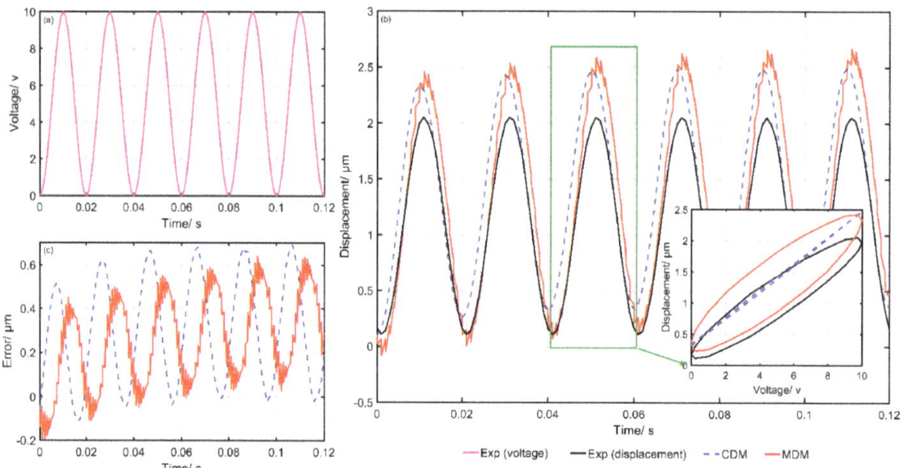

Figure 11. Exp6: Comparison of the experimental and simulation results of the CDM and MDM under $u_5(t) = 5\sin(2\pi \cdot 50t) + 5$: (**a**) Time histories of input voltage, (**b**) time histories of output displacements and (**c**) time histories of errors of the CDM and MDM.

To evaluate further the modeling performance of the MDM, the root mean square error E_{rms}, the relative root mean square error ξ and optimization ratio φ between the CDM and MDM were employed in comparing the errors of two models as follows:

$$E_{rms} = \sqrt{\frac{\sum\limits_{i=1}^{n} \left[Y_{\exp}(i) - Y_{pre}(i) \right]^2}{n}} \tag{10}$$

$$\xi = \frac{E_{rms}}{\max\left[Y_{\exp}(i)\right]} \times 100\% \tag{11}$$

$$\varphi = \frac{\left|E_{rms}^{CDM} - E_{rms}^{MDM}\right|}{E_{rms}^{CDM}} \times 100\% \tag{12}$$

where n is the total number of the sample and i is the i-th value in the sample, Y_{exp} is measured from experiments, Y_{pre} represents the displacements predicted by the hysteresis models, and E_{rms}^{CDM} and E_{rms}^{MDM} represent the root mean square error of the CDM and MDM, respectively. The details modeling errors are shown in Table 3.

Table 3. The simulation errors of the CDM and MDM.

Experiment	CDM		MDM		Optimization Ratio
	E_{rms} (μm)	ξ	E_{rms} (μm)	ξ	φ (%)
Exp1: $u_1(t)$	0.3789	3.1	0.2467	1.96	34.89
Exp2: $u_2(t)$	0.5845	4.6	0.4008	3.1	31.43
Exp3: $u_3(t)$	0.2166	2.84	0.1187	1.56	45.20
Exp4: $u_4(t)$	0.1426	3.6	0.0916	1.2	35.76
Exp5: $u_5(t)$	0.3313	15.94	0.2899	13.95	12.50
Exp6: $u_6(t)$	0.3868	18.86	0.3325	16.21	14.04

It can be seen from Table 3 that in the fifth group of experiments (Exp5) at the single-frequency of 30 Hz, E_{rms} and ξ of the MDM were 0.2899 μm and 13.95%,respectively, while those of the CDM were 0.3313 μm and 15.94%, respectively. Compared with the CDM, the MDM can predict more precisely the output displacements and the optimized ratio was 12.50%. In the last group (Exp6) at the single-frequency of 50 Hz, the optimized ratio was up to 14.04%. In the fourth group of experiments (Exp4) with low frequency and amplitude, E_{rms} and ξ of the MDM were 0.0916 μm and 1.2% respectively, while those of the CDM were 0.1426 μm and 3.6%, respectively. Compared with the CDM, the optimized ratio was 35.76%. With the increasing of frequency and amplitude, E_{rms} and ξ of the CDM in the third group of experiments (Exp3) increased to 0.1181 μm and 1.56%, respectively. The corresponding optimization ratio was up to 45.2%. Compared with the two groups above, the first and second groups of experiments (Exp1 and Exp2) had higher amplitude and frequency. The optimization ratios of Exp1 and Exp2 were 34.89% and 31.43%, respectively. Compared with the other three groups, the second group of experiments (Exp2) had the biggest amplitude and frequency, whose E_{rms} and ξ of MDM were 0.5845 μm and 4.6%, respectively. It was found that the modeling errors of both the CDM and MDM increase with the increasing of frequency and amplitude.

These experimental and simulation results clearly reveal that the MDM can describe more precisely rate-dependent hysteresis behaviors at high-frequency and high-amplitude excitations compared with the CDM.

5. Conclusions

In this paper, a modified Duhem model (MDM) is proposed to describe rate-dependent hysteresis behaviors at high-frequency and high-amplitude excitations. The MDM combines trigonometric functions and derivatives of input signal based on the classical Duhem model (CDM). The MDM parameters can be identified easily by the nonlinear least squares method. Six groups of experiments were conducted and the experimental and simulation results show that the MDM can more precisely describe rate-dependent hysteresis behaviors at high-frequency and high-amplitude excitations than the CDM. It is demonstrated that the MDM is effective and useful.

Author Contributions: Conceptualization, J.G., Z.M., X.C., M.-F.G. and Y.Z.; methodology, J.G. and Z.M.; software, J.G and Z.M.; validation, J.G. and Z.M.; formal analysis, J.G. and Z.M.; investigation, J.G.; resources, J.G. and Z.M.;

data curation, J.G. and Z.M.; writing—original draft preparation, J.G and Z.M.; writing—review and editing, J.G., Z.M., X.C., M.-F.G. and Y.Z.; visualization, J.G. and Z.M.; supervision, J.G. and M.-F.G.; project administration, J.G. and M.-F.G.; funding acquisition, J.G and M.-F.G.

Funding: This work was supported by the National Natural Science Foundation of China (Grant Nos. 51805494 and 61703374), the Fundamental Research Funds for the Central Universities (Grant Nos.CUGL180819), and the Open Foundation of Guangdong Provincial Key Laboratory of Precision Equipment and Manufacturing Technology (PEM201702).

Conflicts of Interest: The authors declare no conflict of interest.

References

1. Aggogeri, F.; Al-Bender, F.; Brunner, B.; Elsaid, M.; Mazzola, M.; Merlo, A.; Ricciardi, D.; Rodriguez, M.D.L.O.; Salvi, E. Design of piezo-based AVC system for machine tool applications. *Mech. Syst. Signal Process.* **2013**, *36*, 53–65. [CrossRef]
2. Grossard, M.; Rotinat-Libersa, C.; Chaillet, N.; Boukallel, M. Mechanical and control-oriented design of a monolithic piezoelectric microgripper using a new topological optimization method. *IEEE/ASME Trans. Mechatron.* **2009**, *14*, 32–45. [CrossRef]
3. Song, C.; Prasad, G.; Chan, K.; V, M. Characterization and optimization of illumination translation vector for dimensional parameters measurement of surface form and feature using DSPI. *Rev. Sci. Instrum.* **2016**, *87*, 63116. [CrossRef] [PubMed]
4. Haridas, A.; Song, C.; Chan, K.; Murukeshan, V.M. Nondestructive characterization of thermal damages and its interactions in carbon fibre composite panels. *Fatigue Fract. Eng. M* **2017**, *40*, 1562–1580. [CrossRef]
5. Gan, J.; Zhang, X. A review of nonlinear hysteresis modeling and control of piezoelectric actuators. *AIP Adv.* **2019**, *9*, 40702. [CrossRef]
6. Mayergoyz, I.D.; Friedman, G. Generalized Preisach model of hysteresis. *IEEE Trans. Magn.* **1988**, *24*, 212–217. [CrossRef]
7. Song, G.; Zhao, J.Q.; Zhou, X.Q.; de Abreu-Garcia, J.A. Tracking control of a piezoceramic actuator with hysteresis compensation using inverse Preisach model. *IEEE/ASME Trans. Mechatron.* **2005**, *10*, 198–209. [CrossRef]
8. U-Xuan, T.; Win, T.L.; Cheng, Y.S.; Riviere, C.N.; Wei, T.A. Feedforward controller of ill-conditioned hysteresis using singularity-free Prandtl–Ishlinskii model. *IEEE/ASME Trans. Mechatron.* **2009**, *14*, 598–605. [CrossRef]
9. Tan, U.X.; Win, T.L.; Ang, W.T. Modeling piezoelectric actuator hysteresis with singularity free Prandtl–Ishlinskii model. In Proceedings of the 2006 IEEE International Conference on Robotics and Biomimetics, Kunming, China, 17–20 December 2006.
10. Goldfarb, M.; Celanovic, N. Modeling piezoelectric stack actuators for control of micromanipulation. *IEEE Control Syst.* **1997**, *17*, 69–79.
11. Tri, V.M.; Tjahjowidodo, T.; Ramon, H.; Van Brussel, H. A new approach to modeling hysteresis in a pneumatic artificial muscle using the Maxwell-slip model. *IEEE/ASME Trans. Mechatron.* **2011**, *16*, 177–186.
12. Ru, C.H.; Sun, L.N. Hysteresis and creep compensation for piezoelectric actuator in open-loop operation. *Sensor Actuat. A Phys.* **2005**, *122*, 124–130.
13. Changhai, R.; Lining, S. Improving positioning accuracy of piezoelectric actuators by feedforward hysteresis compensation based on a new mathematical model. *Rev. Sci. Instrum.* **2005**, *76*, 95111.
14. Zhu, W.; Wang, D.H. Non-symmetrical Bouc–Wen model for piezoelectric ceramic actuators. *Sensor Actuat. A Phys.* **2012**, *181*, 51–60. [CrossRef]
15. Gan, J.; Zhang, X. An enhanced Bouc–Wen model for characterizing rate-dependent hysteresis of piezoelectric actuators. *Rev. Sci. Instrum.* **2018**, *89*, 115002. [CrossRef] [PubMed]
16. Dahl, P.R. A solid friction model. In Proceedings of the Technical Report Tor-0158(3107-18)-1, The Aerospace Corporation, EI Segundo, CA, USA, 1968.
17. Xu, Q.; Li, Y. Dahl model-based hysteresis compensation and precise positioning control of an XY parallel micromanipulator with piezoelectric actuation. *J. Dyn. Syst. Meas. Control* **2010**, *132*, 041011. [CrossRef]
18. Shome, S.K.; Prakash, M.; Mukherjee, A.; Datta, U. Dither control for dahl model based hysteresis compensation. In Proceedings of the 2013 IEEE International Conference on Signal Processing, Computing and Control (ISPCC), Solan, India, 26–28 September 2013.

19. Chen, H.; Tan, Y.; Zhou, X.; Dong, R.; Zhang, Y. Identification of dynamic hysteresis based on Duhem model. In Proceedings of the 2011 International Conference on Intelligent Computation Technology and Automation (ICICTA), Shenzhen, China, 28–29 March 2011.

20. Feng, Y.; Rabbath, C.A.; Chai, T.; Su, C. Robust adaptive control of systems with hysteretic nonlinearities: A Duhem hysteresis modelling approach. In Proceedings of the AFRICON 2009, Nairobi, Kenya, 23–25 September 2009.

21. Padthe, A.K.; Oh, J.; Bernstein, D.S. Counterclockwise dynamics of a rate-independent semilinear Duhem model. In Proceedings of the 44th IEEE Conference on Decision and Control, Seville, Spain, 15 December 2005.

22. Zhou, M.; Wang, J. Research on hysteresis of piezoceramic actuator based on the Duhem model. *Sci. World J.* **2013**, *2013*, 1–6. [CrossRef] [PubMed]

23. Lin, C.; Lin, P. Tracking control of a biaxial piezo-actuated positioning stage using generalized Duhem model. *Comput. Math. Appl.* **2012**, *64*, 766–787. [CrossRef]

24. Wang, X.; Wang, X.; Mao, Y. Hysteresis compensation in GMA actuators using Duhem model. In Proceedings of the 2008 7th World Congress on Intelligent Control and Automaiton, Chongqing, China, 25–27 June 2008.

25. Xie, W.; Fu, J.; Yao, H.; Su, C.Y. Observer based control of piezoelectric actuators with classical Duhem modeled hysteresis. In Proceedings of the 2009 American Control Conference, St. Louis, MO, USA, 10–12 June 2009.

26. Al Janaideh, M.; Su, C.Y.; Rakheja, S. Development of the rate-dependent Prandtl–Ishlinskii model for smart actuators. *Smart Mater. Struct.* **2008**, *17*, 035026. [CrossRef]

27. Gu, G.Y.; Zhu, L.M. Modeling of rate-dependent hysteresis in piezoelectric actuators using a family of ellipses. *Sensor Actuat. A Phys.* **2011**, *165*, 303–309. [CrossRef]

28. Oh, J.; Bernstein, D.S. Semilinear Duhem model for rate-independent and rate-dependent hysteresis. *IEEE Trans. Automat. Contr.* **2005**, *50*, 631–645.

29. Coleman, B.D.; Hodgdon, M.L. A constitutive relation for rate-independent hysteresis in ferromagnetically soft materials. *Int. J. Eng. Sci.* **1986**, *24*, 897–919. [CrossRef]

30. Gu, G.; Yang, M.; Zhu, L. Real-time inverse hysteresis compensation of piezoelectric actuators with a modified Prandtl–Ishlinskii model. *Rev. Sci. Instrum.* **2012**, *83*, 65106. [CrossRef] [PubMed]

31. Ming, M.; Feng, Z.; Ling, J.; Xiao, X. Hysteresis modelling and feedforward compensation of piezoelectric nanopositioning stage with a modified Bouc–Wen model. *Micro Nano Lett.* **2018**, *13*, 1170–1174. [CrossRef]

32. Wang, G.; Chen, G. Identification of piezoelectric hysteresis by a novel Duhem model based neural network. *Sensor Actuat. A Phys.* **2017**, *264*, 282–288. [CrossRef]

33. Gan, J.; Zhang, X. Nonlinear hysteresis modeling of piezoelectric actuators using a generalized Bouc–Wen model. *Micromachines* **2019**, *10*, 183. [CrossRef]

Article

A Low-Frequency MEMS Piezoelectric Energy Harvesting System Based on Frequency Up-Conversion Mechanism

Manjuan Huang [1], Cheng Hou [1], Yunfei Li [1], Huicong Liu [1,*], Fengxia Wang [1,*], Tao Chen [1], Zhan Yang [1], Gang Tang [2] and Lining Sun [1]

[1]   School of Mechanical and Electric Engineering, Jiangsu Provincial Key Laboratory of Advanced Robotics, Soochow University, Suzhou 215123, China; szdxhmj@126.com (M.H.); livehc@126.com (C.H.); liyunfei3321@foxmail.com (Y.L.); chent@suda.edu.cn (T.C.); yangzhan@suda.edu.cn (Z.Y.); lnsun@hit.edu.cn (L.S.)

[2]   Jiangxi Province Key Laboratory of Precision Drive & Control, Department of Mechanical and Electrical Engineering, Nanchang Institute of Technology, Nanchang 330099, China; tanggang@nit.edu.cn

*   Correspondence: hcliu078@suda.edu.cn (H.L.); wangfengxia@suda.edu.cn (F.W.)

Received: 22 August 2019; Accepted: 20 September 2019; Published: 24 September 2019

Abstract: This paper proposes an impact-based micro piezoelectric energy harvesting system (PEHS) working with the frequency up-conversion mechanism. The PEHS consists of a high-frequency straight piezoelectric cantilever (SPC), a low-frequency S-shaped stainless-steel cantilever (SSC), and supporting frames. During the vibration, the frequency up-conversion behavior is realized through the impact between the bottom low-frequency cantilever and the top high-frequency cantilever. The SPC used in the system is fabricated using a new micro electromechanical system (MEMS) fabrication process for a piezoelectric thick film on silicon substrate. The output performances of the single SPC and the PEHS under different excitation accelerations are tested. In the experiment, the normalized power density of the PEHS is 0.216 $\mu W \cdot g^{-1} \cdot Hz^{-1} \cdot cm^{-3}$ at 0.3 g acceleration, which is 34 times higher than that of the SPC at the same acceleration level of 0.3 g. The PEHS can improve the output power under the low frequency and low acceleration scenario.

Keywords: MEMS; piezoelectric vibration energy harvester; frequency up-conversion mechanism; impact; PZT thick film

---

1. Introduction

In recent years, with the rapid development of micro electromechanical systems (MEMSs) and the Internet of things (IoT), various micro wireless sensor nodes (WSNs) have been developed. These nodes are widely used in military surveillance, structural health monitoring, road traffic monitoring, and so on [1–4]. However, the limited lifetime of traditional batteries restricts the application of WSNs in complex environments and increases the working load of changing the batteries periodically. To overcome this restriction, some environmental energy harvesters dedicated to collect solar, thermal, wind, ocean wave, and vibration energies have been developed [5]. Among these, vibration energy is ubiquitous, such as structural vibrations, human activities, and fluid flows. The mechanical vibration energy can be converted into electrical energy through four transduction mechanisms, which are electromagnetic [6,7], piezoelectric [8–10], triboelectric [11,12], and electrostatic [13,14]. Piezoelectric vibration energy harvesters (PVEHs) have received significant attention due to their simple configuration, high energy conversion efficiency, and precision controllability of the mechanical response [15–17].

Some piezoelectric materials are widely used in MEMS energy harvesters, which are aluminum nitride (AlN) [18,19], zinc oxide (ZnO) [20,21], and $Pb(Zr_xTi_{1-x})O_3$ (PZT) [22–27]. Among these,

PZT has a higher electromechanical coupling coefficient compared with AlN and ZnO. Cui et al. [26] developed a multi-beam energy harvester with a PZT thin-film layer using a sol-gel deposition method. The maximum output power of 16.74 nW was obtained under an acceleration of 1 g and resonant frequency of 1400 Hz. Generally, PZT thin-film deposition requires a specific and complicated fabrication recipe, and the output performance of the PZT thin-film is limited. Therefore, PZT thick-film-based energy harvesters were developed. Xu et al. [28] proposed a screen-printed PZT/PZT thick-film bimorph cantilever for energy harvesting. However, the screen-printed PZT thick films are not dense enough, which means their piezoelectricity is low compared with that of bulk PZT. Thus, preparing a high-quality PZT thick film on silicon (Si) substrate through wafer bonding of bulk PZT has been proposed [29–31]. Janphuang et al. [30] demonstrated a wafer-level fabrication process of piezoelectric energy harvester using a spin-on polymetric adhesive WaferBOND as a bonding layer between bulk PZT and Si. The harvester exhibited an average power of 82.4 µW under an excitation of 1 g at 96 Hz. The above studies indicate that the MEMS PVEHs with thinned bulk PZT thick films have the potential for high output performance.

Another challenge for MEMS PVEHs is that the resonant frequencies of piezoelectric cantilevers are higher than most ambient vibration sources. Most of the natural vibration sources are random and at a low-frequency, typically ranging from 30 to 200 Hz [32–34]. In order to effectively utilize the low-frequency environmental vibrations, lowering the resonant frequency and widening the operating bandwidth have been the major target for the small-scale PVEHs. The frequency up-conversion mechanisms provided a good solution to address these issues and have aroused great research interest [35–38]. In general, the frequency up-conversion technologies can be divided into non-impact and impact types. Galchev et al. [36] demonstrated a non-impact piezoelectric generator that utilized a magnetic latching mechanism to convert the ambient low frequency to a higher internal operation frequency. However, the average power of the device was 3.25 µW at 1 g. Improvement of the output power needs to be considered. Jung et al. [39] introduced an energy harvester that uses the snap-through buckling action of a pre-buckled beam for frequency-up conversion instead of magnetic coupling. A maximum output power of 131 µW was generated using a 3 g acceleration. Andò et al. [40] proposed a snap-through buckling based vibrational energy harvester by adopting a flexible buckled beam, which was able to generate power in the excess of 400 µW under an acceleration of 13.35 m/s^2. However, large accelerations are generally required to drive the beam to induce snap-through buckling, and it is difficult to fabricate the buckled beam with standard technologies. In addition to these non-impact frequency up-conversion approaches, Umeda et al. [37] first demonstrated the impact-based frequency up-conversion approach for energy harvesting by investigating the power transformation of a steel ball impacting on a piezoelectric membrane. Halim et al. [38] proposed a mechanical impact-driven PVEH consisting of two series-connected PZT cantilevers and a flexible driving cantilever. A peak power of 734 µW from two series connecting PZT beams was achieved at the resonant frequency of 14.5 Hz. The impact-driven frequency up-conversion technology effectively increases the output power of the energy harvester at low frequency. Liu et al. [8] developed a PZT thin-film MEMS-based frequency up-converted PVEH system by utilizing the periodic impact between an S-shaped, low-frequency driving cantilever and a straight, high-frequency PZT generating cantilever. The PVEH system realized a low operating frequency under 37 Hz and the volume was very small. However, the maximum output power was only 0.12 µW with a 0.8 g acceleration. So far, there have been few studies on silicon-based PVEH fabricated at the scale of MEMS for harvesting energy from low-frequency vibration through an impact-based frequency up-conversion mechanism.

Therefore, this study has carried out research and discussion targeting a low-frequency MEMS PVEH by using a frequency up-conversion mechanism. First, a new wafer-level micromachining process for fabricating the PZT thick-film cantilever energy harvester was put forward. Then, the piezoelectric energy harvesting system (PEHS) with a low-frequency S-shaped stainless-steel cantilever (SSC) and a high-frequency straight piezoelectric cantilever (SPC) was incorporated. The output performances of the system and the single SPC were investigated and compared by using a vibration control and testing system. The experimental results indicated that the impact-based frequency up-conversion mechanism was able to improve the output performance of the harvester under a low-frequency and low-acceleration vibration environment.

2. Design and Simulation

2.1. Device Configuration

In order to harvest the low frequency vibration, a PEHS working with a frequency up-conversion mechanism was designed and is shown in Figure 1a. As can be seen, the PEHS was designed as a parallel-cantilever structure, which consisted of a top high-frequency SPC and a bottom low-frequency S-shaped SSC assembled within a predefined space. Figure 1b,c shows the schematic diagrams of the SPC and the SSC. The surface area of the SPC was 15×14 mm^2 and the dimensions of the whole chip was 22 mm $\times$ 21 mm $\times$ 0.6 mm. The S-shaped structure of the SSC was used to achieve a low-stiffness beam within a small space. At the free end of the SSC, two nickel proof masses were assembled to further reduce the resonant frequency of the cantilever. The SSC was mounted on a piece of printed circuit board (PCB). The SPC was mounted on another piece of PCB, assembled on the top of the SSC. The top and bottom electrode pads of the SPC were connected to the lead interfaces of the PCB using gold thread. A rectangular hollow spacer was fixed between the two PCBs, and the initial gap distance between the SPC and SSC could be adjusted by changing the thickness of the spacer. During the vibration of the PEHS, the frequency up-conversion was realized through the periodic collision between the SSC and SPC. Figure 1d shows the sectional view of the multi-layer SPC. The cantilever consisted of a top Cu electrode layer, a PZT thick film layer, a bottom Cn/Sn electrode layer, a Si supporting layer, and a Si proof mass at the cantilever tip. The thickness of the PZT layer and Si layer were t_p and t_s, respectively. The free end of the cantilever was fixed with a Si mass of thickness t_m and length L_m to reduce the resonance frequency. Figure 1e shows the sectional view of the SSC, where the thickness of the stainless-steel was 100 μm. Table 1 lists the detailed geometric parameters of the SPC and the SSC.

Table 1. Structural parameters and material properties of the straight piezoelectric cantilever.

Parameters	Description	Value
L	Total length of the chip	22 mm
W	Total width of the chip	21 mm
L_b	Length of the cantilever beam	15 mm
L_m	Length of the Si proof mass	5 mm
W_m	Width of the Si proof mass	14 mm
L_n	Length of the nickel proof mass	15 mm
W_n	Width of the nickel proof mass	5 mm
w_s	Width of the S-shaped cantilever	1.5 mm
t_{te}	Thickness of the top Cu electrode	1 μm
t_p	Thickness of the PZT layer	65 μm
t_{be}	Thickness of the bottom electrode	9 μm
t_s	Thickness of the Si substrate	200 μm
t_m	Thickness of the Si proof mass	300 μm
t_n	Thickness of the nickel proof mass	3 mm
t_l	Thickness of the stainless-steel cantilever	100 μm

Figure 2 shows the collision process in one cycle, which can be divided into three states: approaching, impacting, and separating. Assume that the PEHS is excited by a sinusoidal external vibration, and the frequency of the vibration is close to the resonant frequency of the SSC. In the approaching state, the SSC bends and moves upward to the SPC due to the external force. Since the deformation of the SPC is much smaller than that of the SSC, the SPC would hinder the displacement of the SSC. In the impacting state, the SSC impacts on the SPC and then moves upward together with the SPC. After that, the two cantilevers move downward and separate, then vibrate independently at their own resonant frequency until the next collision. As a result, the low-frequency vibration of the SSC is transformed into the high-frequency resonation of the SPC.

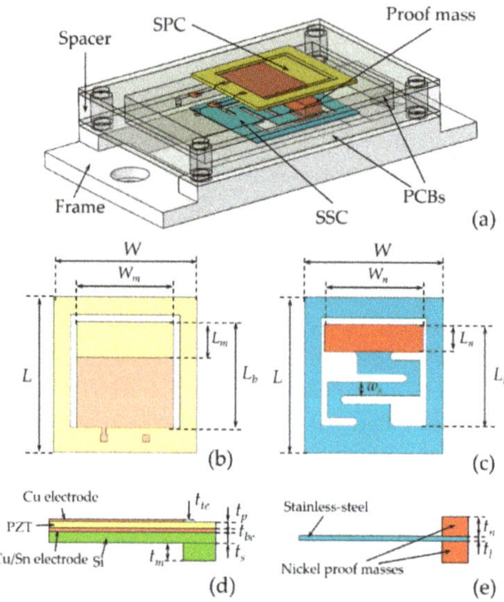

Figure 1. (**a**) 3D schematic of the piezoelectric energy harvesting system. (**b**) Schematic diagrams of the straight piezoelectric cantilever and (**c**) the S-shaped stainless-steel cantilever. (**d**) Sectional views of the piezoelectric cantilever and (**e**) the stainless-steel cantilever.

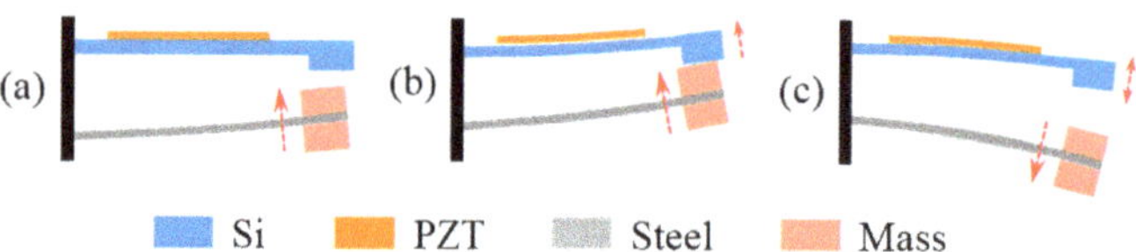

Figure 2. Schematic diagram of the collision process in one cycle: (**a**) approaching, (**b**) impacting, and (**c**) separating states.

2.2. Modal Analysis Using COMSOL

The resonant modes of the SPC and SSC were simulated using the finite element analysis software COMSOL 5.4a (Stockholm, Sweden), as shown in Figure 3. In the simulation, since the top and bottom electrode layers in SPC were too thin, which would lead to an increase in the calculation amount of the mesh division, the simulation model was simplified. The materials of the SPC in the model were defined as Si and PZT-5H. Meanwhile, the materials of the SSC were defined as stainless-steel and nickel. Figure 3a,b show the first-order vibration mode shapes (mode I) and eigenfrequencies of the SPC and SSC, respectively. The simulated resonant frequencies of the high-frequency SPC and the low-frequency SSC were 964.26 and 46.65 Hz, respectively. As can be seen, the maximum y displacements of the SPC and SSC both occurred at the end of the cantilever beam.

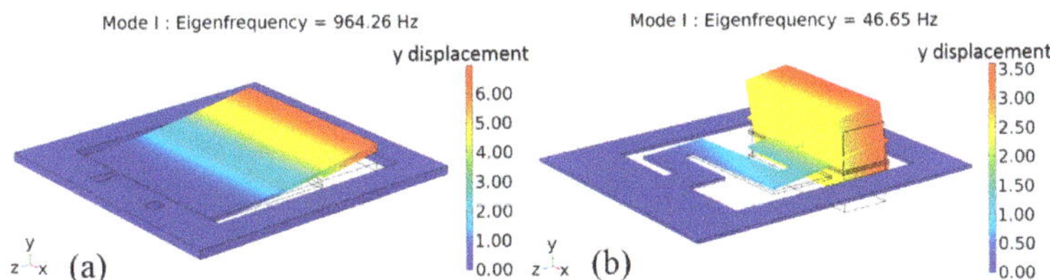

Figure 3. Modal analysis of (**a**) the straight piezoelectric cantilever, and (**b**) the S-shaped stainless-steel cantilever.

3. Micro Fabrication Process

As mentioned above, the PZT thick-film SPC was a PZT/Si composite structure with a Si proof mass on the free end. A schematic illustration of the wafer-level micro fabrication process of the SPC is shown in Figure 4, which mainly included the bonding of bulk PZT and Si wafer, thinning of PZT thick film, electrodes preparation, proof mass etching, and cantilever releasing.

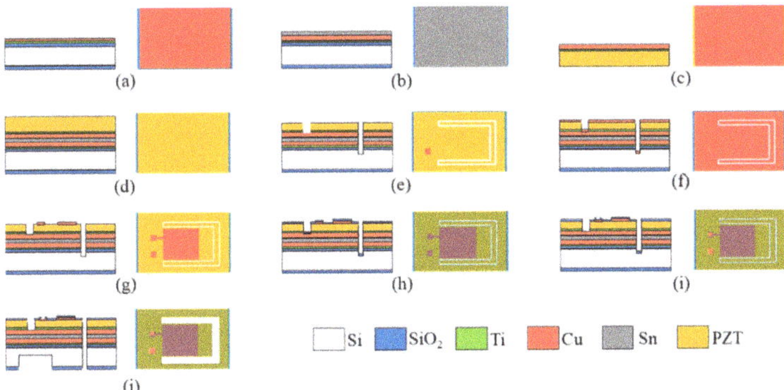

Figure 4. Schematic illustration of the fabrication process of the straight piezoelectric cantilever. (**a**) Sputtering Ti/Cu seed layer and (**b**) electroplating Cu/Sn layer on Si substrate. (**c**) Sputtering Ti/Cu seed layer and electroplating Cu layer on PZT wafer. (**d**) Bonding PZT wafer and Si substrate together. (**e**) Laser cutting the electrode pads and thinning the bulk PZT by mechanical lapping. (**f**) Sputtering Cu top electrode layer. (**g**) Top electrode patterning through lithography and IBE process. (**h**) Depositing SiO_2 by PECVD. (**i**) Welding spots patterning through lithography and RIE process. (**j**) Si proof mass patterning and backside DRIE to release cantilever.

The fabrication process began with a 4-inch double-side-polished bare Si wafer with a thickness of 500 μm. Then, a 500 nm thick silicon dioxide (SiO_2) layer was grown on both sides of the Si wafer using plasma enhanced chemical vapor deposition (PECVD), which served as a mask layer during the etching process. A Ti (20 nm)/Cu (50 nm) seed layer was sputtered on the cleaned Si substrate through magnetron sputtering, as shown in Figure 4a. The seed layer helped to enhance the adhesion of the metal to the Si wafer during the next electroplating. A 5-μm thick Cu layer and a 4.5-μm thick Sn layer were electroplated onto the Si wafer (Figure 4b). Another Ti (20 nm)/Cu (50 nm) seed layer was sputtered on the cleaned 4-inch bulk PZT wafer with a thickness of 400 μm, followed by a 5-μm thick Cu layer electroplated onto the PZT wafer (Figure 4c). Then, the PZT wafer and Si substrate were bonded together by means of Cu-Sn-Cu eutectic bonding (Figure 4d). The metal bonding layer also functioned as the bottom electrode layer. After bonding, the shape of the cantilever and bottom electrode pads were cut on the surface of the PZT wafer using laser cutting. The cutting depth of the bottom electrode pads should be exactly stopped at the metal bonding layer, and the cutting depth of the cantilever boundary should be deeper than the binding layer. Then, the bulk PZT was thinned from 400 μm to 65 μm using mechanical lapping (Figure 4e). Subsequently, a 1-μm thick Cu layer as the top electrode was sputtered onto the polished surface of the PZT using magnetron sputtering (Figure 4f). The top Cu electrode was patterned using ultraviolet (UV) lithography and etched using ion beam etching (IBE) (Figure 4g). A 500-nm thick SiO_2 layer was deposited on the PZT surface using PECVD. This layer was used to prevent the top electrode Cu from being oxidized in the air (Figure 4h). Next, the welding spots of the top and bottom electrodes were patterned using UV lithography and the SiO_2 layer was etched using reactive ton etching (RIE) (Figure 4i). Finally, the structure of the Si proof mass on the backside of the Si wafer were patterned using UV lithography, and then etched through the 500-nm thick SiO_2 using RIE. After the oxide layer was etched, a deep reactive ion etching (DRIE) dry etching process was utilized to ultimately release the cantilever (Figure 4j).

In the above MEMS process, a new bonding method for the bulk PZT and Si wafer was proposed, and the metal bonding layer was employed as the bottom electrode as well, which reduced the step of fabricating the bottom electrode and simplified the process. Preparation of the PZT thick film on the Si substrate was the key technique. It mainly consisted of two steps, which were bonding the bulk PZT wafer with the Si wafer and thinning the bulk PZT to the desired thickness, as shown in Figure 4d,e. Figure 5a shows the photograph of the wafer after being bonded. The Cu-Sn-Cu eutectic bonding method was developed to bond the bulk PZT and Si wafer at 270 °C for 30 min. Since the melting point of Sn is 231.9 °C, in order to ensure that Cu and Sn were sufficiently mutually fused, the bonding temperature should be higher than the melting point of Sn, which produces a high bonding strength. However, the Curie temperature of the PZT material is 295 °C [32]. A high bonding temperature may result in a reduction of the voltage output performance of the PZT thick film, in addition to forming large thermal stresses in the PZT layer. To prevent the PZT layer from cracking due to excessive thermal stress during the thinning process, some grooves were laser-cut on the surface of the PZT layer before the thinning to release the thermal stress in the wafer. In order to further solve these problems, subsequent research should be focused on the development of low-temperature, high-strength bonding methods. The photograph of the wafer after being thinned is shown in Figure 5b. It can be seen that due to the uneven thickness of the bonding layer, the PZT layer on one side of the wafer was completely worn through during the mechanical lapping. Therefore, when placing the device structure on the wafer, it should be placed in the middle as much as possible. Subsequently, in order to facilitate the patterning of the top Cu electrode, the shape of the device was cut out using laser cutting on the surface of the PZT layer. The PZT layer was then polished, and the polished wafer is shown in Figure 5c. Figure 5d shows the photograph of the wafer after being etched, where the shape of the top electrode was patterned using lithography and IBE. The advantage of this new wafer level MEMS process is the ability to simultaneously fabricate PZT thick film energy harvesters of different structures, reducing manufacturing costs and enabling mass production.

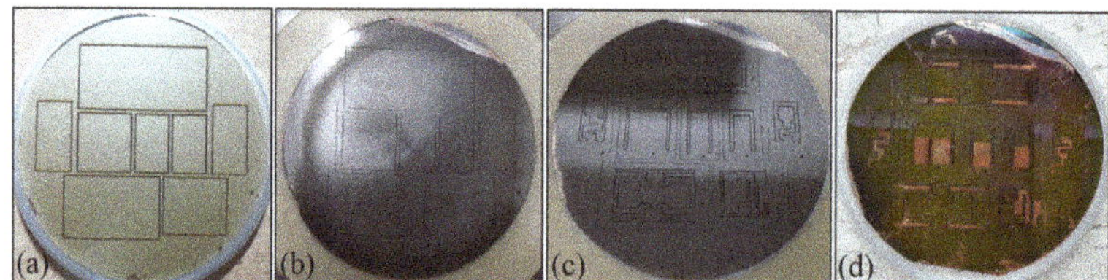

Figure 5. Photographs of the wafer after being (**a**) bonded, (**b**) thinned, (**c**) polished, and (**d**) etched.

The photograph of the PEHS prototype is shown in Figure 6a. Figure 6b shows the cross-section scanning electron microscope (SEM) image of the SPC. The multilayer piezoelectric cantilever consists of a 65-μm PZT layer with a 1-μm Cu electrode layer coated on it, a 9-μm intermediate Cu-Sn-Cu bonding layer, and a 200-μm Si substrate. The thickness of the cantilever was controlled using a DRIE process to about 275 μm. Figure 6c shows the photomicrograph of the top and bottom electrode pads.

The bottom electrode pad was obtained using laser cutting. The area of the electrode pad was 0.5 mm × 0.5 mm.

Figure 6. (**a**) Photograph of the PEHS. (**b**) Cross-section SEM image of the piezoelectric cantilever. (**c**) Photomicrograph of top and bottom electrode pads.

4. Experimental Results and Discussion

Figure 7 shows the experimental setup for the dynamic characterization of the fabricated device. The PVEH prototype (see Figure 6a) was mounted onto a TIRA vibration exciter (TIRA GmbH, Thuringia, Germany) which can generate different external sinusoidal excitations. The sinusoidal excitation signal of the shaker was created using the signal generator and adjusted using the power amplifier. An accelerometer (model 3035BG, DYTRAN, Los Angeles, CA, USA) was fixed on the vibration shaker to monitor the excitation acceleration. The electrical output of the device was recorded via dynamic signal analyzer software on the computer. In this study, in order to verify the effectiveness of the frequency up-conversion mechanism, the output performance of the SPC and the assembled PEHS were tested using a frequency up-sweep method and compared.

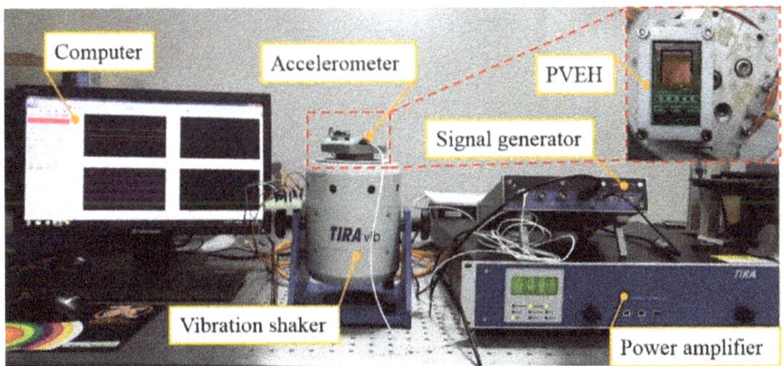

Figure 7. Experimental setup.

4.1. The Output Performance of the SPC

First, Figure 8a shows the open-circuit voltage at various frequencies from 920 Hz to 1100 Hz under different acceleration levels. It can be seen that the resonant frequencies of the SPC gradually decreased as the acceleration increased; when the applied accelerations were 0.1 g, 0.5 g, 1.0 g, and 1.5 g, the resonant frequencies were 1013 Hz, 1011 Hz, 1009 Hz, and 1008 Hz, respectively. This was because of the nonlinear change in the Young's modulus of PZT under a large stress [24,41]. According to the previous modal analysis using COMSOL, the first order resonant frequency of SPC was expected to be

964.26 Hz, which was close to the experimental results. The discrepancy between the simulation and the experimental results may be due to the simplification of the simulation model.

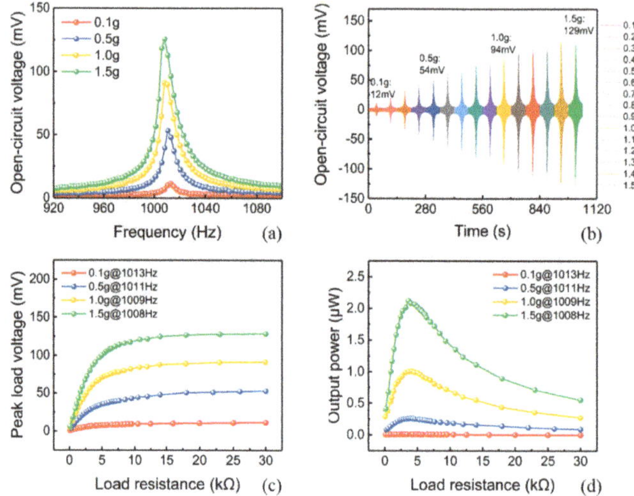

Figure 8. Output performance of the single piezoelectric cantilever. (**a**) The open-circuit voltage versus frequency at different acceleration levels. (**b**) The open-circuit voltage output versus time at various accelerations from 0.1 g to 1.5 g. (**c**) The peak load voltage, and (**d**) the maximum instantaneous output power versus load resistance at accelerations of 0.1 g, 0.5 g, 1.0 g, and 1.5 g.

Figure 8b shows the open-circuit voltage output versus time at accelerations from 0.1 g to 1.5 g. It is clear that the peak open-circuit voltage increased with the increase of the acceleration, which were 12 mV, 54 mV, 94 mV, and 129 mV at the accelerations of 0.1 g, 0.5 g, 1.0 g, and 1.5 g, respectively. To determine the maximum output power of the SPC with the optimal resistance, the voltage output signal was connected to a varying resistor to obtain the relationship between load resistance and output voltage under different vibration conditions. The instantaneous power delivered by the energy harvester can be expressed as:

$$P = V_p^2 / R \qquad (1)$$

where V_p is the voltage across the load, and R is the value of the external load resistance.

Figure 8c shows the peak load voltage (V_p) of the SPC versus the load resistance at different applied acceleration amplitudes of 0.1 g, 0.5 g, 1.0 g, and 1.5 g. Comparing the load voltages under different accelerations, it can be seen that the load voltage increased as the acceleration increased. Furthermore, under a constant acceleration condition, the load voltage clearly increased with the increasing of the load resistance. Based on Equation (1), the maximum output power for different load resistances was calculated and depicted in Figure 8d. A maximum output average power appeared at the optimal matched load resistance, which should be the same as the internal resistance of the device. The value of the optimal matched resistance was related to the acceleration amplitude. For instance, the optimal load resistance under 0.5 g, 1.0 g, and 1.5 g acceleration conditions were 4.2 kΩ, 4.0 kΩ, and 3.6 kΩ, respectively. A conclusion can be drawn that within a certain range of acceleration, the optimal load resistance decreased gradually with the increasing acceleration. As shown in Figure 8d, the maximum output power was 2.12 μW and occurred at the quite high resonance of 1008 Hz and acceleration of 1.5 g.

4.2. The Output Performance of the PEHS

Figure 9a shows the simplified 3D models of the PEHS. The resonant frequency of the SSC was about 40 Hz, obtained using frequency sweep test, which was close to the simulated resonant frequency of 46.65 Hz in COMSOL. Some factors, such as the gap distance between the SPC and the SSC, as well as the vibration acceleration amplitudes, have an influence on the output performance of the PEHS. In order to investigate the effects of the gap distance on the output performance of the PEHS, the output voltages under three gap distance values d_1, d_2, and d_3 were tested using up-sweep. The gap distance should be limited such that the SSC can impact the SPC during low-frequency vibration. However, the gap cannot be equal to zero, because the high-frequency SPC would limit the ability of the low-frequency driving beam SSC to respond to an external low acceleration and low frequency excitation [42]. Here the values of d_1, d_2, and d_3 were set as 0.6 mm, 0.9 mm, and 1.2 mm, respectively.

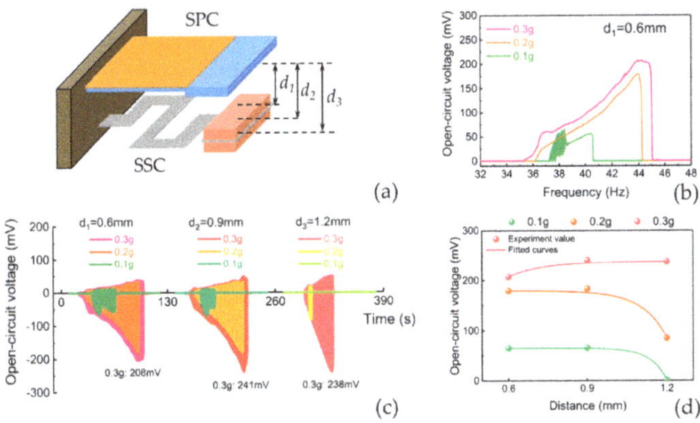

Figure 9. (**a**) The simplified models of the piezoelectric energy harvesting system. (**b**) The open-circuit voltage in the frequency domain at various acceleration levels under the gap of 0.6 mm. (**c**) The open-circuit voltage output in the time domain at various accelerations under different gaps of 0.6 mm, 0.9 mm, and 1.2 mm. (**d**) Relationship between the maximum open-circuit voltage and the initial gap distance.

The measured open-circuit voltage of the PEHS against operating frequencies at various acceleration levels under the gap distance of 0.6 mm is shown in Figure 9b. The maximum open-circuit voltages at the acceleration of 0.1 g, 0.2 g, and 0.3 g were 64 mV, 180 mV, and 208 mV, respectively. It was observed that under a certain gap and a certain acceleration condition, the open-circuit voltage output increased steadily as the operating frequency increased and then fell abruptly. Figure 9b shows that at an acceleration of 0.3 g, the half-bandwidth of the PEHS at a gain of 0.5 was approximately 5 Hz (from 40 Hz to 45 Hz). The reason for the wide operating bandwidth was that the SSC impacted with the SPC, resulting in a hindrance of the motion of the SSC. The frequency response of the SSC deviated from its normal linear behavior and exhibited nonlinearity in the overall stiffness of the SSC [43].

Figure 9c shows the time domain open-circuit voltage output waveforms of the PEHS under three different gaps at accelerations of 0.1 g, 0.2 g, and 0.3 g. The maximum peak voltages were approximately 208 mV, 241 mV, and 238 mV for the distances of 0.6 mm, 0.9 mm, and 1.2 mm, respectively, at an acceleration of 0.3 g. The maximum voltage of the waveforms for a 0.9 mm gap was higher than those under the other gap conditions. However, considering the bandwidth of the voltage waveforms under the three gap conditions, the maximum bandwidth was achieved under a small gap distance of 0.6 mm. It was observed that under the large gap of 1.2 mm, the SSC could not hit the SPC at all at the low vibration acceleration of 0.1 g. Figure 9d shows the fitted curves of the maximum output voltages under three different gap distances at a certain acceleration amplitude, which indicates the relationship between the output performance of the PEHS and the gap distance. Under the condition that the base acceleration was 0.3 g, the voltage continued to increase slightly as the distance increased from 0.6 mm to 1.2 mm. However, the tendency of the voltage curve at 0.1 g and 0.2 g was to increase slightly over a certain distance range and then decrease. The maximum voltage of 238 mV appeared at 0.3 g for a distance of 0.9 mm. It can be inferred that at a certain vibration acceleration, there may exist an optimal distance under which the maximum output voltage can be obtained.

Figure 10a shows the peak load voltage and the maximum output power of the PEHS versus load resistance at a 0.3 g acceleration under the resonant frequency of 40 Hz. The gap distance was 0.9 mm. With the increasing of the load resistance, the load voltage clearly increased, while the corresponding power increased to a maximum value and then decreased. The maximum value of the output power was 0.2 μW at the optimal load resistance of 11 kΩ. The power density of the PEHS normalized by the input acceleration and frequency was 0.216 μW·g^{-1}·Hz^{-1}·cm^{-3}. To verify the effectiveness of the PEHS, the load voltage output of the SPC was also measured at the same acceleration level of 0.3 g under its resonant frequency of 1012 Hz. The peak load voltage and the calculated output power of the SPC for different load resistances is shown in Figure 10b. The tendencies of the voltage and power curves of the SPC were the same as that of the PEHS. However, both the maximum load voltage and the maximum power of the SPC were smaller than those of the PEHS. The maximum output power of the SPC at 0.3 g was 0.15 μW, and the corresponding optimal load resistance was 4.7 kΩ. The normalized power density of the SPC was 0.006 μW·g^{-1}·Hz^{-1}·cm^{-3} at a vibration acceleration of 0.3 g. It can be seen that the normalized power density of the PEHS (0.216 μW·g^{-1}·Hz^{-1}·cm^{-3}) was 34 times higher than the normalized power density of the SPC at the same acceleration level of 0.3 g.

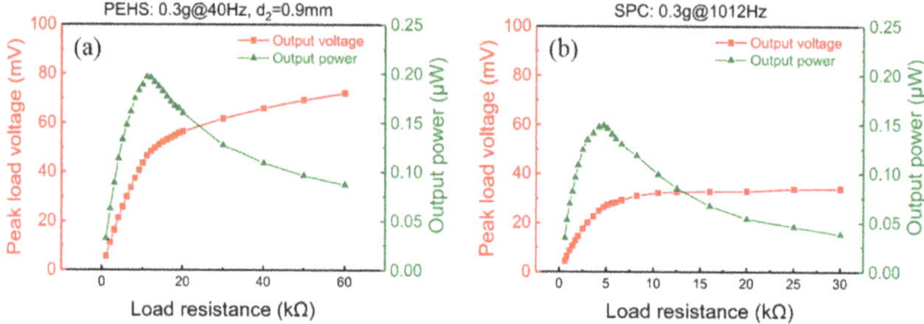

Figure 10. The output load voltage and power versus load resistance of (**a**) the piezoelectric energy harvesting system at 40 Hz, and (**b**) the single straight piezoelectric cantilever at 1012 Hz with a 0.3 g acceleration.

5. Conclusions

In summary, this work presented the design, fabrication, and experimental testing of a MEMS PEHS. As a parallel structure, the PEHS consisted of a piezoelectric cantilever, a stainless-steel S-shaped cantilever with proof mass, and supporting frames. By employing the parallel-cantilever structure, the bottom low-frequency SSC would impact on the top SPC during vibration and realize a frequency up-conversion. The piezoelectric cantilever chip used in the harvester was fabricated using a PZT thick film MEMS fabrication process. Furthermore, the key techniques during fabrication were Cu-Sn-Cu eutectic bonding, mechanical lapping, and electrode layer etching. Experimental results showed that the SPC vibrated at an acceleration of 0.3 g could generate the maximum output power of 0.15 μW at the resonant frequency of 1012 Hz, and the normalized power density was 0.006 μW·g^{-1}·Hz^{-1}·cm^{-3}. The output performances of the PEHS were also investigated under different initial gap distances and accelerations. Under a gap distance of 0.9 mm, the normalized power density of the PEHS was measured to be 0.216 μW·g^{-1}·Hz^{-1}·cm^{-3} at an acceleration of 0.3 g and resonant frequency of 40 Hz, which was much higher than that of the SPC. Moreover, the half-bandwidth of the PEHS broadened to 5 Hz due to the collision between SSC and SPC. It was proven that combining the PEHS with a frequency up-conversion mechanism can increase the output power under low frequency and low acceleration vibrations.

Author Contributions: Conceptualization, M.H. and H.L.; methodology, M.H. and C.H.; software, M.H. and Y.L.; formal analysis, M.H. and T.C.; writing – original draft, M.H., H.L. and C.H.; writing – review & editing, H.L., M.H., F.W. and Z.Y.; project administration, H.L. and F.W.; funding acquisition, H.L., G.T. and L.S.

Funding: This research was funded by the National Natural Science Foundation of China (Grant Nos. 51875377, 41527901 and 51565038).

Conflicts of Interest: The authors declare no conflict of interest.

References

1. Jung, I.; Shin, Y.-H.; Kim, S.; Choi, J.; Kang, C.-Y. Flexible piezoelectric polymer-based energy harvesting system for roadway applications. *Appl. Energy* **2017**, *197*, 222–229. [CrossRef]
2. Hou, C.; Chen, T.; Li, Y.; Huang, M.; Shi, Q.; Liu, H.; Sun, L.; Lee, C. A rotational pendulum based electromagnetic/triboelectric hybrid-generator for ultra-low-frequency vibrations aiming at human motion and blue energy applications. *Nano Energy* **2019**, *63*, 103871. [CrossRef]
3. Wu, J.; Yuan, S.; Zhao, X.; Yin, Y.; Ye, W. A wireless sensor network node designed for exploring a structural health monitoring application. *Smart Mater. Struct.* **2007**, *16*, 1898–1906. [CrossRef]
4. Ahmad, I.; Khan, F.U. Multi-mode vibration based electromagnetic type micro power generator for structural health monitoring of bridges. *Sens. Actuators, A.* **2018**, *275*, 154–161. [CrossRef]
5. Shaikh, F.K.; Zeadally, S. Energy harvesting in wireless sensor networks: A comprehensive review. *Renewable Sustainable Energy Rev.* **2016**, *55*, 1041–1054. [CrossRef]
6. Zhao, X.; Cai, J.; Guo, Y.; Li, C.; Wang, J.; Zheng, H. Modeling and experimental investigation of an AA-sized electromagnetic generator for harvesting energy from human motion. *Smart Mater. Struct.* **2018**, *27*, 085008. [CrossRef]
7. Liu, H.; Hou, C.; Lin, J.; Li, Y.; Shi, Q.; Chen, T.; Sun, L.; Lee, C. A non-resonant rotational electromagnetic energy harvester for low-frequency and irregular human motion. *Appl. Phys. Lett.* **2018**, *113*, 203901. [CrossRef]
8. Liu, H.; Lee, C.; Kobayashi, T.; Tay, C.J.; Quan, C. Piezoelectric MEMS-based wideband energy harvesting systems using a frequency-up-conversion cantilever stopper. *Sens. Actuators A* **2012**, *186*, 242–248. [CrossRef]
9. Fan, K.; Tan, Q.; Zhang, Y.; Liu, S.; Cai, M.; Zhu, Y. A monostable piezoelectric energy harvester for broadband low-level excitations. *Appl. Phys. Lett.* **2018**, *112*, 123901. [CrossRef]
10. Wang, J.; Wen, S.; Zhao, X.; Zhang, M.; Ran, J. Piezoelectric Wind Energy Harvesting from Self-Excited Vibration of Square Cylinder. *J. Sensors* **2016**, *2016*, 2353517. [CrossRef]
11. Zi, Y.; Lin, L.; Wang, J.; Wang, S.; Chen, J.; Fan, X.; Yang, P.K.; Yi, F.; Wang, Z.L. Triboelectric-pyroelectric-piezoelectric hybrid cell for high-efficiency energy-harvesting and self-powered sensing. *Adv. Mater.* **2015**, *27*, 2340–2347. [CrossRef]
12. Yoo, D.; Park, S. C.; Lee, S.; Sim, J. Y.; Song, I.; Choi, D.; Lim, H.; Kim, D.S. Biomimetic anti-reflective triboelectric nanogenerator for concurrent harvesting of solar and raindrop energies. *Nano Energy* **2019**, *57*, 424–431. [CrossRef]
13. Zhang, Y.; Wang, T.; Luo, A.; Hu, Y.; Li, X.; Wang, F. Micro electrostatic energy harvester with both broad bandwidth and high normalized power density. *Appl. Energy* **2018**, *212*, 362–371. [CrossRef]
14. Naito, Y.; Uenishi, K. Electrostatic MEMS Vibration Energy Harvesters inside of Tire Treads. *Sensors* **2019**, *19*, 890. [CrossRef]
15. Han, D.; Yun, K.-S. Piezoelectric energy harvester using mechanical frequency up conversion for operation at low-level accelerations and low-frequency vibration. *Microsyst. Technol.* **2014**, *21*, 1669–1676. [CrossRef]
16. Tang, G.; Yang, B.; Liu, J.; Xu, B.; Zhu, H.; Yang, C. Development of high performance piezoelectric d33 mode MEMS vibration energy harvester based on PMN-PT single crystal thick film. *Sens. Actuators, A* **2014**, *205*, 150–155. [CrossRef]
17. Zawada, T.; Hansen, K.; Lou-Moeller, R.; Ringgaard, E.; Pedersen, T.; Thomsen, E.V. High-performance piezoelectric thick film based energy harvesting micro-generators for MEMS. *Procedia Eng.* **2010**, *5*, 1164–1167. [CrossRef]

18. He, X.; Wen, Q.; Lu, Z.; Shang, Z.; Wen, Z. A micro-electromechanical systems based vibration energy harvester with aluminum nitride piezoelectric thin film deposited by pulsed direct-current magnetron sputtering. *Appl. Energy* **2018**, *228*, 881–890. [CrossRef]

19. Elfrink, R.; Kamel, T.M.; Goedbloed, M.; Matova, S.; Hohlfeld, D.; van Andel, Y.; van Schaijk, R. Vibration energy harvesting with aluminum nitride-based piezoelectric devices. *J. Micromech. Microeng.* **2009**, *19*, 094005. [CrossRef]

20. Joshi, S.; Nayak, M.M.; Rajanna, K. Effect of post-deposition annealing on transverse piezoelectric coefficient and vibration sensing performance of ZnO thin films. *Appl. Surf. Sci.* **2014**, *296*, 169–176. [CrossRef]

21. Md Ralib, A.A.; Nordin, A.N.; Salleh, H.; Othman, R. Fabrication of aluminium doped zinc oxide piezoelectric thin film on a silicon substrate for piezoelectric MEMS energy harvesters. *Microsyst. Technol.* **2012**, *18*, 1761–1769. [CrossRef]

22. Tanaka, K.; Konishi, T.; Ide, M.; Sugiyama, S. Wafer bonding of lead zirconate titanate to Si using an intermediate gold layer for microdevice application. *J. Micromech. Microeng.* **2006**, *16*, 815–820. [CrossRef]

23. Yang, B.; Zhu, Y.B.; Wang, X.Z.; Liu, J.Q.; Chen, X.; Yang, C.S. High performance PZT thick films based on bonding technique for d_{31} mode harvester with integrated proof mass. *Sens. Actuators, A* **2014**, *214*, 88–94. [CrossRef]

24. Tang, G.; Liu, J.; Yang, B.; Luo, J.; Liu, H.; Li, Y.; Yang, C.; He, D.; Dao, V.D.; Tanaka, K.; et al. Fabrication and analysis of high-performance piezoelectric MEMS generators. *J. Micromech. Microeng.* **2012**, *22*, 065017. [CrossRef]

25. Lei, A.; Xu, R.; Thyssen, A.; Stoot, A.C.; Christiansen, T.L.; Hansen, K.; Lou-Moller, R.; Thomsen, E.V.; Birkelund, K. MEMS-based thick film PZT vibrational energy harvester. In Proceedings of the 2011 IEEE 24th International Conference on Micro Electro Mechanical Systems, Cancun, Mexico, 23–27 January 2011; pp. 125–128.

26. Cui, Y.; Zhang, Q.; Yao, M.; Dong, W.; Gao, S. Vibration piezoelectric energy harvester with multi-beam. *AIP Adv.* **2015**, *5*, 041332. [CrossRef]

27. Licheng, D.; Zhiyu, W.; Xingqiang, Z.; Chengwei, Y.; Guoxi, L.; Jike, M. High Voltage Output MEMS Vibration Energy Harvester in d_{31} Mode With PZT Thin Film. *J. Microelectromech. Syst.* **2014**, *23*, 855–861. [CrossRef]

28. Xu, R.; Lei, A.; Dahl-Petersen, C.; Hansen, K.; Guizzetti, M.; Birkelund, K.; Thomsen, E.V.; Hansen, O. Screen printed PZT/PZT thick film bimorph MEMS cantilever device for vibration energy harvesting. *Sens. Actuators, A* **2012**, *188*, 383–388. [CrossRef]

29. Wang, Z.; Miao, J.; Tan, C.W. Acoustic transducers with a perforated damping backplate based on PZT/silicon wafer bonding technique. *Sens. Actuators, A* **2009**, *149*, 277–283. [CrossRef]

30. Janphuang, P.; Lockhart, R.; Uffer, N.; Briand, D.; de Rooij, N.F. Vibrational piezoelectric energy harvesters based on thinned bulk PZT sheets fabricated at the wafer level. *Sens. Actuators, A* **2014**, *210*, 1–9. [CrossRef]

31. Lin, S.-C.; Wu, W.-J. Fabrication of PZT MEMS energy harvester based on silicon and stainless-steel substrates utilizing an aerosol deposition method. *J. Micromech. Microeng.* **2013**, *23*, 125028. [CrossRef]

32. Li, G.; Yi, Z.; Hu, Y.; Liu, J.; Yang, B. High-performance low-frequency MEMS energy harvester via partially covering PZT thick film. *J. Micromech. Microeng.* **2018**, *28*, 095007. [CrossRef]

33. Ju, S.; Ji, C.-H. Impact-based piezoelectric vibration energy harvester. *Appl. Energy* **2018**, *214*, 139–151. [CrossRef]

34. Pillatsch, P.; Yeatman, E.M.; Holmes, A.S. A piezoelectric frequency up-converting energy harvester with rotating proof mass for human body applications. *Sens. Actuators, A* **2014**, *206*, 178–185. [CrossRef]

35. Galchev, T.V. Energy Scavenging From Low Frequency Vibrations. *IEEE Sens. J.* **2008**, *8*, 261–268.

36. Galchev, T.; Aktakka, E.E.; Najafi, K. A Piezoelectric Parametric Frequency Increased Generator for Harvesting Low-Frequency Vibrations. *J. Microelectromech. Syst.* **2012**, *21*, 1311–1320. [CrossRef]

37. Umeda, M.; Nakamura, K.; Ueha, S. Energy Storage Characteristics of a Piezo-Generator using Impact Induced Vibration. *Jpn. J. Appl. Phys.* **1997**, *36*, 3146–3151. [CrossRef]

38. Halim, M.A.; Park, J.Y. Theoretical modeling and analysis of mechanical impact driven and frequency up-converted piezoelectric energy harvester for low-frequency and wide-bandwidth operation. *Sens. Actuators, A* **2014**, *208*, 56–65. [CrossRef]

39. Jung, S.-M.; Yun, K.-S. Energy-harvesting device with mechanical frequency-up conversion mechanism for increased power efficiency and wideband operation. *Appl. Phys. Lett.* **2010**, *96*, 111906. [CrossRef]

40. Andò, B.; Baglio, S.; Marletta, V.; Bulsara, A.R. Modeling a Nonlinear Harvester for Low Energy Vibrations. *IEEE Trans. Instrum. Meas.* **2019**, *68*, 1619–1627. [CrossRef]

41. Yao, L.Q.; Zhang, J.G.; Lu, L.; Lai, M.O. Nonlinear Dynamic Characteristics of Piezoelectric Bending Actuators Under Strong Applied Electric Field. *J. Microelectromech. Syst.* **2004**, *13*, 645–652. [CrossRef]

42. Gu, L.; Livermore, C. Impact-driven, frequency up-converting coupled vibration energy harvesting device for low frequency operation. *Smart Mater. Struct.* **2011**, *20*, 045004. [CrossRef]

43. Chen, S.; Ma, L.; Chen, T.; Liu, H.C.; Sun, L.N.; Wang, J.X. Modeling and verification of a piezoelectric frequency-up-conversion energy harvesting system. *Microsyst. Technol.* **2017**, *23*, 2459–2466. [CrossRef]

Communication

Piezoelectric Impedance-Based Non-Destructive Testing Method for Possible Identification of Composite Debonding Depth

Wongi S. Na [1],* and Jongdae Baek [2]

[1] Sustainable Infrastructure Research Center, Korea Institute of Civil Engineering & Building Technology, Gyeonggi-Do 10223, Korea
[2] Future Infrastructure Research Center, Korea Institute of Civil Engineering & Building Technology, Gyeonggi-Do 10223, Korea
* Correspondence: wongi@kict.re.kr; Tel.: +82-31-910-0155

Received: 3 August 2019; Accepted: 16 September 2019; Published: 17 September 2019

Abstract: Detecting the depth and size of debonding in composite structures is essential for assessing structural safety as it can weaken the structure possibly leading to a failure. As composite materials are used in various fields up to date including aircrafts and bridges, inspections are carried out to maintain structural integrity. Although many inspection methods exist for detection damage of composites, most of the techniques require trained experts or a large equipment that can be time consuming. In this study, the possibility of using the piezoelectric material-based non-destructive method known as the electromechanical impedance (EMI) technique is used to identify the depth of debonding damage of glass epoxy laminates. Laminates with various thicknesses were prepared and tested to seek for the possibility of using the EMI technique for identifying the depth of debonding. Results show promising outcome for bringing the EMI technique a step closer for commercialization.

Keywords: debonding; non-destructive testing; piezoelectric; electromechanical impedance; damage detection; impedance-based technique; damage depth

---

1. Introduction

The high stiffness and strength to weight ratio of composite materials has led to a wide range of applications including the field of aerospace, civil, and mechanical engineering. For this reason, ensuring the safety and reliability of composite structures has always been an important factor. Damage in composite structures such as debonding can cause a serious problem if it goes undetected as it may significantly reduce structural integrity. Thus up to date, many researches have been conducted to detect the damage of composite structures where a well summarized list of contact and non-contact NDT (non-destructive testing) methods can be found in [1]. Thorough review work can be found in previous studies written by various authors including vibration-based model dependent methods for detecting damage in composite structures [2], failure modes that exist in composite materials with a brief discussion on both the destructive and NDT methods [3], applicability of NDT methods on thick walled composites [4], and assessment of debonding subjected to drilling [5].

In this study, a non-destructive testing method known as the electromechanical impedance (EMI) technique is used to study the possibility of detection debonding depth. Up to date, the technique has been proved to be effective for monitoring changes in structural property where some of the recent work can be found in [6–15]. Furthermore, various authors have studied related to debonding and delamination of composites such as modifying the EMI model of piezoelectric (PZT) actuator-sensors to detect debonding of composite patches [16], carbon fiber reinforced polymer (CFRP) and concrete surface debonding detection [17], 2D model and a set of experiments subjected to debonding of RC beams strengthened with FRP strips [18], assessment of three different cases of possible weak CFRP bonding [19], and localization of artificially created delamination in CFRP samples [20].

The technique uses a single PZT patch to act as both actuator and sensor, simultaneously which is a major advantage compared to other contact based non-destructive testing methods. For a successful damage detection process, a frequency range with multiple resonance peaks must be chosen as these peaks will change more when subjected to damage compared to a frequency range without any peaks. Here, a suitable frequency range is selected by the trial and error method by sweeping various frequency ranges. As the 1D EMI model proposed by Liang et al. [21] showed that the electrical impedance of the attached PZT patch is directly related to the mechanical impedance of the host structure, any changes in the structure (i.e., damage) can be detected by monitoring the changes in the measured electrical impedance signature of the PZT patch. Although acquiring impedance signatures can be achieved without too much difficulty, analyzing the data for identifying damage can be a complex task as various factors affect impedance signatures. Factors including change in environment temperature, bonding condition, PZT placement, and sizing can result in different outcomes every time as the EMI technique uses high frequency range (over 20 kHz) making it a sensitive technique subjected to various factors.

Through this work, it was experimentally found that the debonding depth could be found regardless of where the PZT transducer was attached. The difference between signature change due to debonding damage and random PZT transducer attachment was identified in the study.

2. EMI Technique

In this study, experiments were conducted at a room temperature of 24 °C (±0.2 °C) using the AD5933 evaluation board manufactured from Analog Devices Co (Norwood, MA, USA). The equipment allows one to measure impedance up to 100 kHz with over 500 data points. Shown in Figure 1, the AD5933 evaluation board is connected to a computer and the test specimen made of two composite plates (glass fiber epoxy laminate) adhered together using a commercial epoxy adhesive (Loctite Quick Set). The two composite plates of size 300 mm × 100 mm with 0.6 mm thickness (with Poisson ratio 0.3, density of 1550 kg/m^3, Young's modulus 17 Gpa) were purchased from a domestic company (ArtRyx, Gyeonggi-do, Republic of Korea) and the piezoelectric patch model PSI5A4E was purchased from Piezo Systems (Woburn, MA, USA).

The first test for this work was to demonstrate the change in the impedance signature after complete debonding of a composite layer. Since the test specimen shown in Figure 1 consists of two composites bonded together with epoxy, the impedance signatures of the attached 15 mm square PZT were measured before and after removing the bottom composite layer. In general, conducting the EMI technique on metal structures subjected to damage will result in larger signature variations. However, the impedance signature results shown in Figure 1 shows small variation subjected to debonding in the frequency range of 30 kHz to 80 kHz with a small left shift movement. Such reason is due to the composite plate having non-homogenous property and absence of resonance peak in the selected frequency range. However, although the change in the signature indicates the presence of damage, it is difficult to identify the type of damage as a crack damage or debonding damage can both result in change in impedance signatures [22]. In [22], the author found a way to distinguish a crack damage from a debonding damage where it was experimentally proven that only the debonding damage caused the impedance signature peak to increase. Thus, based on this fact, this work will seek for the possibility of finding the debonding depth of a composite structure by analyzing the shape of the impedance signature. The study will attempt to correlate variations of indices obtained from the signature changed to the debonding depth of the composite plates used in this study. Since the EMI technique is strongly affected by small variations in the monitored structure, a composite structure with different dimensions may result in a different outcome. Nevertheless, the results found in this study can be a start to finding the debonding depth using the EMI technique.

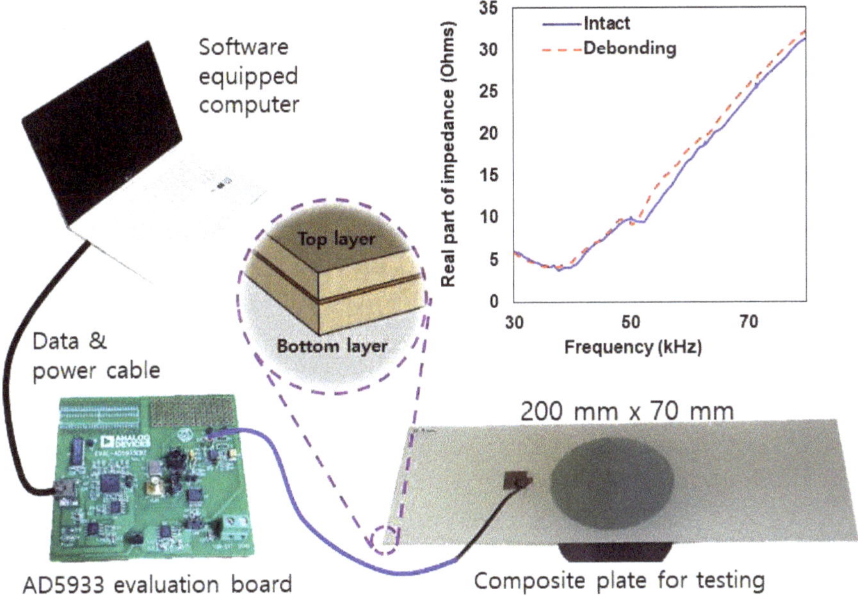

Figure 1. Experiment setup of the electromechanical impedance (EMI) technique for a debonding test.

Since the conventional approach of attaching the PZT patch onto the composite structure, as shown in Figure 1 does not have a large enough impedance peak, the first step was to create one. As previously mentioned, a resonance frequency range with peak(s) is required for a successful damage detection using the EMI technique. Using a 15 mm square PZT patch attached to a 25 mm diameter metal disc with 3 mm thickness, a large impedance signature peak can be created (shown later in Figure 3).

Figure 2 shows the experimental concept for this study where three test specimens are tested with each consisting of a bottom composite plate of 0.6 mm and varying top composite plate thickness of 0.2 mm, 0.4 mm, and 0.6 mm adhered using the epoxy adhesive. Note that only the half of the test specimen is adhered to test the validity of the proposed idea labelled "Area_A" in the figure (the other unattached area is labeled "Area_B"). Thus, Area_A and Area_B are considered as the intact case and damage (debonded) case, respectively. The first test case (shown in Figure 2) involved using the PZT-metal transducer (detailed explanation and research results on this idea can be found in [23]) to be randomly attached to Area_A with the impedance signature being measured then attached onto another spot within Area_A for another measurement until 10 impedance signatures were acquired. The PZT-metal transducer was attached by using a magnet of same size on the opposite side to achieve temporary attachment. Then, another 10 impedance signatures were measured on Area_B for comparison. The second test case was to do the same as the previous test but by using a different top composite plate thickness of 0.4 mm. The third and last test case was to conduct the experiment with 0.6 mm thickness top composite plate. The purpose of the random placement of the PZT-metal transducer is that in reality, it is virtually impossible to exactly attach the sensor onto the same spot every time. Thus, identifying the debonding depth regardless of the sensor attachment would greatly advance the EMI technique for bringing it one step closer for a real use in the (i.e., fault detection during manufacturing process of composite plates at the inspection stage where the PZT-metal can be temporarily be attached to check voids).

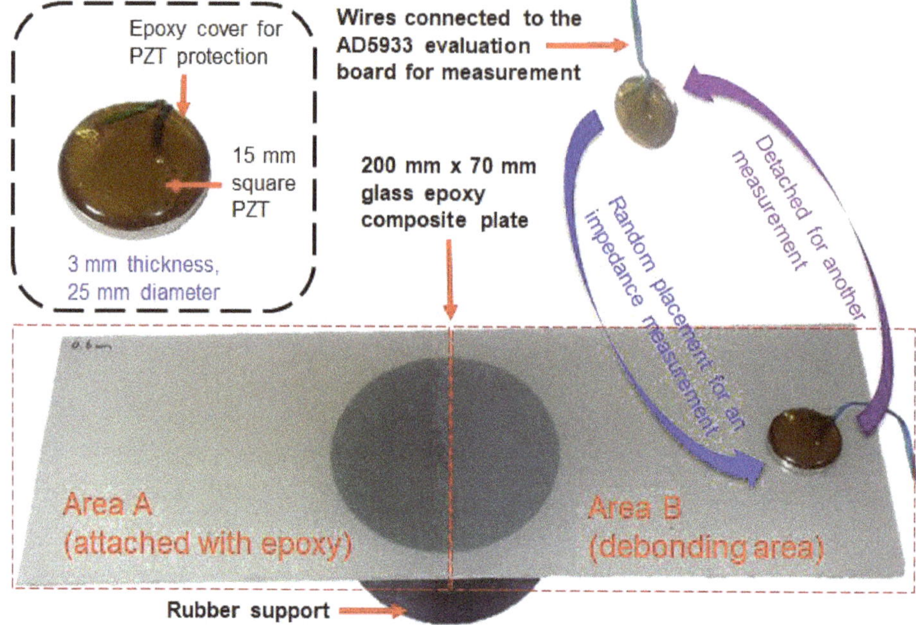

Figure 2. Experiment concept with the piezoelectric (PZT)-metal transducer (c.f. [23]).

Figure 3 shows all the impedance signatures acquired for the three test cases with 20 impedance signatures for Figure 3a–c. First, observing in Figure 3a involving 0.2 mm thick top composite plate, there is a clear difference between the 10 impedance signatures measured before debonding (dark green color on the right side of the figure) and after debonding (light green color on the left side of the figure). Before debonding (intact case), most of the resonance peaks are located near 40 kHz with the peak heights being located around 19 kΩ. With debonding, these peak heights increase to over 30 kΩ with the resonance peak being shifted to the left direction to around 37 kHz. Furthermore, averaging the two set of data clarifies the difference as shown in Figure 3d where the green line labeled "0.2_int" represents the averaged signature obtained from the 10 impedance signatures before debonding and "0.2_del" dotted green line represents the averaged signature from after debonding. With Figure 3b, the difference between the debonding (light blue lines) and intact case (dark blue lines) can easily be visually distinguished. Again, most of the impedance peak heights increase after debonding with a frequency shift in the left direction. In addition, the averaged signature shown in Figure 3d shows a clear difference between the intact case (labeled "0.4_int" blue line) and debonding case (labeled "0.4_del" blue dotted line). With the 0.6 mm thick top composite plate results shown in Figure 3c, the difference between the intact case and debonding case is not as significant compared to the previous two cases. However, it can be visually distinguished as the impedance signatures have shifted to the left after debonding of the bottom composite plate. The averaged signatures before debonding (red line labeled "0.6_int") and after debonding (red line dotted labeled "0.6_del") in Figure 3d once again results in the impedance signature peak increasing and shifting towards the left direction.

Observing in Figure 3d intact cases (0.2_int, 0.4_int, and 0.6_int), there is no frequency shift between 0.2_int and 0.4_int but a frequency shift in the left direction for 0.6_del can be seen. This experimentally proves for this study that the relationship between the thickness of the target structure and the frequency shift movement does not have a simple linear relationship, increasing the complexity when analyzing data. Next, observing at the debonding cases (0.2_del, 0.4_del, and 0.6_del), that the difference between 0.2_del and 0.4_del is small with the 0.4_del impedance peak having the highest value among the three impedance signatures (three dotted lines). This result is interesting when compared to the three intact cases as this shows that the increase in thickness does not always result in higher amplitudes. The higher amplitude for 0.4_del compared to 0.6_del shows that the 0.4 mm thickness may create larger resonances in the 25–45 kHz frequency range using the PZT-metal transducer created for this study. Another possibility of this outcome is that since the impedance signatures were measured 10 times by randomly placing the PZT-metal onto the composite structure, it is possible that this randomness has affected the outcome. Nevertheless, such phenomenon should be further researched in detail in the near future.

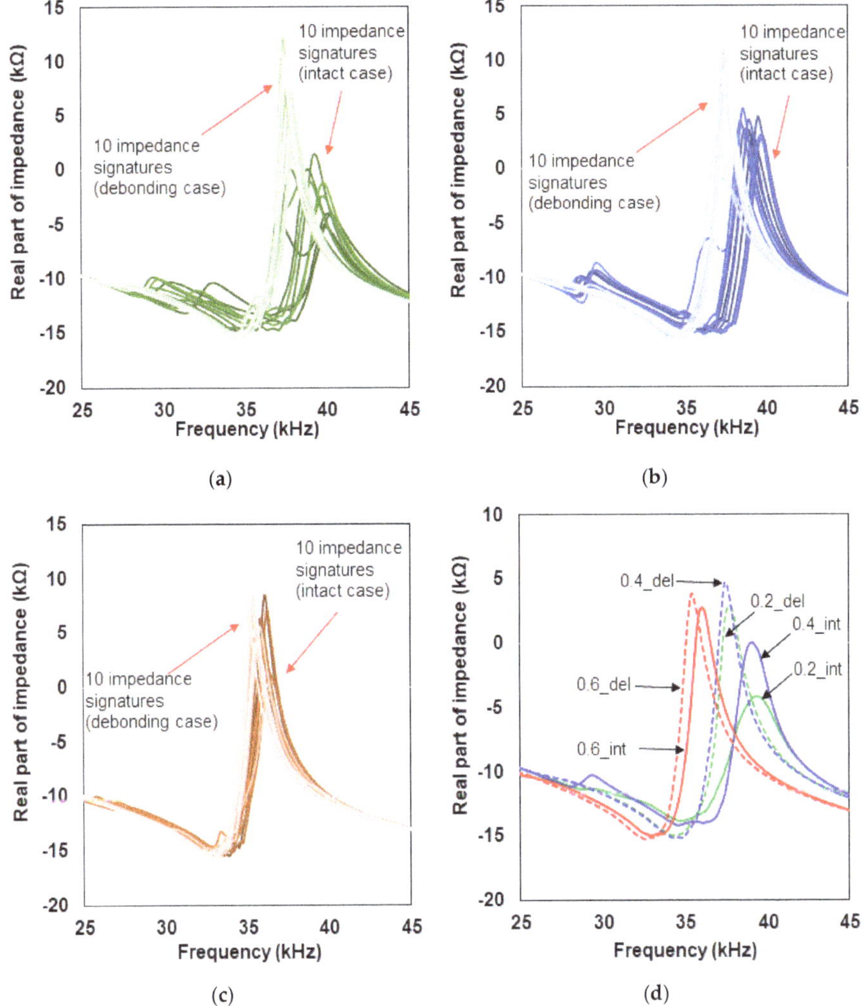

Figure 3. Impedance signatures: (**a**) With 0.2 mm top plate; (**b**) with 0.4 mm top plate; (**c**) with 0.6 mm top plate; (**d**) averaged for all measurements.

3. Possibilities for Identifying Debonding Depth

In the previous section, the changes in impedance signatures subjected to debonding damage were clearly witnessed. Thus, one way of identifying the debonding damage would be to utilize the averaged impedance signatures in Figure 3d. For example, if we have another set of composite structure consisting of two 0.6 mm thick glass epoxy composite plates adhered together, an intact case would result in an impedance signature that nearly matches the impedance signature 0.6_int. With close together and although measuring the impedance signature multiple times for averaging may solve the problem, this idea for solving this problem needs to be more specific. In this section, three statistical equations are used to quantify these changes in the signature to find out which statistical metric is suitable for identifying debonding depth. These three equations are root mean square deviation (RMSD), mean absolute percentage deviation (MAPD), and correlation coefficient deviation (CCD) represented in equations 1 to 3. Here, $Re\left(Z_i^o\right)$ represents the reference impedance signature (real part) and $Re(Z_i)$ the corresponding signature (real part). N is the number of impedance signatures, with the symbols $\overline{Z}$ and σ_Z representing the mean value and standard deviation, respectively.

$$\text{RMSD} = \sqrt{\sum_N \left[Re(Z_i) - Re\left(Z_i^o\right)\right]^2 / \sum_N \left[Re\left(Z_i^o\right)\right]^2} \tag{1}$$

$$\text{MAPD} = \sum_N \left|\left[Re(Z_i) - Re\left(Z_i^o\right)\right]/Re\left(Z_i^o\right)\right| \tag{2}$$

$$\text{CCD} = 1 - \text{CC, where CC} = \frac{1}{\sigma_Z \sigma_{Z^o}} \sum_N \left[Re(Z_i) - Re\left(\overline{Z}\right)\right] \cdot \left[Re\left(Z_i^o\right) - Re\left(\overline{Z}^o\right)\right] \tag{3}$$

For calculating the statistical metrics, the impedance signature of 0.2_int (averaged impedance signature) is used as the reference signature for Figure 3a. Thus, all 20 impedance signatures (10 intact case and 10 debonding case) were used to calculate the RMSD, MAPD, and CCD values for each signature resulting in 60 numbers. For Figure 3b and 3c, the reference signatures used were 0.4_int and 0.6_int, respectively where all the results for applying the three statistical equations can be seen in Table 1 to 3 below. For an easier visualization, all the numbers from the three tables are rearranged and plotted using the surface type plot in Figure 4. From the figure, the left half of the surface plot are all the results from the intact case where the RMSD, MAPD, and CCD values are relatively low compared to the right side of the plot, which represents the values after debonding. Observing at the intact cases (left side of the plot), it is easy to notice that the RMSD and CCD values are higher compared to the MAPD values. Since it is the intact case we are dealing with here, it would be great to have values close to zero to let one know that no damage is experienced. However, the EMI technique is a sensitive technique and randomly placing the PZT-metal transducer has changed the shape of the impedance signatures resulting in relatively high values for some cases (labeled in the figure). Thus, the preferred statistical equation to be used here just by examining the left side of the plot would be to use the MAPD equation since it has the lowest values. However, if we take a look at the debonding case results (right half of the plot), the MAPD has generally low values compared to the other two statistical metrics values. The CCD values result in the highest values here, which is the wanted outcome from the debonding experiment.

The idea of identifying the debonding depth can be referred to the three tables by examining the averaged values for each column. For Table 1, the averaged values (column) of RMSD, MAPD, and CCD are 16.9%, 9.6%, and 18.5% for the intact cases, respectively. The averaged values for the debonding cases are 32.5%, 16.2%, and 32.0% for the RMSD, MAPD, and CCD columns, respectively. Thus, one can propose a guideline by observing only the RMSD and CCD values, a debonding depth at 0.2 mm (in the thickness direction) can be found for the glass epoxy composite plate of the same type used for this study if the averaged RMSD and CCD values are over 30%. Thus, for example, if one was to inspect a composite structure, multiple measurements should be taken regardless of the sensor location placement and averaged values of RMSD and CCD under 20% should be disregarded. However, applying this guideline for Table 2 will result in the same outcome with the 0.4 mm thickness top composite plate with the RMSD and CCD averaged values of less than 20% for intact cases (RMSD of 15.6%, CCD of 11.2%) and over 30% for debonding cases (RMSD of 39.2%, CCD of 51.4%). Thus, the guideline should have more constraints such as averaged RMSD value of over 30% and CCD value of over 50% defined as the debonding damage at 0.4 mm depth (thickness direction). Referring back to Figure 3d, the change in the averaged impedances signatures before and after debonding of the 0.6 mm top composite plate was small compared to the other two cases. The calculated statistical metrics values in Table 3 also show this outcome as the values are considerably smaller compared to Tables 1 and 2. The averaged RMSD, MAPD, and CCD values before debonding are 8.6%, 3.3%, and 2.2%, respectively where these values increase to 20% and 8.6% and 13.2% with debonding. Thus, to identify debonding at 0.6 mm depth, the RMSD and CCD values less than 10% should be the threshold value where obtaining a value higher than this value will mean a debonding damage has occurred at 0.6 mm depth.

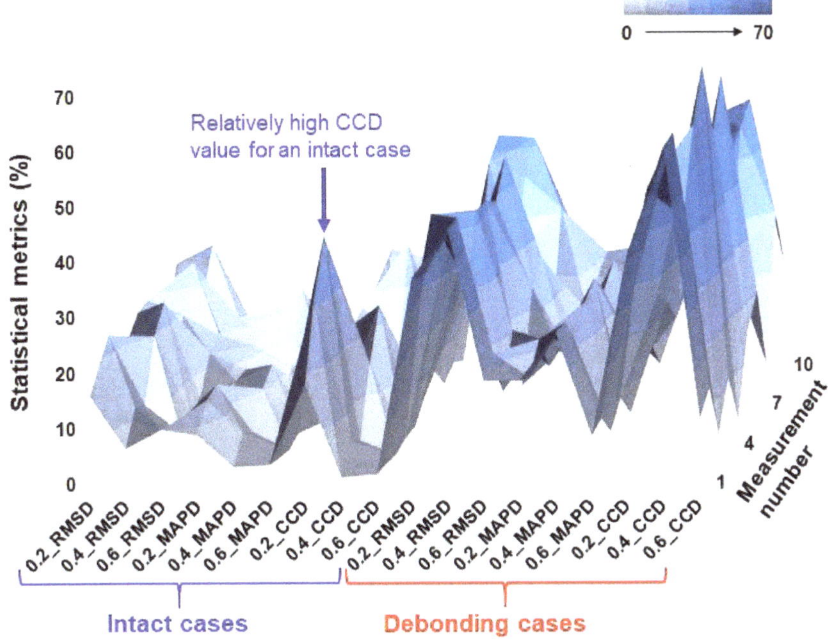

Figure 4. Surface type plot of the statistical metrics obtained from experiments.

Table 1. Results for 0.2 mm thickness top composite plate.

Measurement Number	Intact			Debonding		
	RMSD	MAPD	CCD	RMSD	MAPD	CCD
1	16.1	9.0	18.2	37.2	18.7	41.5
2	19.5	13.0	31.8	34.6	16.9	34.9
3	22.4	13.4	40.2	44.0	21.5	48.3
4	11.4	6.2	5.1	31.2	14.1	26.8
5	17.0	9.3	11.4	24.9	14.4	23.4
6	19.3	9.0	17.1	23.2	11.9	18.1
7	19.2	10.9	17.9	26.0	11.7	20.6
8	14.3	6.6	14.5	23.8	14.2	23.7
9	10.9	7.5	7.5	39.0	18.3	37.8
10	18.6	11	21.1	41.6	20.6	44.8
Averaged	**16.9**	**9.6**	**18.5**	**32.5**	**16.2**	**32.0**

Table 2. Results for 0.4 mm thickness top composite plate.

Measurement Number	Intact			Debonding		
	RMSD	**MAPD**	**CCD**	**RMSD**	**MAPD**	**CCD**
1	6.6	3.3	1.4	46.0	27.3	60.6
2	11.7	7.1	5.5	45.7	26.7	61.0
3	9.8	4.9	3.3	30.9	21.0	43.1
4	19.8	10.4	14.7	42.4	25.7	57.5
5	23.5	12.3	22.9	41.6	27.7	65.9
6	8.3	3.4	2.0	39.1	23.1	49.7
7	16.6	7.6	9.7	45.1	26.0	59.4
8	23.4	19.3	26.9	23.4	11.9	23.2
9	14.3	6.8	6.0	36.3	20.2	45.2
10	22.0	12.8	19.7	41.1	21.4	47.9
Averaged	**15.6**	**8.8**	**11.2**	**39.2**	**23.1**	**51.4**

Table 3. Results for 0.6 mm thickness top composite plate.

Measurement Number	Intact			Debonding		
	RMSD	**MAPD**	**CCD**	**RMSD**	**MAPD**	**CCD**
1	10.0	3.6	1.9	18.7	8.7	11.8
2	7.6	2.5	1.4	21.3	9.4	14.8
3	4.7	1.6	0.7	12.0	4.7	4.2
4	5.5	2.2	1.2	20.7	8.4	11.9
5	6.3	1.8	1.0	8.4	3.2	2.2
6	12.1	4.6	4.1	23.7	10.1	15.6
7	12.9	5.5	6.2	26.6	11.5	20.4
8	6.8	2.7	1.6	11.8	5.1	5.0
9	11.2	4.1	3.1	30.8	14.0	29.6
10	9.1	4.4	1.0	26.1	10.6	16.9
Averaged	**8.6**	**3.3**	**2.2**	**20.0**	**8.6**	**13.2**

To simply summarize the guideline for identifying the possibility of debonding depth of the glass epoxy composite plates used for this study, the conditions are as follows. Averaged RMSD and CCD values both resulting in between 30% and 40% means debonding has occurred at 0.2 mm depth. Averaged RMSD value of over 30% and CCD value of over 50% is that debonding exists at 0.4 mm depth. Lastly, averaged RMSD and CCD values between 10% and 20% states that debonding has occurred at 0.6 mm depth.

Although such guideline may be limited to the experiment conducted in this study, the findings in this study (including change in peak amplitude and left frequency shift movement) brings new possibilities and what needs to be researched in the future. A future study of various composite material types with real debonding cases will be tested with the proposed idea in the near future to bring the EMI technique one step close to real field applications.

4. Conclusions

In this study, the possibility of identifying debonding damage of the glass epoxy composite plates was proposed using the EMI technique. Three experiments were conducted with the 0.6 mm thick composite plate attached to another composite plate with varying thickness of 0.2 mm, 0.4 mm, and 0.6 mm. The PZT transducer was attached to a metal disc before attachment to the composite structure to create a large impedance signature peak, which was necessary for the study. Twenty impedance signatures were acquired for each of the three test specimens where 10 signatures were acquired from the undamaged part of the composite plate and the other 10 signatures from the debonding part of the composite plate. Here, the PZT-metal transducer was randomly placed on the test specimen each time for measuring a signature. The reason for random placement of the sensor was necessary as this is an important issue to be solved for the EMI technique to be used for practical application. For general inspection of composite components or structures, it would be virtually impossible to attach the sensor onto the exact same spot every time. From the experiments in this study, the random placement of the PZT-metal transducer caused the impedance signature to change. However, the change was more severe when subjected to debonding as the resonance peak amplitude increased with a shift towards the left direction.

Using the statistical metrics RMSD and CCD, a simple guideline was proposed based on the findings acquired from the study for identifying the debonding depth of composite structures. By measuring multiple impedance signatures, averaged RMSD and CCD values between 30% and 40% meant debonding has occurred at 0.2 mm depth. Secondly, averaged RMSD over 30% and CCD of 50% indicated debonding damage at 0.4 mm depth and lastly, averaged values of RMSD and CCD between 10% and 20% implied that debonding has occurred at 0.6 mm depth. Since this guideline was proposed based on the experiments conducted in this work, the future work will consist of testing out this idea with real debonding damage case scenarios to bring the EMI technique a step closer for real field applications.

Author Contributions: W.S.N. is the principal author for the research involved in writing the manuscript and analyzing data; J.B. contributed in conducting experiments and acquiring data.

Funding: The research was supported by a grant from "Development of infrastructure technology for hyper speed transportation system (20190140-001)" funded by Korea Institute of Civil engineering and building Technology (KICT), Korea.

Conflicts of Interest: The authors declare no conflict of interest.

References

1. Gholizadeh, S. A review of non-destructive testing methods of composite materials. *Proced. Struct. Integr.* **2016**, *1*, 50–57. [CrossRef]
2. Zou, Y.; Tong, L.P.S.G.; Steven, G.P. Vibration-based model-dependent damage (debonding) identification and health monitoring for composite structures—A review. *J. Sound Vib.* **2000**, *230*, 357–378. [CrossRef]
3. Cantwell, W.J.; Morton, J. The significance of damage and defects and their detection in composite materials: A review. *J. Strain Anal. Eng. Des.* **1992**, *27*, 29–42. [CrossRef]
4. Jolly, M.R.; Prabhakar, A.; Sturzu, B.; Hollstein, K.; Singh, R.; Thomas, S.; Foote, P.; Shaw, A. Review of non-destructive testing (NDT) techniques and their applicability to thick walled composites. *Proced. CIRP* **2015**, *38*, 129–136. [CrossRef]
5. Babu, J.; Sunny, T.; Paul, N.A.; Mohan, K.P.; Philip, J.; Davim, J.P. Assessment of debonding in composite materials: A review. *Proc. Inst. Mech. Eng. Part B J. Eng. Manuf.* **2016**, *230*, 1990–2003. [CrossRef]
6. Tinoco, H.; Cardona, C.; Peña, F.; Gomez, J.; Roldan-Restrepo, S.; Velasco-Mejia, M.; Barco, D. Evaluation of a piezo-actuated sensor for monitoring elastic variations of its support with impedance-based measurements. *Sensors* **2019**, *19*, 184. [CrossRef] [PubMed]
7. Liang, Y.; Feng, Q.; Li, D.; Cai, S. Loosening monitoring of a threaded pipe connection using the electro-mechanical impedance technique—Experimental and numerical studies. *Sensors* **2018**, *18*, 3699. [CrossRef]
8. Tawie, R.; Park, H.B.; Baek, J.; Na, W.S. Damage detection performance of the electromechanical impedance (emi) technique with various attachment methods on glass fibre composite plates. *Sensors* **2019**, *19*, 1000. [CrossRef]
9. Ai, D.; Luo, H.; Zhu, H. Numerical and experimental investigation of flexural performance on pre-stressed concrete structures using electromechanical admittance. *Mech. Syst. Signal Proc.* **2019**, *128*, 244–265. [CrossRef]
10. Na, W.S.; Lee, H. Experimental investigation for an isolation technique on conducting the electromechanical impedance method in high-temperature pipeline facilities. *J. Sound Vib.* **2016**, *383*, 210–220. [CrossRef]
11. Junior, P.; D'Addona, D.M.; Aguiar, P.R. Dressing tool condition monitoring through impedance-based sensors: Part 1—PZT diaphragm transducer response and EMI sensing technique. *Sensors* **2018**, *18*, 4455. [CrossRef] [PubMed]

12. Na, W.S.; Seo, D.W.; Kim, B.C.; Park, K.T. Effects of applying different resonance amplitude on the performance of the impedance-based health monitoring technique subjected to damage. *Sensors* **2018**, *18*, 2267. [CrossRef] [PubMed]

13. Qing, X.; Li, W.; Wang, Y.; Sun, H. Piezoelectric transducer-based structural health monitoring for aircraft applications. *Sensors* **2019**, *19*, 545. [CrossRef] [PubMed]

14. Li, W.; Liu, T.; Zou, D.; Wang, J.; Yi, T.H. PZT based smart corrosion coupon using electromechanical impedance. *Mech. Syst. Signal Proc.* **2019**, *129*, 455–469. [CrossRef]

15. Wang, F.; Ho, S.C.M.; Huo, L.; Song, G. A novel fractal contact-electromechanical impedance model for quantitative monitoring of bolted joint looseness. *IEEE Access* **2018**, *6*, 40212–40220. [CrossRef]

16. Xu, Y.G.; Liu, G.R. A modified electro-mechanical impedance model of piezoelectric actuator-sensors for debonding detection of composite patches. *J. Intell. Mater. Syst. Struct.* **2002**, *13*, 389–396. [CrossRef]

17. Park, S.; Kim, J.W.; Lee, C.; Park, S.K. Impedance-based wireless debonding condition monitoring of CFRP laminated concrete structures. *NDT E Int.* **2011**, *44*, 232–238. [CrossRef]

18. Sun, R.; Sevillano, E.; Perera, R. Debonding detection of FRP strengthened concrete beams by using impedance measurements and an ensemble PSO adaptive spectral model. *Compos. Struct.* **2015**, *125*, 374–387. [CrossRef]

19. Malinowski, P.; Wandowski, T.; Ostachowicz, W. The use of electromechanical impedance conductance signatures for detection of weak adhesive bonds of carbon fibre–reinforced polymer. *Struct. Health Monit.* **2015**, *14*, 332–344. [CrossRef]

20. Wandowski, T.; Malinowski, P.H.; Ostachowicz, W.M. Delamination detection in CFRP panels using EMI method with temperature compensation. *Compos. Struct.* **2016**, *151*, 99–107. [CrossRef]

21. Liang, C.; Sun, F.P.; Rogers, C.A. Coupled electro-mechanical analysis of adaptive material systems—Determination of the actuator power consumption and system energy transfer. *J. Intell. Mater. Syst. Struct.* **1997**, *8*, 335–343. [CrossRef]

22. Na, W.S. Distinguishing crack damage from debonding damage of glass fiber reinforced polymer plate using a piezoelectric transducer based nondestructive testing method. *Compos. Struct.* **2017**, *159*, 517–527. [CrossRef]

23. Na, S.; Lee, H.K. Resonant frequency range utilized electro-mechanical impedance method for damage detection performance enhancement on composite structures. *Compos. Struct.* **2012**, *94*, 2383–2389. [CrossRef]

 *micromachines*

Article

A Resonant Z-Axis Aluminum Nitride Thin-Film Piezoelectric MEMS Accelerometer

Jian Yang [1,2], Meng Zhang [1,2], Yurong He [1,2], Yan Su [1,2], Guowei Han [1,*], Chaowei Si [1,*], Jin Ning [1,2,3], Fuhua Yang [1,2,4] and Xiaodong Wang [1,4,5,6,*]

1 Engineering Research Center for Semiconductor Integrated Technology, Institute of Semiconductors, Chinese Academy of Sciences, Beijing 100083, China
2 Center of Materials Science and Optoelectronics Engineering, University of Chinese Academy of Sciences, Beijing 100049, China
3 State Key Laboratory of Transducer Technology, Chinese Academy of Sciences, Beijing 100083, China
4 Beijing Academy of Quantum Information Science, Beijing 100083, China
5 School of microelectronics, University of Chinese Academy of Sciences, Beijing 100049, China
6 Beijing Engineering Research Center of Semiconductor Micro-Nano Integrated Technology, Beijing 100083, China
* Correspondence: hangw1984@semi.ac.cn (G.H.); schw@semi.ac.cn (C.S.); xdwang@semi.ac.cn (X.W.); Tel.: +86-10-8230-5042 (X.W.)

Received: 1 August 2019; Accepted: 28 August 2019; Published: 6 September 2019

Abstract: In this paper, we report a novel aluminum nitride (AlN) thin-film piezoelectric resonant accelerometer. Different from the ordinary MEMS (micro-electro-mechanical systems) resonant accelerometers, the entire structure of the accelerometer, including the mass and the springs, is excited to resonate in-plane, and the resonance frequency is sensitive to the out-plane acceleration. The structure is centrosymmetrical with serpentine electrodes laid on supporting beams for driving and sensing. The stiffness of the supporting beams changes when an out-plane inertial force is applied on the structure. Therefore, the resonance frequency of the accelerometer will also change under the inertial force. The working principle is analyzed and the properties are simulated in the paper. The proposed AlN accelerometer is fabricated by the MEMS technology, and the structure is released by an ICP isotropic etching. The resonance frequency is 24.66 kHz at a static state. The quality factor is 1868. The relative sensitivity of this accelerometer, defined as the shift in the resonance frequency per gravity unit (1 g = 9.8 m/s^2) is 346 ppm/g. The linearity of the accelerometer is 0.9988. The temperature coefficient of frequency (TCF) of this accelerometer is −2.628 Hz/°C (i.e., −106 ppm/°C), tested from −40 °C to 85 °C.

Keywords: MEMS; AlN thin film; piezoelectric effect; resonant accelerometer; z-axis

1. Introduction

MEMS accelerometers and gyroscopes are widely used in inertial navigation systems [1]. The MEMS accelerometers can be categorized into electrostatic types, piezoelectric, piezoresistive, and pyroelectric ones, based on the working principles [2–8]. Compared with the electrostatic accelerometers (such as the Si accelerometers), the piezoelectric ones have a higher energy transfer efficiency and a simpler structure. In the electrostatic types, a complicated comb structure is a common design, since a capacitor structure is needed for excitation or sensing. The situation tends to be convenient when it comes to the piezoelectric ones, for even a simple electrode could realize the excitation or sensing functions [9,10]. Due to the charge leak, the resonant structure is often used for piezoelectric accelerometers.

A typical MEMS resonant accelerometer contains one or several resonant elements and proof masses, as shown in Figure 1 [7]. The resonant elements as a sensing part work in the resonant state.

When an in-plane inertial force is applied on the MEMS structure, the proof mass will tense or compress the resonant elements. This will lead to a change of the stiffness. Therefore, the resonance frequency will shift with the increase of the in-plane inertial accelerations.

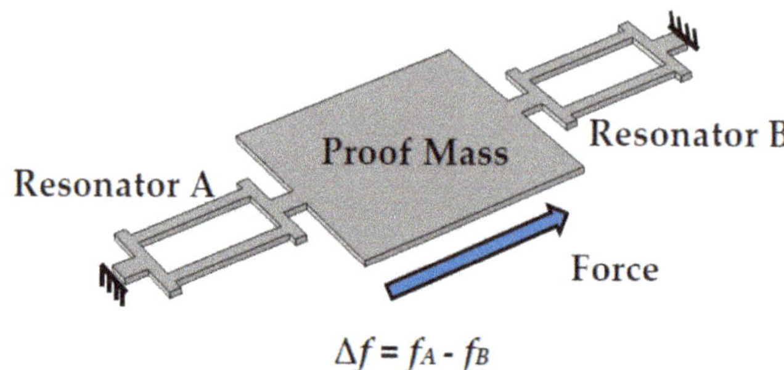

Figure 1. The structure of the resonant micro-electro-mechanical systems (MEMS) accelerometer.

AlN MEMS accelerometers have been studied in recent years [11–14]. However, most of these AlN accelerometers can only detect the in-plane acceleration. This limits the integration of the AlN 3D accelerometer. To detect the out-plane acceleration, a T-shape AlN MEMS accelerometer was designed in my previous work [15]. This accelerometer resonates in-plane but measures the acceleration out-plane. The z-axis sensitivity is 68.9 ppm/g and the quality factor is 1464. In order to improve the sensitivity and Q value, an optimized AlN accelerometer is proposed in this work.

In this paper, a tuning-fork structure is used in this AlN accelerometer. The tuning-fork structure comes from a Si MEMS gyroscope, in the Georgia Institute of Technology [16]. Two proof masses resonate reversely. In addition, the two anchors are designed at the middle of the supporting beams. When resonating, there will be no stress and no displacement at the anchors. Therefore, the anchor loss is zero. This design will benefit for improving the quality factor. This accelerometer is excited in-plane but can measure the out-plane acceleration. When an out-plane inertial force is applied on the AlN thin-film structure, the stiffness of the supporting beams will change. Therefore, the resonance frequency will shift with the increase of the out-plane acceleration. In this work, the z-axis sensitivity is measured at 346 ppm/g with the base frequency of 24.66 kHz. The quality factor is 1868.

2. Design

The AlN MEMS accelerometer is composed of four supported beams and two proof masses. The schematic structure is as shown in Figure 2. It contains four layers, the SiO$_2$ layer, the bottom electrode layer (Mo), the AlN layer, and the top electrode layer (Ti/Pt). The parameters are listed in Table 1. The SiO$_2$ layer works as the support and insulation. The bottom electrode is connected to the ground. The top electrodes, which are serpentine, are placed on the supported beams. They contain the drive electrode and sense electrode. The drive signal is applied on the drive electrode. Due to the piezoelectric effect, the electric field will excite a mechanical strain. This strain will lead to a stress σ, which is the in-plane direction. Moreover, the stress can drive the structure to vibrate.

$$\sigma = E_{ALN} \frac{U}{d_{AlN}} d_{31} \tag{1}$$

where E_{AlN} is the Young's modulus of AlN. d_{AlN} is the thickness of the AlN film. U is the drive voltage. d_{31} is the piezoelectric coefficient of AlN. According to the reported references [17,18], the $d_{31} = -2.6$ pm/V.

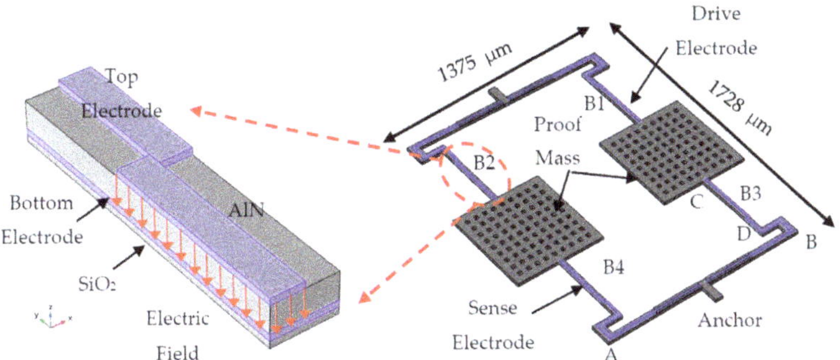

Figure 2. The structure of the aluminum nitride (AlN) MEMS accelerometer. In addition, the detailed analysis of the core beams (B1, B2, B3, B4).

Table 1. The structure parameters of the AlN MEMS accelerometer.

Parameters	Dimensions
Length of the beam (AB)	1150 µm
Width of the beam (AB)	35 µm
Length of the beam (CD)	400 µm
Width of the beam (CD)	25 µm
Width of the electrodes	12 µm
The sides of the proof mass	500 µm
The sides of the releasing hole	20 µm
The thickness of the structure	2.05 µm

To maximize the electromechanical signal, the top electrodes on the core beams are designed as serpentine electrodes. The sizes and the position of the serpentine electrodes come from the stress simulation. Figure 3a shows the stress simulation. The top electrodes are designed according to the stress distribution [11]. ABCDEFG is the line of the top electrode. Figure 3b shows the stress distribution. There are nearly no opposite sign charges on the top electrode. Therefore, a maximized electromechanical signal can be collected by the serpentine electrodes. Table 2 lists the parameters of each material.

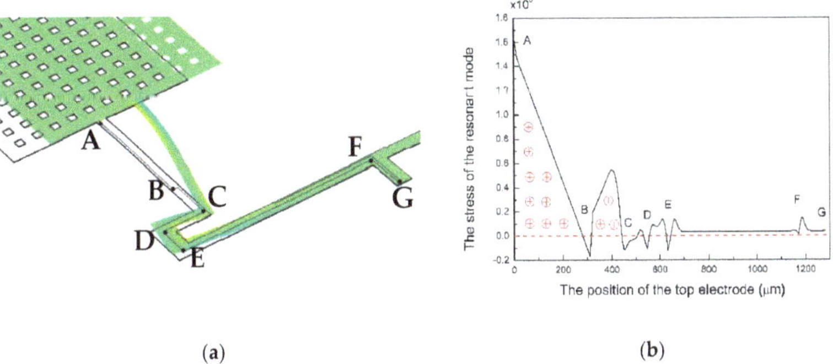

(**a**)   (**b**)

Figure 3. The stress simulation of the AlN MEMS accelerometer (**a**); the stress distribution along the top electrode (**b**).

Table 2. The parameters of each material from COMSOL Multiphysics library.

Parameters	AlN	Mo	SiO$_2$	Si	Pt	Ti
Density (kg/m^3)	3300	10,200	2200	2329	21,450	4506
Young's Modulus (GPa)	410	312	70	170	168	115

Due to the AlN piezoelectric effect, the MEMS structure is excited to resonate in-plane by the drive signal [19]. The resonant mode is simulated by the software COMSOL Multiphysics. The simulation result is shown in Figure 4. The two proof masses vibrate reversely. The positions of the anchors are the nodes. There is no displacement and no stress. This means there will be no anchor loss. The quality factor of this accelerometer will increase [20].

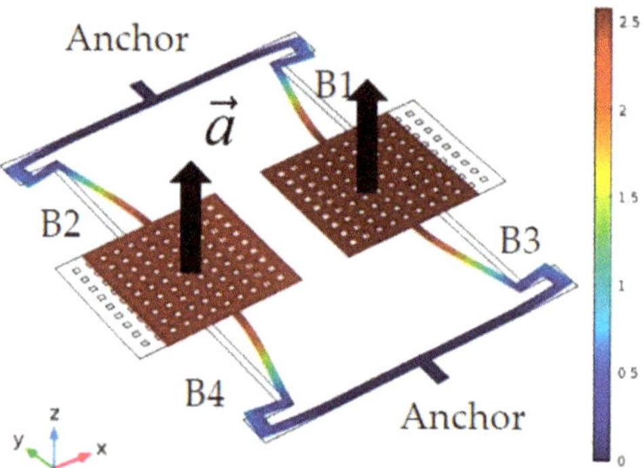

Figure 4. The displacement resonant mode of this AlN MEMS accelerometer.

The proof masses vibrate along the x-axis. The acceleration is along the z-axis. When an acceleration $\vec{a}$ applies on the entire structure, the supporting beams will deform in the z-axis. The stiffness coefficient k_z will change. As a result, the resonance frequency f_x will shift with the z-axis acceleration. The detailed derivation is in reference [15].

$$f_x = f_{x0}\left(1 - \frac{1}{2}\frac{\Delta k_x}{k_{x0}}\right) = f_{x0}\left(1 - \frac{3a}{E_z}\frac{Lm}{wt^3}a\right) = f_{x0}\left(1 - C\frac{Lm}{wt^3}a\right) \tag{2}$$

where f_{x0} is the in-plane resonance frequency at static state. And k_{x0} is the stiffness coefficient along x-axis at static state. C is a constant. L is the length of the beam. w is the width of the beam. t is the thickness of the structure.

The sensitivity of this accelerometer is proportional to $\frac{Lm}{wt^3}$. In order to increase the sensitivity, the mass and the length of the beam should be increased. In this work, the AlN MEMS accelerometer is a thin-film structure. The whole thickness of the structure is 2.05 µm. The size of the proof mass is increased to 500 µm × 500 µm.

Due to the lattice mismatch, there will be a residual stress in the actual accelerometer structure. The mean value of the residual stress is more than 500 MPa. This residual stress will lead the MEMS structure to bend out of plane, as shown in Figure 5a. To simplify the simulation and show the trend of the whole curve, the residual stress is set to 5 MPa, which is less than the actual value. Figure 5b shows the simulation result with the residual stress. The AB part which includes zero g, is monotonically increasing. This part is the range of this accelerometer.

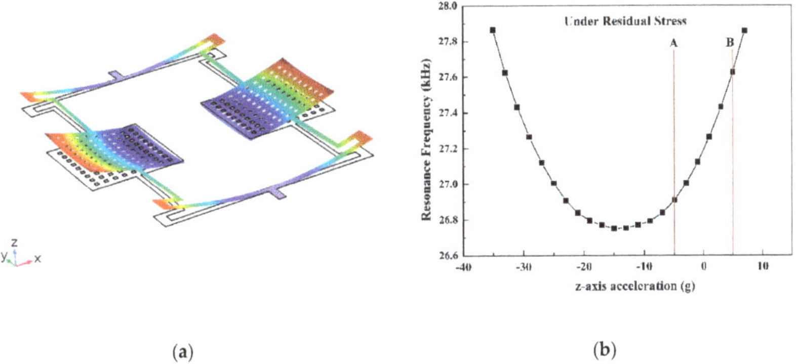

(a)   (b)

Figure 5. The deformation of the structure under the residual stress (**a**); the sensitivity curve of this AlN MEMS accelerometer simulated with the residual stress (**b**).

3. Fabrication

The substrate is the (100) silicon wafer with a 300 nm SiO_2 layer. The bottom electrode is Mo, with 300 nm thick. The (002) oriented AlN film is deposited on the Mo layer by the magnetron sputtering process. It is 1.3 µm thick. The piezoelectric coefficient d_{33} of the AlN film is 5.09 pm/V tested by the piezoresponse force microscopy [21]. A composite top electrode layer 150 nm Ti/Pt is deposited by the electron beam evaporation. The normal MEMS processes have been used to fabricate this accelerometer. The AlN film is etched by the ICP process with $Cl_2/BCl_3/Ar$ [22]. The entire structure is released by the ICP isotropic etching process with SF_6.

Figure 6a,b are the scanning electron microscope (SEM) pictures of the fabricated accelerometer. Figure 6a is the top view of the structure. Figure 6b is the side view of the structure. The structure bends along the +z-axis because of the residual stress.

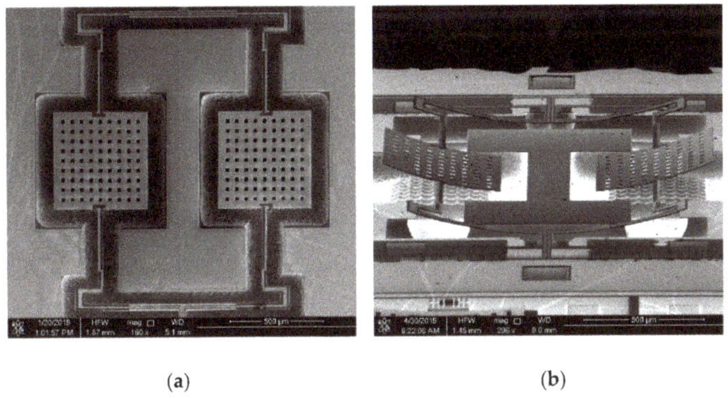

(a)   (b)

Figure 6. The micrographs of the AlN MEMS accelerometer. (**a**) Top view; (**b**) side view.

In order to achieve a better performance, a vacuum package process—the discharge welding—was used. In this process, the vacuum is 0.16 Torr.

4. Results and Discussion

The characteristics of this AlN accelerometer were tested by the dynamic signal analyzer 35,670 A. The source signal, which comes from the 35,670 A, applies on the drive electrode to excite the structure. The amplitude of the input signal is 0.2 Vpk. The GND electrodes are the bottom electrode. It connects the ground. In addition, the resonant signal is detected from the sense electrode. Figure 7 shows the

355

schematics of an electrical test of this accelerometer. The resonance frequency is 24.66 kHz at a static state at room temperature. The quality factor is 1868 as shown in Figure 8.

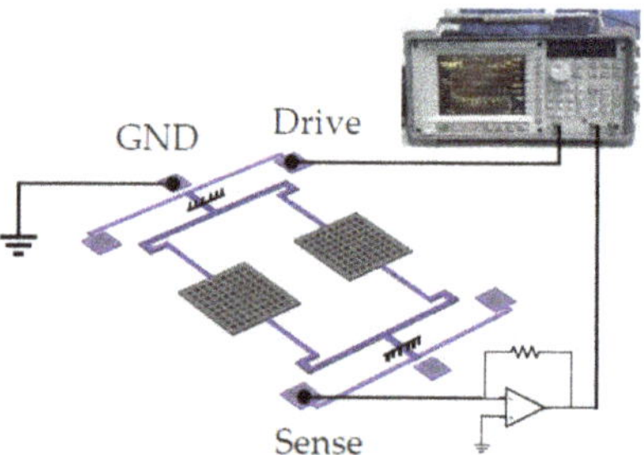

Figure 7. The schematics of the electrical test.

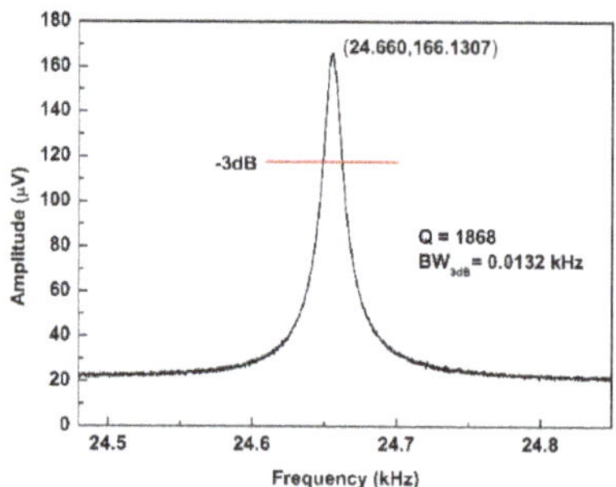

Figure 8. The resonant characteristics of the AlN accelerometer at a static state.

For the sensitivity test, the accelerometer was placed on the rotating platform vertically, as shown in Figure 9. A centripetal acceleration, which comes from the circular motion (i.e., $a = \omega^2 r$), would apply on the AlN resonant accelerometer, and lead to a frequency shift. The rates of the rotating platform range from 0~1000°/s. The accelerometer is placed at the position 20.5 cm away from the center. In this test, the rate was set from zero to 1000°/s with a step of 100°/s. The room temperature is 25 °C.

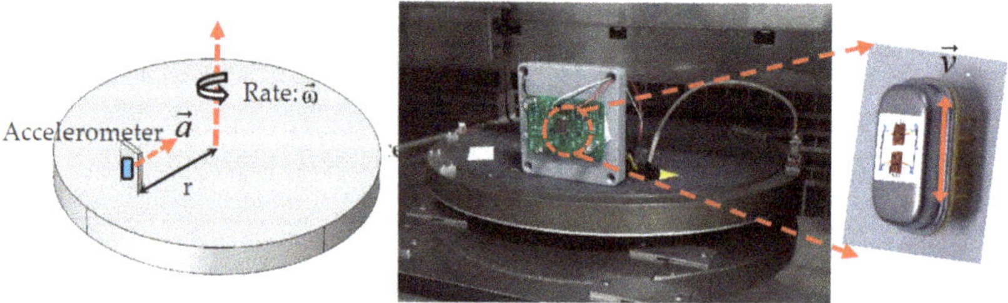

Figure 9. The sensitivity test by the rotating platform.

In the rotation test, the centripetal acceleration is perpendicular to the plane of the accelerometer. To avoid the Coriolis force, the MEMS chip is placed on the support vertically, which the direction of the vibration velocity $\vec{v}$ is parallel with the rate $\vec{\omega}$. Therefore, the Coriolis force is null. Figure 10 shows the experiment results of the relationship between the accelerations and resonance frequencies. The sensitivity of the AlN resonant accelerometer is 8.53 Hz/g, tested from −5 g to +5 g. The relative sensitivity (i.e., $\frac{\Delta f}{f_0}/g$, where f_0 is the resonance frequency) is 346 ppm/g at the base frequency of 24.66 kHz. The linearity of the accelerometer is 0.9988. With the increase of the acceleration, the resonance frequency increases. As shown in Figure 10b, the amplitude of each resonant peak is nearly equivalent.

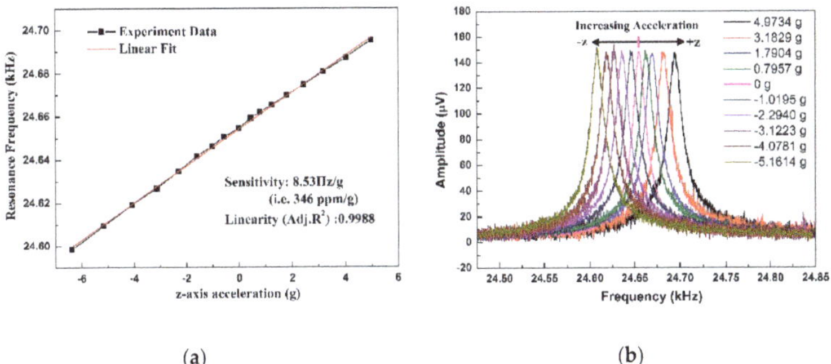

(a)

(b)

Figure 10. (**a**) Sensitivity of the accelerometer; (**b**) resonant peaks at different accelerations.

The cross error of this MEMS accelerometer was characterized by the rotation test. Figure 11a,b show the test results. The output frequencies scatter around 24.66 kHz. The x-axis sensitivity is 0.039 Hz/g, and the y-axis sensitivity is 0.52 Hz/g. Compared with the z-axis sensitivity, the cross error of the x-axis is 0.46% and the y-axis is 6.1%. These results prove that this MEMS accelerometer is insensitive to the in-plane accelerations.

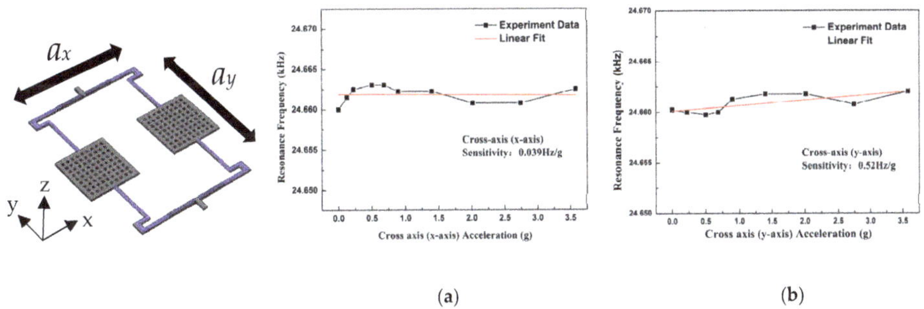

(a)

(b)

Figure 11. Cross axis sensitivities of the AlN MEMS accelerometer. (**a**) X-axis sensitivity; (**b**) Y-axis sensitivity.

The temperature characteristic of the AlN accelerometer is tested. The temperatures range from $-40\,°C$ to $85\,°C$ with a step of $5\,°C$. The temperature coefficient of frequency (TCF) of this accelerometer is -2.628 Hz/°C (i.e., -106 ppm/°C). Figure 12 shows the temperature characteristic of this accelerometer. Due to the negative temperature coefficient of AlN, a higher temperature will lead to a lower Young's modulus. The resonance frequency is proportional to $\sqrt{\frac{E}{\rho}}$. Where E is the Young's modulus. And ρ is the density.

$$f_0 \propto \sqrt{\frac{E}{\rho}} \tag{3}$$

Therefore, the resonance frequency will decrease with the increase of the temperature. In order to reduce the TCF, an optimized temperature compensation layer (SiO_2 layer) or a temperature compensation circuit will be used.

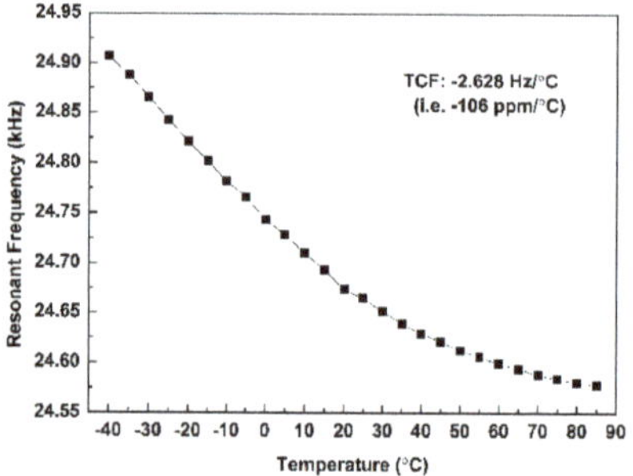

Figure 12. The temperature characteristic of the AlN accelerometer tested from −40 °C to 85 °C.

5. Conclusions

This paper reports an optimized AlN resonant accelerometer with a tuning-fork structure. The accelerometer is excited to vibrate in-plane. A z-axis acceleration, which is vertical to the plane of the accelerometer, can be detected. The z-axis sensitivity is 346 ppm/g, which is five times as much as the T-shape accelerometer. This accelerometer is potential to realize an integrated 3D AlN accelerometer. The Q value is 1868. The temperature coefficient of frequency (TCF) of this accelerometer is −106 ppm/°C, tested from −40 °C to 85 °C. A differential structure and a temperature compensation circuit will be designed to reduce the TCF. In the future work, we will focus on the increase of sensitivity, 3D integration, and reduction of TCF.

Author Contributions: J.Y. and C.S. conceived and designed the MEMS structure; J.Y. performed the simulation, the MEMS fabrication, the test, and wrote the paper; G.H., M.Z., Y.H., Y.S., J.N., F.Y. and X.W. revised the paper.

Funding: This work is supported by the National Natural Science Foundation of China (No.61704165, No. 61474115, No.51720105004, and No.10061704165) and the National Key R&D Program of China (No.2018YFF01010504) and the Science and Technology on Analog Integrated Circuit Laboratory Stability support project (No.6142802WD201806).

Conflicts of Interest: The authors declare no conflict of interest.

References

1. Wang, S.; Deng, Z.; Yin, G. An Accurate GPS-IMU/DR Data Fusion Method for Driverless Car Based on a Set of Predictive Models and Grid Constraints. *Sensors* **2016**, *16*, 280. [CrossRef]
2. Zhao, Y.L.; Li, X.B.; Liang, J.; Jiang, Z.D. Design, fabrication and experiment of a MEMS piezoresistive high-g accelerometer. *J. Mech. Sci. Technol.* **2013**, *27*, 831–836. [CrossRef]
3. Zhao, H.F.; Ren, W.; Liu, X.S. Design and fabrication of micromachined pyroelectric infrared detector array using lead titanate zirconate (PZT) thin film. *Ceram. Int.* **2017**, *43*, S464–S469. [CrossRef]
4. Cooper, E.B.; Post, E.R.; Griffith, S.; Levitan, J.; Manalis, S.R.; Schmidt, M.A.; Quate, C.F. High-resolution micromachined interferometric accelerometer. *Appl. Phys. Lett.* **2000**, *76*, 3316–3318. [CrossRef]
5. Yamane, D.; Konishi, T.; Matsushima, T.; Machida, K.; Toshiyoshi, H.; Masu, K. Design of sub-1g microelectromechanical systems accelerometers. *Appl. Phys. Lett.* **2014**, *104*, 074102. [CrossRef]
6. Chao, M.Y.; Ali, A.; Ghosh, S.; Lee, J.E.Y. An Aluminum Nitride on Silicon resonant MEMS accelerometer operating in ambient pressure. In Proceedings of the 19th International Conference on Solid-State Sensors, Actuators and Microsystems (TRANSDUCERS), Kaohsiung, Taiwan, 18–22 June 2017; pp. 607–610.
7. Wang, Y.; Ding, H.; Le, X.; Wang, W.; Xie, J.A. MEMS piezoelectric in-plane resonant accelerometer based on aluminum nitride with two-stage microleverage mechanism. *Sens. Actuators, A* **2017**, *254*, 126–133. [CrossRef]
8. Vigevani, G. MEMS Aluminum Nitride Technology for Inertial Sensors. Ph.D. Thesis, University of California, Berkeley, CA, USA, 2011.

9. Bao, F.H.; Bao, L.L.; Li, X.Y.; Khan, M.A.; Wu, H.Y.; Qin, F. Multi-stage phononic crystal structure for anchor-loss reduction of thin-film piezoelectric-on-silicon microelectromechanical-system resonator. *Appl. Phys. Express* **2018**, *11*, 067201. [CrossRef]

10. Manzaneque, T.; Hernando-García, J.; Ababneh, A.; Schwarz, P.; Seidel, H.; Schmid, U. Quality-factor amplification in piezoelectric MEMS resonators applying an all-electrical feedback loop. *J. Micromech. Microeng.* **2011**, *21*, 025007. [CrossRef]

11. Ruiz-Díez, V.; Manzaneque, T.; Hernando-García, J.; Ababneh, A.; Kucera, M.; Schmid, U. Design and characterization of AlN-based in-plane microplate resonators. *J. Micromech. Microeng.* **2013**, *23*, 074003. [CrossRef]

12. Ruiz-Díez, V.; Hernando-García, J.; Toledo, J.; Manzaneque, T.; Kucera, M.; Pfusterschmied, G. Modelling and characterization of the roof tile-shaped modes of AlN-based cantilever resonators in liquid media. *J. Micromech. Microeng.* **2016**, *26*, 084008. [CrossRef]

13. OlssonIII, R.H.; Wojciechowski, K.E.; Baker, M.S.; Tuck, M.R.; Fleming, J.G. Post-CMOS-Compatible Aluminum Nitride Resonant MEMS Accelerometers. *J. Microelectromech. Syst.* **2009**, *18*, 671–678. [CrossRef]

14. Vigevani, G.; Goericke, F.T.; Pisano, A.P.; Izyumin, I.I.; Boser, B.E. Microleverage DETF Aluminum Nitride resonating accelerometer. In Proceedings of the 2012 IEEE International Frequency Control Symposium Proceedings, Baltimore, MD, USA, 21–24 May 2012; pp. 1–4.

15. Yang, J.; Zhang, M.; Si, C.; Han, G.; Ning, J.; Yang, F.; Wang, X. A T-Shape Aluminum Nitride Thin-Film Piezoelectric MEMS Resonant Accelerometer. *J. Microelectromech. Syst.* **2019**, 1–6. [CrossRef]

16. Sharma, A.; Zaman, F.M.; Amini, B.V.; Ayazi, F. A high-Q in-plane SOI tuning fork gyroscope. In Proceedings of the Sensors 2004 Proceedings of IEEE, Vienna, Austria, 24–27 October 2004; pp. 467–470.

17. Hernando, J.; Sanchez-Rojas, J.L.; Gonzalez-Castilla, S.; Iborra, E.; Ababneh, A.; Schmid, U. Simulation and laser vibrometry characterization of piezoelectric AlN thin films. *J. Appl. Phys.* **2008**, *104*, 053502. [CrossRef]

18. Martin, F.; Muralt, P.; Dubois, M.-A.; Pezous, A. Thickness dependence of the properties of highly c-axis textured AlN thin films. *J. Vac. Sci. Technol. A* **2004**, *22*, 361–365. [CrossRef]

19. Yang, J.; Si, C.; Yang, F.; Han, G.; Ning, J.; Yang, F. Design and Simulation of A Novel Piezoelectric AlN-Si Cantilever Gyroscope. *Micromachines* **2018**, *9*, 81. [CrossRef]

20. Gil, M.; Manzaneque, T.; Hernando-García, J.; Ababneh, A.; Seidel, H.; Sanchez-Rojas, J.L. Multimodal characterisation of high-Q piezoelectric micro-tuning forks. *IET Circuits Devices Syst.* **2013**, *7*, 361–367. [CrossRef]

21. Zhang, M.; Yang, J.; Si, C.W.; Han, G.W.; Zhao, Y.M.; Ning, J. Research on the Piezoelectric Properties of AlN Thin Films for MEMS Applications. *Micromachines* **2015**, *6*, 1236–1248. [CrossRef]

22. Yang, J.; Si, C.W.; Han, G.W.; Zhang, M.; Ma, L.H.; Zhao, Y.M. Researching the Aluminum Nitride Etching Process for Application in MEMS Resonators. *Micromachines* **2015**, *6*, 281–290. [CrossRef]

Article

Comparative Analysis and Strength Estimation of Fresh Concrete Based on Ultrasonic Wave Propagation and Maturity Using Smart Temperature and PZT Sensors

Najeebullah Tareen [1], Junkyeong Kim [2,*], Won-Kyu Kim [3] and Seunghee Park [3,4,*]

[1] Department of Civil, Architecture & Environmental System Engineering, Sungkyunkwan University, Gyonggi-do 16419, Korea
[2] Research Strategy Team, Advanced Institute of Convergence Technology, Gyonggi-do 16229, Korea
[3] Department of Convergence Engineering for Future City, Sungkyunkwan University, Gyonggi-do 16419, Korea
[4] School of Civil, Architectural Engineering & Landscape Architecture, Sungkyunkwan University, Gyonggi-do 16419, Korea
* Correspondence: junkyeong@snu.ac.kr (J.K.); shparkpc@skku.edu (S.P.)

Received: 24 July 2019; Accepted: 21 August 2019; Published: 23 August 2019

Abstract: Recently, the early-age strength prediction for RC (reinforced concrete) structures has been an important topic in the construction industry, relating to project-time reduction and structural safety. To address this, numerous destructive and NDTs (non-destructive tests) are applied to monitor the early-age strength development of concrete. This study elaborates on the NDT techniques of ultrasonic wave propagation and concrete maturity for the estimation of compressive strength development. The results of these comparative estimation approaches comprise the concrete maturity method, penetration resistance test, and an ultrasonic wave analysis. There is variation of the phase transition in the concrete paste with the changing of boundary limitations of the material in accordance with curing time, so with the formation of phase-transition changes, changes in the velocities of ultrasonic waves occur. As the process of hydration takes place, the maturity method produces a maturity index using the time-feature reflection on the strength-development process of the concrete. Embedded smart temperature sensors (SmartRock) and PZT (piezoelectric) sensors were used for the data acquisition of hydration temperature history and wave propagation. This study suggests a novel relationship between wave propagation, penetration tests, and hydration temperature, and creates a method that relies on the responses of resonant frequency changes with the change of boundary conditions caused by the strength-gain of the concrete specimen. Calculating the changes of these features provides a pattern for estimating concrete strength. The results for the specimens were validated by comparing the strength results with the penetration resistance test by a universal testing machine (UTM). An algorithm used to relate the concrete maturity and ultrasonic wave propagation to the concrete compressive strength. This study leads to a method of acquiring data for forecasting in-situ early-age strength of concrete, used for secure construction of concrete structures, that is fast, cost effective, and comprehensive for SHM (structural health monitoring).

Keywords: nondestructive testing; maturity method; concrete early-age strength; SmartRock; ultrasonic waves; PZT (piezoelectric) sensors; structural health monitoring

---

1. Introduction

Concrete is an extensively used material in construction owing to its rich qualities, including being cost-effective, durable, and moldable into shapes, permitting the construction of safe and reliable structures. By contrast, the complex mixture (cement, sand, aggregates, admixtures, etc.) and field uncertainties make concrete susceptible to imprecise determinations of strength in the early stages of curing; managing these factors is important in preventing structures from unexpected accident or failure due to the premature removal of a framework or immature concrete. It is necessary to understand the materials used for construction to ensure the efficiency and longevity of the structures being built. Several types of structures are built using concrete. It is important to study and determine the properties of cement and concrete at different stages of construction as they directly affect the strength and durability of the structures. In the project schedule for construction, the idle time required for the hardening of concrete is a crucial influencing factor, especially in construction of high rise or multi-staged projects. Unscheduled removal or early removal of the supportive framework sometimes causes huge losses, such as the incident that took place in the construction of 2000 Commonwealth Avenue, Boston [1]; one side of the building progressively fell down, causing the death of four workers and injury to twenty others. Later, it was reported that the collapse was initiated due to a roof slab that was insufficiently cured due to exposure to cold weather. Structures such as high-rise buildings, dams, and bridges need precise monitoring in the early stages of construction because the size of the structures becomes a challenge for safety and quality assurance. For the safe loading of structural elements, applying pre-stressing, removing form work, and limiting unnecessary time between project stages, it is necessary to understand the curing process and be familiar with concrete strength and stiffness. Monitoring of the curing and strength development processes provides enough information for these stages. Consequently, for quality assurance, health monitoring, and safety control, many techniques of destructive and non-destructive testing (NDT) are used [2–13]. The innovations in this field have led the way to smart construction, and utilizing the idea of the IoT (Internet of Things) has upgraded the structural health monitoring process to precisely and remotely monitor the structures and to provide a safe work environment. These innovations tie the IoT to data simplification for destructive and NDT [2]. Conventionally, compressive strength tests of concrete use concrete cylinders or cubes cast along the structure. Even though this technique is simple and economical, various factors cause a difference in the strength of the main structure and of these cubes; one of those factors is spatial variation. On the other hand, NDT is preferred as it is more practical, quick, and cost effective. Some of the NDT techniques that are widely used are the ultrasonic pulse velocity, Schmidt rebound hammer test, radiography, resonant frequency, isothermal calorimeter, maturity method, and some acoustic tests [14–24]. The traditional method for measuring concrete strength may need adjustments of special equipment and to be carried out during construction, which can interrupt the tough construction schedule, while the smart techniques do not delay construction and can provide data for as long it is needed in the process of structural health monitoring (SHM). The accuracy of the results for these tests differs from case-to-case, depending on the technique used. The ultrasonic waves analysis, which is based on a resonant frequency change and uses different features of wave propagation, is the test with enough data availability for analysis. The application of ultrasonic waves propagation is widely used for structural health monitoring, concrete compressive strength prediction, and damage detection in structures [9,25,26]. The technique generously reveals elastic and dynamic properties of the concrete, making this procedure plausible for concrete structural health monitoring using piezoelectric (PZT) materials [6,8,26–28]. In ultrasonic waves analysis, PZT sensors are used to mechanically measure the dynamic responses of the ultrasonic waves that display boundary disparities owing to the development of concrete strength. The velocities of the waves are determined from the arrival time of the signals. Because the PZT materials are used as sensors and actuators, the distance between the actuator and sensor with the arrival of the first return signal is used to modify the propagation process. For fresh concrete, the approach of the ToF (time-of-flight) to the arrival of the first signal gives smooth results; monitoring of velocities for the P-waves and R-waves gives extensive data about the concrete [8,9]. For these approaches, the applied simplification differs for the surface and internal waves. The shared waves may also be used, giving adequate results for damage and crack detection. Simultaneously, embedded, wireless temperature sensors measure the hydration temperature history of the concrete to estimate the concrete early-age strength using the maturity method [16,17,29,30]. The maturity theory states that the graphical area of temperature experienced by a specimen, beyond the datum temperature of the concrete, predicts the concrete strength by using the Nurse–Saul maturity index. In this case, the hydration temperature and the curing age forecast the strength of the concrete on a formulated basis [29,31]. The ASTM (American Society for Testing and Materials) provides two maturity functions for the maturity tests: (i) the Nurse–Saul function and (ii) the Arrhenius function, which uses the temperature history of hydration and the activation energies of the cementation process [16]. In this maturity method study, temperature effects were analyzed for the strength-gaining process and the effects of hydration temperature were determined for early-age concrete strength development. In this procedure, we acquired the data by wireless sensors embedded in the concrete, processed the data for noise reduction, and used the Nurse–Saul method of maturity curve development. For competence of the data and to streamline the work, the results were compared with those formulated by maturity method data, ultrasonic wave propagation results, and the compressive strength gained by a penetration resistance test using a universal testing machine (UTM) and conducted on cylindrical specimens taken from the same concrete mix.

2. Transition of Pattern, Dissimilation of Signals

2.1. Waves Direct Velocities

Ultrasonic wave propagation through the concrete medium is studied widely. The variations of these approaches lay in the methodology used for analyzing the wave propagation. For this purpose, different parameters are analyzed related to concrete stiffness and compressive strength properties. Normally the technique of direct transmission of waves through the concrete medium is used. By using the ultrasonic waves analysis method, three types of data can be acquired when dynamic loads or vibrations are applied to a concrete structure: (i) compressional waves or P-waves, (ii) Rayleigh or surface waves, and (iii) shear waves or S-waves [32,33]. The characteristics of these waves can be determined by the patterns that depend on the velocity, amplitude, and the relationship of these wave parameters. Transmission of compressional waves have high velocity whereas the velocities of shear and surface waves are approximately 60% and 55% of that of compressional waves, respectively. The relationship between the concrete strength and waves is usually given by [34],

$$fc = a \, \exp(b \, V). \tag{1}$$

In this equation fc represents the compressive strength, V is the velocity of the longitudinal waves, and a and b are the parameters determined by P. Turgut [35] using the least squares method. The quality monitoring and assurance by wave propagations follows the theory of kinematics of deformation, which is based on Newton's second law of motion [36]. According to this theory, the total stresses acting on the subjected area of a concrete specimen is the second order differential governing longitudinal wave propagation,

$$\frac{\partial^2 u}{\partial t^2} = S^2 \frac{\partial^2 u}{\partial x^2}. \tag{2}$$

In Equation (2), S is the speed of the wave, u is the displacement, t represents the time, and x is the path length.

For measurement of stressed zones, the waves speed can be used. The wave speed depends on the density and Young's modulus according to,

$$S = \sqrt{\frac{E}{\rho_0}}. \tag{3}$$

In Equation (3), Young's modulus and the density of the material are represented by E and ρ_0 respectively. In stress waves, the Poisson ratio of the material affects the velocity, so the effects are countered using Equation (4) [35],

$$S = \sqrt{\frac{E}{\rho_0}\frac{1-v}{(1+v)(1-2v)}}.$$

(4)

In this equation, v is the Poisson ratio of the material, which is incorporated to give the value for the speed of the waves in the material.

2.2. Molded Variation

Because the stiffness and rigidity grow with the curing time of the concrete, the velocity of the waves increases as a function of time. This variation in stiffness molds the velocity of the waves. The stiffness of the material is related to the Young's modulus of material [37]. Thus, with curing time, the Young's modulus also rises.

According to the wave equation, the relation of wave speed, Young's Modulus, and density of material is defined as,

$$\frac{\partial^2 u}{\partial t^2} = \frac{E}{\rho_0}\frac{\partial^2 u}{\partial x^2} = S_1^2\frac{\partial^2 u}{\partial x^2}.$$

(5)

In Equation (5), E is Young's modulus, S_1 is the speed of the elastic wave, u is the displacement, t is the time, x is the path length, and ρ_0 is the density of the material in its unstrained state.

3. Experimental Setup

This study of concrete comparative strength-development monitoring was prepared using three different results: (i) the maturity method of concrete strength estimation, (ii) ultrasonic wave propagation in concrete, and (iii) UTM-based concrete compressive strength estimation, which was composed of smart PZT sensors and temperature sensors (SmartRock) to monitor the commencement of dynamical changes in the concrete specimens.

3.1. Mixture Composition and Material Gradations

For the experiment, a cubic meter sample composition was prepared according to quantities given in Table 1. For the admixture, silica fume was added for bond strength and compressive-strength gain. The machine mix of cement and aggregates was prepared, and the admixture was added before adding water.

Table 1. Concrete mix details (kg/m^3).

Materials	Cement	Silica Fume	Silica Powder	Sand	Aggregates	Water	Fiber (Vol %)
Quantity	330	0.165 (0.05%)	2.310 (0.70%)	873	916	165	0

Ordinary Portland cement of type ASTM C 150 was used in the preparation of the specimens. The gradation of aggregates, determined by sieve analysis, is given in Figure 1. Sand was used for fine aggregates and crushed stones were used for coarse aggregates. The size of the coarse aggregates used was 18 mm.

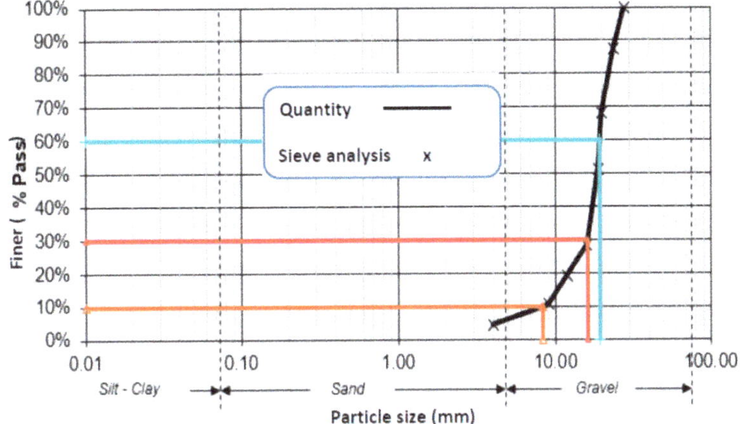

Figure 1. Gradation of gravel size for the concrete mixture.

3.2. Specimen Properties

Concrete specimens of dimensions 400 mm × 100 mm × 100 mm were casted for the test of ultrasonic waves and maturity method strength monitoring. The PZT transducers and temperature sensors were embedded in the specimens during pouring of the concrete. Simultaneously, cylindrical samples were cast for the compressive strength test by penetration resistance using a UTM from the same concrete mixture. Two types of specimens of the same concrete mix were poured for the experiment, the specimens of which dimensions are aforementioned were used for the ultrasonic wave propagation monitoring and SmartRock sensors. The cylindrical samples were poured for the destructive test to be placed as part of the penetration resistance test. The average value of three specimens was considered the value for each specific day during the strength-development monitoring; monitoring was completed on days 1, 2, 3, 7, 14, and 28.

3.3. PZT Material as Sensors and Actuators

The PZT material has the ability to interconvert mechanical and electrical energies and sense the dynamical vibration in paste and solid media. These rich properties enable PZT materials to be used both as sensors and actuators [14,38–40]. As the PZT materials are used as transducers, for the needs of sensing and actuating, the terms of PZT sensors and PZT actuators are used, respectively, according to placement. Researchers have employed these materials using different approaches as sensors. PZT sensors can be used to extract different mechanical properties of the subject samples or structures. Furthermore, using PZT materials for different approaches and to consider various parameters can determine strain, stiffness, and stresses, or other modified features of material. The strain effects and electro-mechanical effects of the PZT sensors are used to investigate the dynamical properties of the host structure. In this study, the property of propagation of ultrasonic waves was utilized to study and analyze the dynamic changes observed with the curing age and development of concrete strength [8–10,41,42].

3.4. Sensors' Preparation

For data acquisition, PZT materials were used to actuate and sense the responses of propagated signals from the medium. The specific PZT material used in this experiment was APC 850 (American Piezo Ceramics), which has high sensitivity characteristics and large displacement effect; these PZT transducers were embedded in the concrete with large damping effects. The material of size 10 mm × 10 mm was attached to 300 mm × 30 mm steel plates such that two PZT specimens were placed on each face of a steel plate separated by 140 mm (a total of four transducers on a single steel plate, PZT specimens were attached on both sides of the plates), as shown in Figure 2. Here, the size of the specimen is smaller than the targeted structure, and further resizing can take place according to the needs and propagation plan. Because the thick transducers generated the ultrasonic waves in the direction of the thickness of the steel plates, thin PZT transducers were used to generate guided waves along the surface of the steel plates. The distance between pairs of the PZT transducers were determined such that signals from the direct path of the PZT transducers could not contaminate the reflected signals from the side and end boundaries of the steel plate.

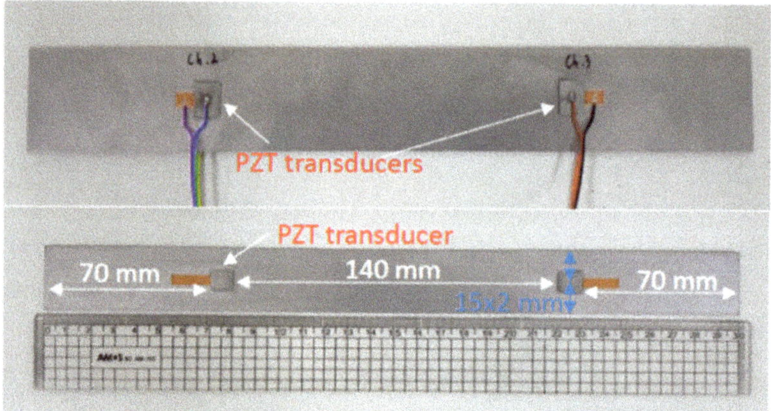

Figure 2. Piezoelectric (PZT) transducer locations on plate, including dimensions.

The PZT transducers were embedded in concrete specimens to actuate and produce guided ultrasonic waves and were considered as a sensor to observe the internal dynamic changes of the subjected material [39]. The sensors were embedded in the specimen during the pouring of the concrete and positioned to keep the connections safe for acquiring the data as shown in Figure 3a,b.

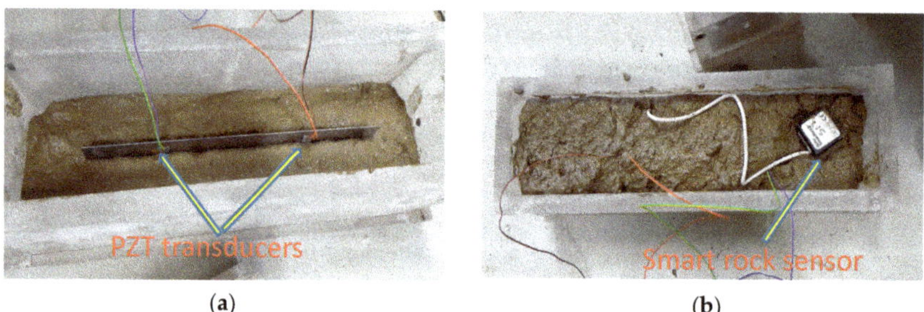

(**a**) (**b**)

Figure 3. (**a**) Embedded PZT transducers in concrete specimen and (**b**) allocation of SmartRock (temperature sensor).

Because the electrical behavior of the material is linear,

$$D_i = \varepsilon_{ij} E_i. \tag{6}$$

In Equation (6), D is the displacement of the electric charge density, E is the electric field strength, and ε is the permittivity of a free body. The factor of proportionality relating the electric field strength and electric displacement is defined as the medium's dielectric constant. The tensor of the displacement of the electric charge density is also represented in Equation (6).

The strain–stress relationship is in accordance with Hooke's Law,

$$S_i = S_{ij}T_j. \tag{7}$$

In this equation, S represents the strain and T represents the stress produced in the body. Equations (6) and (7) can be written as

$$S = s^E T + dE,$$
$$D = dT + \varepsilon^T E. \tag{8}$$

In Equation (8), s^E is the compliance under the electric field and ε^T represents the dielectric constant of the PZT material under constant stress. Equation (8) is used to simplify the stress and strain effects occurring in the body.

3.5. Methodology Analysis for the Longitudinal Wave Variations

The approach for conducting experiments for this study encompassed embedding smart PZT transducers in a concrete specimen to measure the ultrasonic waves variation of center frequencies in the range of 100–300 kHz and thus to monitor the hardening process of concrete by the gradual divergence of phase-transition as the specimens cured. Especially in the liquid phase to the initial sitting time, the strength measuring process is critical, in that it depends on the mixture type and sometimes varies with the measuring approaches. In the initial sitting time, the frequencies of waves indicate the dynamic change in the medium. In this study, we monitored the phase changes of the A-mode and S-mode of signals.

The experiment was carried out with a concrete mixture ratio given in Table 1 and kept at a temperature of 23 °C. The data from the samples were collected continually every thirty minutes, with the aid of a computer setup. For the measurement of data, a setup of an arbitrary waveform generator (AWG), controller, multiplexers (MUX), and a signal digitizer (DIG) was assembled to acquire and process the data. The ultrasonic wave velocity profiles were obtained as a function of time. In this experiment, the velocity variation was studied to estimate the compressive strength of concrete. With the gradual change in the amplitude of the ultrasonic waves, the change of strength became more prominent and was easier to observe and calculate. For this purpose, the changes in the pattern were approached differently from the Roa-Rodriguez method [43].

In the process of ultrasonic wave propagation, to consider the analogy of phase transition, the simplified one-dimensional equation fits adequately because the ultrasonic pulse detector or receiver focuses only on the first arrival of the signal of the longitudinal wave and the subsequent signals are ignored by the system. Furthermore, ultrasonic waves were analyzed through the functionality of the solid-medium wave-transformation analogy. Studies of the phase transition state that for the monitoring of composite structures, the A-mode of Lamb waves gives a clearer and more appropriate result than the S-mode [44,45]. When the propagated wave travels in-phase, it is considered as S-mode and when it travels out-of-phase it is considered A-mode. The relationship can be altered by the placement of the PZT transducers and the polarization of the specimens. This makes it easy to recognize the individual modes of the waves by adding or subtracting one signal from the other. As for the nomination of the PZT transducers, the arrangement of the signals can be made according to its order. If cracks or damage are detected, the signal's phases will be identical. The waves propagation speed, s, in Equation (5) is independent of $\partial u/\partial t$, or that of the local velocity/speed shift of the elements transmitting the wave through sections. The time-of-flight is used for the specification of the velocity change of the waves. The velocity of waves is modified by dynamic variations of the material thickness. The elastic properties of the transmuting medium, density, and Poisson's ratio define the wave speed, s. In this study, all the wave speed values are assumed constant. Hence, the results were compared for all three methods.

4. Hydration Temperature and Maturity Index

The hydration temperature history uses data for the NDT to reveal hydration-temperature-related features to monitor compressive strength of fresh concrete. This method relies on the combined parameters of temperature and time and can be used for in-place concrete during construction [17,46].

When water is added to the mixture, the hydration process commences, and the chemical reaction of disilicate, trisilicate, aluminate, and other admixtures begins, so the hydration temperature of concrete rapidly increases. The temperature rises rapidly during the initial hours of pouring until two days later, after which it stays constant with the development of the material's internal bonding and increased concrete strength [47]. In the maturity method depicted in Figure 4, after pouring concrete, the hydration temperature is the key feature to determine the maturity curve and estimate the maturity function for the prediction of concrete strength. For the strength development, the chemical bonding of individual particles is linear. The related study of N.J. Carino states, in Equation (9), that the shrinkage of chemical bonding is a dependent function of age and gives a clear result for the hydration temperature [17]. The strength can be further clarified by considering the datum temperature as related to the mixture type,

$$S = S_u = \frac{\sqrt{k(t - t_0)}}{1 + \sqrt{k(t - t_0)}}. \tag{9}$$

In this equation, S represents the strength with respect to time t, t_0 is the time at initial sitting, and S_u represents the limiting strength of the sample. In Equation (9), k is a constant rate, which in a molecular system is when the energy is transferred between the molecules due to collisions during their constant motion. The kinetic energy of the molecules increases as the system heats up and molecules surmount the barrier of energy required for lower-energy reaction products. Thus, the rate of reaction increases with increasing temperature observed as a constant rate k [48].

While we can define the datum temperature of the mixture according to the properties of cement and other admixtures, when admixtures are added, the datum temperature can be different. In slack seasons of concrete pouring, freezing-temperature effects on the strength-gaining process of inter-particle bonding must be considered when specifying the datum temperature [17].

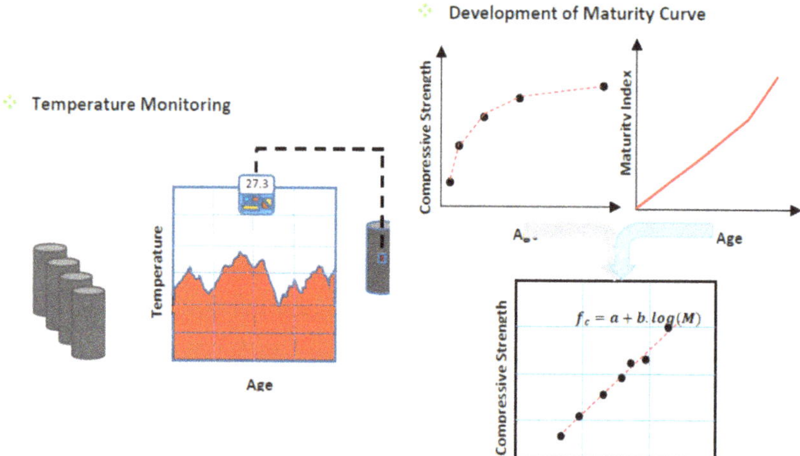

Figure 4. Schematic procedure of temperature history monitoring, maturity index, and compressive strength in the concrete maturity method.

The product of time and hydration temperature of concrete is used for strength estimation. The 1950s Nurse–Saul maturity Function is given in Equation (10) [29]

$$M = \sum_0^t (T - T_0)\Delta t. \tag{10}$$

Equation (10) shows the datum temperature effect is neglected to prominently pursue the strength development.

5. Discussion and Results

5.1. Phase Variation

In monitoring the strength-development process of concrete through ultrasonic wave analysis, the dynamic change of the phase in the material observed for the initial sitting stages used the phase-transition phenomena. At the same time, the hydration-temperature history produced the maturity curve to predict the strength using the maturity index. The comparison of these results was established by measuring the strength of the cylindrical simples by the penetration resistance test using a UTM. The results of the maturity method and the compressive strength were clear and similar to the ultrasonic waves analysis result.

The velocity variations in the wave propagation were observed in the transmuting of phases of the concrete sample. The phase changes were due to the formation of new boundary conditions with curing and strength development, as the strength of the inter-particle bonding caused the expansion in the boundaries of the concrete structure. The velocity change of A-mode compared to S-mode for the specimens is shown in Figure 5. The fast-changing arrival time of signals for the A-mode defines the boundary condition, the S-mode waves change later with the changing of the boundary conditions. The approach of using the first signal observed by the sensor defines the thickness of the material of the transmitted wave, which deals with the signal's travel time between two specific points.

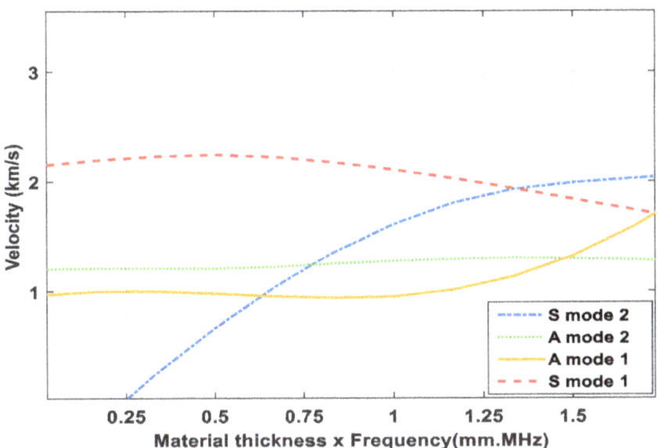

Figure 5. Wave modes for the arrival time of signals.

Pattern irrelevancy in the raw signals can be observed for the amplitude change in Figure 6, which shows the data of three different curing ages. Due to phase regulation in the strength gain, there is a clear difference among the data. The arbitrary waveform generator emits ±10 signals peak-to-peak with a frequency ranging from 100 to 300 kHz; the frequency is exerted by the PZT transducers to form the signals. Figure 6 shows the data of the raw signals for different periods. The peaks of the signals with initial generation time have high amplitudes, which decrease with time. The peak inclination differs for different periods and indicates the dynamic variations.

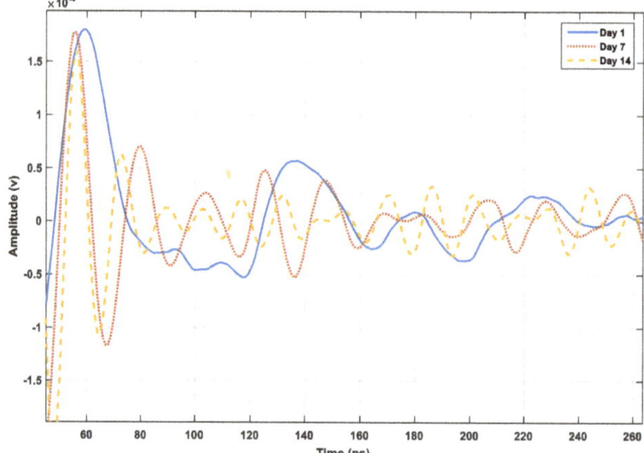

Figure 6. Raw signals amplitude variation on days 1, 7, and 14.

The velocity change in the term of the material stiffness factor occurs prominently in Figure 7; the deepness of the color indicates the rapidity in the velocities and changes of the phases in the material. The wave propagation with curing time changes as the stiffness of the material changes. As the velocities of signals at frequencies near 100 kHz are selected for the A-mode waves, the lower noise data can predict the speed of the waves as a factor for strength monitoring.

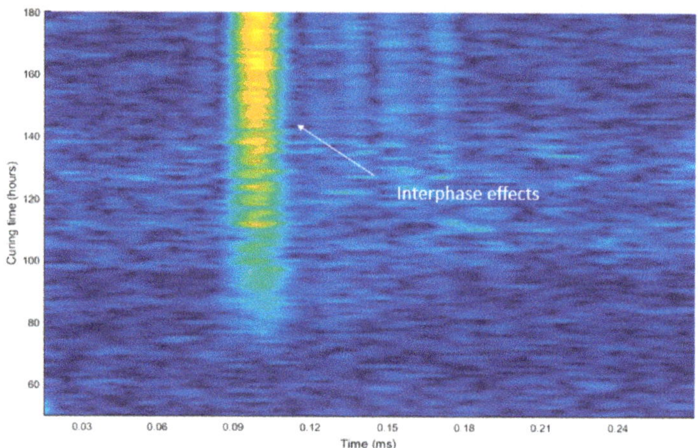

Figure 7. Phase-change response in signals propagation.

5.2. Hydration Temperature and Maturity Index

The hydration temperature was monitored continuously after pouring the concrete and the smart temperature sensors produced the data for the estimation of the maturity of concrete with curing time. The datum temperature was calculated as −2 °C per ASTM [16,17]. The data of the temperature for the hydration heat is shown in Figure 8.

The data of hydration-temperature history was applied to the Nurse–Saul curing function to predict the maturity index. As the datum temperature and time limit were defined, the results showed that hydration temperature was elevated in the initial hours and increased rapidly; that elevation continued to be observed until 28 h after curing, reaching a rise of 20 °C. As time passed, the strength-development improved and the temperature was observed to show lower peaks, as compared with the initial hours, owing to the completion of chemical processes. The experiment was composed of two specimens in the same environmental conditions. Consequently, the temperature data did not show a prominent deviation in curves. The data for the maturity with curing time is shown in Figure 9.

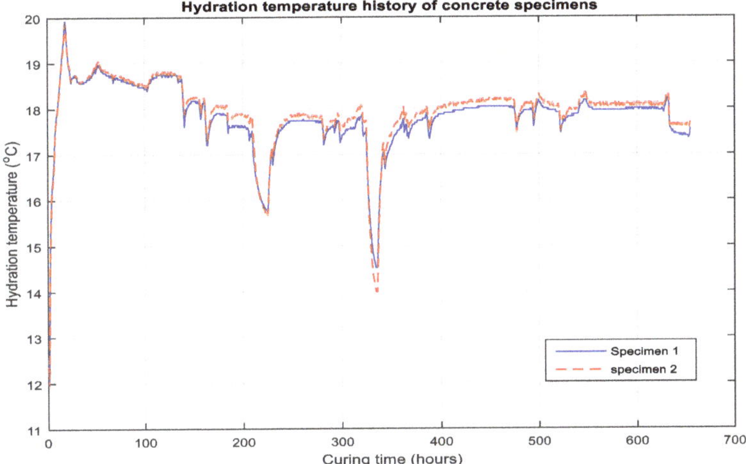

Figure 8. Hydration temperature for concrete specimens 1 and 2.

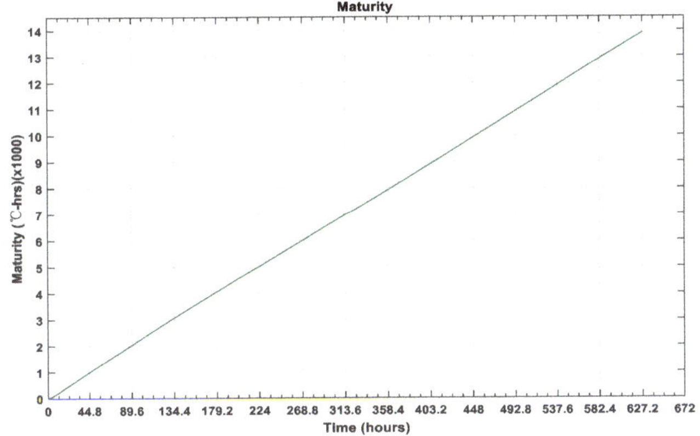

Figure 9. Maturity data for the hydration-temperature monitoring.

Because the temperature data for the concrete specimen 2 appeared slightly higher than that of specimen 1 of concrete, the cured strength was also calculated to be somewhat higher. The strength was calculated from the data shown in Figure 10.

Using the Nurse–Saul function with the datum temperature measured as −2 °C, the maturity index data were calculated as shown in Figure 9, which further simplified the strength of the concrete specimen. The rapid strength development at early ages is prominent from the data results and the compressive strength at the targeted time measured 24.47 MPa, as shown in Figure 10.

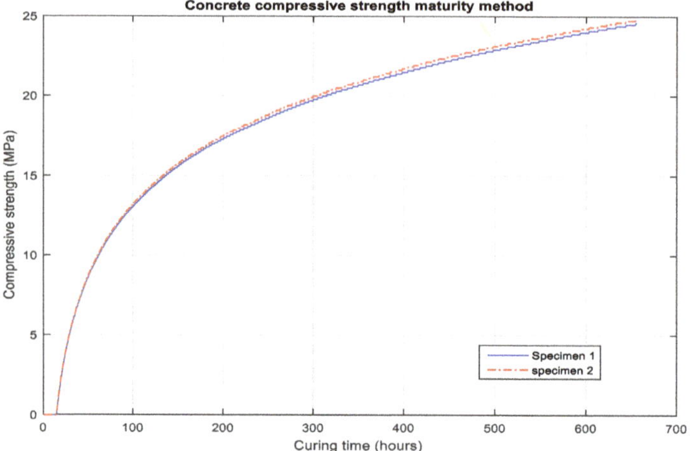

Figure 10. Concrete compressive strength data from temperature sensors.

5.3. Comparative Data for the Strength Prediction

The results of compressive strengths were compared to the penetration test results performed in the laboratory on the specimens for penetration-resistance tests. The data for all indicated dates were the average values of the results for the three cylindrical-specimen tests on each date. After 24 h of curing, the average strength of three cubes for the penetration resistance test was calculated as 3.97 MPa, which rapidly increased until 48 h when the strength value reached 9.58 MPa, as shown in Figure 11. The gain in the strength curve of the ultrasonic waves data and the maturity results was observed to be relatively slow, owing to the specimen size effect. At the third day of curing, the concrete strength approached 11.68 MPa and 14.00 MPa for the ultrasonic waves data and the cylindrical samples penetration test, respectively, while the maturity method yielded a result of 10.92 MPa.

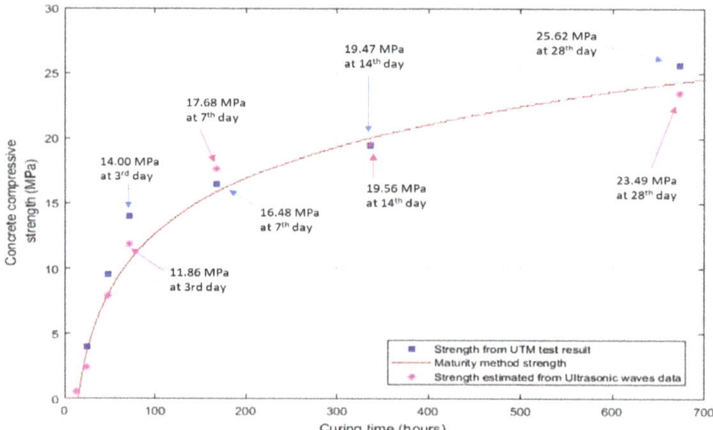

Figure 11. Concrete compressive strength measured by the maturity method, universal testing machine (UTM) test, and ultrasonic wave data.

The variations in the results can be observed by the data, which was due to categorizing the rise of amplitudes in the data of the ultrasonic wave signals in Figure 12. Variation of the waves becomes higher and more prominent with increasing curing time due to the gaining of strength by the concrete specimen and the strengthening of the cohesive power of the concrete particles, which defines the limitations of the concrete material. The propagation of the signals becomes more dominant with time.

The strength prediction for the described methods produced a simultaneous curing process for the gain of strength, and the results were calculated as 16.48 MPa and 17.68 MPa for the UTM test and ultrasonic wave analysis, respectively.

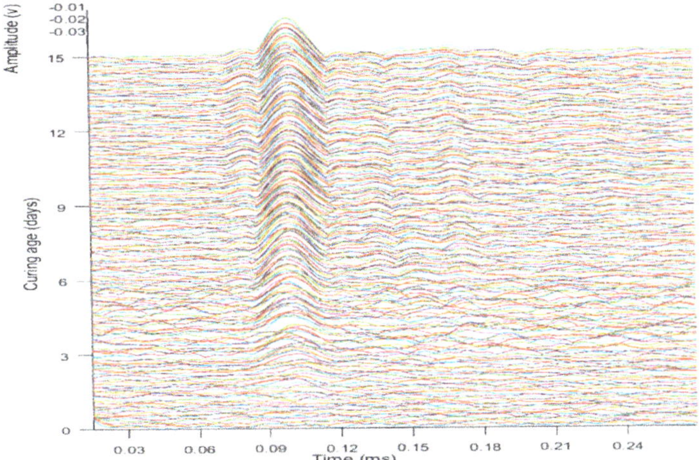

Figure 12. Amplitude rise pattern with curing time for ultrasonic waves data.

The deviation of the data and specification of the strength-development monitoring is compared by fitting a curve to the analyzed data of the ultrasonic waves propagation data and the maturity method results to make the comparison with compressive strength measured by the penetration resistance test. The fitness of regression for the data model was 0.997, implying a valid fit and permitting a novel modified relation for predicting ultrasonic wave velocities in concrete and the maturity of the concrete at an early age. This is specified in Equation (11),

$$C'(v, m) = 9.903v + 0.000599m - 22.72. \tag{11}$$

Here, C' represents the compressive strength calculated from the penetration test data, v represents the wave propagated data, and m is the maturity method result for the estimated compressive strength. The equation provides a promising result for all the data and is in good agreement with the data shown in Figure 13.

The experiments were performed with a controlled temperature, but in addition to temperature, other factors like water/cement ratio, type of bonding agents, aggregates, and age of concrete may affect the concrete-strength process [47,49]. As shown in Figure 11, the strength calculated from the data of hydration temperature differs by 0.86 MPa from the strength gain measured by a UTM test; this is due to the spatial variation effect of the specimens. The effect of concrete compactness was same in the uniform samples.

On day 14, the average strength of the penetration test was calculated as 19.47 MPa and that of the waves analysis was measured as 19.56 MPa, nearly the same for both specimens, while the result of the maturity method was calculated as 20.02 MPa, as shown in Figure 13. The results of the concrete specimens at day 28 were higher: 25.62 MPa for the penetration resistance test of the cylinders, 24.13 MPa for the data of concrete maturity method, and 23.49 MPa for the waves data, which provided a clear demonstration of the test results for the study.

This modified new relationship of maturity and ultrasonic waves data results predict a comparative function for estimation in-situ of concrete early-age strength by NDT methods, which is a novel approach for early-age strength estimation. Concrete strength at day 28 by the destructive test of penetration resistance of concrete cubes through a UTM had the strength of 25.62 MPa, while the result according to the equation calculated the strength as 25.50 MPa at day 28. This relationship of data in the study attempts to predict the compressive strength by different means, leading to safe, promising structural health monitoring.

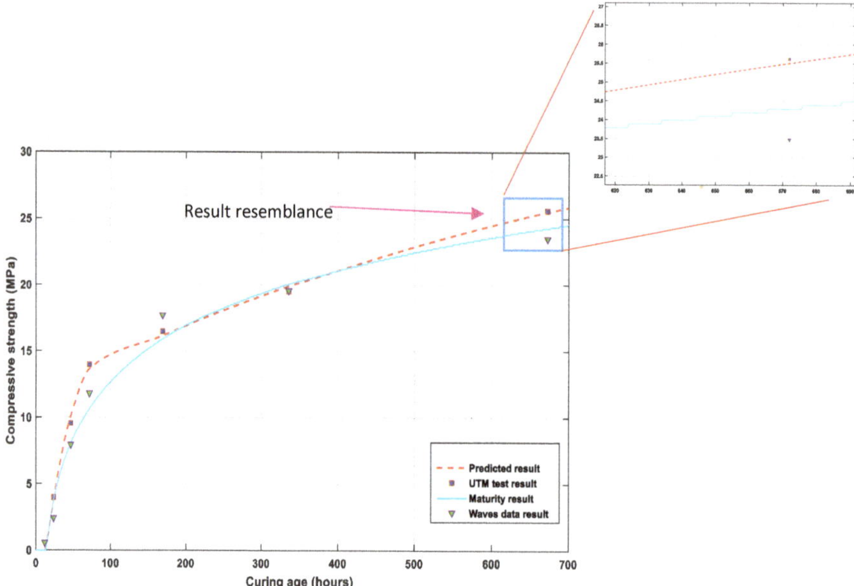

Figure 13. Data comparison for the concrete compressive strength with the maturity- and ultrasonic velocities-modified data.

6. Conclusions

This study covers the limitations of early-age strength development of concrete through an analysis that incorporates different test methodologies. The relationship of the maturity method and the ultrasonic wave propagation is that the hydration temperature results become higher in the initial hours of the hydration process; the wave propagation is slow as the concrete mix paste changes from a liquid paste to a semisolid form, the observed fluctuation of the waves was due to the medium form variation. The results predicted by the data, illustrated in Figure 13, are promising for concrete early-age strength monitoring and strength development estimation. The ultrasonic wave-propagation experiment provided a massive amount of data for predicting the compressive strength of undisturbed field-poured concrete as an NDT function, yielding a modified equation that gives a comparative result for strength estimation, which allows engineers a quick and safe project implementation process.

The data suggest that the behavior of the signals and time of wave propagation to the boundary of the higher rigidity in the concrete sample results in a higher value of the wave propagation velocity.

For the maturity of concrete, the datum temperature plays a basic role in the results of the maturity index, as the maturity method result gave a smooth output and was a little higher than that of the ultrasonic waves. So, for each mix, the datum temperature should be defined according to mix proportions, as the hydration temperature directly affects the strength development. Furthermore, from the results, it has been concluded that high temperature during concrete curing causes rapid strength gains in the structural elements.

The results qualify the proposed method for structural health monitoring in a wide range of industrial multi-stepped project monitoring. The procedure can be advanced in the future for the integration of methodology to the equipped systems for synthesis in the modified polymers.

Author Contributions: N.T. methodology, data collection and analysis, and manuscript writing; J.K. support, data collection, and experimental support; W.-K.K. methodology, data collection, and experimental support; S.P. guidance and support.

Funding: This research was funded by the Korea Agency for Infrastructure Technology Advancement (KAIA) grant funded by the Ministry of Land, Infrastructure and Transport (Grant 19CTAP-C151808-01), the National Research Foundation of Korea (NRF) funded by the Korean government (MSIP; NRF-2017R1A2B3007607, NRF-2017-R1D1A1B03033399), and it was financially supported by the Korea Ministry of Land, Infrastructure and Transport (MOLIT) as the ⌈Innovative Talent Education Program for Smart City⌋.

Acknowledgments: This work was supported by the Korea Agency for Infrastructure Technology Advancement (KAIA) grant funded by the Ministry of Land, Infrastructure, and Transport (Grant 19CTAP-C151808-01), by grants of the National Research Foundation of Korea (NRF) funded by the Korean government (MSIP; NRF-2017R1A2B3007607, NRF-2017-R1D1A1B03033399) and it was financially supported by the Korea Ministry of Land, Infrastructure and Transport (MOLIT) as the ⌈Innovative Talent Education Program for Smart City⌋.

Conflicts of Interest: The authors declared no conflict of interest.

Abbreviations

NDT	non-destructive testing
UTM	universal testing machine
RC	reinforced concrete
AWG	arbitrary waveform generator
MUX	multiplexer
SHM	structural health monitoring
PZT	piezoelectric
ASTM	American Society for Testing and Materials
APC	American Piezo Ceramics
IoT	Internet of Things

References

1. King, S.; Delatte, N.J. Collapse of 2000 Commonwealth Avenue: Punching shear case study. *J. Perform. Constr. Facil.* **2004**, *18*, 54–61. [CrossRef]
2. Zinno, R.; Artese, S.; Clausi, G.; Magaro, F.; Meduri, S.; Miceli, A.; Venneri, A. Structural health monitoring (SHM). In *The Internet of Things for Smart Urban Ecosystem*; Springer: Cham, Switzerland, 2019; pp. 225–249.
3. Sýkora, M.; Diamantidis, D.; Markova, J.; Rózsás, Á. Assessment of compressive strength of historic masonry using non-destructive and destructive techniques. *Constr. Build. Mater.* **2018**, *193*, 196–210. [CrossRef]
4. James, H.; Massoud, S.; Priyan, M. Non-Destructive Testing of Concrete: A Review of Methods. *Electron. J. Struct. Eng.* **2015**, *14*, 97–105.
5. Wahab, A.; Aziz, M.; Rahman, A.; Sam, M.; Qadir, A.; Kassim, K.; You, Y. Review on microwave nondestructive testing techniques and its applications in concrete technology. *Constr. Build. Mater.* **2019**, *203*, 135–146. [CrossRef]
6. Gu, H.; Song, G.; Dhonde, H.; Mo, Y.L.; Yan, S. Concrete early-age strength monitoring using embedded piezoelectric transducers. *Smart Mater. Struct.* **2006**, *15*, 1837–1845. [CrossRef]
7. *Guidebook on Non-Destructive Testing of Concrete Structures*; International Atomic Energy Agency: Vienna, Austria, 2002. Available online: https://www-pub.iaea.org/mtcd/publications/pdf/tcs-17_web.pdf (accessed on 21 August 2019).
8. Lim, Y.Y.; Kwong, K.Z.; Liew, W.Y.H.; Chee, K.S. Non-destructive concrete strength evaluation using smart piezoelectric transducer—A comparative study. *Smart Mater. Struct.* **2016**, *25*, 085021. [CrossRef]
9. Cédric, D.; Grigoris, K.; Jérôme, C.; Stéphanie, S.; Arnaud, D. Monitoring of the ultrasonic P-wave velocity in early-age concrete with embedded piezoelectric transducers. *Smart Mater. Struct.* **2012**, *21*, 047001.

10. Sung, W.S.; Adeel, R.Q.; Jae, Y.L.; Chung, B.Y. Piezoelectric sensor based nondestructive active monitoring of strength gain in concrete. *Smart Mater. Struct.* **2008**, *17*, 055002.

11. Kim, J.W.; Park, S.H. MFL sensing and ANN pattern recognition based automated damage detection and quantification for wire rope NDE. *Sensors* **2018**, *18*, 267–276.

12. Kim, J.; Kim, J.W.; Lee, C.; Park, S. Development of Embedded EM Sensors for Estimating Tensile Forces of PSC Girder Bridges. *Sensors* **2017**, *17*, 1989. [CrossRef]

13. Oh, T.K.; Kim, J.; Lee, C.; Park, S. Nondestructive Concrete Strength Estimation based on Electro-Mechanical Impedance with Artificial Neural Network. *J. Adv. Concr. Technol.* **2017**, *15*, 94–102. [CrossRef]

14. Constantin, E.C.; Chris, G.K.; Georgia, M.A.; Nikos, A.P.; Maria, J.F.; Costas, P.P. Application of smart piezoelectric materials in a wireless admittance monitoring system. (WiAMS) to structure—Test in RC elements. *Case Stud. Constr. Mater.* **2016**, *5*, 1–18.

15. Priya, B.; Thiyagarajan, J.; Monica, B.; Gopalakrishnan, N.; Rao, A. EMI based monitoring of early-age characteristics of concrete and comparison of serial/parallel multi-sensing technique. *Constr. Build. Mater.* **2018**, *191*, 1268–1284. [CrossRef]

16. ASTM International. Standard Practice for Estimation Concrete Strength by the Maturity Method. *ASTM C* **2004**, *1074*, 1074–1093.

17. Carino, N.J.; Lew, H.S. The maturity method: From theory to application. In Proceedings of the 2001 Structures Congress & Exposition, Washington, WA, USA, 21–23 May 2001; p. 19. Available online: https://ws680.nist.gov/publication/get_pdf.cfm?pub_id=860356 (accessed on 21 August 2019).

18. Blaschke, J.H.V.; Torrico, F.A. Estimaing concrete strength using the correlation of the concrete maturity method applied to the mattrials of Cochabamba-Bolivia. *Rev. Investig. Desarro.* **2018**, *18*, 117–127. [CrossRef]

19. Malhotra, V.M.; Carino, N.J. *Handbook on Nondestructive Testing of Concrete Second Edition*; CRC Press: Florida, FL, USA, 2003; pp. 39–381.

20. Bungey, J.H.; Grantham, M.G. *Testing of Concrete in Structures*; CRC Press: London, UK, 2014; pp. 18–331.

21. Park, S.; Kim, J.-W.; Lee, C.; Park, S.-K. Impedance-based wireless Debonding condition monitoring of CFRP Laminated Concrete structure. *NDT E Int.* **2011**, *44*, 232–238. [CrossRef]

22. Costas, P.P.; Stavros, E.T.; Evangelos, V.L. 2-D Statistical Damage Detection of Concrete Structures Combining Smart Piezoelectric Materials and Scanning Laser Doppler Vibrometry. *SDHM* **2018**, *12*, 257–279.

23. Reinhardt, H.W.; Grosse, C.U. Continuous monitoring of setting and hardening of mortar and concrete. *Constr. Build. Mater.* **2004**, *18*, 145–154. [CrossRef]

24. Bhalla, S.; Soh, C.K. Structural health monitoring by piezo-impedance transducers. *J. Aerosp. Eng.* **2004**, *17*, 154–165. [CrossRef]

25. Hyejin, Y.; Kim, J.Y.; Kim, S.H.; Kang, W.J.; Hyun, K.M. Evaluation of Early-Age Concrete Compressive Strength with Ultrasonic Sensors. *Sensors* **2017**, *17*, 1817.

26. Lim, Y.Y.; Tang, Z.S.; Smith, S. Piezoelectric based monitoring of structural adhesives curing: A novel experimental study. *Smart Mater. Struct.* **2018**, *28*, 015016. [CrossRef]

27. Hudson, T.B.; Auwaijan, N.; Yuan, F.G. Guided wave-based system for real-time cure monitoring of composites using piezoelectric discs and phase-shifted fiber Bragg gratings. *J. Compos. Mater.* **2018**, *53*, 969–979. [CrossRef]

28. Tavossi, H.M.; Tittmann, B.R.; Cohen, T.F. Ultrasonic characterization of cement and concrete. In *Review of Progress in Quantitative Nondestructive Evolution*; Springer Science & Business Media: Berlin, Germany, 1999; pp. 1943–1948.

29. Marios, S.; Fragkoulis, K. The modified nurse-saul (MNS) maturity function for improved strength estimates at elevated curing temperatures. *Case Stud. Constr. Mater.* **2018**, *9*, 2214–5095.

30. Marcus, P.; Gregorz, F.; Pawel, N.; Tim, R.; Jack, M. Wireless Concrete Strength Monitoring of Wind Turbine Foundations. *Sensors* **2017**, *17*, 2928.

31. Ramachandran, V.S.; James, J.B. *Handbook of Analytical Techniques in Concrete Science and Technology*; Elsevier: Amsterdam, The Netherlands, 2000; pp. 672–713.

32. Lee, Y.H.; Oh, T. The Measurement of P-, S-, and R-Wave Velocities to Evaluate the Condition of Reinforced and Prestressed Concrete Slabs. *Adv. Mater. Sci. Eng.* **2016**, *2016*. [CrossRef]

33. Hall, K.S. Air-Coupled Ultrasonic Tomographic Imaging of Concrete Element. Ph.D. Thesis, University of Illinois at Urbana Champaign, Champaign, IL, USA, August 2011.

34. Trtnik, G.; Kavcic, F.; Turk, G. Prediction of concrete strength using ultrasonic pulse velocity and artificial neural networks. *Ultrasonics* **2008**, *49*, 53–60. [CrossRef] [PubMed]

35. Turgut, P. Evaluation of the ultrasonic pulse velocity data coming on the field. In Proceedings of the Fourth International Conference on NDE in Relation to Structural Integrity for Nuclear and Pressurized Components 2004, Session B, London, UK, 6–8 December 2004.

36. Jian, H.Y.; Yuan, J.-X.; Liu, M.-G. Assessment of pil integrity by low strain stress wave method. *HKIE Trans.* **1999**, *6*, 42–49.

37. David, R. Mechanical Properties of Materials. 2008. Available online: http://web.mit.edu/course/3/3.225/book.pdf (accessed on 21 August 2019).

38. Kim, J.W.; Kim, J.K.; Park, S.H.; Oh, T.K. Integrating embedded piezoelectric sensors with continuous wavelet transforms for real-time concrete curing strength monitoring. *Struct. Infrastruct. Eng.* **2015**, *11*, 897–903. [CrossRef]

39. Maeder, M.; Damjanovic, D.; Setter, N. Lead Free Piezoelectric Materials. *J. Electroceram.* **2004**, *13*, 385–392. [CrossRef]

40. Elena, A.; Jacob, L.J. Advances in Lead-Free Piezoelectric Materials for Sensors and Actuators. *Sensors* **2010**, *10*, 1935–1954.

41. Yunzhu, C.; Xingwei, X. Advances in the structural health monitoring of bridges using piezoelectric transducers. *Sensors* **2018**, *12*, 4312.

42. Guoqi, Z.; Di, Z.; Lu, Z.; Ben, W. Detection of Defects in Reinforced Concrete Structures Using Ultrasonic Nondestructive Evaluation with Piezoceramic Transducers and the Time Reversal Method. *Sensors* **2018**, *12*, 4176.

43. Roa, R.G.; Aperador, W.; Delgado, A. Simulation of Non-destructive Testing Methods of Ultrasound in Concrete Columns. *Int. J. Electrochem. Sci.* **2013**, *8*, 12226–12237.

44. Zailin, Y.; Hamada, M.E.; Yao, W. Damage Detection Using Lamb Waves (Review). *Adv. Mater. Res.* **2014**, *1028*, 161–166.

45. Marilyne, P.; Constantinos, S.; Matthieu, G.; Kui, Y. Damage Detection in a Composite T-Joint Using Guided Lamb Waves. *Aerospace* **2018**, *5*, 40.

46. Concrete in Practice. *National Ready Mixed Concrete Association.* 2018. Available online: https://www.nrmca.org/aboutconcrete/cips/39p.pdf (accessed on 21 August 2019).

47. Taylor, P.C. *Curing Concrete*; CRC Press: Florida, FL, USA, 2013; pp. 2–22.

48. Brown, T.L.; LeMay, H.E. *Chemistry: The Central Science*, 4th ed.; Prentice Hall: Englewood Cliffs, NJ, USA, 1988; pp. 494–498.

49. Rose, K.; Hope, B.B.; Ip, A.K.C. Factors affecting strength and durability of concrete made with various cements. *Transp. Res. Rec.* **1989**, *1234*, 13–23.

 *micromachines*

Article

Anisotropic Vibration Tactile Model and Human Factor Analysis for a Piezoelectric Tactile Feedback Device

Jichun Xing *, Huajun Li and Dechun Liu

School of Mechanical Engineering, Yanshan University, Qinhuangdao 066004, China
* Correspondence: xingjichun@ysu.edu.cn

Received: 4 June 2019; Accepted: 1 July 2019; Published: 3 July 2019

Abstract: Tactile feedback technology has important development prospects in interactive technology. In order to enrich the tactile sense of haptic devices under simple control, a piezoelectric haptic feedback device is proposed. The piezoelectric tactile feedback device can realize tactile changes in different excitation voltage amplitudes, different excitation frequencies, and different directions through the ciliary body structure. The principle of the anisotropic vibration of the ciliary body structure was analyzed here, and a tactile model was established. The equivalent friction coefficient under full-coverage and local-coverage of the skin of the touch beam was deduced and solved. The effect of system parameters on the friction coefficient was analyzed. The results showed that in the full-coverage, the tactile effect is mainly affected by the proportion of the same directional ciliary bodies and the excitation frequency. The larger the proportion of the same direction ciliary body is, the smaller the coefficient of friction is. The larger the excitation frequency is, the greater the coefficient of friction is. In the local-coverage, the tactile effect is mainly affected by the touch position and voltage amplitude. When changing the touch pressure, it has a certain effect on the change of touch, but it is relatively weak. The experiment on the sliding friction of a cantilever touch beam and the experiment of human factor were conducted. The experimental results of the sliding friction experiment are basically consistent with the theoretical calculations. In the human factor experiment, the effects of haptic regulation are mainly affected by voltage or structure of the ciliary bodies.

Keywords: ciliary bodies touch beam; piezoelectric tactile feedback devices; anisotropic vibration tactile model; human factor experiment

---

1. Introduction

Tactile feedback technology reproduces the tactile sensation for the user through a force, vibration, or other excitation methods [1]. Touch might be the most complex sensing modality compared to sight, hearing, smell, and taste [2]. The technology can be applied to assist the creation and control of virtual scenes and enhance the remote control of machinery and equipment. Tactile feedback technology is usually applied to tactile displays, or touch sensors and other equipment [3]. Some devices are able to reproduce the tactile of the textured surfaces under the finger [4,5]. Tactile feedback technology has been successfully applied in virtual reality gloves, virtual medical, tactile display, and other fields [6]. According to the excitation method, the tactile feedback device can be classified into a pneumatic type, an electromagnetic type, and a piezoelectric type. Among them, the tactile feedback device using piezoelectric materials has received wide attention and application. Piezoelectric tactile feedback devices have stable vibration, slow adaptability to vibration, and low stimulation, etc. [7–9]. Additionally, the response time of piezoelectric materials is relatively short, which is suitable for long-term tactile simulation [10].

The first category uses an ultrasonic vibration of piezoelectric material to modulate the contact effect between a user's fingertip and a vibrating surface [11]. For example, Ma Lu et al. designed a tactile reproduction system based on friction control, which realizes tactile shape reproduction by controlling the frequency of the finger in different positions on the tactile plane [12]. It can be explained either by the squeeze effect or by intermittent contact of the vibrating surface with the finger [13,14]. Other reports focus on tactile displays realized with stimulator arrays, in which mechanical vibrations are generated to stimulate the mechanoreceptors of the skin. For example, in 2002, Yasushi proposed a TextureExplorer that combines tactile and force stimulators to present virtual textures to the user's fingertip. This provided a vibration pin-array excited by piezoelectric plates employed for tactile stimulation in conjunction with the PHANToM, which is a device for force reflection to perform haptic texture rendering [15]. Hayward from McGill University presented a tactile feedback device based on the principle of piezoelectric lateral skin stretching, which was constructed from an array of 64 closely packed piezoelectric actuators connected to a membrane. The deformations of this membrane cause an array of 112 skin contactors to create programmable lateral stress fields in the skin of the finger pad [16].

Tactile feedback devices based on friction control mostly use a combination of sensors and actuators. The virtual touch is realized by detecting the spatial position of the finger to generate a vibration stimulus or to electrostatically stimulate the skin. However, due to a single stimulation method, the effect of tactile stimulation is not optimistic. The tactile effect is relatively simple, and the control system is generally more complicated. The tactile feedback techniques based on stimulator arrays can perform haptic texture rendering, but the resolution is related to the number of used actuators. Therefore, the application of a large number of piezoelectric actuators leads to manufacturing difficulties and high manufacturing costs [17].

To solve the above problems and take advantage of these two technologies, here, we arrange the array tooth structure on ultrasonically vibrating piezoelectric beams, which we call the ciliary body piezoelectric beam. The beam can be any material with elastic properties. So, in this paper, a piezoelectric tactile feedback device with a ciliary body structure is proposed, which has the features of a simple structure and control system. This technique can achieve different touch sensations depending on the direction of finger movement at the same contact position [18]. Firstly, the principle of the anisotropic vibration of the ciliary body structure is analyzed, and a tactile model of the sliding vibration of the anisotropic vibration is established. Secondly, the equivalent friction coefficient of the skin and the touch beam under full-coverage and local-coverage are deduced and solved. The change law of the equivalent friction coefficient is analyzed according to the solution results. Finally, the important parameters, such as the proportion of ciliary bodies in the same direction and ciliary bodies' density, are changed. The equivalent friction coefficient function is solved. The effect of each parameter on the friction coefficient is analyzed, and the tactile control law and scheme are obtained.

2. Principle

The structure of the piezoelectric tactile feedback device is shown in Figure 1, mainly including set screws, piezoelectric ceramics, ciliary body touch beams, bracket, control system, and power supply, of which the control system includes an analog-to-digital conversion module, power amplifier modules, and Bluetooth modules.

When a sinusoidal signal with a frequency close to the natural frequency of the touch beam is supplied to the piezoelectric sheets, the touch beam resonates and produces a bending vibration. The ciliary body at different positions of the touch beam will vibrate in different directions. As shown in Figure 2, the ciliary bodies under the finger are distributed on the right and the left side of the vibration peak. The ciliary bodies of 1, 2, and 3 indicate the ciliary body distributed on the right side of the peak, and 4 and 5 indicate the ciliary body distributed on the left side of the peak. At this time, the ciliary bodies distributed on the right side of the peak will give the finger upward inertial pressure and rightward thrust. The ciliary bodies distributed on the left side of the peak will give the finger upward inertial force and leftward thrust. Therefore, when one finger moves to the right, the vibration direction of the ciliary body on the right side of the peak is the same as the direction of movement of the finger. The vibration direction of the ciliary body on the left side of the peak is opposite to the direction of movement of the finger. When the finger touches the beam, the coefficient of friction of the different parts is constantly changing, which makes the subject feel that the ciliary body on the right side of the peak is smoother than the left side. When the finger moves to the left, the result is reversed. The change in tactile sensation is due to the fact that the equivalent friction coefficient between the finger and the plane is modulated by the vibrating ciliary body beam, and the sliding friction coefficient between the finger and the contact surface depends on the ratio of the ciliary body in both directions covered by the finger. The change in the frequency of the excitation signal causes the vibration mode of the touch beam to change. When the excitation frequency is close to the natural frequency, the acceleration increases, and decreases away from the natural frequency. When the excitation frequency approaches the next natural frequency, the vibration mode of the beam is also changed, and the distribution of the vibration direction of the ciliary bodies are changed as the changing of the vibration mode. The tactile effect is different compared to the previous vibration mode.

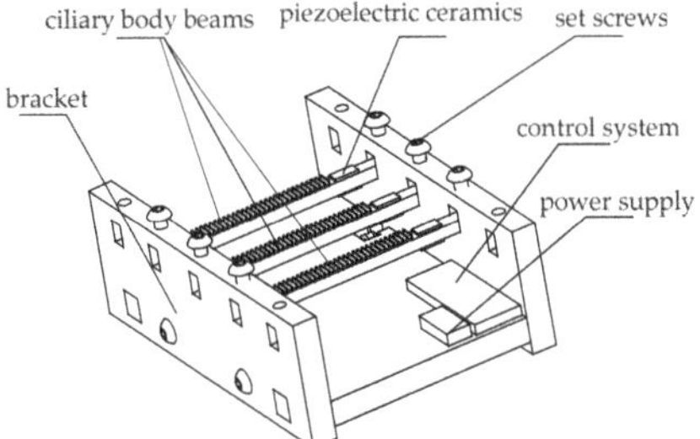

Figure 1. The schematic diagram of the piezoelectric tactile feedback device.

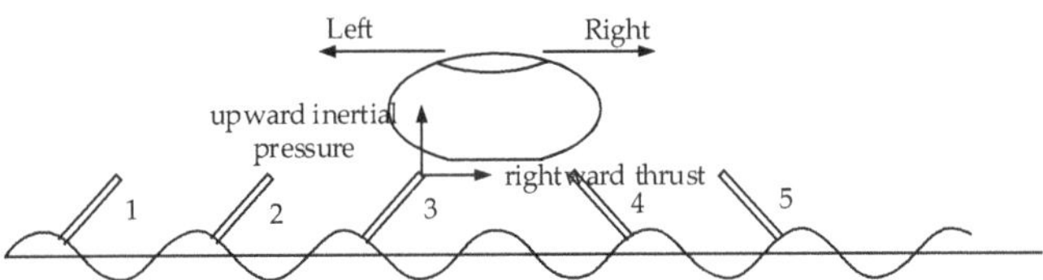

Figure 2. Working principle.

The structure of the ciliary body touch beam is shown in Figure 3. The ciliary bodies' density and number can set multiple sets of data. It can produce a variety of tactile sensations. Piezoelectric sheets are pasted on the upper and lower surfaces of the touch beam, and the position of the paste is one peak of the mode function of the touch beam.

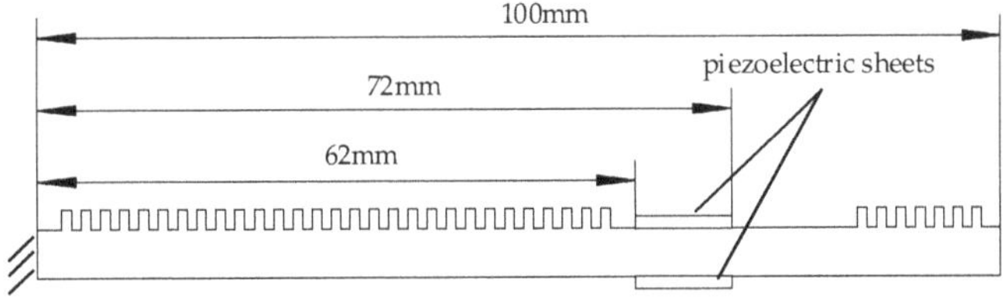

Figure 3. Structure of the ciliary body touch beam.

Since the ciliary body touch beam accomplishes different surface roughness by the anisotropic vibration of ciliary bodies, the number of the ciliary bodies in the two vibrating directions is the main factor in controlling the overall equivalent friction coefficient under the skin-covering length. As shown in Figure 4, the ciliary bodies' anisotropic vibration dynamic model has a vibration of a wavelength, λ. When the ciliary touch beam vibrates between the solid line and the dashed line, the ciliary bodies will squeeze the skin from the bottom to the top. In the process of squeezing the finger skin upwards, the ciliary bodies, a, will move from left to right in the horizontal direction, and the ciliary bodies, b, will move from right to left in the horizontal direction.

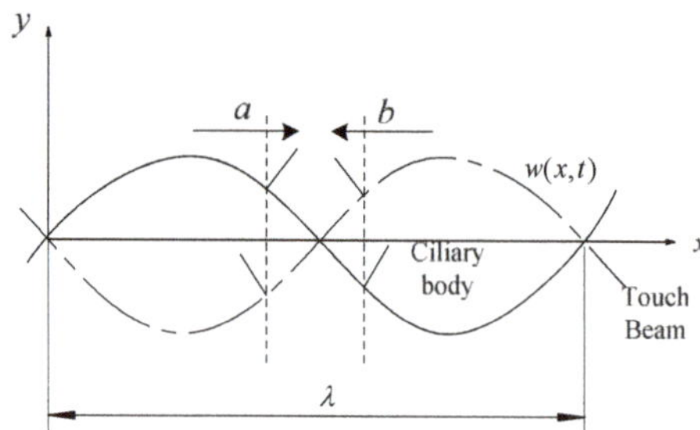

Figure 4. Anisotropic vibration dynamic model of ciliary bodies.

Let the skin move from left to right at a certain speed on the touch beam slowly. When the skin moves above the ciliary body, a, the horizontal component of the vibrational direction of the ciliary body is the same as the direction of movement of the skin, and gives the skin a certain thrust along the direction of motion and reduces the total sliding friction between the skin and the touch beam. Therefore, the equivalent friction coefficient of the skin and the touch beam will be reduced by the total sliding friction of the ciliary bodies. When the skin moves above the ciliary body, b, the horizontal component of the vibration direction of the ciliary body is opposite to the direction of skin movement, which gives the skin the opposite directional resistance. Therefore, the equivalent friction coefficient of the skin and the touch beam affected by the total sliding friction of the ciliary bodies will increase. In the process of moving the skin, the friction coefficient decreases when the ciliary bodies with the same vibration direction as the skin movement direction are touched, and the tactility of the touch beam becomes smooth. The friction coefficient increases when the ciliary bodies with the opposite vibration direction as the skin movement direction are touched and the tactility of the touch beam becomes rough. Additionally, the vibrating direction of ciliary bodies is distributed according to the following Equation (1):

$$\begin{cases} \phi^{(i)}(x) \cdot \phi^{(i)'}(x) < 0, \text{ Right vibration area} \\ \phi^{(i)}(x) \cdot \phi^{(i)'}(x) > 0, \text{ Left vibration area} \end{cases} \tag{1}$$

where $\phi^{(i)}(x)$ is the i-th order mode function of the touch beam [19].

As shown in Figure 5, we assumed that the ciliary bodies' density on the touch beam is sufficiently large. It can be seen that the directions of vibration of the ciliary bodies change at the peak or trough of a wave. Hence, when one finger's skin moves from left to right, the increase and decrease of the friction coefficient shows a periodic variation pattern.

However, if the direction of the vibration of the ciliary body covered by the finger skin is positive and opposite when the finger touches the beam, the sense of touch perceived by the receptor is determined by the ratio of the two vibration directions covered by the skin. In addition, the length of the beam covered by the receptor with a single finger and multi-fingers is not the same. Additionally, the single finger covers less than half of the wavelength of the vibration form. This is called local coverage here. Multi-fingers cover the range of more than one wavelength of the vibration form. This is called full-coverage. Therefore, the tactile model under both full-coverage and local-coverage needs to be analyzed and calculated separately.

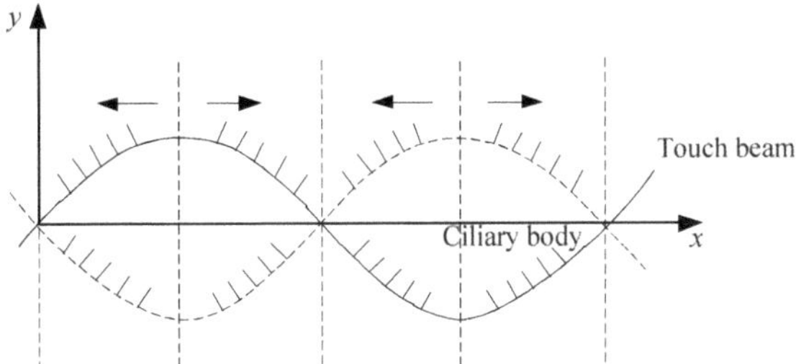

Figure 5. The distribution of the different directional ciliary bodies.

3. Analysis for Anisotropic Vibrating Tactile Models

3.1. Analysis for Full-Coverage Anisotropic Vibration Tactile Model

The full-coverage anisotropic vibration tactile model is shown in Figure 6. The skin covers the full touch beam and moves to the right at a uniform velocity, v. The skin gives pressure to the touch beam, and the pressure between the skin and the touch beam includes the static pressure and the inertia pressure, which are given to the skin during the vibration of the ciliary bodies. Therefore, the total pressure between the skin and the touch beam can be written as:

$$F = (a + b)f + f_p,\qquad(2)$$

where a is the number of vibrating ciliary bodies in the same direction, b is the number of vibrating ciliary bodies in the opposite direction, f is the static pressure of the skin on each ciliary body, and f_p is the ciliary bodies' inertial pressure on the skin of the hand.

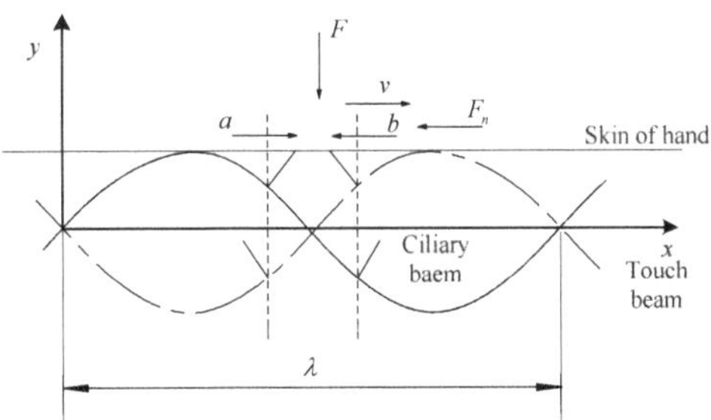

Figure 6. Anisotropic vibration tactile model of full-coverage.

According to the previous analysis, we know the total sliding friction includes the static pressure, the inertial pressure, and the sliding friction caused by lateral vibration of the ciliary bodies. From Equation (2), the total sliding friction force of the hand skin in full-coverage is:

$$\begin{aligned} F_n^{(f)} &= \mu(fa + fb) + \eta(\mu f \cdot b - \mu f \cdot a) + \mu f_p \\ &= \tfrac{3}{2}\mu fb + \tfrac{1}{2}\mu fa + \mu f_p \end{aligned}\qquad(3)$$

where μ is the general sliding friction coefficient between the finger and the touch beam. η is the effective coefficient of the anisotropic vibration of the ciliary bodies. The effect coefficient of the ciliary bodies' anisotropic vibration is the degree of the effect of the skin of the ciliary bodies in the process of touching the beam. The ciliary bodies have a more obvious effect on the skin when moving upwards. However, when the ciliary bodies move away from the skin, the effect is faint. So, here, the effect on the fingers is negligible. Therefore, the vibrating inertial force acting time of the ciliary bodies can be approximated as 1/2 of the total time, so $\eta = 1/2$ is taken.

Let the total number of ciliary bodies remain unchanged on the touch beam. In order to simplify the ciliary bodies' tactile model, it was assumed that the ciliary bodies are dense enough. From Equation (3), the total sliding friction force of the hand skin in full-coverage can be written as:

$$F_n^{(f)} = \mu(\frac{3}{2} - \frac{a}{u} + \frac{f_p}{uf})F. \tag{4}$$

Then, the equivalent friction coefficient can be written as:

$$\mu' = \mu(\frac{3}{2} - \frac{a}{u} + \frac{f_p}{uf}), \tag{5}$$

where u is the total number of ciliary bodies on the touch beam. μ' is the equivalent friction coefficient under the vibration state.

Maintaining the proportion of the same direction of ciliary bodies and the opposite direction of ciliary bodies, let:

$$\frac{a}{b} = \chi, \ \chi \in N. \tag{6}$$

Substituting Equation (6) into (5), we can obtain the relationship between the equivalent friction coefficient and the total number of the ciliary bodies as:

$$\mu' = \mu\left[(\frac{3}{2} - \frac{\chi}{\chi+1}) + \frac{f_p}{uf}\right]. \tag{7}$$

During the contact of the skin with the touch beam, the inertial pressure also affects the sliding friction between the skin and the touch beam. As shown in Figure 7, the touch beam vibrates from the solid line to the dotted line. One vibrating ciliary body produces a normal pressure on the skin that is the inertia force, f_p. The inertial pressure is related to the forced response of the touch beam. The effect coefficient of the inertial pressure, η_p, and the effect coefficient, η, of the anisotropic vibration are the same, taking $\eta_p = 1/2$, so the inertial force of the ciliary bodies can be expressed as:

$$f_p = \eta_p a_c^{(f)} m = \frac{m}{T} \cdot \sum_{j=1}^{u}\left[\int_0^{\frac{T}{2}} \left|w^{(i)''}(x_j, t)\right| dt\right], \tag{8}$$

where $w^{(i)''}$ is the acceleration response of the touch beam, m is the mass of one ciliary body, and $a_c^{(f)}$ is the average acceleration over one vibration period of all ciliary bodies. x_j is the positional coordinate of the j-th ciliary body. T is the vibrational cycle time of the touch beam.

In the case where the density of the ciliary bodies is sufficient, one ciliary body can be approximated as a micro-element, and then the inertial pressure of the full-coverage ciliary bodies can be expressed as:

$$f_p^{(f)} = \frac{1}{2}a_c m = \frac{m}{T} \cdot \int_0^L \int_0^{\frac{T}{2}} \left|w^{(i)''}(x_j, t)\right| dt dx_j. \tag{9}$$

Substituting Equation (9) into (5), the equivalent friction coefficient in full-coverage can be shown by:

$$\mu' = \mu\left(\frac{3}{2} - \frac{a}{u} + \frac{m}{Tuf} \cdot \int_0^L \int_0^{\frac{T}{2}} \left|w^{(i)''}(x_j, t)\right| dt dx_j\right). \tag{10}$$

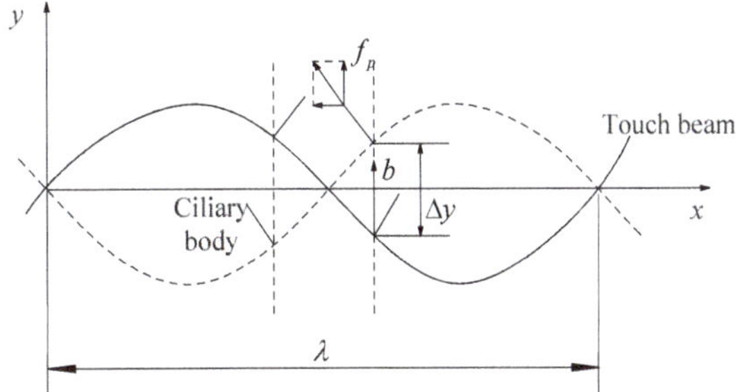

Figure 7. The inertial pressure of the ciliary body.

The cantilever touch beam was chosen as the research object, and the relevant parameters of the piezoelectric sheet and touch beam are shown in Tables 1 and 2. The sliding friction coefficient of the skin and the copper was $\mu = 0.4$, and the parameters related to ciliary bodies are shown in Table 3.

Table 1. Related parameters of the touch beam.

L [m]	h [mm]	W [mm]	E [GPa]	ρ [kg/m^3]
0.1	5	5	105	8500

Table 2. Related parameters of a piezoelectric sheet.

lp [mm]	hp [mm]	Wp [mm]	E [GPa]	ρ_p [kg/m^3]	e_{31} [c/m^2]
10	0.3	5	76.5	7500	2.0

Table 3. Related parameters of the ciliary bodies.

u	ω [Hz]	f [N]	V [V]	m [kg]
20	24,221	20	100	0.01

The relationship between the equivalent friction coefficient of the cantilever touch beam and the number of the vibration direction, *a*, of the ciliary bodies in full-coverage calculated from Equation (10) is shown in Figure 8.

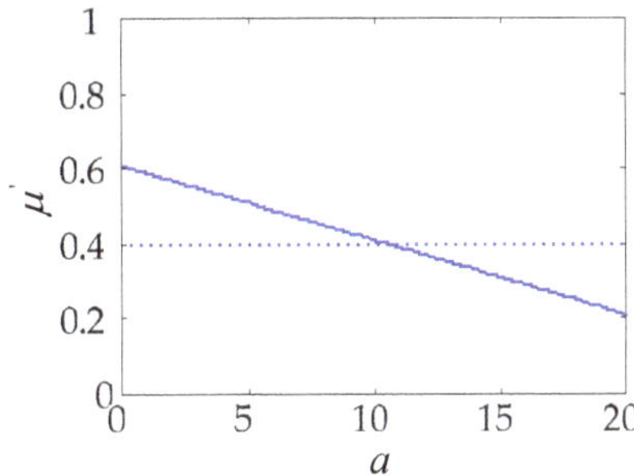

Figure 8. The relationship between μ' and *a*.

Figure 8 shows that if the total number of ciliary bodies on the touch beam is 20, the equivalent friction coefficient between the touch beam and the skin gradually decreases with the increase in the number of vibrating ciliary bodies in the same direction, and the two change linearly. When the number of the same directional ciliary bodies is less than half, the sliding friction force of the touch beam is greater than the general sliding friction force, and it is in the rougher state. When the number of the same directional ciliary bodies is more than half, the sliding friction force of the touch beam is smaller than the general sliding friction force and is in the smoother state. The equivalent friction coefficient of the cantilever touch beam varies from about 0.2 to 0.6.

3.2. Analysis for Anisotropic Vibration Tactile Model in Local-Coverage

The local-coverage anisotropic vibration tactile model is shown in Figure 9. The skin of the finger covers the local length, l_s, of the touch beam and moves to the right at a uniform velocity, v. The finger gives pressure to the touch beam. There are a certain number of the same directional ciliary bodies and the opposite directional ciliary bodies under the skin. The pressure between the skin of the finger and the touch beam also includes the static pressure and the inertia pressure, which is given to the finger during the vibration of the ciliary bodies. Due to the changes of the ciliary bodies' vibration distribution state during the constant movement of the finger, the changeable law of the equivalent friction coefficient with two parts of the pressure at different positions, x, must be analyzed.

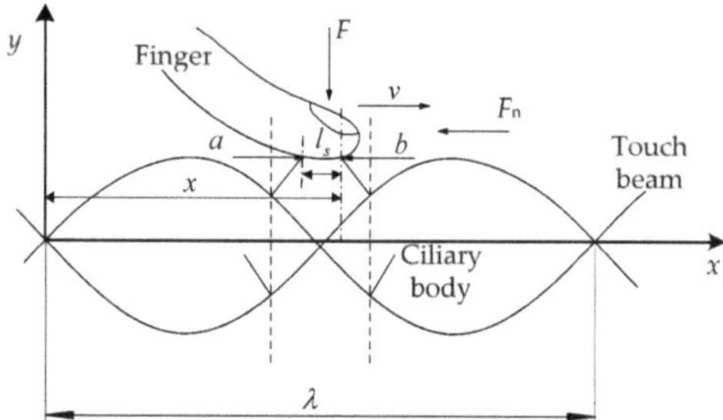

Figure 9. Anisotropic vibrational tactile model of local-coverage.

In the case where the ciliary bodies' density is sufficient, every ciliary body is approximated as a micro-element. One-half vibration cycle of each ciliary body is its effective period on the finger. So, from Equation (9), the inertial pressure in the local-coverage can yield:

$$f_p^{(l)} = \frac{1}{2} a_c^{(l)} m = \frac{m}{T} \cdot \int_{x_j - l_s}^{x_j} \int_0^{\frac{T}{2}} \left| w^{(i)''}(x_j, t) \right| \mathrm{d}t \mathrm{d}x_j, \tag{11}$$

where $a_c^{(l)}$ is the average acceleration within a half vibration period of the locally covering ciliary bodies; ls is the local-coverage length of the finger. Substituting Equation (11) to (5), the equivalent friction coefficient in local-coverage can be shown as:

$$\mu' = \mu \left(\frac{3}{2} - \frac{a}{u} + \frac{m}{Tuf} \cdot \int_{x_j - l_s}^{x_j} \int_0^{\frac{T}{2}} \left| w^{(i)''}(x_j, t) \right| \mathrm{d}t \mathrm{d}x_j \right). \tag{12}$$

From Equation (12), it is apparent that the equivalent friction coefficient under local-coverage is mainly affected by two parts. One is the inertial force, f_p, of the ciliary bodies, and the other is the proportion, a/u, of the total number of ciliary bodies in the same direction of vibrating ciliary bodies. The functional relationship between f_p and the positional coordinate, x, was obtained from Equation (11), and the functional relationship between a/u and the positional coordinate, x, needs to be further analyzed.

One cantilever touch beam was selected as the research object, and the relevant parameters were the same as in Tables 1–3. One operating mode of the touch beam is shown in Figure 10, and the covered length of the finger was considered as $l_s = 0.01$ m. In the course of the uniform movement of the finger on the touch beam, according to the theory of the ciliary body's anisotropic vibration, the ratio of the ciliary bodies of the first whole wave segment, a/u, exists in six stages. Then, seven key positions of the finger exist on the touch beam. The change rules were as follows:

- As shown in Figure 10, when the finger is moved from the position I to the position II, the covered portion of the finger is all in the opposite directional vibrating region, and the ratio, a/u, of the same direction of the vibrating ciliary bodies is maintained at 0.
- When the finger moves from position II to position III, the same directional vibration area of the finger-covering part gradually increases. Also, the ratio of the same-directional ciliary bodies increases and the increased ratio is $(x - 0.0131)/l_s$.
- When the finger moves from position III to position IV, the ratio of the same direction of ciliary bodies, a/u, becomes maximum and remains unchanged. The proportion of the same direction of ciliary bodies in this stage is:

$$\frac{a}{u} = \frac{\lambda}{4l_s}. \tag{13}$$

- When the finger moves from position IV to position V, the opposite directional vibrational area of the finger-covering part gradually increases. Additionally, the ratio of the same directional ciliary bodies decreases from $\lambda/(4l_s)$, and the decreased ratio is $(x - 0.0131 - \lambda/2)/l_s$.
- When the finger moves from position V to position VI, the increasing proportion of the same direction of ciliary bodies is equal to the decrease. So, the proportion remains unchanged, and the proportion of the same direction of ciliary bodies in this stage is:

$$\frac{a}{u} = \left(l_s - \frac{\lambda}{4}\right)/l_s = 1 - \frac{\lambda}{4l_s}. \tag{14}$$

- When the finger moves from position VI to position VII, the same directional vibration area of the finger-covering part gradually increases. The ratio of the same-directional ciliary bodies increases from $1 - \lambda/(4l_s)$, and the increased ratio is $(x - 0.0331 - \lambda/2)/l_s$.

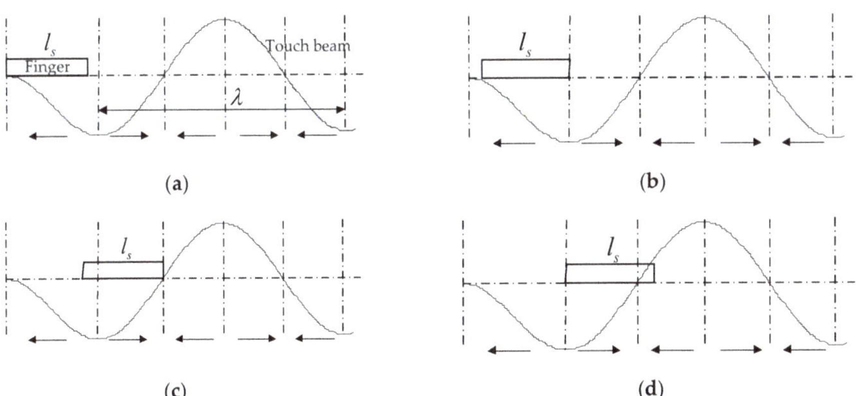

Figure 10. *Cont.*

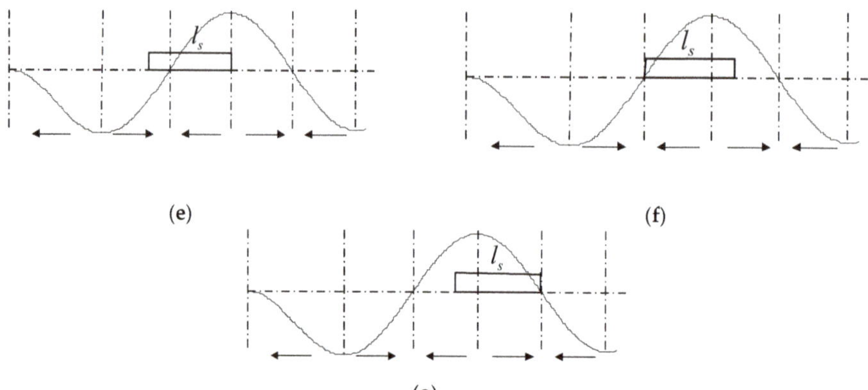

(e)   (f)

(g)

Figure 10. Key positions of the finger on the cantilever beam. (**a**) Position I; (**b**) position II; (**c**) position III; (**d**) position IV; (**e**) position V; (**f**) position VI; (**g**) position VII.

After that, all the whole wave segments met the above change rules 3 to 6. So, the coefficient, γ, was introduced to indicate the number of the whole wave segments after the first. In summary, according to the change rules of the ratios of the same directional ciliary bodies in the five stages, the relationship between a/u and the positional coordinates are as follows:

$$\frac{a}{u} = \begin{cases} 0 & x \in [0, 0.0131) \\ \frac{x-0.0131}{l_s} & x \in [0.0131, 0.0218) \\ \lambda/4l_s & x \in [0.0218+\frac{\lambda\gamma}{2}, 0.0231+\frac{\lambda\gamma}{2}) \\ \frac{\lambda}{4l_s} - \frac{x-0.0231-\frac{\lambda\gamma}{2}}{l_s} & x \in [0.0231+\frac{\lambda\gamma}{2}, 0.0318+\frac{\lambda\gamma}{2}) \\ 1-\lambda/4l_s & x \in [0.0318+\frac{\lambda\gamma}{2}, 0.0331+\frac{\lambda\gamma}{2}) \\ 1-\frac{\lambda}{4l_s} + \frac{x-0.0331-\frac{\lambda\gamma}{2}}{l_s} & x \in [0.0331+\frac{\lambda\gamma}{2}, 0.0418+\frac{\lambda\gamma}{2}) \end{cases} \tag{15}$$

where $\gamma \in Z$, $0 \le \gamma \le 5$, $x \le 0.1$.

From Equation (15), the relationship between the proportion of the same directional ciliary bodies and the positional coordinates is shown in Figure 11. From the image, we know that the ratio, a/u, increases from zero and then decreases when the finger moves from left to right, and periodically alternates in the interval of [0,0.87]. The value of a/u maintains a shorter distance when it is at a maximum or minimum position.

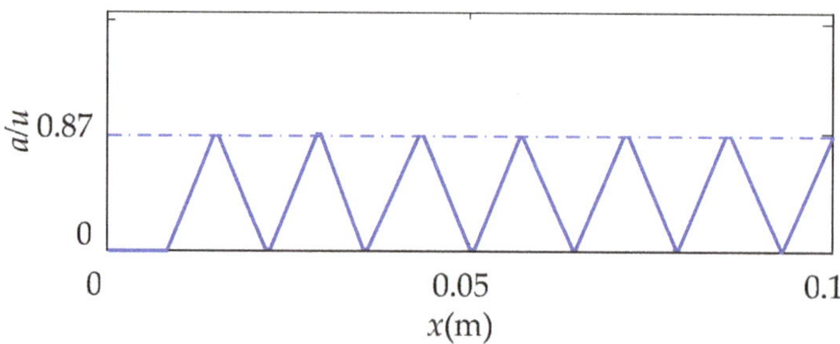

Figure 11. The relationship between the ratio, a/u, and the position coordinates, x.

Substituting Equation (15) into (12), the relationship between the equivalent friction coefficient and the position of the touch beam at one operating mode of the cantilever touch beam can be obtained, as shown in Figure 12. The length of the dotted line, l_b, is the piezoelectric sheet pasted length. From Figure 12, the following observations were worth noting:

The equivalent friction coefficient of the touch beam also alternates periodically and fluctuates around 0.2 to 0.6 and centered at $\mu' = 0.4$. There are four troughs of the equivalent friction coefficient on the left side of the length, l_b, of the piezoelectric sheet, which is the place where the equivalent friction coefficient is relatively low and is also a relatively smooth position. However, the distance between the fourth valley (from left to right) and the segment where the piezoelectric sheet is located is short, and it is difficult to move in the length of this segment during the actual touch process. Therefore, the first three valleys, l_I, l_{II}, and l_{III}, are the three easily perceived smoother positions on the left side of the piezoelectric sheet. Taking the position of the trough at each position as the base point, they are respectively located at 0.0137, 0.0275, and 0.0422 m on the abscissa, x.

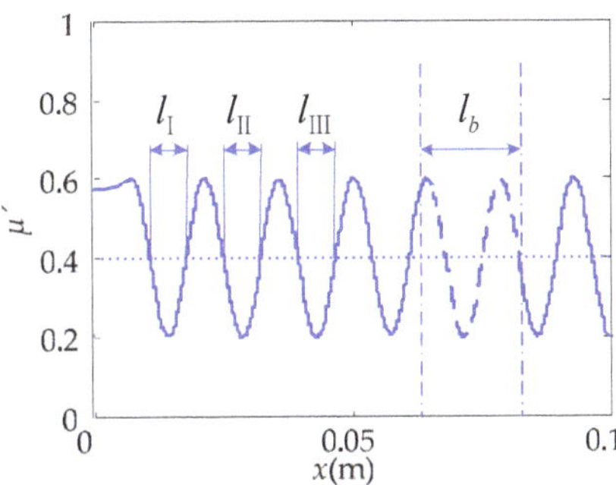

Figure 12. The relationship of μ'-x at the operating frequency of 24,221 Hz.

4. Effect of System Parameters on Tactile Changes

4.1. Effect of System Parameters on Tactile Changes in Full-Coverage

Four parameters were selected to analyze the effect on the equivalent friction coefficient, including the number of the same directional ciliary bodies, a; the total number of ciliary bodies, u; the operating frequency, ω; and the excitation voltage amplitude, V. Figure 13 shows the variation of the equivalent friction coefficient with the parameters changed under the conditions of 2, 10, and 20 N finger pressures. From Figure 13, the following observations were worth noting:

- As a increases, the equivalent friction coefficient, μ', of the full-coverage touch beam gradually becomes smaller, and the two are linearly negatively correlated. As u increases, the equivalent friction coefficient, μ', of the full-coverage touch beam also gradually becomes smaller.
- However, as ω increases, the equivalent friction coefficient, μ', of the full-coverage gradually increases. Additionally, with the increase of V, the equivalent friction coefficient, μ', of the full-coverage touch beam gradually decreases linearly. Comparing the μ' curves under the three touch pressures, the higher the touch pressure, f, is, the smaller the value of μ' becomes.
- Under the full-coverage touch, the change of the parameter, a, and frequency has the greatest influence on the equivalent friction coefficient. Although u has a great influence on the friction coefficient at the beginning, with the increase of u, the effect after 20 is not significantly changed. In order to achieve the best results, the same directional ciliary bodies, a, and voltage, V, should be increased as much as possible, and the parameter u can be kept within 20.

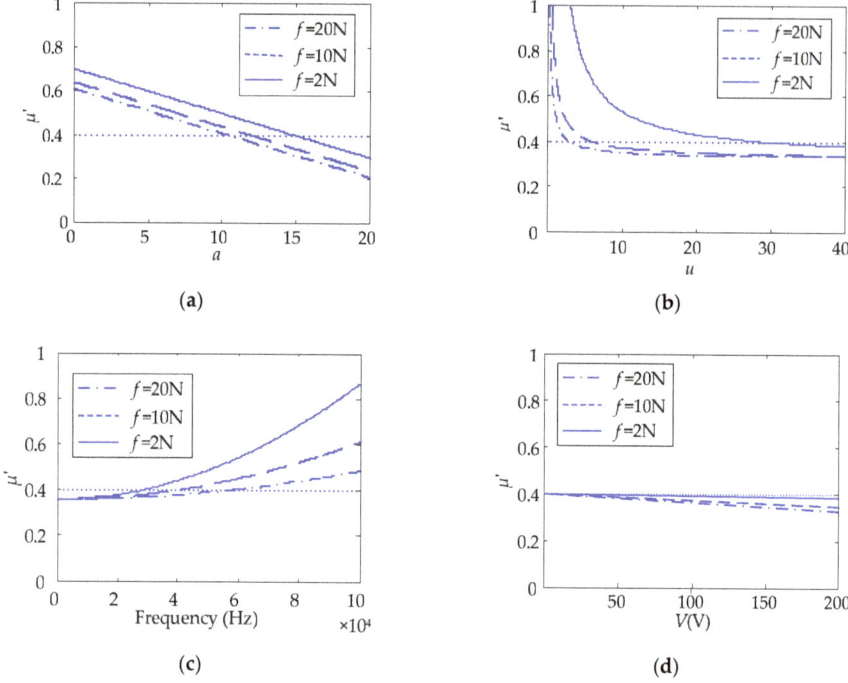

Figure 13. The effect of key parameters on the equivalent friction coefficient in full-coverage. (**a**) μ'-a relational figure; (**b**) μ'-u relational figure; (**c**) μ'-frequency relational figure; (**d**) μ'-V relational figure.

4.2. Effect of System Parameters on Tactile Changes in Local-Coverage

The parameters of the operating voltage, V, and frequency, ω, of the local-coverage cantilever touch beam were changed. The relationship between the equivalent friction coefficient and the position of the touch beam under the effect of the operating parameters was obtained, as shown in Figure 14.

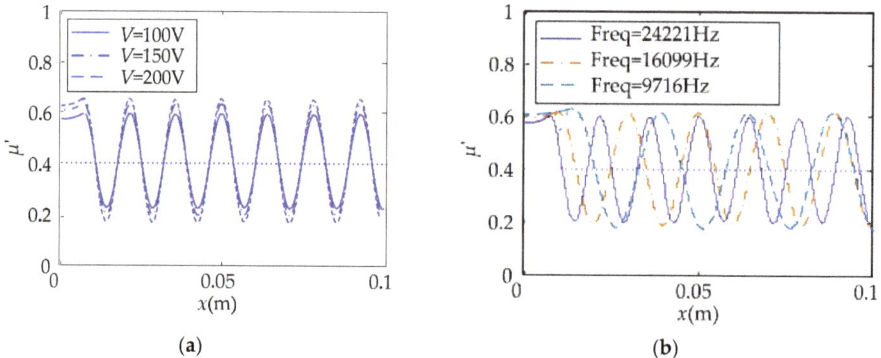

Figure 14. The changes of μ'-x under the effect of operating parameters in local-coverage. (**a**) Operating voltage changed; (**b**) operating frequency changed.

From Figure 14, the following observations were worth noting:

As the operating voltage amplitude, V, increases, the μ' variation amplitude of the touch beam gradually increases with 0.4 as the center, and the corresponding position becomes smoother or rougher. With the decrease of the operating frequency, ω, the μ' variation amplitude of the touch beam slightly increases with 0.4 as the center. However, when the frequency decreases, the number of peaks and troughs gradually decreases.

5. Experiments

5.1. The Experiments on the Sliding Friction of the Cantilever Touch Beam

A cantilever touch beam with a ciliary body spacing of 1 mm was used as the experimental object. One sine signal with a voltage amplitude of 100 V and a frequency of 23,200 Hz was supplied to the touch beam. An acrylic touch block with a weight of 0.56 N was slowly moved on the touch beam from left to right. The real-time friction force change of the block on the touch beam was measured by a digital dynamometer. The test environment is shown in Figure 15.

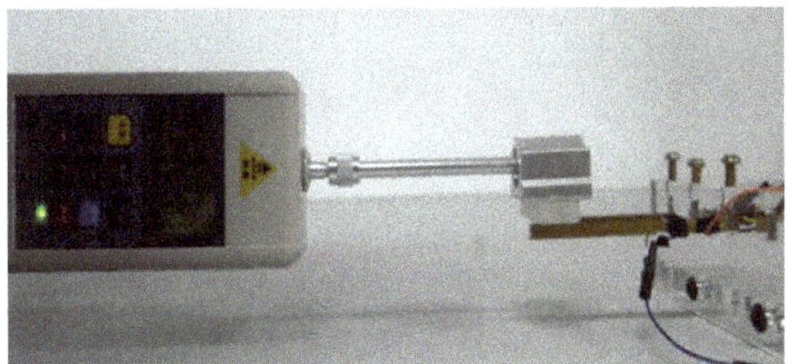

Figure 15. The sliding friction force test.

From Figure 16, the following observations were worth noting:

By comparing the results of local-coverage tactile theory calculations with the experimental test results, the variation trend of the equivalent friction coefficient obtained from the experiment is very close to that of the theoretical calculation. The test results are very similar to the theoretically calculated positions, except that there is a slight error between the theoretical value and the test value at the first trough, while the value of second trough and third trough are very close to the calculated result. However, the error between the test value and the theoretical value is larger at the two peaks of the equivalent friction coefficient, which we think was caused by the jitter of the thrust of the block when the friction coefficient increases. In general, however, the overall trend of change in the test data shows the validity of the theoretical analysis.

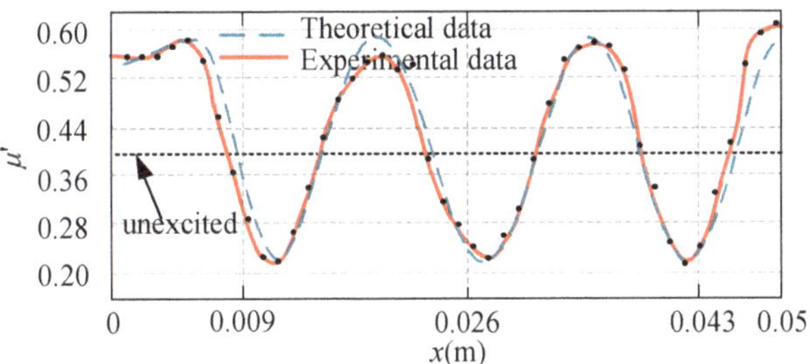

Figure 16. The comparison of the theoretical and experimental data.

5.2. The Experiments of Human Factor

In order to verify the tactile perception effect of the piezoelectric tactile feedback device, nine of the subjects were invited to perform a touch operation on the tactile device, and the tactile perception effect was evaluated. The subjects were random personnel unrelated to the study. The test environment is shown in Figure 17, and the experimental steps were as follows:

- In the case that no signal was supplied to the touch beam, one subject touched the cantilever touch beam with his left index finger and remembered the current tactile sensation.
- A drive signal with a voltage amplitude of 100 V was supplied to the touch beam. Then, the subject was required to touch the ciliary body touch beam again. By comparing this with the tactile feel of the touch beam when no signal was supplied, the subject was questioned to describe the tactile sensory changes about the roughness.
- The operating parameters of the touch beam and the movement direction of the finger were changed. The subject was required to touch the ciliary body touch beam again. The subject needed to describe the changes in the roughness of the touch beam, and we recorded whether the results were consistent with expectations. If the expectations were met, the subject needed to evaluate the degree of tactile perception and we recorded the score.
- Nine subjects were required to perform tests according to steps 1 to 3 and the relevant experimental scores were recorded.

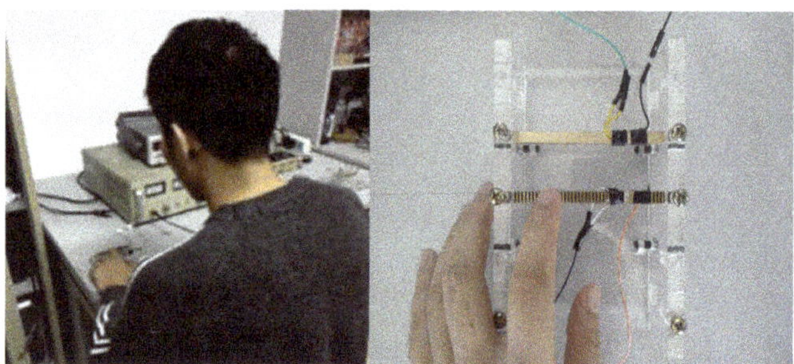

Figure 17. The experiments of human factor.

In order to make the test more accurate and comprehensive, the five parameters of operating frequency, operating voltage peak, beam structure, touch direction, and touch pressure were changed in step 3. The specific test methods for the five parameters were as follows:

- The subjects touched the beam from left to right slowly and the frequency of the operating signal was increased. The subjects were required to describe the changes in tactile sensations after touching. The degree to which the tactile sensations changed as the frequency was close to the resonance point was evaluated, and then scores were recorded.
- The ciliary bodies' structure was changed. Three kinds of ciliary body space, no ciliary body, 1.5 mm, and 1 mm, were selected. The subjects were asked to touch a smooth position of the beam. The subjects were required to describe the changes in tactile sensations after touching. The degree which the tactile sensations changed as the ciliary bodies' density increased was evaluated and then the scores were recorded.
- The voltage amplitude of the operating signal was regulated from 0 to 200 V while touching. The subjects were required to describe how they felt after they finished touching it. The degree to which the tactile sensations changed as the voltage increased was evaluated and then the scores were recorded.

- The subjects touched the beam from left to right slowly, and then the movement direction of the finger was reversed to touch the beam from right to left. The subjects were required to describe the tactile change. Whether the smooth and rough positions alternated in different directions of the finger motion was queried, and the perceived degree of subjects was scored and recorded.
- The subjects were asked to touch a smooth position of the beam and increase the touch pressure slightly. The subjects were required to describe their feeling after they finished touching the beam. Whether the subjects felt smooth with increasing pressure was queried, and the perceived degree of subjects was scored and recorded.

The scoring guidelines were as follows:

The score was 0 to 10 points, with 0 being no effect; 1–4 being very weak; 5–7 being general; and 8–10 being very effective.

After the tests, nine subjects felt that the touch beam became smoother after the signal was applied, and the tactile sensation scores of the five important parameters are shown in Table 4.

Table 4. The result of the human factor behavior effect experimental test scores.

Subjects	Frequency	Ciliary Body Structure	Voltage Amplitude	Movement Direction	Touch Pressure	Comprehensive Score
1	10	8	10	10	5	8.6
2	10	10	10	8	7	9
3	8	8	8	7	4	7
4	10	10	10	10	8	9.6
5	8	8	10	8	5	7.8
6	8	10	10	10	7	9
7	7	10	9	8	7	8.2
8	7	8	10	7	8	8
9	8	8	10	7	5	7.6
average	8.44	8.89	9.67	8.33	6.22	8.31

From Table 4, the following observations were worth noting:

- The given comprehensive scores of the nine subjects ranged from 7 to 10 points. The highest average score was 9.6 points and the lowest average score was 7.2 points. The average value of the comprehensive was 8.31 points. It showed that that the tactile feedback device performs well and the sensation reproduction effect of the tactile feedback device with piezoelectric ciliary body beams is notable.
- Among the five test items, the tactile control effect of changing the voltage amplitude was the best. The tactile perception effect of changing the touch pressure was the weakest. The reason is that the change of touch pressure had little effect on the equivalent friction coefficient in the range of test pressures, while the equivalent friction coefficient was more sensitive to the voltage change in local-coverage.
- By analyzing the lower scores, the third subject's scores were lower. He gave only 4 points for the effect of changing the touch pressure. In other test items, he gave 7 or 8 points, which is approximately the scores given by others. It indicates that there is a difference in the sensitivity of human skin receptors when the perception of tactile changes faintly.

6. Conclusions

The principle of the anisotropic vibration tactile model of the ciliary body touch beam was explored. The equation of the equivalent friction coefficients in full-coverage and local-coverage of the touch beam was established, and the effects of system parameters on the equivalent friction coefficient were analyzed. An experiment on the sliding friction of the touch beam and the experiment of human factors were conducted.

- The full-coverage was mainly affected by the proportion of the same direction of ciliary bodies and the operating frequency. The greater the proportion of the same direction of vibrating ciliary bodies is, the smaller the full-coverage equivalent friction coefficient is. The greater the operating frequency is, the greater the full-coverage equivalent friction coefficient is.

- The local-coverage was mainly affected by the touch position and the amplitude of the operating voltage. The local equivalent friction coefficient at the contact position alternated periodically, and the left side of the piezoelectric sheets, which was 0.0137, 0.0275, and 0.0422 m, respectively, on the abscissa, x, is the relatively smooth position that was easily perceived. The larger the amplitude of the excitation voltage is, the more obvious the tactile change on the touch beam is.

- The experimental results of the sliding friction of the touch beam were basically consistent with the corresponding theoretical calculations. In the human factor experiments, the tactile effect of changing the voltage amplitude and increasing the ciliary body density in the prototype was notable. All the results were consistent with the expectations.

Author Contributions: Conceptualization, J.X.; methodology, J.X.; software, D.L. and J.X.; verification, D.L. and J.X.; formal analysis, D.L.; data management, J.X. writing and manuscript preparation, D.L. and J.X.; project management, J.X.; H.L. improved the manuscript.

Funding: This work was supported in part by a grant from the National Natural Science Foundation of China under grant 51605423 and Youth Projects of the Department of Education of Hebei Province under grant QN2018154.

References

1. Thawani, J.P.; Ramayya, A.G.; Abdullah, K.G.; Hudgins, E.; Vaughan, K.; Piazza, M. Resident simulation training in endoscopic endonasal surgery utilizing haptic feedback technology. *J. Clin. Neurosci.* **2016**, *34*, 112–116. [CrossRef] [PubMed]

2. Lee, M.H.; Nicholls, H.R. Review Article Tactile sensing for mechatronics—A state of the art survey. *Mechatronics* **1999**, *9*, 1–31. [CrossRef]

3. Yoshida, T.; Shimizu, K.; Kurogi, T.; Kamuro, S.; Minamizawa, K.; Nii, H. RePro3D: Full-parallax 3D display with haptic feedback using retro-reflective projection technology. In Proceedings of the IEEE International Symposium on VR Innovation, Melaka, Malaysia, 30 November–2 December 2010.

4. Mélisande, B.; Frédéric, G.; Betty, L. Squeeze film effect for the design of an ultrasonic tactile plate. *IEEE Trans. Ultrason. Ferroelectr. Freq. Control* **2007**, *54*, 2678–2688.

5. Giraud, F.; Amberg, M.; Lemairesemail, B.; Casiez, G. Design of a transparent tactile stimulator. In Proceedings of the 2012 IEEE Haptics Symposium, Vancouver, BC, Canada, 16 April 2012.

6. Özgür, M.; Verena, M.W.; Vincent, H. Contaminant Resistant Pin-Based Tactile Display. In Proceedings of the 2017 World Haptics Conference, Munich, Germany, 6–9 June 2017; pp. 165–170.

7. Yan-Yan, Z.; Chao-Dong, L.I.; Bo-Qian, K. Tactile displays, actuators and their vibroexcitation mode. *Ind. Instrum. Autom.* **2008**, *2*, 82–85.

8. Wong, R.D.P.; Posner, J.D.; Santos, V.J. Flexible microfluidic normal force sensor skin for tactile feedback. *Sens. Actuators A Phys.* **2012**, *179*, 62–69. [CrossRef]

9. Jara, C.A.; Pomares, J.; Candelas, F.A.; Torres, F. Control Framework for Dexterous Manipulation Using Dynamic Visual Servoing and Tactile Sensors' Feedback. *Sensors* **2014**, *14*, 1787–1804. [CrossRef] [PubMed]

10. Fu, Z. Virtual Reality Application Research of Force Tactile Feedback System. Ph.D. Thesis, Shanghai University of Engineering Science, Shanghai, China, 2015.

11. Ghenna, S.; Vezzoli, E.; Giraud-Audine, C.; Giraud, F.; Amberg, M.; Lemaire-Semail, B. Enhancing variable friction tactile display using an ultrasonic travelling wave. *IEEE Trans. Haptics* **2017**, *10*, 296–301. [CrossRef] [PubMed]

12. Ma, L.; Lu, X. Research on tactile rendering system based on friction control. *Comput. Technol. Dev.* **2015**, *1*, 62–65.

13. Vezzoli, E.; Messaoud, W.B.; Amberg, M.; Giraud, F.; Lemaire-Semail, B.; Bueno, M.A. Physical and perceptual independence of ultrasonic vibration and electrovibration for friction modulation. *IEEE Trans. Haptics* **2017**, *8*, 235–239. [CrossRef] [PubMed]

14. Wiertlewski, M.; Fenton Friesen, R.; Colgate, J.E. Partial squeeze film levitation modulates fingertip friction. *Proc. Natl. Acad. Sci. USA* **2016**, *113*, 9210–9215. [CrossRef] [PubMed]

15. Ikei, Y.; Shiratori, M. TextureExplorer: A Tactile and Force Display for Virtual Textures. In Proceedings of the 10th Symposium on Haptic Interfaces for Virtual Environment and Teleoperator Systems, Orlando, FL, USA, 24–25 March 2002; pp. 327–334.

16. Vincent, H.; Juan, M.C. Tactile Display Device Using Distributed Lateral Skin Stretch. In Proceedings of the Symposium on Tactile Interfaces for Virtual Environment and Teleoperator Systems, Osaka, Japan, 25–27 September 2000; pp. 1309–1314.

17. Li, C.; Xing, J.; Xu, L. Coupled vibration of driving sections for an electromechanical integrated harmonic piezodrive system. *AIP Adv.* **2014**, *4*, 031320. [CrossRef]

18. Xing, J.; Liu, D.; Ren, W. Summary of the development of piezoelectric haptic feedback actuators. *Piezoelectr. Acoustoopt.* **2014**, *40*, 146–152.

19. Bashash, S.; Vora, K.; Jalili, N.; Evans, P.G.; Dapino, M.J.; Slaughter, J. Modeling Major and Minor Hysteresis Loops in Galfenol-Driven Micro-Positioning Actuators Using A Memory-Based Hysteresis Framework s. In Proceedings of the 1st Annual Dynamic Systems and Control Conference, Ann Arbor, MI, USA, 20–22 October 2008; p. 1027.

Article

A Shear-Mode Piezoelectric Heterostructure for Electric Current Sensing in Electric Power Grids

Wei He [1],* and Aichao Yang [2]

[1] School of Information Engineering, Baise University, Baise 533000, China
[2] Jiangxi Electric Power Research Institute, Nanchang 330096, China; dkyyac2015@163.com
* Correspondence: weiheky@yeah.net

Received: 1 June 2019; Accepted: 21 June 2019; Published: 23 June 2019

Abstract: This paper presents a shear-mode piezoelectric current sensing device for two-wire power cords in electric power grids. The piezoelectric heterostructure consists of a cymbal structure and a permalloy plate. The cymbal structure is constructed from a permanent magnet, a brass cap, and shear-mode piezoelectric materials. The permalloy plate concentrates the magnetic field generated by the two-wire power cord on the magnet. Under the force amplification effect of the cymbal structure, the response of the device is improved. A prototype has been fabricated to conduct the experiments. The experimental average sensitivity of the device is 12.74 mV/A in the current range of 1–10 A with a separating distance of d = 0 mm, and the resolution reaches 0.04 A. The accuracy is calculated to be ±0.0177 mV at 1.5 A according to the experimental voltage distribution. The current-to-voltage results demonstrate that the proposed heterostructure can also be used as a magnetoelectric device without bias.

Keywords: piezoelectric current sensing device; two-wire power cord; cymbal structure; force amplification effect; sensitivity

---

1. Introduction

Electricity monitoring of power lines based on current sensing devices is of great significance to improve the security and reliability of the electric power systems. The traditional current sensing devices, such as Hall sensors [1], magnetoresistance sensors [2,3], and current transformers [4], have been investigated. But there are some limitations for these devices. The Hall sensors require external power and place high demands on signal conditioners due to the weak Hall voltage. Magnetoresistance sensors also need external power supply and exhibit large thermal drifts. Current transformers have the disadvantage of magnetic saturation and relay maloperation might be induced. Magnetoelectric (ME) structures based on magnetostrictive materials and piezoelectric materials have been reported for current measuring [5–8], but the devices need to encircle the wire in operation, which may limit their real applications. Furthermore, most of the previously proposed ME structures require DC bias magnetic field, and the demand for DC bias magnetic field greatly increases the occupied space. Cantilever-based non-invasive piezoelectric current sensors are proposed [9–11]. The devices are designed to resonate at the power frequency (50 Hz or 60 Hz) to obtain maximal responses. However, due to the nonlinearity of the piezoelectric materials [12], it is difficult to maintain the resonant state of the devices for varying current amplitudes, and non-resonant structures are more suitable for current sensing. Recently, non-resonant piezoelectric current sensing devices were developed for a single wire in electric power grids [13,14], which operate non-invasively and exhibit high linearity, but the proposed structures are not suited for a two-wire power cord carrying identical currents in opposite directions.

In the past several years, the shear effect of the piezoelectric materials has attracted great attention due to the high piezoelectric constant and electromechanical coupling coefficient [15–21]. Ren et al. [15] presented a shear-mode piezoelectric vibration energy harvesting device with a maximum output power of 4.16 mW. Carrera et al. [16,17] analyzed multilayered piezoelectric structures in shear-mode using finite element models (e.g., LW models with different theory approximation orders). Liu et al. [21] theoretically investigated the ME effect of a magnetoelectric laminated composite working in shear–shear (S–S) mode, which exhibits stronger ME coupling coefficients. In this paper, a shear-mode non-resonant piezoelectric heterostructure for current sensing of two-wire power cords is designed. The brass cap, the shear-mode piezoelectric materials, and the permanent magnet constitute a cymbal structure, which results in force amplification and potential enhancement of the response to the currents. Theoretical study was conducted, and a prototype was fabricated to investigate the output characteristics of the device. The least squares method is used to analyze the linearity, and the accuracy is investigated for a given electric current. The prototype exhibits high linearity and sensitivity, which are favorable for current sensing in electric power grids. The experimental results verify the theoretical model and the feasibility of the proposed heterostructure. Meanwhile, a conversion factor of 0.77 V/A is obtained for the current-to-voltage conversion experiment, which indicates the latent advantages of the proposed device in practical applications compared with the magnetoelectric laminated composites using DC bias magnetic field [21–24].

2. Structure and Analysis

Figure 1 shows the schematic diagram and photograph of the proposed piezoelectric heterostructure. The device comprises a cymbal structure and a permalloy plate. The cymbal structure is constructed from a magnet, two shear-mode piezoelectric plates, and a brass cap. The piezoelectric ceramic Pb(Zr,Ti)O$_3$ (PZT5H) is chosen as the material of the piezoelectric plates. The dimension of each PZT5H plate is 1 mm (d_p) × 6 mm (w_p) × 3 mm (l_p). The material of the permanent magnet is NdFeB (N35), and the size of the magnet is 5 mm (d_m) × 6 mm (w_m) × 22 mm (l_m). The magnet also acts as a retaining plate for the cymbal structure. Under the action of the AC magnetic field produced by the two-wire power cord, the magnet is acted upon by an AC magnetic force due to the nonuniform AC magnetic field acting on the magnet. The magnetic force results in amplified shear stress on the piezoelectric plates. Then, a voltage is produced, due to the piezoelectric effect of the piezoelectric material.

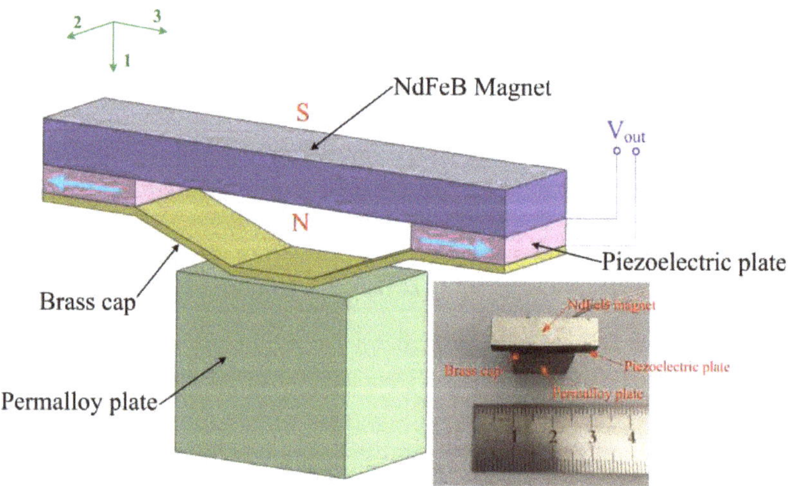

Figure 1. Schematic diagram and photograph of the proposed piezoelectric heterostructure.

The magnetic force on the magnet F_m can be expressed as

$$F_m = \sigma_m \iint_{S_m} \Delta H dS_m,$$ (1)

$$\Delta H = H_{1a} - H_{1b},$$ (2)

where σ_m is the magnetic charge density, $\sigma_m = B_r$ [25], B_r is the remnant flux density, S_m is the surface area of one pole of the magnet, and H_{1a} and H_{1b} are the magnetic fields on the bottom and top surfaces of the magnet, respectively. F_m can be expressed as a power series of the electric current I, and the coefficients of the power series can be determined by fitting curve. Assuming that the vertical force (in 1-direction) acting on one piezoelectric plate by the magnet is $F_m/2$, the vertical force acting on the magnet by one piezoelectric plate is the same in amplitude and opposite in direction ($-F_m/2$).

Figure 2 shows the force F_t exerted by the brass cap on one piezoelectric plate. Based on the decomposition of the force F_t, the following equation is obtained

$$F_v = -\frac{F_m}{2}$$ (3)

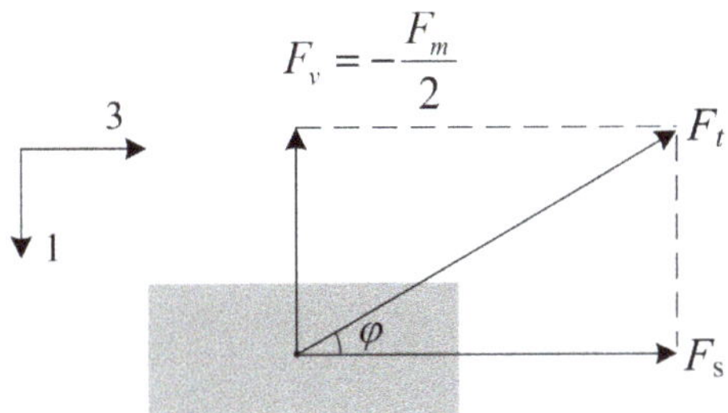

Figure 2. The force F_t exerted by the brass cap on a piezoelectric plate.

The shear force F_s (in 3-direction) can be expressed as

$$F_s = -\frac{F_v}{tg\varphi},$$ (4)

where $tg\varphi$ is determined by

$$tg\varphi = \frac{2(h - d_p)}{l_m - 2l_p - l_t},$$ (5)

where h is the distance between the magnet (N pole) and the top part (inner surface) of the brass cap, d_p and l_p are respectively the thickness (in 1-direction) and length (in 3-direction) of one piezoelectric plate, l_m is the length of the magnet (w_m and d_m are respectively the width and the thickness of the magnet), and l_t is the length of the top part of the brass cap. Then, the shear stress acting on one piezoelectric plate is

$$T_s = \frac{F_s}{w_p l_p} = -\frac{F_v}{w_p l_p tg\varphi},$$ (6)

where w_p is the width of one piezoelectric plate. Based on piezoelectric constitutive equations in shear-mode [26], the electric field in 1-direction of the piezoelectric material is given by

$$E_1 = -h_{15}S_5 = h_{15}\frac{F_v}{w_p l_p tg\varphi c_{55}^D},$$ (7)

where h_{15} is the shear stiffness constant, and c_{55}^D is the elastic stiffness coefficient (at constant D_1) in shear-mode. In open-circuit condition, the electric displacement $D_1 = 0$. Thus, the output voltage of one piezoelectric plate is obtained as

$$V_1 = E_1 d_p = -\frac{F_m h_{15} d_p}{2 w_p l_p t g \varphi c_{55}^D} \tag{8}$$

Based on Equation (8), it is obvious that the voltage is proportional to the magnetic force F_m for determined material and geometric parameters of the device, which depends on the magnetic field magnetic field gradient ΔH on the NdFeB magnet. For current-to-voltage conversion application, the conversion factor can be defined as

$$\gamma = \frac{V_1}{I_{coil}} = -\frac{F_m h_{15} d_p}{2 w_p l_p I_{coil} t g \varphi c_{55}^D} \tag{9}$$

If the output voltage exhibits a linear response to the input current of the coil, the conversion factor λ will show a flat response to the current.

3. Results and Discussions

A prototype was fabricated to study the current sensing performances of the device. The fabrication process is as follows. (1) The brass cap, the piezoelectric plates (PZT5H), the permalloy plate, and the permanent magnet (NdFeB) were dipped in propanone to clean. (2) The brass cap, the piezoelectric plates, and the NdFeB magnet were bonded to constitute the cymbal structure using insulate epoxy adhesive, and the cymbal structure was naturally dried in the air. (3) The permalloy plate was bonded with the cymbal structure. The prototype was then used to current sensing for a two-wire power cord. The configuration and the experimental set-up for the presented device are illustrated in Figure 3 (the power cord carries opposite currents). The electric currents of the two-wire power cord were generated by a current generator, and the output voltages were monitored by a lock-in amplifier. The current sensing device was placed above the two-wire power cord. The permalloy plate concentrates the magnetic field produced by the two wires of the power cord to the NdFeB magnet, which can potentially enhance the response of the device to electric currents. In Figure 3, the parameter d represents the distance between the bottom surface of the permalloy plate and the top surface of the power cord.

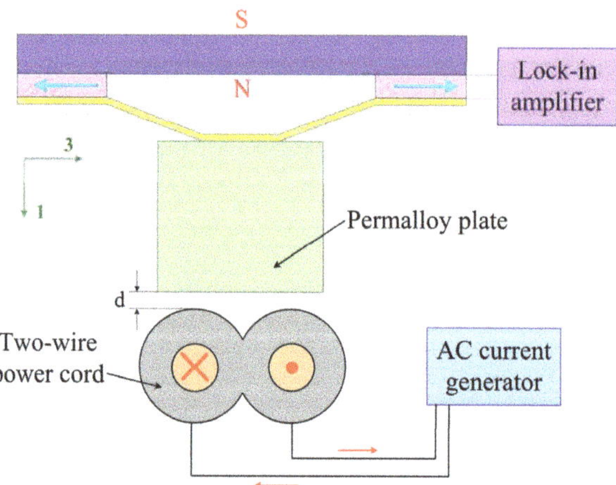

Figure 3. Configuration and experimental set-up for the proposed device.

Figure 4 shows the output peak voltage versus the current in the power cord (f = 50 Hz). It can be seen from Figure 4 that the experimental voltage increases from 13.91 mV to 128.59 mV when the current is increased from 1 A to 10 A for d = 0 mm, and the average sensitivity is 12.74 mV/A in the given current range (1–10 A). For d = 3 mm, there is an obvious drop for the induced voltages, and the voltage varies from 7.23 mV to 59.48 mV with an average sensitivity of 5.81 mV/A. The theoretical voltages obtained from Equation (8) for 4 A, 6 A, and 8 A at d = 0 mm were plotted in Figure 4 (respectively 60.5 mV, 88.1 mV, and 114.2 mV), which validate the developed model. In order to analyze the linearity of the proposed device, the least squares method was adopted. The equation of the fitting curve can be expressed as

$$V = aI + b, \tag{10}$$

where a is the slope and b is the intercept of the equation. Using the experimental data in Figure 4, the slope of the equation a = 12.82077 for d = 0 mm and a = 5.785861 for d = 3 mm. Correspondingly, the intercept b = 1.546273 for d = 0 mm and b = 1.548193 for d = 3 mm. The correlation coefficients are 0.999897 and 0.999945 for d = 0 mm and d = 3 mm, respectively. After plotting the fitting curves in Figure 4, the linearity of the proposed device is calculated by

$$\delta = \frac{\Delta V_{\max}}{V_{\max}} \times 100\%, \tag{11}$$

where $\Delta V_{\max}$ is maximal deviation between the experimental results and the fitting results, and $V_{\max}$ represents the full-scale output of the device. The corresponding results are 0.9% and 0.67% for d = 0 mm and d = 3 mm, respectively. Compared with the resonant structures, the high linearity makes the presented device very suitable for current sensing in electric power systems.

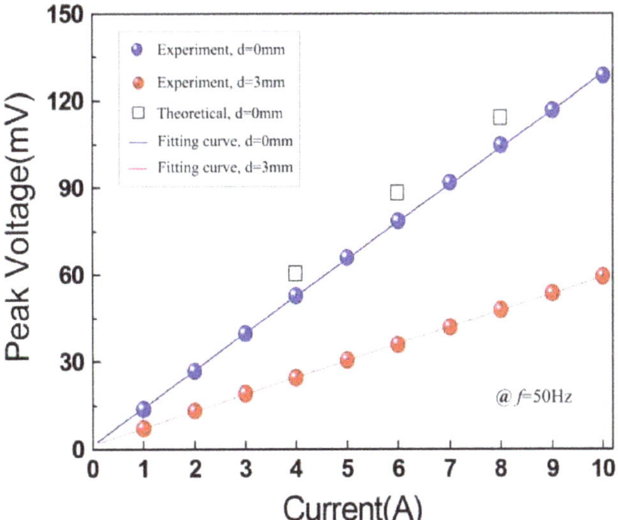

Figure 4. Induced voltage versus the electric current for d = 0 mm and d = 3 mm.

The current remains unchanged at 1.5 A (f = 50 Hz). Figure 5a plots the induced experimental voltage at different times. As can be seen from Figure 5a, the voltage changes with time. The average voltage is 20.4 mA (measurement number n = 120) in the given time range (0–1200 s). The histogram of the voltages is shown in Figure 5b. It can be seen from Figure 5b that the voltages approximately obey normal distribution. Therefore, the accuracy of the device for current sensing can be calculated by

$$A = v \pm \frac{4\sigma}{\sqrt{n}},\tag{12}$$

where v is the systematic error, σ is the standard deviation, and $\pm 4\sigma/\sqrt{n}$ represents the uncertainty. If we do not take into account the systematic error, the accuracy of the device is calculated to be ± 0.0177 mV.

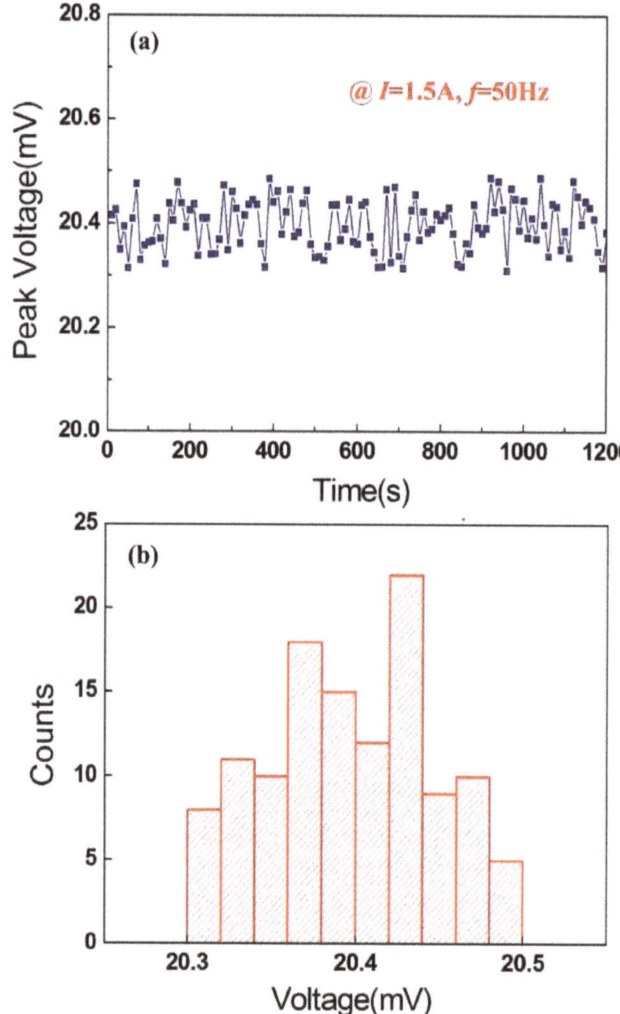

Figure 5. (a) Induced voltage versus time at 50 Hz for $d = 0$ mm and (b) Histogram for the voltage at 1.5 A.

A small current step ($\Delta I = 0.04$ A, $f = 50$ Hz) was applied in the two-wire power cord. Figure 6 shows the induced experimental voltage versus time for $d = 0$ mm. As can be seen from Figure 6, by adjusting the current amplitudes within 140 s, the output voltage exhibits a step change. It is clear that a current change (ΔI) of 0.04 A can be distinguished. We predict further resolution improvement could be achieved by replacing the PZT5H with shear-mode PMN-PT single crystal, which has a higher piezoelectric coefficient d_{15}.

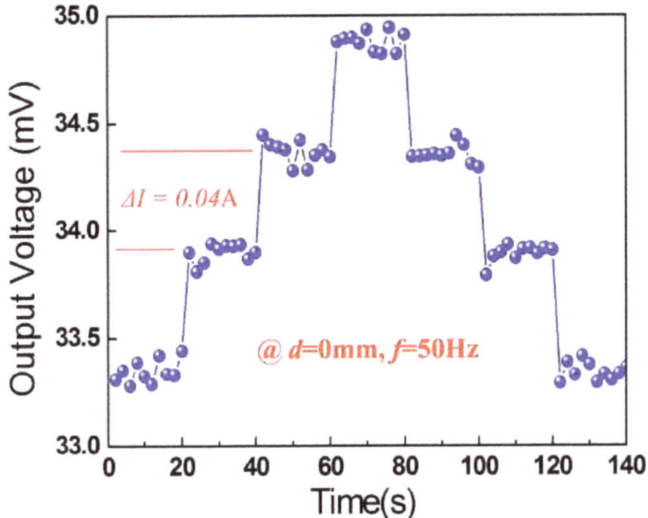

Figure 6. Resolution of the device to the power-frequency current step of 0.04 A at d = 0 mm.

A copper wire coil was wound around the permalloy plate of the prototype, and a current of 40 mA–400 mA (I_{in}) was applied in the coil for current-to-voltage conversion. The experimental output peak voltage increases from 33.18 mV to 270.85 mV with approximately linear response to the current at the low-frequency of 1 kHz, as shown in Figure 7. It also can be seen from the inset of Figure 7 that the factor λ exhibits an approximate flat response. It varies in the range of 0.68 V/A to 0.83 V/A, with an average value of 0.77 V/A. The results show that the heterostructure has the potential to produce large magnetoelectric effect without using magnetostrictive materials and bias magnetic field [27,28].

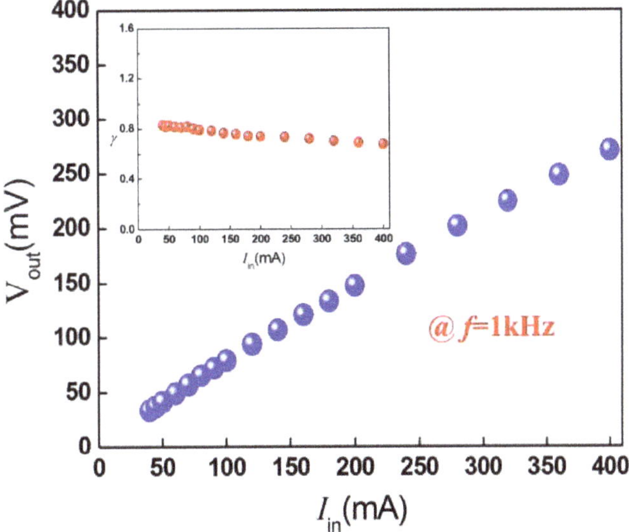

Figure 7. Output open-circuit voltage as a function of the input electric current in the coil at 1 kHz. The inset indicates the conversion factor in the current range of 40 mA to 400 mA.

4. Conclusions

In this paper, a non-resonant piezoelectric current sensing device with high resolution is proposed, which possesses the advantages of passivity, low cost, and simple structure. The device can operate non-invasively for a two-wire power cord carrying opposite currents. The sensitivity of the structure is improved due to the force amplification effect of the cymbal structure and the concentration effect of the permalloy plate. A theoretical model has been developed and validated by the experiments. A large current sensitivity of 12.74 mV/A ($d = 0$ mm) and a high linearity of 0.67% ($d = 3$ mm) are obtained. The experimental current-to-voltage results demonstrate the potential of a large magnetoelectric effect of the proposed piezoelectric device at low-frequency applications.

Author Contributions: W.H. performed the analysis and the experiments; A.Y analyzed the experimental results; W.H. wrote the paper; all authors reviewed the manuscript.

Funding: This work is supported by the National Natural Science Foundation of China (Grant Nos. 61761001 and 61540039) and the Natural Science Foundation of Guangxi Province (Grant No. 2015GXNSFBA139263).

References

1. Cataliotti, A.; Di Cara, D.; Emanuel, A.E.; Nuccio, S. Improvement of Hall effect current transducer metrological performances in the presence of harmonic distortion. *IEEE Trans. Instrum. Meas.* **2010**, *59*, 1091–1097. [CrossRef]

2. Mlejnek, P.; Vopálenský, M.; Ripka, P. AMR current measurement device. *Sens. Actuators A* **2008**, *141*, 649–653. [CrossRef]

3. Jedlicska, I.; Weiss, R.; Weigel, R. Linearizing the output characteristic of GMR current sensors through hysteresis modeling. *IEEE Trans. Ind. Electron.* **2009**, *57*, 1728–1734. [CrossRef]

4. Salach, J.; Hasse, L.; Szewczyk, R.; Smulko, J.; Bienkowski, A.; Frydrych, P.; Kolano-Burian, A. Low current transformer utilizing Co-based amorphous alloys. *IEEE Trans. Magn.* **2012**, *48*, 1493–1496. [CrossRef]

5. Dong, S.; Li, J.F.; Viehland, D. Circumferentially magnetized and circumferentially polarized magnetostrictive/piezoelectric laminated rings. *J. Appl. Phys.* **2004**, *96*, 3382–3387. [CrossRef]

6. Leung, C.M.; Or, S.W.; Zhang, S.; Ho, S.L. Ring-type electric current sensor based on ring-shaped magnetoelectric laminate of epoxy-bonded $Tb_{0.3}Dy_{0.7}Fe_{1.92}$ short-fiber/NdFeB magnet magnetostrictive composite and Pb (Zr, Ti) O_3 piezoelectric ceramic. *J. Appl. Phys.* **2010**, *107*, 09D918.

7. Zhang, J.; Li, P.; Wen, Y.; He, W.; Yang, A.; Lu, C.; Qiu, J.; Wen, J.; Yang, J.; Zhu, Y.; et al. High-resolution current sensor utilizing nanocrystalline alloy and magnetoelectric laminate composite. *Rev. Sci. Instrum.* **2012**, *83*, 115001. [CrossRef]

8. Zhang, J.; He, W.; Zhang, M.; Zhao, H.; Yang, Q.; Guo, S.; Wang, X.; Zheng, X.; Cao, L. Broadband high-sensitivity current-sensing device utilizing nonlinear magnetoelectric medium and nanocrystalline flux concentrator. *Rev. Sci. Instrum.* **2015**, *86*, 095005. [CrossRef]

9. Leland, E.S.; Wright, P.K.; White, R.M. Design of a MEMS passive, proximity-based AC electric current sensor for residential and commercial loads. *Proc. PowerMEMS* **2007**, 77–80.

10. Leland, E.S.; Wright, P.K.; White, R.M. A MEMS AC current sensor for residential and commercial electricity end-use monitoring. *J. Micromech. Microeng.* **2009**, *19*, 094018. [CrossRef]

11. Lu, C.; Li, P.; Wen, Y.; Yang, A.; Yang, C.; Wang, D.; He, W.; Zhang, J. Zero-biased magnetoelectric composite $Fe_{73.5}Cu_1Nb_3Si_{13.5}B_9$/Ni/Pb $(Zr_{1-x}, Ti_x)O_3$ for current sensing. *J. Alloys Compd.* **2014**, *589*, 498–501.

12. Shen, D.; Park, J.H.; Noh, J.H.; Choe, S.Y.; Kim, S.H.; Wikle, H.C., III; Kim, D.J. Micromachined PZT cantilever based on SOI structure for low frequency vibration energy harvesting. *Sens. Actuators A* **2009**, *154*, 103–108.

13. He, W.; Li, P.; Wen, Y.; Zhang, J.; Yang, A.; Lu, C. Note: A high-sensitivity current sensor based on piezoelectric ceramic Pb (Zr, Ti) O_3 and ferromagnetic materials. *Rev. Sci. Instrum.* **2014**, *85*, 026110. [CrossRef]

14. He, W.; Lu, Y.; Qu, C.; Peng, J. A non-invasive electric current sensor employing a modified shear-mode cymbal transducer. *Sens. Actuators A* **2016**, *241*, 120–123. [CrossRef]

15. Ren, B.; Or, S.W.; Zhang, Y.; Zhang, Q.; Li, X.; Jiao, J.; Wang, W.; Liu, D.; Zhao, X.; Luo, H. Piezoelectric energy harvesting using shear mode 0.71Pb (Mg$_{1/3}$Nb$_{2/3}$)O$_3$–0.29 PbTiO$_3$ single crystal cantilever. *Appl. Phys. Lett.* **2010**, *96*, 083502.

16. Carrera, E.; Valvano, S.; Kulikov, G.M. Multilayered plate elements with node-dependent kinematics for electro-mechanical problems. *Int. J. Smart Nano Mater.* **2018**, *9*, 279–317. [CrossRef]

17. Carrera, E.; Valvano, S.; Kulikov, G.M. Electro-mechanical analysis of composite and sandwich multilayered structures by shell elements with node-dependent kinematics. *Int. J. Smart Nano Mater.* **2018**, *9*, 1–33. [CrossRef]

18. Zhu, Y.; Zheng, X.; Li, L.; Yu, Y.; Liu, X.; Chen, J. Evaluation of shear piezoelectric coefficient d$_{15}$ of piezoelectric ceramics by using piezoelectric cantilever beam in dynamic resonance. *Ferroelectrics* **2017**, *520*, 202–211. [CrossRef]

19. Gao, X.; Xin, X.; Wu, J.; Chu, Z.; Dong, S. A multilayered-cylindrical piezoelectric shear actuator operating in shear (d$_{15}$) mode. *Appl. Phys. Lett.* **2018**, *112*, 152902. [CrossRef]

20. Qin, L.; Jia, J.; Choi, M.; Uchino, K. Improvement of electromechanical coupling coefficient in shear-mode of piezoelectric ceramics. *Ceram. Int.* **2019**, *45*, 1496–1502. [CrossRef]

21. Liu, G.; Zhang, C.; Dong, S. Magnetoelectric effect in magnetostrictive/piezoelectric laminated composite operating in shear-shear mode. *J. Appl. Phys.* **2014**, *116*, 074104. [CrossRef]

22. Zhai, J.; Xing, Z.; Dong, S.; Li, J.; Viehland, D. Magnetoelectric laminate composites: an overview. *J. Am. Ceram. Soc.* **2008**, *91*, 351–358. [CrossRef]

23. Yao, Y.P.; Hou, Y.; Dong, S.N.; Li, X.G. Giant magnetodielectric effect in Terfenol-D/PZT magnetoelectric laminate composite. *J. Appl. Phys.* **2011**, *110*, 014508. [CrossRef]

24. Liu, Y.; Xu, G.; Xie, Y.; Lv, H.; Huang, C.; Chen, Y.; Tong, Z.; Shi, J.; Xiong, R. Magnetoelectric behaviors in BaTiO$_3$/CoFe$_2$O$_4$/BaTiO$_3$ laminated ceramic composites prepared by spark plasma sintering. *Ceram. Int.* **2018**, *44*, 9649–9655. [CrossRef]

25. Xu, Y.; Jiang, Z.S.; Wang, Q.; Xu, X.; Sun, D.S.; Zhou, J.; Yang, G. Three dimensional magnetostatic field calculation using equivalent magnetic charge method. *IEEE Trans. Magn.* **1991**, *27*, 5010–5012. [CrossRef]

26. Ikeda, T. *Fundamentals of Piezoelectricity*; Oxford University Press: Oxford, UK, 1990; pp. 38–39.

27. Zeng, M.; Or, S.W.; Chan, H.L.W. Giant resonance frequency tunable magnetoelectric effect in a device of Pb(Zr$_{0.52}$Ti$_{0.48}$)O$_3$ drum transducer, NdFeB magnet, and Fe-core solenoid. *Appl. Phys. Lett.* **2010**, *96*, 203502.

28. Zeng, M.; Or, S.W.; Chan, H.L.W. Giant magnetoelectric effect in magnet-cymbal-solenoid current-to-voltage conversion device. *J. Appl. Phys.* **2010**, *107*, 074509. [CrossRef]

 *micromachines*

Article

Generation of Linear Traveling Waves in Piezoelectric Plates in Air and Liquid

Alex Díaz-Molina [1],*, **Víctor Ruiz-Díez [1]**, **Jorge Hernando-García [1]**, **Abdallah Ababneh [2]**, **Helmut Seidel [3] and José Luis Sánchez-Rojas [1]**

[1] Microsystems, Actuators and Sensors Group, Universidad de Castilla-La Mancha,
   E-13071 Ciudad Real, Spain; victor.ruiz@uclm.es (V.R.-D.); jorge.hernando@uclm.es (J.H.-G.);
   joseluis.saldavero@uclm.es (J.L.S.-R.)
[2] Electronic Engineering Department, Hijjawi Faculty for Engineering Technology, Yarmouk University,
   21163 Irbid, Jordan; a.ababneh@yu.edu.jo
[3] Chair of Micromechanics, Microfluidics/Microactuators, Faculty of Natural Sciences and Technology,
   Saarland University, 66123 Saarbrücken, Germany; seidel@lmm.uni-saarland.de
* Correspondence: alex.diaz@uclm.es; Tel.: +34-926-295-300 (ext. 96667)

Received: 5 April 2019; Accepted: 25 April 2019; Published: 27 April 2019

Abstract: A micro- to milli-sized linear traveling wave (TW) actuator fabricated with microelectromechanical systems (MEMS) technology is demonstrated. The device is a silicon cantilever actuated by piezoelectric aluminum nitride. Specifically designed top electrodes allow the generation of TWs at different frequencies, in air and liquid, by combining two neighboring resonant modes. This approach was supported by analytical calculations, and different TWs were measured on the same plate by laser Doppler vibrometry. Numerical simulations were also carried out and compared with the measurements in air, validating the wave features. A standing wave ratio as low as 1.45 was achieved in air, with a phase velocity of 652 m/s and a peak horizontal velocity on the device surface of 124 μm/s for a driving signal of 1 V at 921.9 kHz. The results show the potential of this kind of actuator for locomotion applications in contact with surfaces or under immersion in liquid.

Keywords: traveling waves; piezoelectric; microactuator; MEMS

---

1. Introduction

Microelectromechanical systems (MEMS) are key components for the progress of miniaturization in disciplines such as consumer electronics, instrumentation, and healthcare [1]. Recently, robotic research on the micro and millimeter scales (i.e., the insect scale) has also benefited greatly from technological advances in MEMS and other enabling technologies, such as additive manufacturing, piezoelectric actuators, and low power sensors [2,3]. In this field, the principles of locomotion are diverse [4,5]. Wave-based locomotion can be adapted to miniature systems, as already happens in nature [6]. The type of wave can be either standing or traveling; however, traveling waves (TWs) are usually preferred, due to the absence of contact tips and the simplicity of bidirectional motion.

Circular motors based on TWs are already well established, with the actuation of two degenerate modes and a proper spatial shift [7]. Similarly, linear motors have also been accomplished by combining two modes at the same frequency, with either two bending modes [8] or a bending mode and a longitudinal mode [9,10]. Most of these approaches are millimeter-sized; however, their assembly is not commonly based on silicon monolithic technology, which would allow for cost reduction, manufacturing precision, and the possibility of system-level integration [11].

It is equally interesting to mention that linear motors can also be implemented by just combining two contiguous bending modes, without the degeneration requirement, which results in less restrictions on the design of the device [12]. Different reports have already demonstrated centimeter-sized devices that move on solid surfaces with this approach [13–15]. Another important field to consider is locomotion within liquids, which emulates the movement of aquatic animals [16]. An example is to mimic the structure of a cuttlefish or a water flatworm, with two moving membranes attached to the length of the body, which has already been considered for macro-scale systems [17–19].

Here we present millimeter-sized linear TW actuators based on piezoelectric aluminum nitride (AlN) on silicon plates, fabricated with MEMS technology. The structure is a cantilever, with the proper top electrode layout to realize progressive waves along its width, mimicking the animal membranes mentioned before. TWs were developed by combining two neighboring modes, and the feasibility of the approach was demonstrated for two different excitation frequencies on the same device, by coupling different pairs of modes. Both the simulations and experimental results were compared, and the investigations were realized in air and liquid. For the TWs in air, a standing wave ratio (SWR) as low as 1.45 was determined, with a phase velocity of 652 m/s and a peak horizontal velocity of 124 μm/s for a 1 V peak amplitude and a frequency near 900 kHz.

2. Materials and Methods

The size of the cantilever under study is 750 μm × 1300 μm (Figure 1). It was fabricated out of a low resistivity silicon p-doped wafer of 520 μm thickness, which served as the bottom electrode. For the piezoelectric layer, 1 μm AlN was sputtered, with a measured d_{33} of 3.22 pm/V. The deposition was realized with a back-pressure level of 4×10^{-3} mbar and a sputtering ratio close to 20 nm/min at 1000 W. A sputtered gold layer was used for the top electrode with a thickness of 400 nm. A silicon membrane was obtained by wet etching with 38% potassium hydroxide (KOH) at 85 °C, previously depositing a 400 nm thick Si_3N_4 layer on the backside using plasma-enhanced chemical vapor deposition (PECVD). This allowed us to pattern the silicon substrate to a residual thickness of about 40 μm in the areas where the suspended beams were fabricated. Finally, the plates were released by a deep reactive-ion etching (DRIE) process. Additional details of the fabrication process can be found in [20].

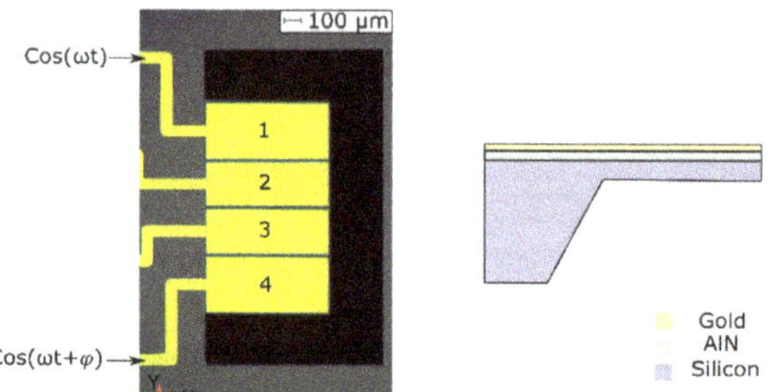

Figure 1. Optical micrograph of the 750 μm × 1300 μm cantilever and cross section view along the length of the device, including the suspended part and the silicon anchor.

The top electrode design comprises 4 isolated strips, which have been previously reported for the actuation of roof tile-shaped modes in liquid sensor applications [21]. Here, only the two outer electrodes (1 and 4 in Figure 1) were used for the generation of TWs.

The displacement of the device versus time and versus frequency was measured by means of a laser Doppler vibrometer (Polytec MSV 400, POLYTEC, Waldbronn, Germany). This instrument allows for the measurement of the out-of-plane displacement through a laser spot, which scans a set of points distributed along the surface of the cantilever.

Finite elements method (FEM) software was used to corroborate the generation of the TWs. The software 'Coventorware' was employed, and a 3D model of the cantilever was analyzed with the help of modal harmonic analysis. In the simulated design, the angle of inclination of the substrate due to the wet etching process (shown in Figure 1) was taken into account, as well as the surrounding bulk material, where the clamped boundary condition was applied. The mesh used was composed of

9255 3D cubic parabolic elements. The dimensions of the device were the nominal values, and the elastic and rest of the material constants were as provided by the software, except for the piezoelectric constant of AlN, which was coincident with the measured value. A good agreement between the measured and calculated resonance frequencies was attained, without including any built-in stress in the structure. This was confirmed with measurements of the plate deformation by an optical profiler, obtaining values close to 773 and 171 nm along the lines perpendicular and parallel to the anchored side of the plate, respectively.

Before the investigation of the TW generation, the static piezoelectric response was measured with the laser Doppler vibrometer, applying an 8 V peak amplitude to the top four electrodes at a frequency of 200 Hz (far from any neighbor resonance). The deflection reached a maximum value of 3.1 nm/V, located at the edge of the tip.

3. Results and Discussion

First, we present the frequency response of the cantilever obtained by means of the laser Doppler vibrometer, by applying a periodic chirp signal to Electrode 1. Figure 2 shows the average displacement versus the frequency from 100 to 300 kHz. Two modes were detected, Modes (11) and (12), following Leissa nomenclature [22]. Mode (11) was antisymmetric, while Mode (12) was symmetric, with respect to the center of the cantilever.

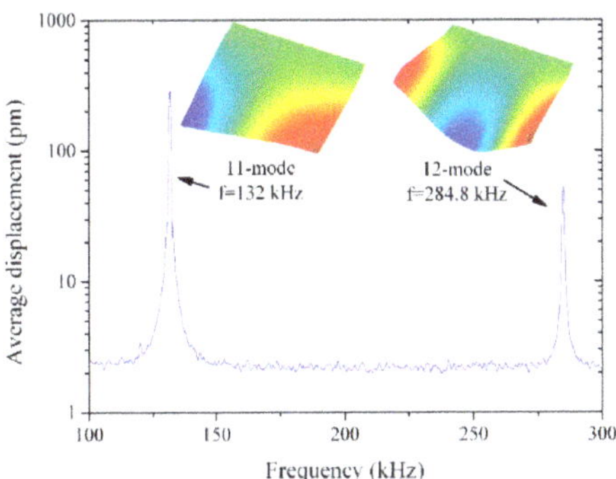

Figure 2. Measured frequency response of the cantilever from 100 to 300 kHz by means of a laser Doppler vibrometer. Modes (11) and (12) were detected. The measured modal shapes are also included.

By applying a sine wave signal to each of the outer electrodes, with a phase difference, φ, and a frequency, ω_f, between the two modes previously mentioned, the displacement of the device, in the two-mode approximation, can be described by

$$w(x,y,t) = [Q_s\Phi_s(x,y) + Q_a\Phi_a(x,y)]\cos(\omega_f t) + [Q_s\Phi_s(x,y) - Q_a\Phi_a(x,y)]\cos(\omega_f t + \varphi), \quad (1)$$

which is the superposition of the displacements associated with the respective sine waves, applied to each of the outer electrodes. Φ_s and Φ_a represent the shapes of the symmetric and antisymmetric modes, respectively and Q_s and Q_a are the weights determining the contribution of each of these modes. As is shown in [23], this equation can be deconstructed into four different progressive wave terms, and, depending on the values of Q_a, Q_s, and φ, the sum of these terms results in either a traveling or standing wave.

This idea is depicted in Figure 3 where we compare the envelope of the maximum displacement of the modes in Figure 2 and that of a progressive wave resulting from the combination of these modes, as in Equation (1). Figure 3a,b correspond to Modes (11) and (12), respectively. Figure 3c is the envelope given by Equation (1), for $Q_a = 1.3$, $Q_s = 1$, $\omega_f = 1.31 \times 10^6$ rad/s, and $\varphi = 90°$. Although these values were chosen only for illustration purposes, while the modes (standing waves) were characterized by nodes along the width of the cantilever edge (dark blue in the figure), the displacement associated with the combination of modes in quadrature resulted in a profile without nodes along the edge, which is characteristic of a progressive wave.

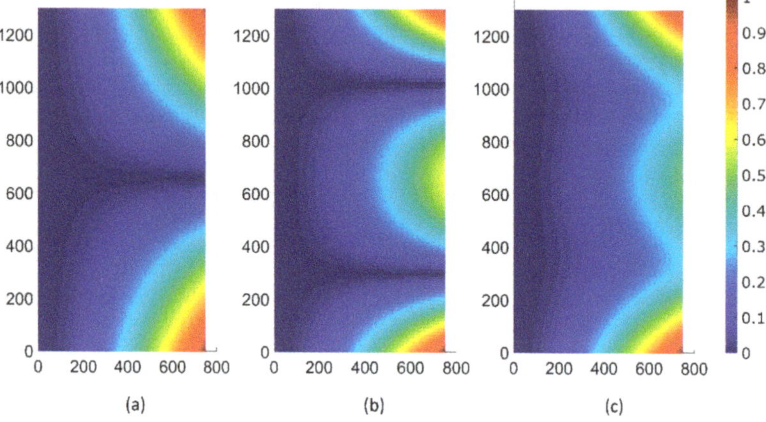

Figure 3. Envelope of the maximum displacement over the cantilever surface of (**a**) Mode (11), (**b**) Mode (12), and (**c**) Equation (1) with the parameters mentioned in the text. The color scale represents the unitary displacement.

This basic analytical calculation suggests that a TW may be generated just by using electrodes that excite the two modes with different weights ($Q_s \neq Q_a$). Next, we present the experimental measurements on the device described above, whose electrodes can be shown to satisfy this condition [24]. The following parameters were chosen for the sine waves applied to the two outer electrodes: a frequency of 208.4 kHz, corresponding to the average of the resonant frequencies of the two modes, a 1 V peak amplitude, and a phase shift, φ of 90°. Measurements were taken in the time domain with the laser Doppler vibrometer. Figure 4a shows the map of the displacement on the device surface (anchored on the left side) at different instants. The plot shows clearly that the peaks and valleys are located in different places depending on the moment in time as the wave travels along the width. This differs from a standing wave-based modal response where the maxima and the minima of the wave maintain their location independently of the instant in time. Figure 4b shows different snapshots of the 3D TW as it propagates from right to left. An animation of the 3D movement can be seen in Video S1. A TW in the opposite direction can be obtained by simply exchanging the sine waves applied to the outer electrodes. The TW envelope reaches a maximum value of 37 pm, and the phase velocity can be estimated as 416 m/s. The horizontal velocity of a point on the surface of the cantilever was estimated to be 3 µm/s for the voltage applied [25]. A pure TW would have a SWR of 1. In our case, due to the free-free boundary condition at the edges, the wave was not ideal. Excluding the edges of the plate, the SWR reached a value of 1.46 for a centered window covering 54% of the total width of the plate. This value is comparable to those reported for centimeter-sized devices [26,27].

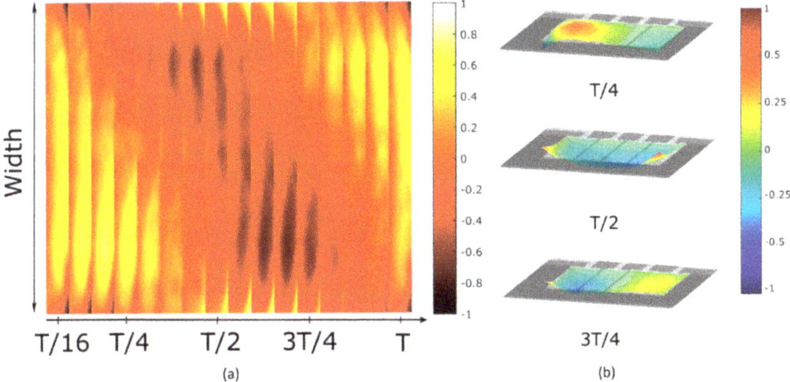

Figure 4. Time frame of the measured traveling wave (TW) by combining Modes (11) and (12): (**a**) 2D representation and (**b**) 3D representation (a video clip is available (Video S1)). The color scale represents the normalized displacement of the cantilever. T corresponds to the period of the TW.

TWs may be generated by combining different couples of modes on the same device. Figure 5a shows the frequency response of the cantilever in a different range between 700 kHz and 1.1 MHz. In this case, the symmetric mode corresponded to Mode (22) and the antisymmetric one to Mode (14). By applying the same excitation as for the previous case but adapting the frequency to the new average (921.9 kHz), we obtained the response of Figure 5b, again characteristic of a TW motion, reaching 122 pm of maximum vertical displacement. The estimated phase velocity was 652 m/s, and the horizontal speed on the surface was 124 μm/s. The SWR reached a value of 1.45 in a central window corresponding to 93% of the total width of the plate. A 3D representation of the TW at different instants in time is shown in Figure 5c. Video S2 shows the 3D movement of this TW.

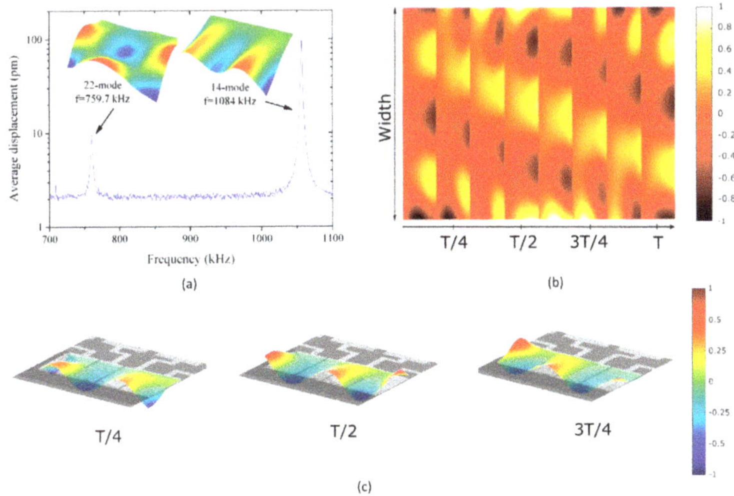

Figure 5. (**a**) Measured frequency response of the cantilever from 700 kHz to 1.1 MHz. Time frame of the measured TW by combining Modes (22) and (14). (**b**) 2D representation and (**c**) 3D representation (a video clip is available (Video S2)). The colors represent the normalized displacement of the cantilever. T corresponds to the period of the TW.

To gain a more complete understanding of the generation of a TW and the effect of the different parameters on both the device and the excitation signals, the experimental results were compared to a FEM model. The FEM analysis provides a more realistic and quantitative description of the device behavior as compared to the basic analytical approach given by Equation (1), as it allows the inclusion of the real shape of the surrounding anchor (see the cross section of Figure 1), the electromechanical coupling for the given geometry of the electrodes, and the contribution of those modes further away from the two modes considered to determine the frequency of the actuation.

Figure 6a,b show a comparison between the measured and the simulated envelopes of the maximum displacement along the width of the plate, close to the edge of the cantilever, for the two cases presented above. It can be seen that there is a reasonable agreement between the experiments and the modeling, which supports our approach. The discrepancy observed may be related to the slight differences between the calculated frequencies and modal shapes and those in the real structure, induced by a non-ideal anchoring. This hypothesis has been checked by the FEM model—varying the portion of the substrate, that is included in the calculation, changed the modal frequencies, affecting the resulting TW envelope. We found that the TW envelope, in shape and amplitude, was very sensitive to the shape and frequency of the neighbor modes.

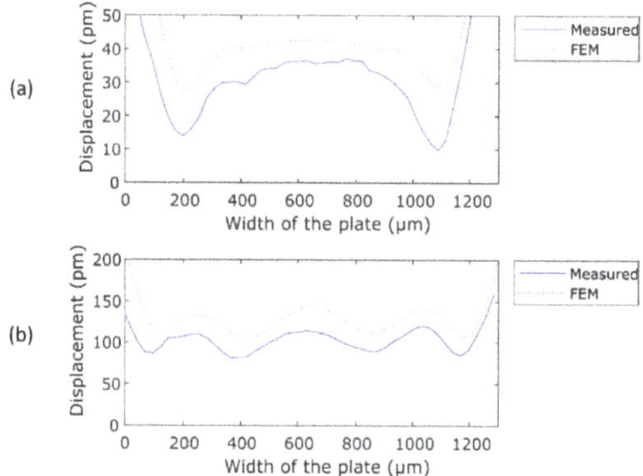

Figure 6. Comparison between the envelope of the measured (blue) and the FEM simulated (red) TW: (**a**) combination of Modes (11) and (12) at 208.4 kHz and (**b**) combination of Modes (22) and (14) at 921.9 kHz.

Generation of TWs in micromechanical plates may be applied to locomotion on surfaces. Here we demonstrate that they can also be generated in liquid, which may be an interesting actuation mechanism for underwater micro-robotics. The measurements were carried out in isopropanol. The plate and its cavity were fully immersed in the liquid and covered with a glass slider to avoid air bubbles and control the volume of liquid [21]. Modes (22) and (14) were chosen again, as they exhibited a quality factor in liquid of 47 and 120, respectively. The scheme for the actuation was similar to that used in the previous measurements in air, except that the applied voltage was increased to 7 V in liquid to improve the signal to noise ratio. Ninety phase-shifted sine waves were applied to the two outer electrodes at a frequency of 583.2 kHz. Figure 7a shows the displacement of the TW at different moments in time, and a selection of them are represented in a 3D view in Figure 7b. The animation of the movement can be seen in Video S3. It can be seen that there was an attenuation in the wave amplitude of 59% with respect to the air measurement, due to the damping induced by the liquid. The estimated phase speed was 441 m/s, and the horizontal velocity was 290 μm/s for the voltage applied. The SWR reached a value of 2.19 in 98% of the total width of the plate.

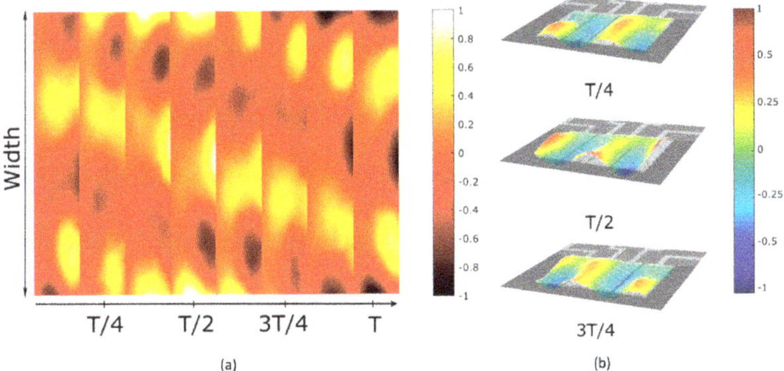

Figure 7. Snapshots of different times of the measured TW in isopropanol at 583.2 kHz: (**a**) 2D representation and (**b**) 3D representation (a video clip is available (Video S3)). The colors represent the normalized displacement of the cantilever. T corresponds to the period of the TW.

4. Conclusions

In summary, this paper demonstrates the feasibility of developing TWs on micro- to millimeter-scaled piezoelectrically actuated silicon-based MEMS structures, by combining two resonant modes with the proper scheme of actuation. Values of the SWR close to the ideal TW support the efficient generation mechanism. In order to validate the experimental results, FEM analysis was performed, and a good agreement between the measurements and the simulations was found. Additionally, TW generation in liquid media was presented, which revealed the potential applicability of this type of structure to locomotion in liquid media, such as miniaturized propulsion systems.

Supplementary Materials: The following are available online at http://www.mdpi.com/2072-666X/10/5/283/s1: Video S1: 3D animation of the TW measured at 208.4 kHz in air; Video S2: 3D animation of the TW measured at 921.9 kHz in air; and Video S3: 3D animation of the TW measured at 583.2 kHz in isopropanol.

Author Contributions: Investigation, A.D.-M.; project administration, J.L.S.-R.; supervision, V.R.-D., J.H.-G., A.A., H.S., and J.L.S.-R.; writing—original draft preparation, A.D.-M.; writing—review and editing, V.R.-D., J.H.-G., A.A., H.S., and J.L.S.-R.

Funding: This work was supported by the European Regional Development Fund, the Spanish Ministerio de Economía y Competitividad project (TEC2015-67470-P), and the regional government (JCCLM) project (SBPLY/17/180501/000139). Abdallah Ababneh and Helmut Seidel acknowledge the funding provided by the German Research Council (DFG) project "Micro resonators for applications in liquids and gases".

Conflicts of Interest: The authors declare no conflict of interest. The funders had no role in the design of the study; in the collection, analyses, or interpretation of data; in the writing of the manuscript; or in the decision to publish the results.

References

1. Bogue, R. Recent developments in MEMS sensors: A review of applications, markets and technologies. *Sens. Rev.* **2013**, *33*, 300–304. [CrossRef]
2. Jager, E.W.H. Microrobots for Micrometer-Size Objects in Aqueous Media: Potential Tools for Single-Cell Manipulation. *Science* **2000**, *288*, 2335–2338. [CrossRef] [PubMed]
3. Palagi, S.; Fischer, P. Bioinspired microrobots. *Nat. Rev. Mater.* **2018**, *3*, 113–124. [CrossRef]
4. Hariri, H.; Bernard, Y.; Razek, A. Locomotion principles for piezoelectric miniature robots. In Proceedings of the 12th International Conference on New Actuators, Bremen, Germany, 14–16 June 2010; pp. 1015–1020.
5. Calisti, M.; Picardi, G.; Laschi, C. Fundamentals of soft robot locomotion. *J. R. Soc. Interface* **2017**, *14*, 20170101. [CrossRef] [PubMed]
6. Taylor, G. Analysis of the swimming of microscopic organisms. *Proc. R. Soc. London. Ser. A Math. Phys. Sci.* **1951**, *209*, 447–461.

7. Hagedorn, P.; Wallaschek, J. Travelling wave ultrasonic motors, Part I: Working principle and mathematical modelling of the stator. *J. Sound Vib.* **1992**, *155*, 31–46. [CrossRef]

8. Chen, Z.; Li, X.; Chenm, J.; Dong, S. A square-plate ultrasonic linear motor operating in two orthogonal first bending modes. *IEEE Trans. Ultrason. Ferroelectr. Freq. Control* **2013**, *60*, 115–120. [CrossRef] [PubMed]

9. Tomikawa, Y.; Takano, T.; Umeda, H. Thin Rotary and Linear Ultrasonic Motors Using a Double-Mode Piezoelectric Vibrator of the First Longitudinal and Second Bending Modes. *Jpn. J. Appl. Phys.* **1992**, *31*, 3073–3076. [CrossRef]

10. Chen, J.; Chen, Z.; Li, X.; Dong, S. A high-temperature piezoelectric linear actuator operating in two orthogonal first bending modes. *Appl. Phys. Lett.* **2013**, *102*, 052902. [CrossRef]

11. Smith, G.L.; Rudy, R.Q.; Polcawich, R.G.; DeVoe, D.L. Integrated thin-film piezoelectric traveling wave ultrasonic motors. *Sens. Actuat. A Phys.* **2012**, *188*, 305–311. [CrossRef]

12. Dehez, B.; Vloebergh, C.; Labrique, F. Study and optimization of traveling wave generation in finite-length beams. *Math. Comput. Simul.* **2010**, *81*, 290–301. [CrossRef]

13. Hariri, H.; Bernard, Y.; Razek, A. A traveling wave piezoelectric beam robot. *Smart Mater. Struct.* **2014**, *23*, 025013. [CrossRef]

14. Malladi, V.V.S.; Albakri, M.; Tarazaga, P.A. An experimental and theoretical study of two-dimensional traveling waves in plates. *J. Intell. Mater. Syst. Struct.* **2017**, *28*, 1803–1815. [CrossRef]

15. Hariri, H.; Bernard, Y.; Razek, A. 2-D Traveling Wave Driven Piezoelectric Plate Robot for Planar Motion. *IEEE/ASME Trans. Mechatron.* **2018**, *23*, 242–251. [CrossRef]

16. Sfakiotakis, M.; Lane, D.M.; Davies, J.B.C. Review of fish swimming modes for aquatic locomotion. *IEEE J. Ocean. Eng.* **1999**, *24*, 237–252. [CrossRef]

17. Kopman, V.; Porfiri, M. Design, Modeling, and Characterization of a Miniature Robotic Fish for Research and Education in Biomimetics and Bioinspiration. *IEEE/ASME Trans. Mechatron.* **2013**, *18*, 471–483. [CrossRef]

18. Wang, S.; Wang, Y.; Wei, Q.; Tan, M.; Yu, J. A Bio-Inspired Robot With Undulatory Fins and Its Control Methods. *IEEE/ASME Trans. Mechatron.* **2017**, *22*, 206–216. [CrossRef]

19. Low, K.H.; Willy, A. Biomimetic Motion Planning of an Undulating Robotic Fish Fin. *J. Vib. Control* **2006**, *12*, 1337–1359. [CrossRef]

20. Ababneh, A.; Al-Omari, A.N.; Dagamseh, A.M.K.; Qiu, H.C.; Feili, D.; Ruiz-Díez, V.; Manzaneque, T.; Hernando, J.; Sánchez-Rojas, J.L.; Bittner, A.; et al. Electrical characterization of micromachined AlN resonators at various back pressures. *Microsyst. Technol.* **2014**, *20*, 663–670. [CrossRef]

21. Ruiz-Díez, V.; Hernando-García, J.; Toledo, J.; Manzaneque, T.; Kucera, M.; Pfusterschmied, G.; Schmid, U.; Sánchez-Rojas, J.L. Modelling and characterization of the roof tile-shaped modes of AlN-based cantilever resonators in liquid media. *J. Micromechan. Microeng.* **2016**, *26*, 084008. [CrossRef]

22. Leissa, A.W. *Vibration of Plates*; Scientific and Technical Information Division, National Aeronautics and Space Administration: Washington, DC, USA, 1969.

23. Malladi, V.V.N.S. Continual Traveling waves in Finite Structures: Theory, Simulations, and Experiments. Ph.D. Thesis, Faculty of the Virginia Polytechnic Institute and State University, Blacksburg, VA, USA, 2016.

24. Sanchez-Rojas, J.L.; Hernando, J.; Donoso, A.; Bellido, J.C.; Manzaneque, T.; Ababneh, A.; Seidel, H.; Schmid, U. Modal optimization and filtering in piezoelectric microplate resonators. *J. Micromechan. Microeng.* **2010**, *20*, 055027. [CrossRef]

25. Inaba, R.; Tokushima, A.; Kawasaki, O.; Ise, Y.; Yoneno, H. Piezoelectric Ultrasonic Motor. In Proceedings of the IEEE 1987 Ultrasonics Symposium, New York, NY, USA, 14–16 October 1987; pp. 747–756.

26. Minikes, A.; Gabay, R.; Bucher, I.; Feldman, M. On the sensing and tuning of progressive structural vibration waves. *IEEE Trans. Ultrason. Ferroelectr. Freq. Control* **2005**, *52*, 1565–1575. [CrossRef] [PubMed]

27. Ghenna, S.; Giraud, F.; Giraud-Audine, C.; Amberg, M.; Lemaire-Semail, B. Modelling and control of a travelling wave in a finite beam, using multi-modal approach and vector control method. In Proceedings of the 2015 Joint Conference of the IEEE International Frequency Control Symposium & the European Frequency and Time Forum, Denver, CO, USA, 12–16 April 2015; pp. 509–514.

Article

3-D Design and Simulation of a Piezoelectric Micropump

Seyed Amir Fouad Farshchi Yazdi, Alberto Corigliano * and Raffaele Ardito

MEMS Modelling and Design Group, Department of Civil and Environmental Engineering,
Politecnico di Milano, 20133 Milan, Italy; seyedamir.farshchi@polimi.it (S.A.F.F.Y.);
raffaele.ardito@polimi.it (R.A.)
* Correspondence: alberto.corigliano@polimi.it; Tel.: +39-0223994244

Received: 14 February 2019; Accepted: 11 April 2019; Published: 18 April 2019

Abstract: The objective of this paper is to carefully study the performances of a new piezoelectric micropump that could be used, e.g., for drug delivery or micro-cooling systems. The proposed micropump is characterized by silicon diaphragms, with a piezoelectric actuation at a 60 V input voltage, and by two passive valves for flow input and output. By means of a 3-D Finite Element (FE) model, the fluid dynamic response during different stages of the working cycle is investigated, together with the fluid–structure interaction. The maximum predicted outflow is 1.62 µL min^{-1}, obtained at 10 Hz working frequency. The computational model enables the optimization of geometrical features, with the goal to improve the pumping efficiency: The outflow is increased until 2.5 µL min^{-1}.

Keywords: piezoelectric material; multiphysics simulation; finite element method (FEM); fluid–structure interaction (FSI); micro electromechanical systems (MEMS)

---

1. Introduction

Pumps are devices that provide momentum to transfer fluids and can be integrated in Micro Electromechanical Systems (MEMS) to develop so-called micro-fluidic systems [1]. As an example, micropumps play an important role in biomedical and drug delivery systems: The micro-dosing feature in such devices has improved the effectiveness of the treatment because the concentration of the drug in the patient's body has been kept constant as well as because toxicity has been prevented [2].

Several actuation systems have been investigated in the literature [3], with particular attention to electrostatic forces and piezoelectric materials. The former case, e.g., studied in Reference [4], is connected to some drawbacks related to the high actuation voltage and the possible occurrence of pull-in instability [5]; on the other hand, piezoelectric actuation has been widely utilized based on advantages such as the small size, low power consumptions, no electromagnetic interference, and an insensitivity of fluid viscosity [6] even though the maximum power density of the piezoelectric materials is dependent on the working frequency [7]. Among different materials with a piezoelectric effect, lead zirconate titanate (PZT) ceramics demonstrated optimal performances in view of the large deflection that can be induced in the pump diaphragm, as recently shown by Cazorla et al. [8]. While Nisar et al. have reported different types of micropumps fabricated for biomedical applications [9], there are some recent studies about the design and modelling of piezoelectric micropumps. Revathi and Padmanabhan designed a valveless micropump with a composite piezoelectric actuator which showed the maximum outflow at an aspect ratio of 15 for a nozzle/diffuser [10]. Sateesh et al. [11] designed and modelled a piezoelectrically actuated micropump with PZT-5h and Polydimethylsiloxane (PDMS) with an outflow rate of 0.029 µL s^{-1}.

In any case, it has been clearly established that, in addition to the actuator, the presence of valves in the device can affect the performance. Despite the fact that the valveless micropumps have the benefit of no risk of wear and fatigue of the valves, it is hard to control the flow to the desired direction at inlet and outlet, causing energy loss and liquid reflux. Consequently, in order to improve the control of the fluid flow, mechanical valves (either active or passive) must be introduced in the system. The main novelty of the present paper is the comprehensive study of a complete pumping system, composed of a piezoelectric actuator and passive valves, using three-dimensional modelling and a simulation of the device performing a complete cycle of pumping with a consideration for the two-way fluid–structure interaction. An innovative layout is considered, starting from a patent [12] that was originally proposed for electrostatic actuation. The purpose of the present paper is to assess the behavior of the new device and to propose some slight modifications in order to improve the performances.

The paper is organized as follows. After the introduction, the layout of the device is shortly discussed. The subsequent section contains a thorough description of the computational model which then leads to some preliminary results. Then, we consider some modifications of the geometrical features of the pump, with the purpose of optimizing its performances.

2. Description of the Proposed Layout

In most cases, a pumping device, composed of a micropump and valves, is of complex realization in view of the difficult integration of the different components of the actuating system on a limited number of wafers. Moreover, the integrated valves may present specific issues related to a lack of tightness, which is the cause of a leakage and backflow.

In order to overcome the aforementioned limits, an innovative device based on the use of two wafers has been recently conceived and patented [12]. In spite of the fact that the invention refers to an electrostatic actuation, the main advantage of using just two wafers for an integrated design can be exploited also for the piezoelectric case. According to the invention, the pumping device comprises (i) a pumping chamber, realized between two silicon wafers bonded to each other; (ii) an inlet valve, with a shutter element in correspondence to the connection with the external reservoir; and (iii) an outlet valve, with a shutter element on the external microfluidic circuit. As shown in Figure 1, when the inlet valve is in the open configuration, the shutter is housed by a recess that is fluidly coupled with the pumping chamber by means of an inlet channel. On the other hand, the outlet shutter is located in a recess that is fluidly decoupled with respect to the pumping chamber. The described configuration of the inlet and outlet valves allows the direction of the processed flow to be controlled in a completely passive way. More precisely, the inlet and outlet valves do not require dedicated actuators, and so, the structure is generally simplified for the benefit of both the overall dimensions and the manufacturing costs. For instance, the micropump, as above defined, may be made from just two semiconductor wafers joined together. Moreover, the micropump control is simplified because it does not have to take into account the synchronization of the valves. Dedicated actuators for the valves, in particular for the output valve, may optionally be provided if specific circumstances make this advisable. However, the micropump is still fully operative even with purely passive valves. The inlet valve and the outlet valve are of the orthoplanar type. The sealing is guaranteed by the presence of an initial stress state that can be easily achieved during the bonding phase by introducing a couple of thick elements on the shutters: If the thickness of such elements is larger than the bonding layer, the orthoplanar valves are forced to close the holes. It is worth noting that the layout of the device allows for the introduction of two opposite pumping diaphragms, with a possible increase of the stroke volume. In the present paper, for the sake of simplicity, just one actuation diaphragm is considered.

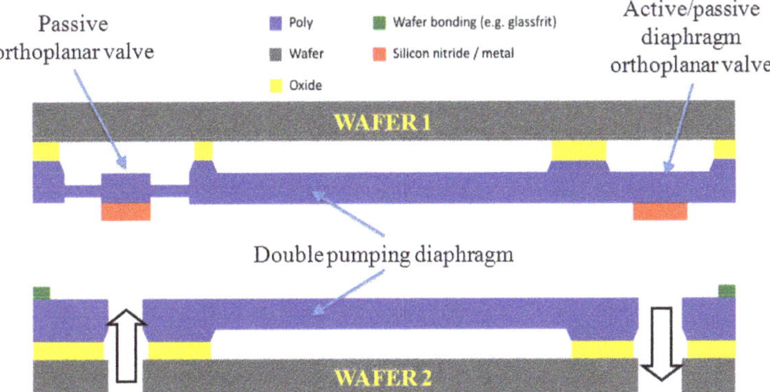

Figure 1. A schematic view of the micropump proposed in Reference [12] and adopted in the present paper with the addition of a piezoelectric actuation.

3. The Three-Dimensional Model

Multiphysics modelling encompasses two main parts: the "deformable" solid, containing the actuation portion and the valves for controlling the fluid flux, and the "fluid domain", which interacts with the solid domain through the interface at the boundary of the solid and fluid domains. Due to the sake of symmetry, half of the micropump is modelled.

In the initial design, the micropump consists of a circular pumping chamber, delimited by lower and lateral fixed surfaces and by an upper silicon diaphragm. As shown in Figures 2 and 3, the fluid domain is completed by two cylindrical spaces from one side connected to the inlet and outlet, which are connected to the pumping chamber by means of two prismatic channels with a rectangular cross section from the other side. It is worth noting that the inlet and outlet are closed by the silicon valves. The radius of the valve's membrane is larger than the hole, so that the fluid flow is prevented when the valve is in the rest configuration.

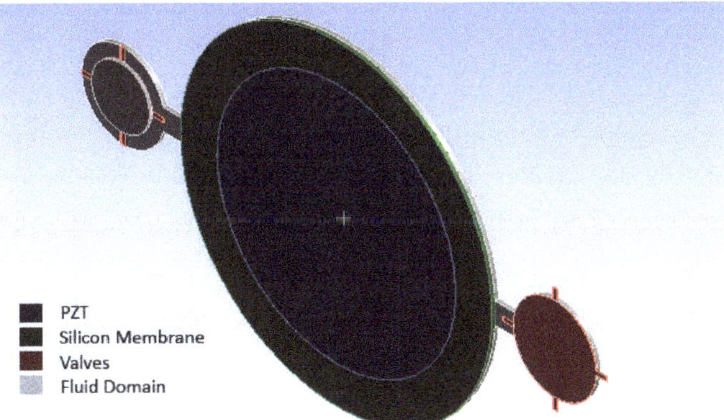

Figure 2. A 3-D scheme of the proposed micropump.

The geometrical specifications of the fluid domain are presented in Table 1. The considered fluid is water, modeled as an incompressible, viscid fluid with the mechanical properties summarized in Table 2.

The model for the solid parts is depicted in Figure 4. The actuation diaphragm is represented by a suspended disc, with a laminate cross section. The actuation is achieved by means of a piezoelectric layer deposited on a silicon plate. Among the various possibilities, lead zirconate titanate (PZT) is chosen as the active material, in the form of a thin film [13]. This is fully compatible with the MEMS production process, through the adoption of one of the available techniques, e.g., sputtering [14], pulsed laser deposition [15], and sol-gel process [16]. The piezoelectric layer has a circular shape, coaxial with the silicon disc. The radius of PZT, see Table 3, is selected according to the results presented in a previous study [6], which showed that the highest stroke volume was obtained when the ratio between the radius of the active layer and the radius of the silicon diaphragm was equal to 0.73. The passive valves consist of silicon discs, that are attached to the rigid frame by means of four rectangular beams. The elastic deformation of the beams allows for the vertical movement of the disc that alternatively opens and closes the inlet hole (the same applies to the outlet). The sealing of the valves in the closed rest configuration is ensured by a prestress in the elastic ligaments, given by an imposed transverse displacement on the edge of the disc.

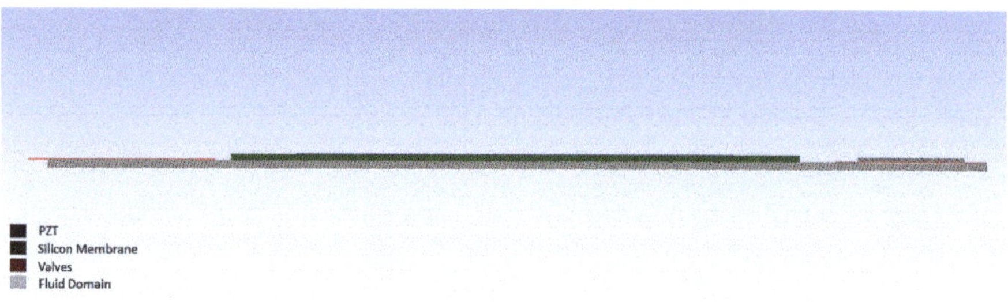

(**a**) A cross-sectional view of the domains.

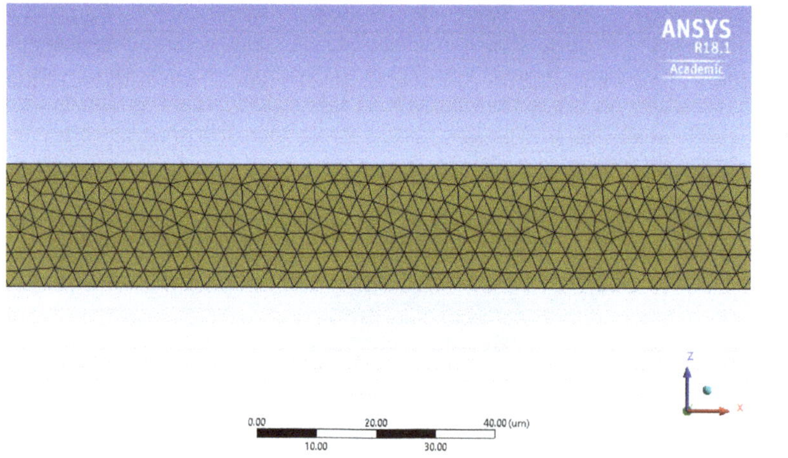

(**b**) The fluid domain mesh in a cross section.

Figure 3. The micropump fluid domain.

Table 1. The geometrical specifications of the fluid domain.

Section	Geometrical Specifications			
	Radius (µm)	Width (µm)	Length (µm)	Thickness (µm)
Chamber	750	-	-	20
Entrance/Exit Cylinders	200	-	-	20
Inlet	140	-	-	
Outlet	250	-	-	
Channels		100	100	20

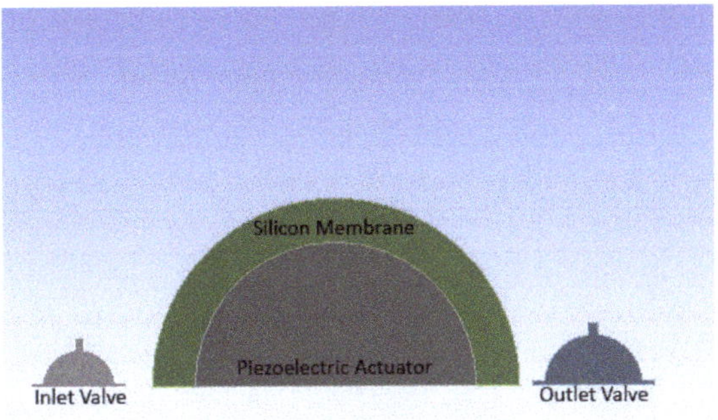

(**a**) The solid domain of the model.

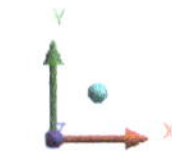

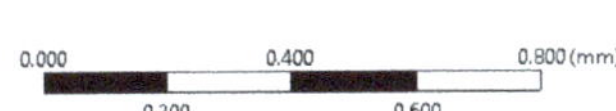

(**b**) The solid domain mesh.

Figure 4. The micropump structural parts.

Table 2. The material properties.

Properties	Materials		
	Silicon	**PZT-5A**	**Water**
E (MPa)	169,000	70,000	-
ν	0.23	0.3	-
ρ (kg/m^3)	2300	7700	999.97
e_{13} (N/V m)	-	-5.04	-
e_{33} (N/V m)	-	19.7	-
$\varepsilon_{11}^s/\varepsilon_0$	-	1320	-
$\varepsilon_{33}^s/\varepsilon_0$	-	1250	-
Viscosity (cP)	-	-	1 (20°C)

Table 3. The geometrical specifications of the solid parts.

Part	Geometrical Specifications	
	Radius (μm)	Thickness (μm)
Membrane	750	16
Piezoelectric	547.5	1
Inlet Valve	150	2
Outlet Valve	250	2

The mechanical properties of polycrystalline silicon and PZT are given in Table 2.

The investigation of the interaction between the moving solid parts and the fluid is a key point, especially in mechanical micropumps. The motion of the solid, i.e., the oscillating displacement field, induces the motion in the fluid. Also, the fluctuating pressure in the fluid acts as a surface load on the fluid/solid interface. By coupling the governing equations of these two domains, the mutual interaction can be thoroughly studied.

For the proposed micropump, a two-way fluid–Structure Interaction (FSI) is modeled with the commercial code ANSYS® 18.1 and 19.2, with the mechanical and CFX solvers. At each staggered loop, the mechanical solver sends the time derivative of the displacement of the interface nodes to the target nodes in the fluid domain. On the other hand, the CFX solver sends back the stress to the solid nodes based on the traction equilibrium at the interface [17].

The valves are modelled by means of quadrilateral 8-node solid elements (SHELL182 in ANSYS nomenclature), which are suitable for modelling thin to moderately thick shell structures. The aim of choosing this kind of element is to reduce the computational cost, taking into account the specific geometrical features of the problem at hand. The 20-node brick element is chosen for the oscillating membrane (SOLID186), and the coupled-field element is chosen for the piezoelectric actuator (SOLID226). In the fluid domain, the tetrahedral elements with 10 nodes (FLUID221) are used. The number of elements for the solid and fluid domains are 33,968 and 1,425,377, respectively. Some views of the adopted mesh are reported in Figures 3 and 4. The piezoelectric actuator is polarized in the out-of-plane direction. The oscillating membrane is clamped at the edge, as well as the valves at the end of connected bars. The applied voltages to the actuator are 20, 40, and 60 V.

For the boundary conditions on the fluid domain, both the inlet and outlet have a zero relative pressure with respect to the outside of the domain. The faces attached to the silicon membrane and valves are set to the fluid–structure interaction sites as well. The other faces of the fluid domain are set to stationary walls, i.e., the velocity of the fluid is equal to zero.

4. Preliminary Results

The proposed micropump is investigated for different values of the input voltage frequency. The most important figure of merit is the outflow of the micropump that is reported in Figure 5 with the actuation voltage equal to 60 V for the range of frequency between 0.01 to 100 Hz. Ideally, the displaced quantity of fluid can be estimated on the basis of the so-called *stroke volume* v_s, i.e., the variation in the volume of the pumping chamber due to the movement of the actuating diaphragm during one cycle. For a given pumping frequency f_p, the nominal outflow would be

$$q = f_p v_s \tag{1}$$

As it is shown by the chart in Figure 5, the computed outflow is strongly different with respect to the nominal quantity. More specifically, the non-monotonic behavior is connected to the inertial force of the fluid so that the maximum outflow is reached at a certain point and a decay is observed after that frequency. The optimal outflow for the designed micropump is achieved at 10 Hz: The numerical results are now thoroughly examined for that specific actuation frequency.

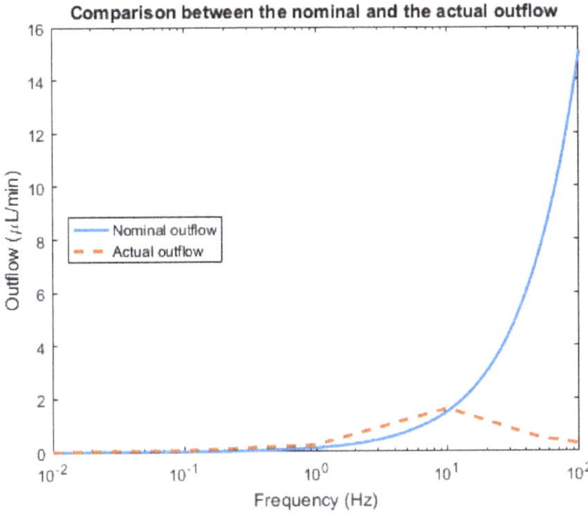

Figure 5. The dependency of the outflow on the pumping frequency.

At the first stage of the micropump working cycle, the silicon membrane is deformed in the upward direction due to the PZT deformation. The pressure gradient causes an inlet valve opening and the fluid enters in the chamber. The computed maximum deflection of the diaphragm for different actuation voltages is demonstrated in Figure 6.

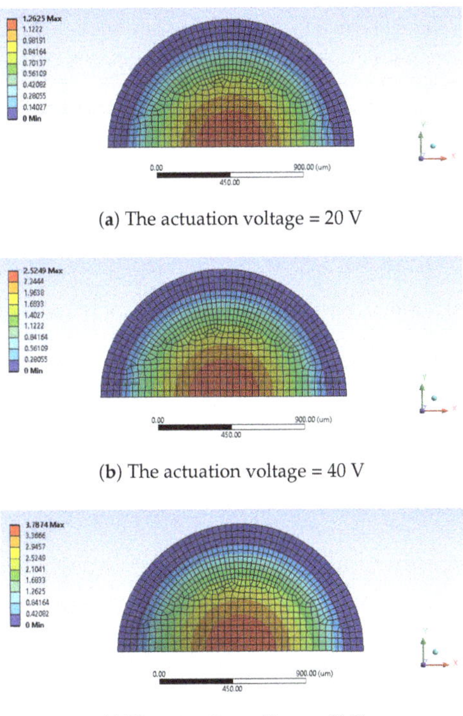

(**a**) The actuation voltage = 20 V

(**b**) The actuation voltage = 40 V

(**c**) The actuation voltage = 60 V

Figure 6. The deformation of the pumping diaphragm: contour plots of the transverse displacement in μm.

Figure 7 represents the outflow of the designed micropump for different actuation voltages at the working frequency of 10 Hz.

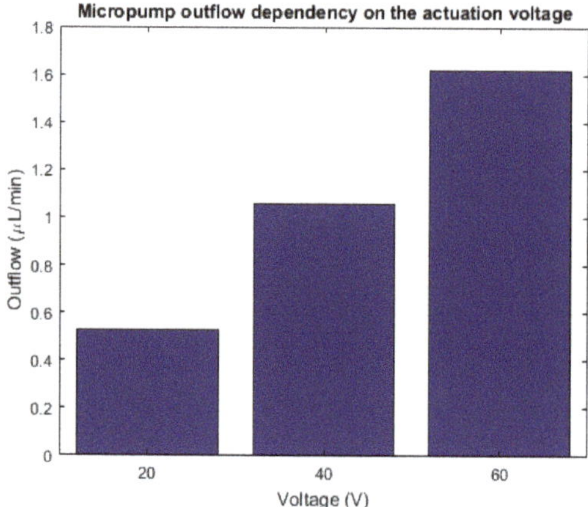

Figure 7. The outflow dependency on the actuation voltage.

Figure 8 presents the variation of the chamber volume during a complete pumping cycle. After the silicon membrane reaches the maximum stroke, as mentioned above, the piezoelectric input is instantaneously dropped and the diaphragm tends to recover its initial configuration. Due to the membrane displacement, the fluid pressure increases, and as a consequence, the outlet valve is pushed in its recess and the hole is opened. At the same time, the elastic recovery enables the closing of the inlet channel. The results of Figure 8 allow one to compute the quantity of fluid that is pushed out from the chamber in a cycle connected to the above defined stroke volume.

The analysis shows that the maximum fluid velocity occurs in the channels during the first and the second half cycles (see Figure 9). The maximum backpressure is equal to 6.8 kPa, and the optimal outflow is 1.62 µL min^{-1}.

The apparent power consumption of the actuator can be approximately computed by assuming harmonic variation [18]:

$$P_A = I_e V_e \tag{2}$$

where I_e and V_e are the root mean square of the current and voltage, respectively. The magnitude of the current generated by the piezoelectric is

$$I = \omega_p \int \int D_3 dx dy \tag{3}$$

where ω_p is the working angular frequency and D_3 is the electric displacement in the piezoelectric layer. The power consumption for the optimal frequency (f_p = 10 Hz) is about 0.29, 0.58, and 0.88 mW for the actuation voltages 20, 40, and 60 V, respectively.

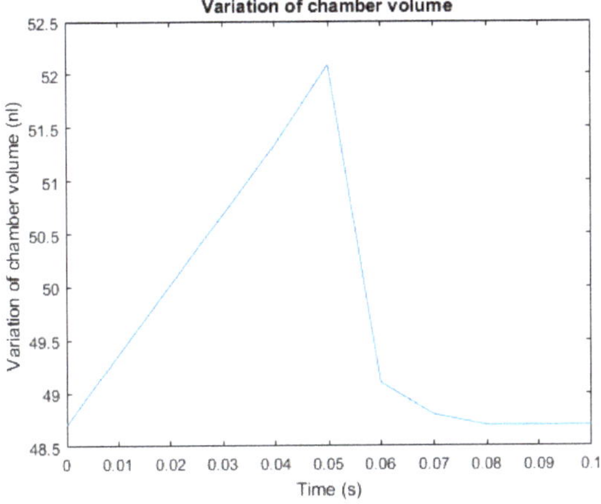

Figure 8. The variation of the chamber volume during a pumping cycle.

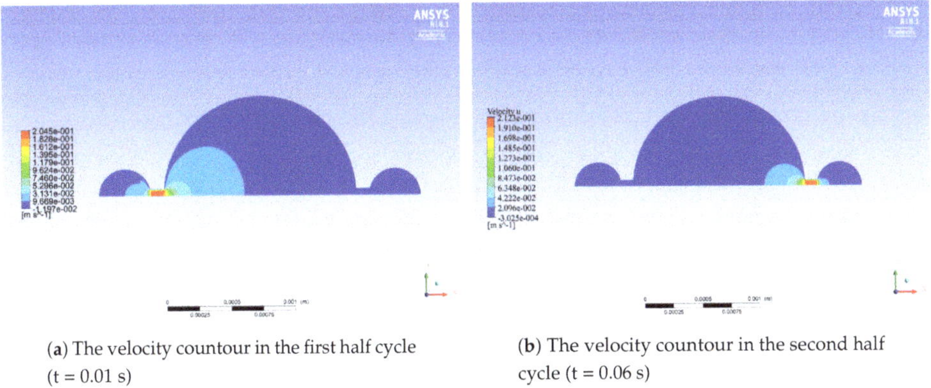

(**a**) The velocity countour in the first half cycle (t = 0.01 s)

(**b**) The velocity countour in the second half cycle (t = 0.06 s)

Figure 9. The velocity profiles in the midplane of the micropump.

5. Geometrical Modifications

Geometry of the Chamber

Figure 10 shows the streamlines, during one pumping cycle, within the planar view of the micropump. There are regions close to the perimeter of the pumping chamber where there is no flux. The presence of such regions, called *dead zones*, affects the efficiency of the micropump, since a certain portion of the volume is not involved in the fluid flux. To decrease the dead zones in the pumping chamber, a geometrical optimization is carried out. The specific shape of the streamlines shown in Figure 10 suggests that a better performance is possibly achieved if the circular geometry is replaced by the elliptical one. As a first attempt, the minor axis of the ellipse is kept equal to the original diameter, i.e., 1500 μm, and the major axis is set to 1800 μm. Of course, the shape of the silicon membrane is also changed according to the modification of the chamber geometry. Conversely, the piezoelectric layer remains unchanged, i.e., a circular disc coaxial with respect to the ellipse. The initial volume of the new pumping chamber is 0.0584 mm^3, whereas the original volume is 0.0487 mm^3 (20% increment).

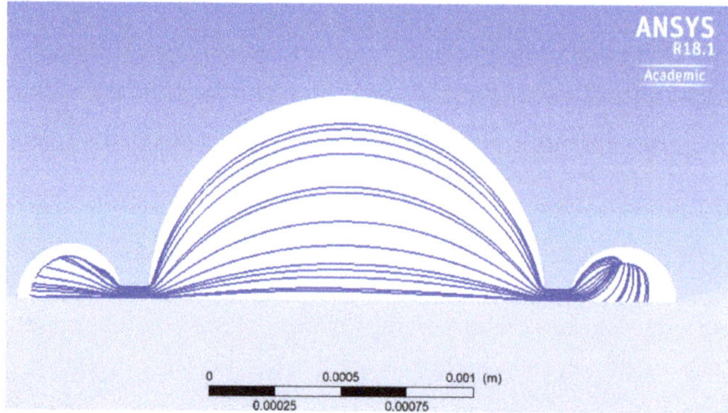

Figure 10. The streamlines of the micropump.

The modified layout is shown in Figure 11. In this modification, the input voltage remains the same as for the circular geometry: In view of the unchanged geometry of the piezoelectric layer, the power consumption does not change significantly. On the other hand, due to the fact that the stiffness of the elliptical membrane is less than the circular one, the same actuation voltage yields larger deflections of the modified membrane. As a matter of fact, the maximum deflection of the circular case is 3.78 μm, whereas for the elliptical case, one obtains 4.57 μm, a 21% increment.

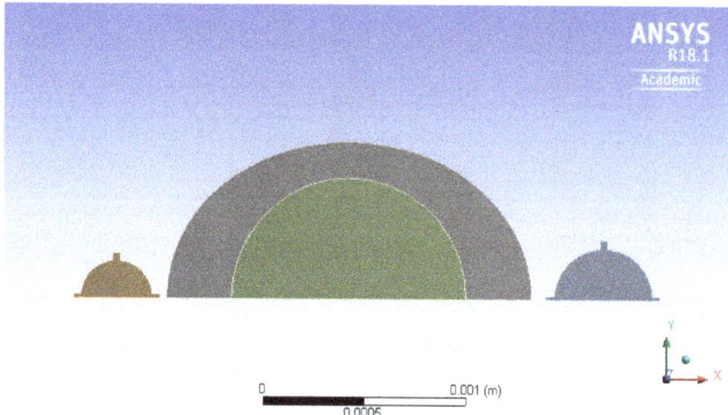

Figure 11. The elliptical chamber micropump.

Figure 12 shows the streamlines for the elliptical chamber: By a comparison with the circular one, it is easily realized that the dead zone is reduced. The obtained outflow of this optimized micropump is 2.11 μL min^{-1}, which indicates that the outflow increases 30%, with the same actuation system and expended power.

The same analysis has been done on similar geometries with different length/width ratios of the ellipse. As noticed in Figure 13, the numerical outcomes indicate a non-monotonic behavior. For aspect ratios close to unity, the outflow steadily increases in view of the optimized flux and of the larger deflection of the diaphragm. Nevertheless, there is a negative effect of the larger volume of fluid to be displaced. As a consequence, after a certain value, the actuator does not provide sufficient power to promote the fluid motion and the outflow decreases. The optimal outflow is achieved for an aspect ratio equal to 2 (namely, length: 3000 μm, width: 1500 μm); in that case, the outflow attains a value of 2.5 μL min^{-1}.

In the manufacturing process of this two-wafer micropump, wafer bonding is one important step. There are several methods for silicon wafer bonding such as anodic bonding [19], metal bonding [20], and glass frit bonding [21]. The residual stresses and possible residual warping coming from the bonding process and due to the coefficients of thermal expansion mismatch can have an influence on the mechanical properties of the micropump components and must be carefully controlled. Another important parameter in the fabrication process which affects the performance of the micropump is the piezoelectric layer thickness. The simulations have shown that, by increasing the piezoelectric layer thickness, the outflow decreases linearly (see Figure 14).

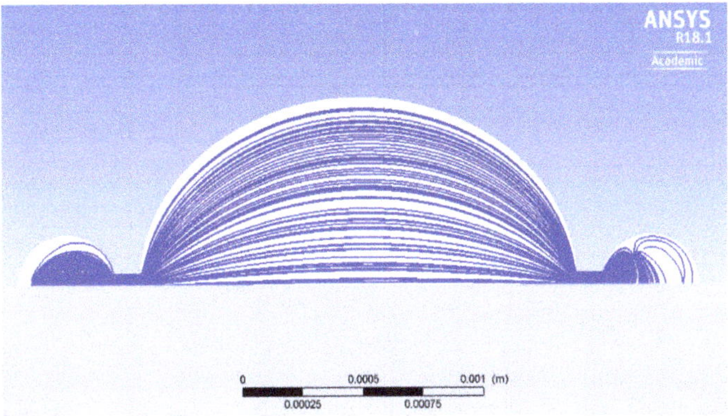

Figure 12. The streamlines in the elliptical geometry.

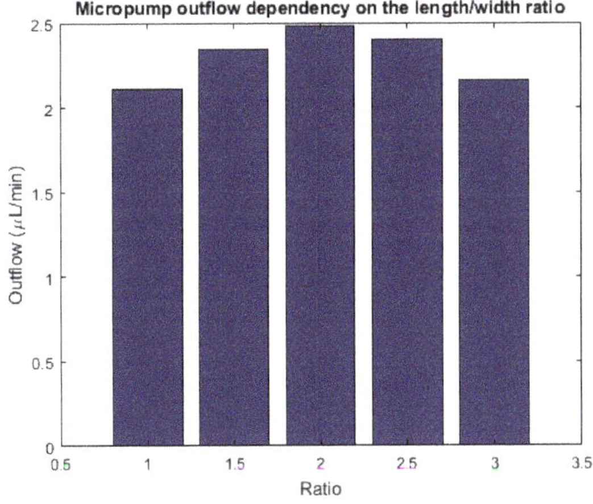

Figure 13. The outflow of the micropump for different length-to-width aspect ratios.

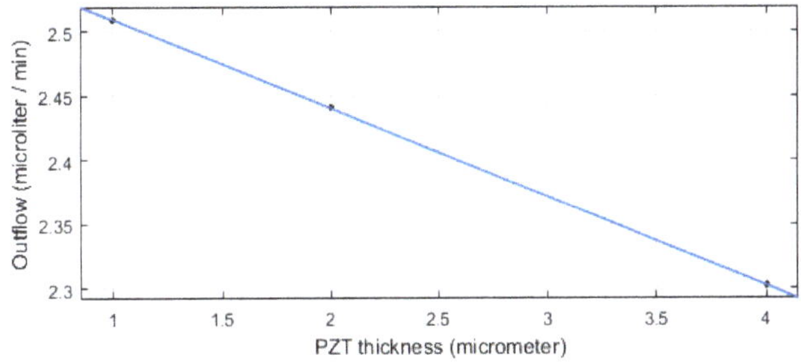

Figure 14. The outflow dependance of the device on the thickness of the lead zirconate titanate (PZT) layer.

6. Comparison with a Commercial, Three-Wafer Micropump

To validate the model, a commercial micropump designed for biomedical applications and fabricated by Debiotech [22] is modelled and compared with the proposed new device. The device proposed in Reference [22] is actuated with a bulk piezoelectric actuator, and has passive valves at the inlet and outlet channels. Figure 15 shows the cross section of the device, while Figure 16 presents the FE mesh.

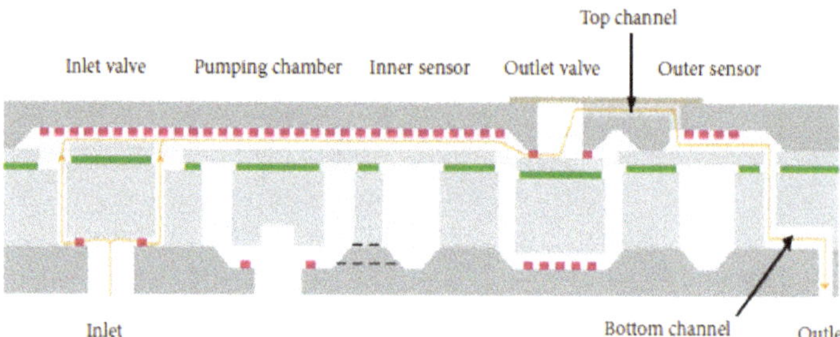

Figure 15. The cross-sectional scheme of the micropump proposed in Reference [22] (Reproduced under Creative Commons Attribution License).

A complete pumping cycle at the frequency of 1 Hz is simulated. Figures 17 and 18 show the streamlines and velocity contours, respectively.

To validate the model, the pressure profiles measured inside and outside the chamber in the Debiotech micropump is compared with the results of the numerical model and shown in Figure 19. In addition, the pressure profile obtained at a 10 Hz frequency of the new device proposed in this paper is shown in Figure 19c in the same time window. The numerical model built in the present work for the simulation of the Debiotech micropump has been obtained with the same approach used for the simulation of the new two-wafers piezoelectric micropump proposed in this paper. As shown from Figure 19, the model shows a good agreement with the experimental and numerical results reported in Reference [22]. Note that the pressure peak for the newly designed micropump is 1.5 mbar, which is almost twice that of the Debiotech device.

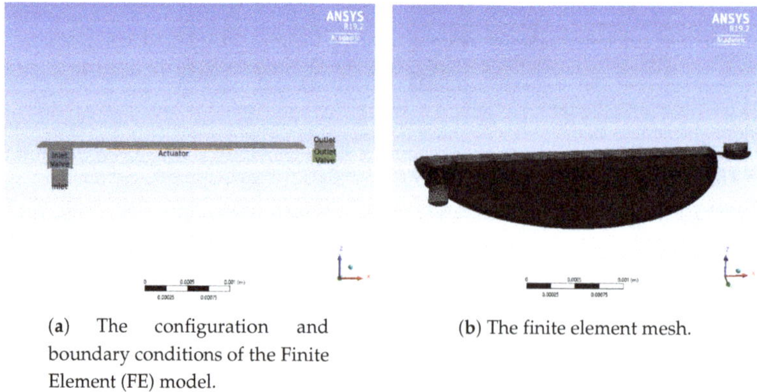

(**a**) The configuration and boundary conditions of the Finite Element (FE) model.

(**b**) The finite element mesh.

Figure 16. The FE model built to simulate the micropump [22].

Figure 17. Computed streamlines for the micropump [22] during the pushing phase.

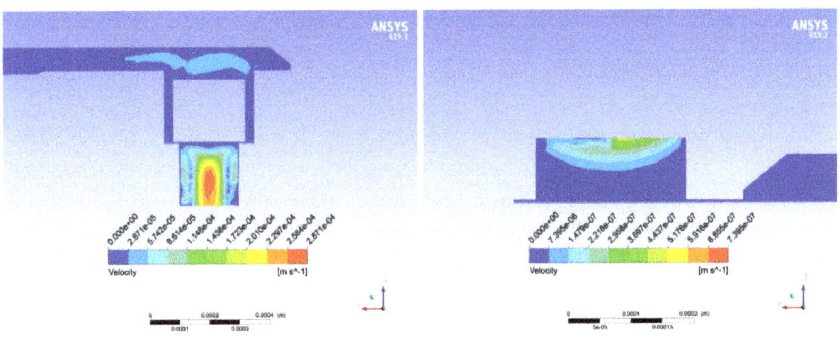

(**a**) The velocity contour at the inlet during a negative stroke.

(**b**) The velocity contour at the outlet during a pushing stroke.

Figure 18. Computed velocity contours at the inlet and outlet of the micropump [22].

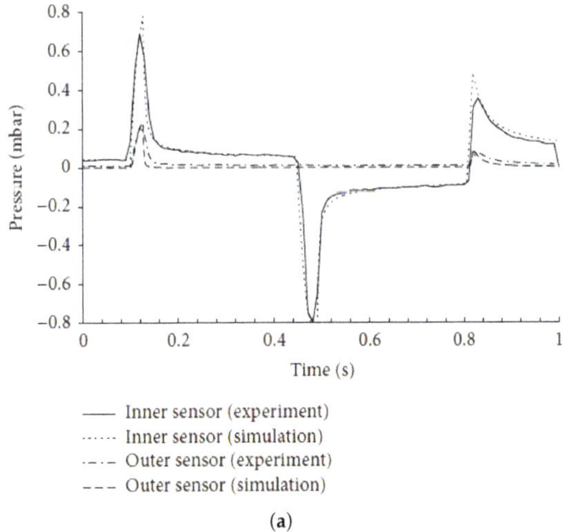

(**a**)

Figure 19. *Cont.*

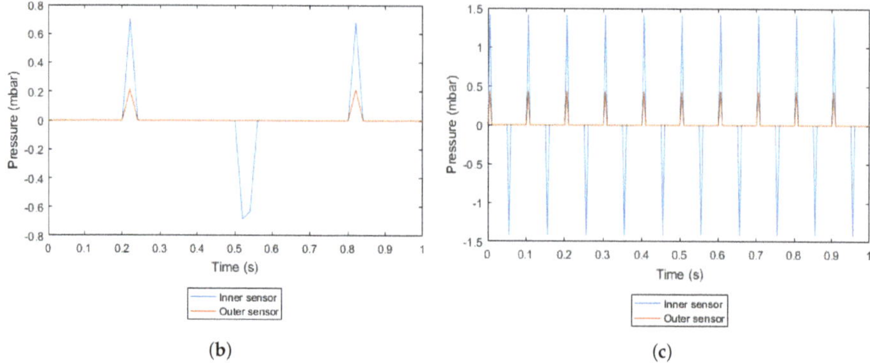

(b)

(c)

Figure 19. A comparison of the pressure profiles for the micropump [23]: The results for the (**a**) experimental, (**b**) numerical, and (**c**) new device proposed in the present paper. (**a**) The experimental results of the inner pressure sensor. From [23] (Reproduced under Creative Commons Attribution License; (**b**) The pressure profile obtained from the simulation; (**c**) The pressure profile obtained for the proposed device.

7. Conclusions

In this paper, the response of a new piezoelectric micropump is investigated by means of a multiphysics computational model. The most important novelty of the proposed design, inspired by a recent patent [12] related to electrostatic actuation, is that the device can be manufactured on two silicon wafers only. In that way, important benefits are introduced with respect to existing devices, such as the simplicity of the process and the absence of a multi-stack wafer bonding and its related defects like misalignments between wafers or damages to the middle wafers during the process. The computational outcomes confirm that the proposed micropump can provide a sufficient outflow for biomedical applications that is in the order of 10 μL min^{-1} [24]. The power consumption appears to be sufficiently limited. The performances of the micropump are boosted by means of simple geometric modifications. By replacing the circular chamber with an elliptical one, the dead zones are decreased significantly and the outflow increases. The analyses suggest that the optimal results are achieved for an aspect ratio equal to 2. Moreover, the presence of rectifiers instead of straight channels is proposed in order to reduce the backflow of the fluid.

Table 4 shows a comparison between the proposed micropump in this research with other devices that have been fabricated up to now. The comparison is satisfactory: The present study paves the way for the development of micropump prototypes that could be tested in order to double check the performance of the new device.

Table 4. A comparison of the proposed micropump with some previously fabricated devices. (* not reported).

Author and Year	Package Size (mm³)	Frequency (Hz)	Max. Back Pressure (kPa)	ΔV (V)	q (μL min^{-1})
Proposed micropump	8	10	10.2	60	2.5
Van Linten 1988 [25]	4100	0.1	24	NR *	0.6
Esashi 1989 [26]	800	30	6.4	90	15
Shoji 1990 [27]	4000	50	NR	100	22
Matsumuto 1999 [28]	NR	5	NR	NR	5.5
Li et al. 2000 [29]	4600	20	35	450	2500
MIP Implantable 2003 [30]	357	0.2	55	150	1.7
Liu et al. 2010 [31]	1413	200	22	36	747.6
Luo et al. 2014 [32]	628	120	NR	60	15030
Debiotech 2014 [22]	84	3.125	NR	200	41.67

Micromachines **2019**, *10*, 259

Author Contributions: Conceptualization, A.C. and R.A.; data curation, S.A.F.F.Y.; funding acquisition, A.C.; methodology, S.A.F.F.Y. and R.A.; software, S.A.F.F.Y.; supervision, A.C.; writing—original draft, S.A.F.F.Y.; writing—review and editing, R.A. and A.C.

Funding: This research received no external funding.

Conflicts of Interest: The authors declare no conflict of interest.

References

1. Corigliano, A.; Ardito, R.; Comi, C.; Frangi, A.; Ghisi, A.; Mariani, S. *Mechanics of Microsystems*; John Wiley & Sons: Hoboken, NJ, USA: 2018; Volume 7646.

2. Shawgo, R.; Grayson, A.R.; Li, Y.; Cima, M. BioMEMS for drug delivery. *Curr. Opin. Solid State Mater. Sci.* **2002**, *6*, 329–334. [CrossRef]

3. Woias, P. Micropumps—Past, progress and future prospects. *Sens. Actuators B Chem.* **2005**, *105*, 28–38. [CrossRef]

4. Bertarelli, E.; Ardito, R.; Bianchi, E.; Laganà, K.; Procopio, F.; Baldo, L.; Corigliano, A.; Dubini, G.; Contro, R. A computational study for design optimization af an electrostatic micropump in stable and pull-in regime. *AES Tech. Rev. Int. J. Ser.* **2010**, *1*, 19–25.

5. Bertarelli, E.; Ardito, R.; Corigliano, A.; Contro, R. A plate model for the evaluation of pull-in instability occurrence in electrostatic micropump diaphragms. *Int. J. Appl. Mech.* **2011**, *3*, 1–19, doi:10.1142/S1758825111000841. [CrossRef]

6. Ardito, R.; Bertarelli, E.; Corigliano, A.; Gafforelli, G. On the application of piezolaminated composites to diaphragm micropumps. *Compos. Struct.* **2013**, *99*, 231–240. [CrossRef]

7. Flynn, A.M.; Sanders, S.R. Fundamental limits on energy transfer and circuit considerations for piezoelectric transformers. *IEEE Trans. Power Electron.* **2002**, *17*, 8–14. [CrossRef]

8. Cazorla, P.H.; Fuchs, O.; Cochet, M.; Maubert, S.; Le Rhun, G.; Fouillet, Y.; Defay, E. A low voltage silicon micro-pump based on piezoelectric thin films. *Sens. Actuators A Phys.* **2016**, *250*, 35–39. [CrossRef]

9. Nisar, A.; Afzulpurkar, N.; Mahisavariya, B.; Tuantranont, A. MEMS-based micropump in drug delivery and biomedical applications. *Sens. Actuators B Chem.* **2008**, *130*, 917–942. [CrossRef]

10. Revathi, S.; Padmapriya, N.; Padmanabhan, R. A design analysis of piezoelectric-polymer composite-based valveless micropump. *Int. J. Model. Simul.* **2018**. [CrossRef]

11. Sateesh, J.; Sravani, K.G.; Kumar, R.A.; Guha, K.; Rao, K.S. Design and flow analysis of MEMS based piezo-electric micro pump. *Microsyst. Technol.* **2018**, *24*, 1609–1614. [CrossRef]

12. Corigliano, A.; Ardito, R.; Bertarelli, E.; Ferrera, M. Micropump with Electrostatic Actuation. Patent PCT/IB2016/053985, 1 July 2016.

13. Trolier-McKinstry, S.; Muralt, P. Thin film piezoelectrics for MEMS. *J. Electroceram.* **2004**, *12*, 7–17. [CrossRef]

14. Jacobsen, H.; Prume, K.; Wagner, B.; Ortner, K.; Jung, T. High-rate sputtering of thick PZT thin films for MEMS. *J. Electroceram.* **2010**, *25*, 198–202. [CrossRef]

15. Horwitz, J.; Grabowski, K.; Chrisey, D.; Leuchtner, R. In situ deposition of epitaxial PbZr$_x$Ti$_{(1-x)}$O$_3$ thin films by pulsed laser deposition. *Appl. Phys. Lett.* **1991**, *59*, 1565–1567. [CrossRef]

16. Schroth, A.; Lee, C.; Matsumoto, S.; Maeda, R. Application of sol-gel deposited thin PZT film for actuation of 1D and 2D scanners. *Sens. Actuators A Phys.* **1999**, *73*, 144–152. [CrossRef]

17. *ANSYS Academic Research Mechanical, Release 15, Help System, System Coupling User's Guide*; ANSYS, Inc.: Canonsburg, PA, USA, 2017.

18. Liang, C.; Sun, F.; Rogers, C. Coupled electro-mechanical analysis of adaptive material systems-determination of the actuator power consumption and system energy transfer. *J. Intell. Mater. Syst. Struct.* **1997**, *8*, 335–343. [CrossRef]

19. Wei, J.; Xie, H.; Nai, M.; Wong, C.; Lee, L. Low temperature wafer anodic bonding. *J. Micromech. Microeng.* **2003**, *13*, 217. [CrossRef]

20. Tan, C.; Reif, R. Silicon multilayer stacking based on copper wafer bonding. *Electrochem. Solid State Lett.* **2005**, *8*, G147–G149. [CrossRef]

21. Knechtel, R. Glass frit bonding: An universal technology for wafer level encapsulation and packaging. *Microsyst. Technol.* **2005**, *12*, 63–68. [CrossRef]

22. Dumont-Fillon, D.; Tahriou, H.; Conan, C.; Chappel, E. Insulin micropump with embedded pressure sensors for failure detection and delivery of accurate monitoring. *Micromachines* **2014**, *5*, 1161–1172. [CrossRef]

23. Fournier, S.; Chappel, E. Modeling of a piezoelectric MEMS micropump dedicated to insulin delivery and experimental validation using integrated pressure sensors: Application to partial occlusion management. *J. Sens.* **2017**, *2017*, 3719853. [CrossRef]

24. Ashraf, M.W.; Tayyaba, S.; Afzulpurkar, N. Micro electromechanical systems (MEMS) based microfluidic devices for biomedical applications. *Int. J. Mol. Sci.* **2011**, *12*, 3648–3704. [CrossRef]

25. Van Lintel, H.; Van de Pol, F.; Bouwstra, S. A piezoelectric micropump based on micromachining of silicon. *Sens. Actuators* **1988**, *15*, 153–167. [CrossRef]

26. Esashi, M.; Shoji, S.; Nakano, A. Normally closed microvalve and mircopump fabricated on a silicon wafer. *Sens. Actuators* **1989**, *20*, 163–169. [CrossRef]

27. Shoji, S.; Nakagawa, S.; Esashi, M. Micropump and sample-injector for integrated chemical analyzing systems. *Sens. Actuators A Phys.* **1990**, *21*, 189–192. [CrossRef]

28. Matsumoto, S.; Klein, A.; Maeda, R. Development of bi-directional valve-less micropump for liquid. In Prodeedings of the Twelfth IEEE International Conference on Micro Electro Mechanical Systems (MEMS'99), Orlando, FL, USA, 21 January 1999; pp. 141–146.

29. Li, H.; Roberts, D.; Steyn, J.; Turner, K.; Carretero, J.; Yaglioglu, O.; Su, Y.; Saggere, L.; Hagood, N.; Spearing, S.; et al. A high frequency high flow rate piezoelectrically driven MEMS micropump. In Prodeedings of the IEEE Solid State Sensors and Actuators Workshop, Hilton Head Island, SC, USA, 4–8 June 2000.

30. MIP Implantable Product Information. Available online: www.debiotech.sa (accessed on 16 April 2019).

31. Liu, G.; Shen, C.; Yang, Z.; Cai, X.; Zhang, H. A disposable piezoelectric micropump with high performance for closed-loop insulin therapy system. *Sens. Actuators A Phys.* **2010**, *163*, 291–296. [CrossRef]

32. Luo, W.F.; Ma, H.K. Development of a Piezoelectric Micropump with Novel Separable Design. *Appl. Mech. Mater.* **2014**, *670–671*, 1426–1429. [CrossRef]

Article

Robust Model-Free Adaptive Iterative Learning Control for Vibration Suppression Based on Evidential Reasoning

Liang Bai [1,*], Yun-Wen Feng [1], Ning Li [2] and Xiao-Feng Xue [1]

[1] School of Aeronautics, Northwestern Polytechnical University, Western Youyi Street 127, Xi'an 710072, China; fengyunwen033616@163.com (Y.-W.F.); xuexiaofeng033616@163.com (X.-F.X.)

[2] College of Sciences, Northeastern University, Shenyang 110819, China; lining80@163.com

* Correspondence: bailiangcom@163.com

Received: 11 February 2019; Accepted: 13 March 2019; Published: 19 March 2019

Abstract: Through combining P-type iterative learning (IL) control, model-free adaptive (MFA) control and sliding mode (SM) control, a robust model-free adaptive iterative learning (MFA-IL) control approach is presented for the active vibration control of piezoelectric smart structures. Considering the uncertainty of the interaction among actuators in the learning control process, MFA control is adopted to adaptively adjust the learning gain of the P-type IL control in order to improve the convergence speed of feedback gain. In order to enhance the robustness of the system and achieve fast response for error tracking, the SM control is integrated with the MFA control to design the appropriate learning gain. Real-time feedback gains which are extracted from controllers construct the basic probability functions (BPFs). The evidence theory is adopted to the design and experimental investigations on a piezoelectric smart cantilever plate are performed to validate the proposed control algorithm. The results demonstrate that the robust MFA-IL control presents a faster learning speed, higher robustness and better control performance in vibration suppression when compared with the P-type IL control.

Keywords: P-type IL; MFA control; SM control; evidence theory; active vibration control; piezoelectric smart structure

---

1. Introduction

As an intelligent control strategy, iterative learning (IL) control has a simple structure and doesn't require accurate system modeling. According to past control experience, this method can improve the current control performance of the system by operating repetitively over a fixed time interval [1]. In 1978, IL control was first proposed by Uchiyama in Japanese [2], which did not receive much attention. After one critical report published by Arimoto in English [3], IL control made significant progress in both theories and applications [4,5]. Practically, IL control has been applied to a wide range of engineering applications, including flexible structures [6], nonholonomic mobile robots [7], flapping wing micro aerial vehicles [8] and the sheet metal forming process [9]. Considering the repeatability of the structural dynamic response in the vibration process, IL control is expected to provide a feasible solution for the control issue here. Several research groups have applied IL control methods to the active vibration control of piezoelectric smart structures. Zhu et al. [10] and Tavakolpour et al. [11] first applied P-type IL control for the attenuation of vibrations in piezoelectric smart structures for the design of the feedback gain, and the efficiency of P-type IL control was proven in their papers. In addition, Fadil et al. [12] proposed a new intelligent proportional-integral-derivative (PID) controller for vibration suppression by using P-type IL control and PID control, in which the P-type IL control was applied to tune the parameters of the PID controller.

In the studies above, although P-type IL controllers can effectively attenuate structural vibrations at some excitation frequencies, the performances of the controllers for vibration suppression are still not obvious when the piezoelectric smart structure is excited by its first natural frequency. Besides, the control effectiveness of the actuators is obvious at the locations of the sensors, and they are not able to effectively compensate for unwanted vibrations at other locations [13]. Moreover, thousands of iterations in P-type IL control are needed for achieving satisfactory control precision, leading to the slow learning speed [11,14]. In addition, the learning process should be accomplished within a limited period, as overlearning may lead to system instability [11]. Therefore, the iterative number should be limited to a predefined value. Finally, the unreasonable selections of learning gains may directly lead to control spillover or system instability [15].

For accelerating the convergence rate and improving the stability of the learning algorithm, adaptive control is needed to modify the learning process of the P-type IL control. Adaptive control has been successfully incorporated into various learning algorithms for the adjustment of parameters of the learning process [16,17]. Effective adaptive control strategies are necessary for the ability to automatically tune parameters to the desired performance at each sampling period, such as an internal model control method with an adaptive algorithm, implemented to reduce fatigue loads and tower vibrations in wind turbines [18]. A real-time control implementation, based on an auto-tuning finite impulse response filter, was applied to active vibration isolation [19]. An online method that tunes the poles of the controller was proposed to adapt to the errors between a real object and its model [20]. An adaptive voltage and frequency control method was proposed for inverter-based distributed generations in a multi-microgrid structure [21]. A characteristic model-based nonlinear golden section adaptive control method was presented for vibration suppression in a flexible Cartesian smart material robot [22].

As mentioned above, most of the adaptive control methods are model-based, in which the dynamic model of system has been already known before the design of the controller. However, for complex practical systems, the mechanism models of the plants are often difficult to establish, and the parameters are also hard to identify, making the design and application of controllers unpractical. Controlling vibrations in plate and shell structures always brings a challenge because of the complexity and density of the vibration modes. The strategy of using piezoelectric actuator-sensor pairs with discrete locations, glued on both surfaces of the plate, realizes a low weight and effective control for structural vibration [23]. The plate-integrated piezoelectric actuator-sensor pairs thus become a multi-input-multi-output (MIMO) system. If an actuator fails to perform as expected, the performance of its neighboring actuators will be negatively affected. In this system, the interaction among all actuators exists in the whole process of active vibration control, and this kind of interaction is always uncertain. The uncertainty caused by this interaction greatly presents a great challenge when designing a controller, and the model-based adaptive controller cannot deal with these conditions. Data-driven control methods, which are designed by directly using input and output (I/O) data of the system, can serve as an efficient alternative. The control problems caused by time-varying parameters and uncertainties of the model are challenging for model-based control, but not with data-driven control approaches [24].

Model free adaptive (MFA) control, as an effective data-driven control method, is an attractive technique which has gained a large amount of interest in recent years. It is easily implemented, with small computational burden for its simple structure and strong robustness. Unlike the neural-network-based adaptive control methods and model-based methods, no additional signal testing or training processes are required during the design of data-driven control methods. Instead of identifying the model of the plant, the MFA method builds an equivalent linearization of the data at each operation by introducing a novel concept named the pseudo-partial derivative (PPD), and the time-varying PPD can be estimated by merely using the I/O measurements of the plant.

In this paper, in order to accelerate the convergence speed of the feedback gain, the learning gain of the P-type IL control is designed by the MFA control. The MFA control can realize adaptive control in parametric and structural manners. And it is also suitable for dealing with system uncertainties [25]. This advantage allows MFA control to be successfully employed in various engineering fields, such as use in the sensing and control of piezoelectrically actuated systems [26], blood pump control [27], multivariable industrial processes control [28], and robotic exoskeleton tracking control [29]. However, the convergence speed of the tracking error may be slow if only MFA control is used to adaptively adjust to the learning gain of the P-type IL control, and external noise in the system will increase the difficulty of vibration control. Various control strategies have been provided and continuously developed for the control of plants with unknown uncertainties and dynamic variations, such as sliding mode (SM) control [30], fuzzy logic control [31], neural networks [32], etc. For nonlinear, time-varying and uncertain systems, neural network approaches have an excellent approximation ability, and fuzzy logic control possesses remarkable robustness and adaptability, nevertheless, the tuning of numerous parameters and complex rules may decrease the efficiency and possibility of these methods [33]. Unlike neural networks and fuzzy logic control, SM control has a simple controller structure and can is easily implemented. Moreover, it has other attractive features, including good transient performance, robustness to parameter variations and insensitivity to disturbances [34]. With the aim of achieving a faster response and better robustness, the SM control was integrated with the MFA control to self-adjust the learning gain of the P-type control. Within the proposed method, the SM control is applied to estimate the parameters by tracking the time-varying PPD, such that the state variables can rapidly converge to the desired trajectory. Additionally, SM control can also be used to compensate for the impact of random disturbance, thereby enabling the system to enhance control effectiveness and maintain superior stability. In the P-type IL algorithm, two parts mainly affect the convergence speed of the feedback gain: The system output error and the learning gain. In the application of vibration suppression, the desired output signal is always zero. Therefore, the measured output signal is the main factor that decides the system output error. In this paper, multi-sensors are used to detect the structural deformation of a cantilever plate. Sensors at various locations generate different measured output signals, and the controllers connected to these sensors may have distinct learning speeds. It is unreasonable to define the same iterative number for all controllers as part of the stopping criteria. After obtaining multi-source information from the controlled plant, it is critical to discover the optimal method of fusing this information. In this paper, in order to solve the multicriteria and multiobjective problems in practical applications, evidence theory was adopted to design the stopping criteria. The evidence theory does not require prior knowledge and has outstanding performance for handling uncertain or inexact information, which makes it an indispensable tool for state diagnosis and defect inspection [35,36]. Applying the combination rules, the evidence theory can carry out reasoning, data fusion or decision making [37]. By using information fusion technology, the learning processes of all controllers can be diagnosed in real-time by the real-time feedback gains obtained from the controllers. On this basis, the stopping criteria were designed for overlearning diagnosis of the robust MFA-IL algorithm.

The finite element (FE) method is a widely accepted and powerful tool to deal with piezoelectric smart structures. Some kinds of efficient and accurate electromechanically coupled dynamic FEs of smart structures have already been developed [38–40]. Among commercial FE analysis codes, ANSYS has the ability to model smart structures with piezoelectric materials, and H Karagülle et al. [41] successfully integrated vibration control actions into ANSYS modeling, where the solution was achieved as well. In this paper, ANSYS parametric design language (APDL) is used to integrate the control law into the ANSYS FE model to perform closed-loop simulations.

In this paper, using the complementary features of P-type IL control, MFA control and SM control, a robust MFA-IL control strategy was developed for the suppression of vibrations in smart structures. Due to its ability to cope with uncertainties in the learning control process, MFA control was applied to adaptively adjust the learning gain of the P-type IL control. By inserting the SM control term into the MFA control, the learning gain can be designed properly and the convergence rate of the tracking error and the robustness of the closed-loop system can be improved. A multi-source information fusion diagnosis method for the overlearning evaluation is presented based on the evidence theory, and the stopping criteria are also be designed. The proposed control method was numerically and experimentally investigated for a clamped plate under various external disturbances, and the results are illustrated and extensively discussed at the end of the present work.

The rest of this paper is organized as follows. In Section 2, based on the FE model of piezoelectric smart structures, the state space model of the equivalent linear system is developed for the purpose of control law design. The P-type IL control is employed for establishing the vibration control equations. Section 3 describes the dynamic transformation and linearization for the vibration control system. Section 4 describes the design of the robust MFA-IL control scheme. Theoretical basis of state diagnosis, based on evidence theory and the design of the stopping criteria, is introduced in Section 5. In Section 6, numerical examples are presented to demonstrate the validity of the proposed method. In Section 7, a complete active vibration control system is set up to conduct an experimental investigation. The conclusions and outlooks are drawn in Section 8.

2. FE Model and P-type IL Control

The linear electro-mechanically coupled dynamic FE equations of the piezoelectric smart structure can be written as [23]:

$$\begin{bmatrix} M_{uu} & 0 \\ 0 & 0 \end{bmatrix} \left\{ \begin{array}{c} \ddot{u} \\ \ddot{\phi} \end{array} \right\} + \begin{bmatrix} C_{uu} & 0 \\ 0 & 0 \end{bmatrix} \left\{ \begin{array}{c} \dot{u} \\ \dot{\phi} \end{array} \right\} + \begin{bmatrix} K_{uu} & K_{u\phi} \\ K_{\phi u} & K_{\phi\phi} \end{bmatrix} \left\{ \begin{array}{c} u \\ \phi \end{array} \right\} = \left\{ \begin{array}{c} F_u \\ F_\phi \end{array} \right\} \tag{1}$$

where u and ϕ are the vectors of nodal displacements and electric potentials; M_{uu}, C_{uu}, K_{uu}, $K_{u\phi}(K_{\phi u})$ and $K_{\phi\phi}$ are the structural mass matrix, the damping matrix, the mechanical stiffness matrix, the piezoelectric coupling matrix and the dielectric stiffness matrix, respectively; F_u and F_ϕ are the vectors of mechanical force and electric load, respectively.

The damping matrix C_{uu} is usually defined as a linear combination of the structural mass matrix M_{uu} and the mechanical stiffness matrix K_{uu}, which is written as follows:

$$C_{uu} = \alpha M_{uu} + \beta K_{uu} \tag{2}$$

where the constants α and β are the Rayleigh damping coefficients.

The piezoelectric sensor generates output electric potential as long as the structure is oscillating. The system (1) can be uncoupled into the following independent equations for the sensor output electric potential:

$$\phi = K_{\phi\phi}^{-1}(F_\phi - K_{\phi u}u) \tag{3}$$

and the structural displacement

$$M_{uu}\ddot{u} + C_{uu}\dot{u} + K^*u = F_u - K_{u\phi}K_{\phi\phi}^{-1}F_\phi \tag{4}$$

where $K^* = K_{uu} - K_{u\phi}K_{\phi\phi}^{-1}K_{\phi u}$.

Note that F_ϕ is usually zero in the sensor, thus, Equation (3) can be rewritten as $\phi = -K_{\phi\phi}^{-1}K_{\phi u}u$. Then, the rate of change of the sensor output electric potential can be written as:

$$\dot{\phi} = -K_{\phi\phi}^{-1}K_{\phi u}\dot{u} \tag{5}$$

The system output error is defined as:

$$e = y_d - y \tag{6}$$

where y_d is the desired output signal and y is measured output signal

In the application of vibration suppression, we require the desired output signal to be zero. The measured output signal in this paper is the rate of change of the sensor output electric potential, namely, $y = \dot{\phi}$. As a discrete-time system, the output error at the kth moment can be given as:

$$e(k) = 0 - \dot{\phi}(k) = K_{\phi\phi}^{-1} K_{\phi u} \dot{u}(k) \tag{7}$$

According to the update rule of the P-type IL control [42], the feedback gain at the kth moment can be given as:

$$G(k) = G(k-1) + \Phi e(k-1) \tag{8}$$

where Φ and $G(k)$ are the learning gain matrix and feedback gain matrix, respectively.

The feedback gain $G(k)$ is stored in memory at the $(k-1)$th moment and applied for the next iteration when the system operates. The input voltage of the actuator is expressed as:

$$V_a = -G\dot{\phi} \tag{9}$$

The electric load vector is defined as:

$$F_\phi = C_a V_a \tag{10}$$

where C_a is the capacitance constant of the piezoelectric actuator.

The control force can be defined as $F_a = -K_{u\phi} K_{\phi\phi}^{-1} F_\phi$ and by combining with Equations (5), (9) and (10), F_a can be rewritten as:

$$F_a = -K_{u\phi} K_{\phi\phi}^{-1} C_a G K_{\phi\phi}^{-1} K_{\phi u} \dot{u} \tag{11}$$

The control force generated by actuator is used to suppress structural vibration. Substituting Equations (5), (9) and (10) into Equation (4), the vibration control equation of the piezoelectric smart structure can be expressed as:

$$M_{uu}\ddot{u} + (C_{uu} + K_{u\phi} K_{\phi\phi}^{-1} C_a G K_{\phi\phi}^{-1} K_{\phi u})\dot{u} + K^* u = F_u \tag{12}$$

3. Dynamic Transformation and Linearization of Vibration Control Equations

As a discrete-time system, the system (12) can be approximated by the following form at the kth moment:

$$M_{uu}\ddot{u}(k) + [C_{uu} + K_{u\phi} K_{\phi\phi}^{-1} C_a G(k) K_{\phi\phi}^{-1} K_{\phi u}]\dot{u}(k) + K^* u(k) = F_u(k) \tag{13}$$

Again, using Equations (5) and (7), the system (13) can be rewritten as:

$$\dot{y}(k) = -(K_M^* C_{uu} K_{\phi u}^{-1} K_{\phi\phi})y(k) - [K_M^* K_C^* G(k)]y(k) + K_M^* F^*(k) \tag{14}$$

where $K_M^* = K_{\phi\phi}^{-1} K_{\phi u} M_{uu}^{-1}$, $K_C^* = K_{u\phi} K_{\phi\phi}^{-1} C_a$, $F^*(k) = K^* u(k) - F_u(k)$.

Similar to Equation (8), a time-varying version for P-type IL updating rule is given as [42]:

$$G(k) = G(k-1) + \Phi(k-1)e(k-1) \tag{15}$$

where $\Phi(k-1)$ is the learning gain matrix, which is now time-varying.

Because $\dot{y}(k) = \frac{y(k+1)-y(k)}{T}$ for the sample period T, substituting Equation (15) into Equation (14), the discrete-time form of the system (14) can be given as:

$$y(k+1) = -T[K_M^* C_{uu} K_{\phi u}^{-1} K_{\phi\phi} + K_M^* K_C^* G(k-1) + K_M^* K_C^* \Phi(k-1)e(k-1) - \tfrac{1}{T}]y(k) + T K_M^* F^*(k) \tag{16}$$

where $\Phi(k-1)$ and $y(k)$ are the system input and the system output, respectively.

It can be known from Equation (16) that the partial derivatives of $y(k+1)$, with respect to output $y(k)$ and input $\Phi(k-1)$, are continuous and the system is generalized *Lipschitz*. For system (16), with $\|[\Delta y(k), \Delta\Phi(k-1)]^T\| \neq 0$ for each fixed k, there must exist $\Psi(k)$, named the PPD matrix, such that Equation (16) can be transformed into the following equivalent full form dynamic linearization model:

$$\Delta y(k+1) = \Psi(k)[\Delta y(k),\ \Delta\Phi(k-1)]^T \tag{17}$$

where $\Psi(k) = [\varphi_1(k), \varphi_2(k)]$, $\varphi_1(k),\ \varphi_2(k) \in R^{N\times N}$ and $\|\Psi(k)\| < b$, b is a positive constant.

Proof. From Equation (16) we have

$$
\begin{aligned}
\Delta y(k+1) &= y(k+1) - y(k)\\
&= \{-T[K_M^* C_{uu} K_{\phi u}^{-1} K_{\phi\phi} + K_M^* K_C^* G(k-1) + K_M^* K_C^* \Phi(k-1)e(k-1) - \tfrac{1}{T}]y(k) + TK_M^* F^*(k)\}\\
&\quad -\{-T[K_M^* C_{uu} K_{\phi u}^{-1} K_{\phi\phi} + K_M^* K_C^* G(k-1) + K_M^* K_C^* \Phi(k-2)e(k-1) - \tfrac{1}{T}]y(k-1) + TK_M^* F^*(k)\}\\
&\quad +\{-T[K_M^* C_{uu} K_{\phi u}^{-1} K_{\phi\phi} + K_M^* K_C^* G(k-1) + K_M^* K_C^* \Phi(k-2)e(k-1) - \tfrac{1}{T}]y(k-1) + TK_M^* F^*(k)\}\\
&\quad +\{-T[K_M^* C_{uu} K_{\phi u}^{-1} K_{\phi\phi} + K_M^* K_C^* G(k-2) + K_M^* K_C^* \Phi(k-1)e(k-2) - \tfrac{1}{T}]y(k-1) + TK_M^* F^*(k-1)\}\\
&\quad -\{-T[K_M^* C_{uu} K_{\phi u}^{-1} K_{\phi\phi} + K_M^* K_C^* G(k-2) + K_M^* K_C^* \Phi(k-2)e(k-2) - \tfrac{1}{T}]y(k-1) + TK_M^* F^*(k-1)\}\\
&\quad -\{-T[K_M^* C_{uu} K_{\phi u}^{-1} K_{\phi\phi} + K_M^* K_C^* G(k-2) + K_M^* K_C^* \Phi(k-1)e(k-2) - \tfrac{1}{T}]y(k-1) + TK_M^* F^*(k-1)\}
\end{aligned}
\tag{18}
$$

$$
\begin{aligned}
\theta(k) &\triangleq \{-T[K_M^* C_{uu} K_{\phi u}^{-1} K_{\phi\phi} + K_M^* K_C^* G(k-1) + K_M^* K_C^* \Phi(k-2)e(k-1) - \tfrac{1}{T}]y(k-1) + TK_M^* F^*(k)\}\\
&\quad -\{-T[K_M^* C_{uu} K_{\phi u}^{-1} K_{\phi\phi} + K_M^* K_C^* G(k-2) + K_M^* K_C^* \Phi(k-1)e(k-2) - \tfrac{1}{T}]y(k-1) + TK_M^* F^*(k-1)\}
\end{aligned}
$$

By virtue of the Cauchy differential mean value theorem, Equation (18) can be rewritten as

$$\Delta y(k+1) = \frac{\partial y(k+1)}{\partial y(k)}\Delta y(k) + \frac{\partial y(k)}{\partial \Phi(k-2)}\Delta\Phi(k-1) + \theta(k) \tag{19}$$

where $\frac{\partial y(k+1)}{\partial y(k)}$ is the partial derivative value of $y(k+1)$, with respect to output $y(k)$, and $\frac{\partial y(k)}{\partial \Phi(k-2)}$ represents the partial derivative value of $y(k)$, with respect to input $\Phi(k-2)$.

For each fixed k, we consider the following equation with the numerical matrix $H(k) \in R^{N\times N}$.

$$\theta(k) = H(k)[\Delta y(k), \Delta\Phi(k-1)]^T \tag{20}$$

Since condition $\|[\Delta y(k), \Delta\Phi(k-1)]^T\| \neq 0$, Equation (20) must have at least one solution $H^*(k)$. In fact, it must have an infinite number of solutions for each k.

Let

$$\Psi(k) = H^*(k) + [\frac{\partial y(k+1)}{\partial y(k)}, \frac{\partial y(k)}{\partial \Phi(k-2)}], \tag{21}$$

Then, we have Equation (17).

Based on Equation (17), the system (16) can be rewritten in the following dynamic linearization form:

$$y(k+1) = \varphi_1(k)\Delta y(k) + \varphi_2(k)\Delta\Phi(k-1) + y(k) \tag{22}$$

where the values of $\varphi_1(k)$ and $\varphi_2(k)$ are dynamic changed. $\quad\square$

4. Controller Design

4.1. MFA Control

Consider the following control input criterion function:

$$J(\Phi(k-1)) = \|y_d(k+1) - y(k+1)\|^2 + \gamma\|\Phi(k-1) - \Phi(k-2)\|^2 \tag{23}$$

where $y_d(k+1)$ is the desired output signal and $\gamma > 0$ is a weighting constant. $\Phi(k-1)$ is the MFA control rate.

Substituting Equation (22) into Equation (23), then differentiating Equation (23) with respect to $\Phi(k-1)$, and letting it be equal zero, gives the following:

$$\Delta\Phi(k-1) = \left(\varphi_2(k)^T\varphi_2(k) + \gamma I\right)^{-1}\varphi_2(k)^T \times [y_d(k+1) - y(k) - \varphi_1(k)\Delta y(k)] \tag{24}$$

Equation (24) includes the calculation of the inverse matrix, which may cause computational burden once the I/O matrixes of the system are of a high dimension. The simplified form of Equation (24) can be expressed as follows:

$$\Phi_{\mathrm{MFA}}(k-1) = \Phi_{\mathrm{MFA}}(k-2) + \frac{\rho\varphi_2(k)^T[y_d(k+1) - y(k) - \varphi_1(k)\Delta y(k)]}{\|\varphi_2(k)\|^2 + \gamma} \tag{25}$$

where $\rho \in (0,1]$ is a step-size constant, which is added to make Equation (25) general.

In this paper, a modified projection algorithm is used to estimate the unknown PPD matrix:

$$J(\Psi(k)) = \|\Delta y(k) - \Psi(k)[\Delta y(k-1), \Delta\Phi(k-2)]^T\|^2 + \mu\|\Psi(k) - \hat{\Psi}(k-1)\|^2 \tag{26}$$

where $\Psi(k) = \hat{\Psi}(k-1) + \Delta\Psi(k) = [\hat{\varphi}_1(k-1), \hat{\varphi}_2(k-1)] + [\Delta\varphi_1(k), \Delta\varphi_2(k)]$, $\mu > 0$ is a weighting constant and $\hat{\Psi}(k-1)$ is an estimated value of $\Psi(k-1)$.

Here, we differentiate Equation (26) with respect to $\Psi(k)$, and letting it be equal to zero. According to the simplified form in Equation (25), we can obtain the parameters of the estimation algorithm as follows:

$$\begin{aligned}
\hat{\varphi}_1(k) &= \hat{\varphi}_1(k-1) + \frac{\eta[\Delta y(k) - (\hat{\varphi}_1(k-1)\hat{\varphi}_2(k-1))(\Delta y(k-1), \Delta\Phi(k-2))^T]\Delta y(k-1)^T}{\mu + \|\Delta y(k-1)\|^2 + \|\Delta\Phi(k-2)\|^2} \\
\hat{\varphi}_2(k) &= \hat{\varphi}_2(k-1) + \frac{\eta[\Delta y(k) - (\hat{\varphi}_1(k-1)\hat{\varphi}_2(k-1))(\Delta y(k-1), \Delta\Phi(k-2))^T]\Delta\Phi(k-2)^T}{\mu + \|\Delta y(k-1)\|^2 + \|\Delta\Phi(k-2)\|^2}
\end{aligned} \tag{27}$$

where $\eta \in (0,1]$ is a step-size constant, which is added to make Equation (27) general.

4.2. Robust MFA-IL Control

The discrete-time SM is used to compensate the external disturbances and guarantee the fast convergence of feedback gain, which can increase the system robustness and control performance. By assuming the discrete sliding surface is

$$s(k) = e(k) \tag{28}$$

and combining Equations (6) and (22), the above equation can be rewritten as follows:

$$s(k+1) = e(k+1) = y_d(k+1) - \varphi_1(k)\Delta y(k) - \varphi_2(k)\Delta\Phi(k-1) - y(k) \tag{29}$$

The sliding reaching law is defined as follows [43]:

$$\begin{cases}
s(k+1) = (1 - qT)s(k) - \varepsilon T\mathrm{fal}(s(k), \sigma, \delta) \\
\delta > \left(\frac{\varepsilon T}{1-qT}\right)^{\frac{1}{1-\sigma}}, \qquad 0 < \frac{\varepsilon T}{1-qT} < 1
\end{cases} \tag{30}$$

where $\mathrm{fal}(s(k), \alpha, \delta) = \begin{cases} |s(k)|^\sigma \mathrm{sign}(s(k)), & |s(k)| \geq \delta \\ \frac{s(k)}{\delta^{1-\sigma}}, & |s(k)| < \delta \end{cases}$, $0 < \sigma < 1, 0 < \delta < 1, \varepsilon > 0, q > 0,$
$1 - qT > 0$, T is the sample period.

Substituting Equation (30) into Equation (29), gives the following:

$$\Delta\boldsymbol{\Phi}(k-1) = \boldsymbol{\varphi}_2(k)^{-1}[\boldsymbol{y}_d(k+1) - \boldsymbol{\varphi}_1(k)\Delta\boldsymbol{y}(k) - \boldsymbol{y}(k) - (1-qT)\boldsymbol{s}(k) + \varepsilon T\mathrm{fal}(s(k),\sigma,\delta)] \tag{31}$$

Because the sliding mode reaching law is based on the transformed dynamic linearization, let $\boldsymbol{\Phi}_{\mathrm{SM}}(k-1) = \Delta\boldsymbol{\Phi}(k-1)$. Then, the final learning gain of the IL controller is equal to Equation (25) plus Equation (31), written as follows:

$$\boldsymbol{\Phi}(k-1) = \boldsymbol{\Phi}_{\mathrm{MFA}}(k-1) + \Gamma\boldsymbol{\Phi}_{\mathrm{SM}}(k-1) \tag{32}$$

where Γ is a weighting factor which is added to make Equation (32) general. $\boldsymbol{\Phi}_{\mathrm{SM}}(k-1)$ is used to compensate for the input disturbance and increase the convergence rate.

For convenience, a block diagram of the robust MFA-IL control approach is presented in Figure 1.

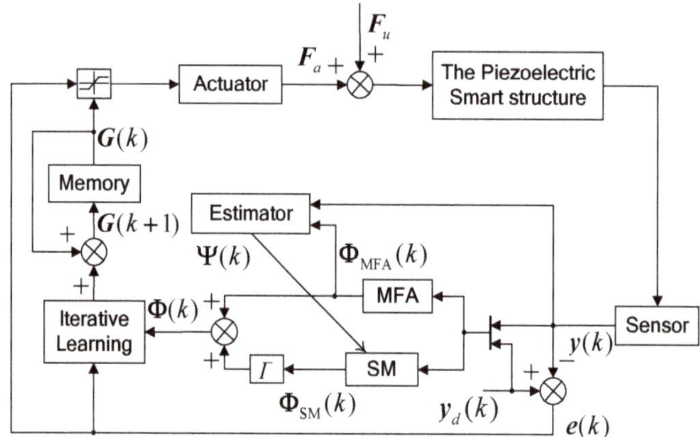

Figure 1. Block diagram of the robust model-free adaptive-iterative learning (MFA-IL) control.

5. The Design of the Stopping Criteria Based on Evidence Theory

5.1. Evidence Theory

The evidence theory is a mathematical theory and general framework for reasoning with uncertainty information in systems, which allows one to combine multiple variables from multiple sources, arriving at a degree of belief. The major definitions and concepts of the theory are briefly introduced as follows [44,45].

Definition 1. *Let a finite set of elements* $\Theta = \{Z_1, Z_2, \cdots, Z_L\}$ *be defined as the frame of discernment. An element can be a hypothesis, object or state. The* 2^Θ*, named the power set, is the set of all subsets of* Θ*. It is composed of each element and multi-subset and can be indicated as* $2^\Theta = \{\varnothing, \{Z_1\}, \{Z_2\}, \cdots \{Z_1, Z_2\}, \cdots, \Theta\}$*, where* $\varnothing$ *is an empty set.*

Definition 2. *A mass function is a mapping of m from* 2^Θ *to* $[0,1]$*, and be formally defined as:*

$$m : 2^\Theta \rightarrow [0,1] \tag{33}$$

Additionally, the function satisfies the following equation:

$$m(\varnothing) = 0, \quad \sum_{A \subseteq 2^\Theta} m(A) = 1 \tag{34}$$

The function m is named basic probability assignment (BPA). m(A) expresses the proportion of all relevant and available evidence. It is claimed that a particular element of Θ belongs to the set A, but to no particular subset of A. If $m(A) > 0$, A is called a focal element of Θ. If $m(A) = 0$, it means that the proposition totally lacks belief.

Definition 3. *Suppose that two BPAs denoted by m_1 and m_2 are obtained from two different information sources in the same frame of discernment Θ. The degree of conflict among the evidence is denoted as follows:*

$$K = \sum_{A \cap B = \varnothing} m_1(A)m_2(B) \tag{35}$$

where if $K = 0$, it means that the two pieces of evidence are fully compatible with each other. On the contrary, if $K = 1$, it means that the two pieces of evidence totally conflict with each other.

Dempster's rule of combination is the most basic and widely used rule for the combination of evidence:

$$\begin{cases} \frac{1}{1-K} \sum\limits_{A \cap B = C} m_1(A)m_2(B) & \forall C \subseteq 2^\Theta, C \neq \varnothing \\ 0 & C = \varnothing \end{cases} \tag{36}$$

Dempster's combination rule is commutative and associates. Thus, the fusion result has nothing to do with the order of the fusion process.

It has two characteristics [46]:

- Pieces of mutually supporting evidence are reinforced.
- Pieces of conflicting evidence weaken each other.

5.2. The Design of Stopping Criterions

Research on stopping criteria based on evidence theory in this paper involves extracting real-time feedback gains from each controller in the vibration control system, constructing the frame of discernment, choosing appropriate feature vectors that describe the learning process of the robust MFA-IL algorithm, calculating the BPAs based on the input signals of actuator, forming the fused BPAs using combination rule and diagnosing the learning states of the control method based on the BPA results.

The real-time feedback gains from each controller can be considered as a piece of evidence for diagnosing the system state, assuming M actuators are glued on the plate and that the feedback gains of the corresponding controllers are measured vectors. For the sake of simplicity, suppose that all types of states are independent of each another. Only one state can occur at any given time. Let S_ω represent the measurement obtained from the ωth controller (information source):

$$S_\omega = [s_{\omega 1} \ s_{\omega 2} \ \cdots \ s_{\omega m_\omega}] \quad \omega = 1, 2, \ldots, M \tag{37}$$

where $s_{\omega i}$ is the ith element of S_ω, $i = 1, 2, \ldots, m_\omega$, m_ω is the number of elements provided by the ωth controller and $\sum\limits_{\omega=1}^{M} m_\omega = n$, n is the number of features.

There are N types of states. The system states matrix can be described as [47]:

$$H = \begin{bmatrix} X_1 \\ X_2 \\ \vdots \\ X_N \end{bmatrix} = \begin{bmatrix} x_{11} & x_{12} & \cdots & x_{1n} \\ x_{21} & x_{22} & \cdots & x_{2n} \\ \vdots & \vdots & \vdots & \vdots \\ x_{N1} & x_{N2} & \cdots & x_{Nn} \end{bmatrix} \tag{38}$$

where X_j is a feature vector describing the jth state, x_{ji} is the ith feature of the jth state, and $i = 1, 2, \ldots, n, j = 1, 2, \ldots, N$.

This paper uses the Minkowski distance for quantifying the objective evaluation of the BPAs. The Minkowski distance between S_ω and X_j is reconstructed as:

$$
d_{\omega j} = \begin{cases} \left[\sum\limits_{i=1}^{m_\omega} \left(\dfrac{s_{\omega i} - x_{ji}}{x_{ji}} \right)^r \right]^{1/r} & \omega = 1 \\[2em] \left[\sum\limits_{i=1}^{m_\omega} \left(\dfrac{s_{\omega i} - x_{j(i+\sum\limits_{l=1}^{\omega-1} m_l)}}{x_{j(i+\sum\limits_{l=1}^{\omega-1} m_l)}} \right)^r \right]^{1/r} & \omega = 2, 3 \ldots, M \end{cases}
\tag{39}
$$

where $d_{\omega j}$ is the distance between S_ω and X_j. r is a constant, such that if $r = 2$, then the distance converges to the Euclidean distance. On the other hand, if $r = 1$, the distance converges to the corner distance. The distances between all measurement vectors S_ω and all state vectors X_j can be obtained in a matrix form:

$$
D = \begin{bmatrix} d_{11} & d_{12} & \cdots & d_{1N} \\ d_{21} & d_{22} & \cdots & d_{2N} \\ \vdots & \vdots & \vdots & \vdots \\ d_{M1} & d_{M2} & \cdots & d_{MN} \end{bmatrix}
\tag{40}
$$

The smaller the distance $d_{\omega j}$, the more probable the jth state, based on the feedback gain acquired from the ωth controller. By defining $p_{\omega j} = 1/d_{\omega j}$, a matrix form after normalizing can be expressed as follows:

$$
P = \begin{bmatrix} p_{11} & p_{12} & \cdots & p_{1N} \\ p_{21} & p_{22} & \cdots & p_{2N} \\ \vdots & \vdots & \vdots & \vdots \\ p_{M1} & p_{M2} & \cdots & p_{MN} \end{bmatrix} = \begin{bmatrix} P_1 \\ P_2 \\ \vdots \\ P_M \end{bmatrix}
\tag{41}
$$

where $P_\omega = \begin{bmatrix} p_{\omega 1} & p_{\omega 2} & \cdots & p_{\omega N} \end{bmatrix}$. P_ω is the BPA assigned by the ωth controller to the set of states, satisfying $\sum\limits_{j=1}^{N} p_{\omega j} = 1$.

According to Equation (35), the degree of conflicting evidence among the various controllers can be calculated. Then, the fused BPAs are computed using the combination rule, denoted in Equation (36). The fused BPAs are the pieces of evidence for state diagnosis. The threshold value was predefined as a stopping criterion. Based on the values of BPAs for a certain state, decision making on a system state can be fulfilled by comparing with the threshold value.

6. Numerical Examples

6.1. FE Model Validation

The purpose of this subsection is to examine the accuracy of the dynamic FE model established by ANSYS, by comparison of the numerical and analytical results in the open literature.

Considering the square laminate plate, which composes of three layers of graphite-epoxy (GE, carbon-fibre reinforced) composite material (0/90/0) covered by PZT-4 piezoelectric layers poled in the z-direction (through-thickness). Then, a five-layer laminate plate (PZT-4/GE $0°$/GE $90°$/GE $0°$/PZT-4) was formed. The geometrical data of the square laminated plate are given in Figure 2. The material properties of the GE composite material and the piezoelectric material are given in Table 1, and the densities of all materials considered are to be considered unitary ($\rho_S = 1 \text{ kg/m}^3$) for the purpose of comparison. The assumed boundary conditions are simply supported. The SOLID46 3-D solid element, which can be used to model the laminated composite beam or plate type structures with various layer orientations, is used for simulating the GE composite plate and the SOLID5 3-D solid element (which has a 3D piezoelectric and structural capability between the fields and is applied for simulating piezoelectric layers).

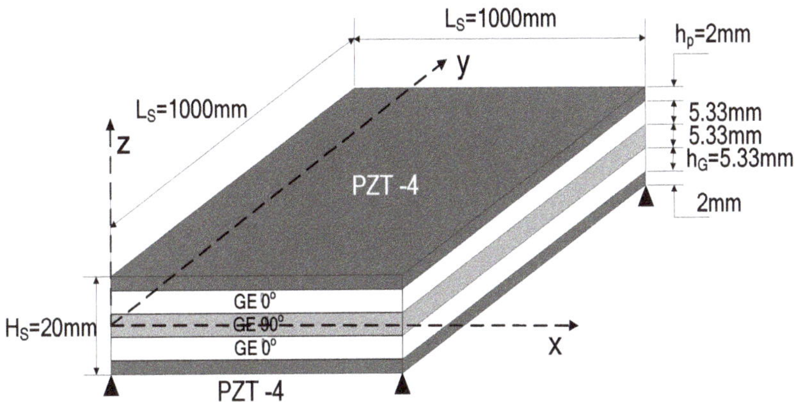

Figure 2. Square laminated plate.

Table 1. Properties of the graphite-epoxy (GE) composite material and the piezoelectric material.

GE [48]	PZT-4 [49]
Yong's modulus (GPa)	Elastic stiffness (GPa)
$E_{11} = 132.38;$	$C_{11} = 138.50; C_{12} = 77.37;$
$E_{22} = E_{33} = 10.76;$	$C_{13} = 73.64; C_{33} = 114.75;$
Shear modulus (GPa)	$C_{44} = 25.60; C_{66} = 30.60;$
$G_{12} = G_{13} = 5.65;$	Piezoelectric stain (C/m^2)
$G_{23} = 3.61;$	$e_{16} = e_{25} = 12.72;$
Poisson's ratio	$e_{31} = e_{32} = -5.2;$
$v_{12} = v_{13} = 0.24;$	$e_{33} = 15.08;$
$v_{23} = 0.49;$	Permittivity (F/m)
-	$\varepsilon_{11} = \varepsilon_{22} = 1.305 \times 10^{-8};$
-	$\varepsilon_{33} = 1.151 \times 10^{-8};$

The first five natural frequencies $\omega_{x,y}$ can be non-dimensionalized by the expression $\lambda_{x,y} = \omega_{x,y}L_S^2/H_S\sqrt{\rho_S} \times 10^3$. The dynamic FE numerical analysis results in this paper are compared to different approaches: (a) The finite element solution FPS (an equivalent single-layer approach and a layerwise representation of the electric potential, using the finite element method with four nodes and five degrees of freedom) by W. Larbi et al. [48]; (b) the finite element solution Q9-HSDT (higher shear deformation theory, using the finite element method with nine nodes and eleven degree of freedom) by Victor M. Franco Correia et al. [50]; (c) the finite element solutions TDST (third order shear deformation theory, using the finite element method with four nodes and seven degrees of freedom) by Tatiane Corrêa de Godoy et al. [49]; and (d) two-dimensional analytical solutions (layerwise first-order shear deformation theory and quadratic electric potential) by Ayech Benjeddou et al. [51]. The present results and their percent errors are relative to the published results ($\Delta\%$) and are shown in Table 2. It was found that a reasonably good approximation to FPS and Q9-HSDT theories was obtained, although the relative errors for some natural frequencies were greater than 5%.

Table 2. First five frequencies parameters.

Mode	Present	FPS [48]	Q9-HSDT [50]	TDST [49]	2D [51]
1	232.62	231.42 (0.52%)	230.46 (0.94%)	225.98 (2.94%)	246.07 (−5.47%)
2	523.90	521.88 (0.39%)	520.38 (0.68%)	542.29 (−3.39%)	559.62 (−6.38%)
3	665.75	667.91 (−0.32%)	662.91 (0.43%)	680.11 (−2.11%)	693.60 (−4.02%)
4	916.35	909.34 (0.77%)	908.46 (0.87%)	906.09 (1.13%)	967.14 (−5.25%)
5	1023.40	1027.22 (−0.37%)	1022.09 (0.13%)	1099.05 (−6.88%)	1091.46 (−6.19%)

6.2. Modeling and Choice of Controller Parameters

In this section, a piezoelectric smart plate is considered for vibration control simulations. The piezoelectric smart plate is made of a host composite plate and six piezoelectric patches bonded in pairs on both sides of the plate. From Figure 3, there are three piezoelectric actuator-sensor pairs marked with *a*, *b* and *c*, respectively. The upper piezoelectric patches work as actuators to control structural vibration, while the lower ones work as sensors to obtain vibration information. The dimensions of the composite plate and the piezoelectric patches are 414 mm × 120 mm × 1 mm and 60 mm × 24 mm × 1 mm, and the configuration of the structure is shown in Figure 3. The localization of the actuator-sensor pairs is selected by referencing to [52]. The host composite plate is made of a GE composite material with five substrate layers, the stacking sequence of which is symmetric angle-ply [0/90/0/90/0]. The properties of the GE composite material and the piezoelectric material are listed in Table 3. The locations of points A, B and C are shown in Figure 3. A clamped boundary condition is assigned in the root of the piezoelectric smart plate.

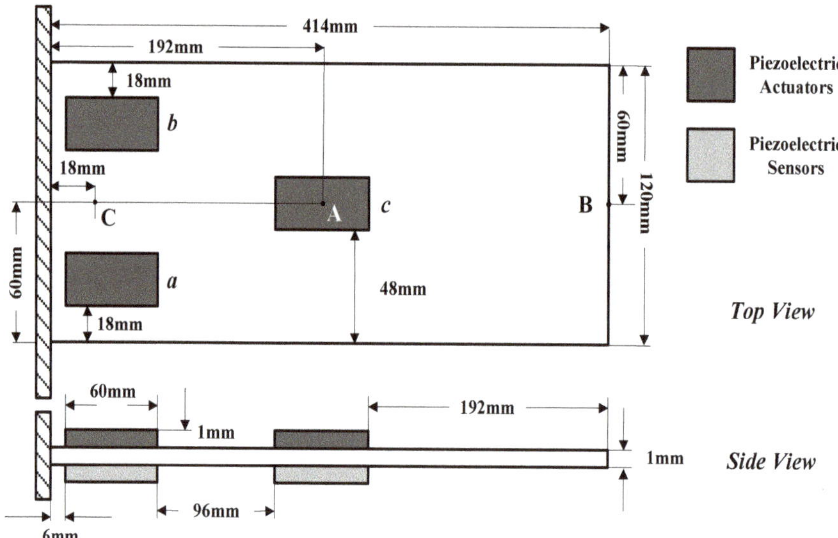

Figure 3. The piezoelectric smart plate.

Table 3. Properties of the GE composite material and the piezoelectric material.

GE	PZT-5H
Yong's modulus (GPa)	Elastic stiffness (GPa)
$E_{11} = 40.51$;	$C_{11} = 126$; $C_{12} = 79.5$;
$E_{22} = E_{33} = 13.96$;	$C_{13} = 84.1$; $C_{33} = 117$;
Shear modulus (GPa)	$C_{44} = 23.3$; $C_{66} = 23$;
$G_{12} = G_{13} = 3.1$;	Piezoelectric stain (C/m^2)
$G_{23} = 1.55$;	$e_{16} = e_{25} = 17$;
Poisson's ratio	$e_{31} = e_{32} = -6.5$;
$v_{12} = v_{13} = 0.22$;	$e_{33} = 23.3$;
$v_{23} = 0.11$;	Permittivity (F/m)
Density (kg/m^3)	$\varepsilon_{11} = \varepsilon_{22} = 1.503 \times 10^{-8}$;
$\rho = 1830$	$\varepsilon_{33} = 1.3 \times 10^{-8}$;
-	Density (kg/m^3)
-	$\rho = 7500$

In this study, the layered solid element SOLID46 was used to model the composite plate, while SOLID5 was used to model the piezoelectric patches. The composite plate was meshed with $69 \times 20 \times 1$ elements, and each piezoelectric patch was meshed with $10 \times 4 \times 1$ elements. It was assumed that the same magnitude but the opposite electric field direction was applied to the upper and lower piezoelectric patches. The degrees of electric freedom for the nodes at the top and bottom surfaces of the piezoelectric patches were coupled by using the ANSYS command CP. Model analysis was implemented to find out the natural frequencies of the plate and determine the sampling period for the closed-loop system simulations [28]. The numerical and experimental results of the first three natural frequencies are given in Table 4. The sampling period was defined as $T = 1/(20f_1)$, where f_1 is the first natural frequency. The Rayleigh damping coefficients were considered as $\alpha = 0.003$ and $\beta = 0.0015$.

Table 4. First three natural frequencies of the piezoelectric smart plate.

Mode	Numerical (Hz)	Experimental (Hz)	Error
1	5.438	5.326	2.1%
2	24.217	21.259	13.9%
3	28.683	31.593	−9.2%

The initial value of feedback gain $G(k)$ in Equation (15) was assumed to be zero. The parameters of the robust MFA-IL controller are given in Table 5. There are three actuator-sensor pairs in the system, and the corresponding controllers offer measurements. The frame of discernment for system state diagnosis was constructed with two types of states, namely the normal learning state and the learning stopping state. The feedback gains of the controllers were chosen as the feature parameters. Let $r = 2$ in Equation (39) and the distance converges to the Euclidean distance. The BPAs can be calculated using the method based on information sources in Section 5.2. The stopping criteria of the robust MFA-IL control algorithm are defined as follows: (a) According to Dempster's rule of combination, each row vector of the BPA matrix P can be fused. The threshold value was set 0.985. If the fused BPA is higher than the threshold value, the learning processes of all controllers should be stopped. (b) Controllers connected to the actuators at different locations may lead to distinct learning speeds. To make all controllers learn sufficiently, the threshold value for the BPAs of each controller should also be proposed. If the BPAs from a certain controller are higher than 0.800, the learning process of the corresponding controller should be stopped. If one of the above conditions is met, the learning process of the controller is then terminated. Otherwise, the controller is in normal learning state. Considering the vibrations generated by various external disturbances, different simulations were investigated to evaluate the effectiveness of the proposed method. In the design of the P-type IL controllers, the maximum number of iterations was limited to 500 as the stopping criterion, and the fixed learning gains were selected as $\Phi_1 = 0.063$ and $\Phi_2 = 0.100$ for various cases.

Table 5. The control parameters.

Parameter	Value	Parameter	Value
λ	1	α	0.5
ρ	0.1	δ	1.5
μ	1	ε	6
η	0.1	q	60
-	-	Γ	0.5

6.3. Harmonic Excitation

In this case, the first mode control was tested by applying the harmonic force $f(t) = 6\cos(\omega_1 t)N$ at the point C, where $\omega_1 = 17.083$ rad/s (5.4377 Hz). Due to the symmetry of the piezoelectric smart structure and the excitation location, sensors *a* and *b* have the same control feedback signals in the process of vibration.

The displacement responses of points A and B are displayed in Figure 4a,b, where it can be seen that both the robust MFA-IL control and the P-type IL control can effectively suppress the first mode vibration. The control effectiveness of the actuators is obvious at the positions with sensors (e.g., point A) and without sensors (e.g., point B). However, it is noteworthy that these results are different from Saleh's [13], which claimed that the P-type IL control was able to compensate for the unwanted vibration at the observation point, while not being effective at other points. In addition, it was also pointed out that the P-type IL method cannot effectively attenuate the amplitude as long as the smart structure is excited by its first natural frequency.

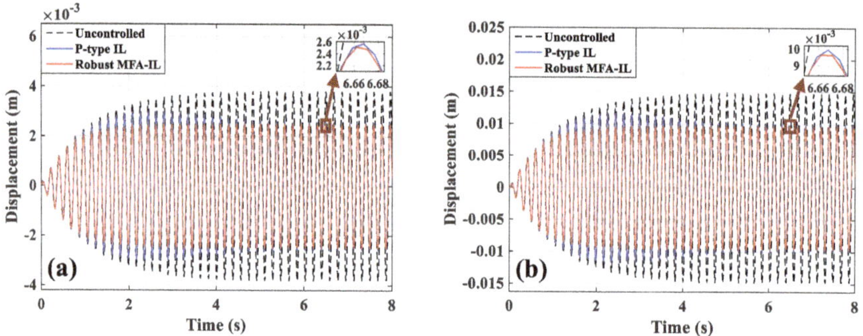

Figure 4. Displacement responses: (**a**) Point A, (**b**) point B.

An effective control system has the ability to suppress the structural vibration of the whole plate rather than a small portion of the plate area. The control strategy plays an important role in designing a vibration control system for obtaining a desired performance. Besides, the locations and sizes of the piezoelectric actuators and sensors should also be seriously considered [53]. The areas of the structure where the mechanical strain is highest are always the best locations for actuators and sensors. To guarantee the actuators generate the desired control forces to suppress structural vibration, the dimensions of the actuators should also be designed appropriately. The sizes of sensors should also be chosen properly, so that accurate information of structural deformation can be obtained. A misread or incorrect sensor measurement signal may lead to unreasonable measurements and inappropriate control force generation, which will deteriorate the dynamic behavior. As long as the locations and the sizes of the actuators and sensors are selected appropriately, the P-type IL control presents good performance on first mode control. Furthermore, both the locations with sensors and the positions without sensors on the smart plate can provide good controllability of structural vibration.

The displacement responses of points A and B are given in Figure 4a,b. The output voltages of sensor *a*/*b* and sensor *c* are shown in Figure 5c,d. By comparing with the P-type IL approach, a smaller amplitude can be obtained when the system is controlled by the robust MFA-IL method. In this paper, the root mean square (RMS) values of the amplitude at points A and B and output electric potential are used to quantitatively evaluate the control performance of the robust MFA-IL control and the P-type IL control. The data used to calculate the RMSs were recorded after all the controllers stopped learning, and the RMSs of amplitude are given in Table 6.

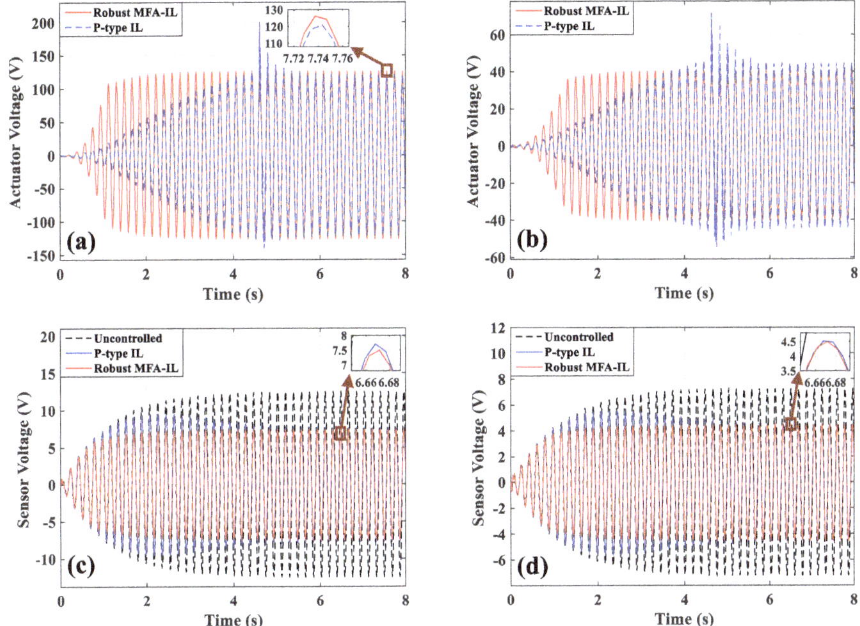

Figure 5. The time-domain responses of actuators/sensors: (**a**) Actuator a/b, (**b**) actuator c, (**c**) sensor a/b, (**d**) sensor c.

Table 6. Root mean square (RMS) values of amplitude.

Algorithm	Case 1				Case 2				Experiment	
	Point A	Point B	Sensor a/b	Sensor c	Point A	Point B	Sensor a/b	Senso c	Sensor a/b	Sensor c
Uncontrolled	0.0027	0.0106	8.9034	5.1768	0.0024	0.0087	8.1071	3.9265	3.7644	2.1672
P-type IL	0.0018	0.0071	5.4976	3.2501	0.0016	0.0055	5.0016	2.3231	2.5570	1.4806
Robust MFA- IL	0.0017	0.0067	5.2075	3.0731	0.0015	0.0053	4.7678	2.2119	2.4876	1.4510

The control voltages applied to actuators a/b and actuator c are shown in Figure 5a,b. In the process of iterative learning, the actuation voltages changed sharply under the control of the P-type IL algorithm at 4.3 s. After the learning was terminated, the amplitude recovered to a smooth value. The feedback gain of the controllers at different locations had a distinct convergence rate, which may cause the control force produced by controllers to mismatch among each other. If an actuator fails to perform as expected, the performance of its neighboring actuators will be negatively affected. In order to avoid this problem, more iterations are necessary to reach satisfying level of control stability. A smaller iterative number may directly cause control spillover or even system instability. The controllers connected with actuators a/b have the same learning processes, as shown in Figure 6a. Figure 6b presents the learning processes of the feedback gains in actuator e. Observing the results of Figures 5 and 6, the robust MFA-IL control has a faster convergence rate when compared with the P-type IL method, and the input voltages of the actuators change smoothly. Therefore, it can be seen that the proposed method enables the system to enhance control effectiveness and maintain superior stability.

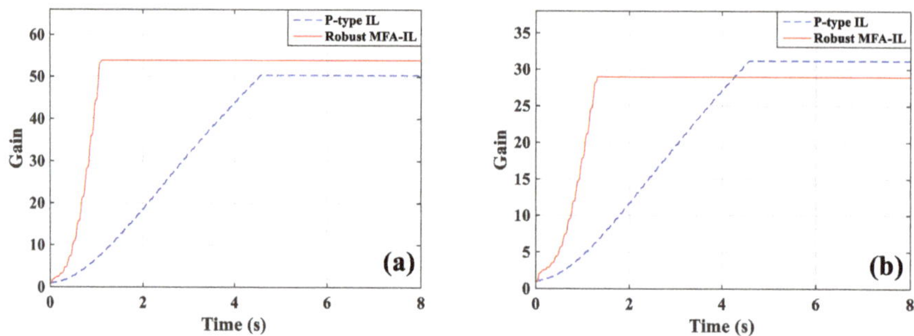

Figure 6. The learning processes of feedback gains: (**a**) Actuator a/b, (**b**) actuator c.

The main contribution of the robust MFA-IL control is the improvement of learning speed. Two parts influence the learning speed of the proposed algorithm: The MFA control and the SM control. Both of these two methods accelerate the convergence speed of the feedback gains in the learning processes. Their contribution percentages to the feedback gains are shown in Figure 7, where it can be seen that the SM control plays a more important role than the MFA control.

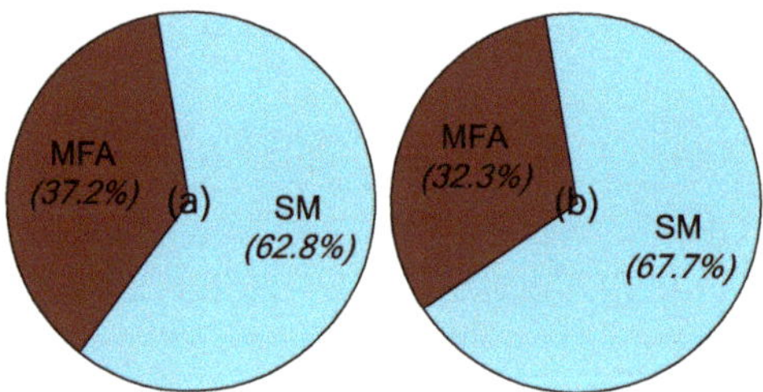

Figure 7. The contribution percentages of various methods to the feedback gains: (**a**) Actuators a/b, (**b**) actuator c.

Real-time monitoring for the fused BPA results and the BPA from each controller was implemented and is shown in Figure 8. The BPAs update at each period and the monitoring curve moves forward over time. First, the BPAs obtained from actuator c meet the stopping criteria. After a short period, the controller connected with actuators a/b stops learning. All controllers can learn sufficiently based on the evidence theory, and thus preferable control performance can be obtained.

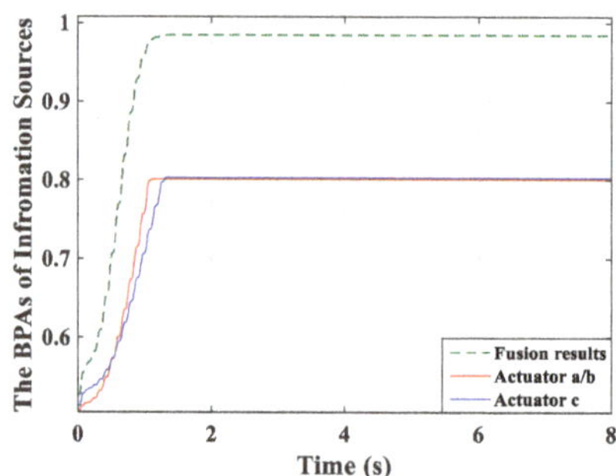

Figure 8. The basic probability assignment (BPA) curves.

6.4. Random Excitation

In the last simulation, a random excitation (shown in Figure 9) was applied to point C to drive the piezoelectric smart plate.

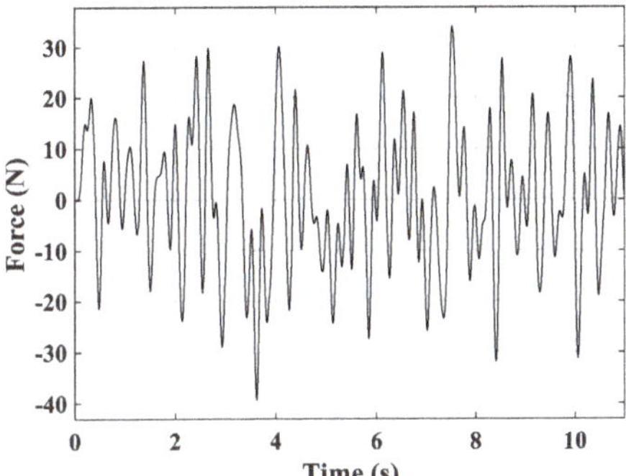

Figure 9. The random excitation.

Figure 10a,b presents the dynamic displacement responses of points A and B, respectively. The control voltages applied on actuators a/b and actuator c are depicted in Figure 11a,b. The output signals of the corresponding sensors are shown in Figure 11c,d. By comparing with the robust MFA-IL method, the actuation voltage amplitudes controlled by the P-type IL method are smaller in the initial stage of simulation. Under the control of the P-type IL algorithm, the actuators cannot work effectively to consume the energy of the vibration system in the initial period. According to Figures 10 and 11, it can be seen that the P-type IL control cannot suppress structural vibration in the short term. Since several hundreds of iterations lead to convergence rate of feedback gain slow. However, the learning speed of the robust MFA-IL algorithm is faster than that of the P-type IL method which has fixed gain. The time-varying learning gain is updated using I/O data, which can reflect the system dynamic behavior in real-time.

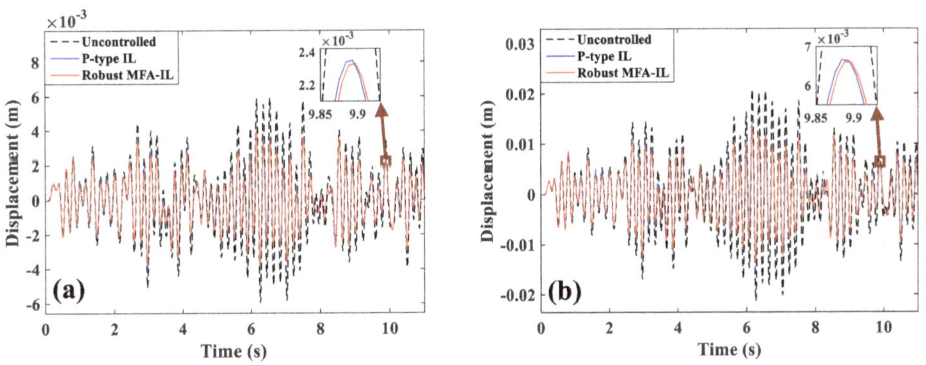

Figure 10. Displacement responses: (**a**) Point A, (**b**) point B.

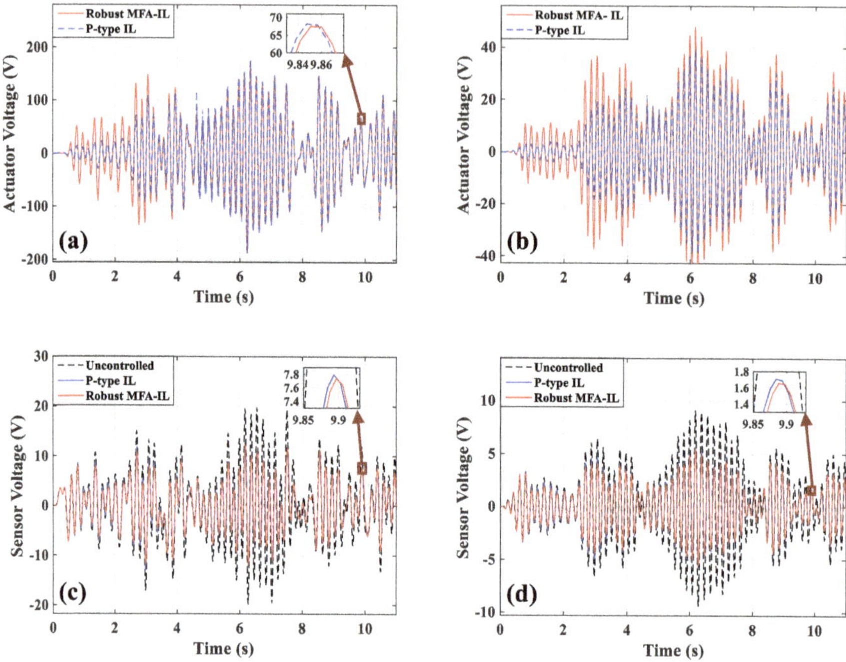

Figure 11. The time-domain responses of actuators/sensors: (**a**) Actuators a/b, (**b**) actuator c, (**c**) sensors a/b, (**d**) sensor c.

The learning processes for feedback gain for the controllers connected actuators a/b and actuator c are shown in Figure 12. The percentages contributed by the MFA control and the SM control are shown in Figure 13, where it can be seen that the SM control has a larger impact on the convergence rate of feedback gain than the MFA control.

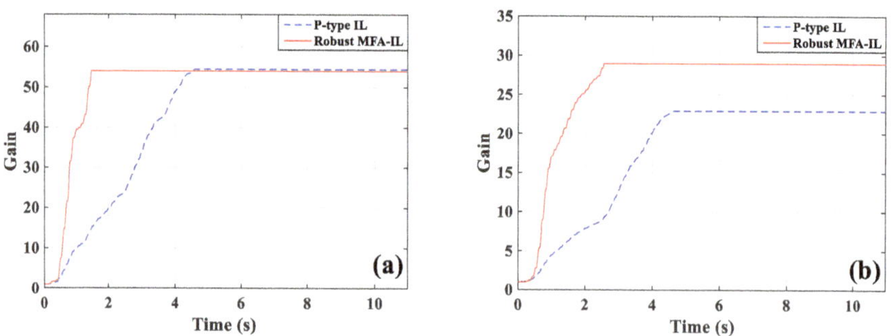

Figure 12. The learning processes of feedback gains: (**a**) Actuators a/b, (**b**) actuator c.

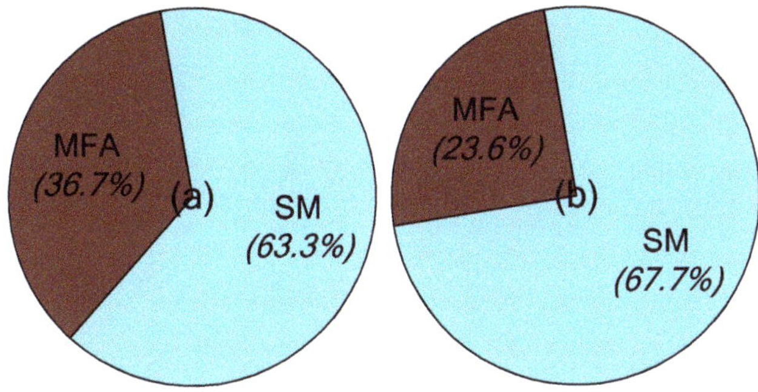

Figure 13. The contribution percentages of various methods to feedback gain: (**a**) Actuators a/b, (**b**) actuator c.

The RMSs of amplitude were calculated and are shown in Table 6. A similar control effect to the previous simulation can be observed. The robust MFA-IL control in this simulation presents a better control performance and makes the learning speed of the controller faster by comparing with the P-type IL control. The real-time monitoring results of the fused BPAs and the BPAs from each controller are depicted in Figure 14.

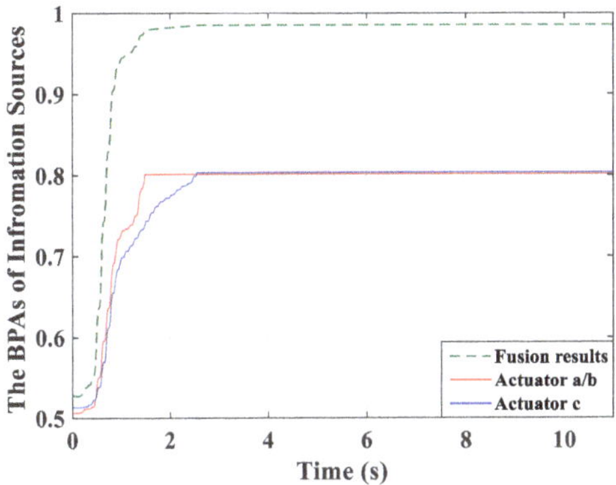

Figure 14. The BPAs curves.

In order to test the robustness of the proposed method, the harmonic noise signal $f(t) = \cos(152 \cdot t)N$ was added to the external excitation at point C. The controller's parameters were set up the same as mentioned above. Figure 15 shows the sensor output signals. The added noise resulted in a decrease of system performance and divergence as long as the system was controlled by the P-type IL method, while the robust MFA-IL control could maintain the stability of the control system. These comparative results validate that the proposed control method possesses excellent control performance and robustness to the noise from external excitation.

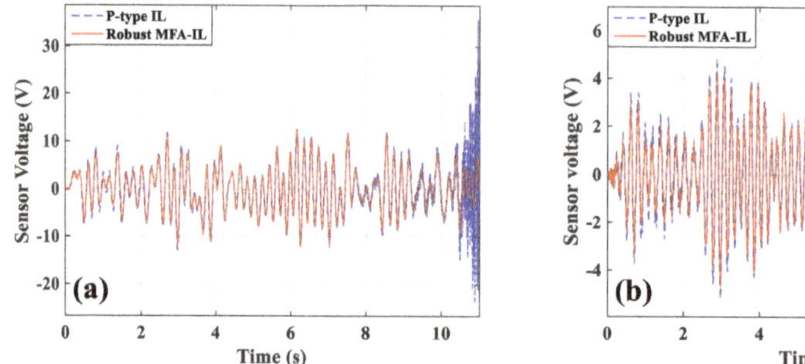

Figure 15. The time-domain responses of sensors: (**a**) Sensors a/b, (**b**) sensor c.

7. Experimental Investigations

7.1. Experimental Model

To verify the feasibility and the performance of the robust MFA-IL algorithm, an active vibration control system was established. The experimental devices are illustrated in Figure 16. The robust MFA-IL control algorithm and the P-type IL control algorithm were implemented in a MATLAB/Simulink environment. The real-time code was automatically operated in the real-time semi-physical simulation system. The dimensions of the specimen were the same as in the numerical simulations above, and exciting position point C was replaced by a metal patch. The experimental setup was composed of a piezoelectric smart plate, an electric-eddy current exciter (JZF-1, Beijing, China), a voltage amplifier (YE5872A, State College, PA, USA), a high voltage amplifier (E70, Harbin, China), a real-time semi-physical simulation system (Quarc, Toronto, ON, Canada) and a personal computer (PC) with a signal acquisition instrument. The piezoelectric patches were glued on the host plate using commercial cyanoacrylate glue.

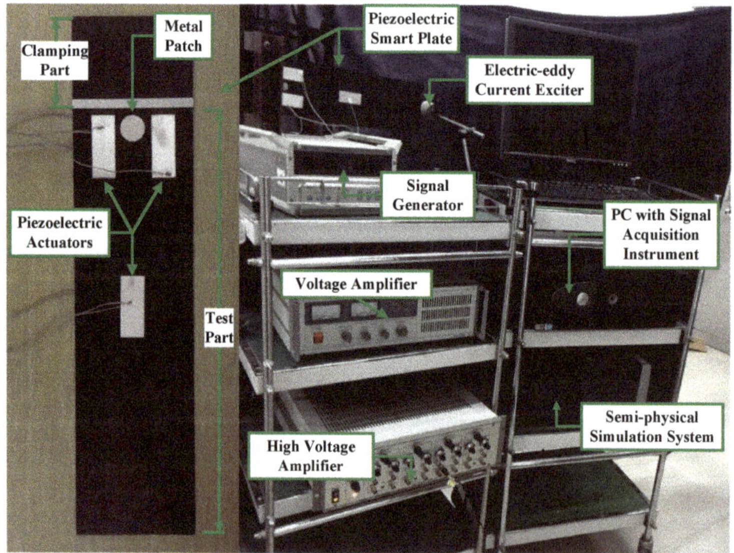

Figure 16. Experimental devices and the specimen.

A block diagram of the experimental system is illustrated in Figure 17. The experimental system includes three subsystems, namely: The vibration exciting system, the signal acquisition system and the vibration control feedback system. The signal transfer paths of these three subsystems are marked by red, green and blue color with different line types.

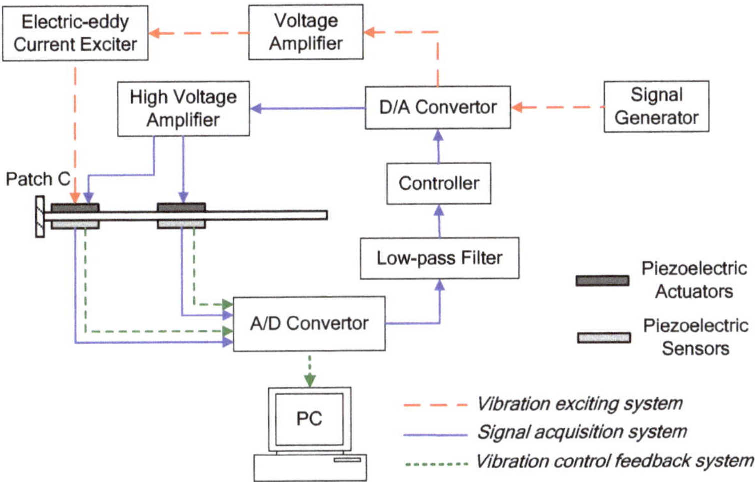

Figure 17. Block diagram of the experimental system.

For the vibration exciting system, the excitation signal was produced by the signal generator, which was used to simulate the external excitation of the piezoelectric smart plate. After digital to analog (D/A) conversion, the excitation signal was amplified by the voltage amplifier. The signal was transmitted to the electric-eddy current exciter, which could then convert the electrical signal into the vibration excitation.

For the signal acquisition system, three piezoelectric sensors (sensors a, b and c) were used to obtain the vibration information of the piezoelectric smart plate. The dynamic signals from the piezoelectric sensors were sent to the PC after analog to digital (A/D) conversion, which were then used to estimate the performance of the investigated control algorithms.

For the vibration control feedback system, the dynamic signals from the piezoelectric sensors were fed to a A/D convertor. To remove high-frequency noise, the input signals were passed through a low-pass filter set to 14 Hz. The controllers conduct both signal processing and controlling design in the MATLAB/Simulink 2014a environment. Running the control algorithms, the control signals were computed and sent through the D/A convertor. After amplification from the high voltage amplifier, the control signals were applied to the actuators for suppressing structural vibration. The experimental sample period was specified as 3 ms.

7.2. Modal Identification

A chirp signal is utilized to determine the natural frequencies of the piezoelectric smart plate. Actuator a was excited by a chirp signal with an amplitude of 100V, while the output voltages of the Sensor a were stored after filtering. The starting frequency was 0.5 Hz, and the stopping frequency was 50 Hz. The sweep time was 100 s. Then, the fast Fourier transform (FFT) of the time response was calculated. Figure 18b depicts the frequency responses of sensor a when applying FFT to the time-domain signal plotted in Figure 18a. The experimental natural frequencies obtained from the FFT plot are given in the same table (Table 4) for comparison. The biggest difference, 13.9%, occurs in the second natural frequency.

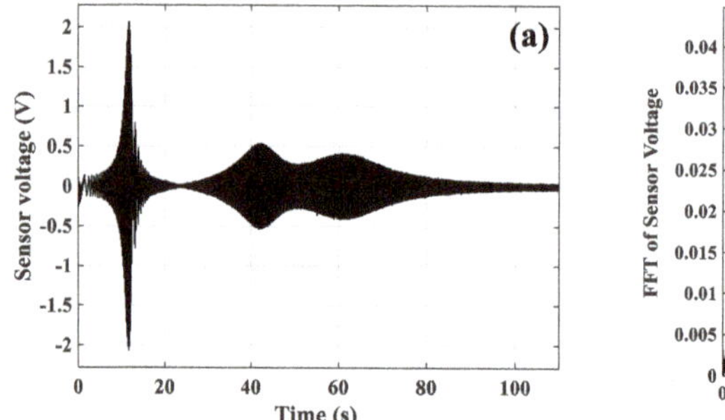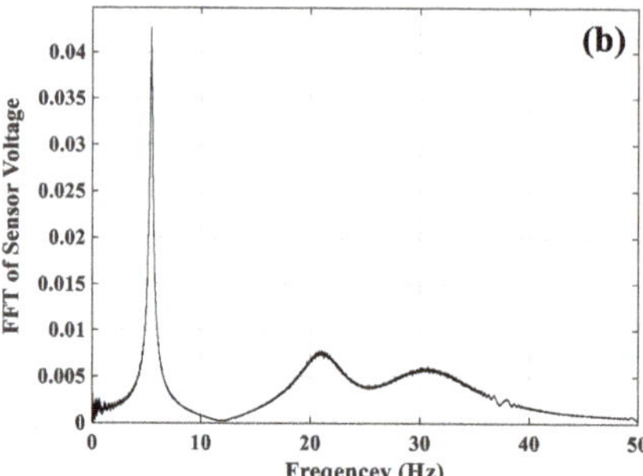

Figure 18. Measured vibration responses excited by actuator *a*: (**a**) Time-domain responses of sensor *a*, (**b**) frequency responses of sensor *a*.

Conditions that may lead to the difference of modal frequencies between the numerical solutions and experimental solutions are considered as follows: (1) The clamp-free boundary condition in the simulations is ideal. While in the experiments, the boundary of the plate may not be totally clamped. (2) The geometric parameters and material properties of the piezoelectric smart plate cannot be known accurately enough. The parameters used in the numerical simulations are not precisely consistent with those of the plate applied in the experimental material. (3) The mass of both the glue and the connected signal wire of the piezoelectric patches are not considered in the simulations. The modal analysis in simulations is used to provide an approximate solution for verifying the feasibility of the control algorithms, the difference of modal frequencies between the numerical results and the experimental results is acceptable. It is clear that the FE model sufficiently predicts the natural frequencies of the piezoelectric smart plate. Filters were designed and utilized in the experiments to deal with high frequency noise. For investigating the control algorithms, the designed low-pass filters were applied. The cutoff frequency of the low-pass filters was specified at 30 Hz.

7.3. Experiments Results and Discussions

During the experiments, the robust MFA-IL control and P-type IL control algorithms were used to suppress the structural vibration of the cantilevered plate. The first mode control was tested in this section, where the parameters of the P-type IL controller are chosen as follows: The fixed learning gain $\Phi = 0.54$. The maximum number of iterations was set to 1500 as to avoid control spillover or system instability. The parameters of the robust MFA-IL controller are listed in Table 7. The stopping criteria in the experiment are the same with those in Section 6.2. The sensor output signals (shown in Figure 19a,b) were moved forward to compensate for the phase lag effect. Five seconds after harmonic excitation driving, the proposed control algorithms were implemented to suppress the structural vibration.

Table 7. The control parameters.

Parameter	Value	Parameter	Value
λ	1	α	0.5
ρ	0.1	δ	1.5
μ	1	ε	16
η	0.05	q	150
-		Γ	0.8

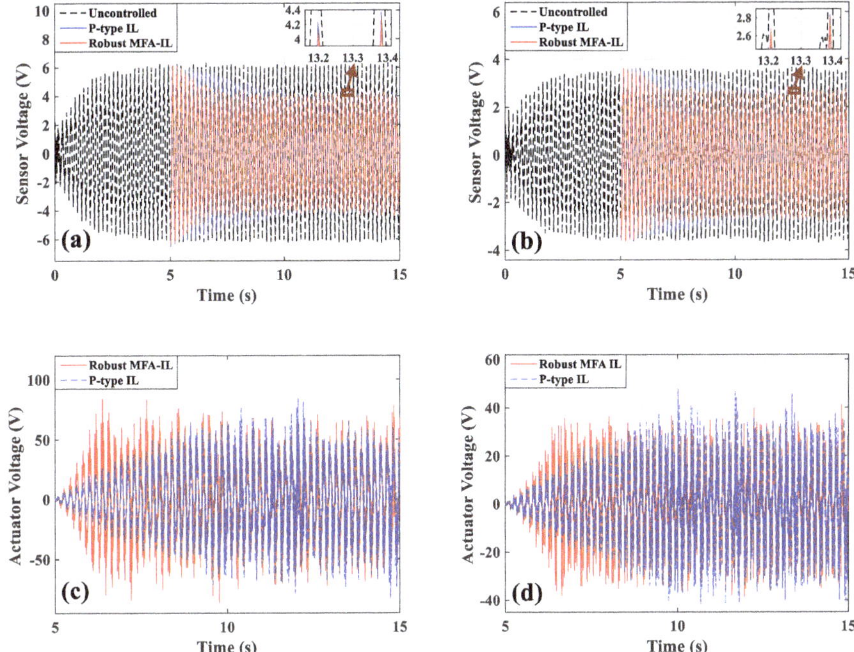

Figure 19. The time-domain response of actuators/sensors: (**a**) Actuators a/b, (**b**) actuator c, (**c**) sensors a/b, (**d**) sensor c.

The measured signals of sensors a/b and sensor c, and the corresponding actuation voltages obtained from the robust MFA-IL controller and the P-type IL controller are shown in Figure 19. The sensor a/b and sensor c output signals used to calculate the RMS values were recorded after learning termination. The RMS values (shown in Table 6) were used to quantitatively evaluate the control performance of the various methods. From the RMS values, it can be seen that the P-type IL control provides a 31.9% reduction and the robust MFA-IL control provides a 33.73% reduction. It is clear that the robust MFA-IL control is more effective at reducing the vibration amplitude of the flexible plate than the P-type IL control. It can be concluded that the proposed control method exhibits excellent performance when integrating MFA control and SM control. The learning processes of the feedback gains are shown in Figure 20. Under the control of the robust MFA-IL method, the structural vibration can be suppressed faster when compared with the robust MFA-IL algorithm.

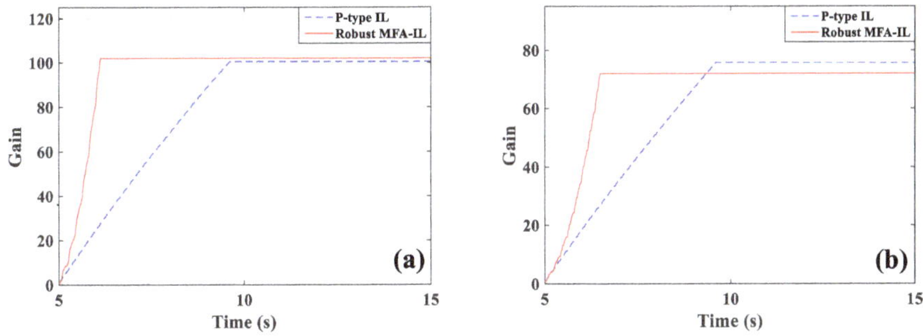

Figure 20. The learning processes of feedback gains: (**a**) Actuators a/b, (**b**) actuator c.

The percentages of contributed feedback gain from the MFA and SM methods are shown in Figure 21, where it can be seen that SM control has a larger impact on the convergence speed of feedback gain than MFA control. The real-time monitoring curves of the fused BPA results and the BPAs from each information source are shown in Figure 22. By applying evidence theory to the design of the stopping criteria, the learning processes of all controllers can be evaluated in real-time. The decision-making based on the appropriate stopping criteria makes all controllers learn sufficiently, so that a better control performance is obtained.

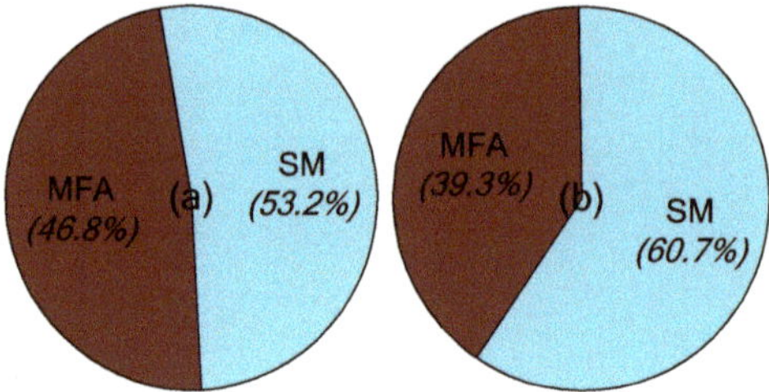

Figure 21. The contribution percentages of various methods to feedback gain: (**a**) Actuators a/b, (**b**) actuator c.

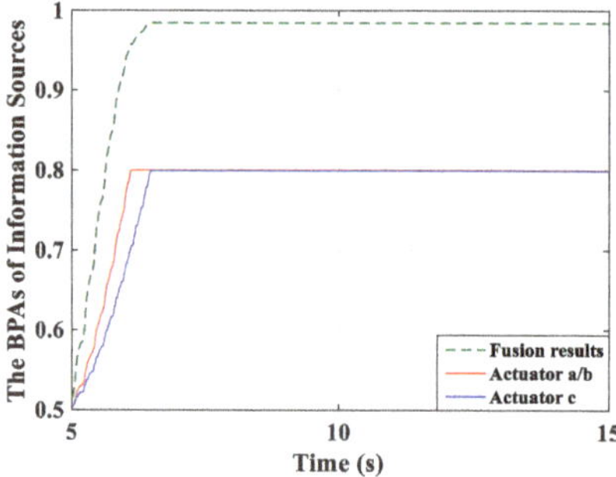

Figure 22. The BPAs curves.

Comparing the experimental results with the numerical results, it can be seen that the vibration suppression trends performed similarly. Due to the difference of the sample period and the unknown nature of the experimental system, the parameters selected in the simulations and experiments are different.

8. Conclusions and Outlooks

A robust MFA-IL control method was developed for the active vibration control of piezoelectric smart structures. Considering the uncertainty of interaction among all actuators in the control process, the MFA control was incorporated into P-type IL control for the adjustment of learning gain. In order to achieve a fast control response and enhance the stability of the system, SM control was adopted to ensure a fast dynamic response and compensate for the influence of uncertain noise. The multi-source information fusion method based on the evidence theory was adopted to design the stopping criteria of the robust MFA-IL method. The vibration control equations of piezoelectric smart structures were derived from the dynamic FE equations of a linear elastic system. The dynamic linear method was applied to transfer the vibration control equations for the design of the robust MFA-IL controller. The simulation and experimental results were presented and compared with the corresponding results using the P-type IL control approach.

As long as the locations and sizes of actuators and sensors are chosen appropriately, both P-type IL control and robust MFA-IL control can effectively suppress structural vibration when the piezoelectric smart plate is excited by its first natural frequency. Furthermore, the whole structure presents good controllability, rather than small portions bonded with piezoelectric sensors, using both the P-type IL method and the robust MFA-IL method.

The robust MFA-IL control method accelerates the learning speed of the controller. Based on the comprehensive comparative analysis of the numerical and experimental results, the proposed control method can achieve better control performance and is more robust to external disturbances when compared with the P-type IL control. In other words, the proposed control method overcomes the inherent drawbacks of P-type IL control and achieves the desired control performance.

Although the robust MFA-IL control method was applied for a plate structure in this paper, considering its advantages presented above, this approach can be applied for other structures, like beam structures, extending the possibilities of engineering and research applications. Dynamic linearization was an effective method in developing the proposed method for nonlinear systems. Dynamic linearization mainly introduces optimal technology as a tool for the controller design and analysis. In future work, it will be expected that the dynamic linearization method will be combined with an online model identification technique to deal with much more practical problems, those typically encountered in industrial applications. This may result in a simpler controller structure and obtain a better degree of control precision.

Author Contributions: L.B. conceived and designed the experiments; X.X. performed the experiment; Y.F. and N.L. contributed the analysis tools; and L.B. analyzed the data and wrote the paper.

Acknowledgments: The work was supported by National Natural Science Foundation of China under Grant 10577015.

References

1. Li, X.F.; Xu, J.X.; Huang, D.Q. An iterative learning control approach for linear systems with randomly varying trial lengths. *IEEE Trans Automat Contr.* **2014**, *59*, 1954–1960. [CrossRef]
2. Uchiyama, M. Formulation of high-speed motion pattern of a mechanical arm by trial (in Japanese). *Trans. Soc. Instrum. Contr. Eng.* **1978**, *14*, 706–712.
3. Arimoto, S.; Kawamura, S.; Miyazaki, F. Bettering operation of dynamic systems by learning: a new control theory for servomechanism or mechatronics systems. In Proceedings of the 23rd Conference on Decision and Control, Las Vegas, NV, USA, 12–14 December 1984; pp. 1064–1069.
4. Xiao, T.F.; Li, X.D.; John, K.L.H. An adaptive discrete-time ILC strategy using fuzzy systems for iteration-varying reference trajectory tracking. *Int. J. Contr. Autom.* **2015**, *13*, 222–230. [CrossRef]
5. Cao, Z.X.; Zhang, R.D.; Yang, Y.; Lu, J.Y.; Gao, F.R. Discrete-time robust iterative learning Kalman filtering for repetitive processes. *IEEE Trans Automat. Contr.* **2016**, *61*, 270–275. [CrossRef]
6. He, W.; Meng, T.T.; He, X.Y.; Ge, S.S.Z. Unified iterative learning control for flexible structures with input constraints. *Automatica* **2018**, *96*, 326–336. [CrossRef]
7. Zhao, Y.; Zhou, F.Y.; Li, Y.; Wang, Y.G. A novel iterative learning path-tracking control for nonholonomic mobile robots against initial shift. *Int. J. Adv. Robot. Syst.* **2017**, *14*, 1–9. [CrossRef]
8. He, W.; Meng, T.T.; He, X.Y.; Sun, C.Y. Iterative learning control for a flapping wing micro aerial vehicle under distributed disturbances. *IEEE Trans. Cybern.* **2019**, *49*, 1524–1535. [CrossRef] [PubMed]
9. Zhang, Q.C.; Liu, Y.Q.; Zhang, Z.B. A new method for automatic optimization of drawbead geometry in the sheet metal forming process based on an iterative learning control model. *Int. J. Adv. Manuf. Technol.* **2017**, *88*, 1845–1861. [CrossRef]
10. Zhu, X.J.; Gao, Z.Y.; Huang, Q.Z.; Yi, J.C. ILC based active vibration control of smart structures. *IEEE Int. Conf. Intell.Comput. Intell. Syst.* **2009**, *2*, 236–240.
11. Tavakolpour, A.R.; Mailah, M.; Intan, Z.; Darus, M.; Tokhi, O. Self-learning active vibration control of a flexible plate structure with piezoelectric actuator. *Simul. Model. Pract. Theory* **2010**, *18*, 516–532. [CrossRef]

12. Fadil, M.A.; Jalil, N.A.; Darus, I.Z.M. Intelligent PID controller using iterative learning algorithm for active vibration controller of flexible beam. In Proceedings of the IEEE Symposium on Computers & Informatics, Langkawi, Malaysia, 7–9 April 2013; pp. 80–85.

13. Saleh, A.R.T.; Mailah, M. Control of resonance phenomenon in flexible structures via active support. *J. Sound Vib.* **2012**, *331*, 3451–3465. [CrossRef]

14. Tayebi, A.; Islam, S. Adaptive iterative learning control for robot manipulators: Experimental results. *Control Eng. Practice* **2006**, *14*, 843–851. [CrossRef]

15. Wang, J.; Li, B. Fuzzy gain p-type iterative learning controller based on FPGA. In Proceedings of the Chinese Control and Decision Conference, Mianyang, China, 23–25 May 2011; pp. 3199–3203.

16. Han, H.G.; Wu, X.L.; Qiao, J.F. Nonlinear systems modeling based on self-organizing fuzzy-neural-network with adaptive computation algorithm. *IEEE Tran. Cybern.* **2014**, *44*, 554–564.

17. He, W.; Meng, T.T.; Huang, D.Q.; Li, X.F. Adaptive boundary iterative learning control for an Euler-Bernoulli beam system with input constraint. *IEEE Trans. Neur. Net. Lear.* **2018**, *29*, 1539–1549. [CrossRef] [PubMed]

18. Mohammadi, E.; Fadaeinedjad, R.; Moschopoulos, G. Implementation of internal model based control and individual pitch control to reduce fatigue loads and tower vibrations in wind turbines. *J. Sound Vib.* **2018**, *421*, 132–152. [CrossRef]

19. Sarban, R.; Jones, R.W. Physical model-based active vibration using a dielectric elastomer actuator. *J. Intel. Mat. Syst. Struct.* **2012**, *23*, 473–483. [CrossRef]

20. Furuya, K.; Ishizuka, S.; Kajiwara, I. Online tuning of a model-based controller by perturbation of its poles. *J. Adv. Mech. Des. Syst.* **2015**, *9*, 15–61. [CrossRef]

21. Amoateng, D.O.; Hosani, M.A.; Elmoursi, M.S.; Turisyn, K.; Jame, L.; Kirtley, J.R. Adaptive voltage and frequency control of islanded multi-microgrids. *IEEE Trans. Power Syst.* **2018**, *33*, 4454–4465. [CrossRef]

22. Qiu, Z.C. Adaptive nonlinear vibration control of a Cartesian flexible manipulator driven by a ballscrew mechanism. *Mech. Syst. Signal Process* **2012**, *30*, 248–266. [CrossRef]

23. Lim, Y.H. Finite-element simulation of closed loop vibration control of a smart plate under transient loading. *Smart Mater. Struct.* **2003**, *12*, 272–286. [CrossRef]

24. Hou, Z.S.; Wang, Z. From model-based control to data-driven control: survey, classification and perspective. *Inf. Sci.* **2013**, *235*, 3–35. [CrossRef]

25. Hou, Z.S.; Chi, R.H.; Gao, H.J. An overview of dynamic-linearization-based data-driven control applications. *IEEE Trans. Ind. Electron.* **2017**, *64*, 4076–4089. [CrossRef]

26. Chen, H.Y.; Liang, J.W. Model-free adaptive sensing and control for a piezoelectrically actuated system. *Sensors* **2010**, *10*, 10545–10559. [CrossRef]

27. Chang, Y.; Gao, B.; Gu, K.Y. A model-free adaptive control to a blood pump based on heart rate. *Asaio. J.* **2011**, *57*, 262–267. [CrossRef]

28. Xu, D.Z.; Jiang, B.; Shi, P. A novel model-free adaptive control design for multivariable industrial processes. *IEEE Tran. Ind. Electron.* **2014**, *61*, 6391–6398. [CrossRef]

29. Wang, X.F.; Li, X.; Wang, J.H.; Fang, X.K.; Zhu, X.F. Data-driven model-free adaptive sliding mode control for multi degree-of-freedom robotic exoskeleton. *Inf. Sci.* **2016**, *327*, 246–257. [CrossRef]

30. Chen, X.Y.; Park, J.H.; Cao, J.D.; Qiu, J.L. Sliding mode synchronization of multiple chaotic systems with uncertainties and disturbances. *Appl. Math. Comput.* **2017**, *308*, 161–173. [CrossRef]

31. Li, Y.M.; Ma, Z.Y.; Tong, S.C. Adaptive fuzzy output-constrained fault-tolerant control of nonlinear stochastic large-scale systems with actuator faults. *IEEE Tran. Cybern.* **2017**, *47*, 2362–2376. [CrossRef]

32. Yang, C.G.; Wang, X.Y.; Cheng, L.; Ma, H.B. Neural-learning-based telerobot control with guaranteed performance. *IEEE Tran. Cybern.* **2017**, *47*, 3148–3159. [CrossRef]

33. Han, S.S.; Wang, H.P.; Tian, Y. Model-free based adaptive nonsingular fast terminal sliding mode control with time-delay estimation for a 12 DOF multi-functional lower limb exoskeleton. *Adv. Eng. Softw.* **2018**, *119*, 38–47. [CrossRef]

34. Boukattaya, M.; Mezghani, N.; Damak, T. Adaptive nonsingular fast terminal sliding-mode control for the tracking problem of uncertain dynamical systems. *ISA Trans.* **2018**, *77*, 1–19. [CrossRef]

35. Lu, C.Q.; Wang, S.P.; Wang, X.J. A multi-source information fusion diagnosis for aviation hydraulic pump based on the new evidence similarity distance. *Aerosp. Sci. Technol.* **2017**, *71*, 392–401. [CrossRef]

36. Li, S.B.; Liu, G.K.; Tang, X.H.; Lu, J.G.; Hu, J.J. An ensemble deep convolutional neural network model with improved D-S evidence fusion for bearing fault diagnosis. *Sensors* **2017**, *17*, 1729. [CrossRef]

37. Zheng, X.Z.; Wang, F.; Zhou, J.L. A hybrid approach for evaluating faulty behavior risk of high-risk operations using ANP and evidence theory. *Math. Probl. Eng.* **2017**, *2017*, 7908737.

38. Zhang, S.Q.; Li, H.N.; Schmidt, R.; Müller, P.C. Disturbance rejection control for vibration suppression of piezoelectric laminated thin-walled structures. *J. Sound Vib.* **2014**, *333*, 1209–1223. [CrossRef]

39. Ray, M.C.; Reddy, J.N. Active damping of laminated cylindrical shells conveying fluid using 1-3 piezoelectric composites. *Compos. Struct.* **2013**, *98*, 261–271. [CrossRef]

40. Kapuria, S.; Yasin, M.Y.; Hagedorn, P. Active vibration control of piezolaminated composite plates considering strong electric field nonlinearity. *AIAA J.* **2015**, *53*, 603–616. [CrossRef]

41. Karagülle, H.; Malgaca, L.; Öktem, H.F. Analysis of active vibration control in smart structures by ANSYS. *Smart Mater. Struct.* **2004**, *13*, 661–667. [CrossRef]

42. Ahn, H.S.; Chen, Y.Q.; Moore, K.L. Iterative learning control: brief survey and categorization. *IEEE Trans. Syst. Man Cybern. Part C Appl. Rev.* **2007**, *37*, 1099–1121. [CrossRef]

43. Zhong, H.; Wang, Q.; Yu, J.H.; Wei, Y.H.; Wang, Y. Discrete sliding mode adaptive vibration control for space frame based on characteristic model. *J. Comput. Nonlin. Dyn.* **2017**, *1*, 011003. [CrossRef]

44. Yin, S.W.; Yu, D.J.; Ma, Z.D.; Xia, B.Z. A unified model approach for probability response analysis of structure-acoustic system with random and epistemic uncertainties. *Mech. Syst. Signal Process* **2018**, *111*, 509–528. [CrossRef]

45. Setiti, H.; Hafezalkotob, A. Developing pessimistic-optimistic risk-based method for multi-sensor fusion: An interval-valued evidence theory approach. *Appl. Soft Comput.* **2018**, *72*, 609–623. [CrossRef]

46. Jiang, W.; Cao, Y.; Yang, L.; He, Z.C. A time-space domain information fusion method for specific emitter identification based on Dempster-Shafer evidence theory. *Sensors* **2017**, *17*, 1972. [CrossRef]

47. Basir, O.; Yuan, X.H. Engine fault diagnosis based on multi-sensor information fusion using Dempster-Shafer evidence theory. *Inf. Fusion* **2007**, *8*, 379–386. [CrossRef]

48. Larbi, W.; Deü, J.F.; Ohayon, R. Finite element formulation of smart piezoelectric composite plates coupled with acoustic fluid. *Compos. Struct.* **2012**, *94*, 501–509. [CrossRef]

49. Godoy, T.C.D.; Trindade, M.A. Modeling and analysis of laminate composite plates with embedded active-passive piezoelectric networks. *J. Sound Vib.* **2011**, *330*, 194–216. [CrossRef]

50. Correia, V.M.F.; Gomes, M.A.A.; Suleman, A.; Soares, C.M.M.; Soares, C.A.M. Modelling and design of adaptive composite structures. *Comput. Method Appl. Mech. Eng.* **2000**, *185*, 325–346. [CrossRef]

51. Benjeddou, A.; Deü, J.F. A two-dimensional closed-form solution for the free-vibrations analysis of piezoelectric sandwich plates. *Int. J. Solids Struct.* **2002**, *39*, 1463–1486. [CrossRef]

52. Dong, X.J.; Peng, Z.K.; Ye, L.; Hua, H.X.; Meng, G. Performance evaluation of vibration controller for piezoelectric smart structures in finite element environment. *J. Vib. Control.* **2014**, *20*, 2146–2161. [CrossRef]

53. Zorić, N.D.; Simonović, A.M.; Mitrović, Z.S.; Stupar, S.N. Optimal vibration control of smart composite beams with optimal size and location of piezoelectric sensing and actuation. *J. Intel. Mat. Syst. Struct.* **2012**, *24*, 499–526. [CrossRef]

 *micromachines*

Article

Nonlinear Hysteresis Modeling of Piezoelectric Actuators Using a Generalized Bouc–Wen Model

Jinqiang Gan [1],* and Xianmin Zhang [2]

[1] School of Mechanical Engineering and Electronic Information, China University of Geosciences, Wuhan 430074, China
[2] Guangdong Provincial Key Laboratory of Precision Equipment and Manufacturing Technology, School of Mechanical and Automotive Engineering, South China University of Technology, Guangzhou 510640, China; zhangxm@scut.edu.cn
* Correspondence: ganjq@cug.edu.cn

Received: 14 February 2019; Accepted: 8 March 2019; Published: 12 March 2019

Abstract: Hysteresis behaviors exist in piezoelectric ceramics actuators (PCAs), which degrade the positioning accuracy badly. The classical Bouc–Wen (CB–W) model is mainly used for describing rate-independent hysteresis behaviors. However, it cannot characterize the rate-dependent hysteresis precisely. In this paper, a generalized Bouc–Wen (GB–W) model with relaxation functions is developed for both rate-independent and rate-dependent hysteresis behaviors of piezoelectric actuators. Meanwhile, the nonlinear least squares method through MATLAB/Simulink is adopted to identify the parameters of hysteresis models. To demonstrate the validity of the developed model, a number of experiments based on a 1-DOF compliant mechanism were conducted to characterize hysteresis behaviors. Comparisons of experiments and simulations show that the developed model can describe rate-dependent and rate-independent hysteresis more accurately than the classical Bouc–Wen model. The results demonstrate that the developed model is effective and useful.

Keywords: piezoelectric ceramics actuators; hysteresis modeling; Bouc–Wen model

1. Introduction

Piezoelectric materials are capable of undergoing reversible phase transitions as a result of voltage and pressure. Due to these special material properties, piezoelectric materials have become increasingly popular in sensors and actuators. Piezoelectric ceramics actuators (PCAs) have been employed in precision positioning systems for their large force generation, high stiffness, high resolution and fast response. However, they commonly exhibit strong hysteresis behaviors, which greatly degrade the overall positioning accuracy.

According to whether the rate of the input is considered or not, hysteresis behaviors can be divided into rate-dependent and rate-independent hysteresis behaviors. The corresponding hysteresis models can be classified into rate-dependent and rate-independent hysteresis models. Over the past few decades, great efforts have been devoted to developing hysteresis models such as the Prandtl–Ishlinskii model [1–3], Preisach model [4–6], Maxwell-Slip model [7,8], Duhem model [9,10], Polynomial-based hysteresis model [11,12], and Bouc–Wen model [13,14]. For modeling of rate-independent hysteresis, it mainly focuses on the nonlinear relationship between the amplitude of input voltage and output displacement at low input frequency or rate. However, for modeling of rate-dependent hysteresis, it needs an analysis of the nonlinear relationship between the rate of input voltage and output displacement at high input frequency or rate. Comparisons of rate-dependent and rate-independent hysteresis models reveal that the rate-independent model is just a special kind of rate-dependent model when the rate of input is low enough. Therefore, it is more difficult to develop the rate-dependent model relatively. Overall, most related literatures focused on developing rate-independent hysteresis models and few literatures paid attention to modeling of rate-dependent hysteresis.

Due to its differential equations and ability to capture an analytical form, the Bouc–Wen model has been widely applied in hysteresis modeling and compensation for piezoelectric ceramics actuators. Based on the classical Bouc–Wen (CB–W) model, Zhu and Wang [15] added a non-symmetrical formula to describe non-symmetrical hysteresis and the corresponding experiments demonstrated its validity. Fujii et al. [16] proposed an extended Bouc–Wen model by introducing a velocity sign sensitivity. To eliminate the influence of nonlinear hysteresis, Li et al. [17] presented an adaptive sliding mode control with perturbation estimation (SMCPE) based on the classical Bouc–Wen model. In addition, Liu et al. [18] proposed an adaptive neural output-feedback control based on a modified Bouc–Wen model. Lin and Yang [19] used a Bouc–Wen model to describe the hysteresis behavior and designed a hysteresis-observer based control to compensate for the piezoelectric actuator.

It should be noted that the piezoelectric actuator possesses a non-symmetrical hysteresis according to a lot of experimental research [20,21]. When the input rate is high, the non-symmetrical characteristic of the piezoelectric actuator is more serious. However, the classical Bouc–Wen model is used to describe a symmetrical hysteresis. When the input frequency or rate is low, the modeling error of the classical Bouc–Wen model is not large. However, when the input frequency or rate is high, its modeling error is large. Therefore, it can be found that the classical Bouc–Wen model is mainly used to characterize the rate-independent hysteresis behavior and modeling, but cannot characterize the rate-dependent hysteresis behavior precisely though it is a rate-dependent hysteresis model according to traditional classifications. The modeling accuracies of published hysteresis models are not high enough. Furthermore, due to the existence of many parameters and differential equations, it is a hard task to identify the parameters of hysteresis models.

In our previous work [22], we have developed an enhanced Bouc–Wen model by introducing input frequency. But there is a limitation that the developed model cannot be applied when the input frequency is unknown. To solve the problems above, this paper proposed a generalized Bouc–Wen (GB–W) model by introducing relaxation functions in the classical Bouc–Wen model, which can characterize both rate-independent and rate-dependent hysteresis behavior for piezoelectric ceramics actuators precisely. A lot of experiments are conducted in advance to characterize hysteresis behaviors and subsequently the relaxation functions are determined cautiously based on these experimental characteristics. The generalized Bouc–Wen (GB–W) model doesn't have the aforementioned limitation and can be widely applied. In addition, the nonlinear least squares method through MATLAB/Simulink is used to identify the corresponding parameters of hysteresis models. Both simulations and experiments finally demonstrate the validity of the developed model. Therein, the classical Bouc–Wen model is set as a comparison. The rest of this paper is arranged as follows: Section 2 gives the descriptions of the classical Bouc–Wen model. In Section 3, the generalized Bouc–Wen model is presented. Section 4 gives the experimental validation of results and discussion. Finally, conclusions are drawn in Section 5.

2. Classical Bouc–Wen Model

The Bouc–Wen model was initially applied for nonlinear vibrational mechanics. With the rapid development of smart actuators, it was gradually used to describe nonlinear hysteresis for piezoelectric actuators. The hysteresis curve can be considered as the superposition of a linear component $X(t)$ and a hysteretic component $h(t)$. The classical hysteretic Bouc–Wen model as a nonlinear system is described as follows:

$$y(t) = X(t) + h(t) = k \cdot u(t) + h(t) \tag{1}$$

$$\dot{h}(t) = \alpha \dot{u}(t) - \beta |\dot{u}(t)||h(t)|^n - \gamma |\dot{u}(t)||h(t)|^{n-1} h(t) \tag{2}$$

where $u(t)$ is the input voltage and $y(t)$ is the output displacement. k, α, β, γ and n are the model parameters, which decide the shape of hysteresis curves. In order to simplify the model, n is usually set as 1 and the hysteretic components is expressed by

$$\dot{h}(t) = \alpha \dot{u}(t) - \beta \dot{u}(t)|h(t)| - \gamma|\dot{u}(t)|h(t) \tag{3}$$

3. Generalized Bouc–Wen Model

To analyze the performance of the classical Bouc–Wen model in detail, some efforts were devoted to research on the characteristics of its parameters. The variations of its parameters k and α at different frequencies of the input are shown in Figures 1 and 2, respectively. Table 1 gives the detailed values of parameters of the classical Bouc–Wen model at different frequencies. The parameters were identified by the nonlinear least squares method through MATLAB/Simulink, which will be introduced in detail in the next part. The identified results based on experimental data clearly reveal that the parameters k and α both decrease with the increase in frequency. Such frequency dependence of the parameters k and α cannot be described by the classical Bouc–Wen model. The parameters of the classical Bouc–Wen model are fixed constants, which cannot characterize their change trend with the increase in frequency. To some degree, thus, it can be concluded that the classical Bouc–Wen model cannot describe rate-dependent hysteresis behaviors.

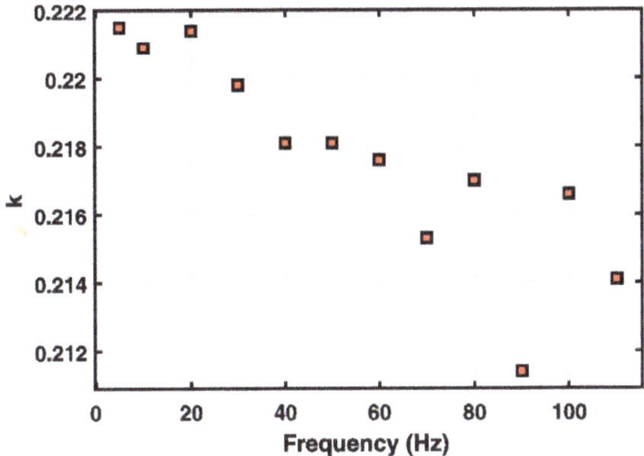

Figure 1. Variations of k under $u(t) = 5\sin(2\pi f t) + 5$ at different frequencies.

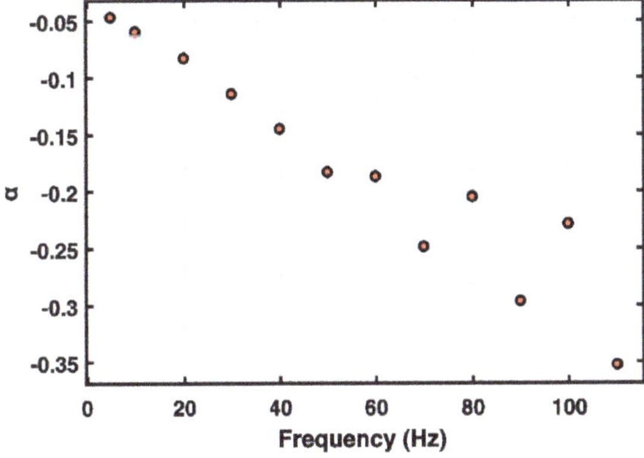

Figure 2. Variations of α under $u(t) = 5\sin(2\pi f t) + 5$ at different frequencies.

Table 1. Identified parameters of the classical Bouc–Wen (CB–W) model at different frequencies.

Frequency (Hz)	k	α	β	γ
5	0.2215	−0.0473	0.0847	0.4477
10	0.2209	−0.0598	0.2629	0.6506
20	0.2214	−0.0832	0.7183	1.1444
30	0.2198	−0.1146	1.5761	2.0662
40	0.2181	−0.1453	2.8847	3.4124
50	0.2181	−0.1838	2.0272	2.6029
60	0.2176	−0.1877	2.4755	3.0932
70	0.2153	−0.2492	4.0013	4.6221
80	0.2170	−0.2055	2.9687	3.5753
90	0.2114	−0.2970	5.9981	6.6169
100	0.2166	−0.2293	3.6008	4.2298
110	0.2141	−0.3534	3.1034	3.7229

3.1. Formulation of the Generalized Bouc–Wen Model

The results above show that the classical Bouc–Wen model cannot describe rate-dependent hysteresis behaviors precisely. It is resulted by a non-symmetrical hysteresis of piezoelectric actuators while the classical Bouc–Wen model is a symmetrical model. Therefore, the classical Bouc–Wen model should be redefined to solve this problem. However, it should be noted that the redefined model should possess the capacity to describe both rate-independent and rate-dependent hysteresis, which is the purpose of this work.

How to formulate a generalized hysteresis model of piezoelectric actuators is still a hard task and some researchers have made some contributions, such as, Al Janaideh et al. [23] presented a generalized P-I model with relaxation functions to characterize rate-dependent hysteresis behaviors. Mayergoyz [24] proposed a generalized Preisach model of hysteresis by introducing a generalized density function. Based on the experimental dynamic characteristics and researchers' experiences above, the generalized Bouc–Wen model is thus formulated upon integrating relaxation functions $k(v(t), \dot{v}(t))$ and $\alpha(v(t), \dot{v}(t))$, such as

$$y(t) = X(t) + h(t) = k(u(t), \dot{u}(t)) \cdot u(t) + h(t) \tag{4}$$

$$\dot{h}(t) = \alpha(u(t), \dot{u}(t)) \cdot \dot{u}(t) - \beta \dot{u}(t)|h(t)| - \gamma|\dot{u}(t)|h(t) \tag{5}$$

where $k(u(t), \dot{u}(t))$ and $\alpha(u(t), \dot{u}(t))$ are relaxation functions of current input $u(t)$ and its rate of $\dot{u}(t)$ and $k(u(t), \dot{u}(t))$ is a positive function. It should be noted that the generalized Bouc–Wen model could describe rate-independent hysteresis behaviors at low input frequency. Therefore, $k(u(t), \dot{u}(t))$ and $\alpha(u(t), \dot{u}(t))$ should converge to fixed constants when the input frequency is low. Based on the characteristics above, the relaxation functions can be defined by inducing the exponential function, such as

$$k(u(t), \dot{u}(t)) = pe^{-q\dot{u}(t)} \tag{6}$$

$$\alpha(u(t), \dot{u}(t)) = \varepsilon e^{\delta|\dot{u}(t)|} \tag{7}$$

where $p \geq 0$, $q \geq 0$, ε, δ, β and γ are constants. From the expressions of the relaxation functions, it can be found that when the input frequency is very low, such that $\dot{u}(t) \cong 0$, the relaxation functions converge to fixed constants, such that $k(u(t), \dot{u}(t)) \cong p$ and $\alpha(u(t), \dot{u}(t)) \cong \varepsilon$. Thus, it can describe the rate-independent hysteresis model the same as the classical Bouc–Wen model.

Compared with the classical Bouc–Wen model, the parameters $k(u(t), \dot{u}(t))$ and $\alpha(u(t), \dot{u}(t))$ of the proposed model values vary with the rate of input, which are not fixed constants any more. Furthermore, the parameters $k(u(t), \dot{u}(t))$ in the rising hysteresis curves at the same input are different from that in the decreasing hysteresis curves. The characteristics above form the non-symmetrical hysteresis of the proposed model.

3.2. Properties of the Generalized Bouc–Wen Model

This section will focus on analyses of the properties of the generalized Bouc–Wen model. First, the characteristics of the hysteretic component $h(t)$ should be analyzed based on simulations. Figure 3 shows the relationship between $h(t)$ and input frequency f under input voltage $u(t) = 5\sin(2\pi f t) + 5$. The corresponding parameters of the generalized Bouc–Wen model are set as $p = 0.2107$, $q = 1.189 \times 10^{-5}$, $\varepsilon = -0.1331$, $\delta = 5.2622 \times 10^{-4}$, $\beta = 5.3743$ and $\gamma = 6.4698$. The results clearly reveal that the width of the hysteretic component $h(t)$ increases monotonically with the increase in the rate of input.

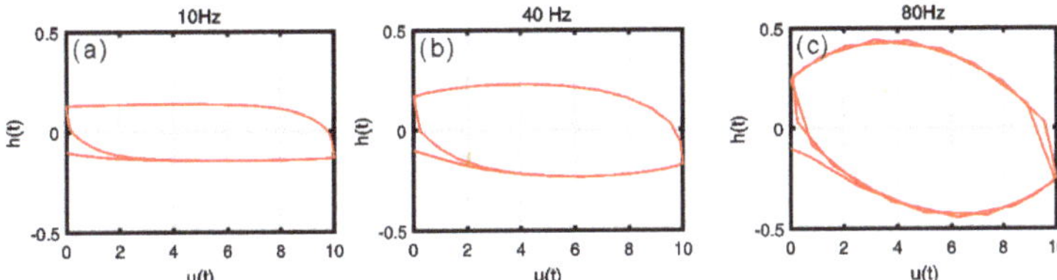

Figure 3. Variations of the component $h(t)$ under $u(t) = 5\sin(2\pi f t) + 5$: (**a**) f = 10 Hz; (**b**) f = 40 Hz; (**c**) f = 80 Hz.

Figure 4 shows the relationship between the component $X(t)$ and input frequency f. The results reveal that the component $X(t)$ is still nearly linear with input voltages at low frequencies. But the curves of the component $X(t)$ have a hysteresis loop with the increase in frequencies, which shows non-symmetrical characteristics.

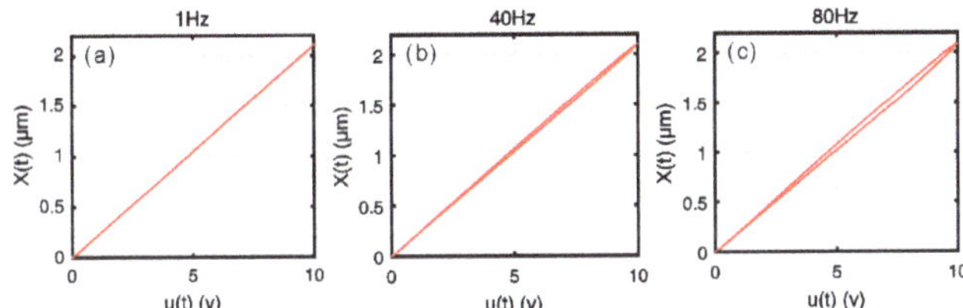

Figure 4. Variations of the component $X(t)$ under $u(t) = 5\sin(2\pi f t) + 5$: (**a**) f = 10 Hz; (**b**) f = 40 Hz; (**c**) f = 80 Hz.

3.3. Parameters Identification

Due to the existence of derivation and more parameters, it is not easy to identify the Bouc–Wen model for most researchers. In this study, the objective function F is expressed by

$$F = Min \sum_{i=1}^{n} f^2(u) \tag{8}$$

with

$$f(u) = y_i - y_i{}^{HM} \tag{9}$$

$$y_i{}^{HM} = X(iT) + h(iT) = pe^{-q\dot{u}(iT)}u(iT) + h(iT) \tag{10}$$

$$\dot{h}(iT) = \varepsilon e^{\delta|\dot{u}(iT)|} \cdot \dot{u}(iT) - \beta\dot{u}(iT)|h(iT)| - \gamma|\dot{u}(iT)|h(iT) \tag{11}$$

where n is the total number of samples, T is the sampling period, $i = 1, 2, 3 \cdots n$ is the ith sampling period, $y_i{}^{HM}$ and y_i are the predicted output by hysteresis model and experimental displacements of piezoelectric actuators, respectively. According to these equations above, it can be concluded that $f(u)$ is a nonlinear function of the parameters p, q, ε, δ, β and γ. It is a nonlinear least squares problem. It is better to choose the nonlinear least squares method instead of the least squares method to identify parameters of the proposed model in our study.

In this paper, the nonlinear least squares method using the Trust-Region-Reflective algorithm is presented to identify the parameters of classical and generalized Bouc–Wen models. This method uses the nonlinear least squares function for optimization through the MATLAB/Simulink Optimization Toolbox. The corresponding identification steps of the nonlinear least squares method is carried out offline as follows:

(1)   Data collection: Experimental data including output displacements and input voltages for piezoelectric actuators are obtained and recorded.

(2)   Model implementation: Classical and generalized Bouc–Wen models are implemented using the MATLAB/Simulink block as shown in Figures 5 and 6, respectively.

(3)   Parameter estimation: The Trust-Region-Reflective algorithm is used to identify the parameters of hysteresis models based on experimental data.

(4)   Validation: Comparison of the measured and simulation results predicted by hysteresis models are shown, and the corresponding modeling errors are plotted.

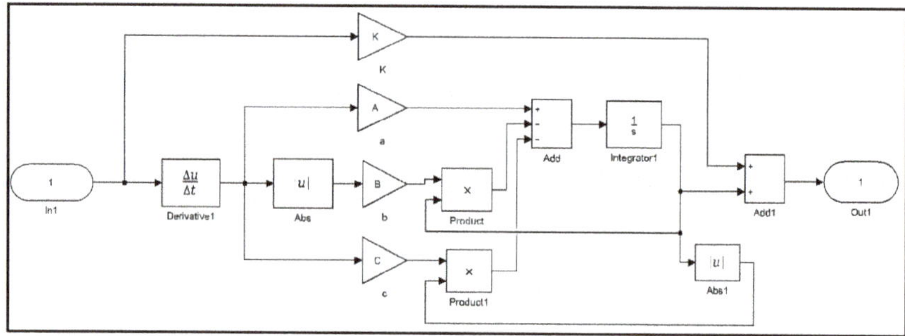

Figure 5. Classical Bouc–Wen (CB–W) model implemented with Matlab/Simulink.

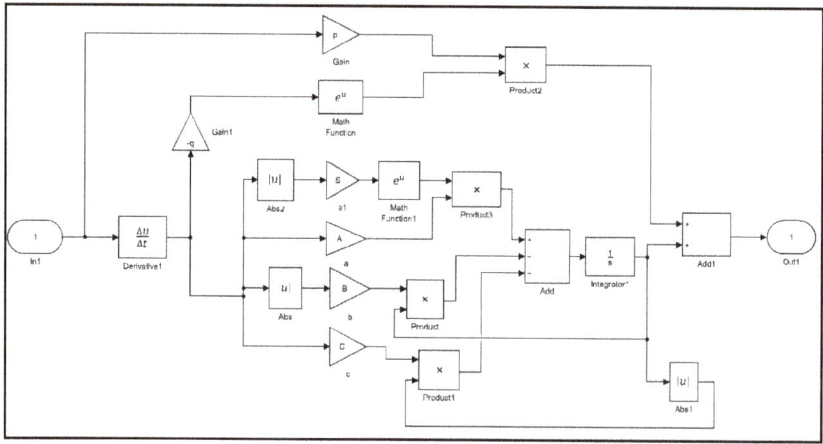

Figure 6. Generalized Bouc–Wen (GB–W) model implemented with MATLAB/Simulink.

In traditional identification methods, the identified procedure is usually a long and complex work with many equations and steps. For example, in reference [15], Zhu and Wang added a non-symmetrical formula based on the classical Bouc–Wen model to describe non-symmetrical hysteresis. The parameters for the linear and hysteresis components are separately identified by utilizing the final value theorem of the Laplace transform and the least squares method, respectively. However, the nonlinear least squares method using the Trust-Region-Reflective algorithm in this study is presented to identify all parameters for the linear and hysteresis components at the same time, which can simplify the identification procedure. In addition, all operations are conducted by using the MATLAB/Simlink optimization tools and there are no complex algorithms needed to write, which can simplify the identification procedure further. Furthermore, there are just four steps above and the whole computing time is controlled in several minutes. Therefore, the nonlinear least squares method can quickly and simply identify the parameters of hysteresis models. Last but not least, this method can be expanded to apply to other complex model identifications, which is useful and meaningful undoubtedly.

4. Experimental Validation Results and Discussion

4.1. Experimental Setup

Experiments were conducted to establish the magnitude of the nonlinearity in the output of the piezoelectric actuator. The experimental setup, as shown in Figure 7, employs a 1-DOF compliant mechanism stage to produce linear motion. The stage is actuated by a stack piezoelectric ceramic actuator (Coremorrow PST 150/7/60VS12, Harbin, China), whose nominal displacement was 60 μm for the maximum input voltage of 150 V. The actuator is a preloaded piezoelectric translator manufactured by Piezomechanik in Germany and made of PZT (Pb-based Lanthanum-doped Zirconate Titanates), whose detailed information is shown in Table 2. This actuator is the superposition of the multilayer piezoelectric ceramic sheets together. The stage was set up on an optical table for vibration isolation and experiments were carried out in a clean room to minimize external noise sources.

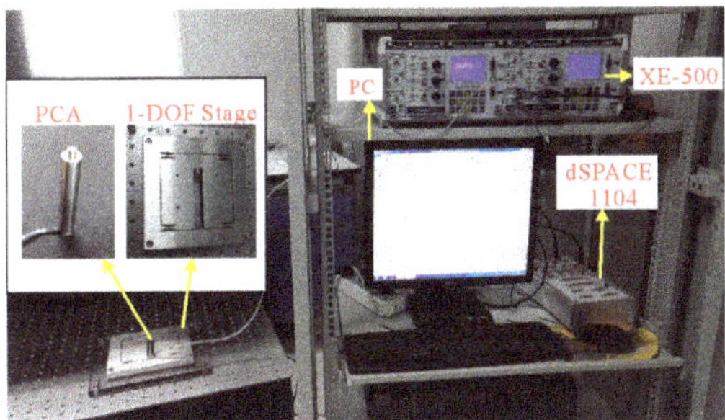

Figure 7. Experimental setup.

Table 2. Information about the piezoelectric ceramics actuator (PCA).

Type	PST 150/7/60VS12
Material	PZT
Length [mm] ±0.3	64
Nominal Thrust/tension [N]	1800/300
Electrical capacitance [μF] ±20%	5.4
Resonant frequency [kHz]	15
Stiffness [N/μm] ±20%	15
Nominal Stroke [μm] ±15%	60

The position of the moving stage was simultaneously measured by the strain gauge position sensor (SGS), which is included in the piezoelectric ceramic actuator. To achieve the real-time control, the dSPACE-DS1104 rapid prototyping controller board equipped with a 16-bit analogue-to-digital converter (ADC) and 16-bit digital-to-analogue converter (DAC) was used as the real-time control system. The output voltage signals were amplified by an amplifier included in XE-500 controller, which provided excitation voltage for the piezoelectric actuator from −20 V to 150 V. The measured displacement signals were obtained by a signal conditioner included in the XE-500 controller. Both the input voltage and output displacement signals were stored by a computer. The control interface is performed by the Control Desk 5.0-dSPACE and MATLAB/Simulink.

4.2. Results

To demonstrate the performance of the generalized Bouc–Wen (GB–W) model for the piezoelectric actuator, hysteresis behaviors in the piezoelectric actuator were measured and the classical Bouc–Wen (CB–W) model was constructed for comparison. In this case, we conducted two groups of experiments to demonstrate the effectiveness of characterizing rate-dependent and rate-independent hysteresis, respectively.

In the first group of experiments, we measured the outputs of the piezoelectric actuators under excitation voltage signals $u(t) = 5\sin(2\pi ft) + 5(f = 5, 10, 20, 40, 60, 80, 90, 100, 110)$ to demonstrate the effectiveness of the GB–W model to characterize rate-dependent hysteresis behaviors. The measured data of excitation signal at 110 Hz is initially adopted to identify the parameters of both generalized and classical Bouc–Wen models. The identified parameters of the generalized Bouc–Wen model are $p = 0.2107$, $q = 1.189e \times 10^{-5}$, $\varepsilon = -0.1331$, $\delta = 5.2622 \times 10^{-4}$, $\beta = 5.3743$ and $\gamma = 6.4698$ Meanwhile, the corresponding parameters of the classical Bouc–Wen model $k = 0.2141$, $\alpha = -0.3534$, $\beta = 3.1034$ and $\gamma = 3.7229$. It must be noted that the input frequency should be controlled under 150 Hz to avoid a high dynamic force of encapsulated stack piezoelectric ceramics for security protection. Therefore, the nine groups of experiments need to follow this rule.

Figure 8 shows comparisons of the measured and simulation results predicted by the generalized and classical Bouc–Wen model. The black, green and red lines represent the experimental data, and the generalized Bouc–Wen model, respectively. It can be found that the simulation results predicted by the generalized Bouc–Wen model agree better with the experimental data than that of the classical Bouc–Wen model. Figure 9 shows the corresponding modeling errors of these models. It is clearly shown that the modeling errors of the generalized Bouc–Wen model is much smaller than that of the classical Bouc–Wen model.

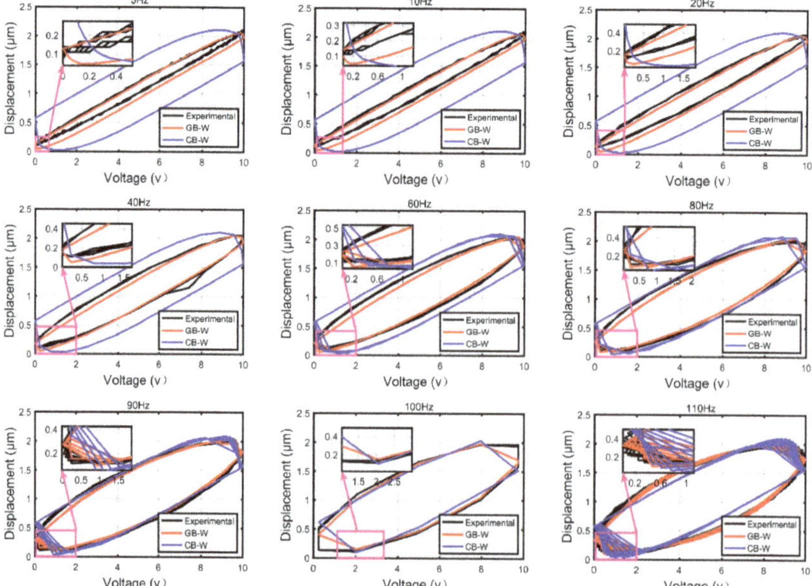

Figure 8. Comparisons of the measured and simulation results predicted by the GB–W model and classical CB–W model.

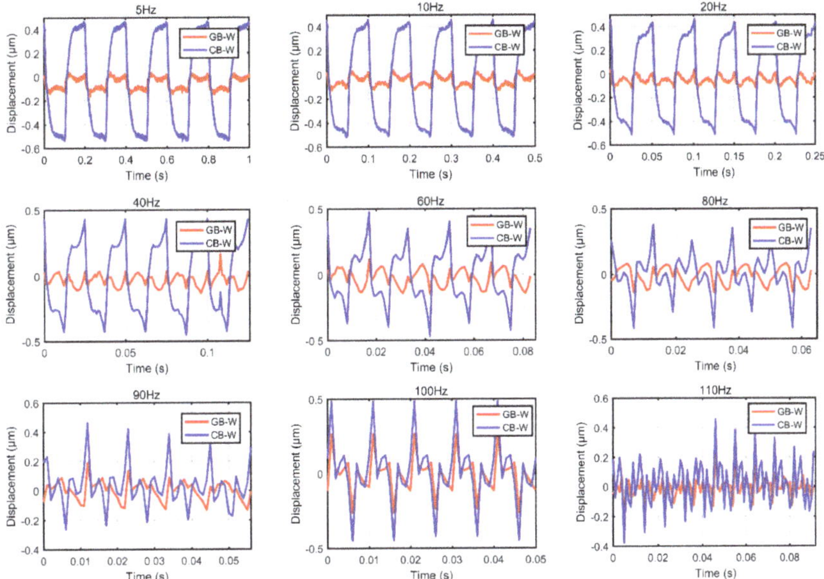

Figure 9. Rate-dependent modeling errors of the GB–W model and classical CB–W model.

In the second group of experiments, two different waveforms of input excitation signals with the amplitude of 10 and 20, respectively, were adopted to actuate the 1-DOF compliant mechanism. These experiments are used to demonstrate the effectiveness of the GB–W model to characterize rate-independent hysteresis behaviors. The parameters of the GB–W model and CB–W model remained the same with the first group of experiments. Figures 10 and 11 show comparisons of the measured and simulation results predicted by the GB–W model and CB–W model. It can be found that the predicted results by the GB–W model agree better with the measured than that by the CB–W model.

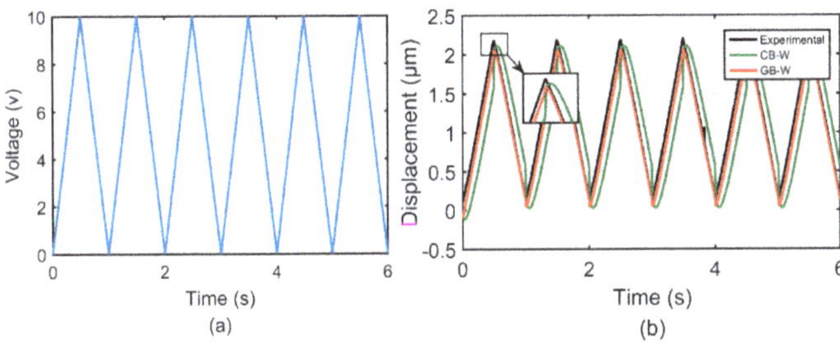

Figure 10. Time histories of: (**a**) A waveform of input excitation signal with the amplitude of 10 and (**b**) the measured and simulation results predicted by the GB–W model and CB–W model.

To qualify modeling errors reasonably, the root-mean-square error e_{rms} (RMSE) and relative root-mean-square error δ (RRMSE) are introduced in this paper as follows:

$$e_{rms} = \sqrt{\frac{1}{T}\int_0^T |y(t) - y_d(t)|^2 dt} \tag{12}$$

$$\delta = \frac{e_{rms}}{\max(y_d(t))} \times 100\% \tag{13}$$

where $y(t)$ and $y_d(t)$ are the simulated and measured displacements, respectively, T is the total time.

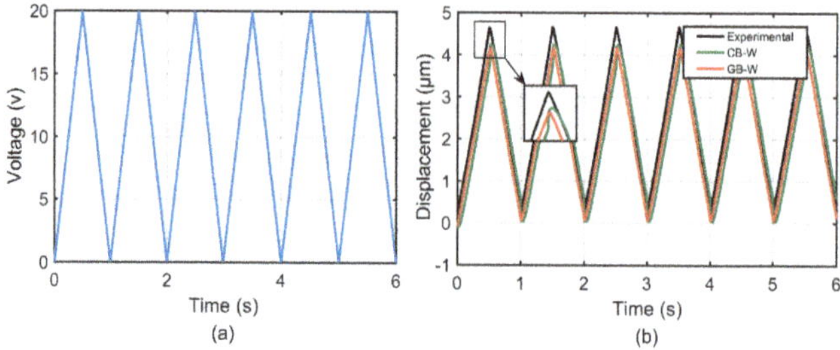

Figure 11. Time histories of: (**a**) A waveform of input excitation signal with the amplitude of 20 and (**b**) the measured and simulation results predicted by the GB–W model and CB–W model.

The detailed modeling errors are shown in Tables 3 and 4. The max displacement is 2.105 μm in the first group of experiments. According to it, in the first group of experiments, RMSE and RRMSE of the GB–W model under excitation signal at 5 Hz are 0.0742 μm and 3.52%, which are reduced by 81.5% compared with that of the CB–W model. When the frequency increases to 100 Hz and 110 Hz, the modeling errors (RMSE and RRMSE) of the GB–W model are reduced by nearly 42.1% and 56.47%, respectively. The results clearly reveal that the generalized Bouc–Wen model can predict the output of rate-dependent hysteresis curves of piezoelectric actuators more precisely. In the second group of experiments, the max measured displacements under two waveforms of input excitation signal are 2.215 μm and 4.676 μm, respectively. RMSE and RRMSE of the GB–W model under a waveform of input excitation signal with the amplitude of 10 are 0.1315 μm and 5.94%, respectively, which are reduced by 70.8% compared with that of the CB–W model. In the other experiment with the amplitude of 20, the rate-independent modeling errors of the GB–W model are still smaller and reduced by 28.9% compared with that of the CB–W model. The results above clearly reveal that the GB–W model can also predict the output of rate-independent hysteresis curves of piezoelectric actuators more precisely. According to the above analyses, it is reasonable to believe that the GB–W model is effective and can characterize both rate-dependent and rate-independent hysteresis behaviors more precisely than the CB–W model.

Table 3. Rate-dependent modeling errors of the GB–W model and CB–W model.

Frequency (Hz)	GB–W Model		CB–W Model	
	RMSE (μm)	RRMSE (%)	RMSE (μm)	RRMSE (%)
5	0.0742	3.52	0.4015	19.07
10	0.0650	3.08	0.3959	18.81
20	0.0700	3.33	0.3420	16.25
40	0.0316	1.50	0.2595	12.32
60	0.0697	3.31	0.2110	10.02
80	0.0718	3.41	0.1608	7.64
90	0.0679	3.23	0.1495	7.10
100	0.1264	6.00	0.2183	10.37
110	0.0673	3.20	0.1546	7.34

Table 4. Rate-independent modeling errors of the GB–W model and CB–W model.

Amplitude	GB–W Model		CB–W Model	
	RMSE (μm)	RRMSE (%)	RMSE (μm)	RRMSE (%)
10	0.1315	5.94	0.4510	20.36
20	0.3869	8.27	0.5441	11.63

4.3. Discussion

The classical Bouc–Wen model is mainly used to characterize the rate-independent hysteresis, but cannot characterize the rate-dependent hysteresis precisely. Compared with the classical Bouc–Wen model, the generalized Bouc–Wen model integrates relaxation functions. $k(u(t), \dot{u}(t))$ and $\alpha(u(t), \dot{u}(t))$ based on the traditional fixed model parameters k and α. The relaxation functions are determined based on experimental characteristics instead of random imaginations and make model parameters closely related to the rate of input $\dot{u}(t)$. Therefore, the proposed model can theoretically describe the rate-dependent hysteresis behaviors more precisely and the experimental and simulation results demonstrate its effectiveness. In addition, the experimental and simulation results also show that the generalized Bouc–Wen model can describe the rate-independent hysteresis behaviors more precisely than the classical Bouc–Wen model. So it can be concluded that the generalized Bouc–Wen model can characterize both rate-dependent and rate-independent hysteresis behaviors.

The enhanced Bouc–Wen model in our previous work [22] is closely related to the input frequency f and cannot be used to describe the rate-dependent hysteresis behaviors when the frequency is unknown. However, the generalized Bouc–Wen model is closely related to the rate of input $\dot{u}(t)$ and can be widely used to describe the rate-dependent hysteresis behaviors without limitations. This is the main advantage of the developed model.

Compared with other existing models such as the Prandtl–Ishlinskii model and Preisach model, which are rate-independent models, the proposed model is a rate-dependent model. In addition, the proposed model has differential equations and the ability to capture an analytical form, which can provide more convenience for hysteresis compensation control. So the developed model has a broader application prospect in hysteresis modeling and compensation controls.

5. Conclusions

In this paper, a generalized Bouc–Wen model is established to characterize both rate-independent and rate-dependent hysteresis behaviors by introducing relaxation functions in the classical Bouc–Wen model. The corresponding parameter was identified by the nonlinear least squares method through Matlab/Simulink. The validity of the developed model is demonstrated by a number of experiments. Comparing the predicted data of the generalized Bouc–Wen model with the experimental data revealed reasonably good agreements. The results showed that the developed model can describe rate-dependent and rate-independent hysteresis behaviors more precisely than the classical Bouc–Wen model. The modeling errors of the generalized Bouc–Wen model can be reduced greatly.

Author Contributions: Conceptualization, J.G. and X.Z.; methodology, J.G. and X.Z.; software, J.G.; validation, J.G.; formal analysis, J.G.; investigation, J.G.; resources, J.G. and X.Z.; data curation, J.G.; writing—original draft preparation, J.G.; writing—review and editing, J.G.; visualization, J.G.; supervision, X.Z.; project administration, J.G. and X.Z.; funding acquisition, J.G.

Funding: This work was supported by the National Natural Science Foundation of China (Grant Nos. 51805494, U1501247 and U1609206), the Fundamental Research Funds for the Central Universities (Grant Nos.CUGL180819), and Open Foundation of Guangdong Provincial Key Laboratory of Precision Equipment and Manufacturing Technology (PEM201702).

Conflicts of Interest: The authors declare no conflict of interest.

References

1. Dong, R.; Tan, Y.; Chen, H.; Xie, Y. A neural networks based model for rate-dependent hysteresis for piezoceramic actuators. *Sens. Actuators A Phys.* **2008**, *143*, 370–376. [CrossRef]
2. Al Janaideh, M.; Krejčí, P. Prandtl-Ishlinskii hysteresis models for complex time dependent hysteresis nonlinearities. *Phys. B Condens. Matter* **2012**, *407*, 1365–1367. [CrossRef]
3. Gan, J.; Zhang, X.; Wu, H. A generalized Prandtl-Ishlinskii model for characterizing the rate-independent and rate-dependent hysteresis of piezoelectric actuators. *Rev. Sci. Instrum.* **2016**, *87*, 35002. [CrossRef] [PubMed]

4.	Bernard, Y.; Maalej, A.H.; Lebrun, L.; Ducharne, B. Preisach modelling of ferroelectric behaviour. *Int. J. Appl. Electromagn.* **2007**, *25*, 729–733. [CrossRef]

5.	Ge, P.; Jouaneh, M. Generalized Preisach model for hysteresis nonlinearity of piezoceramic actuators. *Precis. Eng. J. Am. Soc. Precis. Eng.* **1997**, *20*, 99–111. [CrossRef]

6.	Nguyen, P.; Choi, S.; Song, B. A new approach to hysteresis modelling for a piezoelectric actuator using Preisach model and recursive method with an application to open-loop position tracking control. *Sens. Actuators A Phys.* **2018**, *270*, 136–152. [CrossRef]

7.	Tri, V.M.; Tjahjowidodo, T.; Ramon, H.; Van Brussel, H. A new approach to modeling hysteresis in a pneumatic artificial muscle using the Maxwell-slip model. *IEEE ASME Trans. Mech.* **2011**, *16*, 177–186.

8.	Goldfarb, M.; Celanovic, N. Modeling piezoelectric stack actuators for control of micromanipulation. *Control Syst. IEEE* **1997**, *17*, 69–79.

9.	Oh, J.; Bernstein, D.S. Semilinear Duhem model for rate-independent and rate-dependent hysteresis. *IEEE Trans. Automat. Contr.* **2005**, *50*, 631–645.

10.	Oh, J.; Bernstein, D.S. Piecewise linear identification for the rate-independent and rate-dependent DUHEM hysteresis models. *IEEE Trans. Automat. Contr.* **2007**, *52*, 576–582. [CrossRef]

11.	Sun, L.N.; Ru, C.H.; Rong, W.B.; Chen, L.G.; Kong, M.X. Tracking control of piezoelectric actuator based on a new mathematical model. *J. Micromech. Microeng.* **2004**, *14*, 1439–1444.

12.	Gan, J.; Zhang, X.; Wu, H. Tracking control of piezoelectric actuators using a polynomial-based hysteresis model. *AIP ADV* **2016**, *6*, 65204. [CrossRef]

13.	Zhu, W.; Rui, X. Hysteresis modeling and displacement control of piezoelectric actuators with the frequency-dependent behavior using a generalized Bouc Wen model. *Precis. Eng.* **2016**, *43*, 299–307. [CrossRef]

14.	Rakotondrabe, M. Bouc–Wen modeling and inverse multiplicative structure to compensate hysteresis nonlinearity in piezoelectric actuators. *IEEE Trans. Autom. Sci. Eng.* **2011**, *8*, 428–431. [CrossRef]

15.	Zhu, W.; Wang, D.H. Non-symmetrical Bouc-Wen model for piezoelectric ceramic actuators. *Sens. Actuat. A Phys.* **2012**, *181*, 51–60. [CrossRef]

16.	Fujii, F.; Tatebatake, K.I.; Morita, K.; Shiinoki, T. A Bouc–Wen model-based compensation of the frequency-dependent hysteresis of a piezoelectric actuator exhibiting odd harmonic oscillation. *Actuators* **2018**, *7*, 37. [CrossRef]

17.	Li, Y.; Xu, Q. Adaptive sliding mode control with perturbation estimation and PID sliding surface for motion tracking of a piezo-driven micromanipulator. *IEEE Trans. Contr. Syst. Tech.* **2010**, *18*, 798–810. [CrossRef]

18.	Liu, Z.; Lai, G.; Zhang, Y.; Chen, C.L.P. Adaptive neural output feedback control of output-constrained nonlinear systems with unknown output nonlinearity. *IEEE Trans. Control Syst. Tech.* **2015**, *26*, 1789–1802.

19.	Lin, C.; Yang, S. Precise positioning of piezo-actuated stages using hysteresis-observer based control. *Mechatronics* **2006**, *16*, 417–426. [CrossRef]

20.	Qin, Y.; Zhao, X.; Zhou, L. Modeling and Identification of the rate-dependent hysteresis of piezoelectric actuator using a modified Prandtl-Ishlinskii model. *Micromachines* **2017**, *8*, 114. [CrossRef]

21.	Gu, G.; Yang, M.; Zhu, L. Real-time inverse hysteresis compensation of piezoelectric actuators with a modified Prandtl-Ishlinskii model. *Rev. Sci. Instrum.* **2012**, *83*, 65106. [CrossRef]

22.	Gan, J.; Zhang, X. An enhanced Bouc-Wen model for characterizing rate-dependent hysteresis of piezoelectric actuators. *Rev. Sci. Instrum.* **2018**, *89*, 115002. [CrossRef]

23.	Al Janaideh, M.; Su, C.Y.; Rakheja, S. Development of the rate-dependent Prandtl-Ishlinskii model for smart actuators. *Smart Mater. Struct.* **2008**, *17*. [CrossRef]

24.	Mayergoyz, I.D.; Friedman, G. Generalized Preisach model of hysteresis. *IEEE Trans. Magn.* **1988**, *24*, 212–217. [CrossRef]

Article

A Piezoelectric Resonance Pump Based on a Flexible Support

Jiantao Wang [1,2,*], **Xiaolong Zhao** [3], **Xiafei Chen** [3] **and Haoren Yang** [3]

1   Postdoctoral Mobile Station of Mechanical Engineering, Yanshan University, Qinhuangdao 066004, China
2   School of Vehicle and Energy, Yanshan University, Qinhuangdao 066004, China
3   School of Mechanical Engineering, Yanshan University, Qinhuangdao 066004, China;
    xlzhao@stumail.ysu.edu.cn (X.Z.); chenxiafi@163.com (X.C.); yanghaoren9@163.com (H.Y.)
*   Correspondence: wjt@ysu.edu.cn; Tel.: +86-335-8074682

Received: 12 February 2019; Accepted: 25 February 2019; Published: 28 February 2019

Abstract: Small volume changes are important factors that restrict the improvement of the performance of a piezoelectric diaphragm pump. In order to increase the volume change of the pump chamber, a square piezoelectric vibrator with a flexible support is proposed in this paper and used as the driving unit of the pump. The pump chamber diaphragm was separated from the driving unit, and the resonance principle was used to amplify the amplitude of the pump diaphragm. After analyzing the working principle of the piezoelectric resonance pump and establishing the motion differential equation of the vibration system, prototypes with different structural parameters were made and tested. The results show that the piezoelectric resonance pump resonated at 236 Hz when pumping air. When the peak-to-peak voltage of the driving power was 220 V, the amplitude of the diaphragm reached a maximum value of 0.43933 mm, and the volume change of the pump was correspondingly improved. When the pump chamber height was 0.25 mm, the output flow rate of pumping water reached a maximum value of 213.5 mL/min. When the chamber height was 0.15 mm, the output pressure reached a maximum value of 85.2 kPa.

Keywords: square piezoelectric vibrator; resonance; piezoelectric diaphragm pump; flexible support; piezoelectric resonance pump

---

1. Introduction

The piezoelectric diaphragm pump, which uses a piezoelectric actuator as a drive unit, is a common form of positive displacement pump. This type of pump has the advantages of simple structure, easy miniaturization, zero electromagnetic interference and low noise, so it is widely used in many fields such as biomedicine [1–3], fuel cells [4–6], cooling systems [7–9], and household appliances [10].

A typical piezoelectric diaphragm pump uses a circular piezoelectric vibrator to directly drive fluid. In order to construct a closed volume pump chamber, the vibrator must be sealed to the periphery of the pump body. This installation limits the deformation of the vibrator, making it difficult for the pump chamber to obtain a large volume change. To eliminate the limitations on the structure and installation, researchers have separated the piezoelectric actuator from the diaphragm of the pump chamber to make an indirect drive piezoelectric diaphragm pump [9,11,12]. Meanwhile, the resonance principle is used to amplify the vibration displacement of the piezoelectric actuator to drive the diaphragm—and to increase the volume change of the pump [12]—in order to form a piezoelectric resonance diaphragm pump (hereafter referred to as a piezoelectric resonance pump).

The driving device of the early piezoelectric resonance pump was mainly composed of a piezoelectric stack [5,9,13–15]. It had a large driving force and high precision, but the manufacturing process was complex, with the features of high cost, large volume and high replacement cost if damaged. In recent years we have seen an increase in studies of piezoelectric resonance pumps. These studies mainly used piezoelectric vibrators in different shapes, such as circular, annular and rectangular vibrators, as the driving device of the pump.

In 2012, Jilin University developed a gas piezoelectric resonance pump with a large pump chamber compression ratio, using an annular bimorph piezoelectric vibrator as the driving device [11]. In 2014, for the precise transportation of chemical fuel cells, the University of Science and Technology of China developed a piezoelectric resonance pump with a flexible fluid buffer cavity at the outlet and inlet of the pump chamber. The pump used the rectangular piezoelectric vibrator with free ends as its driving device [16] and used the vibration inertia of the vibrator to drive the movement of the diaphragm in the pump chamber. In 2015, the University of Science and Technology of China also designed a piezoelectric resonance pump with double pump chamber units [17]. It consisted of a U-shaped piezoelectric actuator and two diaphragm pump chamber units that were symmetrically arranged. The U-shaped actuator had very good displacement driving capability in the resonant state. In 2016, National Taiwan University developed a split-type piezoelectric resonance pump used in the medical industry [1]. The circular bimorph piezoelectric vibrator drove the diaphragm through the middle strut. The pump chamber unit could be freely disassembled and replaced, reducing the cost of disposals.

Piezoelectric actuators can be divided into several types according to their structures, such as unimorph/bimorph piezoelectric vibrators [10,11], piezoelectric stacks [5], cylindrical piezoelectric actuators [18], cymbal-shaped piezoelectric actuators [19] and so on. These piezoelectric actuators are based on the strain induced in the longitudinal or transverse directions. The unimorph/bimorph piezoelectric vibrator has the characteristics of simple structure, low cost and various forms. Specifically, the annular and the circular piezoelectric vibrators mainly adopt a periphery fixed installation method, which suits a compact piezoelectric resonance pump system [20]. In contrast, the rectangular piezoelectric vibrator is more flexible in installation. For example, it can be fixed at both ends, at one end, or with both free ends. This method is suitable for a piezoelectric resonance pump system with a large displacement output [12]. After analyzing the advantages and characteristics of common piezoelectric vibrators, a square vibrator with a flexible support is proposed in this paper. The vibrator's substrate adopts a hollow design, with only four corners fixed, to form the flexible support. This structure can reduce the constraint of the support method on the vibration of the square vibrator, thereby improving the vibrator's performance on the displacement output. In this study, the abovementioned piezoelectric vibrator was used as the driving device to construct the piezoelectric resonance pump. We also analyzed and calculated a dynamic model of the pump's vibration and conducted experimental research on the prototype.

2. Development of the Piezoelectric Resonance Pump

2.1. Design Concept

Figure 1a shows the overall structure of the piezoelectric resonance pump, which includes a piezoelectric actuator unit and a pump chamber unit. A square piezoelectric vibrator is the core of the actuator unit, and can periodically vibrate when driven by the sine wave signal. Under it is the elastic vibration transfer block, which is used to amplify the vibrator's amplitude to drive the pump chamber diaphragm. The block is also the stiffness and mass adjusting element of the pump vibration system.

The pump body, processed by polymethyl methacrylate (PMMA), and the circular elastic diaphragm, made from beryllium bronze, were glued together to form an airtight pump chamber unit. The inlet and outlet of the pump employed wheeled check valves. The valve piece was cut from beryllium bronze by ultraviolet laser and bonded to the inlet and outlet seats. Due to the one side limit, the wheeled valve at the inlet only opened to the inside of the pump chamber, and the outlet wheeled valve only to the outside. Figure 1b is the assembly view of the piezoelectric resonance pump.

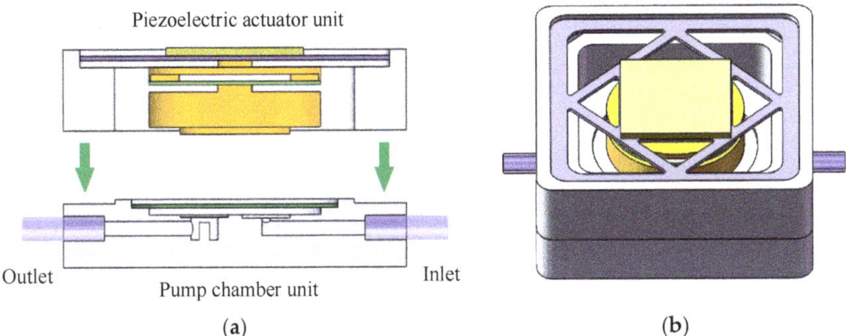

Figure 1. Structure of the piezoelectric resonance pump; (**a**) cross-sectional view; (**b**) assembly view.

Figure 2 is the schematic of the piezoelectric resonance pump when pumping liquid. The square piezoelectric vibrator periodically vibrates when it is driven by the sine wave signal. In resonance, the center amplitude of the vibrator can be amplified by the elastic vibration transfer block. When the vibrator moves upward, the pump chamber diaphragm moves upwards as well, under the driving force of the block. The pressure in the pump chamber is lowered with the inlet valve open and the outlet valve closed, at which point liquid flows into the pump chamber through the inlet. Conversely, when the vibrator moves downward, the pump chamber diaphragm moves downwards under the action of the elastic vibration transfer block. The pressure in the pump chamber increases with the inlet valve closed and the outlet valve open, at which point liquid flows out through the outlet.

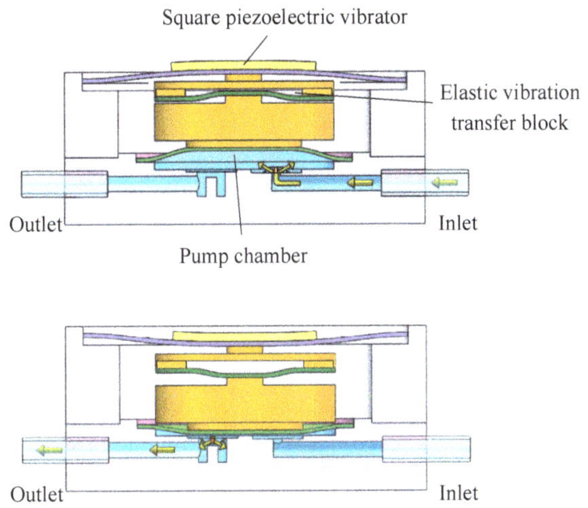

Figure 2. Schematic diagram of pump operation.

2.2. Design of the Piezoelectric Vibrator with Flexible Support

Figure 3 is a structural diagram of the square piezoelectric vibrator designed in this paper. It consists of a 60Si2Mn substrate and a square piezoelectric layer, bonded together. The metal substrate adopts a hollow design, and its four corners are fixed on the vibrator holder, thereby forming a structure with flexible support for the square piezoelectric vibrator. This structure can reduce the vibration constraint of the vibrator and concentrate the deformation in the hollow area during the vibration. The metal substrate with flexible support can improve the mechanical properties of the piezoelectric vibrator and protect the piezoelectric layer.

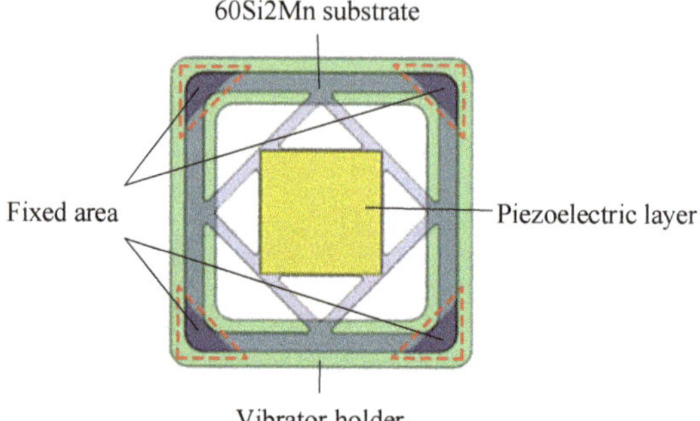

Figure 3. Piezoelectric vibrator based on a flexible support.

3. Dynamic Model

The piezoelectric resonance pump uses system resonance to amplify the amplitude of the piezoelectric vibrator and drive the pump chamber diaphragm to improve the volume change of the pump chamber and optimize the pump's output performance. The dynamic model of the vibration system is established in this section to study the influence of system parameters on the natural frequency and amplitude amplification coefficient. Figure 4 shows the simplified dynamic model of the piezoelectric resonance pump. M_{act} is the equivalent mass of the square piezoelectric vibrator; M_{dia} is the equivalent mass of elastic vibration transfer block and pump chamber diaphragm; K_{act} is the equivalent stiffness of the square piezoelectric vibrator; K_{tra} is the equivalent stiffness of the elastic vibration transfer block; and K_{dia} is the equivalent stiffness of the pump chamber diaphragm. The interaction between the fluid and the flow channel is equivalent to damping C, ignoring the material damping of the elastic element in the vibration system.

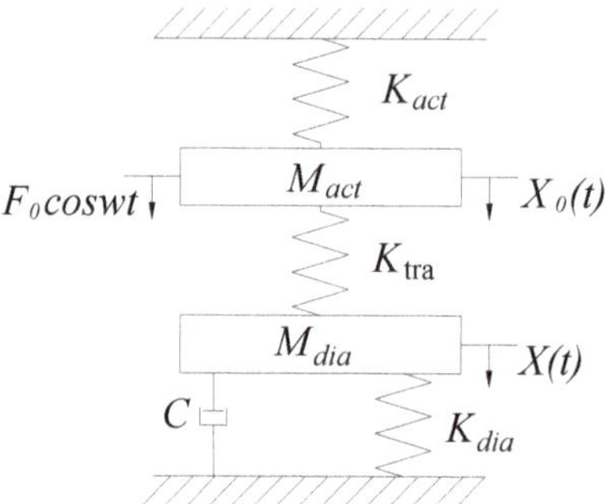

Figure 4. Dynamic model of the piezoelectric resonance pump.

The square piezoelectric vibrator uses a sine wave driving power, and w is the frequency of the driving power. If F_0 is the amplitude of the output force of the square piezoelectric vibrator, then assume that the vibration displacement and output force at time t are $X_0(t)$ and F_0coswt respectively, and $X(t)$ is the vibration displacement of the pump chamber diaphragm. Thus, the motion differential equation of the system is

$$\begin{cases} M_{act}\ddot{X}_0 + K_{act}X_0 + K_{tra}(X_0 - X) = F_0 \cos \omega t \\ M_{dia}\ddot{X} + C\dot{X} - K_{tra}(X_0 - X) + K_{dia}X = 0 \end{cases} \tag{1}$$

In Equation (1), $F_0 \cos \omega t$ is the output force from the center of the square piezoelectric vibrator. If the vibrator is considered as an ideal spring that provides vibration power, the mass of M_{act} can be ignored, so the motion differential equation can be converted to

$$M_{dia}\ddot{X} + C\dot{X} + \frac{K_{act}K_{tra} + K_{act}K_{dia} + K_{tra}K_{dia}}{K_{act} + K_{tra}} X = \frac{K_{tra}F_0 \cos \omega t}{K_{act} + K_{tra}}. \tag{2}$$

The natural frequency of the system can be obtained from Equation (2):

$$\omega_n = \sqrt{\frac{K_{act}K_{tra} + K_{act}K_{dia} + K_{act}K_{dia}}{M_{dia}(K_{act} + K_{tra})}}. \tag{3}$$

$\zeta = \frac{C}{2M_{dia}\omega_n}$ is the damping ratio. The steady-state response amplitude of the system is obtained by using Laplace transform to Equation (2):

$$X = \frac{K_{tra}}{K_{act}K_{tra} + K_{act}K_{dia} + K_{act}K_{dia}} \cdot \frac{F_0}{\left[1 - \left(\frac{\omega}{\omega_n}\right)^2\right]^2 + \left(2\zeta\frac{\omega}{\omega_n}\right)^2}. \tag{4}$$

The deformation under the static force F_0 is

$$\delta_{st} = \frac{K_{tra}F_0}{K_{act}K_{tra} + K_{act}K_{dia} + K_{act}K_{dia}}. \tag{5}$$

Then, the amplitude amplification coefficient of the system is calculated by Equations (4) and (5):

$$\frac{X}{\delta_{st}} = \frac{1}{\sqrt{\left[1 - \left(\frac{\omega}{\omega_n}\right)^2\right]^2 + \left(2\zeta\frac{\omega}{\omega_n}\right)^2}}. \tag{6}$$

According to Equation (6), when the driving frequency fulfills $\omega = \omega_n$, the system resonates, with the amplitude amplification coefficient reaching maximum. Due to the damping of the vibration system, the amplification coefficient is limited.

Wheeled check valves with different structural parameters all have the feature of response hysteresis. The higher the driving frequency of the system, the more obvious the hysteresis, which will reduce the working efficiency of the wheeled check valve and lower the performance of the pump. Therefore, it is vitally important to select the appropriate system operating frequency to improve output performance. We can make a preliminary calculation on the resonance frequency, according to Equation (3). On this basis, considering the feature of hysteresis, the appropriate structural parameters for the wheeled check valve can be determined to improve the output performance.

4. Prototype Fabrication and Experimental Device

4.1. Experimental Prototypes

Figure 5a shows the key structural parameters of the piezoelectric resonance pump. D_t is the diameter of the fixed connection part between the lower vibration transfer block and the pump chamber diaphragm, hereinafter referred to as "fixed connection diameter". D_c is the diameter of the pump chamber, h_r is the height of the chamber and d_o is the diameter of the flow channel. D_i and D_o are the inner and outer diameters of the annular elastic gasket in the middle of the elastic vibration transfer block, and h_k is the thickness of the annular elastic gasket. Figure 5b shows the key structural parameters of the wheeled check valves at the inlet and outlet of the pump. d_s is the outer diameter of the valve, d_m is the outer diameter of the moving disc, and k_v indicates the stiffness of the valve in the opening direction. These structural parameters, which have a direct impact on the output performance of the piezoelectric resonance pump, were tested experimentally in the following section. The main parameters for the prototype designed are shown below in Table 1.

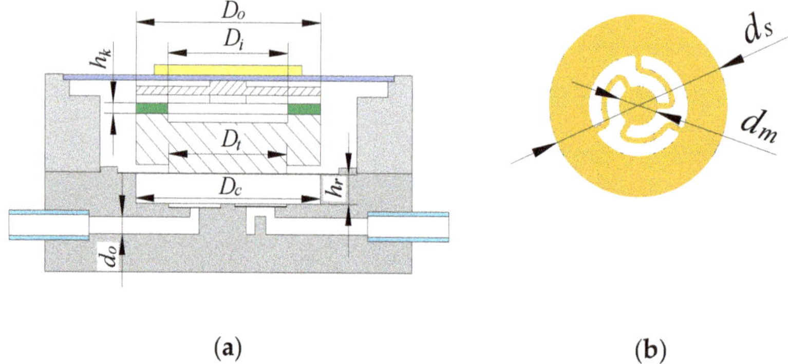

(**a**)   (**b**)

Figure 5. Key structural parameters of the piezoelectric resonance pump. (**a**) Main structure of the pump; (**b**) Structure of the wheeled check valve.

Table 1. The main structural parameters of the prototype.

Structural Parameters	Values
Fixed connection diameter D_t (mm)	5, 10, 15
Diameter of the pump chamber D_C (mm)	25
Height of the pump chamber h_r (mm)	0.15, 0.2, 0.25, 0.3, 0.35
Diameter of the flow channel d_o (mm)	2.1
Inner diameter of the annular elastic gasket D_i (mm)	17
Outer diameter of the annular elastic gasket D_o (mm)	25
Thickness of the annular elastic gasket h_k (mm)	0.3
Outer diameter of the wheeled check valve d_s (mm)	7
Outer diameter of the moving disc d_m (mm)	1.5

Figure 6 is a photo of the piezoelectric resonance pump designed in this paper. Its overall dimensions were 50 mm × 50 mm × 20 mm. The material of the pump body was polymethyl methacrylate (PMMA), which is highly transparent and convenient to observe the working status of the wheeled check valve. The diaphragm and the valve were made of beryllium bronze sheets, and the annular elastic gasket was made of 60Si2Mn. The 60Si2Mn is a kind of silicomanganese alloy spring steel and commonly used as the material for piezoelectric vibrator substrates. Beryllium bronze has features such as high strength, hardness, elastic limit and fatigue limit, and has a small elastic hysteresis.

Figure 6. Photo of the piezoelectric resonance pump.

4.2. Experimental Device

Figure 7 shows the performance testing device for the piezoelectric resonance pump. The SDVC-40 driving power source could generate sine wave driving signals, with a peak-to-peak voltage ranging from 0 to 220 V and a driving frequency ranging from 40 to 400 Hz. Water, as the pumped medium, was heated to 60 °C and the temperature was kept constant by a thermostatic water bath. The inlet and outlet tubes were placed horizontally on the bench, so the back pressure of the pump was zero. The flow rate of the piezoelectric resonance pump was measured by the weighing method, and the output pressure was measured by the digital manometer. The laser micrometer was used to measure the amplitudes of the piezoelectric vibrator and the diaphragm. Figure 8 is a photograph of the experimental device of the piezoelectric resonance pump. In order to lower the measurement error of the experimental data, each prototype was measured four times and the average value calculated.

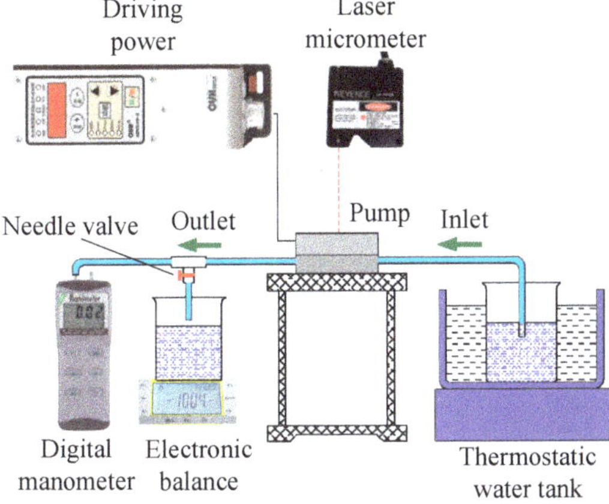

Figure 7. Schematic of the experimental device.

Figure 8. Photograph of the experimental device.

5. Results and Analysis

5.1. Experiments on Amplitude–Frequency Characteristics

Experimental prototypes were made for the performance tests. The amplitude–frequency characteristics of the pump vibration system were studied without water pumping. To begin with, we set the peak-to-peak voltage of the driving power as 220 V. Then we changed the driving frequency and measured the amplitude of the square piezoelectric vibrator and the pump chamber diaphragm by laser micrometer. The test results are shown in Figure 9; as the driving frequency increased, the amplitudes of the vibrator and the diaphragm both initially increased, but then declined. When the driving frequency was 236 Hz, the amplitudes of the vibrator and the diaphragm reached the maximum values of 0.21808 mm and 0.43933 mm, respectively. At this point, the system resonated.

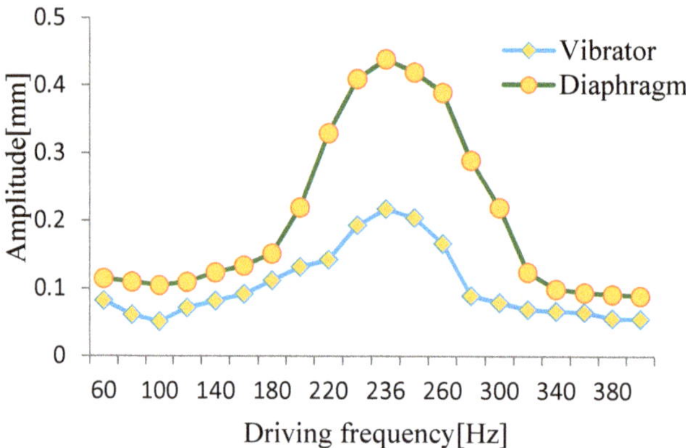

Figure 9. Amplitude–frequency characteristics of the pump vibration system.

5.2. Optimization Test of the Fixed Connection Diameter

When the diameter of the pump chamber diaphragm was constant, the fixed connection diameter (D_t) between the diaphragm and the transfer block can make a difference in the pump's output capability. We chose the fixed connection diameter D_t as 5 mm, 10 mm and 15 mm, respectively, to make prototypes to carry out three tests. The test results of the output flow rate and output pressure were recorded, as shown in Figure 10. The results showed that the piezoelectric resonance pump had the best output performance when the fixed connection diameter was 10 mm, with a flow rate of 189 mL/min and an output pressure of 63 kPa at this point.

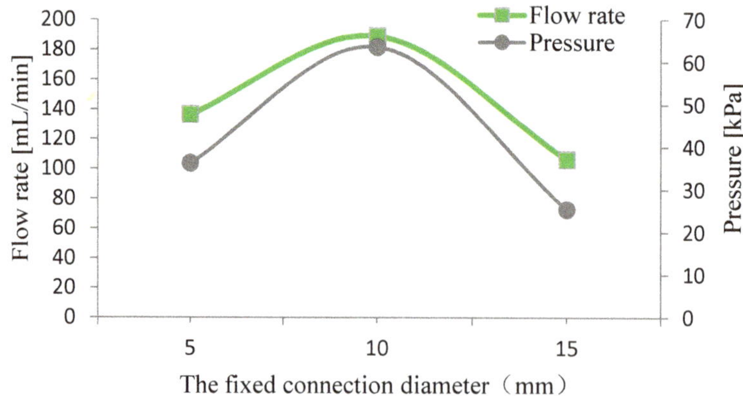

Figure 10. Output performance under different fixed connection diameters.

5.3. Optimization Test of the Pump Chamber Height

Another important parameter of the pump is chamber height. If the chamber is too high, it will reduce the liquid compression ratio, while being too low will increase the flow resistance of the pump chamber. Therefore, we made five prototypes with different chamber heights to conduct tests, with the heights 0.15 mm, 0.20 mm, 0.25 mm, 0.30 mm and 0.35 mm. First, we opened the needle valve and tested the output flow rate of the prototypes without load. It was found that the resonance frequencies of the five prototypes differed little, ranging from 190 Hz to 213 Hz. When closing the needle valve and testing the output pressure, the resonance frequencies of the five prototypes also differed little, within the range of 170 Hz to 186 Hz. However, this was quite different from the situation when the prototypes were unloaded. The reason for this phenomenon is that the working load affects the stiffness and damping of the vibration system, thereby changing its resonance frequency.

The test results of the five prototypes with different chamber heights are shown in Figure 11. Comparing the trends between curves in the chart, it can be seen that the flow rate of the resonance pump first increased and then fell with the increase of the pump chamber height. When the height was 0.25 mm, the pump's flow rate reached a maximum of 213.5 mL/min. Conversely, the output pressure of the resonance pump decreased as the chamber height increased, achieving the maximum of 85.2 kPa with a height of 0.15 mm. Increasing the chamber height meant lowering the liquid compression ratio in the pump chamber, and thus the output pressure was reduced.

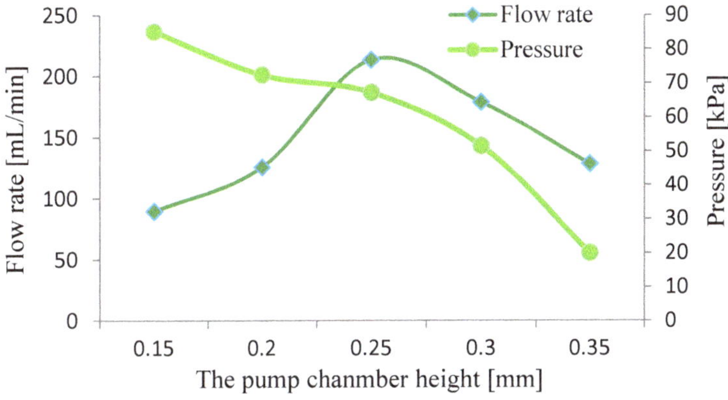

Figure 11. Output performance under different chamber heights.

5.4. Optimization Test of the Wheeled Check Valve

The response of the wheeled check valve has hysteresis, and the key factors affecting this hysteresis include the working frequency of the pump and the stiffness of the check valve itself. For a piezoelectric resonance pump working at a specific resonance frequency, it is essential to select a wheeled check valve with appropriate stiffness to improve the pump's output performance.

We used wheeled check valves with different stiffnesses to make prototypes, and found the appropriate stiffness by testing. Two valves, with a stiffness of 153 N/m and 359 N/m, were chosen for comparison, and the results are shown in Figures 12 and 13. For the prototype with the valve stiffness of 153 N/m, the output flow rate and output pressure changed with the driving frequency. Two peaks were formed, one near the optimal working frequency of the wheeled check valve and the other near the resonance frequency of the pump. For the other prototype of 359 N/m, the flow rate and output pressure also changed with the driving frequency, but only one peak appeared near the pump's resonance frequency. The above tests show that selecting a wheeled check valve with appropriate stiffness can improve the output performance of the piezoelectric resonance pump.

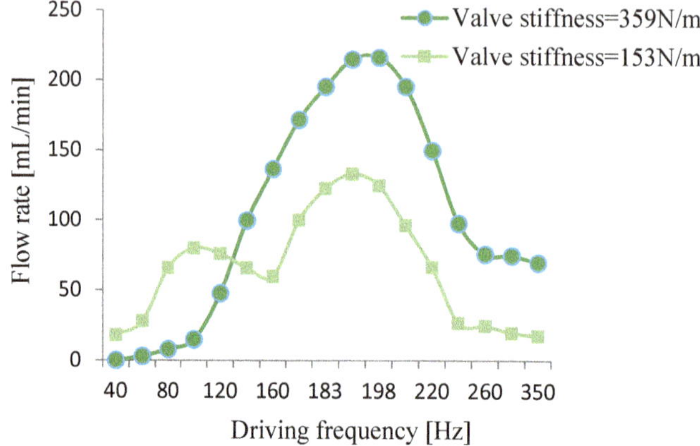

Figure 12. Relation curve between flow rate and valve stiffness.

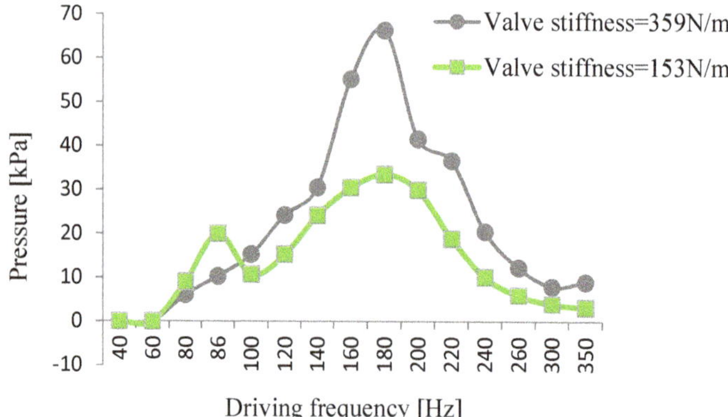

Figure 13. Relation curve between pressure and valve stiffness.

6. Conclusions

This paper discussed a piezoelectric resonance pump with a flexible support, which took a square piezoelectric vibrator as the driving unit. After analyzing the working principle and establishing a dynamic model of the vibration system, prototypes with different structural parameters were made and tested. Finally, the experimental results were provided to show several performance indexes to optimize the pump system.

A vibration test of the piezoelectric resonance pump was carried out. The results show that the pump resonated at 236 Hz when pumping air. When the peak-to-peak voltage of the driving power was 220 V, the amplitude of the diaphragm reached a maximum value of 0.43933 mm, and the volume change of the pump improved. This suggests that the flexible supported square piezoelectric vibrator can effectively reduce the constraints of fixed installation with its own vibration.

For the results of the performance test of the piezoelectric resonance pump show that when the pump chamber height was 0.25 mm, the output flow rate reached the maximum value of 213.5 mL/min. When the chamber height was 0.15 mm, the output pressure reached the maximum value of 85.2 kPa. This indicates that the chamber height directly affected the liquid compression ratio in the pump chamber. Increasing the chamber height means that the liquid compression ratio decreased, thereby reducing the output pressure.

Finally, the check valve comparison test showed that it is vital to select a wheeled check valve with suitable stiffness according to the frequency for a piezoelectric resonance pump working at a specific resonance frequency. An appropriate check valve can improve the pump's output performance.

Author Contributions: J.W. designed the concept and organized the study. X.Z. conducted the experiment and analyzed the experiment results. X.C. completed the assembly and production of the prototype and assisted in the prototype test. H.Y. assisted in the manuscript preparation. All of the authors contributed in writing the manuscript.

Funding: This research was funded by Doctoral Fund of Yanshan University, grant number BL18052, and China Postdoctoral Science Foundation, grant number 2018M631765.

Acknowledgments: This study was supported by the Doctoral Fund of Yanshan University (Grant No.BL18052).

Conflicts of Interest: The authors declare no conflict of interest.

References

1. Ma, H.K.; Chen, R.H.; Yu, N.S.; Hsu, Y.H. A miniature circular pump with a piezoelectric bimorph and a disposable chamber for biomedical applications. *Sensors Actuators A—Phys.* **2016**, *251*, 108–118. [CrossRef]
2. Liu, G.J.; Shen, C.L.; Yang, Z.G.; Cai, X.X.; Zhang, H.H. A disposable piezoelectric micropump with high performance for closed-loop insulin therapy system. *Sensors Actuators A—Phys.* **2010**, *163*, 291–296. [CrossRef]
3. Jenke, C.; Rubio, J.P.; Kibler, S.; Haefner, J.; Richter, M.; Kutter, C. The Combination of Micro Diaphragm Pumps and Flow Sensors for Single Stroke Based Liquid Flow Control. *Sensors* **2017**, *17*, 755. [CrossRef] [PubMed]
4. Yang, X.; Zhou, Z.Y.; Cho, H.J.; Luo, X.B. Study on a PZT-actuated diaphragm pump for air supply for micro fuel cells. *Sensors Actuators A—Phys.* **2006**, *130*, 531–536. [CrossRef]
5. Park, J.H.; Seo, M.Y.; Ham, Y.B.; Yun, S.N.; Kim, D.I. A study on high-output piezoelectric micropumps for application in DMFC. *J. Electroceram.* **2013**, *30*, 102–107. [CrossRef]
6. Ma, H.K.; Huang, S.H.; Kuo, Y.Z. A novel ribbed cathode polar plate design in piezoelectric proton exchange membrane fuel cells. *J. Power Sources* **2008**, *185*, 1154–1161. [CrossRef]
7. Ma, H.K.; Hou, B.R.; Lin, C.Y.; Gao, J.J. The improved performance of one-side actuating diaphragm micropump for a liquid cooling system. *Int. Commun. Heat Mass Transf.* **2008**, *35*, 957–966. [CrossRef]
8. Ma, H.K.; Chen, B.R.; Gao, J.J.; Lin, C.Y. Development of an OAPCP-micropump liquid cooling system in a laptop. *Int. Commun. Heat Mass Transf.* **2009**, *36*, 225–232. [CrossRef]
9. Lyndon, B. Diaphragm pump with resonant piezoelectric drive, 2007. Available online: http://www.techbriefs.com/component/content/article/2182 (accessed on 25 February 2019).
10. Wang, C.-P.; Wang, G.-B.; Ting, Y. Analysis of dynamic characteristics and cooling performance of ultrasonic micro-blower. *Microelectron. Eng.* **2018**, *195*, 1–6. [CrossRef]
11. Wu, Y.; Liu, Y.; Liu, J.; Wang, L.; Jiao, X.; Yang, Z. An improved resonantly driven piezoelectric gas pump. *J. Mechan. Sci. Technol.* **2013**, *27*, 793–798. [CrossRef]
12. Wang, J.T.; Liu, Y.; Shen, Y.H.; Chen, S.; Yang, Z.G. A Resonant Piezoelectric Diaphragm Pump Transferring Gas with Compact Structure. *Micromachines* **2016**, *7*, 219. [CrossRef] [PubMed]
13. Lee, D.G.; Or, S.W.; Carman, G.P. Design of a piezoelectric-hydraulic pump with active valves. *Intell. Mater. Syst. Struct.* **2004**, *15*, 107–115. [CrossRef]
14. Park, J.H.; Yoshida, K.; Yokota, S. Resonantly driven piezoelectric micropump—Fabrication of a micropump having high power density. *Mechatronics* **1999**, *9*, 687–702. [CrossRef]
15. Park, J.H.; Yoshida, K.; Yokota, S.; Seto, T.; Takagi, K. Development of Micromachines Using Improved Resonantly-Driven Piezoelectric Micropumps. In Proceedings of the Fourth International Symposium on Fluid Power Transmission and Control, Wuhan, China, 8–10 April 2003; pp. 536–541.
16. Wang, X.Y.; Ma, Y.T.; Yan, G.Y.; Huang, D.; Feng, Z.H. High flow-rate piezoelectric micropump with two fixed ends polydimethylsiloxane valves and compressible spaces. *Sensors Actuators A-Phys.* **2014**, *218*, 94–104. [CrossRef]
17. Chen, J.; Huang, D.; Feng, Z.H. A U-shaped piezoelectric resonator for a compact and high-performance pump system. *Smart Mater. Struct.* **2015**, *24*, 105009. [CrossRef]

18. Putra, A.S.; Huang, S.; Tan, K.K.; Panda, S.K.; Lee, T.H. Design, modeling, and control of piezoelectric actuators for intracytoplasmic sperm injection. *IEEE Trans. Control Syst. Technol.* **2007**, *15*, 879–890. [CrossRef]

19. Poikselka, K.; Leinonen, M.; Palosaari, J.; Vallivaara, I.; Haverinen, J.; Roning, J.; Juuti, J. End cap profile optimization of a piezoelectric Cymbal actuator for quasi-static operation by using a genetic algorithm. *J. Intell. Mater. Syst. Struct.* **2016**, *27*, 444–452. [CrossRef]

20. Ma, H.K.; Chen, R.H.; Hsu, Y.H. Development of a piezoelectric-driven miniature pump for biomedical applications. *Sensors Actuators A-Phys.* **2015**, *234*, 23–33. [CrossRef]

 *micromachines*

Article

Single Cylinder-Type Piezoelectric Actuator with Two Active Kinematic Pairs

Ramutis Bansevicius [1,†], Jurate Janutenaite-Bogdaniene [2,*,†], Vytautas Jurenas [1,†], Genadijus Kulvietis [2,†], Dalius Mazeika [2,†] and Asta Drukteiniene [2,†]

[1]   Institute of Mechatronics, Kaunas University of Technology, K. Donelaicio st. 73, LT-44249 Kaunas, Lithuania; ramutis.bansevicius@ktu.lt (R.B.); vytautas.jurenas@ktu.lt (V.J.)
[2]   Department of Information Systems, Vilnius Gediminas Technical University, Sauletekio al. 11, LT-10223 Vilnius, Lithuania; genadijus.kulvietis@vgtu.lt (G.K.); dalius.mazeika@vgtu.lt (D.M.); asta.drukteiniene@gmail.com (A.D.)
*   Correspondence: jurate.janutenaite@gmail.com; Tel.: +370-662-35465
†   These authors contributed equally to this work.

Received: 27 September 2018; Accepted: 2 November 2018; Published: 15 November 2018

Abstract: There is an ever-increasing demand for small-size, low-cost, and high-precision positioning systems. Therefore, investigation in this field is performed to search for various solutions that can meet technical requirements of precise multi-degree-of-freedom (DOF) positioning systems. This paper presents a new design of a piezoelectric cylindrical actuator with two active kinematic pairs. This means that a single actuator is used to create vibrations that are transformed into the rotation of the sphere located on the top of the cylinder and at the same time ensure movement of the piezoelectric cylinder on the plane. Numerical and experimental investigations of the piezoelectric cylinder have been performed. A mathematical model of contacting force control was developed to solve the problem of positioning of the rotor when it needs to be rotated or moved according to a specific motion trajectory. The numerical simulation included harmonic response analysis of the actuator to analyze the trajectories of the contact points motion. A prototype actuator has been manufactured and tested. Obtained results confirmed that such a device is suitable for both positioning and movement of the actuator in the plane.

Keywords: piezoelectric actuators; positioning; trajectory control; numerical analysis; trajectory planning

---

1. Introduction

Over recent decades, there has been an ever-increasing demand for the positioning systems with several degrees-of-freedom that would be faster, more accurate, and reliable, yet more compact at the same time. Piezoelectric devices are preferable when solving this type of problems because of the high precision, quick response, and low cost. Such advantages of piezoelectric devices are applicable in different fields, where nanometric scale resolution is required as, for example, in microscopes, laser systems, precision positioning systems [1–4], etc.

Precise positioning is a complicated problem because it affects the accuracy and proper functioning of the mechanism. Various precision positioning systems are proposed that employ piezoelectric actuators and flexible hinges to move or rotate the platform and to achieve the desired motion [1,2]. However, in most cases piezoelectric actuators are used as single degree-of-freedom (DOF) devices despite that there is a high demand for multiple DOF systems. Multiple DOF piezoelectric systems usually require one actuator for each degree-of-freedom, where each actuator requires its own piezoelectric transducer to transform the electrical input signal into a mechanical output. It is difficult to achieve high-resolution systems applying such design principle.

Cai et al. proposed a 6-DOF system, which is a combination of two 3-DOF precision positioning stages [1]. The first stage operates on the plane and ensures translation in X and Y directions and rotation about Z axis. The second stage is out-of-plane and has the first stage mounted on it. Both stages use piezoelectric actuators in combination with flexure hinges to achieve high accuracy. However, the structure of the proposed system is quite complicated. Lu et al. investigated the positioning stage comprised of the ball-screw stage and 3-DOF piezoelectric stage [3]. In this system, the first stage ensures a large translation range and fast motion, while the second stage is responsible for the high-precision positioning. The stage uses piezoelectric actuators, as well as translation and rotation mechanisms to ensure proper operation. The system design, where high accuracy is achieved by using components based on flexure hinges and piezoelectric actuators, is efficient enough [4–7]. However, the main drawback of such systems is a complex structure and many components that must be controlled. Besides, piezoelectric stack actuators used in such systems are affected by hysteresis of piezoelectric material that must be evaluated by the control system of the device.

There are several multi-DOF ultrasonic actuators developed that are capable to position an object in two or three directions [7–13]. A 3-DOF Langevin type piezoelectric actuator was developed by Zhao et al. [9]. It is composed of a cylindrical stator and a spherical rotor. The actuator operates employing the superposition of bending and longitudinal vibration modes, when the excitation voltages with certain frequency, phase and amplitude are applied to the three groups of piezoelectric ceramic rings. A 3-DOF piezoelectric actuator generating rotation of the sphere about three axes has been developed by Vasiljev et al. [10]. The operation of the actuator is based on a shaking beam principle. The proposed actuator consists of a vibrating frame, four piezoceramic stacks and four overlays. Four driving tips are mounted on the top of the vibrating frame and are in contact with the sphere. Four electric signals with shifted phases by $\pi/2$ are used for excitation. Multiple degrees-of-freedom ultrasonic motor consisting of a bar-shaped stator and a spherical rotor has been reported [11]. Motor can generate 3-DOF rotation of the rotor around perpendicular axes using the bending and longitudinal vibration of the stator. A similar small-size cylinder-type multi-DOF actuator was developed by Gouda et al. [12]. The actuator consists of a small cylinder fixed on a substrate and a thin PZT ring that is glued to the substrate. The PZT electrodes are divided into four parts and are driven by shifted phase signals. A combination of the first longitudinal and the second bending mode is used to obtain elliptical vibrations on the top surface of the cylinder and to rotate a ball about three axes independently.

A multi-DOF cylindrical resonant-type piezoelectric actuator is presented and analyzed in this paper. This actuator has a triangular shape configuration of the electrodes and two active kinematic pairs. Compared to our previous research, an additional kinematic pair has been introduced and two additional degrees-of-freedom have been achieved, respectively [14,15]. This piezoelectric actuator can rotate the sphere about three axes and simultaneously move itself on the plane. The positioning of the sphere can be achieved through the direct impact of the sphere when the contacting point hits the sphere, or by moving it to the needed position on the plane. A combination of different voltages applied to electrodes creates an elliptical movement of the contact points. The mathematical model of contacting force control was developed for sphere trajectory planning. A numerical investigation of the actuator was performed, and a prototype actuator was manufactured and tested. The multi-DOF positioning system was realized using a single piezoelectric actuator that ensures higher accuracy. The proposed actuator can be applied for laser beam positioning in space, adaptive focusing of optical lenses in optoelectronics and laser systems, or optic beam stabilization.

2. Description of the Actuator and Modeling Results

Several principle design schemes of the piezoelectric actuator with two kinematic pairs are shown in Figure 1, where w is the total number of the degrees-of-freedom of moving links *1* and *1a*; *2* is the piezoelectric cylinder with special sectioning of electrodes, and *3* is a frame, on which additional displacements occur. Thus, the displacements of the moving link are quite evident: link *1* rotates around axes *x*, *y*, *z* and affects the translational motion in directions *x*, *y* and rotation around axis *z* (Figure 1a); link *1* rotates around axis *z*, affects translational motion in the direction of *z* axis and moves on the plane *x*, *y* (Figure 1b); links *1* and *1a* rotate around axis x and move along it (Figure 1c); link *1* rotates around axes *x*, *y*, *z* and affects translational motion in the direction of *z* axis and rotation around the same axis (Figure 1d).

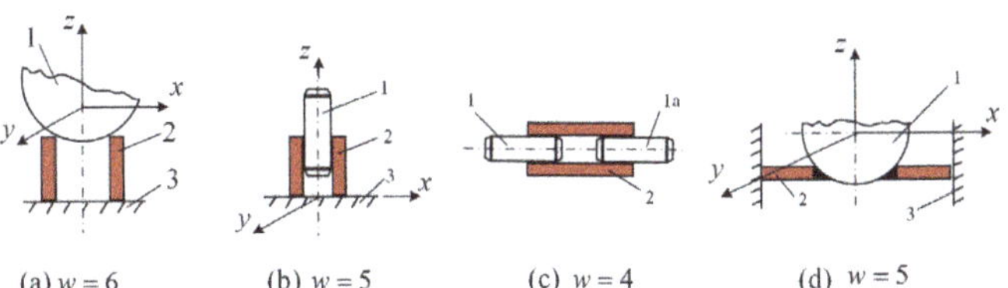

Figure 1. General schematics of the piezoelectric actuator with two active kinematic pairs.

The number of degrees-of-freedom could be increased: a schematic of the piezoelectric actuator with three active kinematic pairs ($w = 8$) is shown in Figure 2. To obtain 8 degrees-of-freedom, it is necessary to solve the problem of electrode sectioning of the piezoelectric actuator with the aim to exclude the unwanted oscillations of some specific forms.

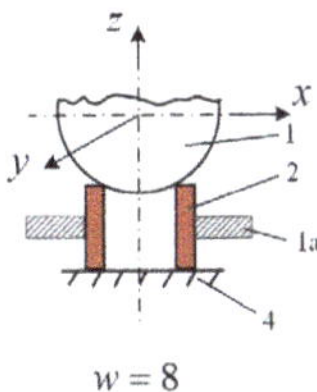

Figure 2. The schematic of the piezoelectric actuator with three active kinematic pairs.

Two types of motion (the sphere rotating on the cylinder and the sphere moving on the plane) can be controlled separately by:

- controlling the forces, acting in the contact zones of the sphere, piezoelectric cylinder and plane by introducing controllable brakes with the help of permanent magnets (Figure 3) or using layers of electro-rheological suspensions, controlled by high voltage
- connecting the same frequency voltage to specific electrodes, but with a different phase (180°) and amplitude, as in the main driving source
- using additional elements related to the position of the nodes of oscillation

We will further analyze the piezoelectric cylinder-type actuator with two kinematic pairs as shown in Figure 4a. A triangular electrode configuration is used to excite vibrations of the actuator (Figure 4b). The brakes of active kinematic pairs are not included. The piezoelectric cylinder has the following dimensions: 28 mm (inner diameter) × 33 mm (outer diameter) × 20 mm (height) and is made from PZT-8 material. A steel sphere is positioned on top of the cylinder (Figure 4a). The outer surface of the cylinder is covered with triangular electrodes as shown in Figure 4b, while the inner surface of the cylinder has a single electrode and is grounded.

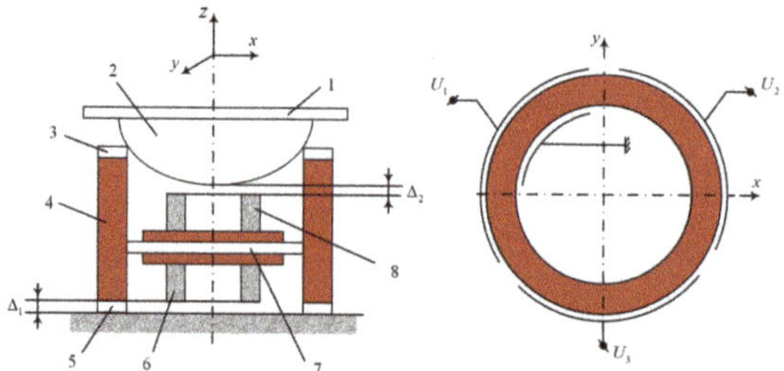

Figure 3. The laser deflector made on the basis of a single piezoelectric actuator with two active kinematic pairs and two brakes, activated with the help of a bimorph piezoelectric actuator, generating static bending displacements: *1*—mirror, *2*—spherical element, *3*—three contacts made of high friction material, *4*—piezoelectric cylinder with radial poling, *5*—three contacts made of high friction material, *6* and *8*— permanent magnets - brakes, fixed to bimorph actuator *7*, working in static displacement mode; $\Delta_1 = 0$, when the displacement of bimorph actuator *7* is negative (the brake is applied to the lower kinematic pair— displacements on the plane); $\Delta_2 = 0$, when the displacement of bimorph actuator *7* is positive (the brake is applied to the upper kinematic pair—angular displacements on the sphere).

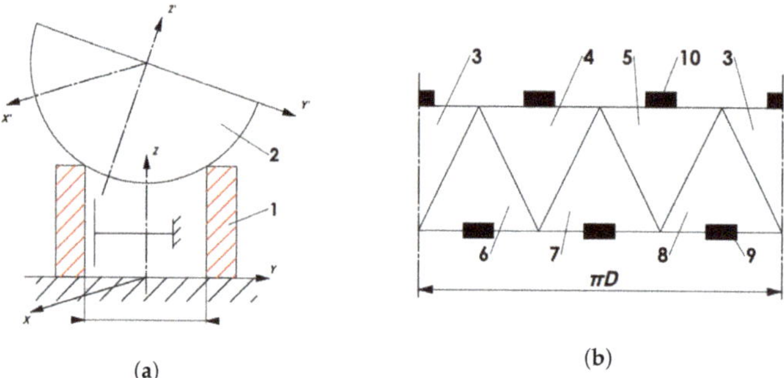

(a)

(b)

Figure 4. (**a**) Analyzed kinematic pair; (**b**) Proposed triangular electrode scheme.

Two electrodes on the outer surface are excited at the same time to achieve rotation of the rotor. Different voltages are applied to achieve the desired motion—larger voltage results in higher amplitude but not necessarily in the proper direction of the contact point. Therefore, improper voltages may result in elliptical trajectories of the contact points moving in opposite directions. It was chosen to apply a voltage of 100 V to the first electrode and 50 V to the second. Harmonic excitation signal was used. To determine proper resonance frequencies, the amplitudes of contact point vibration were investigated, when excitation frequency was changing in the range 0 Hz–150,000 Hz.

The amplitudes of the contact point in the middle of electrode 4 (Figure 4b) are shown at Figure 5. The modal analysis shows that the required eigenform is at 88,371 Hz (Figure 6); therefore, a resonant frequency at 88,600 Hz will be further analyzed. This frequency resonates with the correct eigenform. The calculations and analysis were performed using ANSYS software.

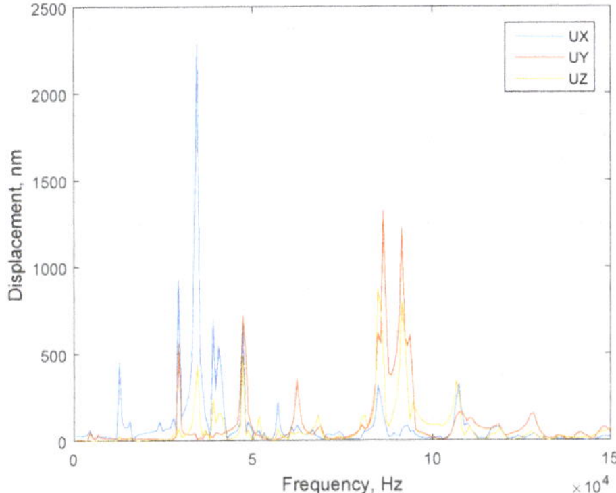

Figure 5. Harmonic analysis: The amplitudes of the contact point in the middle of electrode 4.

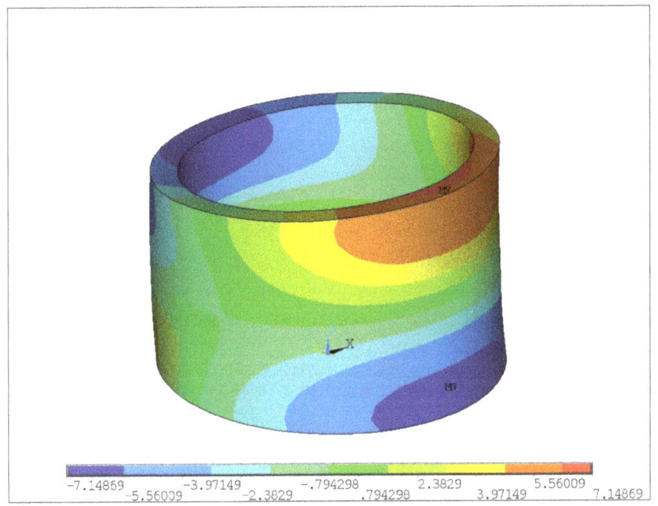

Figure 6. Modal analysis: The suitable eigenform at 88,371 Hz, deformations in Z direction.

The proposed electrode configuration and the excitation of two electrodes with different voltages result in rotation of the sphere by two contact points. The contact point trajectory was calculated. The analysis shows that the contact point between electrodes 4 and 5 on XZ plane (Figure 1b) moves in ellipsis that has angle $\phi = \deg 88$ (angle ϕ describes rotation of ellipse on the plane). The contact points that are positioned in the middle of electrodes 4 and 5 produce elliptical motion as well, but the angles are $\phi = 74°$ and $\phi = 82°$ respectively. The movement of the second contact point (Figure 4b—node 10) is shown in Figure 7. The direction and angles of the contact points can be effectively controlled by changing the applied voltage. When electrodes 6, 7 or 8 are excited, the biggest deformations occur in the bottom nodes. These are used to move the actuator on the plane. Exciting the top and bottom electrodes at the same time results in both rotations of the sphere and movement of the actuator.

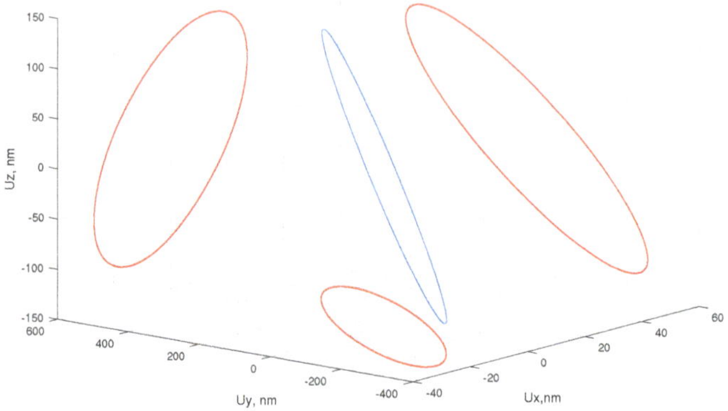

Figure 7. Trajectories of the contact point.

3. Mathematical Model of the Contact Force Control

A mathematical model for sphere positioning in space through contacting force control was developed. One active kinematic pair of the actuator was used for sphere positioning on the plane. This kind of positioning is performed by moving the piezoelectric cylinder on the plane. Another active kinematic pair is used to rotate the sphere about three axes. All previously studied positioning systems had only one kinematic pair, i.e., the active cylinder was moved on the plane or the sphere was rotated about different axes. Combining two kinematic pairs and applying the brakes, as shown in Figure 3, we obtain one active cylinder with two kinematic pairs (Figure 4a). The configuration of electrodes should be as follows (Figure 4b):

- the sphere rotates on the cylinder and the cylinder does not move, when electrodes 3, 4, 5 are excited
- the cylinder moves on the plane and the sphere does not rotate, when electrodes 6, 7, 8 are excited

Different electrode switching and trajectory planning algorithms of the cylinder movement on the plane and sphere rotation are developed [16,17]. However, there is no research performed and trajectory planning algorithm developed, when two active kinematic pairs are used. Therefore, a new algorithm must be created. Based on the previous investigation of trajectory planning algorithms, it can be concluded that only a fine trajectory planning algorithm is suitable for this task [14,15]. The fine trajectory planning algorithm allows exciting two or more electrodes of the cylinder at the same time for obtaining a linear motion of the cylinder or rotation of the sphere. This method has the advantage of achieving the maximum available speed of movement, when certain contact forces are used. Therefore, the main task of the trajectory planning algorithm is to calculate the function of the contacting forces that can be obtained by exciting a particular electrode section of the piezoelectric cylinder.

The cylinder moves on the plane by the trajectory described as function $C(t)$ (Figure 8a), the sphere rotates around its central axis by the trajectory described as function $S(\tau) = \vec{r}\left(\left|\vec{R}\right|, \theta(\tau), \phi(\tau)\right)$ (Figure 8b), where $\theta(\tau) \in [0, \pi]$ is an angle between positive z axis and $\vec{r}$; $\phi(\tau) \in [0, 2\pi]$ is an angle between positive x axis and projection in xy plane of $\vec{r}$ (Figure 8); t and τ are parameters of the trajectories. Each kinematic pair is controlled at different value intervals:

$$\left|\vec{f}_{Jt}\right| \in \left[f_{min}^t, f_{max}^t\right]$$
$$\left|\vec{f}_{J\tau}\right| \in \left[f_{min}^\tau, f_{max}^\tau\right]$$

$$(1)$$

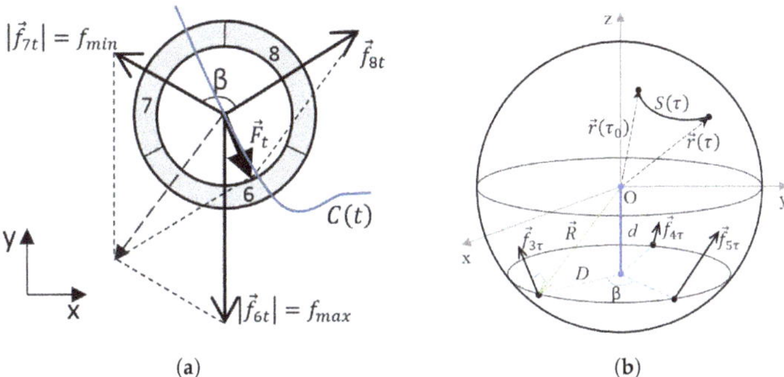

Figure 8. (**a**) Trajectory of the cylinder; (**b**) Sphere rotation trajectory.

The orientation of contact forces $J = 3, 4, 5$ of the sphere does not change throughout the movement time:

$$\vec{f_J}^{\,\circ} = (\cos(\alpha_{J\tau}), \cos(\beta_{J\tau}), \cos(\gamma_{J\tau})) \tag{2}$$

where $\alpha_J, \beta_J, \gamma_J$ is an angle between x, y, z axis and forces $\vec{f}_{J\tau}$. The rotation direction of the sphere is the same as the direction of the unit tangent vector at point τ of function $S(\tau)$. The unit tangent vector also shows the direction, in which the total force must be active. The unit tangent vector $\vec{T}(\tau) = \langle T_x | T_y | T_z \rangle$ is calculated by the formula [14]:

$$\vec{T}(\tau) = \frac{\vec{r}'(\tau)}{|\vec{r}'(\tau)|} \tag{3}$$

The distance from the center of the sphere to the plane, where the contacts of the cylinder are located, are equal to $d = \sqrt{R^2 - D^2}$, where R is the radius of the sphere; D is the radius of the cylinder and $d < R$. Then orientation of forces $\vec{f}_{3,4,5}$ of the sphere can be calculated by formula [15]:

$$\left[\vec{f}_J(\alpha_J, \beta_J, \gamma_J) \right] = \cos^{-1} \begin{bmatrix} \frac{\sqrt{R^2 - D^2}}{R} & 0 & \sqrt{D^2 - R^2 + 1} \\ -\frac{\sqrt{R^2 - D^2}}{2R} & \frac{\sqrt{3}}{2}\frac{\sqrt{R^2 - D^2}}{R} & \sqrt{D^2 - R^2 + 1} \\ -\frac{\sqrt{R^2 - D^2}}{2R} & \cos\left(180° - \cos^{-1}\left(\frac{\sqrt{3}}{2}\frac{\sqrt{R^2 - D^2}}{R}\right)\right) & \sqrt{D^2 - R^2 + 1} \end{bmatrix} \tag{4}$$

To choose an appropriate electrode segment for excitation, it is necessary to know the total force $\vec{F}$ of all segments:

$$\vec{F}^C(t) = \sum_{J=6}^{n} \vec{f}_J(t), \text{ in case } J = 6, 7, 8 \; n = 3;$$

$$\vec{F}^C(\tau) = \sum_{J=3}^{n} \vec{f}_J(\tau), \text{ in case } J = 3, 4, 5 \; n = 3; \tag{5}$$

The electrode segment that is the nearest to the total force $\vec{F}$ vector must be selected for excitation and its numeric value must be equal to maximum f_{max} (Figure 8a). To obtain precision of the motion, a force that is cardinally in the opposite orientation than the total force must be activated. Its value must be equal to the minimum value of the force f_{min}. The value of the third electrode is calculated after solving a system of equations [14]:

$$\begin{cases} \left|\overrightarrow{F}\right|\cos\mu = \sum_{j}^{n}\left|\overrightarrow{f}_{j}\right|\cos\phi_{J} \\ \left|\overrightarrow{F}\right|\sin\mu = \sum_{j}^{n}\left|\overrightarrow{f}_{j}\right|\sin\phi_{J} \end{cases} \tag{6}$$

$$\text{where } \left|\overrightarrow{F}\right| = \tan^{-1}\frac{T_y}{T_x}; \phi_J = \widehat{\overrightarrow{F}\,\overrightarrow{f}}_J \text{ and } \left|\overrightarrow{f}\right| = \begin{cases} f_{min}, \text{ if } \left|\overrightarrow{f}_{J}\right| = 0, \\ f_{max}, \text{ if } \phi_J = min([\phi_J]) \\ \left|\overrightarrow{f}_{J}^{*}\right| \end{cases}$$

The integration scheme of the fine trajectory planning algorithm for one active element with two kinematic pairs, when parallel calculations are used, is presented in Figure 9.

The sphere of the radius R = 20 mm and the cylinder of the radius D = 16.5 mm were selected for the numerical experiment.

The trajectory of the sphere rotation around different axes was described by formulas (Figure 10):

$$S(\tau) = \begin{cases} \theta(\tau) = 4\tau^3 + 10 \\ \phi(\tau) = 15\tau^2 + 2\tau \end{cases}, \text{ where } t \in [0,3] \tag{7}$$

$$C(t) = \begin{cases} x(t) = t\cos(t) \\ y(t) = 2t\sin(t) \end{cases}, \text{ where } t \in [0,3] \tag{8}$$

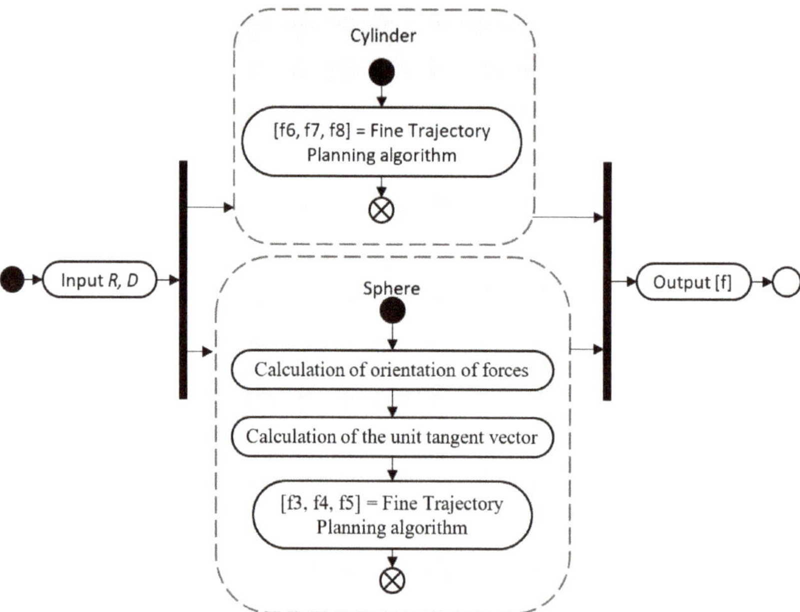

Figure 9. The integration system of the fine trajectory planning algorithm.

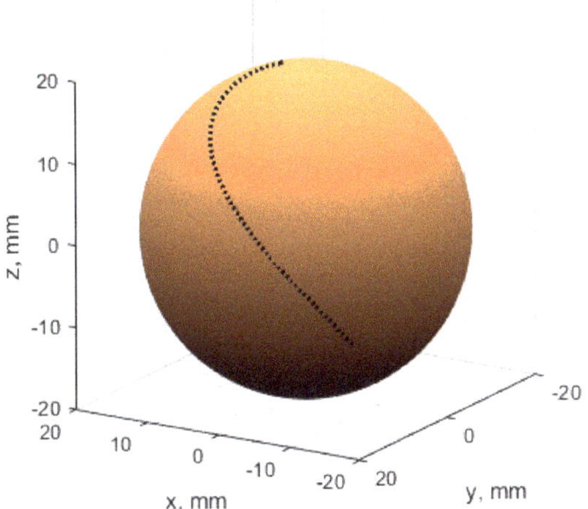

Figure 10. The sphere rotation trajectory (in dots).

The trajetory of the cylinder is given in Figure 11. The forces of the sphere must be in the interval $\left|\vec{f}_{6,7,8}\right| \in [0.5, 1]$ mN, and the forces of the cylinder in the interval $\left|\vec{f}_{3,4,5}\right| \in [3, 5]$ mN. The results of the calculation of the cylinder forces and the sphere electrode segments are presented in Figure 12a,b.

The numerical experiment confirmed that the fine trajectory planning algorithm is successfully integrated and is suitable for both movements: the movement on the plane and sphere rotation on certain points of the contact. This means that kinematic pairs of the actuator can be managed separately.

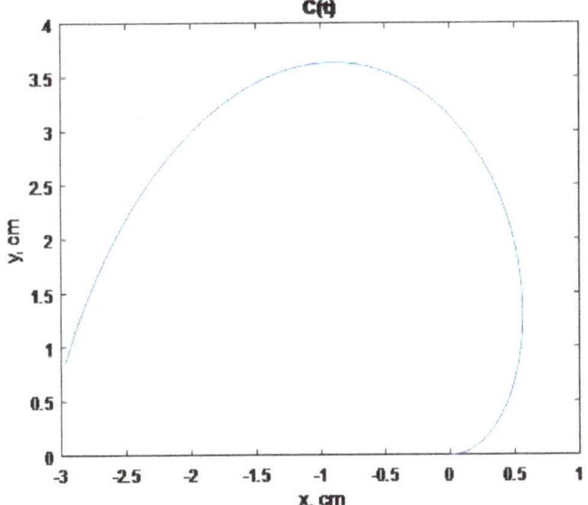

Figure 11. The cylinder motion trajectory.

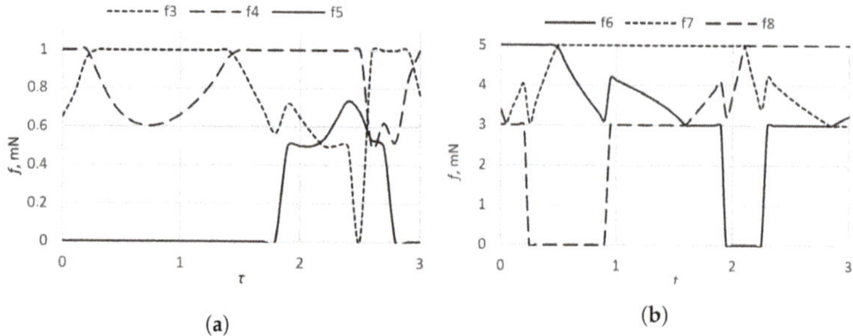

Figure 12. Forces of electrode segments: (**a**) the sphere; (**b**) the cylinder.

4. Experimental Results

A radially poled piezoelectric cylinder with six triangle-shaped electrodes on the outer surface was produced based on the results of numerical simulations. The dimensions of the piezoelectric cylinder were 33 mm(outer diameter) × 28 mm (inner diameter) × 20 mm (height). The material was the hard type piezo ceramic of type PZT-8. Triangular electrodes on the outer surface of the piezoceramic cylinder were formed by etching using nitric acid.

Measurements using a 3D scanning vibrometer and an impedance analyzer (Figure 13) are conducted to investigate the vibration modes of the piezoelectric actuator and to identify resonance frequencies. Equipment used for the experiment included a 3D scanning vibrometer PSV-500-3D-HV (Polytec GmbH, Springe, Germany), a linear amplifier P200 (FLC Electronics AB, Sweden) for the actuator driving and an impedance analyzer 6500B (Wayne Kerr Electronics Ltd., West Sussex, UK). The driving voltage of amplitude 5 V was a harmonic signal within the scanning frequency range of 10–150 kHz.

The vibrations of a cylinder-shaped transducer visualized with 3-D laser scanning vibrometer at a driving frequency of 93.6 kHz are shown in Figure 14. It can be noticed that the area of the deflections of the top surface is about 2 times larger, when driving voltage is connected to the triangle-shaped electrode with the tip directed backwards to the scanned surface. It confirms that the position of the 3 points of contact on the end surfaces of each side of the piezo cylinder should be by the middle of the side of the triangular electrode as shown in Figure 4b. Figure 15 shows dependence of the amplitude of the axial point of the contact point versus applied voltage. The dashed line shows the case when the triangle-shaped electrode with the tip directed to the active contacting element is excited; while solid line represents the case when the triangle-shaped electrode with the tip directed backwards to the active contacting element is excited. It can be noticed that almost linear dependence was obtained. Moreover, it can be seen that the vibration amplitude depends on electrode configuration, and larger amplitude value is obtained when the triangle-shaped electrode with the tip directed backwards to the active contacting element is excited. The results of the experiment show that three-dimensional resonant vibrations of contact points of the piezo cylinder are excited at a driving frequency of 93.6 kHz (Figure 16). These vibrations generate the lateral motion of the piezo cylinder placed on the horizontal plane, or the angular rotation of the sphere placed on 3 contact points of the piezo cylinder. The results of the rotation angle measurement about x axis of the sphere are shown in Figure 17. One electrode of the piezoelectric actuator was excited with the harmonic electric signal containing 10 periods. The excitation frequency was 93.6 kHz, while voltage amplitude was 40 V. Such electric signal was applied in steps every second. A minimal resolution of 50 μrad was obtained, whereas average rotation speed was 42.5 μrad/s. Compared to the results of FEM analysis, the difference of driving frequency differed only by 5.5% (88,361 Hz compared to 93,600 Hz). The direction of the lateral motion of the cylinder and the direction of the angular rotation of the sphere can be controlled by the algorithm presented in Section 3. A working prototype is shown in a Supplementary Video.

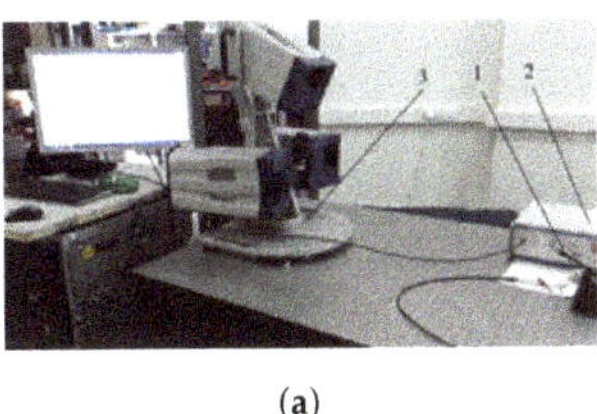
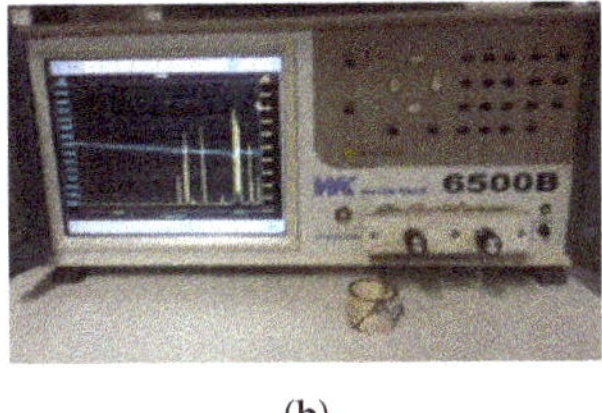

(**a**)   (**b**)

Figure 13. Experimental setup for investigating the modes of vibration (**a**) and impedance measurement of the piezoelectric cylinder (**b**): 1—piezoelectric cylinder; 2—linear amplifier; 3—3D scanning vibrometer; 4—impedance analyzer.

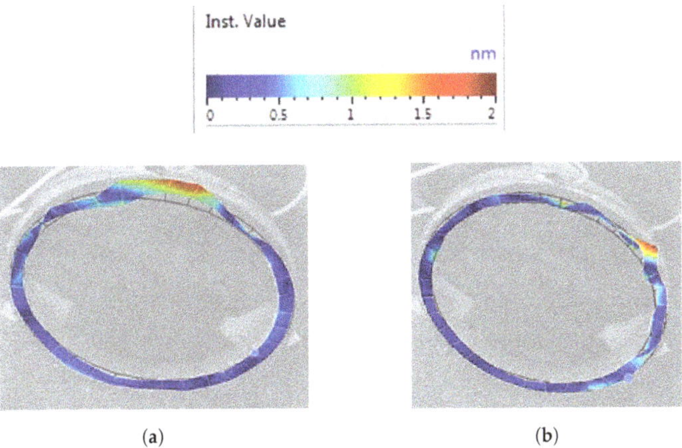

(**a**)   (**b**)

Figure 14. Deformation modes of the end side of piezoelectric cylinder, when an electric signal of the voltage amplitude of 5 V and a driving frequency of 93.6 kHz are applied into the triangle-shaped electrode: (**a**) the driving voltage is connected to the triangle-shaped electrode with the tip directed to the scanned surface; (**b**) the driving voltage is connected to the triangle-shaped electrode with the tip directed backwards to the scanned surface.

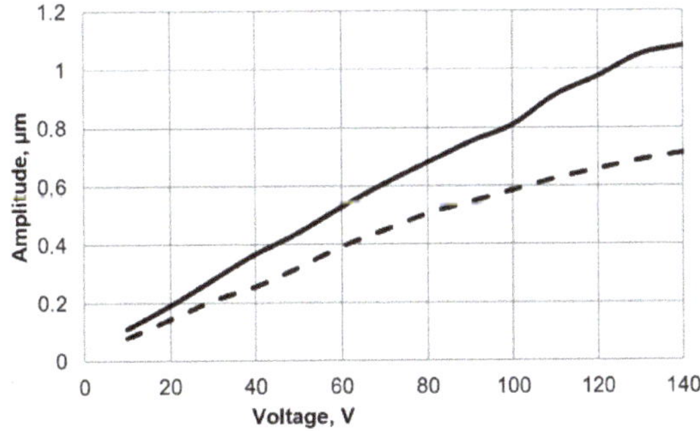

Figure 15. Amplitude vs applied voltage of the axial vibrations of contacting element at the operational frequency 93.6 kHz: dashed line—when the triangle-shaped electrode tip is directed to an active contacting element; solid line—when the triangle-shaped electrode tip is directed backwards to the active contacting element.

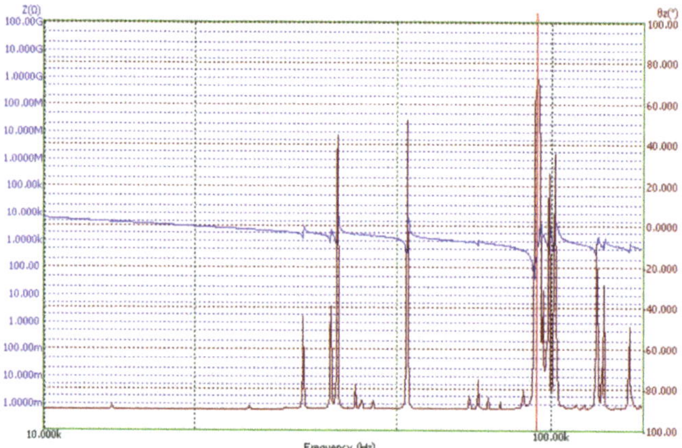

Figure 16. Impedance/phase diagrams of the piezoelectric cylinder. A resonant frequency for driving a transducer is 93.6 kHz.

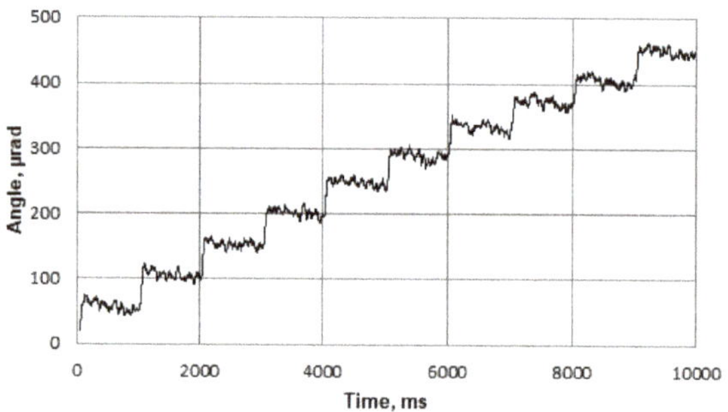

Figure 17. Measured stepped rotation of the sphere.

5. Conclusions

A new piezoelectric actuator with two active kinematic pairs was proposed and investigated. A novel electrode configuration scheme was used to excite vibrations of the actuator. Besides, a mathematical model of the contact force control and a trajectory planning algorithm were proposed and evaluated. Exciting different electrodes and controlling the contact force in kinematic pairs with brakes can create rotation of the sphere or movement on the plane. The trajectory planning method was validated by the numerical analysis. A prototype actuator was made, and mechanical and electrical characteristics were measured. The results of the numerical simulation and experimental measurements were in a good agreement.

Supplementary Materials: The following are available online at http://www.mdpi.com/2072-666X/9/11/597/s1. Video S1: A motion of the sphere when driving voltage is applied to the actuator.

Author Contributions: J.J.-B. and D.M. were responsible for FEM analysis; A.D. and G.K. designed the trajectory planning algorithm; R.B. conceived the idea, supervised the experiments, and analyzed the data; V.J. produced the prototype and performed experimental measurements. All the authors contributed to the manuscript preparation.

Funding: This research was funded by European Regional Development Fund, grant number DOTSUT-234.

Acknowledgments: This research was funded by the European Regional Development Fund according to the supported activity No. 01.2.2-LMT-K-718 under the project No. DOTSUT-234.

Conflicts of Interest: The authors declare no conflict of interest.

References

1. Cai, K.; Tian, Y.; Wang, F.; Zhang, D.; Liu, X.; Shirinzadeh, B. Design and control of a 6-degree-of-freedom precision positioning system. *Robot. Comput. Integr. Manuf.* **2017**, *44*, 77–96. [CrossRef] [CrossRef]
2. Zhu, W.; Yang, F.; Rui, X. Robust independent modal space control of a coupled nano-positioning piezo-stage. *Mech. Syst. Signal Process.* **2018**, *106*, 466–478. [CrossRef] [CrossRef]
3. Liu, C.H.; Jywe, W.Y.; Jeng, Y.R.; Hsu, T.H.; Li, Y.T. Design and control of a long-traveling nano-positioning stage. *Precis. Eng.* **2010**, *34*, 497–506. [CrossRef] [CrossRef]
4. Guo, Z.; Tian, Y.; Liu, C.; Wang, F.; Liu, X.; Shirinzadeh, B.; Zhang, D. Design and control methodology of a 3-DOF flexure-based mechanism for micro/nano-positioning. *Robot. Comput.-Integr. Manuf.* **2015**, *32*, 93–105. [CrossRef] [CrossRef]
5. Chen, J.; Xu, M.L.; Feng, B.; Tian, Q.G. A novel positioning stage using piezoelectric actuator for antenna pointing. In *Recent Advances in Structural Integrity Analysis, Proceedings of the International Congress APCFS/SIF 2014*; Woodhead Publishing: Delhi , India, 2014; pp. 437–441. [CrossRef]
6. Liu, P.; Yan, P.; Zhang, Z.; Leng, T. Modeling and control of a novel X-Y parallel piezoelectric-actuator driven nanopositioner. *ISA Trans.* **2015**, *56*, 145–154. [CrossRef] [CrossRef] [PubMed]
7. Liaw, H.C.; Shirinzadeh, B. Constrained motion tracking control of piezo-actuated flexure-based four-bar mechanisms for micro/nano manipulation. *IEEE Trans. Autom. Sci. Eng.* **2010**, *7*, 699–705. [CrossRef] [CrossRef]
8. Gozen, B.A.; Ozdoganlar, B. A method for open-loop control of dynamic motions of piezo-stack actuators. *Sens. Actuators A Phys.* **2012**, *184*, 160–172. [CrossRef] [CrossRef]
9. Zhao, C.; Li, Z.; Huang, W. Optimal design of the stator of a three-DOF ultrasonic motor. *Sens. Actuator A Phys.* **2005**, *121*, 494–499. [CrossRef] [CrossRef]
10. Vasiljev, P.; Mazeika, D.; Kulvietis, G.; Vaiciuliene, S. Piezoelectric Actuator Generating 3D-Rotations of the Sphere. *Solid State Phenom.* **2006**, *113*, 173–178. [CrossRef] [CrossRef]
11. Aoyagi, M.; Beeby, S.P.; White, N.M. A novel multi-degree-of-freedom thick-film ultrasonic motor. *IEEE Trans. Ultrason. Ferroelectr. Freq. Control* **2002**, *49*, 151–158. [CrossRef] [CrossRef] [PubMed]
12. Gouda, Y.; Nakamura, K.; Ueha, S. A miniaturization of the multi-degree-of-freedom ultrasonic actuator using a small cylinder fixed on a substrate. *Ultrasonics* **2006**, *44*, e617–e620. [CrossRef] [CrossRef] [PubMed]
13. Zhao, C. *Ultrasonic Motors: Technologies and Applications*; Science Press: Beijing, China; Springer: Berlin/Heidelberg, Germany, 2011; p. 494, ISBN 978-3-642-15304-4. [CrossRef]
14. Bansevicius, R.P.; Drukteiniene, A.; Kulvietis, G.; Macerauskas, E.; Janutenaite, J.; Mazeika, D. Fine trajectory planning method for mobile piezorobots. *J. Vibroeng.* **2016**, *18*, 2043–2052. [CrossRef]
15. Bansevicius, R.P.; Genadijus, K.; Macerauskas, E.; Janutenaite, J.; Drukteiniene, A.; Mazeika, D. Trajectory planning for stabilization system of nanosatellite. In Proceedings of the Mechanika, 21th International Scientific Conference, Kaunas, Lithuania, 12–13 May 2016; pp. 22–28.
16. Janutnaite-Bogdaniene, J.; Macerauskas, E.; Drukteiniene, A.; Kulvietis, G.; Bansevicius, R.P. Cylindrical piezorobot's trajectory planning and control. *J. Vibroeng.* **2017**, *19*, 2670–2679. [CrossRef] [CrossRef]
17. Bansevicius R.P.; Mazeika, D.; Kulvietis, G.; Tumasoniene, I.; Drukteiniene, A.; Jurenas, V.; Bakanauskas, V. Investigation of sphere trajectories of a rotational type piezoelectric deflector. *Mech. Syst. Signal Process.* **2018**, 1–8. [CrossRef] [CrossRef]

Article

Parametric Analysis and Optimization of Radially Layered Cylindrical Piezoceramic/Epoxy Composite Transducers

Jianjun Wang [1,*, Lei Qin [2], Weijie Li [3] and Weibin Song [4]**

[1]   Department of Applied Mechanics, University of Science and Technology Beijing, Beijing 100083, China
[2]   Beijing Key Laboratory for Sensors, Beijing Information Science & Technology University,
       Beijing 100101, China; qinlei@bistu.edu.cn
[3]   School of Civil and Environmental Engineering, Harbin Institute of Technology, Shenzhen 518055, China;
       wli27@uh.edu
[4]   Department of Mechanics, Beijing Jiaotong University, Beijing 100044, China; 10115246@bjtu.edu.cn
*   Correspondence: jianjunwang168@ustb.edu.cn; Tel.: +86-010-6233-2901

Received: 8 October 2018; Accepted: 6 November 2018; Published: 9 November 2018

Abstract: Radially layered cylindrical piezoceramic/epoxy composite transducers have been designed by integrating the excellent performance of piezoelectric/polymer composites and the radial radiation ability of cylindrical configurations, which are promising in developing novel ultrasonic and underwater sound techniques. Our previous study has explored the effects of the external resistance on the electromechanical characteristics of the transducer, and obtained some valuable findings. To clearly understand the electromechanical characteristics of the transducer and to guide the device design, in this paper, parametric analysis was performed to reveal the effects of multiple key factors on the electromechanical characteristics. These factors include material parameters of epoxy layers, piezoceramic material types, and locations of piezoceramic rings. In order to better analyze the influence of these factors, a modified theoretical model, in which every layer has different geometric and material parameters, was developed based on the model given in the previous work. Furthermore, the reliability of the model was validated by the ANSYS simulation results and the experimental results. The present investigation provides some helpful guidelines to design and optimize the radially layered cylindrical piezoceramic/epoxy composite transducers.

Keywords: cylindrical composite; piezoceramic/epoxy composite; electromechanical characteristics; transducer

---

1. Introduction

Piezoelectric ultrasonic and underwater sound transducers, by virtue of their excellent electromechanical properties, have been widely applied in medical imaging [1–3], non-destructive testing [4,5], underwater communications [6,7], and so on. In recent years, researchers have been trying to improve their design in hope of obtaining more excellent abilities to perform related tasks. Many attempts have been made and this research topic has received more and more attention.

One effective attempt is to add a flexible polymer phase when fabricating the piezoelectric/polymer composites. These composites can overcome the shortcomings of single-phase piezoelectric materials, such as brittleness, high acoustic impedance, and at the same time possess the advantages of both the piezoceramics and the polymers, including low acoustic impedance, high electromechanical coupling coefficient, low mechanical loss and large dielectric constant [8,9]. Through designing and optimizing various connectivity patterns, such as 1-3 type [10–18], 2-2 type [3,19–21], these transducers can be used in high-frequency, high-power, and high-temperature applications [9]. In addition, an addition of modified epoxy, i.e., aluminum load epoxy, can also enhance their dielectric, piezoelectric and acoustic dampening [22–24], which is very useful for structural health monitoring, energy harvesting, and acoustic liners [22].

Another approach is to adopt new structural forms to realize multiple functionalities. The earlier transducers mainly adopted the approximated one-dimensional structures, which can excite one-dimensional longitudinal waves. The representative devices are the longitudinally sandwiched transducers [25–29] that are composed of axially polarized piezoceramic rings and end metal masses. Recently, two dimensional structures, i.e., cylindrical configurations, were proposed to increase the wave coverage area and output power. These structures include tubes, shells, disks and rings [30–47]. They can realize the radial wave radiations by utilizing their radial vibrations, and can be used as the omni-directional emitter and receiver in underwater sound and ultrasonic applications [48].

The above two methods can improve the performance of piezoelectric transducers to meet different requirements. However, both of these methods have certain limitations. In addition, in some special applications, such as endoscopic ultrasound (EUS), it is required that the transducer is able to realize 360° imaging [49]. To obtain a comprehensive performance, some efforts have been made to focus on combining the advantages of piezoelectric/polymer composites and those of cylindrical configurations. One attempt was adopting the 1-3 type piezocomposite to design the cylindrical EUS transducers, which can acquire high-resolution EUS imaging [50]. Another attempt was using the 2-2 type piezocomposite to develop the cylindrical underwater acoustic transducers, which can achieve high-frequency wideband ability [51]. In the latest work, a new type of radially layered cylindrical piezoceramic/epoxy composite transducer was developed by integrating two concentric axially polarized piezoceramic rings into cylindrical epoxy matrixes, which is expected to be utilized in developing novel ultrasonic and underwater sound techniques [52]. This preliminary study mainly focused on studying the effects of the external resistance on the electromechanical characteristics of the transducer, which lacked of clear understanding on the influences of other key factors, including material parameters of epoxy layers, piezoceramic material types, and locations of piezoceramic rings. To clearly understand the electromechanical characteristics of this type of transducer and to guide the device design, in this paper, parametric analysis was performed to reveal the effects of these multiple key factors on the electromechanical characteristics of the transducer.

The remainder of this paper is organized as follows. Section 2 exhibits a schematic representation of the radially layered cylindrical piezoceramic/epoxy composite transducer and gives its modified theoretical model based on the model developed in previous work. Section 3 validates the theoretical solution by comparing it to solutions from both ANSYS numerical simulation and experimental investigation. Section 4 discusses the effects of material parameters of epoxy layers, piezoceramic material types, and locations of piezoceramic rings on the electromechanical characteristics of the transducer through numerical analysis. Section 5 draws the conclusions of the paper.

2. Modeling

2.1. Basic Equations

Figure 1 exhibits the schematic representation of the radially layered cylindrical piezoceramic/epoxy composite transducer. It consists of a solid epoxy disk, two epoxy rings, and two axially polarized piezoceramic rings. These components are arranged alternatively in the radial direction. The two piezoceramic rings are connected in parallel electrically, and are denoted as piezoceramic ring #1 and piezoceramic ring #2, respectively. Three epoxy layers are denoted as epoxy disk #1, epoxy ring #2 and epoxy ring #3, respectively. The geometric and material parameters of each layer are different. The radial location of the interface between each layer and the axial height of the transducer are defined as R_i $(i = 1, 2, 3, 4, 5)$ and h, respectively.

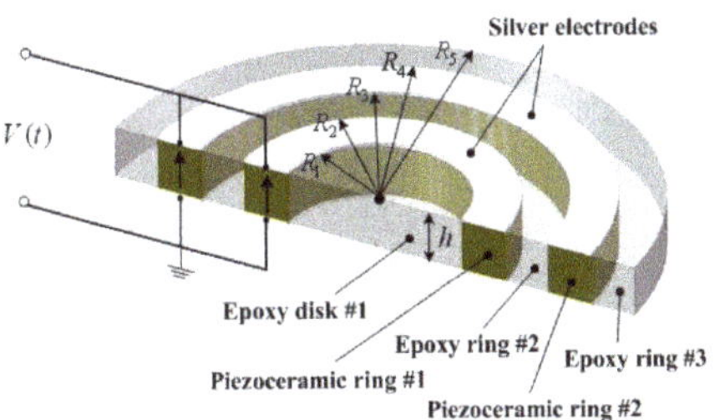

Figure 1. Schematic representation of the radially layered cylindrical piezoceramic/epoxy composite transducer.

A harmonic form of voltage is used as the excitation source, which is expressed as:

$$V(t) = V_0 e^{j\omega t}, \tag{1}$$

where V_0, $j = \sqrt{-1}$, $\omega = 2\pi f$, f and t are the excitation amplitude, the imaginary unit, the circular frequency, the excitation frequency and the time, respectively.

Under the assumption of plane stress, the harmonic radial displacement $u_{rP(i)}$, radial stress $\sigma_{rP(i)}$, electric potential $\phi_{(i)}$ and electric displacement $D_{z(i)}$ of the i-th piezoelectric layer ($i = 1, 2$) are expressed as follows [52,53]:

$$u_{rP(i)} = [A_{P(i)} f_1(r,i) + B_{P(i)} f_2(r,i)] e^{j\omega t}, \tag{2}$$

$$\sigma_{rP(i)} = [A_{P(i)} f_3(r,i) + B_{P(i)} f_4(r,i) + e_{31(i)}(V_0/h)] e^{j\omega t}, \tag{3}$$

$$\phi_{(i)} = z(V_0/h) e^{j\omega t}, \tag{4}$$

$$D_{z(i)}(z) = [A_{P(i)} e_{31(i)} k_{P(i)} J_0(k_{P(i)}r) + B_{P(i)} e_{31(i)} k_{P(i)} Y_0(k_{P(i)}r) - \kappa^\varepsilon_{33(i)}(V_0/h)] e^{j\omega t}, \tag{5}$$

In Equations (2)–(5), the functions from $f_1(r,i)$ to $f_4(r,i)$ can be expressed as following [52,53]:

$$f_1(r,i) = J_1(k_{P(i)}r), \tag{6}$$

$$f_2(r,i) = Y_1(k_{P(i)}r), \tag{7}$$

$$f_3(r,i) = c^E_{11(i)} k_{P(i)} J_0(k_{P(i)}r) + [(c^E_{12(i)} - c^E_{11(i)})/r] J_1(k_{P(i)}r), \tag{8}$$

$$f_4(r,i) = c^E_{11(i)} k_{P(i)} Y_0(k_{P(i)}r) + [(c^E_{12(i)} - c^E_{11(i)})/r] Y_1(k_{P(i)}r), \tag{9}$$

where $c^E_{11(i)} = s^E_{11(i)}/(s^E_{11(i)} s^E_{11(i)} - s^E_{12(i)} s^E_{12(i)})$, $c^E_{12(i)} = -s^E_{12(i)}/(s^E_{11(i)} s^E_{11(i)} - s^E_{12(i)} s^E_{12(i)})$, $e_{31(i)} = d_{31(i)}/(s^E_{11(i)} + s^E_{12(i)})$, $\kappa^\varepsilon_{33(i)} = \kappa^\sigma_{33(i)} - 2d^2_{31(i)}/(s^E_{11(i)} + s^E_{12(i)})$; $c^E_{11(i)}$, $c^E_{12(i)}$, $e_{31(i)}$, and $\kappa^\varepsilon_{33(i)}$ are the effective elastic, piezoelectric and dielectric constants of the i-th piezoceramic layer, respectively. $k_{P(i)} = \omega/V_{rP(i)}$ and $V_{rP(i)} = \sqrt{c^E_{11(i)}/\rho_{P(i)}}$ are the radial wave number and sound speed, respectively; $\rho_{P(i)}$ is the density of the piezoceramic. $J_0(k_{P(i)}r)$ is the Bessel function of the first kind, and $Y_0(k_{P(i)}r)$ is the Bessel function of the second kind.

Similarly, the harmonic radial displacement $u_{rE(i)}$ and radial stress $\sigma_{rE(i)}$ of the i-th elastic layers ($i = 1, 2, 3$) are expressed as follows [52–54]:

$$u_{rE(i)}(r) = [A_{E(i)} f_5(r,i) + B_{E(i)} f_6(r,i)] e^{j\omega t}, \tag{10}$$

$$\sigma_{rE(i)}(r) = [A_{E(i)} f_7(r,i) + B_{E(i)} f_8(r,i)] e^{j\omega t}. \tag{11}$$

In Equations (10) and (11), the functions from $f_5(r,i)$ to $f_8(r,i)$ can be expressed as following [52–54]:

$$f_5(r,i) = J_1(k_{E(i)}r),\tag{12}$$

$$f_6(r,i) = Y_1(k_{E(i)}r),\tag{13}$$

$$f_7(r,i) = [(\overline{E}_{(i)}k_{E(i)})/(1 - \mu_{(i)}^2)]\{J_0(k_{E(i)}r) + [(\mu_{(i)} - 1)/(k_{E(i)}r)]J_1(k_{E(i)}r)\},\tag{14}$$

$$f_8(r,i) = [(\overline{E}_{(i)}k_{E(i)})/(1 - \mu_{(i)}^2)]\{Y_0(k_{E(i)}r) + [(\mu_{(i)} - 1)/(k_{E(i)}r)]Y_1(k_{E(i)}r)\},\tag{15}$$

where $k_{E(i)} = \omega/V_{rE(i)}$, $V_{rE(i)}^2 = \overline{E}_{(i)}/[\rho_{E(i)}(1 - \mu_{(i)}^2)]$; $\rho_{E(i)}$, $\overline{E}_{(i)}$ and $\mu_{(i)}$ are the density, Young's modulus and Poisson's ratio of epoxy, respectively.

2.2. Solution

As shown in Figure 1, the boundary and continuity conditions of the cylindrical transducer consist of one innermost displacement boundary condition, one outermost stress boundary condition and eight continuous conditions, which are given as follows:

$$\left\{ \begin{array}{l} u_{rE(1)}\big|_{r=0} = 0 \\ \sigma_{rE(3)}\big|_{r=R_5} = 0 \end{array} \right.,\tag{16}$$

$$\left\{ \begin{array}{l} u_{rP(i)}\big|_{r=R_{2i-1}} = u_{rE(i)}\big|_{r=R_{2i-1}} \\ \sigma_{rP(i)}\big|_{r=R_{2i-1}} = \sigma_{rE(i)}\big|_{r=R_{2i-1}} \end{array} \right. \ (i = 1,\ 2),\tag{17}$$

$$\left\{ \begin{array}{l} u_{rP(i)}\big|_{r=R_{2i}} = u_{rE(i+1)}\big|_{r=R_{2i}} \\ \sigma_{rP(i)}\big|_{r=R_{2i}} = \sigma_{rE(i+1)}\big|_{r=R_{2i}} \end{array} \right. \ (i = 1,\ 2).\tag{18}$$

Combining Equations (2), (3), (10), (11), and (16)–(18), 10 coefficients can be derived, which are listed in Equations (A1)–(A5) (Appendix A). Further, the total electrical charge $Q_{total}(t)$ and the total current $I_{total}(t)$ can be expressed by the following expressions:

$$Q_{total}(t) = Q_{(1)}(t) + Q_{(2)}(t) = \int_0^{2\pi}\int_{R_1}^{R_2} D_{z(1)}rd\theta\,dr + \int_0^{2\pi}\int_{R_3}^{R_4} D_{z(2)}rd\theta\,dr = (\widetilde{C}_1 + \widetilde{C}_2)V_0e^{j\omega t},\tag{19}$$

$$I_{total}(t) = -dQ_{total}(t)/dt = -j\omega(\widetilde{C}_1 + \widetilde{C}_2)V_0e^{j\omega t},\tag{20}$$

where

$$\widetilde{C}_1 = 2\pi\{(a_1v_9 + v_1)[f_9(R_2,1) - f_9(R_1,1)] + (a_2v_9 + v_2)[f_{10}(R_2,1) - f_{10}(R_1,1)]\} - C_1,\tag{21}$$

$$\widetilde{C}_2 = 2\pi\{(a_5v_9 + v_5)[f_9(R_4,2) - f_9(R_3,2)] + (a_6v_9 + v_6)[f_{10}(R_4,2) - f_{10}(R_3,2)]\} - C_2,\tag{22}$$

$$f_9(r,i) = e_{31(i)}rJ_1(k_{P(i)}r)(i = 1,2),\tag{23}$$

$$f_{10}(r,i) = e_{31(i)}rY_1(k_{P(i)}r)(i = 1,2).\tag{24}$$

In Equations (23) and (24), $C_1 = \pi(\kappa_{33(1)}^\varepsilon/h)(R_2^2 - R_1^2)$ and $C_2 = \pi(\kappa_{33(2)}^\varepsilon/h)(R_4^2 - R_3^2)$ are the clamped electric capacitances of piezoceramic rings #1 and #2 in radial vibration, respectively. $\widetilde{C}_1$ and $\widetilde{C}_2$ are the effective electric capacitances of piezoceramic rings #1 and #2 in radial vibration, respectively. Then, the electrical impedance of the transducer Z can be given as:

$$Z = V(t)/I_{total}(t) = -1/[j\omega(\widetilde{C}_1 + \widetilde{C}_2)],\tag{25}$$

Subsequently, by letting $|Z| = 0$ and $|Z| = \infty$, we can obtain the resonance frequency f_r and anti-resonance frequency f_a, respectively. Based on these frequencies, the electromechanical coupling factor of the transducer is obtained as [55]:

$$k_d^2 = (f_a^2 - f_r^2)/f_a^2, \tag{26}$$

3. Validation

In this section, an ANSYS numerical simulation and an experimental study were conducted to validate the reliability of the theoretical solution. The geometric dimensions of the transducer are given as: $R_1 = 10$ mm, $R_2 = 15$ mm, $R_3 = 20$ mm, $R_4 = 25$ mm, $R_5 = 30$ mm, and $h = 5.63$ mm. Materials of the piezoceramic layers were selected as piezoceramic material type (PZT-5H), of which material parameters are listed in Table 1. Three different epoxy materials were chosen for epoxy layers, which have the same Poisson's ratio, approximately equal density and a certain difference in their Young's modulus. The material parameters of these three different epoxy materials can be found in Table 2, which are numbered as ①, ②, and ③, respectively. In the following analysis, these geometric dimensions and material parameters will be adopted, unless otherwise stated.

Table 1. Material parameters of piezoceramic materials [52,56–58].

Material Types	Elastic Constant ($\times 10^{-12}$ m²/N)		Piezoelectric Constant ($\times 10^{-12}$ C/N)	Dielectric Constant	Density (kg/m³)	Radial Sound Speed (m/s)	Plane Electromechanical Coupling Factor
	s_{11}^E	s_{12}^E	d_{31}	$\kappa_{33}^\sigma/\varepsilon_0$	ρ_P	V_{rP}	$k_{P(p)}$
PZT-5H	13	−4.29	−186	4500	7450	3404	0.45
PZT-4	12.3	−4.05	−123	1300	7500	3487	0.56
EC-64	12.8	−4.2	−127	1300	7500	3417	0.57
PZT-5A	16.4	−5.74	−171	1700	7750	2994	0.60
BaTiO₃	8.55	−2.61	−79	1900	5700	4757	0.35

Permittivity of free space: $\varepsilon_0 = 8.85 \times 10^{-12}$ F/m.

Table 2. Material parameters of three types of epoxy materials.

Epoxy Types	Young's Modulus ($\times 10^6$ N/m²)	Poisson's Ratio	Density (kg/m³)
	$\bar{E}$	μ	ρ_E
①	2301.47	0.43	1186
②	2862.17	0.43	1197
③	2930	0.43	1205

3.1. ANSYS Numerical Simulation

In this section, a finite element analysis based on the software ANSYS R17.1 was performed to compare with the theoretical results. A three-dimensional model of one-twelfth of the transducer was created because of the structural symmetry, as shown in Figure 2. In the simulation, the elements, Solid 185 and Solid 5, were used for the epoxy parts and the piezoelectric parts, respectively. The total amounts of elements and nodes were set as 2460 and 3258, respectively, to guarantee the computational precision. All the voltage degrees of freedom (DOFs) of the positive electrodes were coupled together, and the electrical condition $V_0 = 1$ V was applied. All the voltage DOFs of the negative electrodes were also coupled together, and the electrical condition $V_0 = 0$ V was applied. The harmonic analysis type was selected, and the frequency range was from 10 kHz to 40 kHz. The simulated impedance–frequency curve is plotted in Figure 3. In addition, the theoretical impedance–frequency relation is also plotted in Figure 3. It can be found that the results from theoretical analysis and finite element analysis agree reasonably well with each other. Further, the theoretical and simulated first resonance and anti-resonance frequencies are compared in Table 3. The relative errors between the theoretical values and the simulated ones for the first resonance frequency and the first anti-resonance frequency are −1.31% and −1.89%, respectively. The above comparative results validate the reliability of the theoretical solution.

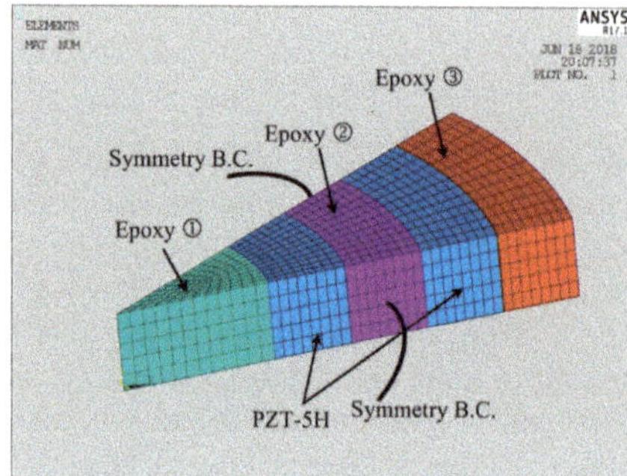

Figure 2. Three-dimensional finite element model of one-twelfth of the transducer.

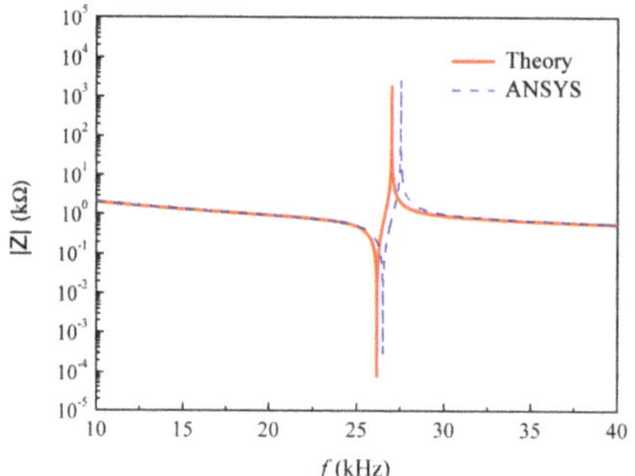

Figure 3. Theoretical and simulated impedance spectra.

Table 3. Comparisons between the calculated, simulated and experimental frequencies.

First Resonance and Anti-Resonance Frequencies	Theory (kHz)	ANSYS (kHz)	Error 1 (%)	Experiment (kHz)	Error 2 (%)
f_r	26.154	26.497	−1.31	23.179	11.37
f_a	26.995	27.504	−1.89	23.780	11.91

Error 1 = (Theory − ANSYS)/Theory; Error 2 = (Theory − Experiment)/Theory.

3.2. Experimental Validation

In this section, a test specimen of the radially layered cylindrical piezoceramic/epoxy composite transducer was fabricated by utilizing the mold-filling technique [52], as shown in Figure 4a. Similar to the previous work [52], the specimen fabrication mainly includes 8 steps: (1) mold and piezoceramic rings preparation, (2) epoxy preparation, (3) pouring epoxy into mould, (4) curing, (5) demolding, (6) polishing, (7) silvering and (8) final specimen. A difference is that in the step (3), three different epoxy materials were poured into the mold in this experiment. Three different epoxy materials, shown in Table 2, were prepared by mixing curing agents of 4,4′-methylenedianiline and bisphenol-A epoxy resin (E-51 with an epoxy value of 0.51 mol/100 g) at a mass ratio of 15/100, 17/100, 19/100, respectively. The curing agents of 4,4′-methylenedianiline was provided by Acros Organics Co. (Geel, Belgium). The E-51 was supplied by Nantong Xingchen Synthetic Material Co., Ltd. (Nantong, China). PZT-5H was selected as the piezoceramic material, shown in Table 1, which was provided by Baoding Hongsheng Acoustics Electron Apparatus Co., Ltd. (Baoding, China).

The specimen sizes were same as the given ones. The impedance test system is shown in Figure 4b, which included an Agilent 4294A Precision Impedance Analyzer for measurement and a computer for data acquisition. The electrical parallel connection was realized by using two conductive copper foil tapes. The measured impedance spectra and the phase of the impedance over the frequency range between 10 kHz and 40 kHz are shown in Figure 5. From the spectra, the first resonance and anti-resonance frequencies can be obtained as 23.179 kHz and 23.780 kHz, respectively. These two frequencies are also addressed in Table 3 to compare with the theoretical values. As can be seen, the calculated values are larger than the experimental ones; however, they agree reasonably well with each other. The relative errors between the theoretical values and experimental values for the first resonance frequency and the first anti-resonance frequency are 11.37% and 11.91%, respectively. There are two main factors accounting for the errors. Firstly, the theoretical model was established based on plane stress assumption, which is not the case for the fabricated composite. Secondly, the material parameters provided by the manufacturer were used here and the provided values may not be the exact values of the components used.

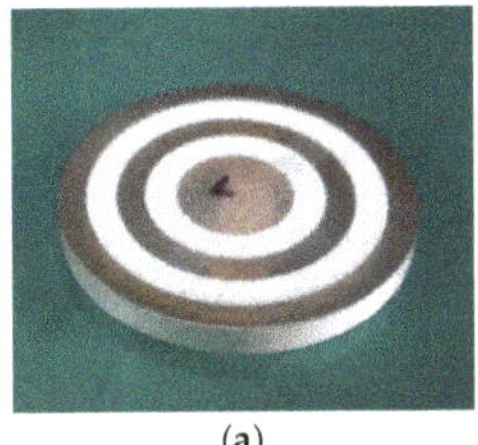

(a)

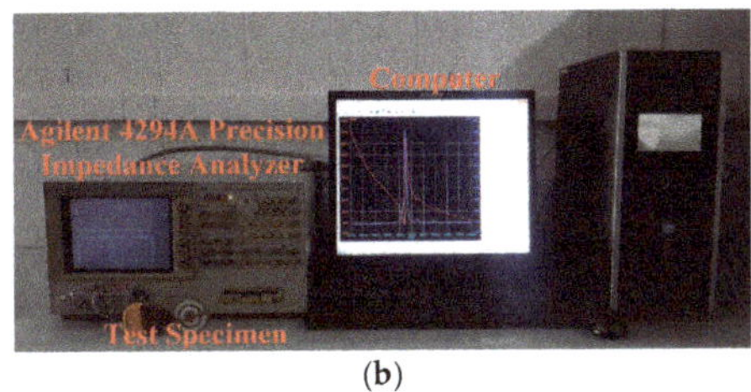

(b)

Figure 4. Experimental setup: (**a**) the fabricated specimen; (**b**) the impedance test system.

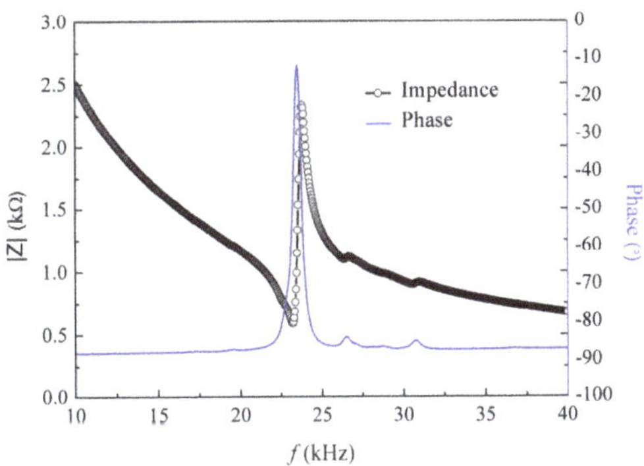

Figure 5. Measured impedance spectrum and its phase.

4. Results and Discussion

In this section, the effects of material parameters of epoxy layers, piezoceramic material types, and locations of piezoceramic rings on the electromechanical characteristics will be analyzed and discussed.

4.1. Effect of Material Parameters of Epoxy Layers

In the above experiment, the transducer with a sequence of material parameters ①-②-③ for epoxy layers #1, #2 and #3 was fabricated and tested. Here, the sequence ①-②-③ denotes that the material parameters for epoxy layers #1, #2, and #3 are materials ①, ②, and ③, respectively. Keeping the PZT-5H and geometric dimensions of the transducer unchanged, 27 sequences can be formulated according to different material arrangements of these three epoxy layers. Figure 6 plots the electromechanical characteristics for these 27 sequences. These electromechanical characteristics are the first resonance and anti-resonance frequencies and the corresponding electromechanical coupling factors. It can be seen that these 27 different sequences present 27 sets of electromechanical characteristics, which enable the multi-frequency characteristics of the transducer. In addition, transducer with the sequence ③-③-② has the maximum first resonance and anti-resonance frequencies, while the one with the sequence ①-①-① has the minimum frequencies. The transducer with the sequence ①-③-③ has the maximum electromechanical coupling factor, while the one with the sequence ③-①-② has the minimum value. As can be seen from Table 1, since Poisson's ratios are the same for all these three epoxy layers, the Young's modulus and density are the contributing factors to the variation in the electromechanical characteristics. The following analysis will discuss their effects on the electromechanical characteristics in order to distinguish the dominant factor.

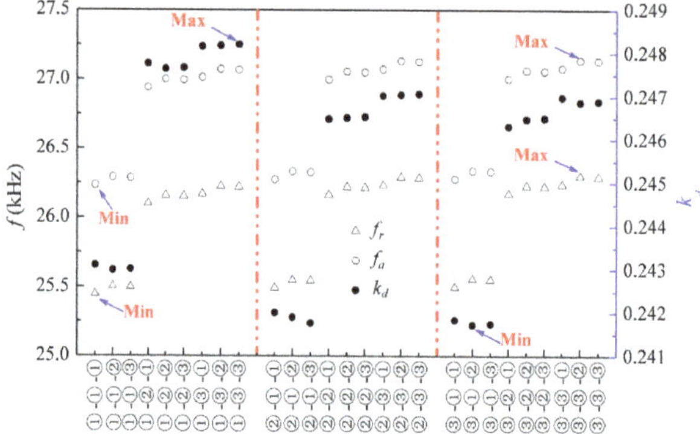

Figure 6. Electromechanical characteristics of the transducers with different material sequences of the epoxy layers.

Keeping other parameters unchanged, Figure 7 presents the effects of variation in density on the electromechanical characteristics. We had one reference group and three comparison groups. Here, the ①-②-③ combination was selected as the reference group, and three special cases with the same density within the group were selected as the comparison groups. From Figure 7, it can be observed that all of the first resonance and anti-resonance frequencies, as well as the corresponding electromechanical coupling factors, are very close to each other. A maximum relative error is −0.14%, which indicates that the effect of density on the electromechanical characteristics is very small and even negligible.

Similarly, Figure 8 shows the effects of variation in Young's modulus on the electromechanical characteristics of the transducer. For this case, three special cases with the same Young's modulus within each group were selected as the comparison groups. The differences between the electromechanical characteristics of these examples can be seen from Figure 8, where the maximum relative error is −3.24%. It is worth noting that this maximum error is 23 times more than that for density, which proves that the Young's modulus is the dominant factor for the electromechanical characteristics of the transducer. These results can serve as a good reference for designing the transducer.

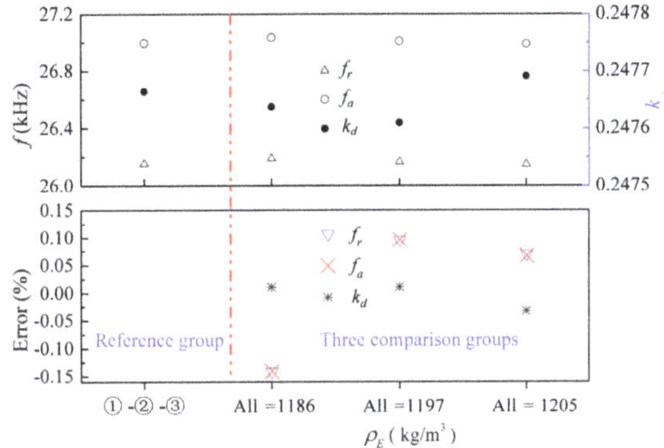

Figure 7. Influence of density of the epoxy layers on the electromechanical characteristics of the transducer. Error = (Reference group − Comparison group)/Reference group.

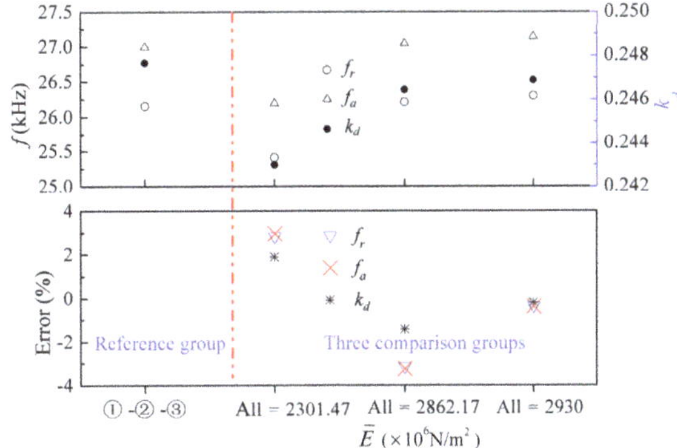

Figure 8. Influence of Young's modulus of the epoxy layers on the electromechanical characteristics of the transducer. Error = (Reference group − Comparison group)/Reference group.

Further, keeping the density of all epoxy layers as 1186 kg/m³, Figure 9 plots the variation of the electromechanical characteristics of the transducer when Young's modulus of epoxy layers changes from 2300×10^6 N/m² to 2940×10^6 N/m². Four cases are presented, i.e., the case of changing all epoxy layers, the case with only epoxy disk #1 changing, the case with only epoxy ring #2 changing, and the case with only epoxy ring #3 changing. It can be seen that the first resonance and anti-resonance frequencies increase with the increase of the Young's modulus of the epoxy layers. This is because larger Young's modulus will increase the stiffness of the transducer, which leads to higher resonant frequencies. Furthermore, it can be seen that changing Young's modulus of epoxy disk #1 and epoxy ring #3 has negligible effects on these two frequencies as compared to the case of changing the Young's modulus of epoxy ring #2. Therefore, in the transducer design, adjusting the Young's modulus of epoxy ring #2 can only realize frequency control of the proposed radial layered cylindrical piezoceramic/epoxy composite transducer. From Figure 9, it can also be found that for every case, the variation of Young's modulus of the epoxy layers has almost no effect on the corresponding electromechanical coupling factors. Here, it should be pointed out that the Poisson's ratio also greatly influences the electromechanical characteristics of the piezoelectric composites, which has been proved by the previous works [59–61]. However, in the present work, the main focus is to design a type of new transducers controlled by Young's modulus of the epoxy layers. Therefore, three different epoxy materials were chosen for the epoxy layers, which have the same Poisson's ratio, approximately equal density and a certain difference in their Young's moduli. The results also demonstrate the feasibility of this design.

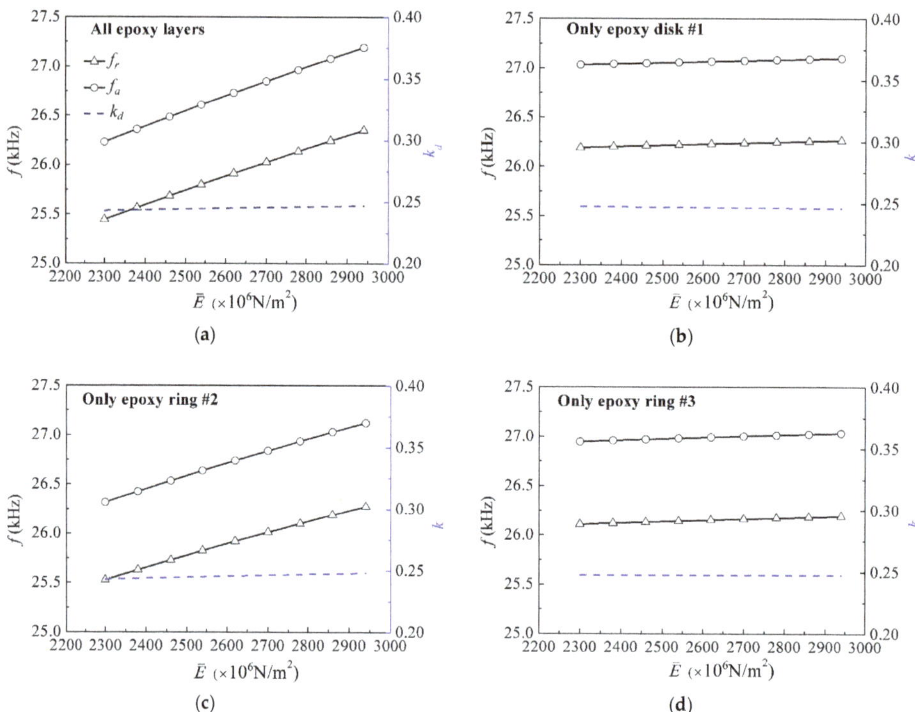

Figure 9. Electromechanical characteristics versus Young's moduli of the epoxy layers: (**a**) all epoxy layers; (**b**) only epoxy disk #1; (**c**) only epoxy ring #2; (**d**) only epoxy ring #3.

4.2. Effect of Piezoceramic Material Types

Selecting the material parameters of epoxy layers as the sequence ①-①-① and keeping geometric dimensions of the transducer unchanged, Figure 10 gives the effect of combinations of five commonly used piezoceramic materials on the electromechanical characteristics. These piezoceramic materials include PZT-5H, PZT-4, EC-64, PZT-5A and BaTiO$_3$, of which material parameters are shown in Table 3. The piezoceramic material types of PZT ring #1 are marked in the abscissa. The piezoceramic material types of PZT ring #2 are listed in the graph. From Figure 10, it can be found that when the PZT ring #2 is chosen as PZT-5A, the transducer has the minimum first resonance and anti-resonance frequencies, but the maximum first electromechanical coupling factor. When the PZT ring #2 is chosen as BaTiO$_3$, the transducer has the maximum first resonance and anti-resonance frequencies, but the minimum first electromechanical coupling factor. When the PZT ring #2 are chosen as PZT-5H, PZT-4, EC-64, the transducer has the similar first resonance and anti-resonance frequencies. The reasons are as follows. For a piezoelectric circular ring in radial vibration, when keeping its geometric sizes unchanged, its resonance frequency depends on the radial sound speed $V_{rP} = \sqrt{c_{11}^E/\rho_P}$ [36]. The radial sound speed reflects its stiffness–mass ratio, of which values are listed in Table 1. A larger V_{rP} for PZT ring #2 means its stiffness is enhanced, which further induces the stiffness increase of the transducer. In addition, its electromechanical coupling effect depends on the plane coupling factor $k_{P(p)} = \sqrt{2d_{31}^2/[\kappa_{33}^\sigma(s_{11}^E + s_{12}^E)]}$ [36], as shown in Table 1. A larger $k_{P(p)}$ for PZT ring #2 means its electromechanical coupling effect is better, which further improves the whole coupling effect. The PZT-5A has the minimum V_{rP} and maximum $k_{P(p)}$; therefore, the transducer with PZT-5A ring #2 has the smallest resonance frequency and best electromechanical coupling effect than the other types.

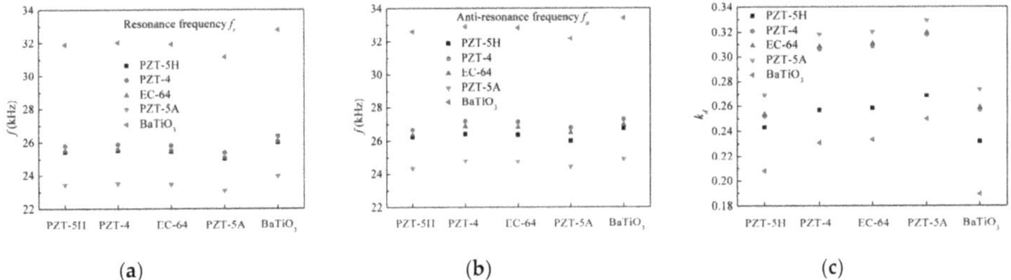

(a) (b) (c)

Figure 10. Electromechanical characteristics: (**a**) resonance frequency f_r; (**b**) anti-resonance frequency f_a; (**c**) electromechanical coupling factor k_d versus piezoceramic material types.

4.3. Effect of Locations of Piezoceramic Rings

Selecting the material parameters of epoxy layers as the sequence ①-①-①, piezoceramic material types of two PZT rings as PZT-5H, and keeping the area of one pizeoceramic ring unchanged, Figure 11 shows the relations between the electromechanical characteristics and locations of the other piezoceramic rings. Here, the inner radii R_1 and R_3 of the PZT rings #1 and #2 are used to denote their locations, respectively. The corresponding outer radii R_2 and R_4 of the PZT rings #1 and #2 also need to be changed to maintain the same areas, which are defined as $R_2 = \sqrt{S_1/\pi + R_1^2}$ and $R_4 = \sqrt{S_2/\pi + R_3^2}$, respectively. Symbols S_1 and S_2 are the areas of the PZT rings #1 and #2, respectively. When fixing the location of PZT ring #1 and varying the location of PZT ring #2, the inner and outer radii R_3 and R_4 of the PZT ring #2, and the areas of the epoxy rings #2 and #3 also vary. When fixing the location of PZT ring #2 and varying the location of PZT ring #1, the inner and outer radii R_1 and R_2 of the PZT ring #1, and the areas of the epoxy disk #1 and ring #2 also vary. From Figure 11, it is indicated that when the location of PZT ring #1 is fixed, the first resonance and anti-resonance frequencies, as well as the first electromechanical coupling factor, decrease with the increase of location of PZT ring #2. When the location of PZT ring #2 is fixed, the first resonance and anti-resonance frequencies, as well as the first electromechanical coupling factor, firstly increase to the maximum values, and then decrease. That is because different locations of one piezoelectric ring relative to the other will change the geometric sizes itself and those of the adjacent epoxy layers. When the material parameters are unchanged, these variations in the geometric sizes will vary their stiffness and mass, which lead to the change in the electromechanical coupling effect of the transducer In Figure 11b, three maximum values are f_r = 25.685 kHz, f_a = 26.578 kHz, k_d = 0.26, respectively. The corresponding locations are R_1 = 12 mm, 12.5 mm, and 14 mm, respectively. This rule can be used to design the improved transducer that has the maximum first resonance and anti-resonance frequencies as well as the first electromechanical coupling factor.

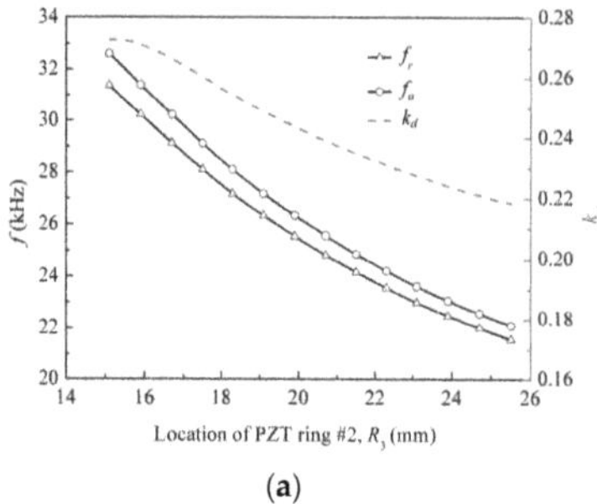 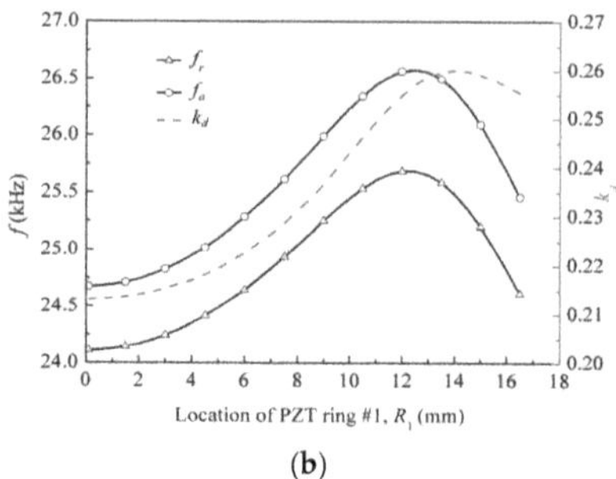

(a)   (b)

Figure 11. Electromechanical characteristics versus locations of piezoceramic rings: (**a**) location of PZT ring #2; (**b**) location of PZT ring #1.

5. Conclusions

A parametric analysis was performed to study the effects of multiple key factors, including material parameters of epoxy layers, piezoceramic material types, and locations of piezoceramic rings, on the electromechanical characteristics of the radially layered cylindrical piezoceramic/epoxy composite transducer. The main results can be concluded as follow.

(1) Based on the presented three different epoxy materials that have the same Poisson's ratio, approximately equal density and the certain difference in Young's modulus, the transducer can present 27 sets of electromechanical characteristics by utilizing different material sequences. Furthermore, these electromechanical characteristics are mainly controlled by the Young's moduli of the epoxy layers, especially for that of the epoxy ring #2. This result demonstrates that only regulating the Young's modulus of the epoxy layers can realize the design and optimization of the electromechanical characteristics of the transducer.

(2) Among five commonly used piezoceramic materials (PZT-5H, PZT-4, EC-64, PZT-5A and BaTiO$_3$), the transducer with the PZT-5A ring #2 has the minimum first resonance and anti-resonance frequencies as well as the maximum first electromechanical coupling factor; the transducer with the BaTiO$_3$ ring #2 has the maximum first resonance and anti-resonance frequencies as well as the minimum first electromechanical coupling factor; the transducer with PZT-5H, PZT-4, EC-64 ring #2 has the similar first resonance and anti-resonance frequencies. That is to say, the selections of piezoceramic material types in the ring #2, the piezoceramic materials with the lager radial sound speed and plane electromechanical coupling factor can optimize the electromechanical characteristics of the transducer.

(3) The locations of piezoceramic rings have great effects on the electromechanical characteristics of the transducer, in particular, an appropriate location can be used to optimize the transducer design, making it have the maximum first resonance and anti-resonance frequencies as well as the first electromechanical coupling factor.

Author Contributions: J.W. established the modelling and performed the numerical analysis; J.W., L.Q. and W.S. fabricated the transducer; J.W. and L.Q. tested the transducer; J.W. and W.L. wrote the paper; and all the authors discussed the results.

Funding: This research was funded by the National Natural Science Foundation of China (51708025, 61671068) and the Fundamental Research Funds for the Central Universities (FRF-TP-16-069A1).

Conflicts of Interest: The authors declare no conflicts of interest.

Appendix A

$$\begin{cases} A_{E(1)} = v_9 V_0 \\ B_{E(1)} = 0 \end{cases}, \tag{A1}$$

$$\left\{ \begin{array}{l} A_{P(1)} = (a_1 v_9 + v_1) V_0 \\ B_{P(1)} = (a_2 v_9 + v_2) V_0 \end{array} \right. , \tag{A2}$$

$$\left\{ \begin{array}{l} A_{E(2)} = (a_3 v_9 + v_3) V_0 \\ B_{E(2)} = (a_4 v_9 + v_4) V_0 \end{array} \right. , \tag{A3}$$

$$\left\{ \begin{array}{l} A_{P(2)} = (a_5 v_9 + v_5) V_0 \\ B_{P(2)} = (a_6 v_9 + v_6) V_0 \end{array} \right. , \tag{A4}$$

$$\left\{ \begin{array}{l} A_{E(3)} = (a_7 v_9 + v_7) V_0 \\ B_{E(3)} = (a_8 v_9 + v_8) V_0 \end{array} \right. , \tag{A5}$$

where

$$\left\{ \begin{array}{l} H_1 = f_1(R_1,1) f_4(R_1,1) - f_3(R_1,1) f_2(R_1,1) \\ a_1 = [f_5(R_1,1) f_4(R_1,1) - f_7(R_1,1) f_2(R_1,1)]/H_1 \\ v_1 = [(e_{31(1)}/h) f_2(R_1,1)]/H_1 \\ a_2 = -[f_5(R_1,1) f_3(R_1,1) - f_7(R_1,1) f_1(R_1,1)]/H_1 \\ v_2 = -[(e_{31(1)}/h) f_1(R_1,1)]/H_1 \end{array} \right. , \tag{A6}$$

$$\left\{ \begin{array}{l} H_2 = f_5(R_2,2) f_8(R_2,2) - f_7(R_2,2) f_6(R_2,2) \\ a_3 = \{ f_8(R_2,2)[a_1 f_1(R_2,1) + a_2 f_2(R_2,1)] - f_6(R_2,2)[a_1 f_3(R_2,1) + a_2 f_4(R_2,1)] \}/H_2 \\ v_3 = \left\{ f_8(R_2,2)[v_1 f_1(R_2,1) + v_2 f_2(R_2,1)] - f_6(R_2,2)[v_1 f_3(R_2,1) + v_2 f_4(R_2,1) + e_{31(1)}/h] \right\}/H_2 \\ a_4 = -\{ f_7(R_2,2)[a_1 f_1(R_2,1) + a_2 f_2(R_2,1)] - f_5(R_2,2)[a_1 f_3(R_2,1) + a_2 f_4(R_2,1)] \}/H_2 \\ v_4 = -\left\{ f_7(R_2,2)[v_1 f_1(R_2,1) + v_2 f_2(R_2,1)] - f_5(R_2,2)[v_1 f_3(R_2,1) + v_2 f_4(R_2,1) + e_{31(1)}/h] \right\}/H_2 \end{array} \right. , \tag{A7}$$

$$\left\{ \begin{array}{l} H_3 = f_1(R_3,2) f_4(R_3,2) - f_3(R_3,2) f_2(R_3,2) \\ a_5 = \{ f_4(R_3,2)[a_3 f_5(R_3,2) + a_4 f_6(R_3,2)] - f_2(R_3,2)[a_3 f_7(R_3,2) + a_4 f_8(R_3,2)] \}/H_3 \\ v_5 = \left\{ f_4(R_3,2)[v_3 f_5(R_3,2) + v_4 f_6(R_3,2)] - f_2(R_3,2)[v_3 f_7(R_3,2) + v_4 f_8(R_3,2) - e_{31(2)}/h] \right\}/H_3 \\ a_6 = -\{ f_3(R_3,2)[a_3 f_5(R_3,2) + a_4 f_6(R_3,2)] - f_1(R_3,2)[a_3 f_7(R_3,2) + a_4 f_8(R_3,2)] \}/H_3 \\ v_6 = -\left\{ f_3(R_3,2)[v_3 f_5(R_3,2) + v_4 f_6(R_3,2)] - f_1(R_3,2)[v_3 f_7(R_3,2) + v_4 f_8(R_3,2) - e_{31(2)}/h] \right\}/H_3 \end{array} \right. , \tag{A8}$$

$$\left\{ \begin{array}{l} H_4 = f_5(R_4,3) f_8(R_4,3) - f_7(R_4,3) f_6(R_4,3) \\ a_7 = \{ f_8(R_4,3)[a_5 f_1(R_4,2) + a_6 f_2(R_4,2)] - f_6(R_4,3)[a_5 f_3(R_4,2) + a_6 f_4(R_4,2)] \}/H_4 \\ v_7 = \left\{ f_8(R_4,3)[v_5 f_1(R_4,2) + v_6 f_2(R_4,2)] - f_6(R_4,3)[v_5 f_3(R_4,2) + v_6 f_4(R_4,2) - e_{31(2)}/h] \right\}/H_4 \\ a_8 = -\{ f_7(R_4,3)[a_5 f_1(R_4,2) + a_6 f_2(R_4,2)] - f_5(R_4,3)[a_5 f_3(R_4,2) + a_6 f_4(R_4,2)] \}/H_4 \\ v_8 = -\left\{ f_7(R_4,3)[v_5 f_1(R_4,2) + v_6 f_2(R_4,2)] - f_5(R_4,3)[v_5 f_3(R_4,2) + v_6 f_4(R_4,2) - e_{31(2)}/h] \right\}/H_4 \end{array} \right. , \tag{A9}$$

$$v_9 = -[v_7 f_7(R_5,3) + v_8 f_8(R_5,3)]/[a_7 f_7(R_5,3) + a_8 f_8(R_5,3)]. \tag{A10}$$

References

1. Maréchal, P.; Levassort, F.; Holc, J.; Tran-Huu-Hue, L.-P.; Kosec, M.; Lethiecq, M. High-frequency transducers based on integrated piezoelectric thick films for medical imaging. *IEEE Trans. Ultrason. Ferroelectr. Freq. Control* **2006**, *53*, 1524–1533. [CrossRef] [PubMed]
2. Shung, K.K.; Cannata, J.; Zhou, Q. Piezoelectric materials for high frequency medical imaging applications: A review. *J. Electroceram.* **2007**, *19*, 141–147. [CrossRef]
3. Cannata, J.M.; Williams, J.A.; Zhou, Q.; Ritter, T.A.; Shung, K.K. Development of a 35-MHz piezo-composite ultrasound array for medical imaging. *IEEE Trans. Ultrason. Ferroelectr. Freq. Control* **2006**, *53*, 224–236. [CrossRef] [PubMed]
4. Tawie, R.; Lee, H.-K.; Park, S. Non-destructive evaluation of concrete quality using PZT transducers. *Smart Struct. Syst.* **2010**, *6*, 851–866. [CrossRef]
5. Lim, Y.Y.; Kwong, K.Z.; Liew, W.Y.H.; Soh, C.K. Non-destructive concrete strength evaluation using smart piezoelectric transducer—A comparative study. *Smart Mater. Struct.* **2016**, *25*, 085021. [CrossRef]
6. Martins, M.; Correia, V.; Cabral, J.; Lanceros-Mendez, S.; Rocha, J. Optimization of piezoelectric ultrasound emitter transducers for underwater communications. *Sens. Actuators A Phys.* **2012**, *184*, 141–148. [CrossRef]

7. Mosca, F.; Matte, G.; Shimura, T. Low-frequency source for very long-range underwater communication. *J. Acoust. Soc. Am.* **2013**, *133*, EL61–EL67. [CrossRef] [PubMed]

8. Sun, J.; Ngernchuklin, P.; Vittadello, M.; Akdoğan, E.; Safari, A. Development of 2-2 piezoelectric ceramic/polymer composites by direct-write technique. *J. Electroceram.* **2010**, *24*, 219–225. [CrossRef]

9. Lee, H.J.; Zhang, S.; Bar-Cohen, Y.; Sherrit, S. High temperature, high power piezoelectric composite transducers. *Sensors* **2014**, *14*, 14526–14552. [CrossRef] [PubMed]

10. Li, L.; Zhang, S.; Xu, Z.; Wen, F.; Geng, X.; Lee, H.J.; Shrout, T.R. 1-3 piezoelectric composites for high-temperature transducer applications. *J. Phys. D Appl. Phys.* **2013**, *46*, 165306. [CrossRef] [PubMed]

11. Chabok, H.R.; Cannata, J.M.; Kim, H.H.; Williams, J.A.; Park, J.; Shung, K.K. A high-frequency annular-array transducer using an interdigital bonded 1-3 composite. *IEEE Trans. Ultrason. Ferroelectr. Freq. Control* **2011**, *58*, 206–214. [CrossRef] [PubMed]

12. Garcia-Gancedo, L.; Olhero, S.M.; Alves, F.J.; Ferreira, J.M.F.; Demore, C.E.M.; Cochran, S.; Button, T.W. Application of gel-casting to the fabrication of 1-3 piezoelectric ceramic-polymer composites for high-frequency ultrasound devices. *J. Micromech. Microeng.* **2012**, *22*, 125001. [CrossRef]

13. Lee, H.J.; Zhang, S.; Geng, X.; Shrout, T.R. Electroacoustic response of 1-3 piezocomposite transducers for high power applications. *Appl. Phys. Lett.* **2012**, *101*, 253504. [CrossRef] [PubMed]

14. Lee, H.J.; Zhang, S. Design of low-loss 1-3 piezoelectric composites for high-power transducer applications. *IEEE Trans. Ultrason. Ferroelectr. Freq. Control* **2012**, *59*, 1969–1975. [PubMed]

15. Lee, H.J.; Zhang, S.; Meyer, R.J., Jr.; Sherlock, N.P.; Shrout, T.R. Characterization of piezoelectric ceramics and 1-3 composites for high power transducers. *Appl. Phys. Lett.* **2012**, *101*, 032902. [CrossRef] [PubMed]

16. Abrar, A.; Zhang, D.; Su, B.; Button, T.W.; Kirk, K.J.; Cochran, S. 1-3 connectivity piezoelectric ceramic-polymer composite transducers made with viscous polymer processing for high frequency ultrasound. *Ultrasonics* **2004**, *42*, 479–484. [CrossRef] [PubMed]

17. Wang, W.; Or, S.W.; Yue, Q.; Zhang, Y.; Jiao, J.; Ren, B.; Lin, D.; Leung, C.M.; Zhao, X.; Luo, H. Cylindrically shaped ultrasonic linear array fabricated using PIMNT/epoxy 1-3 piezoelectric composite. *Sens. Actuators A Phys.* **2013**, *192*, 69–75. [CrossRef]

18. Zhang, Y.; Wang, L.; Qin, L. Equivalent parameter model of 1-3 piezocomposite with a sandwich polymer. *Results Phys.* **2018**, *9*, 1256–1261. [CrossRef]

19. Zhang, Y.; Jiang, Y.; Lin, X.; Xie, R.; Zhou, K.; Button, T.W.; Zhang, D. Fine-scaled piezoelectric ceramic/polymer 2-2 composites for high-frequency transducer. *J. Am. Ceram. Soc.* **2014**, *97*, 1060–1064. [CrossRef]

20. Cannata, J.M.; Williams, J.A.; Zhang, L.; Hu, C.-H.; Shung, K.K. A high-frequency linear ultrasonic array utilizing an interdigitally bonded 2-2 piezo-composite. *IEEE Trans. Ultrason. Ferroelectr. Freq. Control* **2011**, *58*, 2202–2212. [CrossRef] [PubMed]

21. Ritter, T.A.; Shrout, T.R.; Tutwiler, R.; Shung, K.K. A 30-MHz piezo-composite ultrasound array for medical imaging applications. *IEEE Trans. Ultrason. Ferroelectr. Freq. Control* **2002**, *49*, 217–230. [CrossRef] [PubMed]

22. Banerjee, S.; Cook-Chennault, K.A. Influence of al particle size and lead zirconate titanate (PZT) volume fraction on the dielectric properties of pzt-epoxy-aluminum composites. *J. Eng. Mater. Technol.* **2011**, *133*, 041016. [CrossRef]

23. Banerjee, S.; Cook-Chennault, K.A. An investigation into the influence of electrically conductive particle size on electromechanical coupling and effective dielectric strain coefficients in three phase composite piezoelectric polymers. *Compos. Part A Appl. Sci. Manuf.* **2012**, *43*, 1612–1619. [CrossRef]

24. Sundar, U.; Cook-Chennault, K.A.; Banerjee, S.; Refour, E. Dielectric and piezoelectric properties of percolative three-phase piezoelectric polymer composites. *J. Vac. Sci. Technol. B* **2016**, *34*, 041232. [CrossRef]

25. Lin, S. Effect of electric load impedances on the performance of sandwich piezoelectric transducers. *IEEE Trans. Ultrason. Ferroelectr. Freq. Control* **2004**, *51*, 1280–1286. [PubMed]

26. Lin, S. Analysis of multifrequency langevin composite ultrasonic transducers. *IEEE Trans. Ultrason. Ferroelectr. Freq. Control* **2009**, *56*, 1990–1998. [PubMed]

27. Lin, S.; Xu, C. Analysis of the sandwich ultrasonic transducer with two sets of piezoelectric elements. *Smart Mater. Struct.* **2008**, *17*, 065008. [CrossRef]

28. Arnold, F.J.; Bravo-Roger, L.L.; Gonçalves, M.S.; Grilo, M. Characterization of sandwiched piezoelectric transducers-a complement for teaching electric circuits. *Lat.-Am. J. Phys. Educ.* **2012**, *6*, 216–220.

29. Arnold, F.J.; Mühlen, S.S. The resonance frequencies on mechanically pre-stressed ultrasonic piezotransducers. *Ultrasonics* **2001**, *39*, 1–5. [CrossRef]
30. Adelman, N.T.; Stavsky, Y.; Segal, E. Axisymmetric vibrations of radially polarized piezoelectric ceramic cylinders. *J. Sound Vib.* **1975**, *38*, 245–254. [CrossRef]
31. Ebenezer, D.D.; Abraham, P. Eigenfunction analysis of radially polarized piezoelectric cylindrical shells of finite length. *J. Acoust. Soc. Am.* **1997**, *102*, 1549–1558. [CrossRef]
32. Hussein, M.; Heyliger, P. Discrete layer analysis of axisymmetric vibrations of laminated piezoelectric cylinders. *J. Sound Vib.* **1996**, *192*, 995–1013. [CrossRef]
33. Kim, J.O.; Hwang, K.K.; Jeong, H.G. Radial vibration characteristics of piezoelectric cylindrical transducers. *J. Sound Vib.* **2004**, *276*, 1135–1144. [CrossRef]
34. Kim, J.O.; Lee, J.G. Dynamic characteristics of piezoelectric cylindrical transducers with radial polarization. *J. Sound Vib.* **2007**, *300*, 241–249. [CrossRef]
35. Li, X.F.; Peng, X.L.; Lee, K.Y. Radially polarized functionally graded piezoelectric hollow cylinders as sensors and actuators. *Eur. J. Mech. A Solids* **2010**, *29*, 704–713. [CrossRef]
36. Lin, S. Electro-mechanical equivalent circuit of a piezoelectric ceramic thin circular ring in radial vibration. *Sens. Actuators A Phys.* **2007**, *134*, 505–512. [CrossRef]
37. Lin, S.; Wang, S.J.; Fu, Z.Q. Electro-mechanical equivalent circuit for the radial vibration of the radially poled piezoelectric ceramic long tubes with arbitrary wall thickness. *Sens. Actuators A Phys.* **2012**, *180*, 87–96. [CrossRef]
38. Piao, C.; Kim, J.O. Vibration characteristics of a piezoelectric transducer laminated with piezoelectric disks. *Noise Control Eng. J.* **2016**, *64*, 444–458. [CrossRef]
39. Piao, C.; Kim, J.O. Vibration characteristics of an ultrasonic transducer of two piezoelectric discs. *Ultrasonics* **2017**, *74*, 72–80. [CrossRef] [PubMed]
40. Zhang, T.T.; Shi, Z.F. Exact analyses for two kinds of piezoelectric hollow cylinders with graded properties. *Smart Struct. Syst.* **2010**, *6*, 975–989. [CrossRef]
41. Zhang, T.T.; Shi, Z.F.; Spencer, B.F., Jr. Vibration analysis of a functionally graded piezoelectric cylindrical actuator. *Smart Mater. Struct.* **2008**, *17*, 025018. [CrossRef]
42. Ebenezer, D.D. Determination of complex coefficients of radially polarized piezoelectric ceramic cylindrical shells using thin shell theory. *IEEE Trans. Ultrason. Ferroelectr. Freq. Control* **2004**, *51*, 1209–1215. [CrossRef] [PubMed]
43. Ebenezer, D.D.; Joseph, L. Frequency-dependent open-circuit acoustic sensitivity of fluid-filled, coated, radially polarized piezoelectric ceramic cylindrical shells of arbitrary thickness and infinite length. *IEEE Trans. Ultrason. Ferroelectr. Freq. Control* **2001**, *48*, 914–921. [CrossRef] [PubMed]
44. Huan, Q.; Miao, H.; Li, F. A uniform-sensitivity omnidirectional shear-horizontal (SH) wave transducer based on a thickness poled, thickness-shear (d15) piezoelectric ring. *Smart Mater. Struct.* **2017**, *26*, 08LT01. [CrossRef]
45. Huan, Q.; Miao, H.; Li, F. A variable-frequency structural health monitoring system based on omnidirectional shear horizontal wave piezoelectric transducers. *Smart Mater. Struct.* **2018**, *27*, 025008. [CrossRef]
46. Gao, W.; Huo, L.; Li, H.; Song, G. Smart concrete slabs with embedded tubular pzt transducers for damage detection. *Smart Mater. Struct.* **2018**, *27*, 025002. [CrossRef]
47. Gao, W.; Huo, L.; Li, H.; Song, G. An embedded tubular PZT transducer based damage imaging method for two-dimensional concrete structures. *IEEE Access* **2018**, *6*, 30100–30109. [CrossRef]
48. Hu, J.; Lin, S.; Zhang, X.; Wang, Y. Radially sandwiched composite transducers composed of the radially polarized piezoelectric ceramic circular ring and metal rings. *Acta Acust. United Acust.* **2014**, *100*, 418–426. [CrossRef]
49. Cui, M.-T.; Xue, S.; Guo, R.-B.; Su, M.; Li, Y.-C.; Qian, M. Design and fabrication of cylindrical transducer based on 2-2 piezoelectric composite. In Proceedings of the 2015 Symposium on Piezoelectricity, Acoustic Waves, and Device Applications (SPAWDA), Jinan, China, 30 October–2 November 2015; pp. 277–281.
50. Zhou, D.; Cheung, K.F.; Chen, Y.; Lau, S.T.; Zhou, Q.; Shung, K.K.; Luo, H.S.; Dai, J.; Chan, H.L.W. Fabrication and performance of endoscopic ultrasound radial arrays based on PMN-PT single crystal/epoxy 1-3 composite. *IEEE Trans. Ultrason. Ferroelectr. Freq. Control* **2011**, *58*, 477–484. [CrossRef] [PubMed]

51. Wang, H.; Wang, L. High frequency wide band underwater acoustic transducer for ring shaped composite material. *Acta Acust.* **2017**, *42*, 53–59.

52. Wang, J.J.; Qin, L.; Song, W.B.; Shi, Z.F.; Song, G. Electromechanical characteristics of radially layered piezoceramic/epoxy cylindrical composite transducers: Theoretical solution, numerical simulation and experimental verification. *IEEE Trans. Ultrason. Ferroelectr. Freq. Control* **2018**, *65*, 1643–1656. [CrossRef] [PubMed]

53. Wang, J.J.; Shi, Z.F. Dynamic characteristics of an axially polarized multilayer piezoelectric/elastic composite cylindrical transducer. *IEEE Trans. Ultrason. Ferroelectr. Freq. Control* **2013**, *60*, 2196–2203. [CrossRef] [PubMed]

54. Wang, J.J.; Shi, Z.F. Models for designing radially polarized multilayer piezoelectric/elastic composite cylindrical transducers. *J. Intell. Mater. Syst. Struct.* **2016**, *27*, 500–511. [CrossRef]

55. Mason, W.P. *Piezoelectric Crystals and Their Application to Ultrasonics*; Van Nostrand Reinhold: New York, NY, USA, 1950.

56. Kim, D.; Kim, J.O.; Il Jung, S. Vibration characteristics of a piezoelectric open-shell transducer. *J. Sound Vib.* **2012**, *331*, 2038–2054. [CrossRef]

57. Butler, J.L.; Sherman, C.H. *Transducers and Arrays for Underwater Sound*; Springer: Cham, Switzerland, 2016.

58. Berlincourt, D.; Cmolik, C.; Jaffe, H. Piezoelectric properties of polycrystalline lead titanate zirconate compositions. *Proc. IRE* **1960**, *48*, 220–229. [CrossRef]

59. Avellaneda, M.; Swart, P.J. Calculating the performance of 1–3 piezoelectric composites for hydrophone applications: An effective medium approach. *J. Acoust. Soc. Am.* **1998**, *103*, 1449–1467. [CrossRef]

60. Hayward, G.; Bennett, J.; Hamilton, R. A theoretical study on the influence of some constituent material properties on the behavior of 1-3 connectivity composite transducers. *J. Acoust. Soc. Am.* **1995**, *98*, 2187–2196. [CrossRef]

61. Smith, W.A. Optimizing electromechanical coupling in piezocomposites using polymers with negative poisson's ratio. In Proceedings of the IEEE 1991 Ultrasonics Symposium, Orlando, FL, USA, 8–11 December 1991; pp. 661–666.

MDPI

St. Alban-Anlage 66

4052 Basel

Switzerland

Tel. +41 61 683 77 34

Fax +41 61 302 89 18

www.mdpi.com

Micromachines Editorial Office

E-mail: micromachines@mdpi.com

www.mdpi.com/journal/micromachines